# Introductory Solid State Physics with MATLAB® Applications

# Introductory Solid State Physics with MATLAB® Applications

**Javier E. Hasbun**
**Trinanjan Datta**

CRC Press
Taylor & Francis Group
Boca Raton London New York

CRC Press is an imprint of the
Taylor & Francis Group, an **informa** business

CRC Press
Taylor & Francis Group
6000 Broken Sound Parkway NW, Suite 300
Boca Raton, FL 33487-2742



Printed on acid-free paper

International Standard Book Number-13: 978-1-4665-1230-6 (Hardback)



**Visit the Taylor & Francis Web site at**
**http://www.taylorandfrancis.com**

**and the CRC Press Web site at**
**http://www.crcpress.com**

To my wife Margie and my son Ernest who have, spiritually and steadfastly, accompanied me for the most part of this journey.
—JEH

To my mother Subhra,
and
my wife Joya.
—TD

# Contents

# Preface

Solid state physics is a branch of condensed matter physics that is mainly concerned with the study of crystals. Crystals are considered to be systems of atoms arranged in particular and infinitely repeatable patterns. These patterns are responsible for much of the electrical and magnetic characteristics of many materials. Furthermore, the concepts learned in solid state physics and the way such concepts are used play an important role in treating systems that deviate away from the solid state, namely, glasses and perhaps even liquids for those who pursue such condensed matter topics. Thus the subject of solid state physics is a significant tool in the physicist's arsenal to help tackle many different problems. As such, solid state physics deals with the properties of materials. These properties can be in the electrical as well as in the magnetic regimes. Many problems in solid state physics begin with simple physical ideas for which whatever analytical expressions can be obtained under special circumstances help to gain insight into the problem of interest. However, it is the use of computer programs, as applied to solid state problems, that enable much of the subject in helping us learn about nature's secrets. Furthermore, the efficient use of computers has led to breakthroughs in the studies of materials which, in turn, has led us to a deeper understanding of materials. In this text, we use computers to calculate, visualize, animate, and simulate electronic and magnetic properties of the solid state. In that regard, throughout the text, we make use of MATLAB® (https://www.mathworks.com/) and, as part of the various concepts and their applications, we present programs or scripts to perform sophisticated calculations. The scripts are basically computer code that is interpreted by MATLAB to perform computations. The usefulness of MATLAB is realized quickly in the speed and simplicity of the code. Also open source software that is compatible with the MATLAB code is widely available, such as, for example, OCTAVE (https://www.gnu.org/software/octave/) and SCILAB (https://www.scilab.org/). The text's reader will benefit from the code that is part of this text and be able to modify it for further purposes of interest.

The text contains 14 chapters. Chapter 1 concentrates on the overall periodicity in the atomic arrangement found in crystals. It is in this chapter that we begin to learn that it is the interactions of the atomic electrons with the ions in a crystal that make the crystal act the way it does. In Chapter 2 we learn that the information gathered from the scattered particles can be analyzed to learn about the crystal plane spacing as well as the kind of atoms that compose the crystal system. Also we learn about the difference between standard light diffraction and crystal diffraction. Chapter 3 is concerned with what holds a crystal together. Energy is what makes crystal binding possible. In this regard, two kinds of energy terms are in general use to describe the binding energy. One is the cohesive energy, and the other is the lattice energy. Chapter 4 deals with lattice vibrations or collective excitations of a crystal. The atoms in crystals oscillate about their equilibrium positions in a similar way that a mass at the end of a spring oscillates about its equilibrium position. In a crystal, however, its periodicity plays a role in the behavior of the vibration. In Chapter 5 we discuss the free electron gas, which refers to the electrons in a crystal whose behavior is treated as if they were free from the binding forces that keep them confined to the crystal. These are the same as the valence electrons associated with the atoms in a crystal. In Chapter 6 we go beyond the free electron picture in order to begin in earnest the process of accounting for the difference between different materials such as metals, semimetals, semiconductors, and insulators; and various properties such as hall coefficients, the relationships between conduction electrons in metals and valence electrons in free atoms, etc.

Here energy bands is a term used to describe the electronic energies of electrons in crystalline materials. The energy bands in crystals are the analogue of electronic energy levels in atoms. Chapter 7 deals with semiconductor materials characterized by electronic conductivity values between metals and insulators and which have carrier concentrations that depend on temperature. We also talk about the effect of impurities on their electronic properties. Chapter 8 deals with band structure calculations in a simple way; that is, using the tight binding method. This method is based on the idea of a collection of isolated atoms which are slowly brought closer together in order to form a crystal. The wave function is a linear combination of atomic orbitals, but the only orbitals considered are those of the valence electrons. In Chapter 9 we build on previous chapters to discuss impurities in a deeper way, such as, the conductivity, and also discuss some disordered systems' properties. For example, random alloys in which at least two atomic species take random positions in a crystal's lattice sites fall within the realm of disordered systems. In Chapter 10 we begin by introducing the basic ideas of magnetism. The discussion begins at an elementary level such that readers with knowledge of electricity and magnetism from calculus based physics can follow. Wherever possible every term and definition has been stated along with the discussion. The main focus of this chapter is to build the foundation towards understanding the macroscopic physical properties of magnets. Throughout the chapter we have provided computational examples including the associated MATLAB codes. The chapter ends by providing an exposure to the very useful Monte Carlo simulation technique along with a MATLAB GUI. The GUI is a very handy tool for initiating student exploration activities to learn about classical Monte Carlo simulation techniques. It can also serve as a starting point for creating code for more complicated lattices and studying their behavior. The code could also be used as part of a thermal physics course. In Chapter 11 we introduce the quantum theory of magnetism. We discuss in detail the building blocks of magnetism and electron spin. We present the associated spin algebra and its consequence on the arrangement of electrons within an electronic shell. We discuss the origin and effect of spin-orbit interaction in a magnetic crystal, the concept of Hund's rule, the crystal field effect, diamagnetism, paramagnetism, and ferromagnetism. The chapter concludes by discussing the quantum mechanical origins of exchange interaction, magnetic resonance (primarily NMR), and Pauli paramagnetism. This chapter supplies codes that can be utilized to visualize atomic orbitals. These codes should be useful for chemists also. The problems in this chapter are more challenging compared to those in Chapter 10. In Chapter 12 we deal with the concept of superconductivity. The emphasis in this chapter has been on the qualitative aspects of superconductivity as opposed to a quantitative treatment based on BCS theory. The focus has been on the explanation of physical properties (at a qualitative level) such as zero electrical resistance, persistent current, Meissner and London effects, and the thermodynamic properties of superconductors. Finally, the chapter ends by motivating the various technological uses of superconductivity to the reader. Since this textbook is for undergraduates, we feel one of our obligations is not just to show case complex theory but also to expose the students to the areas where these ideas can be applied. We believe our book can be used not only by senior physics majors, but also by chemists, engineers, and other practitioners of science who want to gain a basic yet fundamental understanding of solid state physics. In Chapter 13 we highlight the optical properties of solids. While undergraduates are routinely exposed to various optical properties such as reflection, refraction, and transmission in their courses, they are barely exposed to the solid state applications of those fundamental concepts. The goal of this chapter has been to bring a synergy between their understanding of basic E&M concepts and optical properties. Beginning with the concept of a complex refractive index and a dielectric constant, the reader is initiated into the free electron Drude theory, the Drude-Lorentz dipole oscillator theory, and a qualitative description of the optical properties of glass, metal, and semiconductors. The chapter ends by providing a summary of the various commonly utilized optical spectroscopy techniques that are currently available to the scientific community. The last section of this chapter discusses the advanced concept of the Kramers-Kronig relationship, which may be skipped on a first reading or for those students who have not been exposed to the ideas of complex analysis. In Chapter 14 we discuss the transport properties of solids. The discussion in this chapter utilizes the

Boltzmann transport equation formalism. This approach is slightly more advanced than any previous exposition on transport properties in earlier chapters. The theme of this chapter has been to provide an overall comprehension and application of the various transport formalisms that are routinely utilized in modern condensed matter applications. The emphasis is more on the Boltzmann transport formulation and its application to the Drude conductivity, effects of electric and temperature gradients, drift and diffusion currents, thermal conductivity, and the thermoelectric effect. We finish the chapter with a brief discussion on the Landauer theory of transport. Finally, Appendix A contains a MATLAB tutorial that should be useful in understanding, modifying, and developing scripts associated with the computational code found throughout the text. Appendix B includes a discussion on the Boltzmann, the Fermi-Dirac, and the Bose-Einstein distribution functions. Finally, the MATLAB code listed throughout the text will be made available at the publisher's website (https://www.crcpress.com/), the author's website (https://www.westga.edu/ jhasbun/osp/osp.htm), and The MathWorks website (https://www.mathworks.com/academia/books).

MATLAB® is a registered trademark of The MathWorks, Inc. For product information, please contact:

The MathWorks, Inc.
3 Apple Hill Drive
Natick, MA 01760-2098 USA
Tel: 508-647-7000
Fax: 508-647-7001
E-mail: info@mathworks.com
Web: www.mathworks.com

# *Authors*

**Javier E. Hasbun** is a Professor of Physics in the Department of Physics at the University of West Georgia, Carrollton, GA. He obtained his B.S in Physics from the Massachusetts College of Liberal Arts, North Adams, MA. He obtained his M.S. and Ph.D. from the University of New York at Albany, Albany, NY in theoretical solid state physics. His post-doctoral experience was obtained from the Naval Air Warfare Center - Weapons Division, China Lake, CA. He has been at the University of West Georgia since 1990 and has taught courses in lower and upper level physics. He is a lifetime member of the American Physical Society, the American Association of Physics Teacher, the Sigma Xi Scientific Research Society, and the Georgia Academy of Science. His specialty is theoretical Solid State Physics, and conducts research in the electronic properties of solids. He is also interested in developing methods that use computers to enhance physics learning and teaching. He has published research in professional journals and presentations dealing with teaching and/or research in physics. To learn more, visit http://www.westga.edu/ jhasbun/.

**Trinanjan Datta** is a Professor of Physics in the Department of Chemistry and Physics at Augusta University, Augusta, GA. He obtained his B.S in Physics from the University of Calcutta (St. Xavier's College, Kolkata, India), M.S. from the Indian Institute of Technology (Kanpur, India), and his doctoral degree in theoretical condensed matter physics from Purdue University (West Lafayette, IN, USA). His research interests include theoretical and computational studies of magnetic systems and the scholarship of teaching and learning (SoTL). He has published numerous research articles on magnetism and delivered several invited national and international research presentations. Dr. Datta actively mentors undergraduate research students at Augusta University and masters and doctoral students at Sun Yat-Sen University. He is the recipient of the Augusta State University Louis K. Bell Alumni Research Award, the Kavli Institute for Theoretical Physics (KITP) scholar award, the University System of Georgia SoTL fellowship, and Augusta University Individual Teaching Excellence Award. He is a member of the Anacapa Society and the American Physical Society. To learn more, visit http://spots.gru.edu/tdatta/index.html.

# Acknowledgments

This text would not have been possible without the efforts of many professors throughout my years of study at the Massachusetts College of Liberal Arts and the State University of New York at Albany. I am especially grateful to Carl Wolf and Laura Roth whose words of wisdom I have carried in my mind for so many years. I thank the department of physics at the University of West Georgia for all the support afforded to me throughout the development of this text.

—Javier Hasbun, Carrollton, Georgia

I would like to thank my co-author Javier Hasbun for inviting me to write this textbook and for offering patient guidance, over the last six years, on my book chapters. Over the last twelve years, I have collaborated with a very talented group of undergraduate (physics and math) research students from Augusta University whom I would like to acknowledge: Billy Baez, Philip Javernick, Alexander Price, CurtisLee Thornton, Kenny Stiwinter, Sean Mongan, and Greg Price. Chapters 10 and 11 in this book are a direct outcome of my numerous discussions with them on how to explain magnetism to an undergraduate research student. I would like to acknowledge my longstanding research collaborator Prof. Dao-Xin Yao and graduate students Zewei Chen, Luo Cheng, Zengye Huang, Zijian Xiong, Meiyu He, Jia-Zheng Ma, Shangjian Jin, and Jun Li from Sun Yat-Sen University, Guangzhou, China. I would like to thank Prof. Kingshuk Majumdar of Grand Valley State University for initiating me into research on frustrated quantum magnetism. I would like to thank Prof. Per Rikvold (Florida State University and University of Oslo) and Prof. Mark Novotny (Mississippi State University) for helping me gain a deeper understanding of the various aspects of classical Monte Carlo simulation in non-equilibrium magnetic systems. I would like to thank the National Science Foundation, Cottrell College Science Award, State Key Laboratory of Optoelectronic Materials and Technologies at Sun Yat-Sen University, Savannah River National Laboratory, Augusta University Center for Undergraduate Research and Scholarship, and Augusta University for funding my research on magnetism. I am grateful to my senior departmental colleagues (Tom Colbert, Andy Hauger, and Tom Crute) for being supportive of my teaching and research pursuits. I would like to thank my middle school tutor Mr. Balakrishnan in Zambia and my high school teacher Prof. Dr. Anindya Ghose-Choudhury for teaching me the fundamentals of mathematics and physics. I would like to thank my father (Utpal), my mother (Subhra), and my sister (Aparajita) for being understanding of my passion to pursue physics. I would like to especially thank my mother Subhra for always being supportive of my aspiration to pursue physics as a profession and constantly encouraging me towards that goal. Finally, I would like to thank my wife Joya for sacrificing innumerable weekends and weeknights over the last six years as I toiled through the chapters in this book. Her constant support of my research activities during my doctoral degree and over the last sixteen years has been invaluable towards the completion of this book and for my own professional development.

—Trinanjan Datta, Augusta, Georgia

# 1

# *Introduction*

**Contents**



## 1.1 What Is Solid State Physics?

Solid state physics is mainly concerned with the study of crystals. Crystals are considered to be systems of atoms arranged in particular and infinitely repeatable patterns. The arrangement of the atoms can be in one, two, or three dimensions as conceptually shown in Figure 1.1.1 for finite size examples. It is the overall periodicity of the atomic arrangement that enables the understanding of crystals. The interaction of the atomic electrons with the ions in a crystal is of primary interest. Whereas Figure 1.1.1(c) is a three-dimensional example of a few atoms in a simple cubic system (eg., polonium), a quantum wire is a system that closely resembles a one-dimensional system of atoms (Ref. [1]), and a hexagonal lattice of atoms, or graphene sheet, is an example of a two-dimensional crystal (Ref. [2]). Images of both the quantum wire and a graphene sheet are shown in Figure 1.1.2.

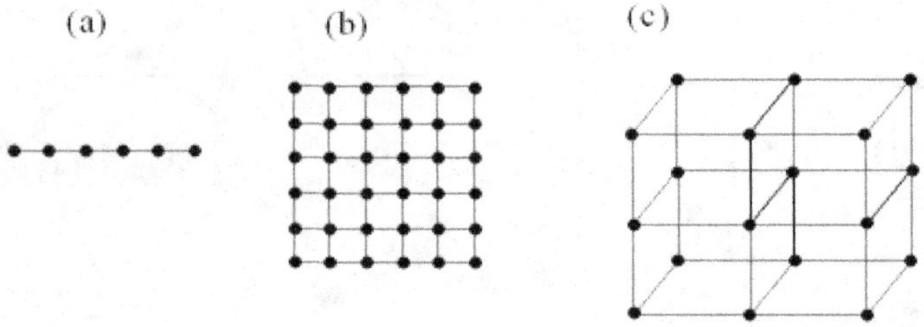


Figure 1.1.1: Examples of crystals in (a) one, (b) two, and (c) three dimensions. Here the black dots represent atoms.

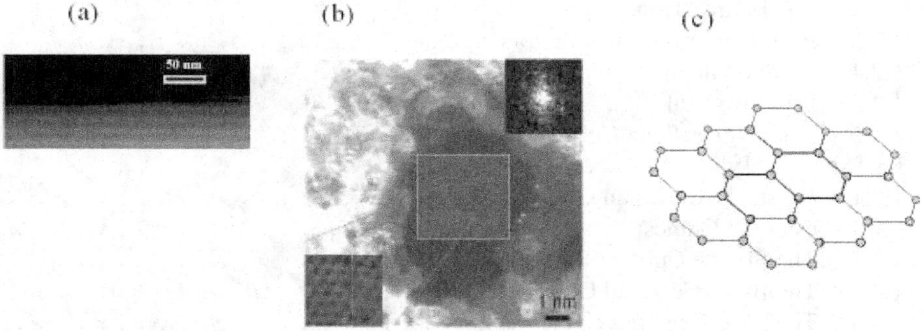


Figure 1.1.2: (a) A transmission electron microscope (TEM) image of a quantum wire (adapted from Ref. [1], reprinted with permission), (b) a scanning transmission electron microscope (STEM) image of a finite layer of graphene (adapted from Ref. [2]), and (c) a drawing of a finite size graphene lattice with the filled circles representing the atoms.

As the understanding of crystal structures grew, physicists could later extend learned techniques and begin to recognize the importance of amorphous systems, such as glass and liquids. In such systems, however, the best one can hope for is the presence of short range order, as shown for diamond in Figure 1.1.3, rather than the long range order that characterizes crystals. In Figure 1.1.3(a) the carbon atoms are placed at standard crystalline positions, while in Figure 1.1.3(b), the atoms are displaced by a small random deviation from the standard crystalline positions in order to simulate amorphous diamond. One can imagine going from the amorphous state to the liquid state by adding more randomness to simulate a liquid's atomic motion and thus averaging techniques are developed to understand this state. It is therefore crucial to understand crystal structures and the methods used to study them to be able to extend them to the more complex regimes. Studies of crystals began when in 1912 Max von Laue pondered the behavior of short wavelength electromagnetic radiation that interacts with crystals. He had the insight that x-rays, having wavelengths shorter than the atomic spacing within crystals, would experience interference. Laue and coworkers sent an x-ray beam through a copper sulfate crystal and observed a diffraction pattern on a photographic plate as if the crystal behaved as a three-dimensional diffraction grating (Figure 1.1.4). The plate recorded a large number of bright spots spread in a pattern of intersecting circles around the intense central beam spot. This represented direct evidence that the atoms in crystals organize in regular periodic structural units. Each structural unit contains one or more atoms in a particular arrangement. The Bragg Law (discussed later) also proposed in 1913 can be applied to the diffracted waves to obtain the interplanar crystal spacings. These advances were further aided by the serendipitous

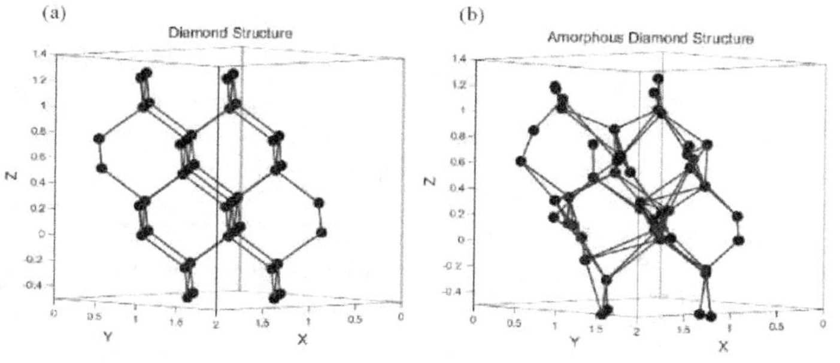


Figure 1.1.3: Examples of (a) crystalline diamond (Exercise 1.9.23) and (b) the simulated amorphous diamond for the same number of carbon atoms (Exercise 1.9.24).

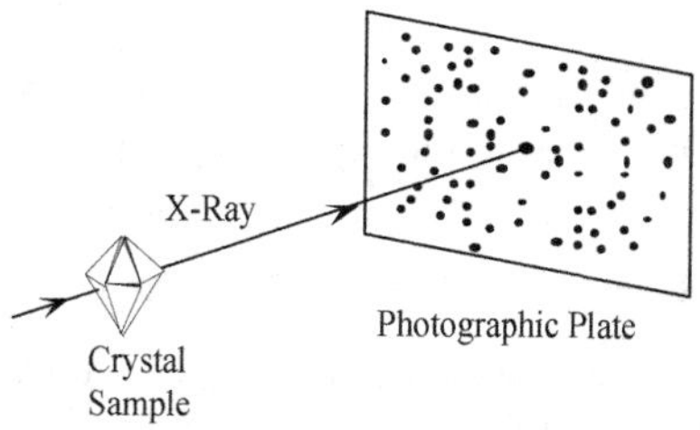


Figure 1.1.4: A simple x-ray Laue diffraction experiment.

Davisson-Germer (1927) experiment in which electrons showed diffraction peaks at certain scattering angles indicative of crystalline patterns in a nickel metal crystal (Figure 1.1.5). Diffraction peaks are absent in a disordered system, such as a liquid. The electrons' kinetic energy was high enough to enable them to have x-ray size de Broglie wavelength. Crystals are available naturally or can be artificially made in several different ways. In the case of diamonds, for example, they are formed of carbon atoms under high pressure and temperature deep inside Earth's mantle and brought to the surface through volcanic vents. They can be replicated in the laboratory under such conditions in a constant environment. Crystals can also be grown an atom at a time in ultra-high vacuum chambers or they can be chemically formed through saturated solutions. When a crystal grows in a constant environment, a form or shape develops as if identical blocks were added. These blocks or groups of atoms, arranged in a particular pattern, when periodically and indefinitely assembled, are responsible for the crystal structure; e.g., Figures 1.1.1(c) and 1.1.3(a).

## 1.2 Crystal Structure Basics

Following the above discussion, an ideal crystal is understood to be an infinite repetition of structural units in space. In the simplest crystals, the structural unit is a geometric equivalent arrangement of a single atom as in the metal crystals of aluminum, copper, silver, and gold. In iron, a different structural unit, also with a single atom, leads to a metal crystal of different geometry. In diamond, the structural unit is the single carbon atom arranged in as yet another different geometry. In a crystal

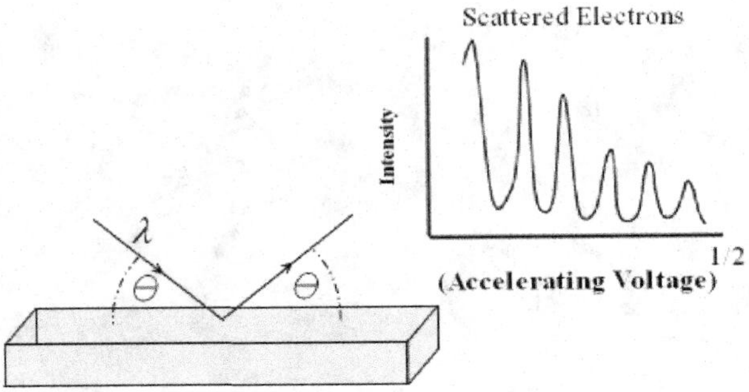


Figure 1.1.5: Davisson and Germer (1927) showed that an annealed sample of nickel metal showed electron diffraction.

of table salt, the structural unit is the diatomic molecule of NaCl organized in the same geometry found in a copper crystal, but making a different crystal structure, as will be seen later.

### 1.2.1 The Lattice and the Basis

The structure of all crystals can be described by a **lattice**, with a **group of atoms** attached to every lattice point. We make the following definitions. The **lattice**, in a particular dimension, is an indefinitely extended array of points, each of which is surrounded in an identical way by its neighbors. By definition, the lattice is a mathematical abstraction. A **lattice point** is referred to as each point that makes up the lattice. The **group of atoms** is called the **basis**. The basis can be one, two, three, etc. atoms. When the basis is repeated in space according to the lattice geometry, the crystal is formed. As an example, consider the simple two-dimensional square lattice of equally spaced points shown in Figure 1.2.6(a). Also consider the basis of a single atom shown in Figure 1.2.6(b). Finally, after locating the atom basis onto every lattice point one gets Figure 1.2.6(c) which together make up the crystal. For the same two-dimensional lattice, choosing a basis of two different atoms results in a different crystal structure as shown in Figure 1.2.7.

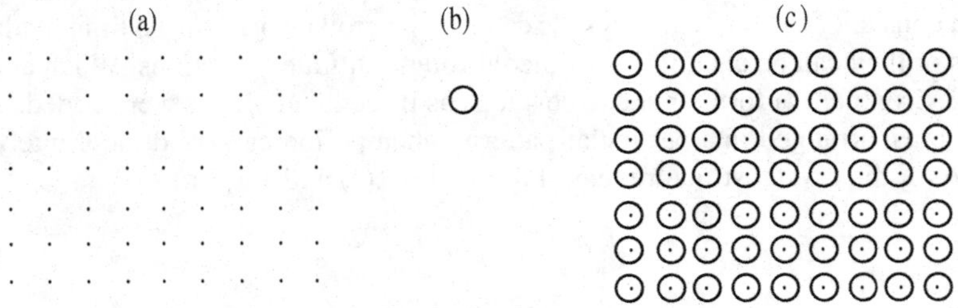


Figure 1.2.6: (a) A two-dimensional square lattice, (b) a basis of one atom, and (c) the crystal formed when a replica of the basis is placed at every lattice point.

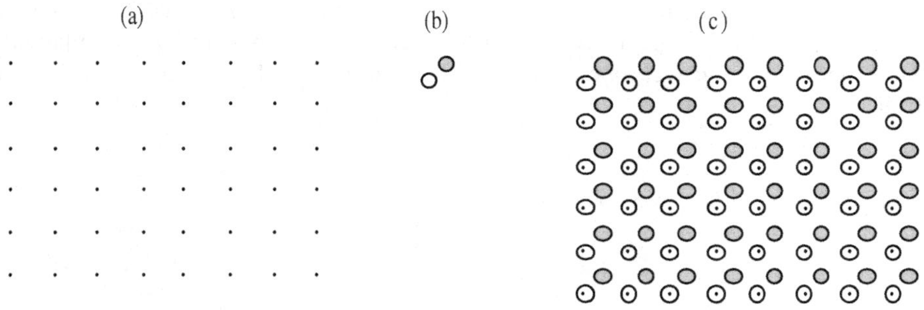


Figure 1.2.7: (a) The two-dimensional square lattice, (b) a basis of two atoms, and (c) the crystal formed when a replica of the basis is placed at every lattice point.

### 1.2.2 The Lattice Translation Vector

In mathematical terms, a crystal lattice is constructed according to lattice translation vectors. A general vector in a crystal is written as

$$\vec{r}' = \vec{r} + \vec{T}, \tag{1.2.1}$$

such that the atomic arrangement looks the same when viewed from $\vec{r}'$ as when viewed from $\vec{r}$. Here $\vec{T}$ is a **translation vector**

$$\vec{T} = u_1\vec{a}_1 + u_2\vec{a}_2 + u_3\vec{a}_3, \tag{1.2.2}$$

or **lattice translation vector**, where $u_1$, $u_2$, and $u_3$ are arbitrary integers and $\vec{a}_1$, $\vec{a}_2$, $\vec{a}_3$ are fundamental translation vectors. An example of fundamental vectors are the primitive translation vectors discussed next.

### 1.2.3 Primitive Translation Vectors

Consider two points $\vec{r}$ and $\vec{r}'$ in a crystal from which the atomic arrangement looks the same and which satisfy Equation (1.2.1) with a suitable choice of $u_1$, $u_2$, and $u_3$ (integers), then $\vec{a}_1$, $\vec{a}_2$, $\vec{a}_3$ (and therefore the unit cell so created) are primitive. This also means that $\vec{T}$ is a true lattice translation vector. It is possible to think of situations where one can use any set of vectors other than the primitive vectors such that Equation (1.2.1) is also true. However, the volume enclosed by the primitive lattice vectors is the smallest volume (building bock) that can be repeated in space to form the crystal structure. Here we use primitive translation vectors to define axes; i.e., primitive crystal axes, instead of Cartesian, for example, and which do not have to be orthogonal. Sometimes physicists use non-primitive crystal axes when they have a simpler relation to the symmetry of the crystal structure. The non-primitive crystal axes can also reproduce the crystal but the volume enclosed by these axes may not necessarily be the smallest volume that can be repeated to form the crystal structure.

**Example 1.2.3.1**
Consider a two-dimensional crystal on a rectangular lattice and let's pick two points from the origin described by $\vec{r}$ and $\vec{r}'$ as shown in Figure 1.2.8(a) where the primitive vectors $\vec{a}$ and $\vec{b}$ are also depicted. As we shall see, these vectors are not picked at random; they are actually related. Let's obtain the expression for $\vec{T}$ so that $\vec{r}'$ obeys Equation (1.2.1). We proceed as shown in Figure 1.2.8(b), an observer at $\vec{r}$ sees the same surroundings from $\vec{r}'$ as seen from $\vec{r}$ and according to Equation (1.2.1),

if $\vec{T}$ satisfies Equation (1.2.2), then for this two-dimensional example, $\vec{T} = u_1\vec{a} + u_2\vec{b}$, with the $u_i$'s having integer values of $u_1 = -3$ and $u_2 = 2$. Finally, according to the definition of a lattice translation vector, with $\vec{T} = -3\vec{a} + 2\vec{b}$, the atomic arrangement viewed from $\vec{r}$ looks the same as that viewed from $\vec{r}'$.

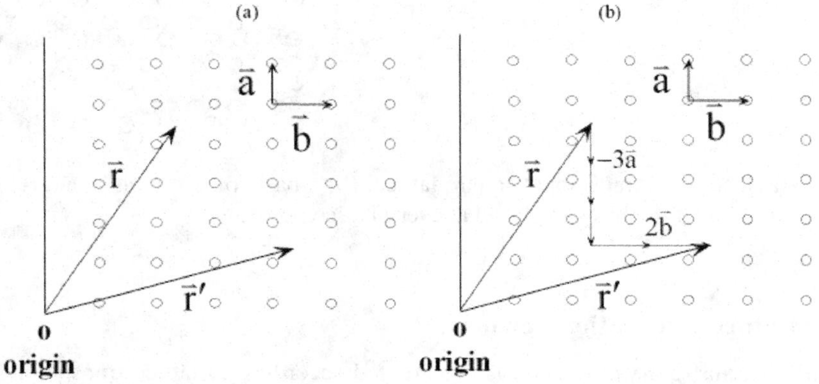


Figure 1.2.8: (a) A two-dimensional rectangular lattice showing position vectors $\vec{r}$ and $\vec{r}'$. (b) $\vec{r}' = \vec{r} + \vec{T}$ in this two-dimensional rectangular lattice (see Example 1.2.3.1).

In summary, to describe a crystal structure, we need the following:

1. What is the lattice?
2. Determine the best choice of $\vec{a}_1$, $\vec{a}_2$, and $\vec{a}_3$, whether they are primitive or non-primitive.
3. What is the basis?

The above-mentioned translation operation $\vec{T}$ belongs to a class of operations called symmetry operations. There are several symmetry operations: (a) translation, (b) rotation, (c) reflections, and (d) compound operations of two or more of these. The symmetry operations are significant because when performed they carry the crystal structure onto itself leaving it unmodified.

### 1.2.4 More on the Basis and the Crystal Structure

As pointed out before, a basis of atoms is attached to every lattice point, with every basis identical in composition, arrangement, and orientation. Let's do another but slightly different example with a rectangular lattice.

**Example 1.2.4.1**
Here we wish to see the effect of symmetry on the rectangular lattice of Figure 1.2.9(a), where every four adjacent lattice points form a rectangle. Consider the two atom basis of Figure 1.2.9(b). The basis is such that every basis atom lies at exactly $1/2$ this rectangle's diagonal. Let the filled circles be the A atoms and the empty circles be the B atoms. When this basis is added to the lattice, we get the two-dimensional hypothetical crystal of Figure 1.2.9(c). Notice that in this case, the symmetry is such that the A atoms are immediately and identically surrounded by four B atoms and that each B atom is surrounded in an identical way by four A atoms as in Figure 1.2.9(d). Due to the perfect order, this is an example of a compound. Were the A atoms surrounded by either A or B atoms in a random way, the resulting crystal would be called a random alloy, a subject discussed much later in the text. In that case, the B atoms would, in turn, be surrounded randomly by either A or B as well. The basis would consist of a random basis of two atoms. Thus the basis is very important in

determining the kind of crystal structure we end up dealing with. There is one caveat in what we have said here; that is, alloys are not considered to have long range order and do not necessarily follow the same rules that crystals do. We will come back to this issue in a later chapter.

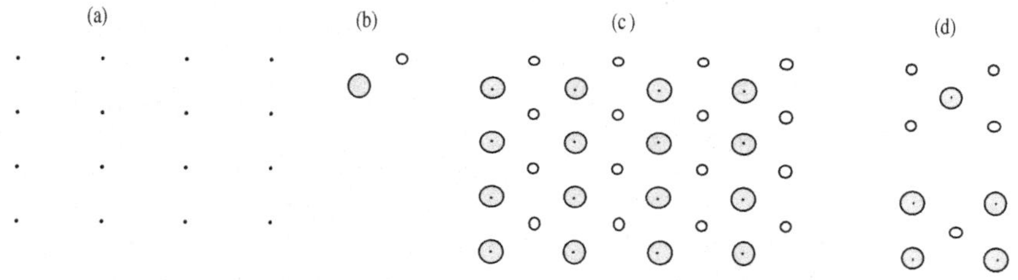


Figure 1.2.9: (a) The two-dimensional rectangular lattice, (b) a 2-atom basis with a separation of 1/2 the smallest lattice's rectangle diagonal, (c) the crystal formed, and (d) the immediate surroundings seen by each type of atom (see Example 1.2.4.1).

As we have seen, the number of atoms in the basis may be more than one. The position of the center atom $j$ of the basis relative to the associated lattice point is given by

$$\vec{r}_j = x_j\vec{a}_1 + y_j\vec{a}_2 + z_j\vec{a}_3, \tag{1.2.3}$$

where $j = 1, 2, ...$ is an integer that runs over the number of atoms in the basis. The positions $x_j$, $y_j$, and $z_j$ may be non-integers and are different from the quantities $u_1$, $u_2$, and $u_3$ in Equation (1.2.1), which take on integer values. The quantities $\vec{r}_j$ give the positions of all the atoms in a crystal, as we show later. For simple structures, the origin can be rearranged so that the coefficients of the lattice vectors in Equation 1.2.3 are fractions; that is, $0 \le x_j, y_j, z_j \le 1$. We can have a translation vector to atom $j$ in the form $\vec{T}_j = \vec{T} + \vec{r}_j$ with $\vec{T}$ as in Equation 1.2.2. This is done in the following example.

**Example 1.2.4.2**
Let's work with Figure 1.2.10(a), where $\vec{r} = \frac{1}{2}\vec{a} + \frac{1}{2}\vec{b}$ for the 'x' atom in the basis (Figure 1.2.10(b)) and where $\vec{a}$, $\vec{b}$ are the lattice vectors. If A is the origin atom, then atom B is located at $\vec{r}_B = \vec{b}$, atom C at $\vec{r}_C = \vec{a} + 2\vec{b}$, and atom D at $\vec{r}_D = \vec{a} + 2\vec{b} + \vec{r} = \frac{3}{2}\vec{a} + \frac{5}{2}\vec{b}$ and is equivalent to the atom at $\vec{r}$. Thus, here $\vec{r}_D = \vec{r}'$ and so $\vec{r}' = \vec{T} + \vec{r}$ where $\vec{T} = \vec{a} + 2\vec{b}$ as in Equation 1.2.1, and one also identifies $\vec{T}_D \equiv \vec{r}'$ as the translation vector to atom D from the origin.

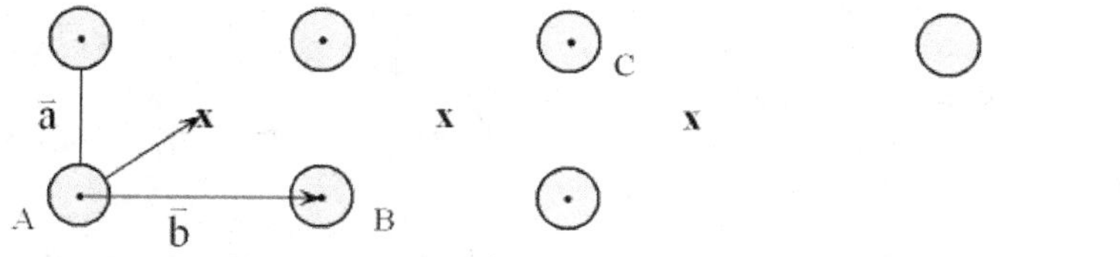


Figure 1.2.10: (a) Rectangular crystal lattice with (b) a two atom basis (see Example 1.2.4.2).

### 1.2.5 Primitive Cell

The primitive cell is the minimum volume (or area in two dimensions and length in one dimension) cell composed of primitive lattice vectors. It is a type of cell or a unit cell. A cell can fill all space by the repetition of itself by suitable crystal symmetry operations. An example of a unit cell is shown in Figure 1.2.11.

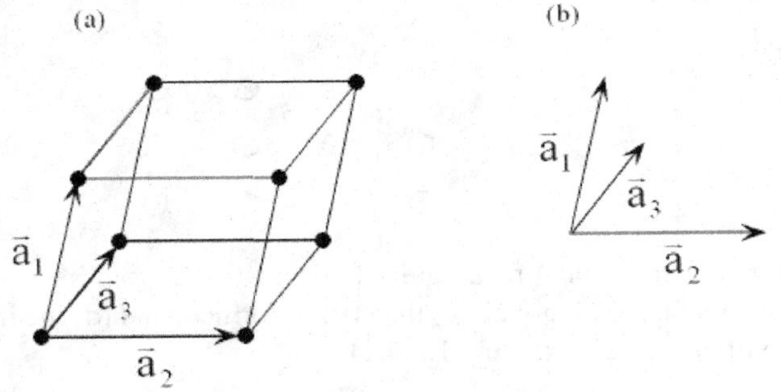


Figure 1.2.11: (a) Primitive unit cell with (b) primitive lattice vectors.

There are many ways of choosing the primitive axes and primitive cell of a given lattice. While the number of atoms in a primitive cell or primitive basis can be more than one, it is always the same for a given crystal structure. The areas or volumes defined by primitive axes are equal because they all define the unit cell from which the crystal is created. As shown in Figure 1.2.12(a), in two dimensions, the area enclosed by the primitive axes is

$$A = |\vec{a}_1 \times \vec{a}_2| = a_1 a_2 \sin\theta. \tag{1.2.4}$$

Note that for a different set of primitive axes, the magnitudes of $\vec{a}_1$ and $\vec{a}_2$ as well as the angle $\theta$ between them may change to keep the primitive area constant. In three dimensions, the volume of a parallelepiped (see Figure 1.2.12(b)) is

$$V = |(\vec{a}_1 \times \vec{a}_2) \cdot \vec{a}_3|. \tag{1.2.5}$$

For numerical calculation purposes, the MATLAB appendix describes ways to carry out the '×' and '·' operations in that environment.

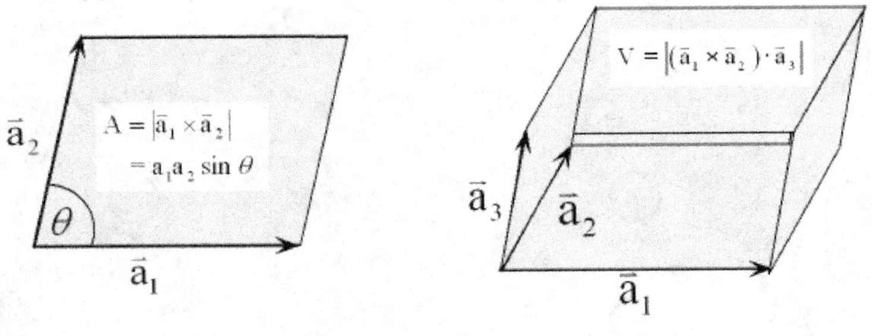


Figure 1.2.12: (a) Area of a two-dimensional unit cell and (b) the volume of a three-dimensional unit cell.

Let's do an example of different primitive axes in a two dimensional lattice with a single atom basis as shown in Figure 1.2.13.

**Example 1.2.5.1**
Referring to Figure 1.2.13, in (a) we see that we can make translation vectors $\vec{T}_1 = \vec{a} + \vec{b}$, $\vec{T}_2 = \vec{a} + 2\vec{b}$, and the pair of vectors $\vec{a}$, $\vec{b}$ are primitive lattice vectors. The shaded area is the primitive cell. In (b) $\vec{T}_1 = \vec{a}' + \vec{b}'$, $\vec{T}_2 = \vec{a}' + 2\vec{b}'$, and $\vec{T}_3 = \vec{a}' - \vec{b}'$. The lattice vectors $\vec{a}'$, $\vec{b}'$ may be primitive if their encompassed shaded area shown is equal to that of (a). In (c) the translation vector $\vec{T}_2 = \vec{a}'' + \vec{b}''$, but $\vec{T}_1 = \vec{a}'' + \frac{1}{2}\vec{b}''$, thus since $\vec{T}_1$ does not obey Equation 1.2.2, then the lattice vectors $\vec{a}''$, $\vec{b}''$, are not primitive.

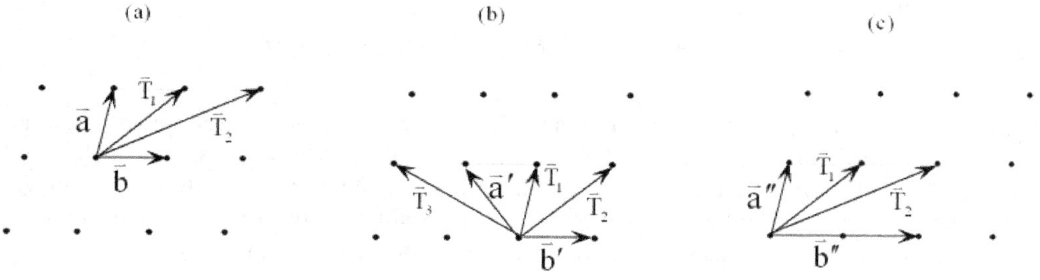


Figure 1.2.13: (a) Primitive cell with translation vectors that can be written in terms of the primitive axes. The primitive axes make up the primitive cell shaded area. (b) Here the translation vectors can be expressed in terms of the axes $\vec{a}'$ and $\vec{b}'$, and the shaded area may be a primitive cell if its area is equal to that shaded in (a), in which case the axes may also be primitive. (c) Here the axes $\vec{a}''$, $\vec{b}''$ are not primitive because the translation vectors cannot be expressed as in Equation 1.2.2 (see Example 1.2.5.1).

There is always one lattice point per primitive cell. To illustrate this, let's do the following example.

**Example 1.2.5.2**
In Figure 1.2.14, suppose $\vec{a}$ and $\vec{b}$ in (a) are the primitive vectors associated with the starred primitive cell shown in (b). Assuming a one atom basis then each point in the cell contains an atom. Let's count how many lattice points there are in this unit cell. Notice the cell is surrounded by 4 other unit cells. This means that each of the 4 lattice points is shared with a total of 4 unit cells. Therefore, we have $(4\ lattice\ points)/4\ cells = total\ number\ of\ lattice\ points\ per\ cell = 1\ lattice\ point\ per\ cell.$ This, of course, is not a general proof so that, formally, one uses the Wigner-Seitz method (discussed below) to do it.

Every primitive cell has one lattice point. The reason the Wigner-Seitz method for finding unit cells is important is that it clearly identifies the primitive cell without having to discuss the sharing of the atoms.

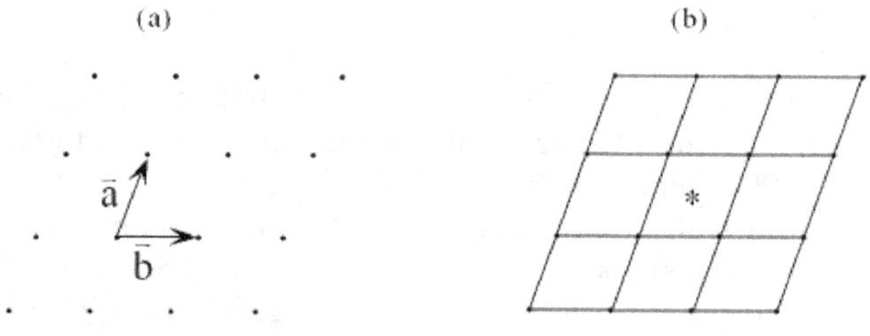


Figure 1.2.14: (a) A two-dimensional crystal lattice with one atom basis, and (b) the starred cell whose lattice points per cell we count in Example 1.2.5.2.

### 1.2.6 The Wigner-Seitz Cell

The Wigner-Seitz method is a very accurate way of obtaining a primitive cell (Wigner-Seitz cell). The method is as follows:

1. draw lines to connect a given lattice point to all nearby lattice points;
2. at the midpoint and normal to these lines, draw new lines (two-dimensions) or planes (three-dimensions);
3. the smallest area (two-dimensions) or volume (three-dimensions) enclosed this way is the Wigner-Seitz primitive cell.

As an example, let's find the Wigner-Seitz cell for a two-dimensional rectangular lattice.

**Example 1.2.6.1**
In Figure 1.2.15, the dashed lines in (a) are drawn from the circled atom to every other atom in the lattice. We then draw perpendicular solid bisectors lines as shown. The smallest area enclosing the centered atom works out to be a rectangle as well. This is the Wigner-Seitz cell for this lattice. It contains one lattice point. Further, this lattice has a one-atom basis but the same results apply to a basis with more than one atom as shown in (b), for a two-atom basis, for instance.

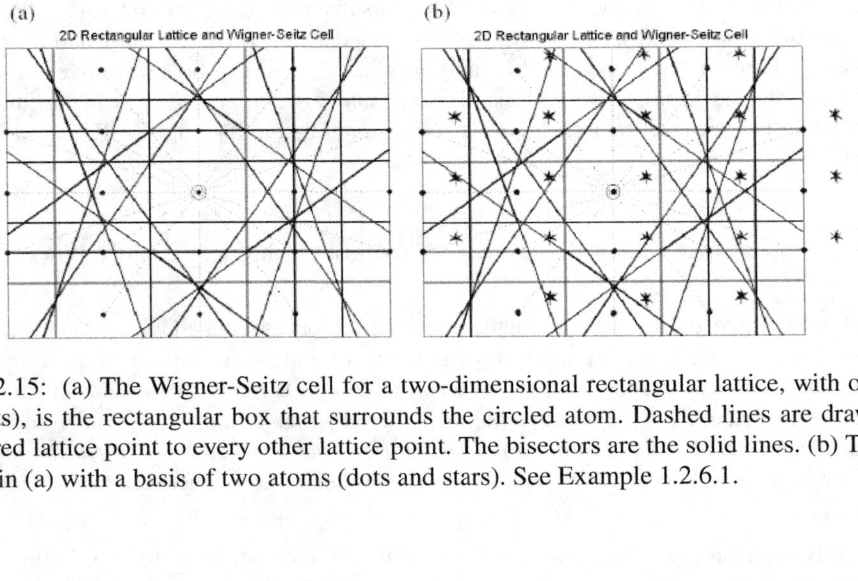


Figure 1.2.15: (a) The Wigner-Seitz cell for a two-dimensional rectangular lattice, with one atom basis (dots), is the rectangular box that surrounds the circled atom. Dashed lines are drawn from the centered lattice point to every other lattice point. The bisectors are the solid lines. (b) The same lattice as in (a) with a basis of two atoms (dots and stars). See Example 1.2.6.1.

## 1.3 Basic Lattice Types

Crystal lattices can be carried or mapped into themselves by lattice symmetry operations. A common operation is that of rotation about a lattice point. Lattices can be found to have one (360°), two (180°), three (120°), four (90°), and six (60°) fold rotations corresponding to rotations by $2m\pi/n$, where $n = 1$; 2, 3, 4, and 6 with m an integral multiple. The rotation axes are denoted by the index $n$. Pentagon ($n = 5$) or heptagon ($n = 7$) lattices have not been found. There is, however, a class of structures that contain some pentagons with hexagons mixed in; they are carbon-based structures known as Buckminster Fullerines with soccer ball shapes composed of 20 hexagons and 12 pentagons. A carbon atom is at the vertex of each polygon. A single molecule can possibly have $x$-fold rotations ($x = 1,2,\ldots$), but an infinite crystal of these cannot. In the following example, we show why a crystal of pentagons is not possible.

**Example 1.3.0.1**

Referring to Figure 1.3.16, in (a) we see $\theta = \frac{2\pi}{5} = 72°$, and $2\phi + \theta = \pi$ so that $\phi = \frac{(\pi-\theta)}{2} = 3\pi/10 = 54°$. Now, $\alpha = 2\phi = 3\pi/5 = 108°$, and no matter how many rotations by amount $\alpha$ we make about point A the result does not equal 360°. In fact $360/108 = 3.33$, which is not an integer. Thus, in general, for a polygon lattice with internal angle $\alpha$, one needs $2\pi/\alpha = n$, where $n$ is an integer, to be able to get an infinite crystal with no empty spaces. In (b), there is no waste of any space and an infinite lattice of hexagons can quite well ensue.

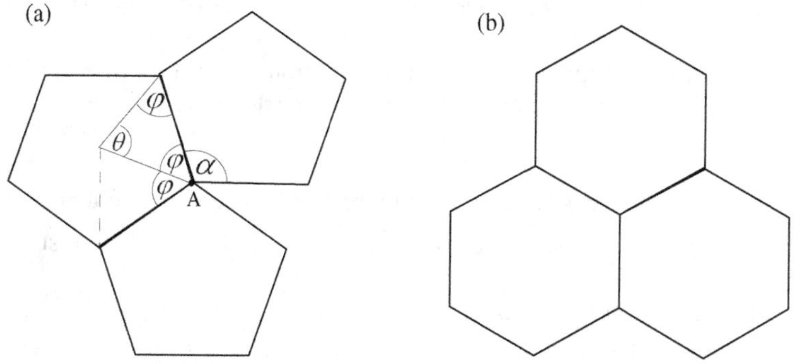


Figure 1.3.16: (a) A lattice of pentagons is impossible because $2\pi/\alpha$ is not an integer. (b) A lattice of hexagons has no empty spaces. See Example 1.3.0.1.

A *lattice point group* is the collection of symmetry operations which, applied to a lattice point, carry the crystal into itself. We can have:

a) possible rotations ($2\pi/n$ where $n = 1, 2, 3, 6$) about an axis and labeled $C_n$ (C for cyclic) rotations (e.g., Figure 1.3.17(a));
b) mirror reflections about a plane through a lattice point. Here the axis perpendicular to the mirror plane changes sign (i.e., $z \to -z$ if the mirror plane is perpendicular to the $z$-axis). The operation is labeled $m$ or $\sigma$ operations (e.g., Figure 1.3.17(b));
c) inversion, composed of rotations by $\pi$ followed by reflection in a plane normal to the rotation axis (here the rotation changes the sign of two axes, say $x \to -x$, $y \to -y$, if rotating about $z$, then the mirror plane perpendicular to $z$ causes $z \to -z$), with the net effect that $\vec{r} \to -\vec{r}$. The label used is $i$ for inversion or $s_4$ for the combination of operations (e.g., Figure 1.3.17(c)).

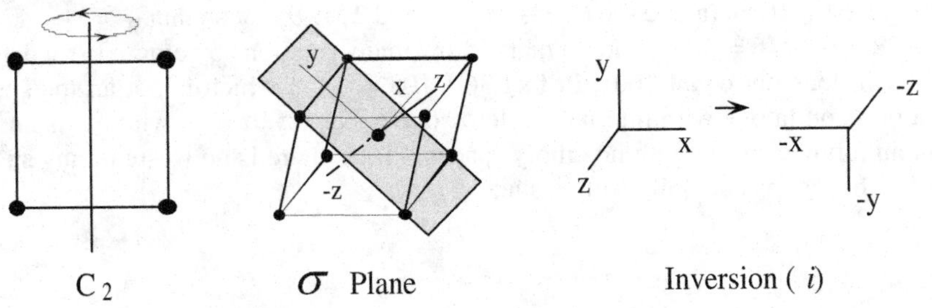


Figure 1.3.17: (a) Example of operation $C_2$ or 180° rotation about the symmetry axis shown. (b) Example of a mirror reflection about a plane. (c) Example of the inversion operation.

In two dimensions, the lattice types are listed in Figure 1.3.18 where we see that

1. in the oblique lattice, there is no restriction on the angle $\phi$ and $\vec{a} \neq \vec{b}$, the rest are special cases of this type;
2. the square lattice has $\phi = 90°$, and $|\vec{a}| = |\vec{b}|$;
3. the rectangular lattice has $\phi = 90°$, and $|\vec{a}| \neq |\vec{b}|$;
4. the hexagonal lattice has $\phi = 120°$ and $|\vec{a}| = |\vec{b}|$;
5. the centered rectangular also has $\phi = 90°$, and $|\vec{a}| \neq |\vec{b}|$ with an added lattice point at the rectangle's center.

These are also called Bravais lattices named after the French physicist Auguste Bravais (1811-1863).

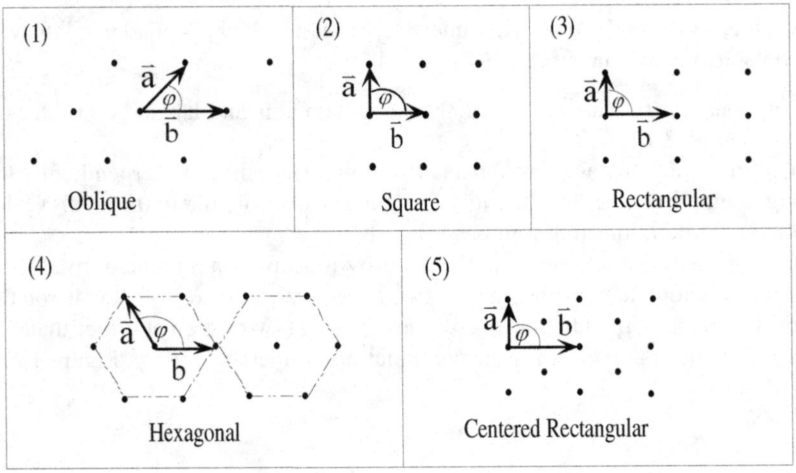


Figure 1.3.18: The two-dimensional Bravais lattices.

There are fourteen three dimensional Bravais lattice types, the triclinic is the general one and the remaining thirteen are special cases of this type as shown in example cases of Figure 1.3.19. Notice that the crystal axes $\vec{a}$, $\vec{b}$, and $\vec{c}$ are in general not orthogonal. Their associated angles with each other, $\alpha$, $\beta$, and $\gamma$ are shown in (a). The crystal lattices are:

(b) one triclinic with $\vec{a} \neq \vec{b} \neq \vec{c}$, and $\alpha \neq \beta \neq \gamma$;

(c-d) two monoclinic (simple and base-centered) with $\vec{a} \neq \vec{b} \neq \vec{c}$, and $\alpha = \gamma \neq \beta$;

(e-h) four orthorhombic (simple, body-centered, base-centered, and face-centered) with $\vec{a} \neq \vec{b} \neq \vec{c}$, and $\alpha = \beta = \gamma = 90°$;

(i-j) two tetragonal (simple and body-centered) with $\vec{a} = \vec{b} \neq \vec{c}$ and $\alpha = \beta = \gamma = 90°$;

(k) one trigonal with $\vec{a} = \vec{b} = \vec{c}$ and $\alpha = \beta = \gamma < 120° \neq 90°$;

(l) one hexagonal with $\vec{a} = \vec{b} \neq \vec{c}$ and $\alpha = \beta = 90°$, $\gamma = 120°$;

(m-o) and three cubic (simple, body-centered, and face-centered) with $\vec{a} = \vec{b} = \vec{c}$ and $\alpha = \beta = \gamma = 90°$.

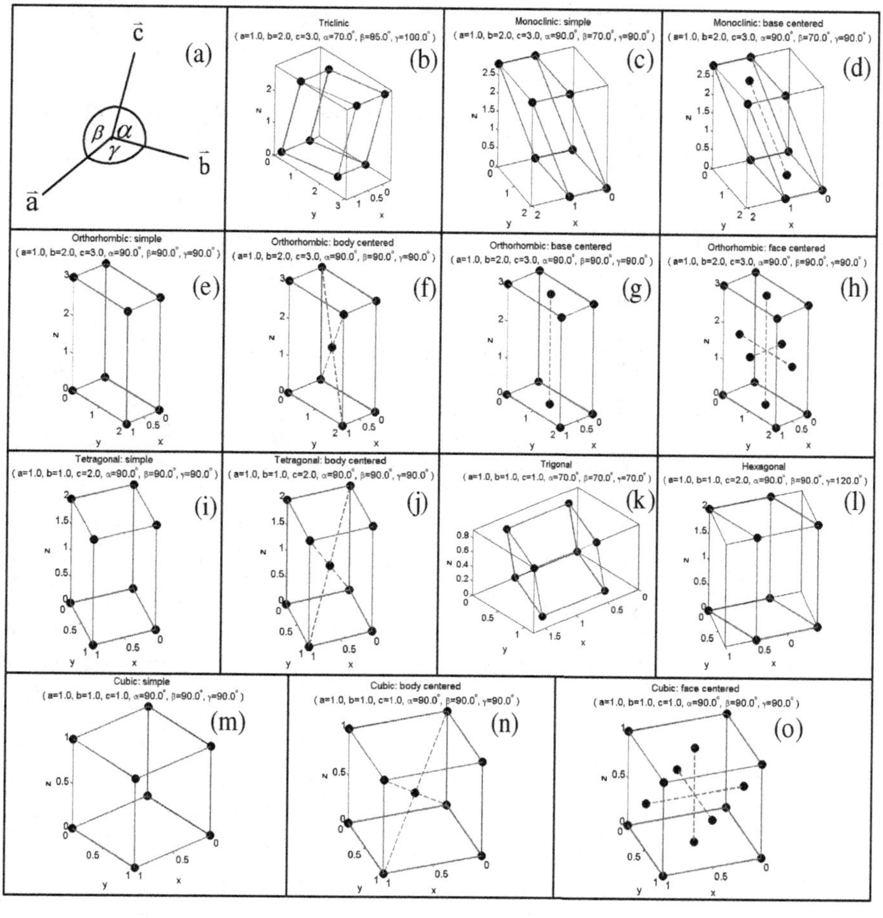


Figure 1.3.19: Examples of the three-dimensional Bravais lattice types.

### 1.3.1 Crystal to Cartesian Coordinates

We have learned that the location of atoms in crystals is based on the crystal lattice vectors, which are not necessarily orthogonal and which do not coincide with the x-y-z Cartesian coordinate system. The examples of three-dimensional lattices of Figure 1.3.19 were made possible by the transformation from crystal axes to Cartesian axes. We next demonstrate the process of locating the $x-y-z$ Cartesian atomic coordinates if we have their positions in terms of the crystal axes (Figure 1.3.19(a)). Consider the $x-y-z$ orthogonal coordinate system, and let's pick the vector $\vec{a}$ to be along x so that

$$\vec{a} = a\hat{x}; \tag{1.3.6}$$

additionally, let $\vec{b}$ lie on the x-y plane. Both situations are shown in Figure 1.3.20.

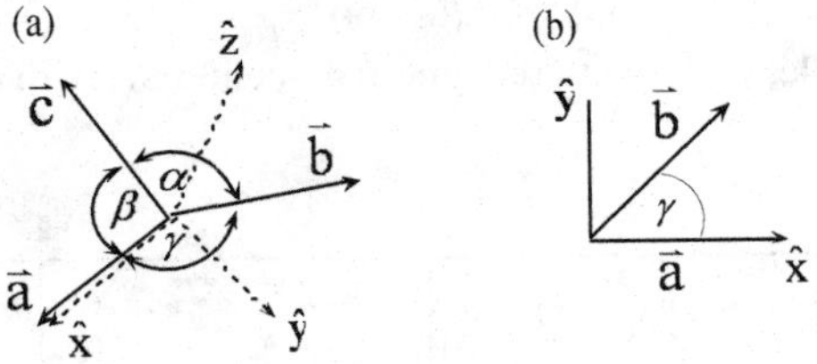


Figure 1.3.20: (a) Superimposed on the Cartesian coordinate systems, the general $\vec{a}$ is chosen along the $\hat{x}$ direction and $\vec{b}$ lies on the plane as in (b).

We can write $\vec{b}$ in the form

$$\vec{b} = b_1\hat{x} + b_2\hat{y} = b\cos\gamma\hat{x} + b\sin\gamma\hat{y}. \tag{1.3.7}$$

The choice left is for $\vec{c}$ to lie in a general direction; that is,

$$\vec{c} = c_1\hat{x} + c_2\hat{y} + c_3\hat{z} = c\cos\beta\hat{x} + c_2\hat{y} + c_3\hat{z}. \tag{1.3.8}$$

So that $\vec{a}\cdot\vec{c} = ac\cos\beta = a\hat{x}\cdot(c_1\hat{x} + c_2\hat{y} + c_3\hat{z}) = ac_1$, or

$$c_1 = c\cos\beta. \tag{1.3.9}$$

Similarly, we obtain $\vec{b}\cdot\vec{c} = bc\cos\alpha = (b\cos\gamma\hat{x} + b\sin\gamma\hat{y})\cdot(c_1\hat{x} + c_2\hat{y} + c_3\hat{z}) = bc_1\cos\gamma + bc_2\sin\gamma$, which, using Equation 1.3.9 gives

$$\vec{b}\cdot\vec{c} = bc\cos\alpha = bc\cos\beta\cos\gamma + bc_2\sin\gamma, \tag{1.3.10}$$

or

$$c_2 = \frac{c(\cos\alpha - \cos\beta\cos\gamma)}{\sin\gamma}. \tag{1.3.11}$$

Finally,

$$|\vec{c}|^2 = c^2 = c_1^2 + c_2^2 + c_3^2, \tag{1.3.12a}$$

and using Equations (1.3.9) and (1.3.11), we have

$$c_3 = \sqrt{c^2 - c_2^2 - c_3^2} = c\sqrt{1 - \cos^2\beta - (\frac{\cos\alpha - \cos\beta\cos\gamma}{\cos\gamma})}. \tag{1.3.12b}$$

Notice that, from Equations (1.3.6) to (1.3.8), we can write the following

$$(\ \vec{a}\ \ \vec{b}\ \ \vec{c}\ ) = (\ \hat{x}\ \ \hat{y}\ \ \hat{z}\ )\begin{pmatrix} a & b\cos\gamma & c_1 \\ 0 & b\sin\gamma & c_2 \\ 0 & 0 & c_3 \end{pmatrix} = (\ \hat{x}\ \ \hat{y}\ \ \hat{z}\ )C, \tag{1.3.13a}$$

where

$$C \equiv \begin{pmatrix} a & b\cos\gamma & c_1 \\ 0 & b\sin\gamma & c_2 \\ 0 & 0 & c_3 \end{pmatrix}, \tag{1.3.13b}$$

with $c_1$, $c_2$, and $c_3$ given by Equations (1.3.9) to (1.3.12). Next, consider a vector $\vec{t}$ which gives the position of an atom in crystal coordinates, such as

$$\vec{t} = u\vec{a} + v\vec{b} + w\vec{c}, \tag{1.3.14}$$

where $u$, $v$, and $w$ are integers. The question arises, what are the $x$, $y$, and $z$ Cartesian coordinates of this atom relative to the origin? The answer is found in rewriting $\vec{t}$ as

$$\vec{t} = x\hat{x} + y\hat{y} + z\hat{z} = u\vec{a} + v\vec{b} + w\vec{c}. \tag{1.3.15}$$

This, with the help of expression (1.3.13a) for the row vector $(\ \vec{a}\ \ \vec{b}\ \ \vec{c}\ )$, becomes

$$\vec{t} = (\ \hat{x}\ \ \hat{y}\ \ \hat{z}\ )\begin{pmatrix} x \\ y \\ z \end{pmatrix} = (\ \hat{x}\ \ \hat{y}\ \ \hat{z}\ )C\begin{pmatrix} u \\ v \\ w \end{pmatrix}, \tag{1.3.16}$$

or

$$\begin{pmatrix} x \\ y \\ z \end{pmatrix} = C\begin{pmatrix} u \\ v \\ w \end{pmatrix}, \tag{1.3.17}$$

which are the sought Cartesian coordinate magnitudes in terms of the crystal axes magnitudes. Here the matrix $C$ is known as the conversion matrix. Let's do an example.

**Example 1.3.1.1**

Consider an orthorhombic system as shown in Figure 1.3.21 in which $\alpha = 90° = \beta = \gamma$; in addition, given the vector magnitudes, $a = 1\text{Å}$, $b = 1.5\text{Å}$, and $c = 2\text{Å}$, find an atom's $x$, $y$, and $z$ coordinates if its position is given by the crystal vector $\vec{t} = 3\vec{a} + 4\vec{b} - 3\vec{c}$.

**Solution**

In this example, the conversion matrix $C$ works out to be diagonal with elements $a = |\vec{a}|$, $b = |\vec{a}|$, and $c = |\vec{c}|$, where $a = 1$, $b = 3/2$, and $c = 2$. From the given $\vec{t}$, $u = 3$, $v = 4$, and $w = -3$ so that, from Equations (1.3.16) and (1.3.17), we find $\vec{t} = ua\hat{x} + vb\hat{y} + wc\hat{z} = 3\hat{x} + 4(3/2)\hat{y} - 3(2\hat{z}) = 3\hat{x} + 6\hat{y} - 6\hat{z}$.

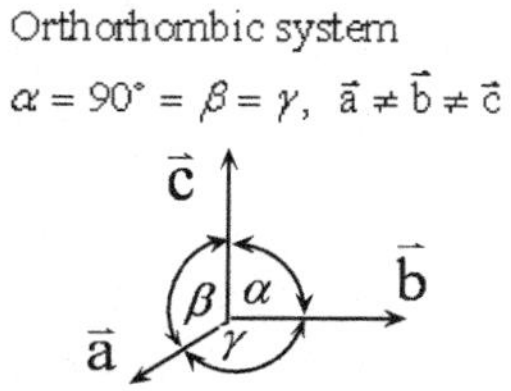


Figure 1.3.21: Orthogonal vectors $\vec{a}$, $\vec{b}$, and $\vec{c}$ in an orthorhombic system.

It is helpful to create MATLAB code to make the conversion from crystal to Cartesian axes tasks easier to perform. In the following example, useful code is listed for such purpose.

**Example 1.3.1.2**
Below follows sample code for the purpose of obtaining the Cartesian axes coordinates from crystal axes information and for which the input values have not yet been specified.

```
%Example code to carry out the conversion process from crystal
%axes to Cartesian axes.
clear                       %It is a good idea to clear the memory
a=                          %axes in angstroms
b=
c=
alpha=                      %angles
beta=
gamma=
%Define c1, c2, and c3 to compose the C matrix
c1= c*cos(beta)
c2= c*(cos(alpha)-cos(gamma)*cos(beta))/(sin(gamma))
c3= + sqrt((c^2 - (c1)^2 - (c2)^2))
C =[[a b*cos(gamma) c1] [0 b*sin(gamma) c2] [0 0 c3 ]]   %C matrix
%The coefficients of the atoms in the crystal representation
u=
v=
w=
%Print the vector in the Cartesian representation
C*[u;v;w]
```

If we needed to plot the atoms' locations in three-dimensions, we could create a two-dimensional matrix, say, `loc` whose rows contain the atoms' crystal's u's, v's, and w's and whose columns correspond to different atoms. Let the result of the conversion be the matrix `p=C*loc`, and let `x=p(1,:)`, `y=p(2,:)`, and `z=p(3,:)` be the coordinates for, say `n` atoms, where the colon `':'` refers to all the columns in the array; we then place dots at their positions using the command `plot3(x,y,z,'k.')`. If desired, lines can be added from each atom to its neighbors by using a loop that runs over the atoms' positions, the following loop, for example, will connect the first atom to its neighbors with straight lines.

```
i=1;  % current atom
n=3;  % neighbors
for j=i+1:i+n
  line([p(1,i),p(1,j)],[p(2,i),p(2,j)],[p(3,i),p(3,j)]);
end
```

## 1.4 Properties of the Cubics

The cubic system is the simplest three-dimensional lattice, and it is worth learning some details about them. In general, the conventional cell of a lattice may not necessarily be the same as the primitive cell. For the cubics, the conventional cell of each lattice is shown in Figures 1.3.19(m-o). As we see below, for the simple cubic (SC) the conventional cell is the same as the primitive cell, but that is neither the case for the body-centered cubic (BCC) nor the face-centered cubic (FCC). The lattice constant defines the size of the cubic conventional cell and here we refer to it as just *a*.

### 1.4.1 The Simple Cubic

In this lattice, the atoms' primitive basis vectors are expressed in terms of the lattice constant $a$; that is, $\vec{a}_1 = a\hat{i}$, $\vec{a}_2 = a\hat{j}$, and $\vec{a}_3 = a\hat{k}$. The primitive cell has volume $V = |(\vec{a}_1 \times \vec{a}_2) \cdot \vec{a}_3)| = a^3$. This is more clearly shown in Figure 1.4.22(a).

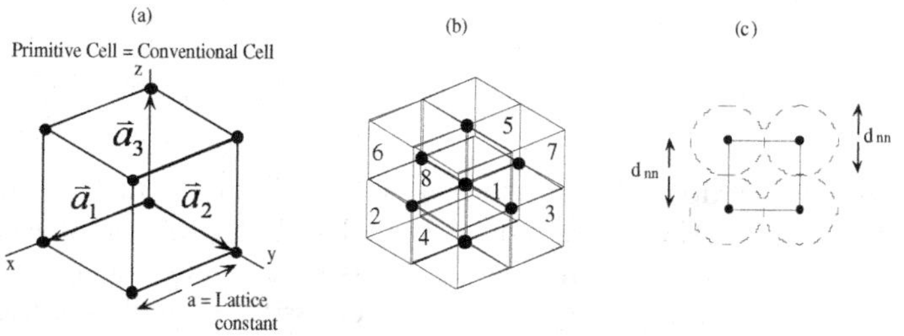


Figure 1.4.22: (a) The conventional cell for the simple cubic is the same as its primitive cell. The lattice constant is the length of a side. (b) The central atom is shared by 8 cells (numbered 1-8) yielding one atom per cell in the simple cubic. Each atom has six nearest neighbors as seen here for the central atom. (c) The packing fraction's sphere diameter associated with an atom equals the nearest neighbor distance $d_{nn}$ as depicted here.

As shown in Figure 1.4.22(b), the simple cubic (SC) has 1 atom at each of its 8 corners, but each corner is shared by 8 cubes (numbered 1-8, 4 below and 4 above the central atom) to yield one atom per cell. Each atom can be thought to be immediately surrounded by 6 nearest neighbors ($\pm x$, $\pm y$, and $\pm z$). The packing fraction is defined as the ratio of a sphere volume to that of the cell volume. The sphere volume is taken as the volume of an atom whose radius is equal to one half the nearest neighbor distance, $d_{nn}$, as shown for the cube's bottom plane in Figure 1.4.22(c). For the SC system, $d_{nn} = a$, so the sphere volume is $\frac{4}{3}\pi(a/2)^3$, and the cell volume obtained above is $a^3$ to get a packing fraction of $\frac{4}{3}\pi(a/2)^3/a^3 = \pi/6 = 0.5236$. It is useful to use the notation $(u, v, w)a$ to indicate the atomic positions, and sometimes even the lattice constant is omitted. For example, the six SC nearest neighbors to, say the atom (000) at the origin, are (100), (010), (001), $(\bar{1}00)$, $(0\bar{1}0)$, and $(00\bar{1})$, in units of $a$ of course. By making linear combinations of these (including self-combinations) but keeping only those that lead to new but distinct positions, and whose distance is the smallest possible, we can get the 2nd neighbors of the (000) atom. Doing this, we get the positions (110), (101), (011), $(\bar{1}\bar{1}0)$, $(\bar{1}0\bar{1})$, $(0\bar{1}\bar{1})$, $(1\bar{1}0)$, $(10\bar{1})$, $(\bar{1}10)$, $(01\bar{1})$, $(\bar{1}01)$, and $(0\bar{1}1)$. Thus there are twelve 2nd neighbors. They are so recognized because they are equidistantly located at a distance of $d_{2nd} = \sqrt{1^2+1^2+0^2}a = \sqrt{2}a$ from the origin. Repeating this process, using all previously known distinct positions, and keeping those whose distance is the smallest possible we can get the 3rd neighbors, and so on. All the atoms in the crystal lattice can be identified this way. Finally, refer to Table 1.8.3 in Section 1.8 for systems that crystallize in this structure.

### 1.4.2 The Body-Centered Cubic

The body-centered cubic (BCC)'s conventional cell is shown in Figure 1.4.23(a), while (b) shows it contains two atoms per cell and each atom has eight nearest neighbors (seen here for the central atom). (c) Shows the BCC's primitive unit cell's rhombohedral shape based on its primitive vectors shown in (a).

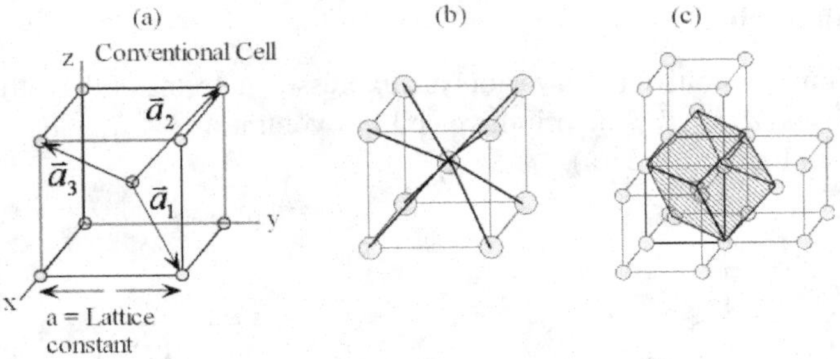


Figure 1.4.23: (a) The conventional cell of the body-centered cubic with the lattice constant the length of a side as in the SC. (b) The central atom is shared by 1 cell and corner atoms are shared by 8 cells each, yielding two atoms per cell for the BCC. Also each atom has eight nearest neighbors as seen here for the central atom. (c) The BCC primitive cell is the shaded rhombohedron.

The primitive lattice vectors are $\vec{a}_1 = \frac{1}{2}a(\hat{\imath}+\hat{\jmath}-k)$, $\vec{a}_2 = \frac{1}{2}a(-\hat{\imath}+\hat{\jmath}+k)$, and $\vec{a}_3 = \frac{1}{2}a(\hat{\imath}-\hat{\jmath}+k)$ as measured from the body-centered atom position at (000). Its volume is $V = |(\vec{a}_1 \times \vec{a}_2)\cdot\vec{a}_3)| = a^3/2$. Notice that the inverse of this number is the same as the number of atoms (and lattice points) per conventional cell. The eight nearest neighbors to the central atom are located at positions $(11\bar{1})a/2$, $(1\bar{1}1)a/2$, $(\bar{1}11)a/2$, $(111)a/2$, $(1\bar{1}\bar{1})a/2$, $(\bar{1}1\bar{1})a/2$, $(\bar{1}\bar{1}1)a/2$, and $(\bar{1}\bar{1}\bar{1})a/2$ with the nearest neighbor distance of $d = a\sqrt{(\frac{1}{2})^2+(\frac{1}{2})^2+(\frac{1}{2})^2} = \frac{\sqrt{3}}{2}a$. The BCC's packing fraction is $\frac{4}{3}\pi(\frac{\sqrt{3}}{2}a/2)^3/(\frac{a^3}{2}) = \pi\sqrt{3}/8 = 0.6802$. It is possible to run a script specifically designed to visualize the conventional cell, in addition to the primitive cell, of the BCC. The code follows in the example below.

**Example 1.4.2.1**
In this example, when run, the script unit_cell_BCC.m requires the user to enter the number of atoms as input. It builds the BCC translation vectors, and makes linear combinations of them (some are repeated). Large dots are placed at the atom positions to show the structure, and the nearest neighbors are connected with straight lines to simulate bonds. The conventional cell and primitive cell are drawn separately based on the lattice vectors as indicated by the comments within. Figure 1.4.24 is the output produced after running the script with a default input of 52 for the number of atoms used (some repeat).

```
%copyright by J. E Hasbun and T. Datta
%unit_cell_BCC.m
%Here we plot the atomic positions for the BCC crystal
%and then plot its conventional and primitive unit cells.
clc
close all
clear all
a=1.0;   %lattice constant
%The BCC smallest possible translation vectors
a0=[0;0;0]; a1=a*[1/2;1/2;-1/2]; a2=a*[1/2;-1/2;1/2]; a3=a*[-1/2;1/2;1/2];
a4=a1+a2; a5=a1+a3; a6=a2+a3; a7=a4+a3;  %unit cell can be built now
                                          %with 8 corner positions
%and also make some new ones for structure purposes
a8=a1-a2; a9=a1-a3; a10=a2-a3; a11=a4-a3;
a12=2*a1; a13=2*a2; a14=2*a3;
%First eight in T are to construct the primitive cell
```

```
T=[a0,a1,a2,a3,a4,a5,a6,a7,-a1,-a2,-a3,a8,a9,a10,a11];
T=[T,-a4,-a5,-a6,-a7,-a8,-a9,-a10,-a11,a12,a13,a14,-a12,-a13,-a14];
%We should identify the cube positions too
b0=[0;0;0]; b1=a*[1;0;0]; b2=a*[0;1;0]; b3=a*[0;0;1];
b4=b1+b2; b5=b1+b3; b6=b2+b3; b7=b4+b3;  %To locate cube position atoms
%plus new ones
b8=b1-b2; b9=b1-b3; b10=b2-b3; b11=b4-b3;
b12=2*b1; b13=2*b2; b14=2*b3;
Tb=[b0,b1,b2,b3,b4,b5,b6,b7,-b1,-b2,-b3,b8,b9,b10,b11];
Tb=[Tb,-b4,-b5,-b6,-b7,-b8,-b9,-b10,-b11,b12,b13,b14,-b12,-b13,-b14];
%final T
T=[T,Tb];
%
M=length(T);              %number of possible smallest translation vectors
%N=52;                     %number of atoms desired some are repeated
N=input('Enter the number of atoms to plot (some are repeated) [52] -> ');
if(isempty(N)), N=52; end
%p= atomic positions in the diamond lattice
p0=a*[0;0;0];             %the atoms at the origin
p=zeros(3,N);             %define the dimensions of atomic position matrix
jc=0;
for i=0:N-1
  j=mod(i,M);                 %j can only go from 1 to M
  if(j==0); jc=jc+1; end      %counts number of times j goes through 0
  %jc allows for reusage of previously obtained position vectors to
  %which T can be added to obtain new position vectors, etc
  p(:,i+1)=p0+T(:,j+1);
  p0=p(:,jc);
end
rnn=norm(T(:,1)-T(:,2));        %bcc nn distance
x=p(1,:); y=p(2,:); z=p(3,:); %x, y, z coords of all atoms
%draw atoms
plot3(x,y,z,'ko','MarkerSize',6,'MarkerFaceColor','k')
hold on
%Draw bonds between near neighbors only
for i=1:N
  for j=1:N
    if(j ~= i)
      rd=norm(p(:,i)-p(:,j)); %calculate the distance between atoms
                              %but draw lines between nn only
      if(rd <= rnn)
        line([p(1,i),p(1,j)],[p(2,i),p(2,j)],[p(3,i),p(3,j)],...
          'Color','b','LineWidth',1);
      end
    end
  end
end
%----------
%The lines below work better here for the primitive cell
line([a0(1),a1(1)],[a0(2),a1(2)],[a0(3),a1(3)],'Color','b','LineWidth',3);
line([a0(1),a2(1)],[a0(2),a2(2)],[a0(3),a2(3)],'Color','b','LineWidth',3);
```

```
line([a0(1),a3(1)],[a0(2),a3(2)],[a0(3),a3(3)],'Color','b','LineWidth',3);
line([a1(1),a4(1)],[a1(2),a4(2)],[a1(3),a4(3)],'Color','b','LineWidth',3);
line([a1(1),a5(1)],[a1(2),a5(2)],[a1(3),a5(3)],'Color','b','LineWidth',3);
line([a2(1),a4(1)],[a2(2),a4(2)],[a2(3),a4(3)],'Color','b','LineWidth',3);
line([a2(1),a6(1)],[a2(2),a6(2)],[a2(3),a6(3)],'Color','b','LineWidth',3);
line([a3(1),a5(1)],[a3(2),a5(2)],[a3(3),a5(3)],'Color','b','LineWidth',3);
line([a3(1),a6(1)],[a3(2),a6(2)],[a3(3),a6(3)],'Color','b','LineWidth',3);
line([a4(1),a7(1)],[a4(2),a7(2)],[a4(3),a7(3)],'Color','b','LineWidth',3);
line([a5(1),a7(1)],[a5(2),a7(2)],[a5(3),a7(3)],'Color','b','LineWidth',3);
line([a6(1),a7(1)],[a6(2),a7(2)],[a6(3),a7(3)],'Color','b','LineWidth',3);
%----------
%To show the cube positions too - not so thick lines
rnnb=norm(Tb(:,1)-Tb(:,2));      %cube position nn distances
for i=1:8
  for j=1:8
    if(j ~= i)
      rd=norm(Tb(:,i)-Tb(:,j)); %calculate the distance between atoms
                                %but draw lines between nn only
      if(rd <= rnnb)
        line([Tb(1,i),Tb(1,j)],[Tb(2,i),Tb(2,j)],[Tb(3,i),Tb(3,j)],...
          'Color','k','LineStyle','-.','LineWidth',2);
      end
    end
  end
end
%Make sure the cube corner atoms appear too
plot3(Tb(1,1:8),Tb(2,1:8),Tb(3,1:8),'ko','MarkerSize',6,'MarkerFaceColor',
'k')
%----------
view(-156,12)         %Angle for viewing purposes
box on
axis equal
xlabel('X'), ylabel('Y'), zlabel('Z')
title('BCC, conventional cell (dashed black), primitive cell (thick blue)')
```

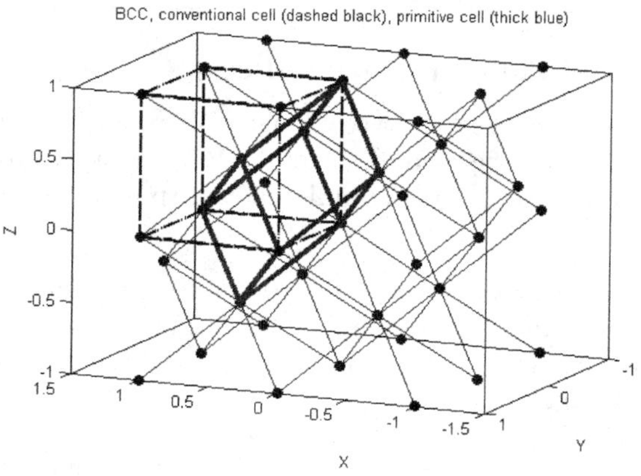


Figure 1.4.24: The output figure from the script unit_cell_BCC.m of Example 1.4.2.1 for the BCC structure with the default input number of atoms (not all show).

You may also refer to Table 1.8.3 of Section 1.8 for systems that crystallize in this structure.

### 1.4.3 The Face-Centered Cubic

The face-centered cubic (FCC)'s conventional cell of Figure 1.4.25(a) shows it contains four atoms per cell, while (b) shows each atom has twelve nearest neighbor (seen here for the central atom). (c) Shows the FCC's primitive unit cell's rhombohedral shape based on its primitive vectors shown in (a).

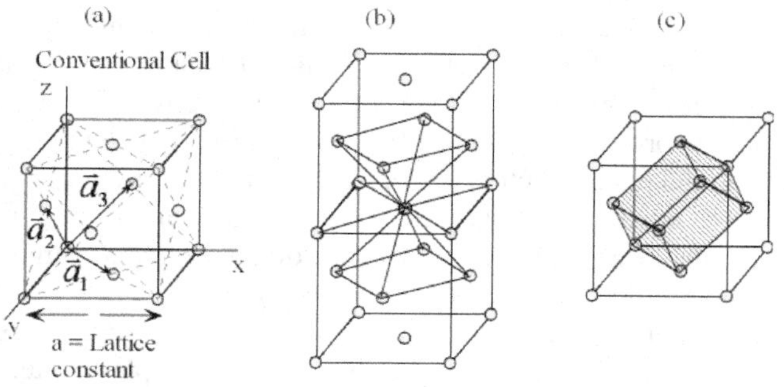


Figure 1.4.25: (a) The conventional cell of the face-centered cubic with the lattice constant the length of a side as in the SC. Each of the eight corner atoms is shared by 8 cells, each of the six face atoms is shared by 2 cells, yielding $1+3=4$ atoms per cell for the FCC. (b) Each atom has twelve nearest neighbors as seen here for the central atom. (c) The FCC primitive cell is the shaded rhombohedron.

The primitive lattice vectors are $\vec{a}_1 = \frac{1}{2}a(\hat{\imath}+\hat{\jmath})$, $\vec{a}_2 = \frac{1}{2}a(\hat{\jmath}+k)$, and $\vec{a}_3 = \frac{1}{2}a(\hat{\imath}+k)$ as measured from the origin atom at (000). Its volume (see exercises) works out to be $V = a^3/4$. The nearest neighbor distance is $d = \frac{a}{\sqrt{2}}$. Finally, the FCC's packing fraction is $\frac{\sqrt{2}}{6}\pi = 0.7405$. Again, you may also refer to Table 1.8.3 of Section 1.8 for systems that crystallize in this structure.

## 1.5 Indexing of Crystal Planes (Miller Indices)

The Miller indices refer to the indices of a vector that is perpendicular to a plane. The plane's intercepts with the crystal axes can be used to obtain the Miller indices, commonly written in the notation $(hkl)$ (often no spaces are used). The Miller indices specify the orientation of a plane by the following rules:

a) Find the intercepts of the plane with the axes in terms of the lattice constants $\vec{a}_1$, $\vec{a}_2$, and $\vec{a}_3$. The axes may be those of a primitive or non-primitive cell. The integer coefficients of each intercept, $u$, $v$, and $w$ are written with braces in the form $\{u,v,w\}$ to indicate lattice point plane intercepts $u\vec{a}_1$, $v\vec{a}_2$, and $w\vec{a}_3$, respectively.
b) Take the reciprocals of the intercept coefficients ($\{u,v,w\}$) and reduce them to three integers having the same ratio, usually the smallest three integers. What results, expressed in parenthesis form, $(hkl)$ are the crystallographic plane indices or Miller indices. That is,

$$(hkl) = \texttt{(smallest number envenly divisible by u,v,w)} \left(\frac{1}{u}, \frac{1}{v}, \frac{1}{w}\right). \tag{1.5.18}$$

The Miller indices are useful because parallel planes will have the same set of indices. The set of planes so described is referred to as a crystallographic plane. Let's follow with a simple example.

**Example 1.5.0.1**
Consider a crystal plane with intercepts $\{1,4,3\}$; obtain the plane's Miller indices.

**Solution**
The smallest number evenly divisible by 4 and 3 is 12, we have $12(1/1, 1/4, 1/3) = (12\,3\,4)$.

Associated with a crystal is the crystallographic direction. This is a vector between two lattice points in which the direction is indicated in the bracketed form $[uvw]$. The integers $u$, $v$, and $w$ (without common factors) are the indices of the crystallographic direction and specify an infinite set of parallel vectors. If we consider the origin and a lattice point whose position is given by (for example) the sum of the position vectors of Example 1.5.0.1 or $\vec{t} = u\vec{a}_1 + v\vec{a}_2 + w\vec{a}_3$, since the $u$, $v$, and $w$ given do not contain common integer factors (except 1), the crystal direction is $[1\,4\,3]$. In crystals for which $\vec{a}_1$, $\vec{a}_2$, and $\vec{a}_3$ are all equal, as in the case of cubic crystals, and if we consider a lattice point at location $\{u,v,w\}$ from the origin, with $u = v = w$, the direction to that lattice point $[uvw]$ is perpendicular to the plane with Miller indices $(hkl)$ where $u = h$, $v = k$, and $w = l$. Negative directions have a bar over the index; that is, if the direction from the origin is in the direction of the vector $\vec{t} = -u\vec{a}_1 + v\vec{a}_2 + w\vec{a}_3$, the direction is $[\bar{u}vw]$. Further, if we have plane intercepts of $\{-1,4,3\}$, the plane's Miller indices are $(\bar{12}\,3\,4)$. A two-dimensional plane with intercepts $\{2,1,\infty\}$ has Miller indices of $(1\,2\,0)$, indicating that it's an infinite plane that does not touch the third dimension. The plane with Miller indices $(1\,0\,0)$ has plane intercepts $\{1,\infty,\infty\}$ because it is an infinite plane that only touches the first dimension. Certain crystal directions are special and are in common use. The direction from the origin to a lattice point at $\vec{t} = \vec{a}_1$ is denoted as $[1\,0\,0]$ which represents a vector perpendicular to a plane with intercepts $\{1,\infty,\infty\}$ and Miller indices $(1\,0\,0)$. In a similar way, the direction $[1\,1\,0]$ is perpendicular to a plane with intercepts $\{1,1,\infty\}$ and whose Miller indices are $(1\,1\,0)$. Some of these and other examples are illustrated in Figure 1.5.26.

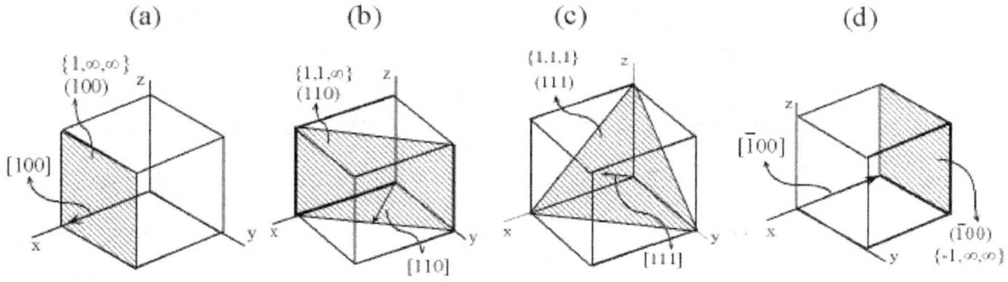


Figure 1.5.26: (a)-(d) Examples of special crystal directions. The plane intercepts are given in curly brackets ({}), the Miller indices in parenthesis, and the directions in square brackets ([]).

One of the interesting aspects of the Miller indices can be seen by looking at Figure 1.5.27. Consider the plane shown in (a) with three intersecting points $p_1$, $p_2$, and $p_3$ whose associated vectors are $\vec{p}_1 =$, $\vec{p}_2$, and $\vec{p}_3$. We next construct vectors along the edges of the plane given by $\vec{v}_1 = \vec{p}_1 - \vec{p}_3$, $\vec{v}_2 = \vec{p}_2 - \vec{p}_1$, and $\vec{v}_3 = \vec{p}_3 - \vec{p}_2$. If we cross any two of these edge vectors in such a way as to obtain a perpendicular vector that points away from the origin (right-hand rule [RHR]), its coefficients (when reduced to the smallest integers with the same ratio) correspond to the plane's Miller indices. The following example illustrates this concept.

**Example 1.5.0.2**
Let's obtain the vector that lies perpendicular to the plane shown in Figure 1.5.27(b) with intersecting points $p_1 = (6,0,0)$, $p_2 = (0,5,0)$, and $p_3 = (0,0,4)$. The edge vectors discussed above become $\vec{v}_1 = 6\hat{\imath} - 4\hat{k}$, $\vec{v}_2 = 5\hat{\jmath} - 6\hat{\imath}$, and $\vec{v}_3 = 4\hat{k} - 5\hat{\jmath}$. Crossing any two of these a la RHR, say $\vec{v}_1 \times \vec{v}_2 = 20\hat{\imath} + 24\hat{\jmath} + 30\hat{k}$, gives a vector that lies perpendicular to the given plane. Its coefficients $(20\,24\,30)$, when reduced to the smallest integers with the same ratio, become $(10\,12\,15) = (hkl)$, which are the Miller indices corresponding to the plane with intercepts $\{6,5,4\}$. (b) also shows the direction vector $[6\,5\,4]$ for illustration purposes.

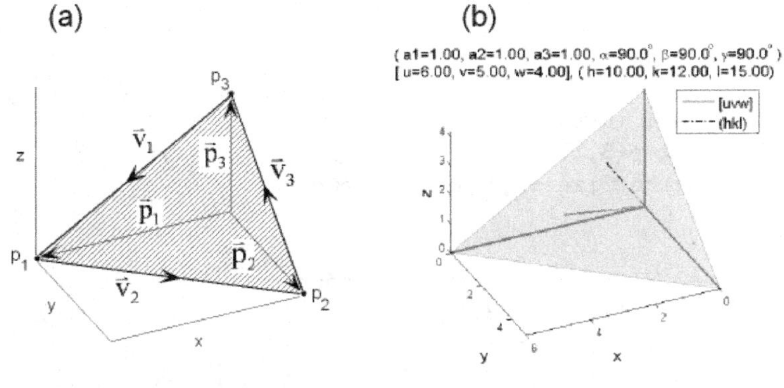


Figure 1.5.27: (a) Shows the plane intercepts and the edge vectors which are crossed to obtain the plane's perpendicular. (b) Illustrates that the coefficients of the perpendicular vector, when reduced to the smallest integers with the same ratio, correspond to the Miller indices of the plane with intercepts $\{6,5,4\}$. The direction vector $[6\,5\,4]$ is also shown (see Example 1.5.0.2). Code plane_indices.m was used in the calculation of (b).

The MATLAB code, `plane_indices.m`, used to create Figure 1.5.27(b) is listed below for reproducibility purposes. The run was made with the default input parameters.

```
%copyright by J. E Hasbun and T. Datta
%plane_indices.m
%The script's purpose is to visualize a plane whose intercepts
%are provided and shows their relatiosnship to the Miller indices
function plane_indices
clear
a1=1.0;               %axes in angstroms
a2=1.0;
a3=1.0;
alph=90;              %angles in degrees
bet =90;
gamm=90;
alpha=(pi/180)*alph; %angles in radians
beta =(pi/180)*bet;
gamma=(pi/180)*gamm;
%Define c1, c2, and c3 to compose the C matrix
c1= a3*cos(beta);
c2= a3*(cos(alpha)-cos(gamma)*cos(beta))/(sin(gamma));
c3= + sqrt((a3^2 - (c1)^2 - (c2)^2));
%plane's intercepts are u*a1, v*a2, w*a3; u, v, w are the coefficients.
%u=6; v=5; w=4; %examples
cI=input('Enter [u,v,w] as a row vector [6,5,4] -> ');
if isempty(cI), cI=[6,5,4]; end
u=cI(1); v=cI(2); w=cI(3);
loc=[u 0 0
     0 v 0
     0 0 w];
%Conversion matrix
cM =[a1 a2*cos(gamma) c1
     0  a2*sin(gamma) c2
     0   0 c3 ];
%Transformed intercepts
p=cM*loc;
%x, y, z coords of all intercepts
x=p(1,:); y=p(2,:); z=p(3,:);
plot3(x,y,z,'ko','MarkerSize',2,'MarkerFaceColor','w')
view(155,32)
hold on
%The Axes a1, a2, a3 plotted in Cartesian coords
line([0,p(1,1)],[0,p(2,1)],[0,p(3,1)],'Color','b','LineWidth',2)
line([0,p(1,2)],[0,p(2,2)],[0,p(3,2)],'Color','b','LineWidth',2)
line([0,p(1,3)],[0,p(2,3)],[0,p(3,3)],'Color','b','LineWidth',2)
%Vector u*a1+v*a2+w*a3 in Cartesian coords (crystal direction [uvw] from 0)
pmax=max([norm(p(:,1)),norm(p(:,2)),norm(p(:,3))]);
vC=p(:,1)+p(:,2)+p(:,3);
vC=bestDiv(vC,53,10); %vector with smallest integers
vCm=vC*pmax/norm(vC); %For plotting purposes, its size is moderated
h1=line([0,vCm(1)],[0,vCm(2)],[0,vCm(3)],'Color',[0.6 0.3 0],
    'LineWidth',1.5);
%Now the vector vA perpendicular to the plane: Miller indices (hkl)
sg=sgn(u)*sgn(v)*sgn(w); %tricky trick to get the (hkl) sign right!
```

```
v1=p(:,1)-p(:,3);        %plane border vectors
v2=p(:,2)-p(:,1);
vA=sg*cross(v1,v2);      %v1 X v2 gives a plane perpendicular
%v3=p(:,3)-p(:,2);       %can create another border vector for some checking
%vB=sg*cross(v2,v3)      %another perpendicular to the plane
%dot(v1,vA)              %check: if 0, v1 & vB are perpendicular to each other
%dot(v1,vB)              %check: if 0 also for v1 & vB
%We next use prime numbers to find a common divisor, up to number 53,
%to simplify further and plot the perpendicular
vA=bestDiv(vA,53,10); %vector with smallest integers (Miller indices)
pmax=max([norm(p(:,1)),norm(p(:,2)),norm(p(:,3))]);
vAm=vA*pmax/norm(vA); %moderate the size of vA for plotting purposes
h2=line([0,vAm(1)],[0,vAm(2)],[0,vAm(3)],'lineStyle','-.','Color','k',
    'LineWidth',1.5);
%Make the 3-D polygon defined by the vector intercepts with color
h=fill3(p(1,:),p(2,:),p(3,:),[0.75 0.75 0.75]);
set(h,'EdgeAlpha',[0.3],'FaceAlpha',[0.5]) %edges, transparent]
axis equal, hb(1)=xlabel('x'); hb(2)=ylabel('y'); hb(3)=zlabel('z');
hl=legend([h1,h2],'[uvw]','(hkl)',1);
set(hl,'FontSize',14), set(hb,'FontSize',14)
ti=' ';
alx=cat(2,', \alpha=',num2str(alph,'%3.1f'),'^\circ');
bex=cat(2,', \beta=' ,num2str(bet ,'%3.1f'),'^\circ');
gax=cat(2,', \gamma=',num2str(gamm,'%3.1f'),'^\circ');
ax=cat(2, ' a1=',num2str(a1,'%4.2f'));
bx=cat(2,', a2=',num2str(a2,'%4.2f'));
cx=cat(2,', a3=',num2str(a3,'%4.2f'));
str1=cat(2,ti,' (',ax,bx,cx,alx,bex,gax,' )');
% ux=cat(2, ' u=',num2str(u,'%4.2f'));
% vx=cat(2,', v=',num2str(v,'%4.2f'));
% wx=cat(2,', w=',num2str(w,'%4.2f'));
ux=cat(2, ' u=',num2str(vC(1),'%4.2f'));
vx=cat(2,', v=',num2str(vC(2),'%4.2f'));
wx=cat(2,', w=',num2str(vC(3),'%4.2f'));
hx=cat(2, ' h=',num2str(vA(1),'%4.2f'));
kx=cat(2,', k=',num2str(vA(2),'%4.2f'));
lx=cat(2,', l=',num2str(vA(3),'%4.2f'));
str2=cat(2,'[',ux,vx,wx,']',', (',hx,kx,lx,')');
title({str1;str2},'FontSize',14)

function y=sgn(x)
%Returns the sign of x. If x=0, the result is > 0.
if(x >= 0.0), y=1.0; else y=-1.0; end

function V=bestDiv(V,Np,ipasses)
%This finds the largest common divisor among the numbers in V
%and returns the simplified V. The idea is based on dividing by
%prime integers until we have such as divisor. Then V is simplified.
%Prime numbers up to NP are used and several passes can be made.
%ipasses=number of passes to make for the simplification in case
%more simplification is possible and is needed
```

```
lV=length(V);
p=primes(Np);
p0=1.0;
for ip=1:ipasses
  p1=p0;               %reset p1 and pp
  pp=ones(1,lV);
  for i=1:length(p)  %go through the primes
    iflag=1;
    for j=1:lV
      a=round(V(j)*1.e12)/1.e12; %to avoid small decimals
      b=p(i);
      c=mod(a,b);
      if(c==0), pp(j)=p(i); end  %keep a divisor if it works
    end
    %If we have a common divisor, all pp should be equal, let's check
    for m=2:lV
      if(pp(m)~=pp(1)), iflag=0; break; end  %not found, go to next
    end
    if(pp(1) > p1 & iflag==1), p1=pp(1); end %keep the largest divisor
  end
  V=V/p1;
end
```

## 1.6 Examples of Crystal Structures

Below, examples of commonly known simple crystal structures are illustrated. These are sodium chloride, cesium chloride, close-packed, diamond, and zinc sulfide or zinc-blende.

### 1.6.1 Sodium Chloride (Salt)

The sodium chloride (NaCl) structure is an example of a FCC lattice. The lattice actually consists of two FCCs interlocked. The Na atoms are located on one sublattice and the Cl atoms on the other. Imagine creating a FCC lattice of Cl and another of Na, then bringing them together in such a way that the Na sublattice ends up displaced from the Cl sublattice by $1/2$ of the body diagonal of the unit cube ($\sqrt{a^2+a^2+a^2} = a\sqrt{3}$). This is an example of an ionic system. The chlorine accepts an electron from sodium, which is happy to give it up, and both reach the electronic configuration of Neon. Thus, the system is held together by ionic bonding of which more will be said later in the text. In Figure 1.6.28 notice that the Na atoms are surrounded by Cl atoms and vice versa; each has 6 nearest neighbors of the opposite type. The basis consists of one Cl atom and one Na atom separated by the vector $(1/2, 1/2, 1/2)a$. The Cl atoms are located at $(0,0,0)$, $(1/2,1/2,0)$, $(1/2,0,1/2)$, and $(0,1/2,1/2)$ while the Na atoms are at $(1/2,1/2,1/2)$, $(0,0,1/2)$, $(0,1/2,0)$, and $(1/2,0,0)$ in units of the lattice constant, $a$.

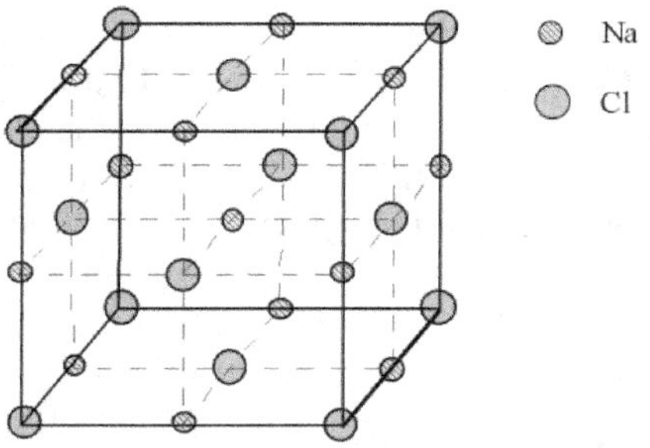


Figure 1.6.28: The NaCl structure consists of two FCC sublattices displaced by $1/2$ a cube body diagonal.

Table 1.6.1 contains system examples that crystallize in the NaCl structure.

### 1.6.2 Cesium Chloride

The cesium chloride (CsCl) structure is an example of a BCC lattice. This crystal has basis with the molecule's Cs atom located at $(0,0,0)$ (cube center), and the Cl atom at $(1/2,1/2,1/2)$ (cube corner) in units of the lattice constant $a$. This is also an example of an ionic system as is NaCl, albeit with a different structure because the Cs positive ion is larger than the positive Na ion. Each atom may be viewed at the center of a cube of atoms of opposite type as, for example, Figure 1.6.29, shows for the Cs atom at the body-center position.

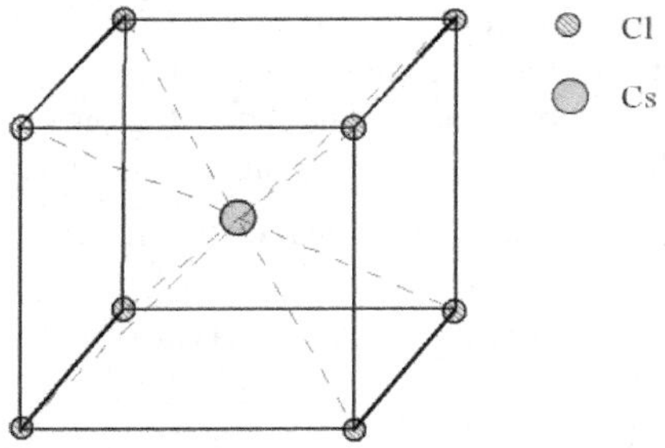


Figure 1.6.29: The CsCl structure consists of the BCC lattice.

Table 1.6.1 contains system examples that crystallize in the CsCl structure.

### 1.6.3 Close-Packed: Hexagonal and Cubic

There are various ways to arrange atoms so as to minimize the packing fraction (how well they are packed). One is the hexagonal close packed (HCP) and the other is the face-centered cubic close packed (CCP). The fraction of the total volume occupied by the atomic spheres is the same as that quoted in Section 1.4.3 with a value of about 0.74 (see also Exercise 1.9.11). Close packing occurs for spherical atoms and it happens in two ways. Consider Figure 1.6.30. As indicated, we refer to the solid spheres those on the 1st layer whose centers are located at $A$ positions. The 2nd layer of close packed spheres (cross-hatched) are placed over the $B$ void positions of the first layer (small dark circles). When a third atomic layer (crisscross-hatched) is added, there are two choices where the atoms are placed, as shown in the middle part of the figure. If they are located over the $A$ atomic positions, the layer arrangement is referred to as the *ABABAB*... stacking and leads to the HCP structure shown in the upper right side of the figure. If the third layer of atoms is instead placed over the *C* void positions of the first layer (small open circles), it leads to the *ABCABC*... stacking associated with the CCP structure which is face centered cubic (FCC) and is shown at the lower right side of the figure. The number of nearest neighbors for both structures is the same as the FCC; i.e., twelve.

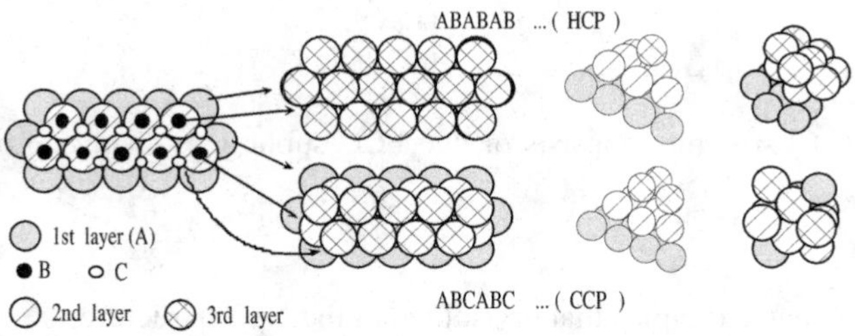


Figure 1.6.30: Close packing starts with the first layer of solid atomic spheres, whose centers are referred to as *A* positons, and a 2nd layer of spheres (cross-hatched) over the *B* void positions of the first layer (small filled circles). There are two choices for the third layer of spheres (crisscrossed-hatched). One is over the original *A* atoms which leads to the hexagonal close packed (HCP) of the upper right; the other choice is for the third layer to go over the *C* void positions of the first layer (small unfilled circles) and leads to the cubic close packed (CCP) of the lower right.

The FCC structure was discussed earlier in Section 1.4.3. For the HCP structure, Figure 1.6.31 shows the positions of the two basis atoms *A* and *B* along with the primitive lattice vectors. The position of the *B* atom with respect to the *A* atom at the origin is given by $\vec{r} = \frac{2}{3}\vec{a} + \frac{1}{3}\vec{b} + \frac{1}{2}\vec{c}$. On the right side of the figure, notice that the positions of the *A* and *B* atoms are such as to be located at the corners of a tetrahedron with sides $|\vec{a}| = |\vec{b}| = |\vec{r}|$. The ratio of the magnitudes between the $c$ and the $a$ axes is given by $\frac{c}{a} = \sqrt{\frac{8}{3}} = 1.633$, which is the ideal or theoretical ratio limit for close packing.

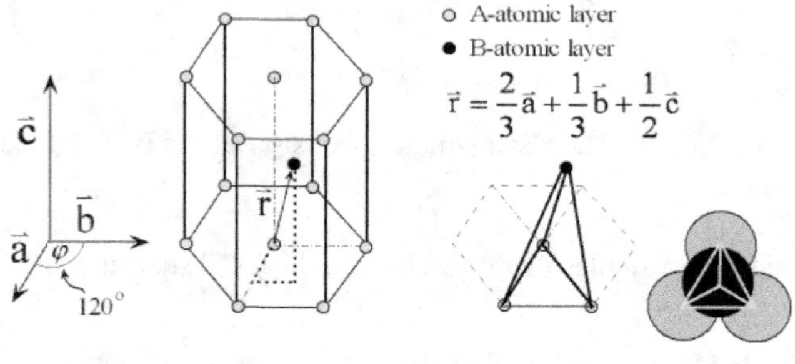


Figure 1.6.31: The two basis atoms *A* and *B* along with the primitive lattice vectors in the HCP structure are shown. The positions of the *A* and *B* atoms are such that the four atoms are located at each of the corners of a tetrahedron with sides $|\vec{a}| = |\vec{b}| = |\vec{r}|$. When the spheres are made to occupy 100% of their space, the tetrahedron structure is seen on the lower right.

Table 1.6.1 contains system examples of close packed (HCP and CCP) structures. Notice some materials that crystalize in the HCP structure; i.e., He, have $c/a$ ratios that are close to the ideal value.

Table 1.6.1: Examples of material systems that crystallize in the sodium chloride (NaCl), cesium chloride (CsCl), cubic close packed (CCP) and hexagonal close packed (HCP) and their lattice constants (in Angstroms) [3].

| **NaCl** | $a$(Å) | **CsCl** | $a$(Å) | **CCP** | $a$(Å) | **HCP** | $a$(Å) , $c$(Å), $c/a$ |
|---|---|---|---|---|---|---|---|
| LiH | 4.08 | BeCu | 2.70 | Cu | 3.61 | He | 3.50, 5.72, 1.63 |
| MgO | 4.20 | AlNi | 2.88 | Ag | 4.09 | Be | 2.29, 3.58, 1.56 |
| MnO | 4.43 | CuPd | 2.99 | Au | 4.08 | Mg | 3.21, 5.21, 1.62 |
| NaCl | 5.63 | AgMg | 3.28 | Al | 4.05 | Ti | 2.95, 4.68, 1.59 |
| AgBr | 5.77 | LiHg | 3.29 | Ni | 3.52 | Zn | 2.66, 4.95, 1.86 |
| PbS | 5.92 | TlBr | 3.97 | Pd | 3.89 | Cd | 2.98, 5.62, 1.89 |
| KCl | 6.29 | CsCl | 4.11 | Pt | 3.92 | Zr | 3.23, 5.15, 1.59 |
| KBr | 6.59 | TlI | 4.20 | Pb | 4.95 | Os | 2.74, 4.32, 1.58 |

### 1.6.4 Diamond

Materials that crystallize in the diamond structure are mainly insulators and semiconductors. Diamond's lattice has a primitive basis of two identical atoms at (000), and $(1/4, 1/4, 1/4)$, in units of the lattice constant $a$, associated with each point of an FCC space lattice. This structure has eight atoms/cell because it has the conventional FCC with 4 atoms/cell and then there are 4 more atoms inside the large cube that are not shared, as seen in Figure 1.6.32. In (a) one can see that each atom can be thought to be surrounded by four nearest neighbors located at tetrahedral corners. The nearest neighbor distance being $\frac{\sqrt{3}}{4}a$. Four of the basic blocks in (a) can be combined to make up the larger block shown in (b) with four additional atoms to occupy the cube corners (shown unbonded).

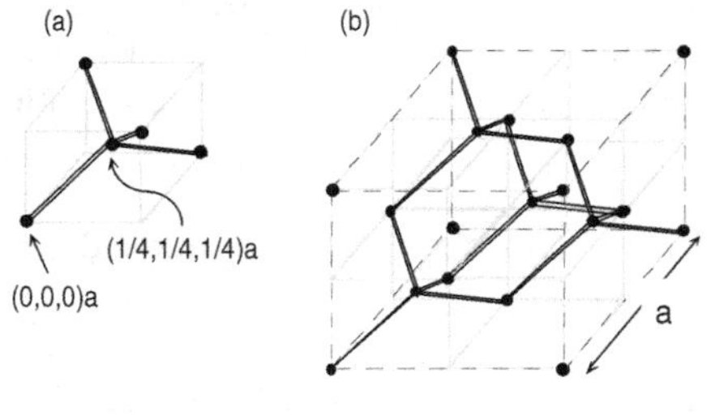


Figure 1.6.32: (a) The basic block of the diamond structure which can be used to make up the larger block shown in (b) with four additional atoms (shown unconnected) at the large cube corners.

The atoms at the tetrahedral corners in (a) form the standard FCC structure; the tetrahedron center atom is characteristic to diamond and is part of another FCC. The diamond structure is thought of two FCCs interlaced and displaced from each other by $1/4$ of a body diagonal, which is somewhat similar to the NaCl structure, albeit with a different displacement; also, in diamond, all the atoms are of the same type. The diamond lattice is the result of the directional covalent bonding associated with column IV of the periodic table. Table 1.6.2 contains examples of systems that crystallize in the diamond structure. Typical diamond (carbon) is an insulator, silicon and germanium are typical semiconductors, as is also tin ($\alpha$-tin or grey tin).

### 1.6.5 Zinc-Sulfide or Zinc-Blende

This structure is similar to the diamond structure, but instead of the atoms being all the same, the zinc sulfide structure (or zinc blende) allows for two different species of atoms. The basis is again two atoms displaced from each other by $1/4$ a body diagonal. Thus the *Zn* atoms lie on one FCC with coordinates $(0,0,0)$, $(0,1/2,1/2)$, $(1/2,0,1/2)$, and $(1/2,1/2,0)$ while the *S* atoms lie on the other FCC with coordinates $(1/4,1/4,1/4)$, $(1/4,3/4,3/4)$, $(3/4,1/4,3/4)$, and $(3/4,3/4,1/4)$ in units of $a$. Thus there are four molecules of ZnS per cell similar to diamond. Also, there are four atoms surrounding the opposite atom at the corners of a tetrahedron as shown in Figure 1.6.33.

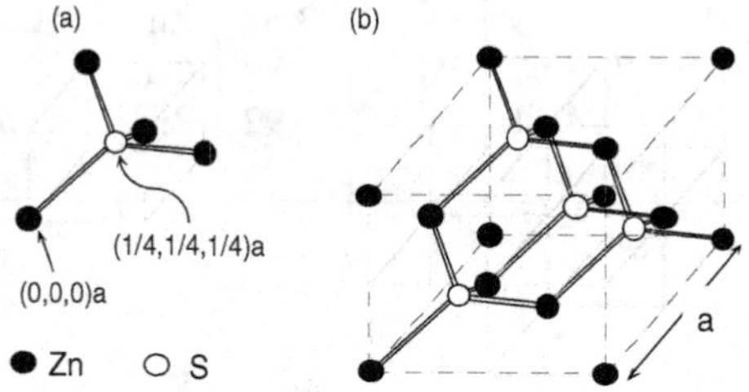


Figure 1.6.33: (a) The basic block of the zinc blende structure which can be used to make up the larger block shown in (b) with four additional atoms (shown unconnected) at the large cube corners. The structure is similar to diamond but with a basis of two different atomic species as indicated by the filled and unfilled circles.

The zinc blende structure allows the formation of semiconducting and insulating compounds and alloys. The compounds always have one atomic species as nearest neighbors to the other, while in alloys the arrangement is such that the different atomic species can occupy either sublattice with a certain probability that depends on the concentration. More will be said later in the text about alloys. Table 1.6.2 contains examples of systems that crystallize in the zinc blende structure.

Table 1.6.2: Examples of material systems that crystallize in the diamond and the zinc blende and their lattice constants (in Angstroms) [3].

| Diamond | $a$(Å) | Zinc Blende | $a$(Å) |
|---|---|---|---|
| C | 3.57 | SiC | 4.35 |
| Si | 5.43 | ZnS | 5.41 |
| Ge | 5.67 | AlP | 5.45 |
| Sn | 6.49 | GaP | 5.45 |
| | | ZnSe | 5.65 |
| | | GaAs | 5.65 |
| | | AsAl | 5.66 |
| | | InSb | 6.46 |

Remember also to refer to Table 1.8.3 for systems that crystallize in various structures as well as to see other properties listed.

## 1.7 Atomic Surface Microscopes

It is possible to image the atomic surface of a solid, atom by atom, by methods such as atomic force microscopy (AFM) and scanning tunneling microscopy (STM). The AFM was developed by Gerd Binnig and Christopher Gerber in 1985. It uses a feedback signal from the detection of a laser beam that reflects off a mirror. The mirror is mounted on a cantilever which is in contact with the sample. The feedback signal is used to move the sample up or down to keep the cantilever force constant. The sample movement traces the atomic sample contours thus providing an image of the surface atoms as shown in Figure 1.7.34.

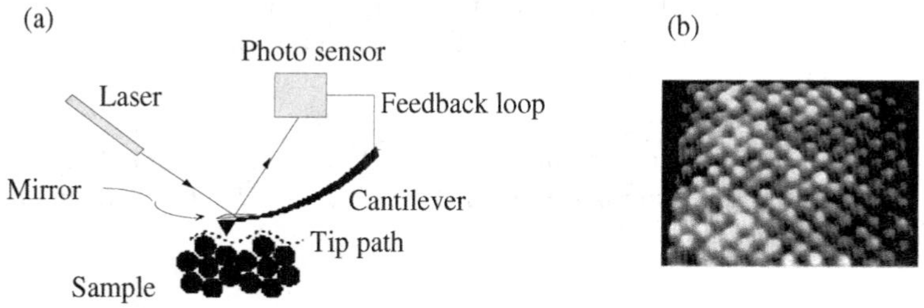


Figure 1.7.34: (a) The basic atomic force microscopy (AFM) experimental setup. (b) A NaCl crystal surface example of an AFM image from [4] (reprinted with permission).

The STM technique was developed also by Gerd Binnig in collaboration with Heinrich Rohrer who were awarded the Nobel Prize in 1986 for their work. Here a constant bias voltage is applied between a metal tip and the sample of interest. As the needle's tip moves over the surface, sample electrons tunnel across the gap (of width $d$) from the sample to the needle. The tunneling current is proportional to the quantum mechanical tunneling probability as $I \sim |T|$, where $T$ is the ratio of the absolute value squared of each of the transmitted electronic wavefunction through the barrier, $\psi_{transmitted}$, and the incident wavefunction on the barrier, $\psi_{incident}$, or

$$T = \frac{|\psi_{transmitted}|^2}{|\psi_{incident}|^2} \sim \exp(-2kd), \tag{1.7.19a}$$

where

$$k = \sqrt{\frac{2m}{\hbar^2}(W - E)}, \tag{1.7.19b}$$

with $m$, the electron mass; $E$ the electron's kinetic energy; and $W$, the sample electrons' work function (minimum energy to jump the gap). The electron's kinetic energy can be related to the applied voltage as $E = e\phi_{app}$; and by analyzing the tunneling current, the crystal sample's surface structure can thus be imaged. This is shown in Figure 1.7.35.

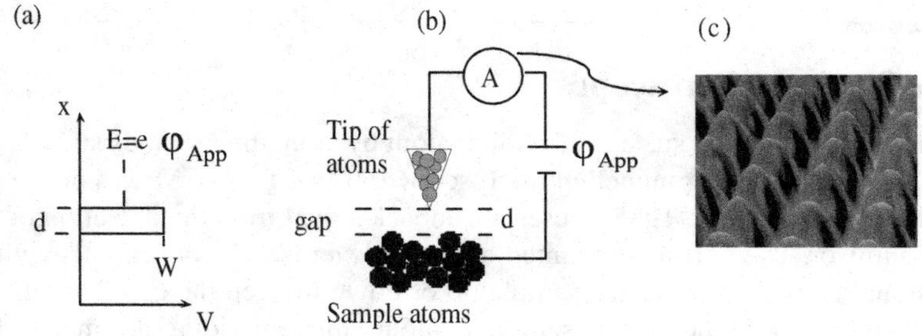


Figure 1.7.35: (a) The quantum mechanical picture model of the scanning tunneling microscope (STM). (b) The basic STM experimental setup. (b) A crystal surface example of an STM image. Shown is Nickel FCC (110) surface (image originally created by IBM [5]).

For this model, electrons experience quantum mechanical tunneling. In one dimension, if we have a quantum mechanical barrier of height $V = V_0$ and width $a$, an electron carrying energy $E$ has a transmission probability given by

$$T = \frac{|\psi_{transmitted}|^2}{|\psi_{incident}|^2} = \frac{1}{1 + D\sinh^2(\alpha a)}, \tag{1.7.20a}$$

where

$$D = \frac{V_0^2}{4E(V_0 - E)}, \tag{1.7.20b}$$

and

$$\alpha = \sqrt{\frac{2m}{\hbar^2}(V_0 - E)}. \tag{1.7.20c}$$

In the limit of $\alpha a \gg 1$, and taking $V_0 = W$. Equations 1.7.20 become

$$T = \frac{|\psi_{transmitted}|^2}{|\psi_{incident}|^2} \approx T_0 \exp(-2kd), \tag{1.7.21a}$$

where

$$T_0 = \frac{16E(W - E)}{W^2}, \tag{1.7.21b}$$

which, by roughly approximating $T_0 \sim 1$, become equivalent to Equations 1.7.19. Let's do a numerical example using Equations 1.7.19.

**Example 1.7.0.1**
Consider a material whose atoms we wish to image has a work function of $W = 4.3\,eV$; if the STM probing needle is under a bias voltage of $3.0\,volts$ and is located at about $0.3\,nm$ from the sample, use Equations 1.7.19 to estimate the percentage of electrons that tunnel.

**Solution**
We will use units of $eV$ for energy and units of $nm$ for distance. We write $kd = \sqrt{\frac{2mc^2}{(\hbar c)^2}(W - E)d^2}$ where $c$ is the speed of light, and use the fact that $mc^2 = 0.511 \times 10^6\,eV$ for electrons, $\hbar c = 197\,eV \cdot nm$, and since electronic charge times voltage is already in $eV$'s, we have

$kd = \sqrt{\frac{2\times0.511\times10^6\,eV}{(197\,eV\cdot nm)^2}(4.3\,eV - 3.0\,eV)(0.3nm)^2} = 1.7553,$
and $\exp(-2kd) = \exp(-2 * 1.7553) = 0.0299$. That is, about 3% of the total electrons will be tunneling.

It is also useful to get a feeling for what this simple approximation (Equations 1.7.19) gives versus the important parameters such as energy and gap width. We do this in Figure 1.7.36. (a) Shows the tunneling probability versus applied voltage (in eV), while keeping the gap width constant, and (b) shows it versus gap width while keeping the energy constant.

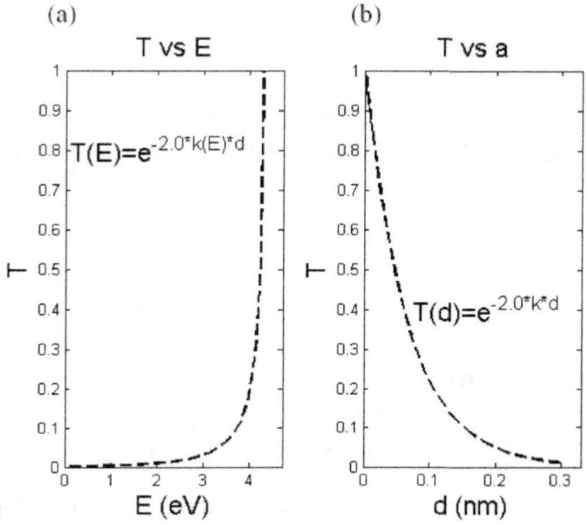


Figure 1.7.36: (a) Tunneling probability versus applied voltage for a gap width of $0.3nm$ and a work function of $W = 4.3\,volts$. (b) Tunneling probability versus gap width for an energy of $0.5W$ with $W$ and in (a). Code tunnel0.m was used to do the calculation.

In Figure 1.7.36(a) notice that as the energy $E$ increases, the higher does the tunneling probability become. In (b), as the gap width increases, the tunneling probability decreases. The code used in Figure 1.7.36's calculation, `tunnel0.m`, is listed below. Be sure to run the code to reproduce the results obtained.

```
%copyright by J. E Hasbun and T. Datta
%tunnel0.m
%Here we use the approximate tunneling formula with maximum
%amplitude of unity and plot it versus energy and spacing.
clear
hbC=197.0;      %hbar*C in eV nm
V0=4.3;         %example work function in eV
% ****** varying the applied voltage *****************
Va=0:V0/100.0:V0;  %applied voltage in electron volts
E=Va;                              %energy in eV
mc2=0.511e6;                       %rest mass of the electon mc^2 in eV
k=sqrt(2*mc2*(V0-E)/hbC^2);
d=0.3;                             %gap width in nm
T1_approx=exp(-2.0*k*d);           %approximation
subplot(1,2,1)
plot(E,T1_approx,'k--','LineWidth',2)
```

```
axis([0 1.1*V0 0 1])
text(0.1,0.8,'T(E)=e^{(-2.0*k(E)*d)}','FontSize',16)
xlabel('E (eV)','FontSize',16), ylabel ('T','FontSize',16)
title('T vs E','FontSize',16)
% ****** varying the gap width *****************
s=0:d/25:d;
EE=0.5*V0;
k=sqrt(2*mc2*(V0-EE)/hbC^2);
T2_approx=exp(-2*k*s);          %approximation
subplot(1,2,2)
plot(s,T2_approx,'k--','LineWidth',2)
axis([0 1.1*d 0 1])
text(d/4,0.4,'T(d)=e^{(-2.0*k*d)}','FontSize',16)
xlabel('d (nm)','FontSize',16), ylabel ('T','FontSize',16)
title('T vs a','FontSize',16)
```

## 1.8 Element Properties Table

In this section, a table is given that lists some properties of the elements that may become useful throughout the text. This is followed by a standard periodic table.

Table 1.8.3: Some properties of the elements that may become useful throughout the text.

```
ElementalProperties0.txt
Z:      Atomic number
Sy:     Symbol
a,c:    Lattice constant and c axis in Angstroms (A)
Str:    Structure
D:      Density in gr/cc (or g/l for gases) at a temperature of 20degC at 1 atm (p)
        unless stated.
nn:     Nearest neighbor distance
MP:     Melting point in Kelvin (K)
BP:     Boiling point in Kelvin (K)
EC:     Electronic Configuration
CUB:    Cubic
MCL:    monoclinic
FCC:    face centered cubic
DIA:    diamond
BCC:    body centered cubic
ORC:    orthorhombic
HCP:    HCPagonal close packed
RHL:    rhombohedral
TET:    tetragonal
SCB:    simple cubic
Other data at room temperature or at the stated temperature (K)
Sources
1) Introduction to Solid State, Charles Kittel, 8th Ed. 13-19 (John Wiley, NY 2005).
2) Inorganic Crystal Structure Database (ICSD) online.
3) Curtesy of Matpack C++ Numerics and Graphics Library
   (online http://www.matpack.de/Info/Nuclear/Elements/lattice.html)
4) http://physics.nist.gov/data, http://www.nist.gov/pml/data/periodic.cfm
```

| Z | Sy | Name | a[A] | K | Str | c/a | D (gr/cc) | nn [A] | MP [K] | BP [K] | EC |
|---|---|---|---|---|---|---|---|---|---|---|---|
| 1 | H | Hydrogen | 3.75 | 4 | HCP | 1.63 | 0.086 g/l | | 14.01 | 20.28 | 1s |
| 2 | He | Helium | 3.57 | 2 | HCP | 1.63 | 0.205(37p) | | 0.95 | 4.216 | 1s2 |
| 3 | Li | Lithium | 3.49 | 78 | BCC | - | 0.542 | 3.023 | 453.69 | 1590 | [He] 2s |
| 4 | Be | Beryllium | 2.27 | | HCP | 1.58 | 1.82 | 2.22 | 1551 | 3243 | [He] 2s2 |
| 5 | B | Boron | 8.73 | | TET | 0.58 | 2.47 | | 2573 | 2823 | [He] 2s2 2p |
| 6 | C | Carbon | 3.57 | | DIA | - | 3.516 | 1.54 | 3823 | 5100 | [He] 2s2 2p2 |
| 7 | N | Nitrogen(N2) | 5.66 | 20 | CUB | - | 1.03 | | 63.29 | 77.4 | [He] 2s2 2p3 |
| 8 | O | Oxygen | 6.83 | | CUB | - | 1.33 g/l | | 54.75 | 90.188 | [He] 2s2 2p4 |
| 9 | F | Fluorine | | | MCL | - | 1.58 g/l | | 53.53 | 85.01 | [He] 2s2 2p5 |
| 10 | Ne | Neon | 4.46 | 4 | FCC | - | 1.51 | 3.16 | 24.48 | 27.1 | [He] 2s2 2p6 |
| 11 | Na | Sodium | 4.23 | 5 | BCC | - | 1.013 | 3.659 | 370.95 | 1165 | [Ne] 3s |
| 12 | Mg | Magnesium | 3.21 | | HCP | 1.62 | 1.74 | 3.20 | 921.95 | 1380 | [Ne] 3s2 |
| 13 | Al | Aluminum | 4.05 | | FCC | - | 2.70 | 2.86 | 933.52 | 2740 | [Ne] 3s2 3p |
| 14 | Si | Silicon | 5.43 | | DIA | - | 2.33 | 2.35 | 1683 | 2628 | [Ne] 3s2 3p2 |
| 15 | P | Phosphorus | 7.17 | | CUB | - | 1.82 | | 317.3 | 553 | [Ne] 3s2 3p3 |
| 16 | S | Sulfur | 10.47 | | ORC | 2.34 | 2.06 | | 386 | 717.824 | [Ne] 3s2 3p4 |
| 17 | Cl | Chlorine | 6.24 | | ORC | 1.32 | 2.03 g/l | | 172.17 | 238.55 | [Ne] 3s2 3p5 |
| 18 | Ar | Argon | 5.31 | 4 | FCC | - | 1.77 | 3.76 | 83.78 | 87.29 | [Ne] 3s2 3p6 |
| 19 | K | Potassium | 5.23 | 5 | BCC | - | 0.910 | 4.525 | 336.8 | 1047 | [Ar] 4s |
| 20 | Ca | Calcium | 5.58 | | FCC | - | 1.53 | 3.95 | 1112 | 1760 | [Ar] 4s2 |
| 21 | Sc | Scandium | 3.31 | | HCP | 1.59 | 2.99 | 3.25 | 1812 | 3105 | [Ar] 3d 4s2 |
| 22 | Ti | Titanium | 2.95 | | HCP | 1.59 | 4.51 | 2.89 | 1933 | 3533 | [Ar] 3d2 4s2 |
| 23 | V | Vanadium | 3.03 | | BCC | - | 6.09 | 2.62 | 2163 | 3653 | [Ar] 3d3 4s2 |
| 24 | Cr | Chromium | 2.88 | | BCC | - | 7.19 | 2.50 | 2130 | 2755 | [Ar] 3d5 4s |
| 25 | Mn | Manganese | 8.89 | | CUB | - | 7.47 | 2.24 | 1517 | 2370 | [Ar] 3d5 4s2 |
| 26 | Fe | Iron | 2.87 | | BCC | - | 7.87 | 2.48 | 1808 | 3023 | [Ar] 3d6 4s2 |
| 27 | Co | Cobalt | 2.51 | | HCP | 1.62 | 8.89 | 2.50 | 1768 | 3143 | [Ar] 3d7 4s2 |
| 28 | Ni | Nickel | 3.52 | | FCC | - | 8.91 | 2.49 | 1726 | 3005 | [Ar] 3d8 4s2 |
| 29 | Cu | Copper | 3.61 | | FCC | - | 8.93 | 2.56 | 1356.6 | 2868 | [Ar] 3d10 4s |
| 30 | Zn | Zinc | 2.66 | | HCP | 1.86 | 7.13 | 2.66 | 692.73 | 1180 | [Ar] 3d10 4s2 |
| 31 | Ga | Gallium | 4.51 | | ORC | 1.70 | 5.91 | 2.44 | 302.93 | 2676 | [Ar] 3d10 4s2 4p |
| 32 | Ge | Germanium | 5.66 | | DIA | - | 5.32 | 2.45 | 1210.55 | 3103 | [Ar] 3d10 4s2 4p2 |
| 33 | As | Arsenic | 4.13 | | RHL | - | 5.77 | 3.16 | 886 | sublime | [Ar] 3d10 4s2 4p3 |
| 34 | Se | Selenium | 4.36 | | HCP | 1.14 | 4.81 | 2.32 | 490 | 958.1 | [Ar] 3d10 4s2 4p4 |
| 35 | Br | Bromine | 6.67 | | ORC | 0.67 | 3.14 | | 265.9 | 331.93 | [Ar] 3d10 4s2 4p5 |
| 36 | Kr | Krypton | 5.64 | 4 | FCC | - | 3.09 | 4.00 | 116.55 | 120.85 | [Ar] 3d10 4s2 4p6 |
| 37 | Rb | Rubidium | 5.59 | 5 | BCC | - | 1.629 | 4.837 | 312.2 | 961 | [Kr] 5s |
| 38 | Sr | Strontium | 6.08 | | FCC | - | 2.58 | 4.30 | 1042 | 1657 | [Kr] 5s2 |
| 39 | Y | Yttrium | 3.65 | | HCP | 1.57 | 4.48 | 3.55 | 1796 | 3610 | [Kr] 4d 5s2 |
| 40 | Zr | Zirconium | 3.23 | | HCP | 1.59 | 6.51 | 3.17 | 2125 | 4650 | [Kr] 4d2 5s2 |
| 41 | Nb | Niobium | 3.30 | | BCC | - | 8.58 | 2.86 | 2741 | 5200 | [Kr] 4d4 5s |
| 42 | Mo | Molybdenum | 3.15 | | BCC | - | 10.22 | 2.72 | 2890 | 5833 | [Kr] 4d5 5s |
| 43 | Tc | Technetium | 2.74 | | HCP | 1.61 | 11.50 | 2.71 | 2445 | 5303 | [Kr] 4d6 5s |
| 44 | Ru | Ruthenium | 2.71 | | HCP | 1.58 | 12.36 | 2.65 | 2583 | 4173 | [Kr] 4d7 5s |
| 45 | Rh | Rhodium | 3.80 | | FCC | - | 12.42 | 2.69 | 2239 | 4000 | [Kr] 4d8 5s |
| 46 | Pd | Palladium | 3.89 | | FCC | - | 12.00 | 2.75 | 1825 | 3413 | [Kr] 4d10 |
| 47 | Ag | Silver | 4.09 | | FCC | - | 10.50 | 2.89 | 1235.08 | 2485 | [Kr] 4d10 5s |
| 48 | Cd | Cadmium | 2.98 | | HCP | 1.89 | 8.65 | 2.98 | 594.1 | 1038 | [Kr] 4d10 5s2 |
| 49 | In | Indium | 3.25 | | TET | 1.52 | 7.29 | 3.25 | 429.32 | 2353 | [Kr] 4d10 5s2 5p |
| 50 | Sn | Tin(\alpha) | 6.49 | | DIA | - | 5.76 | 2.81 | 505.118 | 2543 | [Kr] 4d10 5s2 5p2 |
| 51 | Sb | Antimony | 4.51 | | RHL | - | 6.69 | 2.91 | 903.89 | 2023 | [Kr] 4d10 5s2 5p3 |
| 52 | Te | Tellurium | 4.45 | | HCP | 1.33 | 6.25 | 2.86 | 722.7 | 1263 | [Kr] 4d10 5s2 5p4 |
| 53 | I | Iodine | 7.27 | | ORC | 0.66 | 4.95 | 3.54 | 386.65 | 457.55 | [Kr] 4d10 5s2 5p5 |
| 54 | Xe | Xenon | 6.13 | 4 | FCC | - | 3.78 | 4.34 | 161.3 | 166.1 | [Kr] 4d10 5s2 5p6 |
| 55 | Cs | Cesium | 6.05 | 5 | BCC | - | 1.997 | 5.235 | 301.55 | 963 | [Xe] 6s |
| 56 | Ba | Barium | 5.02 | | BCC | - | 3.59 | 4.35 | 998 | 1913 | [Xe] 6s2 |
| 57 | La | Lanthanum | 3.77 | | HCP | 1.62 | 6.17 | 3.73 | 1193 | 3727 | [Xe] 5d 6s2 |
| 58 | Ce | Cerium | 5.16 | | FCC | - | 6.77 | 3.65 | 1071 | 3530 | [Xe] 4f2 6s2 |
| 59 | Pr | Praseodymium | 3.67 | | HCP | 1.61 | 6.78 | 3.63 | 1204 | 3485 | [Xe] 4f3 6s2 |
| 60 | Nd | Neodymium | 3.66 | | HCP | 1.61 | 7.00 | 3.66 | 1283 | 3400 | [Xe] 4f4 6s2 |
| 61 | Pm | Promethium | - | | - | - | 7.22 | | 1353 | 3000 | [Xe] 4f5 6s2 |
| 62 | Sm | Samarium | 9.00 | | RHL | - | 7.54 | 3.59 | 1345 | 2051 | [Xe] 4f6 6s2 |
| 63 | Eu | Europium | 4.58 | | BCC | - | 5.25 | 3.96 | 1095 | 1870 | [Xe] 4f7 6s2 |

| | | | | | | | | | | | |
|---|---|---|---|---|---|---|---|---|---|---|---|
| 64 | Gd | Gadolinium | 3.63 | HCP | 1.59 | 7.89 | | 3.58 | 1584 | 3506 | [Xe] 4f7 5d 6s2 |
| 65 | Tb | Terbium | 3.60 | HCP | 1.58 | 8.27 | | 3.52 | 1633 | 3314 | [Xe] 4f9 6s2 |
| 66 | Dy | Dysprosium | 3.59 | HCP | 1.57 | 8.53 | | 3.51 | 1682 | 2608 | [Xe] 4f10 6s2 |
| 67 | Ho | Holmium | 3.58 | HCP | 1.57 | 8.80 | | 3.49 | 1743 | 2993 | [Xe] 4f11 6s2 |
| 68 | Er | Erbium | 3.56 | HCP | 1.57 | 9.04 | | 3.47 | 1795 | 2783 | [Xe] 4f12 6s2 |
| 69 | Tm | Thulium | 3.54 | HCP | 1.57 | 9.32 | | 3.54 | 1818 | 2000 | [Xe] 4f13 6s2 |
| 70 | Yb | Ytterbium | 5.48 | FCC | - | 6.97 | | 3.88 | 1097 | 1466 | [Xe] 4f14 6s2 |
| 71 | Lu | Lutetium | 3.50 | HCP | 1.59 | 9.84 | | 3.43 | 1929 | 3588 | [Xe] 4f14 5d 6s2 |
| 72 | Hf | Hafnium | 3.19 | HCP | 1.58 | 13.20 | | 3.13 | 2423 | 5673 | [Xe] 4f14 5d2 6s2 |
| 73 | Ta | Tantalum | 3.30 | BCC | - | 16.66 | | 2.86 | 3269 | 5698 | [Xe] 4f14 5d3 6s2 |
| 74 | W | Tungsten | 3.16 | BCC | - | 19.25 | | 2.74 | 3680 | 6200 | [Xe] 4f14 5d4 6s2 |
| 75 | Re | Rhenium | 2.76 | HCP | 1.62 | 21.03 | | 2.74 | 3453 | 5900 | [Xe] 4f14 5d5 6s2 |
| 76 | Os | Osmium | 2.74 | HCP | 1.58 | 22.58 | | 2.68 | 3318 | 5300 | [Xe] 4f14 5d6 6s2 |
| 77 | Ir | Iridium | 3.84 | FCC | - | 22.55 | | 2.71 | 2683 | 4403 | [Xe] 4f14 5d7 6s2 |
| 78 | Pt | Platinum | 3.92 | FCC | - | 21.47 | | 2.77 | 2045 | 4100 | [Xe] 4f14 5d9 6s |
| 79 | Au | Gold | 4.08 | FCC | - | 19.28 | | 2.88 | 1337.58 | 3213 | [Xe] 4f14 5d10 6s |
| 80 | Hg | Mercury | 2.99 | RHL | - | 13.55 | | 3.01 | 234.28 | 629.73 | [Xe] 4f14 5d10 6s2 |
| 81 | Tl | Thallium | 3.46 | HCP | 1.60 | 11.87 | | 3.46 | 576.7 | 1730 | [Xe] 4f14 5d10 6s2 6p |
| 82 | Pb | Lead | 4.95 | FCC | - | 11.34 | | 3.50 | 600.65 | 2013 | [Xe] 4f14 5d10 6s2 6p2 |
| 83 | Bi | Bismuth | 4.75 | RHL | - | 9.80 | | 3.07 | 544.5 | 1833 | [Xe] 4f14 5d10 6s2 6p3 |
| 84 | Po | Polonium | 3.34 | SCB | - | 9.31 | | 3.34 | 527 | 1235 | [Xe] 4f14 5d10 6s2 6p4 |
| 85 | At | Astatine | - | - | - | | | | 575 | 610 | [Xe] 4f14 5d10 6s2 6p5 |
| 86 | Rn | Radon | - | FCC | - | 9.23 | g/l | | 202 | 211.4 | [Xe] 4f14 5d10 6s2 6p6 |
| 87 | Fr | Francium | - | BCC | - | | | | 300 | 950 | [Rn] 7s |
| 88 | Ra | Radium | - | - | - | 5.50 | | | 973 | 1413 | [Rn] 7s2 |
| 89 | Ac | Actinium | 5.31 | FCC | - | 10.07 | | 3.76 | 1320 | 3470 | [Rn] 6d 7s2 |
| 90 | Th | Thorium | 5.08 | FCC | - | 11.72 | | 3.60 | 2023 | 5060 | [Rn] 6d2 7s2 |
| 91 | Pa | Protactinium | 3.92 | TET | 0.83 | 15.37 | | 3.21 | 1827 | 4300 | [Rn] 5f2 6d 7s2 |
| 92 | U | Uranium | 2.85 | ORC | 1.74 | 18.97 | | 2.75 | 1405.5 | 4091 | [Rn] 5f3 6d 7s2 |
| 93 | Np | Neptunium | 4.72 | ORC | 1.04 | 20.48 | | 2.62 | 913 | 4175 | [Rn] 5f4 6d 7s2 |
| 94 | Pu | Plutonium | | MCL | - | 19.74 | | 3.1 | 914 | 3600 | [Rn] 5f6 7s2 |
| 95 | Am | Americium | 3.64 | HCP | - | 11.87 | | 3.61 | 1267 | 2880 | [Rn] 5f7 7s2 |
| 96 | Cm | Curium | - | - | - | 13.51 | | | 1613 | | [Rn] 5f7 6d 7s2 |
| 97 | Bk | Berkelium | - | - | - | 13.25 | | | 1259 | | [Rn] 5f9 7s2 |
| 98 | Cf | Californium | - | - | - | 15.1 | | | 1173 | | [Rn] 5f10 7s2 |
| 99 | Es | Einsteinium | - | - | - | | | | 1133 | | [Rn] 5f11 7s2 |
| 100 | Fm | Fermium | - | - | - | | | | | | [Rn] 5f12 7s2 |
| 101 | Md | Mendelevium | - | - | - | | | | | | [Rn] 5f13 7s2 |
| 102 | No | Nobelium | - | - | - | | | | | | [Rn] 5f14 7s2 |
| 103 | Lr | Lawrencium | - | - | - | | | | | | [Rn] 5f14 6d 7s2 |
| 104 | Rf | Rutherfordium | - | - | - | | | | | | [Rn] 5f14 6d2 7s2 |
| 105 | Db | Dubnium | - | - | - | | | | | | [Rn] 5f14 6d3 7s2 |
| 106 | Sg | Seaborgium | - | - | - | | | | | | [Rn] 5f14 6d4 7s2 |
| 107 | Bh | Bohrium | - | - | - | | | | | | [Rn] 5f14 6d5 7s2 |
| 108 | Hs | Hassium | - | - | - | | | | | | [Rn] 5f14 6d6 7s2 |
| 109 | Mt | Meitnerium | - | - | - | | | | | | [Rn] 5f14 6d7 7s2 |

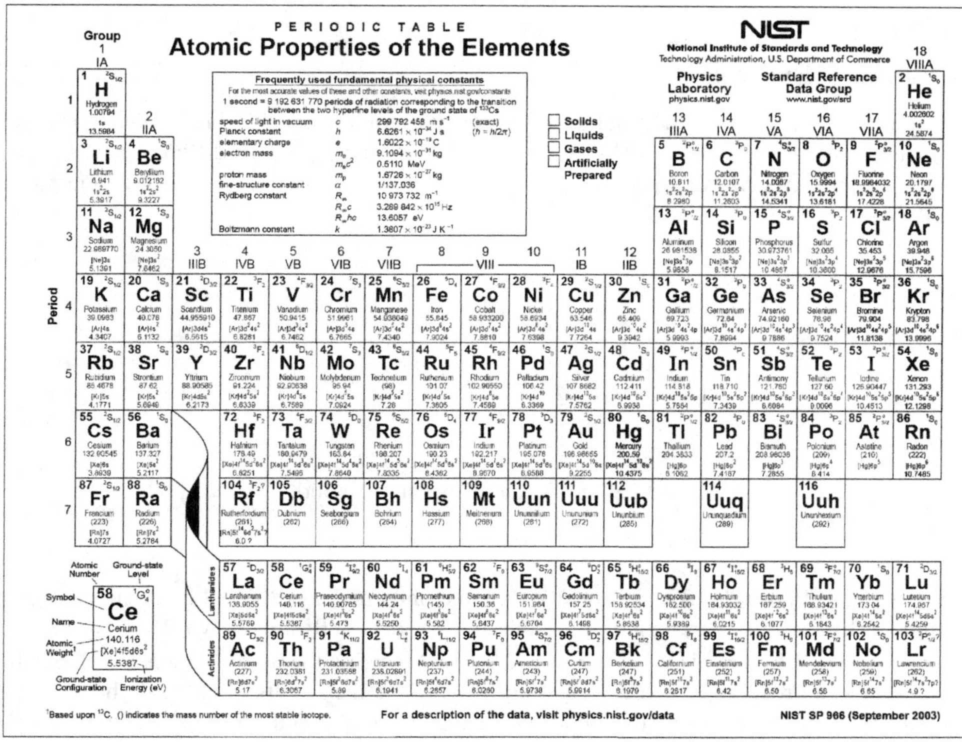


Figure 1.8.37: The standard periodic table (Courtesy of NIST - `http://www.nist.gov`)

## 1.9 Chapter 1 Exercises

1.9.1. Write an expression for the total surface area and volume for the three-dimensional solid displayed in Figure 1.9.38. (a) Assume that , $\vec{a}_1$, $\vec{a}_2$, and $\vec{a}_3$ are all different, and that $\alpha \neq \beta \neq \gamma \neq 90^\circ$. (b) Choosing your own values for each vector and each angle, give a numerical answer.

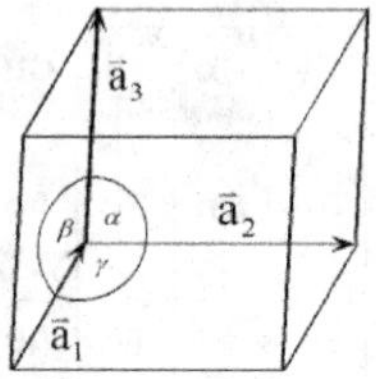


Figure 1.9.38: Three-dimensional solid.

1.9.2. Referring to Figure 1.9.39, in (a) the lattice vectors $\vec{a}$ and $\vec{b}$ of the oblique lattice are at the angle $\theta$ from each other, where $|\vec{a}| \neq |\vec{b}|$ and $0 \leq \theta \leq 180^\circ$ are shown. Here we chose an angle of $\theta = 70^\circ$, and $|\vec{a}| = 1.75a$, $|\vec{b}| = 2.0a$, with $a$ being the unit distance. For this lattice, (b) shows the resulting Wigner-Seitz cell. Reproduce the results of (b) and choosing your own values of $\vec{a}$ and $\vec{b}$ as well as their respective angle, obtain the Wigner-Seitz cell of your oblique lattice.

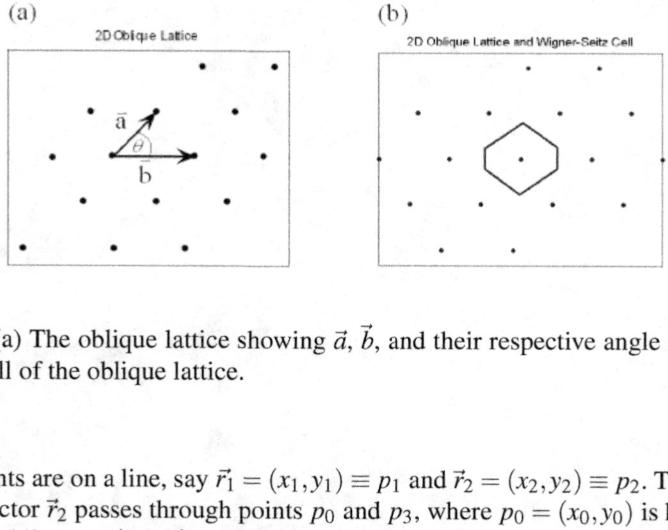


Figure 1.9.39: (a) The oblique lattice showing $\vec{a}$, $\vec{b}$, and their respective angle $\theta$. (b) The resulting Wigner-Seitz cell of the oblique lattice.

1.9.3. Two points are on a line, say $\vec{r}_1 = (x_1, y_1) \equiv p_1$ and $\vec{r}_2 = (x_2, y_2) \equiv p_2$. The line's perpendicular bisector $\vec{r}_2$ passes through points $p_0$ and $p_3$, where $p_0 = (x_0, y_0)$ is halfway between $p_1$ and $p_2$, while $p_3 = (x_3, y_3)$ is located at $d = |\vec{r}_2|$ away from $p_0$. (a) Show that the coordinates of point $p_3$ are given by $x_3 = x_0 \pm d\frac{(y_2-y_1)}{b}$ and $y_3 = y_0 \mp d\frac{(x_2-x_1)}{b}$ with $b \equiv |\vec{r}_1|$. (b) Using these results write the code needed to make possible graphs such as those of Figure 1.2.15. Below is a MATLAB code snippet to get you started.

```
clear
x1=0.0; y1=0.0; x2=1.0; y2=0.0; %inputs
p1=[x1;y1]; p2=[x2;y2];          %original points
```

```
p0=(p1+p2)/2.;                    %point midway between p1, p2
r1=norm(p2+p1);                   %length of r1
l2=2.0*r1;                        %perpendicular line length desired
r2_hat=[-(p2(2)-p1(2)),+(p2(1)-p1(1))]/r1; %solved direction of r2
p3=p0+l2*r2_hat';                 %output
line([p1(1) p2(1)],[p1(2) p2(2)],'Color','k','LineStyle','-',
     'LineWidth', 2.0)
```

1.9.4. Refer to Example 1.3.0.1 and Figure 1.3.16. (a) Show why it is possible to have an infinite lattice of hexagons. (b) What about heptagons? (c) Apply these concepts to your own polygon and explain your conclusions.

1.9.5. Graphene forms a two-dimensional honeycomb lattice with atoms at the corners of a hexagon separated by distance $d$ ($= 1.42$Å). The primitive lattice vectors are shown in Figure 1.9.40. (a) Find the lattice vectors' magnitude in terms of $d$. Call this magnitude $a$. (b) Rewrite $\vec{a}_1$ and $\vec{a}_2$ in terms of $a$, and express them in Cartesian component form with unit vectors $\hat{i}$ and $\hat{j}$. (c) Obtain the number of atoms/cell in graphene and justify your counting.

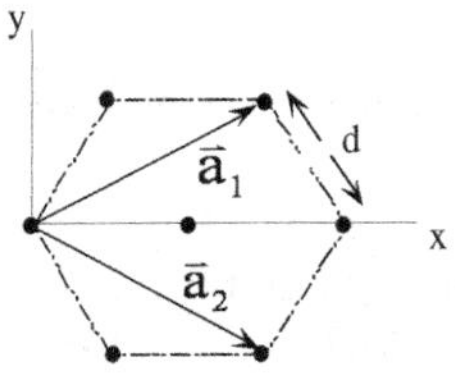


Figure 1.9.40: Graphene's honeycomb lattice and primitive lattice vectors.

1.9.6. Consider a monoclinic system similar to that shown in Figure 1.3.19(c) but for which $\alpha = \gamma = 90°$ and $\beta = 75°$ with vector magnitudes, $a = 1.50$Å, $b = 1.25$Å, and $c = 2.50$Å, find an atom's $x$, $y$, and $z$ coordinates if its position is given by the crystal vector $\vec{t} = 3.5\vec{a} - 2.25\vec{b} + 3.25\vec{c}$. Hint: see Example 1.3.1.2.

1.9.7. Consider the monoclinic system of Exercise 1.9.6. Produce a structure plot of all the atoms associated with its simple Bravais lattice, which is similar to that of Figure 1.3.19(c), albeit with different parameters.

1.9.8. (a) Obtain all the third neighbor positions for the simple cubic lattice, how many are there? (b) Identify them graphically. (c) What is the 3rd neighbor distance from the origin? (d) Can you predict what is the 4th neighbor distance? What about the nth neighbor distance?

1.9.9. Find the angle between two body diagonals of the simple cubic system.

1.9.10. (a) Obtain the 2nd neighbor positions from the central atom in the BCC structure, how many are there? (b) What is the 2nd neighbor distance from the central atom?

1.9.11. Work out the Section 1.4.3 details for the FCC's (a) volume, (b) nearest neighbor distance, (c) packing fraction, and (d) confirm that the number of nearest neighbors in the FCC structure is twelve, what are their positions with respect to the origin?

1.9.12. (a) Obtain the 2nd neighbor positions from the central atom in the FCC structure, how many are there? (b) What is the 2nd neighbor distance from the central atom?

1.9.13. Run the BCC code provided in Example 1.4.2.1 to reproduce Figure 1.4.24.

1.9.14. Modify the BCC code provided in Example 1.4.2.1 in order to repeat the process of drawing the conventional cell as well as the primitive cell, and structure of the FCC lattice. Refer to Section 1.4.3 for other details.

1.9.15. Find the Miller indices for a plane with intercepts $\{12,6,4\}$.

1.9.16. What is the crystallographic direction from the lattice point with coordinates $\{1,0,1\}$ to one with coordinates $\{2,3,5\}$?

1.9.17. Refer to Example 1.5.0.2, show that crossing vectors $\vec{v}_1$ with $\vec{v}_3$ and $\vec{v}_2$ with $\vec{v}_3$ according to the RHR, as explained there, produces the same results.

1.9.18. Run the code associated with Example 1.5.0.2 and reproduce Figure 1.5.27(b).

1.9.19. Run the code associated with Example 1.5.0.2 using the plane intercepts of Exercise 1.9.15. Comment on the results.

1.9.20. In the NaCl crystal structure, (a) what are the nearest neighbor and (b) second neighbor distances?

1.9.21. Derive the expression for the theoretical packing limit of $\frac{c}{a} = \sqrt{\frac{8}{3}} = 1.633$ in the close pack systems. Hint: refer to Section 1.6.3.

1.9.22. Show that the nearest neighbor distance in the diamond structure is given by $\frac{\sqrt{3}}{4}a$ and obtain the tetrahedral angles of the basic block shown in Figure 1.6.32(a).

1.9.23. Apply the concepts of Section 1.6.4 and reproduce the diamond structure of Figure 1.1.3(a). Hint: it might help to start with the FCC's smallest translation vectors.

1.9.24. After you have done Exercise 1.9.23, apply a random displacement to the C atomic positions in the diamond structure in order to simulate the amorphous diamond similar to Figure 1.1.3(b). Hint: in MATLAB random numbers between, say $a$ and $b$, can be generated by the command $a+(b-a)*rand$.

1.9.25. Repeat Example 1.7.0.1 but instead of using Equations 1.7.19 to estimate the percentage of electrons that tunnel, use Equations 1.7.21; i.e., without approximating $T_0$. Comment on your answer.

1.9.26. (a) Show the details of how in the limit of $\alpha a \gg 1$ as well as taking $V_0 = W$, Equations 1.7.20 become those of Equations 1.7.21. (b) When is $T_0 \sim 1$ a good approximation? (c) For large barrier heights, what could be a reasonable expression for the tunneling probability?

1.9.27. Modify the code associated with Figure 1.7.36; that is, tunnel0.m, in order to plot the three forms of the tunneling probability of Equations 1.7.19, 1.7.20, and 1.7.21. Use a value of $d = 0.3\,nm$ when plotting versus energy, and a value of $E = 0.5W$ when plotting versus gap width. For the barrier height, use the value $W = 4.3\,eV$. Comment on your results.

# 2

# *The Reciprocal Lattice*

**Contents**



## 2.1 Introduction

Physicists study crystal structures using various forms of radiation, such as photons, neutrons, and electrons. In this chapter we learn that the information gathered from the scattered particles can be analyzed to learn about the crystal plane spacing as well as the kind of atoms that compose the crystal system. What makes crystal diffraction different from ordinary light ray reflection/refraction theory is the wavelength used. If the wavelength associated with the radiation is large compared to the lattice constant, the result is simply standard reflection and refraction. In the case of standard reflection, the reflected ray leaves at an angle that is equal to the angle of incidence or $\theta_i = \theta_r$, where $\theta_i$ is the angle of incidence and $\theta_r$ is the angle of reflection, both measured from the sample's normal to the surface. In the case of refraction, the refracted ray obeys Snell's law; that is, $n_i \sin\theta_i = n_t \sin\theta_t$ where $n_i$ and $n_t$ are the indices of refraction of the incident and transmission media, respectively, and $\theta_t$ is the angle of transmission as measure from the normal. Both of these situations for ordinary reflection and refraction are shown in Figure 2.1.1 as in the case of optical wavelengths (3900 – 7500 Å).

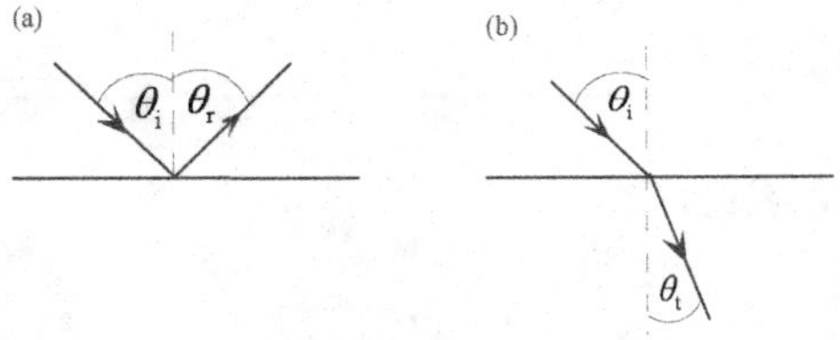


Figure 2.1.1: (a) Ordinary wave reflection and (b) refraction occur when the wavelength of radiation is much greater than the lattice spacing ($\lambda >> a$).

When radiation wavelengths, on the order of the crystal lattice constant, undergo elastic scattering from the crystal atoms, diffraction occurs in directions that are quite different from ordinary reflection. To see how this happens, we will look at the simple approach presented by William L. Bragg (1890-1971) in Section 2.2. For now, let's see what are the physical properties of the radiation that affect the size of the wavelength. To this end, let's recall the relativistic expression for the kinetic energy of a particle,

$$E_k = E - mc^2 = \sqrt{(pc)^2 + (mc^2)^2} - mc^2, \tag{2.1.1}$$

and let's use de Broglie's relation for the momentum, $p = h/\lambda$, to solve for $\lambda$ in terms of $E_k$. We obtain

$$\lambda = hc/\sqrt{(E_k + mc^2)^2 - (mc^2)^2}. \tag{2.1.2}$$

From this result, we see that the wavelength of the radiation $\lambda \propto 1/E_k$ for a photon; that is, increasing a particle's kinetic energy decreases its wavelength. It is convenient to write this expression in units that are more appropriate to the energy scale of the problem. If we use energy units of $eV's$, then we can use the value $hc = 12398\,eV \cdot \text{Å}$. Furthermore, keeping in mind that $mc^2$ is to be expressed in $eV's$ as well, then the above equation for lambda becomes

$$\lambda(\text{Å}) = 12398eV \cdot \text{Å}/\sqrt{(E_k(eV) + mc^2(eV))^2 - (mc^2(eV))^2}. \tag{2.1.3}$$

In the case of a photon with $E_k = E$, $\lambda = hc/E$ and taking the natural log of both sides of this equation we get $\ln(\lambda) = \ln(hc) - \ln(E)$, which, on a logarithmic scale, describes a straight line with a negative slope for $\lambda$ versus $E$. This explains the photon wavelength behavior shown in the log-log plot of Figure 2.1.2. Thus, in order to produce photons with wavelengths on the order of the crystal spacing (about Angstrom size) we need energies such that $\lambda(\text{Å}) = 12398eV\text{Å}/E(eV) = 1\text{Å}$ or $E \approx 1.2 \times 10^4 eV$, which lies in the X-ray regime. Figure 2.1.2 also contains the results from Equation 2.1.3 for electrons ($mc^2 \approx 0.51MeV$) and neutrons ($mc^2 \approx 939.6MeV$) with the energy units shown on the figure.

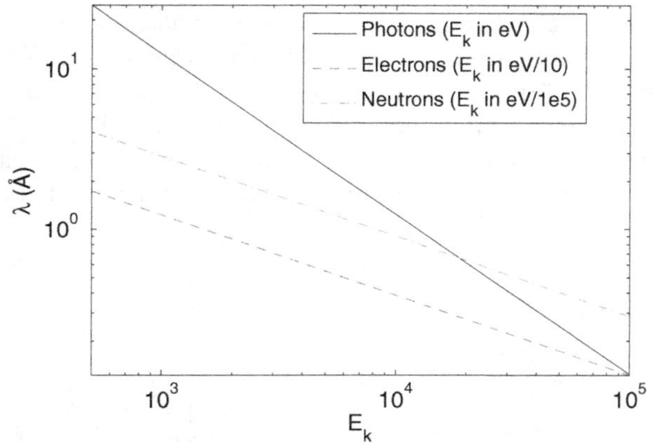


Figure 2.1.2: A log-log plot of Equation 2.1.3 for photons, electrons and neutrons. The energy scale is in units of $eV's$ for photons, tenths of $eV's$ for electrons, and tens of $\mu eV's$ for neutrons as shown in the legend.

**Example 2.1.0.1**
Listed below is a code example that will reproduce the photon line shown in Figure 2.1.2 and, with a slight modification, it can be used to reproduce the electron as well as the neutron lines.

```
%Example for the calculation of a photon's wavelength versus
%photon energy: a log-log plot
clear;
hc=12398;               %h*c in eV-angstroms
N=100;                  %steps to use in the plot
%Ek = particles' relativistic kinetic energy
Eki=5e2; Ekf=1e5;       %photon E range (eV): initial, final
Eks=(Ekf-Eki)/(N-1);    %step size
Ek=Eki:Eks:Ekf;
lambda=hc./Ek;
loglog(Ek,lambda,'k')   %log-log plot
axis tight
legend('Photons (E_k in eV)',0);
xlabel('E_k')
ylabel('\lambda ()')
```

## 2.2 Bragg's Law

The English physicists William Lawrence Bragg (1890-1971) and William Henry Bragg (1862-1942), a son-father team, exploited the wave nature of x-rays in 1913. They simplified Max von Laue's earlier proposed analysis of the scattering of x-rays by crystals. The Bragg team found that at certain specific wavelengths and incident angles, the crystals' reflected radiation produced diffraction peaks or Bragg peaks. Diffraction is an example of interference; whereas interference is the result of the superposition of only a few waves, diffraction is the result of the superposition of a large number of waves. The diffraction in a crystal is thus similar to the concept of the Young's double slit interference experiment where waves passing through two slits, separated by a distance $d$, cause light waves to interfere constructively if they arrive in phase at a point $y$ from the center of a screen located at a distance $L$ away from the slits, as shown in Figure 2.2.3. The two rays reaching point $y$ on the screen produce a bright point if $\ell_2 - \ell_1 = d\sin\theta = n\lambda$ where $n$ is the fringe order integer (1, 2, ...). There is always a bright point at the center of the screen. For small angles, the position of the bright points can be found using $y = L\tan\theta$. From these two equations and using the small angle approximation $\tan\theta \approx \sin\theta$, the position of the *nth* fringe is obtained as $y_n \approx n\lambda L/d$.

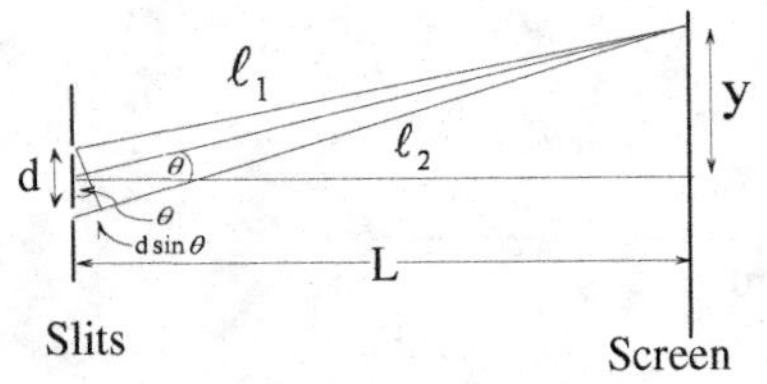


Figure 2.2.3: Young's double slit diffraction experiment, where $\ell_2 - \ell_1 = d\sin\theta = n\lambda$. The position of the *nth* fringe is $y_n \approx n\lambda L/d$. Shown is the case when $n = 1$, or $y = y_1$.

In a similar manner, crystals are responsible for X-ray, neutron, and electron diffraction. William L. Bragg imagined the crystal to be composed of parallel planes spaced from each other by amount $d$. The crystal acts as a three-dimensional diffraction grating which causes diffraction peaks to occur wherever the crystal reflected waves interfere constructively as shown in Figure 2.2.4.

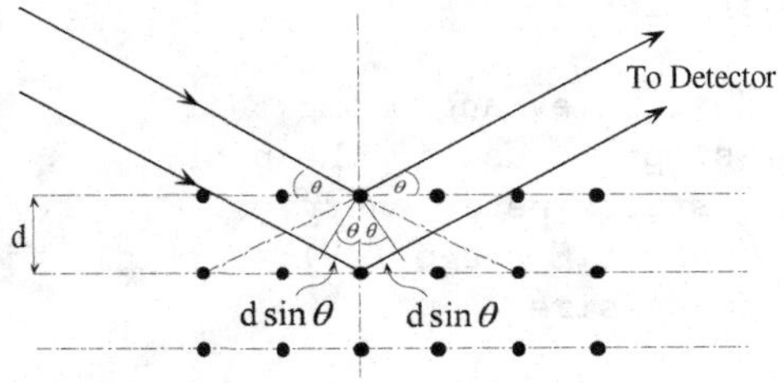


Figure 2.2.4: Experimental setup leading to crystal diffraction. Whenever Bragg's Law is satisfied, the detected rays give rise to diffraction intensity maxima known as Bragg peaks.

Thus, with $d$ the spacing between planes, if $\lambda \lesssim d$; that is, small enough so as not to produce standard reflection or refraction in a crystal, we suppose that the incident beams are reflected by the atomic planes. The outgoing rays experience constructive interference if their path difference is a multiple of the beam's wavelength. From Figure 2.2.4, we have

$$2d\sin\theta = n\lambda, \tag{2.2.4}$$

and diffraction peaks are obtained whenever this relation is satisfied. This is known as Bragg's Law of diffraction of waves by crystals. The angle $\theta$ is the angle between the incident radiation and the atomic planes within the crystal; it is known as the Bragg angle. Twice the Bragg angle or $2\theta$ is known as the diffraction angle; that is, the angle between the incident and the diffracted radiation. From Equation 2.2.4, since the maximum angle is 90°, we notice that $2d < n\lambda_{max}$, so that, for $n = 1$, one has to choose the wavelength of the beam wisely. In the present case we need $\lambda \leq 2d$ but one has to search for the best angle that produces the maximum intensity of the bright points in order to obtain $d$ correctly. In the experimental setup, the intensity of the reflected beam is plotted as a function of angle. Example calculations of the Bragg intensity peaks for an Fe crystal with the BCC structure is shown in Figures 2.2.5 (a) and (b).

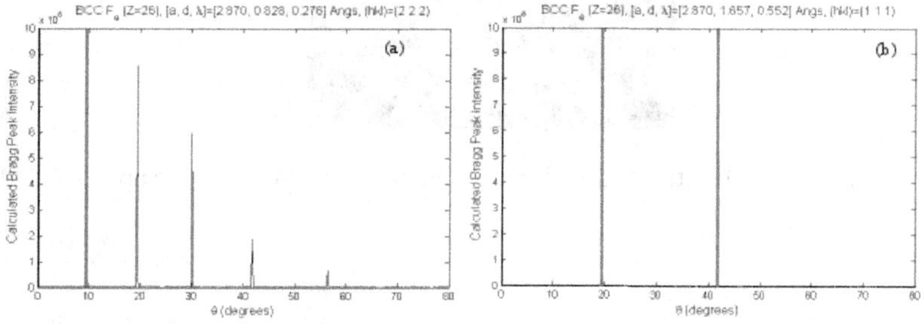


Figure 2.2.5: Example calculations of Bragg peaks for Fe in the BCC structure using the (a) (222) and (b) (111) planes. The lattice constant used is 2.87 Å, and the interplanar distance is taken as $a/\sqrt{h^2+k^2+l^2}$ (see Example 2.4.4.1 discussed later), where $(hkl)$ are the Miller indices, and where we have used $\lambda = d/3$. The method used will be developed later in Section 2.4.2.

In (a) the peaks shown correspond to high intensity diffraction peaks from the $(222)$ plane. Six peaks (some with very low intensity) are discerned in the figure for each value of $n = 0, 1, 2, 3, 4, 5$ whose angles are obtained from the solution of Equation 2.2.4 for a lattice constant of 2.87 Å using a plane separation of $d = \frac{a}{2\sqrt{3}} = 0.828$ Å. A similar calculation is shown in (b) for the $(111)$ plane with $d = \frac{a}{\sqrt{3}} = 1.657$ Å, but notice that only three Bragg peaks are shown here (including the $n = 0$ case). Both calculations assumed a radiation wavelength of $\lambda = d/3$ with $d$ being the corresponding interplanar distance value associated with each set of Miller indices. The method used in obtaining the figures will be explained in Section 2.3. From the Bragg reflections, by analyzing the intensity as a function of angle, we obtain information about the lattice points and, in particular, the distances between the planes of atoms, thus gaining an understanding of the crystal structure. However, we need to learn more about the beam's scattered intensity and its relationship to the atoms in the crystal, in particular their electron distribution. It is the electron distribution that identifies the atomic species which the radiation interacts with.

## 2.3 Reciprocal Lattice Vectors

We will learn in this section that just as a crystal has unique lattice vectors or real space lattice vectors which identify the unique crystal structure, there are reciprocal lattice vectors associated with every set of real space lattice vectors. The reciprocal lattice vectors are vectors in the abstract or the momentum space (k-space) of the crystal. When an electron diffraction image of a crystal is made, a series of bright dots is seen as shown, for example, in Figure 2.3.6.

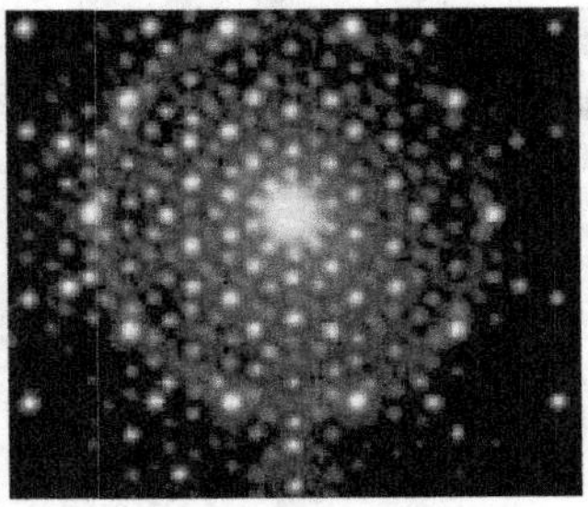

Figure 2.3.6: An electron diffraction pattern obtained from the $Al_{78}Mn_{22}$ rapidly solidified alloy (Ref. [9]).

The dots are thought of as the reciprocal space version of the real crystal. Reciprocal space is also referred to as Fourier space. Each dot in reciprocal lattice space is connected to each other by a reciprocal lattice translation vector denoted by $\vec{G}$. Just as a real space translation vector $\vec{T}$ of Chapter 1 is represented in terms of real space fundamental lattice vectors; i.e.,

$$\vec{T} = u_1\vec{a}_1 + u_2\vec{a}_2 + u_3\vec{a}_3, \tag{2.3.5}$$

the reciprocal lattice translation vector also has a similar representation. It is written in the form

$$\vec{G} = v_1\vec{b}_1 + v_2\vec{b}_2 + v_3\vec{b}_3, \tag{2.3.6}$$

where the $v_i's$ are integers and $\vec{b}_1$, $\vec{b}_2$, $\vec{b}_3$ are fundamental reciprocal lattice vectors which are related to the fundamental lattice vectors $\vec{a}_1$, $\vec{a}_2$, $\vec{a}_3$ of the real crystal as we will see later. We now proceed to study how the reciprocal lattice vectors play an important role in exploiting crystal periodic properties. In particular, we will employ the periodic property of the atomic electron concentration in a crystal.

### 2.3.1 Electron Concentration and Reciprocal Lattice Vectors in 1-D

A crystal is invariant under any translation of the form of Equation 2.3.5. Similarly any local property of the crystal is invariant under $T$. One example of this is the charge concentration or electron number density. Let the electron number density be $n(\vec{r})$, which is a periodic function of $\vec{r}$ with periods $\vec{a}_1$, $\vec{a}_2$, $\vec{a}_3$ in the direction of the crystal axes. We have

$$n(\vec{r} + \vec{T}) = n(\vec{r}); \tag{2.3.7}$$

that is, the electron concentration at position $\vec{r} + \vec{T}$ is the same as that at position $\vec{r}$. Let's work in one-dimension first, later we will extend the results to three dimensions. Consider $n(x)$ to be of period $a$ in the $x$-direction as shown in Figure 2.3.7.

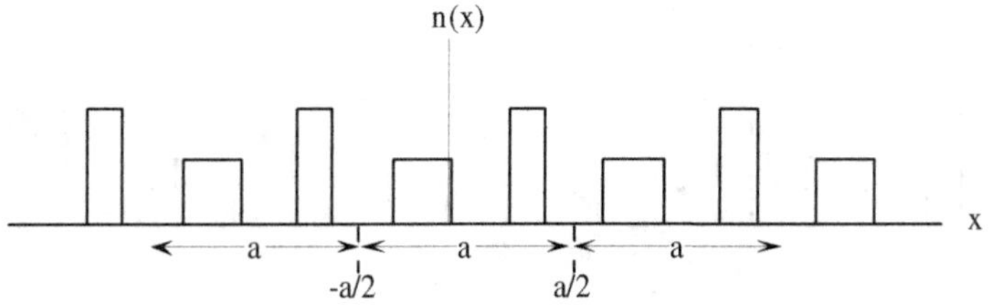


Figure 2.3.7: An anharmonic function of $x$ is shown as a one-dimensional model of a crystal's electron distribution with period $a$. The electron distribution follows the periodic atomic distribution. Here the model represents two different atomic species.

An anharmonic function of period $a$ can be expanded in terms of a Fourier series. We do this using the complex representation as follows,

$$n(x) = \sum_p n_p e^{ikpx}, \tag{2.3.8}$$

where $p$ is an integer in the range $-\infty < x < \infty$, and $n_p$ are the Fourier coefficients. The $n_p$'s are obtained by inverting the Fourier series with the help of the Kronecker delta function representation $a\delta_{pp'} = \int_0^a e^{ik(p-p')x}\,dx$, where $\delta_{pp'}$ takes on the value 1 if $p = p'$, else it is zero. In this way one obtains

$$n_p = \frac{1}{a}\int_0^a n(x)e^{-ikpx}\,dx, \tag{2.3.9}$$

We next probe to see what is the condition necessary for periodicity to occur in $n(x)$; that is, we need $n(x+a) = n(x)$, which is the one-dimensional version of Equation 2.3.7. From the one-dimensional Equation 2.3.8, let's make the displacement $x \to x+a$ to write

$$n(x+a) = \sum_p n_p e^{ikp(x+a)} = \sum_p n_p e^{ikpx} e^{ikpa} = \sum_p n_p e^{ikpx} = n(x), \tag{2.3.10}$$

where we see that in order for $n(x+a) = n(x)$ we have taken $e^{ikpa} = 1$ or that $kpa = 2p\pi$, which implies that $k = 2\pi/a$. In one dimension, from this we conclude that in order for the electron concentration to be periodic with period $a$, $k$ must take on the special value

$$k = 2\pi/a. \tag{2.3.11}$$

It is this special value of $k$ that ensures the periodicity of the crystal properties, such as $n(x)$ in the present case. In the above, we think of the products $kp = \frac{2\pi}{a}p$, for integer $p$, as special points in Fourier space; i.e., points in reciprocal lattice space as shown in Figure 2.3.8.

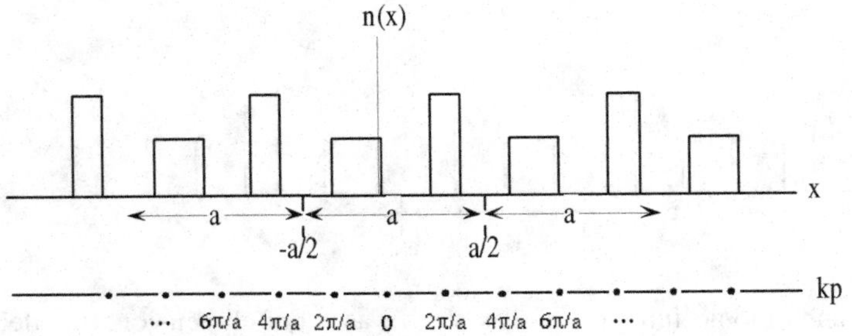


Figure 2.3.8: The anharmonic function of $x$ of Figure 2.3.7 with the associated reciprocal lattice points, $kp$, shown below the real space $n(x)$ distribution. Later in Equation 2.3.13c, for one dimension, we define the product $G \equiv kp$.

While on this subject and before moving any further, let's discuss the properties of the Fourier coefficients $n_p$ in Equation 2.3.8. The electron number density is real, and so must be the case for its complex Fourier representation; that is, $n(x) = n^*(x)$. This implies that $n^*(x) = \sum_p n_p^* e^{-ikpx} = \sum_p n_p e^{ikpx}$. Since the sum is over all p (positive and negative), then if in the first of these sums we replace $p$ with $-p$ we have

$$n^*(x) = \sum_p n_{-p}^* e^{ikpx} = \sum_p n_p e^{ikpx}, \tag{2.3.12a}$$

from which we conclude that in order for $n(x)$ to remain real, the Fourier coefficients, $n_p$, must satisfy the relation

$$n_{-p}^* = n_p. \tag{2.3.12b}$$

Returning to Equation 2.3.8, and making use of Equation 2.3.11, we can write

$$n(x) = \sum_G n_G e^{iGx}, \tag{2.3.13a}$$

whose Fourier components (instead of Equation 2.3.9) are now written as

$$n_G = \frac{1}{a}\int_0^a n(x)e^{-iGx}\,dx, \tag{2.3.13b}$$

where we have defined the one-dimensional $G$ in the form

$$G \equiv kp = p\frac{2\pi}{a}, \tag{2.3.13c}$$

with the integer $p$ absorbed in the definition. Also notice that the $n_G$ now replaces $n_p$, and the sum over the $p$'s has been replaced with a sum over the $G$'s. We next compare this $G$ with the one-dimensional version of Equation 2.3.6. If in one dimension we write $\vec{G_{1D}} = v_1\vec{b_1} = G_{1D}\hat{b_1}$, we see that, $\hat{b_1}$ must be $\hat{x}$, $G_{1D}$ is simply our $G$, $b_1$ is our $k = 2\pi/a$, and $v_1$ is our $p$. Thus, we have obtained the one-dimensional version of the reciprocal lattice vector, $\vec{b_1} = (2\pi/a)\hat{x}$, as well as the one-dimensional version of the reciprocal lattice translation vector, $\vec{G_{1D}} = p(2\pi/a)\hat{x}$. In one dimension, the $G$s (Equation 2.3.13c) correspond to a lattice of points in the one-dimensional reciprocal lattice space of the one-dimensional real crystal as illustrated in Figure 2.3.8. With these findings we notice that

$$Ga = 2p\pi, \tag{2.3.14a}$$

and

$$b_1 a = 2\pi. \tag{2.3.14b}$$

As we will see later, the behavior of the electron density, the reciprocal lattice translation vector $\vec{G}$, and the reciprocal lattice vectors $\vec{b_1}$, $\vec{b_2}$, $\vec{b_3}$ obey similar analogical properties in three dimensions.

### 2.3.2 Electron Concentration and Reciprocal Lattice Vectors in 3-D

By analogy with the one-dimensional Equations 2.3.13 for the electron density, the Fourier coefficients, and the reciprocal lattice points, we simply extend the expressions to three dimensions and write

$$n(\vec{r}) = \sum_{\vec{G}} n_G e^{i\vec{G}\cdot\vec{r}}, \tag{2.3.15a}$$

whose Fourier components are now given by

$$n_G = \frac{1}{V_c}\int_{\text{cell}} n(\vec{r}) e^{-i\vec{G}\cdot\vec{r}}\, dV, \tag{2.3.15b}$$

where, because the integral is now over the crystal cell volume, the denominator is the volume of the cell. In addition, we have made use of the three-dimensional $\vec{G}$ defined as

$$\vec{G} \equiv \vec{k}p, \tag{2.3.15c}$$

with $p$ still an integer, and $\vec{k}$ is a three-dimensional reciprocal lattice vector. The $\vec{G}$'s so created span all the special points of the three-dimensional reciprocal lattice space.
If in Equation 2.3.15b, we take $\vec{G} = 0$, then $n_{G=0} = \frac{1}{V_c}\int_{\text{cell}} n(\vec{r})\, dV$ which corresponds to the number of electrons per cell volume. In general $n_G$ is a measure of the density of electrons as a function of $\vec{G}$. We will also see later that $n_G$ is related to the scattering amplitude of an electromagnetic wave in a crystal, which through Equation 2.3.15b tells us that a diffraction experiment provides indirect information about a crystal's $n(\vec{r})$.
In analogy with Equation 2.3.10, in three dimensions, after a translation of the vectors $\vec{a}_1$, $\vec{a}_2$, or $\vec{a}_3$ any crystal property is to remain invariant. For example, in the case of the electron concentration if in Equation 2.3.15a we make the displacement of $\vec{r} \to \vec{r} + \vec{a}_i$ for $i = 1, 2, 3$, we also require that

$$n(\vec{r} + \vec{a}_i) = \sum_{\vec{G}} n_G e^{i\vec{G}\cdot(\vec{r}+\vec{a}_i)} = \sum_{\vec{G}} n_G e^{i\vec{G}\cdot\vec{r}} e^{i\vec{G}\cdot\vec{a}_i} = n(\vec{r}), \tag{2.3.16}$$

which can only happen if $e^{i\vec{G}\cdot\vec{a}_i} = 1$. In fact, if for any factor of the form

$$f(\vec{r}) = e^{i\vec{G}\cdot\vec{r}}, \tag{2.3.17a}$$

we replace $\vec{r}$ with $\vec{r} + \vec{a}_i$ we require that the condition

$$f(\vec{r} + \vec{a}_i) = e^{i\vec{G}\cdot(\vec{r}+\vec{a}_i)} = e^{i\vec{G}\cdot\vec{r}} e^{i\vec{G}\cdot\vec{a}_i} = f(\vec{r}) e^{i\vec{G}\cdot\vec{a}_i} = f(\vec{r}), \tag{2.3.17b}$$

be obeyed. Again, this can only happen if the phase factor $e^{i\vec{G}\cdot\vec{a}_i}$ takes on the value

$$e^{i\vec{G}\cdot\vec{a}_i} = 1, \tag{2.3.17c}$$

which guarantees that the crystal property under consideration remains invariant under a displacement through a crystal translation vector. The condition of Equation 2.3.17c implies that for each of the crystal lattice vectors $a_{i=1,2,3}$ we must have

$$\vec{G} \cdot \vec{a}_i = 2m\pi, \tag{2.3.18}$$

where $m$ is an integer. In a similar fashion this relation also applies in the case of a general translation vector $\vec{T}$ of the form of Equation 2.3.5 such that $e^{i\vec{G}\cdot\vec{T}} = 1$.
Recall that the vector Equation 2.3.6 is the general form of $\vec{G}$. The $\vec{G}$'s are special reciprocal lattice vectors. Points in reciprocal lattice space are described by the various possible values of $\vec{G}$ through the $v_i$ integers along with the crystal specific fundamental reciprocal lattice vectors $\vec{b}_1$, $\vec{b}_2$, and $\vec{b}_3$. This brings us directly to the root of the matter; i.e., which are the $\vec{b}_i$'s. Since we know $\vec{G}$ from Equation 2.3.6, then if Equation 2.3.18 is to be true, it must also be that

$$\vec{b}_i \cdot \vec{a}_j = 2\pi\delta_{ij}; \tag{2.3.19}$$

that is, $\vec{b}_1 \cdot \vec{a}_1 = \vec{b}_2 \cdot \vec{a}_2 = \vec{b}_3 \cdot \vec{a}_3 = 2\pi$ and $\vec{b}_1 \cdot \vec{a}_2 = \vec{b}_1 \cdot \vec{a}_3 = \vec{b}_2 \cdot \vec{a}_1 = \vec{b}_2 \cdot \vec{a}_3 = \vec{b}_3 \cdot \vec{a}_1 = \vec{b}_3 \cdot \vec{a}_2 = 0$. In three dimensions, given the $\vec{a}_i$'s, Equation 2.3.19 corresponds to 9 equations and 9 unknowns for the components of the $\vec{b}_i$'s. We can, however, make an educated guess at their form. For example, in order to have the vector equations $\vec{b}_1 \cdot \vec{a}_1 = 2\pi$ and $\vec{b}_1 \cdot \vec{a}_2 = \vec{b}_1 \cdot \vec{a}_3 = 0$, we could make the ansatz $\vec{b}_1 \propto \vec{a}_2 \times \vec{a}_3$. In this way $\vec{b}_1$ is perpendicular to both $\vec{a}_2$ and , $\vec{a}_3$, which will produce a zero dot product and solve two of the vector equations associated with $\vec{b}_1$. The last vector equation could be used to solve for the constant of proportionality. If we know the crystal lattice vectors, $\vec{a}_{i=1,2,3}$, the final result for the crystal reciprocal lattice vectors is

$$\vec{b}_1 = 2\pi\frac{\vec{a}_2 \times \vec{a}_3}{V_c}, \qquad \vec{b}_2 = 2\pi\frac{\vec{a}_3 \times \vec{a}_1}{V_c}, \qquad \vec{b}_3 = 2\pi\frac{\vec{a}_1 \times \vec{a}_2}{V_c}, \tag{2.3.20a}$$

where the real cell volume is written as

$$V_c \equiv \vec{a}_1 \cdot (\vec{a}_2 \times \vec{a}_3). \tag{2.3.20b}$$

We, therefore, find that every crystal structure has two lattices associated with it: the crystal lattice (or real space lattice) and the reciprocal lattice (or the Fourier space lattice). Let's do a simple example.

**Example 2.3.2.1**
Use Equations 2.3.20 to obtain the simple orthorhombic system's reciprocal lattice vectors.

**Solution**
From Chapter 1, the real space lattice vectors of the simple orthorhombic system are $\vec{a}_1 = a\hat{i}$, $\vec{a}_2 = b\hat{j}$, and $\vec{a}_3 = c\hat{k}$, so that $V_c = abc$. Furthermore, $\vec{a}_1 \times \vec{a}_2 = ab\hat{i} \times \hat{j} = ab\hat{k}$, and similarly $\vec{a}_2 \times \vec{a}_3 = bc\hat{i}$ and $\vec{a}_3 \times \vec{a}_1 = ac\hat{j}$, to obtain for the simple orthorhombic reciprocal lattice vectors

$$\vec{b}_1 = \frac{2\pi}{a}\hat{i}, \qquad \vec{b}_2 = \frac{2\pi}{b}\hat{j}, \qquad \vec{b}_3 = \frac{2\pi}{c}\hat{k}. \tag{2.3.21}$$

We will come back to this subject later in the chapter. The reciprocal lattice is crucial in understanding a crystal structure because the diffraction pattern of a crystal is a map of its reciprocal lattice. Equations 2.3.20 express the relationship between both of these lattices. It is worth remembering that, working with wavelengths in the optical region of the electromagnetic spectrum, a microscope image is a map of a structure in real space, since $\lambda >> d$, where $d$ is the lattice spacing. However, for $\lambda \lesssim d$ the map is that of the crystal's reciprocal lattice space.
Notice that while the real space lattice vectors have units of *length*, the reciprocal lattice vectors have units of $1/length$. Also, weavevectors as in, for example, $e^{\vec{k}\cdot\vec{r}}$ are always drawn in Fourier space, and carry momentum information in the description of a wave (e.g., $\hbar\vec{k}$), but the reciprocal lattice vector $\vec{G}$ has a special significance. That is so because of the definitions in Equations 2.3.6 and 2.3.20, the electron number density is invariant under a crystal translation of $\vec{T}$; that is $n(\vec{r}+\vec{T}) = n(\vec{r})$.

## 2.4 Revisiting Bragg's Law

In this section we revisit Bragg's Law. In particular we show that the radiation scattering intensity from a crystal is related to the reciprocal lattice vectors $\vec{G}$. We will show that the diffraction peaks occur because Bragg's Law is naturally embedded in the description of the scattering intensity from a crystal. We begin with a discussion on an important property of the vector $\vec{G}$.
Start

### 2.4.1 $\vec{G}$ and its perpendicularity to the (*hkl*) plane

One of the interesting aspects of the reciprocal lattice vector, Equation 2.3.6, is that if we consider a crystal plane with Miller indices $(hkl)$ and we use these integers to create the vector $\vec{G}$; that is, we pick $v_1 = h$, $v_2 = k$, and $v_3 = l$, and write

$$\vec{G}_{(hkl)} = h\vec{b}_1 + k\vec{b}_2 + l\vec{b}_3, \tag{2.4.22}$$

where $\vec{b}_1$, $\vec{b}_2$, and $\vec{b}_3$ are the crystal's reciprocal lattice vectors, then it can be shown that this vector $\vec{G}_{(hkl)}$ is perpendicular to the the crystal plane $(hkl)$ as illustrated in Figure 2.4.9 (see Exercise 2.6.10).

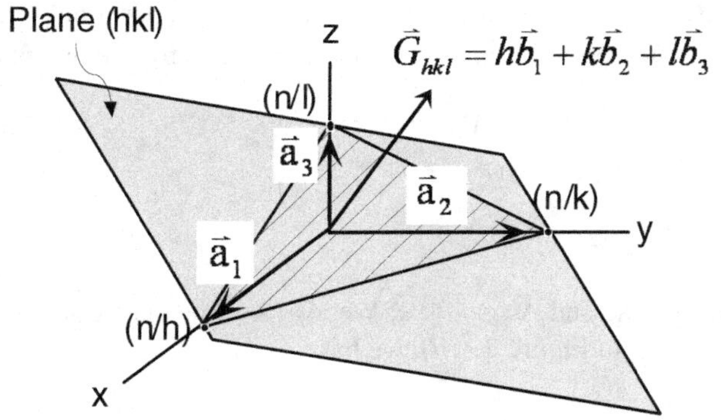


Figure 2.4.9: Shown is the vector $\vec{G}_{(hkl)} = h\vec{b}_1 + k\vec{b}_2 + l\vec{b}_3$, which is perpendicular to the the plane with Miller indices $(hkl)$, where $\vec{b}_1$, $\vec{b}_2$, and $\vec{b}_3$ are the crystal's reciprocal lattice vectors, and $\vec{a}_1, \vec{a}_2$, and $\vec{a}_3$ are the real space lattice vectors. Here the plane intercepts, $n/h$, $n/k$, and $n/l$, are also shown.

This is an important property of vector $\vec{G}$ which we will exploit shortly below.

### 2.4.2 X-Ray Scattering Intensity from Crystals

In this Section we consider the scattering of waves from a crystal. We assign the wavevector $\vec{k}$ to the incident wave and the wavevector $\vec{k}'$ to the scattered wave. We think of a crystal volume element $dV'$ located at position $\vec{r}'$, as in Figure 2.4.10, and a detector of the scattered wave to be located at a constant distance $|\vec{r} - \vec{r}'|$ from the atom.

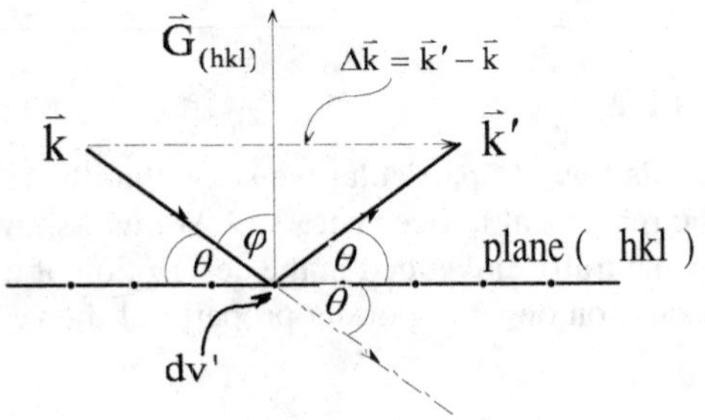


Figure 2.4.10: The scattering from volume element $dV'$ of a crystal plane $(hkl)$ is shown. The angle of the incoming wave with weavevector $\vec{k}$ is $\theta$. Similarly the angle of the scattered wave with weavevector $\vec{k}'$ is angle $\theta$ or Bragg angle and $2\theta$ is the diffraction angle. The $\vec{G}_{(hkl)}$ vector, associated with the $hkl$ plane is shown $\perp$ to it and at an angle $\phi$ from the incident wavevector $\vec{k}$.

Let the propagation vector of the scattered wave be $\vec{k}'$ and let's write the scattered wave as a plane wave of the form

$$\psi(\vec{r},t) = A_s e^{i[\vec{k}'\cdot(\vec{r}'-\vec{r})-\omega t]}. \tag{2.4.23}$$

Here $A_s$ is the amplitude of the scattered wave which is taken proportional to the incident wave or $A_s = Ae^{i\vec{k}\cdot\vec{r}'}$, we then write the contribution to the scattered wave from element of volume $dV'$ as

$$\begin{aligned}\psi(\vec{r},\vec{r}',t)dV' &= Ae^{i\vec{k}\cdot\vec{r}'}e^{i[\vec{k}'\cdot(\vec{r}-\vec{r}')-\omega t]}dV' \\ &= Ae^{i(\vec{k}'\cdot\vec{r}-\omega t)}e^{-i(\vec{k}'-\vec{k})\cdot\vec{r}'}dV' \\ &= Ae^{i(\vec{k}'\cdot\vec{r}-\omega t)}e^{-i\Delta\vec{k}\cdot\vec{r}'}dV',\end{aligned} \tag{2.4.24}$$

where $V'$ is the sample volume and $\Delta\vec{k} \equiv \vec{k}' - \vec{k}$. We also assume elastic scattering so that $k = k' = 2\pi/\lambda$; furthermore, referring to Figure 2.4.10, we have

$$\begin{aligned}|\Delta\vec{k}| = |\vec{k}'-\vec{k}| &= \sqrt{k'^2+k^2-2kk'\cos 2\theta} \\ &= \sqrt{2k^2-2k^2\cos 2\theta} \\ &= \sqrt{2}k\sqrt{1-\cos 2\theta} \\ &= 2k\sin\theta = \frac{4\pi}{\lambda}\sin\theta,\end{aligned} \tag{2.4.25}$$

where we have used $2\sin^2\theta = 1-\cos 2\theta$. We next consider the total scattering of X-rays by the electrons in a crystal. We take $n(r')$ as the concentration of electrons at position $\vec{r}'$ and define the total scattering as $\psi_{T,\Delta k}(\vec{r},t) \equiv \int \psi(\vec{r},\vec{r}',t)n(r')dV'$, which with Equation 2.4.24 becomes

$$\psi_{T,\Delta k}(\vec{r},t) = Ae^{i(\vec{k}'\cdot\vec{r}-\omega t)}\int e^{-i\Delta\vec{k}\cdot\vec{r}'}n(r')dV'. \tag{2.4.26}$$

At this point, it is worth showing a particular property of this expression that involves the relationship of $\Delta\vec{k}$ and the reciprocal lattice vector $\vec{G}$. To see this, we expand $n(r')$ as in Equation 2.3.15a, and rewrite Equation 2.4.26 as

$$\psi_{T,\Delta k}(\vec{r},t) = Ae^{i(\vec{k}'\cdot\vec{r}-\omega t)}F, \tag{2.4.27a}$$

where the following definition has been made

$$F \equiv \sum_{\vec{G}} n_G \int e^{i(\vec{G}-\vec{\Delta k})\cdot\vec{r}'} dV'. \tag{2.4.27b}$$

We notice that the function $F$ in Equation 2.4.27a determines the strength of the scattering. The largest term in the sum over $G$ in $F$ itself is the term for which

$$\vec{\Delta k} = \vec{G}, \tag{2.4.28}$$

see, for example, Exercise 2.6.11. As we will see below, this is a formal expression of Bragg's Law. It says that the lattice vectors $\vec{G}$ determine the possible x-ray reflections. The idea is that Equation 2.4.26 gives the Bragg peaks whenever Equation 2.4.28 is satisfied.

We now go back to Equation 2.4.26 and incorporate the Bragg condition as expressed in Equation 2.4.28; that is,

$$\psi_{T,G}(\vec{r},t) = Ae^{i(\vec{k}'\cdot\vec{r}-\omega t)} \int e^{-i\vec{G}\cdot\vec{r}'} n(r')dV'. \tag{2.4.29}$$

If we let the nucleus of atom $j$ be located at $\vec{r}_j$ and let the electron concentration around it be $n_j$, then we also let $\vec{r}''$ be the displacement vector from the nucleus to the electron concentration (see Figure 2.4.11(a)) or

$$\vec{r}' = \vec{r}_j + \vec{r}'', \tag{2.4.30}$$

so that Equation 2.4.29 becomes

$$\psi_{T,G}(\vec{r},t) = Ae^{i(\vec{k}'\cdot\vec{r}-\omega t)} \sum_j^{\text{all atoms}} e^{-\vec{G}\cdot\vec{r}_j} \int_{\text{atom}} e^{-i\vec{G}\cdot\vec{r}''} n(r'')dV''. \tag{2.4.31}$$

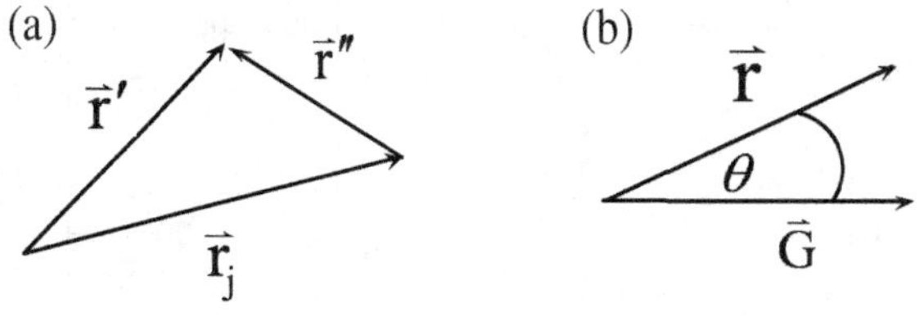


Figure 2.4.11: (a) The $j$th atomic nucleus is located at $\vec{r}_j$ and $\vec{r}'\prime$ sweeps over its electron concentration. (b) The vector $\vec{r}$ is shown making an angle $\theta$ with the vector $\vec{G}$.

The quantity with the integral is known as the form factor or

$$f_j \equiv \int_{\text{atom}} e^{-i\vec{G}\cdot\vec{r}''} n(r'')dV''. \tag{2.4.32}$$

The contribution to the total scattering due to the electron concentration surrounding atom $j$ in the crystal finally becomes

$$\psi_{T,G}(\vec{r},t) = Ae^{i(\vec{k}'\cdot\vec{r}-\omega t)} \sum_j^{\text{all atoms}} e^{-i\vec{G}\cdot\vec{r}_j} f_j. \tag{2.4.33}$$

As an example, we will do a calculation of the form factor under the approximation that the electron concentration is constant or $n(r) = n$.

**Example 2.4.2.1**

With constant $n$ Equation 2.4.32 is $f_j = n\int_{\mathrm{atom}} e^{-i\vec{G}\cdot\vec{r}}dV$. Using the volume element in spherical coordinates $dV = r^2\sin\theta dr d\theta d\phi$, and carrying out the integration over $\phi$ get $f_j = 2n\pi\int_0^R r^2 dr I_r$, where $I_r \equiv \int_0^\pi e^{-iGr\cos\theta}\sin\theta d\theta$ and where we have taken $\vec{G}\cdot\vec{r} = Gr\cos\theta$ as shown in Figure 2.4.11(b). Performing the $\theta$ integration, find $I_r = e^{-iGr\cos\theta}|_0^\pi/(iGr) = \frac{2}{Gr}\sin Gr$ and $f_j$ becomes $f_j = \frac{4n\pi}{G}\int_0^R r\sin(Gr)dr$, where $R$ is taken as the atomic radius. Finally, using $\int x\sin ax = \sin ax/a^2 - x\cos ax/a$, we get

$$f_j = \frac{4n\pi}{G^3}(\sin GR - GR\cos GR) = \frac{3Z}{(GR)^3}(\sin GR - GR\cos GR), \tag{2.4.34a}$$

where we have also used $n = Z/(4\pi R^3/3)$ with $Z$ the atomic number associated with the $j$'th crystalline atom. This expression for $f_j$ becomes more meaningful when we make use of Equations 2.4.25 and the Bragg condition, Equation 2.4.28, to obtain the reciprocal lattice vector magnitude in the form

$$G = |\Delta k| = \frac{4\pi}{\lambda}\sin\theta. \tag{2.4.34b}$$

In the limit as $GR \to 0$, $f_j \to Z$. From this we see that the form factor bears information about the atomic composition of the crystal. It actually measures the scattering contribution of the $j$'th atom of the crystal as a function of $G$. Listed below is the script form_factor_n_const.m which was used to produce the plot of Equation 2.4.34a versus $\theta$ and which uses the above Equation 2.4.34b for the Fe crystal with a BCC structure. We have used the atomic radius $R = d_{nn}/2$ as in Chapter 1, where the nearest neighbor distance in the BCC is $d_{nn} = \sqrt{(3)}a/2$ with a lattice constant $a = 2.87$Å. Finally, Figure 2.4.12 shows the decaying nature of the form factor away from the $G = 0$ value. The oscillatory behavior depends on the electron concentration used and may not necessarily be there for a more realistic model.

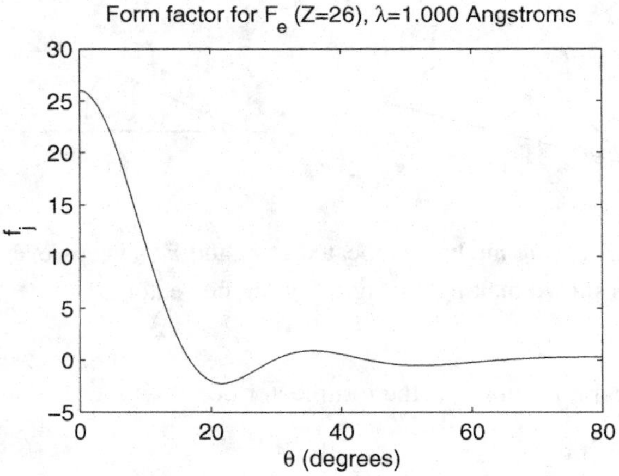


Figure 2.4.12: The form factor of Equation 2.4.34a versus $\theta$ for the Fe BCC structure, assuming a constant electron concentration.

The code used the create Figure 2.4.12 follows.

```
%copyright by J. Hasbun and Trinanjan Datta
%form_factor_n_const.m
```

```
%We use the constant n approximation:
%fj=3Zj*(sin(GR)-GR*cos(GR))/G^3 where Z=atomic number
%We do iron, with a BCC structure
clear
Z = 26;       %Fe number of electrons for the j^th basis atom
lambda=1.0;   %lambda in angstroms
a = 2.87;     %Lattice constant (Angstroms) Iron (Fe),
%Radius of atom is neareast neighbor distance divided by 2
dnn=sqrt(3)*a/2;            %BCC nearest neighbor distance
R=dnn/2;                    %atom Radius close packing model
thes=(80-0)/400;            %angle range
thet =0:thes:80;            %angle in degrees
theta = thet*pi/180;        %angle in radians
for i = 1:length(theta)
  thv=theta(i); if(thv==0), thv=1.e-6; end %prevent theta=0 problems
  G = (4.0*pi/lambda)*sin(thv); %reciprocal lattice vector magnitude
  fj(i) = 3*Z*(sin(G*R)-(G*R).*cos(G*R))./(G*R).^3; %constant n approx
end
str=cat(2,'Form factor for F_e (Z=26), \lambda=',...
  num2str(lambda,'%6.3f'),' Angstroms');
plot(thet,fj,'k')
title (str)
xlabel ('\theta (degrees)')
ylabel ('f_j')
```

We will next make use of Equations 2.4.31, and 2.4.33 to obtain an expression for the scattered intensity. In crystals, the atomic positions are given by $\vec{r}_j = n_1\vec{a}_1 + n_2\vec{a}_2 + n_3\vec{a}_3 + \vec{\rho}_m$ where $\vec{a}_1, \vec{a}_2, \vec{a}_3$ are fundamental lattice vectors, $\vec{\rho}_m$ is a basis vector, and $n_1, n_2, n_3$ are integers that run over the lattice points over N cells. With this, Equation 2.4.33 becomes

$$\begin{aligned}\psi_{T,G}(\vec{r},t) &= Ae^{i(\vec{k}'\cdot\vec{r}-\omega t)} \sum_{n_1=1}^{N}\sum_{n_2=1}^{N}\sum_{n_3=1}^{N}\sum_{m}^{s} f_m e^{-i\vec{G}\cdot(n_1\vec{a}_1+n_2\vec{a}_2+n_3\vec{a}_3+\vec{\rho}_m)} \\ &= Ae^{i(\vec{k}'\cdot\vec{r}-\omega t)} S_G \sum_{n_1=1}^{N} e^{-in_1\vec{G}\cdot\vec{a}_1} \sum_{n_2=1}^{N} e^{-in_2\vec{G}\cdot\vec{a}_2} \sum_{n_3=1}^{N} e^{-in_3\vec{G}\cdot\vec{a}_3},\end{aligned} \tag{2.4.35a}$$

where we have defined the structure factor

$$S_G \equiv \sum_{m}^{s} f_m e^{-i\vec{G}\cdot\vec{\rho}_m}, \tag{2.4.35b}$$

which gives the contribution to the intensity amplitude from electrons in a single cell due to $s$ atoms in the basis. Noticing that

$$\sum_{n=1}^{N} x^n = \sum_{n=0}^{N-1} x^n = \frac{1-x^N}{1-x}, \tag{2.4.36a}$$

so that with $x = e^{-i\vec{G}\cdot\vec{a}_1}$ we find

$$\sum_{n_1=0}^{N-1} e^{-in_1\vec{G}\cdot\vec{a}_1} = \frac{1-e^{-iN\vec{G}\cdot\vec{a}_1}}{1-e^{-i\vec{G}\cdot\vec{a}_1}} = \frac{e^{-iN\frac{\vec{G}\cdot\vec{a}_1}{2}}}{e^{-i\frac{\vec{G}\cdot\vec{a}_1}{2}}}\frac{\sin(N\frac{\vec{G}\cdot\vec{a}_1}{2})}{\sin(\frac{\vec{G}\cdot\vec{a}_1}{2})}, \tag{2.4.36b}$$

where we have factored out $e^{-iN\vec{G}\cdot\vec{a}_1/2}$ from the numerator and $e^{-i\vec{G}\cdot\vec{a}_1/2}$ from the denominator. Continuing with the process for the other two sums in Equation 2.4.35a, the result is

$$\psi_{T,G}(\vec{r},t) = Ae^{i(\vec{k}'\cdot\vec{r}-\omega t)}S_G \frac{\sin(N\frac{\vec{G}\cdot\vec{a}_1}{2})}{\sin(\frac{\vec{G}\cdot\vec{a}_1}{2})} \frac{\sin(N\frac{\vec{G}\cdot\vec{a}_2}{2})}{\sin(\frac{\vec{G}\cdot\vec{a}_2}{2})} \frac{\sin(N\frac{\vec{G}\cdot\vec{a}_3}{2})}{\sin(\frac{\vec{G}\cdot\vec{a}_3}{2})} \frac{e^{-iN\frac{\vec{G}\cdot\vec{a}_1}{2}}}{e^{-i\frac{\vec{G}\cdot\vec{a}_1}{2}}} \frac{e^{-iN\frac{\vec{G}\cdot\vec{a}_2}{2}}}{e^{-i\frac{\vec{G}\cdot\vec{a}_2}{2}}} \frac{e^{-iN\frac{\vec{G}\cdot\vec{a}_3}{2}}}{e^{-i\frac{\vec{G}\cdot\vec{a}_3}{2}}}. \quad (2.4.37)$$

Finally, the intensity at the detector is given by $I = |\psi_{T,G}|^2$ or

$$I = |A|^2|S_G|^2 \left(\frac{\sin(N\frac{\vec{G}\cdot\vec{a}_1}{2})}{\sin(\frac{\vec{G}\cdot\vec{a}_1}{2})}\right)^2 \left(\frac{\sin(N\frac{\vec{G}\cdot\vec{a}_2}{2})}{\sin(\frac{\vec{G}\cdot\vec{a}_2}{2})}\right)^2 \left(\frac{\sin(N\frac{\vec{G}\cdot\vec{a}_3}{2})}{\sin(\frac{\vec{G}\cdot\vec{a}_3}{2})}\right)^2. \quad (2.4.38)$$

This intensity when plot versus $G$ or equivalently, by Equation 2.4.34b, versus angle, gives rise to large peaks whenever

$$\frac{\vec{G}\cdot\vec{a}_i}{2} = n\pi, \quad (2.4.39a)$$

where $n$ is an integer and $i = 1, 2, 3$. The reason for the large peaks is that

$$\lim_{\frac{\vec{G}\cdot\vec{a}_i}{2}\to n\pi} \left(\frac{\sin(N\frac{\vec{G}\cdot\vec{a}_i}{2})}{\sin(\frac{\vec{G}\cdot\vec{a}_i}{2})}\right)^2 \to N^2. \quad (2.4.39b)$$

### 2.4.3 Laue Equations

We have seen that the scattered intensity, Equation 2.4.38, peaks at special values as indicated by Equations 2.4.39. Thus in the process of varying $\vec{G}$, if it should happen that $\vec{G}$ acquires the special value of $\vec{G}_{(hkl)}$ of Equation 2.4.22, where $h$, $k$, $l$ are integers, then by Equation 2.3.19

$$\vec{G}\cdot\vec{a}_1 = 2h\pi, \qquad \vec{G}\cdot\vec{a}_2 = 2k\pi, \qquad \vec{G}\cdot\vec{a}_2 = 2l\pi, \quad (2.4.40a)$$

in which case Equations 2.4.39 become relevant and peaks in the intensity are obtained. Together, Equations 2.4.40a are known as the Laue condition. If we let $s_i = h, k, l$ for $i = 1, 2, 3$ then the Laue condition can be written as

$$\vec{G}\cdot\vec{a}_i = 2s_i\pi. \quad (2.4.40b)$$

These equations are in fact another formal expression of Bragg's Law. To demonstrate this, it becomes useful to show the relationship between the interplanar spacing and the magnitude of the reciprocal lattice vector. Using Equation 2.2.4 for the spacing between $(h, k, l)$ planes we have $2\sin\theta/\lambda = n/d_{(h,k,l)}$, and from Equation 2.4.34b the same ratio is given by $2\sin\theta/\lambda = |\vec{G}_{(h,k,l)}|/2\pi$. Equating these two expressions we see that

$$d_{(h,k,l)} = \frac{2n\pi}{|\vec{G}_{(h,k,l)}|}, \quad (2.4.41)$$

for integer $n$, which, when it equals 1, gives the shortest distance between two adjacent planes. For a simple orthorhombic system, this expression, for the distance between two adjacent $(h, k, l)$ planes gives (see Exercise 2.6.15)

$$d_{(h,k,l)} = \frac{1}{\sqrt{(h/a)^2 + (k/b)^2 + (l/c)^2}}, \quad (2.4.42)$$

where $a$ is the lattice constant. If we now use Equation 2.4.41 and write $\hat{G} = \vec{G}/G = (d/2n\pi)\vec{G}$, then with the use of the Laue condition Equation 2.4.40b, we find

$$\hat{G}\cdot\vec{a}_i = \frac{d}{2n\pi}2s_i\pi = \frac{ds_i}{n}. \tag{2.4.43a}$$

From this, and if we make use of the second form of G; that is, Equation 2.4.34b, as well as the Laue condition again, Equation 2.4.40b, we see that

$$\vec{G}\cdot\vec{a}_i = G\hat{G}\cdot\vec{a}_i = \left(\frac{4\pi}{\lambda}\sin\theta\right)\left(\frac{ds_i}{n}\right) = 2s_i\pi, \tag{2.4.43b}$$

or $2d\sin\theta = n\lambda$; i.e., the Bragg Law, Equation 2.2.4, as obtained from the Laue condition, and shows the consistency between the two pictures.

### 2.4.4 Scattering Intensity Calculation

In this section we will perform the scattering intensity calculation alluded to before when we first introduced Figure 2.2.5 in Section 2.2. We have all the necessary ingredients for the calculation, except one; that is, the details regarding the structure factor Equation 2.4.35b. We will do an example of the structure factor for the case of the BCC structure, which is the case of interest here.

**Example 2.4.4.1**

In this example, we obtain the structure factor for $Fe$ with a BCC structure (see Chapter 1) with a basis of $s = 2$ atoms and using a cubic symmetry. We work with the expression of Equation 2.4.35b, which we rewrite as $S_G = \sum_{j=1}^{s=2} f_j e^{-i\vec{G}\cdot\vec{r}_j}$ where we take $\vec{r}_j = x_j\vec{a}_1 + y_j\vec{a}_2 + z_j\vec{a}_3$ and $\vec{G} = \vec{G}_{(hkl)} = h\vec{b}_1 + k\vec{b}_2 + l\vec{b}_3$ as in Equation 2.4.22. This leads to $S_G = \sum_{j=1}^{2} f_j e^{-i2\pi(hx_j+ky_j+lz_j)}$, where we have used Equation 2.3.19. Letting $(x_1, y_1, z_1) = (0,0,0)$ for the first atom in the basis and $(x_2, y_2, z_2) = (1/2, 1/2, 1/2)$ for the second basis atom, the final structure factor expression is

$$SG = f\left(1 + e^{-i\pi(h+k+l)}\right), \tag{2.4.44}$$

where we have let the form factor be $f = f_1 = f_2$ because the atoms are identical. The form factor, itself a function of $\theta$, is given in Equation 2.4.32, and a model in which the electron conentration is constant is shown in Figure 2.4.12 of Example 2.4.2.1.

In Equation 2.4.44, the value 1 is due to the atom at the origin, and the second term inside the parenthesis is due to the second basis atom. Notice that if the sum of indices $h+k+l$ is an even integer, $SG = 2f$, but if the sum is an odd integer, $SG = 0$. Thus, looking at the reflected intensity Equation 2.4.38 versus angle $\theta$, for a particular direction $(hkl)$ in the BCC lattice, as $G(\theta)$ (see Equation 2.4.34b) varies and acquires the special value $|\vec{G}_{(hkl)}|$ there will be reflection from planes whose Miller indices add up to an even integer, else Bragg peaks will be absent from planes whose Miller indices add up to an odd integer. This happens to be the situation described in Figures 2.2.5(a) and (b), for the (222) and (111) planes, respectively. In (a) all the six expected peaks are shown versus angle (Exercise 2.6.3), and in (b) only three peaks show up. The other three at angles of about $\theta = 9.59$, 30, and, 56.44 degrees have been suppressed. Since for a general $\vec{G}$, our BCC structure factor is $SG = f\left(1 + e^{-i\vec{G}\cdot\vec{r}_2}\right)$, we must have that $e^{-i\vec{G}\cdot\vec{r}_2} = -1$ or $\vec{G}\cdot\vec{r}_2 = p\pi$, where $p$ is an odd integer. This must be so in order to produce $SG = 0$. We now show this is happening in our case for specific angles whenever $(h+k+l)$ is odd as expected from Equation 2.4.44. Let's write $\vec{G} = G\hat{G}$ whose direction is the same as the cubic system's $\vec{G}_{(hkl)} = (h,k,l)2\pi/a$ or

$\hat{G} = (h,k,l)/\sqrt{(h^2+k^2+l^2)}$ and with $\vec{r}_2 = (1/2,1/2,1/2)a$ we have

$$p\pi = \vec{G}\cdot\vec{r}_2 = (4\pi\sin\theta/\lambda)\hat{G}\cdot\vec{r}_2 = \frac{(4\pi\sin\theta/\lambda)a(h+k+l)}{2\sqrt{(h^2+k^2+l^2)}}. \tag{2.4.45}$$

However, from Equation 2.4.41, the smallest interplanar distance in the BCC is

$$d = 2\pi/|\vec{G}_{(hkl)}| = a/\sqrt{h^2+k^2+l^2}, \tag{2.4.46a}$$

then Equation 2.4.45 becomes

$$p\pi = (2d\pi\sin\theta/\lambda)(h+k+l). \tag{2.4.46b}$$

Also, by Bragg's Law we know that $2d\sin\theta/\lambda = n$, then we finally have

$$p = n(h+k+l), \tag{2.4.46c}$$

as the condition for the value of $n$ used in Bragg's Law to get the angles that produce no peaks given odd $(h+k+l)$. Since we know $p$ is also odd, then we conclude $n$ must also be odd. Putting this in the above Bragg's Law and using $\lambda = d/3$, as used in obtaining Figure 2.2.5(b), the angles for which no peaks are seen in the figure for odd $h+k+l$ are, therefore, given by

$$6\sin\theta = n_{\text{odd}}, \tag{2.4.47}$$

where $n_{\text{odd}}$ is an odd integer 1,3, ....

In the calculation of Figure 2.2.5, we have used $N = 50$ cells. The code employed in creating the figure is scat_intensity.m and is listed below.

```
%copyright by J. Hasbun and Trinanjan Datta
%scat_intensity.m
%We calculate the scattered intensity for an Fe BCC system
%For the form factor the approximation:
%fj=3Zj*(sin(GR)-GR*cos(GR))/G^3 where Zj=number, is used.
%Here, G=k*sin(theta), k=2*pi/lambda.
%The atom radius is the neareast neighbor distance divided by 2.
%For the BCC dnn=sqrt(3)a/2 with a the lattice constant in angstroms.
%We include the structure factor here as well as the total
%intensity from a crystal of N cells and we employ a cubic BCC crystal
%of Fe. The calculation of G.a, G.b, and G.c is done carefully
%for the BCC structure.
clear
%BCC - cubic crystal vectors used here
a = 2.87;          %Lattice constant (Angstroms) Iron (Fe),
a1=a*[1,0,0];      %using the cubic cell vectors
a2=a*[0,1,0];
a3=a*[0,0,1];
%A=[h,k,l];        %the Miller indices, avoid zeros and infinities
A=input('Enter [h,k,l] as a row vector [2,2,2] -> ');
if isempty(A), A=[2,2,2]; end
h=A(1); k=A(2); l=A(3);
fprintf('[h,k,l]=[%6.3f,%6.3f,%6.3f ]\n',h,k,l)
%Reciprocal lattice vectors follow
Vt=dot(a1,cross(a2,a3)); %system's unit cell volume
```

```
b1=2*pi*cross(a2,a3)/Vt;
b2=2*pi*cross(a3,a1)/Vt;
b3=2*pi*cross(a1,a2)/Vt;
G_v=h*b1+k*b2+l*b3;         %reciprocal lattice vector
G_m=norm(G_v);              %magnitude of G
G_hat=G_v/G_m;              %getting the G unit vector
%fprintf('G_hat=[%6.3f,%6.3f,%6.3f ]\n',G_hat)
d=2*pi/G_m;                 %plane distance
lambda=d/3;                 %lambda in angstroms as used here
themin=0; themax=80;        %angle range
%
thes=(themax-themin)/400;
thet =themin:thes:themax; %angle in degrees
theta = thet*pi/180;        %angle in radians
Z = 26;     %Fe number of electrons for the j^th basis atom
%Radius of atom is neareast neighbor distance divided by 2
%for the BCC dnn=sqrt(3)a/2.
dnn=sqrt(3)*a/2;            %BCC nearest neighbor distance
R =dnn/2;                   %atom Radius close packing model
N=50;      %number of cells used in getting Rh, Rk, Rl below
Ga1=dot(G_hat,a1);
Ga2=dot(G_hat,a2);
Ga3=dot(G_hat,a3);
%basis atoms (BCC)
nb=2;
rb(1,:)=a*[0 0 0];
rb(2,:)=a*[1 1 1]/2;
%dot products of G_hat with the basis vectors
for j=1:nb
  Gb(j)=dot(G_hat,rb(j,:));
end
for i = 1:length(theta)
  %G = 2*k*sin(theta) where k is equal to 2*pi/lambda
  thv=theta(i); if(thv==0), thv=1.e-6; end %prevent theta=0 problems
  G = (4.0*pi/lambda)*sin(thv);
  fj(i) = 3*Z*(sin(G*R)-(G*R).*cos(G*R))./(G*R).^3; %constant n approx
  %G.a1/2, G.a2/2, G.a3/2 follow. The Laue condition is
  %G.a_i/2=pi*s_i. where a_i=a1, a2, a3, and s_i=h,k,l, for i=1,2,3.
  %Thus we need G*Ga_i/2/s_i=pi. This means that whenever
  %G*Ga_i/2/s_i=pi we get Bragg peaks. We divide by that factor to
  %work with the Laue condition.
  betA=G*Ga1/2.0/h;
  betB=G*Ga2/2.0/k;
  betC=G*Ga3/2.0/l;
  denoA=sin(betA);
  denoB=sin(betB);
  denoC=sin(betC);
  %If the denominator=0, sin(N*0)/sin(0)->N
  Rh(i)=N; Rk(i)=N; Rl(i)=N;
  if(abs(denoA) >= 1.e-6), Rh(i)=sin(N*betA)/denoA; end
  if(abs(denoB) >= 1.e-6), Rk(i)=sin(N*betB)/denoB; end
```

```
  if(abs(denoC) >= 1.e-6), Rl(i)=sin(N*betC)/denoC; end
  %structure factor
  %This way of calculating SG, it takes out reflections
  %from BCC planes for which h+k+l odd. However, for this to happen
  %G has to be the right magnitude such that G*Gb(j)=pi*(h+k+l)
  %in which case exp(-zim*G*Gb(j))=-1 which cancels the the
  %exp(0)=1 term producing zero reflection at that angle.
  SG(i)=0.0;
  zim=complex(0,1);
  for j=1:nb
    SG(i)=SG(i)+fj(i)*exp(-zim*G*Gb(j));
  end
end
str=cat(2,'Form factor for F_e (Z=26), \lambda=',...
  num2str(lambda,'%6.3f'),' Angstroms');
plot(thet,fj,'k')
title (str)
xlabel ('\theta (degrees)')
ylabel ('f_j')
%Structure Factor & Intensity BCC lattice - Note for iron fj is the same
%for each atom (BCC: 2 atom basis)
I0=abs(Rh.*Rk.*Rl).^2;     %lattice sum intensity
SG=abs(SG).^2;
figure
plot(thet,SG,'k')          %structure factor vs theta
axis([0 max(thet) 0 1])
str=cat(2,'       BCC F_e (Z=26), [a, d, \lambda]=[',...
  num2str(a,' %6.3f'),', ',num2str(d,'%6.3f'),', ',...
  num2str(lambda,'%6.3f'),'] Angs, (hkl)=(',num2str(h,'%4.0f'),...
  ' ',num2str(k,'%4.0f'),' ',num2str(l,'%4.0f'),')');
title (str)
xlabel ('\theta (degrees)')
ylabel ('Structure Factor')
%
figure
I=SG.*I0;                  %intensity at detector vs theta
plot(thet,I,'k')
axis([0 max(thet) 0 1E9])
str=cat(2,'       BCC F_e (Z=26), [a, d, \lambda]=[',...
  num2str(a,' %6.3f'),', ',num2str(d,'%6.3f'),', ',...
  num2str(lambda,'%6.3f'),'] Angs, (hkl)=(',num2str(h,'%4.0f'),...
  ' ',num2str(k,'%4.0f'),' ',num2str(l,'%4.0f'),')');
title (str)
xlabel ('\theta (degrees)')
ylabel ('Calculated Bragg Peak Intensity')
```

Notice the code, in addition to producing the scattered intensity graph, also produces a plot of the form factor as well as the structure factor employed in the calculation. These extra figures are not shown here.

## 2.5 Brillouin Zones

At this point, it is important to have a visual representation of the scattering condition indicated by Equation 2.4.28; i.e., $\vec{\Delta k} = \vec{G}$, which we rewrite as

$$\vec{k}' = \vec{k} + \vec{G}, \tag{2.5.48}$$

which says that whenever $\vec{k}'$ obeys this condition, an x-ray beam experiences diffraction. This brings us directly into contact with the work of P. P. Ewald, namely, the Ewald construction.

### 2.5.1 The Ewald Sphere

The Ewald construction refers to a graphical representation of the conditions that lead to crystal diffraction. This is represented by the so-called Ewald sphere shown in Figure 2.5.13 and illustrates how Equation 2.5.48 is to be satisfied. For elastic scattering $k = k' = 2\pi/\lambda$, so that the sphere radius is determined through the radiation wavelength. When the Bragg diffraction condition $\vec{\Delta k} = \vec{G}$ is met, two reciprocal lattice points intersect the sphere's perimeter so that $\vec{k}'$ and $\vec{k}$ become connected by a reciprocal lattice vector $\vec{G}$ as shown.

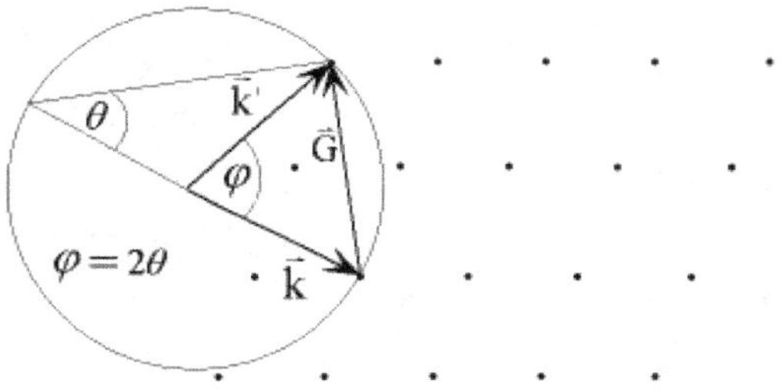


Figure 2.5.13: The Ewald construction shows how two reciprocal lattice points, intersecting the sphere's perimeter, are connected by a reciprocal lattice vector $\vec{G}$ while simultaneously satisfying the Bragg diffraction condition Equation 2.5.48. The angle between $\vec{k}'$ and $\vec{k}$ is the diffraction angle, here labeled $\phi$, which is twice the Bragg angle $\theta$ (see also Figure 2.4.10).

From Equation 2.5.48 and assuming elastic scattering, we also have that $k'^2 = k^2 = (\vec{k}+\vec{G})\cdot(\vec{k}+\vec{G})$, which results in $0 = 2\vec{k}\cdot\vec{G} + \vec{G}\cdot\vec{G}$, or since if $\vec{G}$ is a reciprocal lattice vector, so is $-\vec{G}$, to write

$$\vec{k}\cdot\left(\frac{\vec{G}}{2}\right) = \left(\frac{|\vec{G}|}{2}\right)^2. \tag{2.5.49}$$

The significance of this result is that we can now ask about all the possible wavevectors that give rise to Bragg reflections. The answer lies in regions of reciprocal lattice space known as Brillouin zones (BZs).

### 2.5.2 The First Brillouin Zone

BZs are regions of reciprocal lattice space that exhibit all the wavevectors $\vec{k}$ which can be Bragg reflected by the crystal. They correspond to boundaries from which Bragg reflections take place. While there are many BZs, the first BZ is the smallest region of reciprocal lattice space centered

at the origin and bounded by perpendicular bisectors such as to satisfy the condition indicated by Equation 2.5.49. This is similar to what we did in Chapter 1 when discussing real space primitive cells or the Wigner-Seitz cell. In reciprocal lattice space, the first BZ is defined as the Wigner-Seitz cell in reciprocal lattice space.
The Wigner-Seitz BZ construction method is as follows:

1. select a vector $\vec{G}$ from the origin to a reciprocal lattice point;
2. construct a plane normal to this vector $\vec{G}$ at its midpoint;
3. the plane so formed is part of the BZ;
4. the smallest geometric space, enclosed entirely by planes that are perpendicular bisectors of the reciprocal lattice vectors drawn from the origin, is the first BZ.

An x-ray beam in the crystal will be diffracted if its wavector $\vec{k}$ has a magnitude and direction required by Equation 2.5.49. The diffracted beam will be in the direction $\vec{k}' = \vec{k} \pm \vec{G}$. In this way we find all vectors $\vec{k}$ of the beam which will be Bragg-reflected; such vectors map the regions of reciprocal lattice which we call BZs.
A simple two-dimensional example of a BZ is shown in Figure 2.5.14. Several vectors to various reciprocal lattice points have been drawn, but the smallest enclosed area is the first BZ. Beyond the first BZ, there will be a 2nd, a 3rd, etc., BZs.

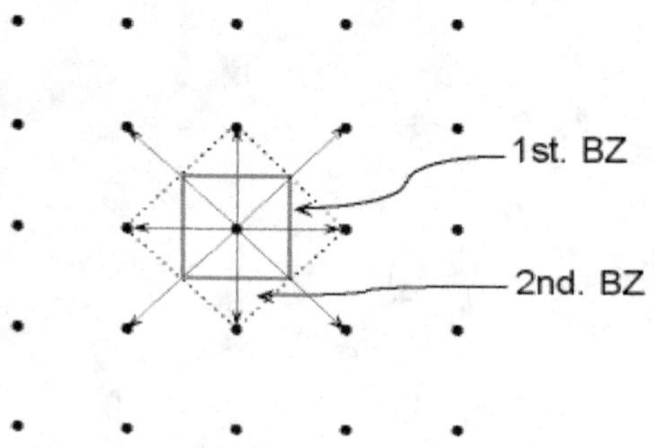


Figure 2.5.14: Example of the first (square shape) and second (triangle shapes) Brillouin zones of a two-dimensional square reciprocal lattice. More zones can be obtained by drawing more vectors from the origin to the rest of the reciprocal lattice points and obtaining their perpendicular bisectors.

As another example, Figure 2.5.15 shows the case of two planes drawn perpendicular to $\vec{G}/2$ as well as how the Bragg condition, as written in Equation 2.5.49, is satisfied.

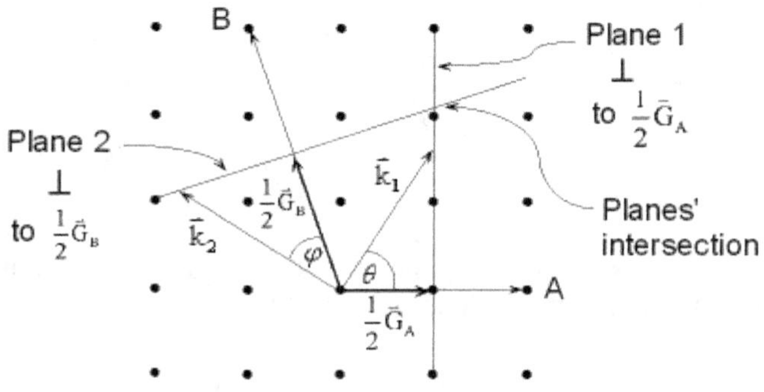


Figure 2.5.15: Two Brillouin zone planes associated with the perpendicular bisectors of reciprocal lattice vectors $\vec{G}_A$ and $\vec{G}_B$ are shown. In each case, the Bragg condition Equation 2.5.49 is obeyed; that is, $\vec{k}_1 \cdot \left(\frac{\vec{G}_A}{2}\right) = \left(\frac{|\vec{G}_A|}{2}\right)^2$ and $\vec{k}_2 \cdot \left(\frac{\vec{G}_B}{2}\right) = \left(\frac{|\vec{G}_B|}{2}\right)^2$.

In Figure 2.5.15, notice that $k_1 \cos\theta = \frac{G_A}{2}$ so that $\vec{k}_1 \cdot \left(\frac{\vec{G}_A}{2}\right) = k_1 \cos\theta \frac{G_A}{2} = \left(\frac{|\vec{G}_A|}{2}\right)^2$, consistent with Equation 2.5.49, and similarly for the wavector $\vec{k}_2 \cdot \left(\frac{\vec{G}_B}{2}\right) = \left(\frac{|\vec{G}_B|}{2}\right)^2$. At the point where the two planes intersect, we have $\vec{k}_1 = \vec{k}_2 \equiv \vec{k}$ so that $\vec{k} \cdot \left(\frac{\vec{G}_B}{2}\right) = \vec{k} \cdot \left(\frac{\vec{G}_A}{2}\right) = \left(\frac{|\vec{G}_A|}{2}\right)^2 = \left(\frac{|\vec{G}_B|}{2}\right)^2$, and are, therefore, special points in the Brillouin zone.

**Example 2.5.2.1**
Consider a one-dimensional lattice with primitive lattice vector $\vec{a}$. (a) Obtain the associated reciprocal lattice vector $\vec{b}$. (b) Obtain the first Brillouin zone (BZ) of this one-dimensional lattice. (c) What is the size of the BZ? (d) What is the value of the wavevector that makes possible Brag reflections in this lattice?

**Solution**
(a) To obtain the reciprocal lattice vector $\vec{b}$ we employ the one-dimensional analogue of Equations 2.3.17c and 2.3.18 to write $\vec{G} = \vec{b}$ and require that $\vec{G} \cdot \vec{a} = 2\pi$, which means that they are parallel in one dimension or $\hat{b} = \hat{a}$. Additionally, $\vec{b} \cdot \vec{a} = 2\pi = ba$. This gives the magnitude $b = 2\pi/a$. We can now write $\vec{b} = (2\pi/a)\hat{b}$.
(b) Writing the reciprocal lattice translation vector $\vec{G} = \vec{b} = (2\pi/a)\hat{b}$, then $|\vec{G}|/2 = \pi/a$. This means that we draw lines that are perpendicular bisectors at $\pm\pi/a$. These lines become the Brillouin zone boundaries in reciprocal lattice space. Figures 2.5.16 (a) and (b) contain the real and reciprocal lattice spaces along with the first BZ boundaries (dashed lines), respectively.
(c) The BZ has length $2\pi/a$, as shown in the figure.
(d) The Bragg wavevector is given by Equation 2.5.49 or since the only possible values of $\vec{k}$ are parallel to $\vec{b}$ to write $|\vec{k}||\vec{G}|/2 = (|\vec{G}|/2)^2$ or $|k| = |\vec{G}|/2 \Rightarrow k = \pm\pi/2$.

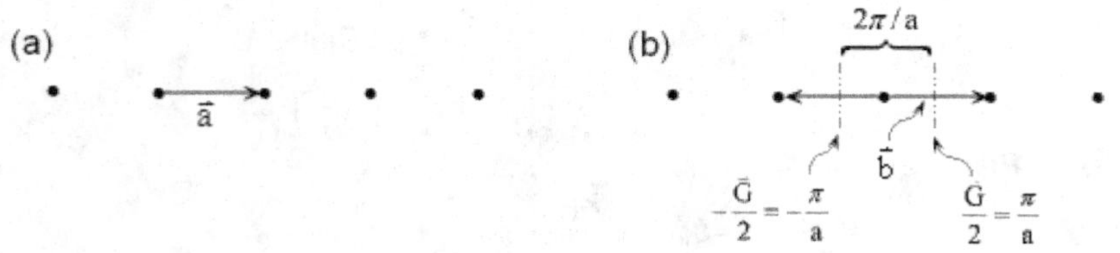


Figure 2.5.16: Example 2.5.2.1's (a) one-dimensional real space lattice, and (b) the corresponding reciprocal lattice showing the size of the Brillouin zone is $2\pi/a$.

### 2.5.3 Simple Cubic Reciprocal Lattice Vectors and BZ

The simple cubic's primitive real lattice vectors of Chapter 1 are given by $\vec{a}_1 = a\hat{i}$, $\vec{a}_2 = a\hat{j}$, and $\vec{a}_3 = a\hat{k}$. The reciprocal lattice vectors can be obtained by the use of Equations 2.3.20 and the result is similar to that of the simple orthorhombic system of Example 2.3.2.1, where we take $a = b = c$ to obtain

$$\vec{b}_1 = \frac{2\pi}{a}\hat{i}, \qquad \vec{b}_2 = \frac{2\pi}{a}\hat{j}, \qquad \vec{b}_3 = \frac{2\pi}{a}\hat{k}. \tag{2.5.50}$$

The reciprocal lattice is, therefore, itself a simple cubic and the reciprocal lattice unit cell volume is $V_{SC-rc} = (2\pi/a)^3$. A general reciprocal lattice translation vector is given by $\vec{G} = v_1\vec{b}_1 + v_2\vec{b}_2 + v_3\vec{b}_3 = (2\pi/a)(v_1\hat{i} + v_2\hat{j} + v_3\hat{k})$, where the $v_i's$ are $\pm$ integers. From the origin of reciprocal lattice space we can draw $\vec{G}$ vectors to the nearest reciprocal lattice points and find bisector planes at positions $\pm\vec{b}_1/2 = \pm(\pi/a)\hat{i}$, $\pm\vec{b}_2/2 = \pm(\pi/a)\hat{j}$, and $\pm\vec{b}_3/2 = \pm(\pi/a)\hat{k}$. This corresponds to a Brillouin zone bounded by six planes with a simple cubic shape of sides $2\pi/a$ and volume $V_{SC-rc}$ as above. Figure 2.5.17 shows the simple cubic's BZ in units of $2\pi/a$. There are special points in the cubic's BZ that are of much significance in computations. These are the high symmetry points $\Gamma = [0,0,0]$, $X = [1/2,0,0]$, $R = [1/2,1/2,1/2]$, and $M = [1/2,1/2,0]$ also in units of $2\pi/a$. These coordinates form a tetrahedron whose volume is the $(1/48)$th part of the total cube's volume. This tetrahedron is the smallest irreducible part of the simple cubic BZ from which the whole cube can be reproduced with proper symmetry operations. The electronic properties of this irreducible part of the zone bears the properties of the entire cube. Thus, performing computations on the irreducible tetrahedron is much less time-consuming than performing them over the entire cube. Furthermore, the results for the entire cube are obtained by multiplying the tetrahedron's results by the factor of 48 in the present case. We will make use of this symmetric property of the cubic system in a later chapter.

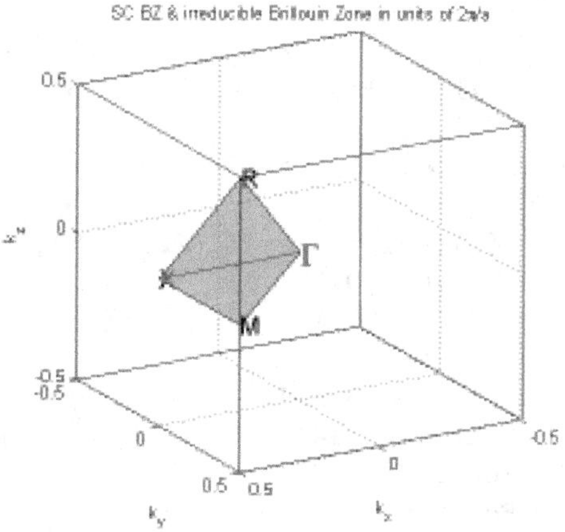


Figure 2.5.17: The simple cubic BZ with plane boundaries at $\pm(\pi/a)\hat{i}$, $\pm(\pi/a)\hat{j}$, $\pm(\pi/a)\hat{k}$. Also shown is the irreducible part of the cube's BZ; i.e., the tetrahedron, 48 of which make up the cube and whose coordinates are high symmetry points in the cube's BZ. These high symmetry points, in units of $2\pi/a$, are $\Gamma = [0,0,0], X = [1/2,0,0], R = [1/2,1/2,1/2], M = [1/2,1/2,0]$, and are also shown. The plotted coordinates are also in units of $2\pi/a$.

Below the function code ScBZ.m employed to produce the BZ plot of Figure 2.5.17 is listed. A "function" is used because it can make use of other functions within it which are needed in the script.

```
%copyright by J. E Hasbun and T. Datta
%ScBZ.m
%Draws the Brillouin zone for the simple cubic and its irreducible
%tetrahedron and calculates the volume.
%Vectors in units of 2*pi/a
%
function ScBZ
clear
clc
Z=[0,0,0];           %zero point (gamma) - origin
X=[1/2,0,0];         %high symmetry points X, R, M
R=[1/2,1/2,1/2];
M=[1/2,1/2,0];
VSC=abs(dot(X,cross(R,M))/6); %tetrahedron volume
disp('============ Simple Cubic ==============')
fprintf('SC: irreducible V=%s%s\n',rats(VSC),' of BZ volume')
disp('SC symmetry points (in units of 2*pi/a):')
disp('X=[1/2,0,0], R=[1/2,1/2,1/2], M=[1/2,1/2,0]')
disp('Known SC BZ volume: (2*pi/a)^3')
view(150,20)                           %viewpoint longitude, latitude
grid on                                %show a grid
%Begin the tetrahedron
liner(Z,X,'-','k',1.0)                 %lines to symmetry points
liner(Z,R,'-','k',1.0)
liner(Z,M,'-','k',1.0)
liner(X,R,'-','k',1.0)
liner(X,M,'-','k',1.0)
```

```
liner(R,M,'-','k',1.0)
hold on                                   %Begin to fill tetrahedron faces
x = [0;X(1);R(1);]; y=[0;X(2);R(2);]; z=[0;X(3);R(3);];
setter(x,y,z,[0.70 0.60 0.70],0.3,0.5)
x = [0;X(1);M(1);]; y=[0;X(2);M(2);]; z=[0;X(3);M(3);];
setter(x,y,z,[0.70 0.60 0.70],0.3,0.5)
x = [0;R(1);M(1);]; y=[0;R(2);M(2);]; z=[0;R(3);M(3);];
setter(x,y,z,[0.70 0.60 0.70],0.3,0.5)
x = [X(1);R(1);M(1);]; y=[X(2);R(2);M(2);]; z=[X(3);R(3);M(3);];
setter(x,y,z,[0.70 0.60 0.70],0.3,0.5)
text(Z(1),Z(2),Z(3),'\Gamma','FontSize',18)
text(X(1),X(2),X(3),'X','FontSize',14)
text(R(1),R(2),R(3),'R','FontSize',14)
text(M(1),M(2),M(3),'M','FontSize',14)
title('SC BZ & irreducible Brillouin Zone in units of 2\pi/a')
%SC BZ next.
%Corners to which lines will be drawn
c1=[1/2,1/2,1/2]; c2=[-1/2,1/2,1/2]; c3=[-1/2,-1/2,1/2]; c4=[1/2,-1/2,1/2];
c5=[1/2,1/2,-1/2]; c6=[-1/2,1/2,-1/2]; c7=[-1/2,-1/2,-1/2];
    c8=[1/2,-1/2,-1/2];
liner(c1,c2,'-','b',1.0)    %c1 top corner connectors
liner(c1,c4,'-','b',1.0)
liner(c1,c5,'-','b',1.0)
liner(c2,c6,'-','b',1.0)    %c2 top corner connectors
liner(c2,c3,'-','b',1.0)
liner(c3,c4,'-','b',1.0)    %c3 top corner connectors
liner(c3,c7,'-','b',1.0)
liner(c5,c6,'-','b',1.0)    %c5 bottom corner connectors
liner(c5,c8,'-','b',1.0)
liner(c7,c6,'-','b',1.0)    %c7 bottom corner connectors
liner(c7,c8,'-','b',1.0)
liner(c8,c4,'-','b',1.0)    %c8 bottom corner connectors
hold off
axis equal                  %helps to make the cube look like it
xlabel('k_x'), ylabel('k_y'), zlabel('k_z')

function liner(v1,v2,lin_style_txt,lin_color_txt,lin_width_num)
%Draws a line given initial vector v1 and final vector v2
%lin_style_txt: line style text format
%lin_color_txt: line color text format
%lin_width_num: line width number format
line([v1(1),v2(1)],[v1(2),v2(2)],[v1(3),v2(3)],...
    'LineStyle',lin_style_txt,'color',...
    lin_color_txt,'linewidth',lin_width_num)

function setter(x,y,z,v,Edge_num,Face_num)
%fills a polygon according to coords x,y,z vectors
%v is a color vector like v=[0.70 0.70 0.40] for example
%Edge_num is a number like 0.3, and Face_num is also a number like 0.5
h=fill3(x,y,z,v);                                   %fill face
set(h,'EdgeAlpha',Edge_num,'FaceAlpha',Face_num) %edges, transparent
```

Notice that the code also takes advantage of the calculation to find and print the volume of the tetrahedron which makes up the 1/48th part of the SC BZ.

### 2.5.4 Body Centered Cubic Reciprocal Lattice Vectors and BZ

The body centered cubic's primitive lattice vectors of Chapter 1 are given by $\vec{a}_1 = a(\hat{\imath}+\hat{\jmath}-\hat{k})/2$, $\vec{a}_2 = a(-\hat{\imath}+\hat{\jmath}+\hat{k})/2$, and $\vec{a}_3 = a(\hat{\imath}-\hat{\jmath}+\hat{k})/2$. The reciprocal lattice vectors can be obtained by the use of Equations 2.3.20 and the result is

$$\vec{b}_1 = \frac{2\pi}{a}(\hat{\imath}+\hat{\jmath}), \qquad \vec{b}_2 = \frac{2\pi}{a}(\hat{\jmath}+\hat{k}), \qquad \vec{b}_3 = \frac{2\pi}{a}(\hat{\imath}+\hat{k}). \tag{2.5.51}$$

The BCC BZ volume is, therefore, $V_{BCC-rc} = 2(2\pi/a)^3$. A general reciprocal lattice translation vector here is given by $\vec{G} = v_1\vec{b}_1 + v_2\vec{b}_2 + v_3\vec{b}_3 = (2\pi/a)\left(v_1(\hat{\imath}+\hat{\jmath}) + v_2(\hat{\jmath}+\hat{k}) + v_3(\hat{\imath}+\hat{k})\right)$. Since the $v_i$'s can take on $\pm$ integers, the first BZ is obtained by drawing planes that are the perpendicular bisectors of $\pm\vec{G}$ or $\pm\vec{b}_1, \pm\vec{b}_2, \pm\vec{b}_3$; i.e., at the reciprocal lattice space locations $\pi(\pm\hat{\imath}\pm\hat{\jmath})/a, \pi(\pm\hat{\jmath}\pm\hat{k})/a, \pi(\pm\hat{\imath}\pm\hat{k})/a$. This corresponds to a BZ bounded by $4\times4\times4 = 12$ planes. This has a rhombic dodecahedron shape with volume $V_{BCC-rc}$ as above. Figure 2.5.18 shows the body centered cubic's BZ in units of $2\pi/a$. As in the simple cubic, there are special points in the BCC's BZ that are of much significance in computations. These are the high symmetry points $\Gamma = [0,0,0], P = [1/2,1/2,1/2], H = [1,0,0]$, and $N = [1/2,1/2,0]$ also in units of $2\pi/a$. These coordinates form a tetrahedron whose volume is the $(1/24)$th part of the total BZ volume. This tetrahedron is the smallest irreducible part of the BCC's BZ from which the whole dodecahedron can be reproduced with proper symmetry operations. The electronic properties of this irreducible zone bear the properties of the entire BZ. Again, performing computations on the irreducible tetrahedron is much less time consuming than performing them over the entire BZ. Furthermore, the results for the entire BZ are obtained by multiplying the tetrahedron's results by the factor of 24. This symmetric property of the BCC system is exploited in a later chapter.

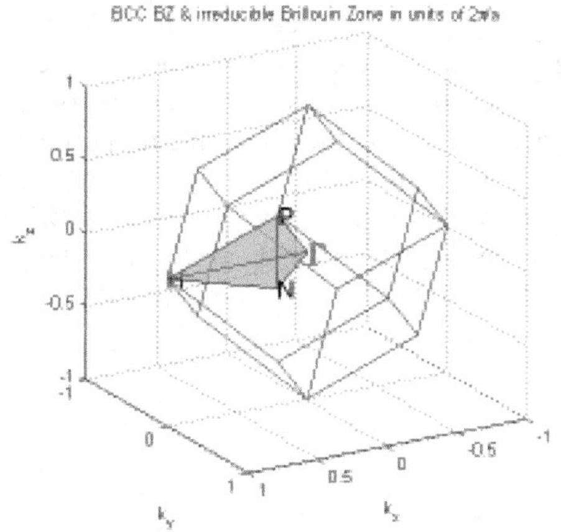


Figure 2.5.18: The body centered cubic BZ with plane boundaries at $\pi(\pm\hat{\imath}\pm\hat{\jmath})/a, \pi(\pm\hat{\jmath}\pm\hat{k})/a, \pi(\pm\hat{\imath}\pm\hat{k})/a$. Also shown is the irreducible part of the BCC's BZ; i.e., the tetrahedron, 24 of which make up the dodecahedron and whose coordinates are high symmetry points in the BCC BZ. These high symmetry points, in units of $2\pi/a$ are $\Gamma = [0,0,0]$, $P = [1/2,1/2,1/2]$, $H = [1,0,0]$, and $N = [1/2,1/2,0]$ and are so labeled. The plotted coordinates are also in units $2\pi/a$.

### 2.5.5 Face Centered Cubic Reciprocal Lattice Vectors and BZ

The face centered cubic's primitive lattice vectors of Chapter 1 are given by $\vec{a}_1 = a(\hat{\imath}+\hat{\jmath})/2$, $\vec{a}_2 = a(\hat{\jmath}+\hat{k})/2$, and $\vec{a}_3 = a(\hat{\imath}+\hat{k})/2$. The reciprocal lattice vectors can be obtained by the use of Equations 2.3.20 and the result is

$$\vec{b}_1 = \frac{2\pi}{a}(\hat{\imath}+\hat{\jmath}-\hat{k}), \qquad \vec{b}_2 = \frac{2\pi}{a}(-\hat{\imath}+\hat{\jmath}+\hat{k}), \qquad \vec{b}_3 = \frac{2\pi}{a}(\hat{\imath}-\hat{\jmath}+\hat{k}). \tag{2.5.52}$$

Notice that these FCC reciprocal lattice vectors look like the real space BCC primitive vectors. This tells us that the BCC lattice is a reciprocal lattice to the FCC. The FCC reciprocal cell volume is, therefore, $V_{FCC-rc} = 4(2\pi/a)^3$. A general reciprocal lattice translation vector here is given by $\vec{G} = v_1\vec{b}_1 + v_2\vec{b}_2 + v_3\vec{b}_3 = (2\pi/a)\left(v_1(\hat{\imath}+\hat{\jmath}-\hat{k}) + v_2(-\hat{\imath}+\hat{\jmath}+\hat{k}) + v_3(\hat{\imath}-\hat{\jmath}+\hat{k})\right) = (2\pi/a)\left(\hat{\imath}(v_1-v_2+v_3)+\hat{\jmath}(v_1+v_2-v_3)+\hat{k}(-v_1+v_2+v_3)\right)$. Since the $v_i$'s can take on $\pm$ integers, then their sum takes on $\pm$ integers as well. We, therefore, find BZ bisector planes at the reciprocal lattice positions $(\vec{G}_\alpha/2) = (\pi/a)(\pm\hat{\imath}\pm\hat{\jmath}\pm\hat{k})$ with magnitude $(G_\alpha/2) = \sqrt{3}\pi/a$. These amount to a total of $2\times 2\times 2 = 8$ planes. However, it has been noticed that, if one makes a linear combination of any pair of the above reciprocal lattice vectors, it is possible to obtain $\vec{G}$ bisectors whose magnitudes are slightly shorter than the intersection of two octahedron faces. These are $\vec{b}_1+\vec{b}_2 = 2\pi(2\hat{\jmath})/a$, $\vec{b}_1+\vec{b}_3 = 2\pi(2\hat{\imath})/a$, $\vec{b}_2+\vec{b}_3 = 2\pi(2\hat{k})/a$ so that we have six extra bisector planes located at positions $(\vec{G}_\beta/2) = \pm\pi(2\hat{\jmath})/a$, $\pm\pi(2\hat{\imath})/a$, $\pm\pi(2\hat{k})/a$ with magnitude $(G_\beta/2) = 2\pi/a$ as described. Adding all these planes we get $8+6 = 14$ planes. The first eight planes form the faces of an octahedron, the second six planes cut through six of its corners and thus we find that the FCC BZ is a 14-sided polyhedron (an octahedron truncated by a cube) as shown in Figure 2.5.19 with the above volume $V_{FCC-rc}$. There are several special points in the FCC BZ that are of interest. These are, in units of $2\pi/a$, $\Gamma = [0,0,0]$, $L = [1/2,1/2,1/2]$, $K = [3/4,3/4,0]$, $W = [1,1/2,0]$, $U = [1,1/4,1/4]$, and $X = [1,0,0]$. These symmetry points are encompassed by three different irreducible tetrahedrons (here labeled $T_1, T_2, T_3$) with coordinates $\Gamma, L, K, W$, for $T_1$; $\Gamma, L, U, W$, for $T_2$; and $\Gamma, X, U, W$, for $T_3$. Each of the $T_1$, $T_2$, and $T_3$ occupy the corresponding fraction of $1/32, 1/32, 1/48$ part of the BZ, respectively. However, it is all three together, which by proper symmetry operations, combine to make up the FCC BZ. Adding their volumes, we see they make up 1/12th of the entire BZ. The three tetrahedrons correspond to the irreducible zone of the FCC BZ. They are shown as they fit snugly together in their fractional part of the BZ in Figure 2.5.19. As mentioned before, the electronic properties of this irreducible zone bear the properties of the entire BZ. Computation on the irreducible zone is much less time-consuming than performing them over the entire BZ. The results for the entire BZ are obtained by multiplying the results for the three tetrahedrons by the factor of 12. As with the other cubics, this symmetric property of the FCC system is exploited in a later chapter.

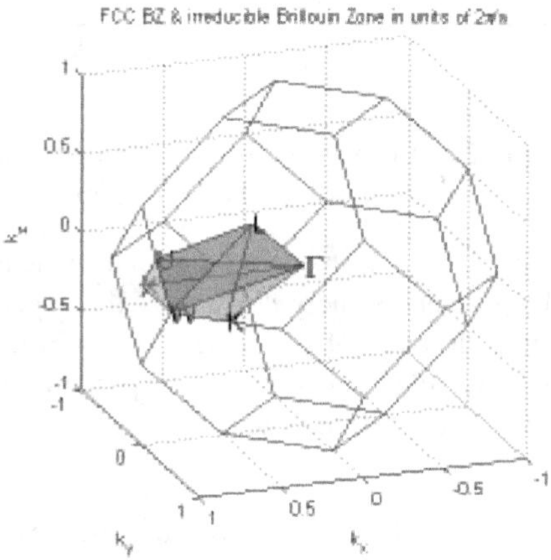


Figure 2.5.19: The face centered cubic BZ with plane boundaries at $(\pi/a)(\pm\hat{\imath} \pm \hat{\jmath} \pm k)$ as well as $\pm\pi(2\hat{\jmath})/a$, $\pm\pi(2\hat{\imath})/a$, $\pm\pi(2\hat{k})/a$ is shown. Here the irreducible part of the FCC's BZ is made up of a group of three irreducible tetrahedrons, 12 of which (together) make up the 14-sided polyhedron. The coordinates of each tetrahedron are high symmetry points in the FCC BZ. These high symmetry points, in units $2\pi/a$, are $\Gamma = [0,0,0]$, $L = [1/2,1/2,1/2]$, $K = [3/4,3/4,0]$, $W = [1,1/2,0]$, $U = [1,1/4,1/4]$, and $X = [1,0,0]$ and are shown along with the tetrahedrons they share. The plotted coordinates are in units $2\pi/a$.

## 2.6 Chapter 2 Exercises

2.6.1. Modify the code of Example 2.1.0.1 in order to incorporate the calculation of Equation 2.1.3 for a particle of finite mass. Use the modified code to reproduce Figure 2.1.2.

2.6.2. (a) Is it necessary to perform calculations using the relativistic expression of Equation 2.1.2 for the behavior of wavelength versus energy? Analyze the expression and obtain its classical limit version. (b) Modify your code of Exercise 2.6.1 in order to perform a comparison between the classical and the relativistic results of part (a) for a particle of finite mass. Apply your results to electrons as well as neutrons and show a comparison similar to that of Figure 2.1.2.

2.6.3. Using the information provided in Figure 2.2.5 (a), use the Bragg Law to produce the values of the angles at which the peaks from the (222) plane occur, thus confirming the positions of the calculated reflection peaks.

2.6.4. Prove the Kronecker delta function relation $a\delta_{pp'} = \int_0^a e^{ik(p-p')x}dx$ where $\delta_{pp'}$ is defined such that it takes on the value 1 if $p = p'$, else it is zero.

2.6.5. Starting from Equation 2.3.8 for $n(x)$, show the steps necessary in order to arrive at Equation 2.3.9 for the Fourier coefficients $n_p$.

2.6.6. Show that Equation 2.3.12b is condition enough for the Fourier representation of $n(x)$ in Equation 2.3.8 to be real.

2.6.7. Show that for a vector of the form of Equation 2.3.5, similar to Equation 2.3.18, $\vec{G}\cdot\vec{T} = 2m\pi$ so that, as in Equation 2.3.17c, $e^{i\vec{G}\cdot\vec{T}} = 1$ is also obeyed.

2.6.8. Use the non-rigorous procedure described above Equations 2.3.20 in order to derive the expresions for the $\vec{b}_i$'s.

2.6.9. Show that the electron number concentration is invariant under a crystal translation of the vector $\vec{T}$.

2.6.10. Referring to Figure 2.4.9, consider a crystal plane with Miller indices $(hkl)$. Show that $\vec{G}_{(hkl)} = h\vec{b}_1 + k\vec{b}_2 + l\vec{b}_3$, where $\vec{b}_1$, $\vec{b}_2$, and $\vec{b}_3$ are the crystal's reciprocal lattice vectors, corresponds to a vector that lies perpendicular to the plane.

2.6.11. The largest term in the sum over $G$ in the function $F$ of Equation 2.4.27b occurs whenever Equation 2.4.28 is satisfied. Show that a way to demonstrate this is to assume a spherical cell geometry and write $\left(\vec{G}-\vec{\Delta k}\right)\cdot\vec{r}' = |\vec{G}-\vec{\Delta k}|r'\cos\theta'$ and to integrate over all space. While doing so, notice that an integral over $r'$; i.e., $\int_0^\infty dr'$, can be approximated by $N\int_0^R dr'$, where $R$ is the radius of a crystal cell and $N$ is the number of crystal cells.

2.6.12. Show the form factor, Equation 2.4.34a, of Example 2.4.2.1 does indeed give the atomic number $Z$ in the limit as $GR \to 0$.

2.6.13. It is possible to assume an exponential behavior for the electron concentration in Equation 2.4.32 rather than the constant value used in Example 2.4.2.1. Imagine a model for which $n(r) = n_0 e^{-br}$ where $n_0 = \frac{A_0 b^3}{8\pi}$ and $b$ to be thought of as a parameter in such a way that its value can be varied so as to moderate the oscillatory behavior of $f_j$ (for the constant $n$ case) shown in Figure 2.4.12. (a) Show that $\int_0^\infty n(r)d^3r = A_0$. (b) Show that in order to have $f_j \to Z$ in the limit of $G \to 0$ the constant $A_0 = \frac{2Z}{b^3 I}$ where $I \equiv \int_0^R r^2 e^{-br} dr$. (c) Perform the calculation of the form factor so described and modify the code of Example 2.4.2.1 to do a comparison between both Fe form factors; i.e., the constant $n$ case of Example 2.4.2.1 and the form factor that results when $n(r) = \frac{A_0 b^3}{8\pi} e^{-br}$ is used. Assume that $b = p/R$ where $p$ is a dimensionless parameter in the range $0.01 \cdots 5.0$. For Fe, use other needed information as provided in Example 2.4.2.1. In your comparison, discuss what happens when $p$ is varied from a small to a large value.

2.6.14. Show the correctness of Equation 2.4.39b.

2.6.15. (a) Obtain an expression for the reciprocal lattice vector magnitude $G_{(h,k,l)}$, associated with planes with Miller indices $(h, k, l)$, for a simple orthorhombic system. (b) Show that the resulting interplanar distance is given by Equation 2.4.42. (c) If the lattice constant for Polonium, with simple cubic system lattice, is 3.34 Å, what is the distance between two adjacent $(134)$ planes?

2.6.16. Use the ideas of Example 2.4.4.1 in order to obtain (a) the interplanar distance (in Å) for the $(222)$ and $(111)$ *Fe* planes associated with Figures 2.2.5; and (b) obtain the angles at which no reflections occur from the $(111)$ plane, thus confirming their absence in the (b) part of the figure.

2.6.17. Run the scat_intensity.m code listed in Section 2.4.4 and reproduce the results shown in Figures 2.2.5 (a) and (b).

2.6.18. Modify the scat_intensity.m code listed in Section 2.4.4 and, rather than using the constant electron concentration form factor, incorporate the form factor that results by the use of the exponential model, $n(r) = n_0 e^{-br}$, discussed in Exercise 2.6.13 for a value of $p \approx 5$ where $b = p/R$ and $R$ is the approximate atom radius. Give plots of the resulting form factor and scattering intensity for the $(222)$ plane. Comment on your results, especially as you contrast with the results shown in Figure 2.2.5(a).

2.6.19. Referring to Figure 2.5.15, prove that the vector $\vec{k}_2$ obeys the Bragg diffraction condition as written in Equation 2.5.49.

2.6.20. Obtain the BCC reciprocal lattice vectors shown in Equations 2.5.51 and obtain the BZ volume.

2.6.21. (a) Make the necessary modification to the simple cubic BZ code ScBZ.m of Section 2.5.3 in order to obtain the BCC BZ as described in Section 2.5.4 and as shown in Figure 2.5.18. (b) When modifying the code, be sure to show that the volume of the involved tetrahedron is 1/24th the volume of the BCC BZ.

2.6.22. In Section 2.5.5 it is mentioned that planes at bisector positions of magnitude $(G_\alpha/2) = \sqrt{3}\pi/a$ create the faces of an octahedron in the FCC BZ; furthermore, that corners of the octahedron are cut by six planes located at bisector positions of magnitude $(G_\beta/2) = 2\pi/a$. If this is the case, the distance from the center of the BZ to the intersection of two octahedron faces must be greater than $(G_\beta/2)$. Show that this is in fact true.

2.6.23. (a) Make the necessary modification to the simple cubic BZ code ScBZ.m of Section 2.5.3 in order to obtain the FCC BZ as described in Section 2.5.5 and as shown in Figure 2.5.19. (b) When modifying the code, be sure to show that the volume of the three involved tetrahedrons add up to 1/12th the volume of the FCC BZ as discussed in that section.

# 3

# *Crystal Binding*

**Contents**



## 3.1 Introduction

In this chapter we are interested in what holds a crystal together. Energy is what makes crystal binding possible. In this regard, two kinds of energy terms are in general use to describe the binding energy. One is the cohesive energy and the other is the lattice energy. The cohesive energy refers to the total energy it takes to disassemble the crystal into its constituent neutral atoms at an infinite separation, at rest, while retaining the same electronic configuration. The lattice energy term, used in a discussion of ionic crystals, is analogous to the cohesive energy with the difference that the component atoms are replaced with the component's ions. In Table 3.1.1, the melting and boiling points as well the bulk modulus, cohesive energies, and the first two ionization potentials are listed for the elements of the periodic table.

Table 3.1.1: [3]. The elements' physical properties. Where known, columns 1-8 list the elements atomic number (Z), symbol, melting point (MP in Kelvin), boiling point (BP in Kelvin), bulk modulus (BM in $10^{11}$ $N/m^2$), cohesive energy (CE in $eV/atom$), first ionization energy (IP1 in $eV$), and second ionization energy (IP2 in $eV$), respectively.

```
Z:     Atomic number
Sy:    Symbol
MP:    Melting point in Kelvin (K)
BP:    Boiling point in Kelvin (K)
BM:    Isothermal bulk modulus 10^11 (N/m^2) Source (1)
CE:    Cohesive energy (eV/atom) at zero K. Note: 1 eV/atom=23.05 kcal/mole
IP1:   First ionization energy (eV)
IP2:   Second ionization energy (eV)
```

Sources

(1) Introduction to Solid State, Charles Kittel, 8th Ed. 13-19 (John Wiley, NY 2005).

(2) Inorganic Crystal Structure Database (ICSD) online.

(3) Curtesy of Matpack C++ Numerics and Graphics Library
(online http://www.matpack.de/Info/Nuclear/Elements/lattice.html)

(4) http://physics.nist.gov/data, http://www.nist.gov/pml/data/periodic.cfm

| Z | Sy | Name | MP | BP | BM | CE | IP1 | IP2 |
|---|---|---|---|---|---|---|---|---|
| 1 | H | Hydrogen | 14.01 | 20.28 | 0.002 | | 13.595 | |
| 2 | He | Helium | 0.95 | 4.216 | 0.00 | | 24.58 | 78.98 |
| 3 | Li | Lithium | 453.69 | 1590 | 0.116 | 1.63 | 5.39 | 81.01 |
| 4 | Be | Beryllium | 1551 | 3243 | 1.003 | 3.32 | 9.32 | 27.53 |
| 5 | B | Boron | 2573 | 2823 | 1.78 | 5.81 | 8.30 | 33.45 |
| 6 | C | Carbon | 3823 | 5100 | 4.43 | 7.37 | 11.26 | 35.64 |
| 7 | N | Nitrogen(N2) | 63.29 | 77.4 | 0.012 | 4.92 | 14.54 | 44.14 |
| 8 | O | Oxygen | 54.75 | 90.188 | | 2.60 | 13.61 | 48.76 |
| 9 | F | Fluorine | 53.53 | 85.01 | | 0.84 | 17.42 | 52.40 |
| 10 | Ne | Neon | 24.48 | 27.1 | 0.010 | 0.020 | 21.56 | 62.63 |
| 11 | Na | Sodium | 370.95 | 1165 | 0.068 | 1.113 | 5.14 | 52.43 |
| 12 | Mg | Magnesium | 921.95 | 1380 | 0.354 | 1.51 | 7.64 | 22.67 |
| 13 | Al | Aluminum | 933.52 | 2740 | 0.722 | 3.39 | 5.98 | 24.80 |
| 14 | Si | Silicon | 1683 | 2628 | 0.988 | 4.63 | 8.15 | 24.49 |
| 15 | P | Phosphorus | 317.3 | 553 | 0.304 | 3.43 | 10.55 | 30.20 |
| 16 | S | Sulfur | 386 | 717.824 | 0.178 | 2.85 | 10.36 | 34.0 |
| 17 | Cl | Chlorine | 172.17 | 238.55 | | 1.40 | 13.01 | 36.81 |
| 18 | Ar | Argon | 83.78 | 87.29 | 0.013 | 0.080 | 15.76 | 43.38 |
| 19 | K | Potassium | 336.8 | 1047 | 0.032 | 0.934 | 4.34 | 36.15 |
| 20 | Ca | Calcium | 1112 | 1760 | 0.152 | 1.84 | 6.11 | 17.98 |
| 21 | Sc | Scandium | 1812 | 3105 | 0.435 | 3.90 | 6.56 | 19.45 |
| 22 | Ti | Titanium | 1933 | 3533 | 1.051 | 4.85 | 6.83 | 20.46 |
| 23 | V | Vanadium | 2163 | 3653 | 1.619 | 5.31 | 6.74 | 21.39 |
| 24 | Cr | Chromium | 2130 | 2755 | 1.901 | 4.10 | 6.76 | 23.25 |
| 25 | Mn | Manganese | 1517 | 2370 | 0.596 | 2.92 | 7.43 | 23.07 |
| 26 | Fe | Iron | 1808 | 3023 | 1.683 | 4.28 | 7.90 | 24.08 |
| 27 | Co | Cobalt | 1768 | 3143 | 1.914 | 4.39 | 7.86 | 24.91 |
| 28 | Ni | Nickel | 1726 | 3005 | 1.86 | 4.44 | 7.63 | 25.78 |
| 29 | Cu | Copper | 1356.6 | 2868 | 1.37 | 3.49 | 7.72 | 27.93 |
| 30 | Zn | Zinc | 692.73 | 1180 | 0.598 | 1.35 | 9.39 | 27.35 |
| 31 | Ga | Gallium | 302.93 | 2676 | 0.569 | 2.81 | 6.00 | 26.51 |
| 32 | Ge | Germanium | 1210.55 | 3103 | 0.772 | 3.85 | 7.88 | 23.81 |
| 33 | As | Arsenic | 886 | sublime | 0.394 | 2.96 | 9.81 | 30.0 |
| 34 | Se | Selenium | 490 | 958.1 | 0.091 | 2.46 | 9.75 | 31.2 |
| 35 | Br | Bromine | 265.9 | 331.93 | | 1.22 | 11.84 | 33.4 |
| 36 | Kr | Krypton | 116.55 | 120.85 | 0.018 | 0.116 | 14.00 | 38.56 |
| 37 | Rb | Rubidium | 312.2 | 961 | 0.031 | 0.852 | 4.18 | 31.7 |
| 38 | Sr | Strontium | 1042 | 1657 | 0.116 | 1.72 | 5.69 | 16.72 |
| 39 | Y | Yttrium | 1796 | 3610 | 0.366 | 4.37 | 6.5 | 18.9 |
| 40 | Zr | Zirconium | 2125 | 4650 | 0.833 | 6.25 | 6.95 | 20.98 |
| 41 | Nb | Niobium | 2741 | 5200 | 1.702 | 7.57 | 6.77 | 21.22 |
| 42 | Mo | Molybdenum | 2890 | 5833 | 2.725 | 6.82 | 7.18 | 23.25 |
| 43 | Tc | Technetium | 2445 | 5303 | 2.97 | 6.85 | 7.28 | 22.54 |
| 44 | Ru | Ruthenium | 2583 | 4173 | 3.208 | 6.74 | 7.36 | 24.12 |
| 45 | Rh | Rhodium | 2239 | 4000 | 2.704 | 5.75 | 7.46 | 25.53 |
| 46 | Pd | Palladium | 1825 | 3413 | 1.808 | 3.89 | 8.33 | 27.75 |
| 47 | Ag | Silver | 1235.08 | 2485 | 1.007 | 2.95 | 7.57 | 29.05 |
| 48 | Cd | Cadmium | 594.1 | 1038 | 0.467 | 1.16 | 8.99 | 25.89 |
| 49 | In | Indium | 429.32 | 2353 | 0.411 | 2.52 | 5.78 | 24.64 |
| 50 | Sn | Tin(\alpha) | 505.118 | 2543 | 1.11 | 3.14 | 7.34 | 21.97 |
| 51 | Sb | Antimony | 903.89 | 2023 | 0.383 | 2.75 | 8.64 | 25.1 |
| 52 | Te | Tellurium | 722.7 | 1263 | 0.230 | 2.19 | 9.01 | 27.6 |
| 53 | I | Iodine | 386.65 | 457.55 | | 1.11 | 10.45 | 29.54 |
| 54 | Xe | Xenon | 161.3 | 166.1 | | 0.16 | 12.13 | 33.3 |
| 55 | Cs | Cesium | 301.55 | 963 | 0.020 | 0.804 | 3.89 | 29.0 |
| 56 | Ba | Barium | 998 | 1913 | 0.103 | 1.90 | 5.21 | 15.21 |

| | | | | | | | | |
|---|---|---|---|---|---|---|---|---|
| 57 | La | Lanthanum | 1193 | 3727 | 0.243 | 4.47 | 5.61 | 17.04 |
| 58 | Ce | Cerium | 1071 | 3530 | 0.239 | 4.32 | 6.91 | |
| 59 | Pr | Praseodymium | 1204 | 3485 | 0.306 | 3.70 | 5.76 | |
| 60 | Nd | Neodymium | 1283 | 3400 | 0.327 | 3.40 | 6.31 | |
| 61 | Pm | Promethium | 1353 | 3000 | 0.35 | | | |
| 62 | Sm | Samarium | 1345 | 2051 | 0.294 | 2.14 | 5.6 | |
| 63 | Eu | Europium | 1095 | 1870 | 0.147 | 1.86 | 5.67 | |
| 64 | Gd | Gadolinium | 1584 | 3506 | 0.383 | 4.14 | 6.16 | |
| 65 | Tb | Terbium | 1633 | 3314 | 0.399 | 4.05 | 6.74 | |
| 66 | Dy | Dysprosium | 1682 | 2608 | 0.384 | 3.04 | 6.82 | |
| 67 | Ho | Holmium | 1743 | 2993 | 0.397 | 3.14 | | |
| 68 | Er | Erbium | 1795 | 2783 | 0.411 | 3.29 | | |
| 69 | Tm | Thulium | 1818 | 2000 | 0.397 | 2.42 | | |
| 70 | Yb | Ytterbium | 1097 | 1466 | 0.133 | 1.60 | 6.2 | |
| 71 | Lu | Lutetium | 1929 | 3588 | 0.411 | 4.43 | 5.0 | |
| 72 | Hf | Hafnium | 2423 | 5673 | 1.09 | 6.44 | 7. | 22. |
| 73 | Ta | Tantalum | 3269 | 5698 | 2.00 | 8.10 | 7.88 | 24.1 |
| 74 | W | Tungsten | 3680 | 6200 | 3.232 | 8.90 | 7.98 | 25.7 |
| 75 | Re | Rhenium | 3453 | 5900 | 3.72 | 8.03 | 7.87 | 24.5 |
| 76 | Os | Osmium | 3318 | 5300 | 4.18 | 8.17 | 8.7 | 26. |
| 77 | Ir | Iridium | 2683 | 4403 | 3.55 | 6.94 | 9. | |
| 78 | Pt | Platinum | 2045 | 4100 | 2.783 | 5.84 | 8.96 | 27.52 |
| 79 | Au | Gold | 1337.58 | 3213 | 1.732 | 3.81 | 9.22 | 29.7 |
| 80 | Hg | Mercury | 234.28 | 629.73 | 0.382 | 0.67 | 10.43 | 29.18 |
| 81 | Tl | Thallium | 576.7 | 1730 | 0.359 | 1.88 | 6.11 | 26.53 |
| 82 | Pb | Lead | 600.65 | 2013 | 0.430 | 2.03 | 7.41 | 22.44 |
| 83 | Bi | Bismuth | 544.5 | 1833 | 0.315 | 2.18 | 7.29 | 23.97 |
| 84 | Po | Polonium | 527 | 1235 | 0.26 | 1.50 | 8.43 | |
| 85 | At | Astatine | 575 | 610 | | | | |
| 86 | Rn | Radon | 202 | 211.4 | | 0.202 | 10.74 | |
| 87 | Fr | Francium | 300 | 950 | 0.020 | | | |
| 88 | Ra | Radium | 973 | 1413 | 0.132 | 1.66 | 5.28 | 15.42 |
| 89 | Ac | Actinium | 1320 | 3470 | 0.25 | 4.25 | 6.9 | 19.0 |
| 90 | Th | Thorium | 2023 | 5060 | 0.543 | 6.20 | | |
| 91 | Pa | Protactinium | 1827 | 4300 | 0.76 | | | |
| 92 | U | Uranium | 1405.5 | 4091 | 0.987 | 5.55 | | |
| 93 | Np | Neptunium | 913 | 4175 | 0.68 | 4.73 | | |
| 94 | Pu | Plutonium | 914 | 3600 | 0.54 | 3.60 | | |
| 95 | Am | Americium | 1267 | 2880 | | 2.73 | | |
| 96 | Cm | Curium | 1613 | | | 3.99 | | |
| 97 | Bk | Berkelium | 1259 | | | | | |
| 98 | Cf | Californium | 1173 | | | | | |
| 99 | Es | Einsteinium | 1133 | | | | | |
| 100 | Fm | Fermium | | | | | | |
| 101 | Md | Mendelevium | | | | | | |
| 102 | No | Nobelium | | | | | | |
| 103 | Lr | Lawrencium | | | | | | |
| 104 | Rf | Rutherfordium | | | | | | |
| 105 | Db | Dubnium | | | | | | |
| 106 | Sg | Seaborgium | | | | | | |
| 107 | Bh | Bohrium | | | | | | |
| 108 | Hs | Hassium | | | | | | |
| 109 | Mt | Meitnerium | | | | | | |

From the table we notice that the inert gas crystals have the lowest melting temperatures, bulk moduli, and cohesive energies. The alkali atoms have intermediate values of these properties between the noble gasses and the transition metals. In fact, these behaviors are more clearly seen in Figure 3.1.1 where the elements' melting temperatures, their bulk moduli, and their cohesive energies are plotted versus atomic number $Z$. We notice that the graphs of these properties bear similar shapes. The inert gas crystals have the lowest values, the alkali metals are next, followed by the transition metals. Also, notice that the group IV elements C, Si, and Ge, have respectable values for these properties as well.

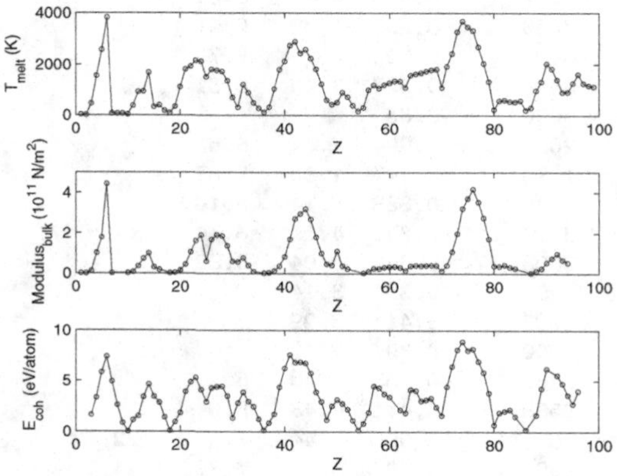


Figure 3.1.1: The elements' melting temperatures (top in Kelvin), the bulk moduli (middle in $10^{11}$ $N/m^2$), and the cohesive energies (bottom in $eV/atom$) are plotted versus atomic number $Z$.

As mentioned, the above properties are correlated. For example, if we sort the elements according to their cohesive energies, and we plot their melting temperatures and bulk moduli versus their energies, we find the results shown in Figure 3.1.2. A general trend emerges, and while there are large deviations, as the cohesive energy increases so do the melting temperatures and the bulk moduli. The dashed line is a simple 2nd-order polynomial fit to convey this understanding.

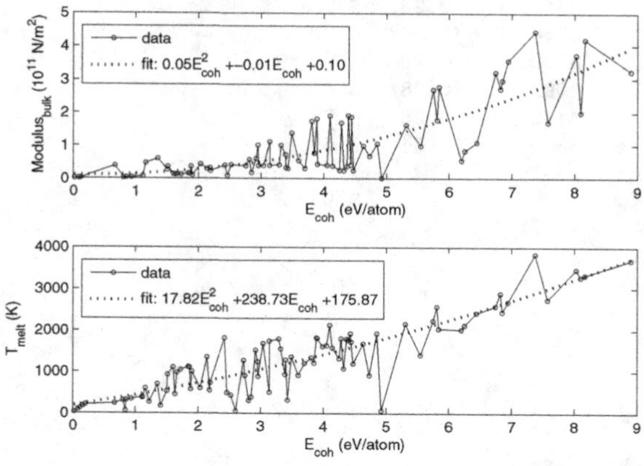


Figure 3.1.2: Elements' melting temperatures (top in Kelvin) and the bulk moduli (bottom in $10^{11}$ $N/m^2$) versus cohesive energy ($E_{coh}$ in $eV/atom$). The dashed lines in both graphs are simple 2nd-order polynomial fits that show a general increasing trend of the melting temperatures and the bulk moduli with cohesive energies.

The attractive interaction between the positive ion cores and the negatively charged electrons is responsible for the cohesion in solids. The cohesion is different for different elements in the periodic table. The noble gas atoms bind through van der Walls forces which are associated with fluctuations in their charge distributions. The alkali-halogen compounds are held together by ionic bonds, in which atoms transfer electrons and electrostatic forces play a dominant role. The groups II, III, IV, V, and VI atoms and compounds bind through sharing of electrons. A pair of atoms, for example, share electrons so that each atom fills its electronic subshells. The electronic distribution around the atoms overlap with a high concentration of charge in between the atoms and thus producing the binding between them. Alkali, alkaline, and transition metal atoms form metallic bonds. Here the valence electrons form a Fermi sea (a uniform electron background) wherein the ions are embedded. The attraction between the fixed ions and the uniformly distributed electrons is essentially responsible for the bonding in these metals.

## 3.2 Inert Gas Solids

The inert gas crystals are held together by van der Walls forces. This interaction is responsible for their low cohesive energies. In fact, the ionization potential of their free atoms is very high compared to their cohesive energies in crystal form as can be seen for Table 3.2.2. In the solid form, this means that the atomic electronic distributions are not significantly changed compared to that of their free atom distributions. Crystals of these elements are possible because of the attractive dipole-dipole interactions between the atoms. A general observation is that the noble gas crystals, at low temperatures, are weakly bound transparent insulators, and they form cubic closed packed (FCC) structures with lattice constants given by $a = \sqrt{2}d_{nn}$ (see Chapter 1).

Table 3.2.2: Inert gas crystal properties showing values for the nearest neighbor distance, $d_{nn}$; the experimental cohesive energy, $E_{coh}$; the melting temperature, $T_{melt}$; the first ionization potential, $I_p$. The parameters are all extrapolated to zero temperature and pressure [3]. †$\varepsilon$ and $\sigma$ are the parameters for the Lennard-Jones potential of Equation 3.2.19a.

| | $d_{nn}$(Å) | $E_{coh}$(eV/atom) | $T_{melt}$(K) | $I_p$ (eV) | $\varepsilon$ ($10^{-22}$ J)† | $\sigma$ (Å)† |
|---|---|---|---|---|---|---|
| He | | | | 24.58 | 1.4 | 2.56 |
| Ne | 3.13 | 0.02 | 24.56 | 21.56 | 5.0 | 2.74 |
| Ar | 3.76 | 0.080 | 83.81 | 15.76 | 16.7 | 3.40 |
| Kr | 4.01 | 0.116 | 115.8 | 14.00 | 22.5 | 3.65 |
| Xe | 4.35 | 0.17 | 161.4 | 12.13 | 32.0 | 3.98 |

### 3.2.1 van der Walls Interaction

To get a flavor of this interaction, a two-springs system is commonly used to model two interacting atoms and to study what happens as they are brought closer to each other. At the ends of each spring is a positive and a negative charge, respectively. The charges on the same atom are themselves held together by a spring of constant $C$. Referring to Figure 3.2.3, the idea being that when the atoms are far away from each other, the negative charge is distributed fairly spherically around the positive charges as shown in (a). In (b), as the atoms are brought closer to an equilibrium distance, the charges interact, thereby creating a different charge distribution, one that is similar to that of two interacting dipoles. Finally, in (c) the two-springs model replaces the interacting dipoles.

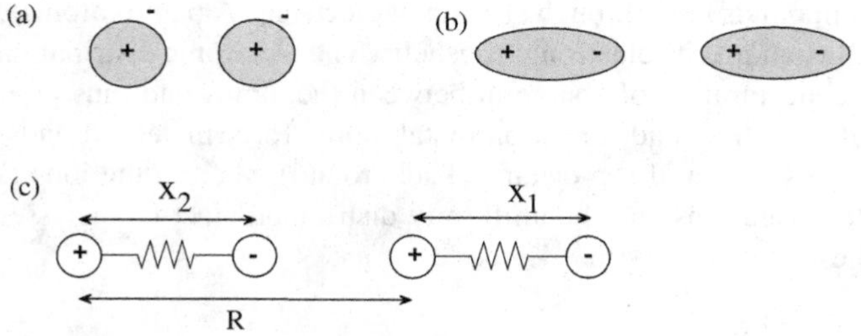


Figure 3.2.3: (a) Shows two non-interacting atoms located far apart from each other. (b) Shows that when the atoms are brought closer to each other, each develops a charge distribution much as that of two interacting dipoles. (c) Shows the two-springs model of the interacting dipoles.

The one-dimensional unperturbed Hamiltonian for this uncoupled system consists of the free atoms, modeled as two harmonic oscillators of equal energy, or

$$H_0 = \frac{p_1^2}{2m} + \frac{1}{2}Cx_1^2 + \frac{p_2^2}{2m} + \frac{1}{2}Cx_2^2, \tag{3.2.1a}$$

each oscillator with frequency and ground state quantum mechanical energy of

$$\omega_0 \equiv \sqrt{\frac{C}{m}}, \qquad E_1 = E_2 = \frac{\hbar\omega_0}{2} \equiv E_0, \tag{3.2.1b}$$

respectively, where $m$ is the electronic mass and $C$ is the spring constant associated with each atom. The total system's initial energy is

$$E_{Ti} = E_1 + E_2 = \hbar\omega_0 = 2E_0. \tag{3.2.1c}$$

The above spring constant can possibly be estimated if we use a hydrogenic model and express the potential between the atom's $+$ and $-$ charges in the form of an effective potential that includes the coulomb contribution as well as the electron's rotational kinetic energy; i.e., $V_{eff}(r) = -Zke^2/r + L^2/2mr^2$, where $k = 1/4\pi\varepsilon_0$, $L$ is the angular momentum, and $Z$ is the atomic number. Using a Taylor expansion of this potential about an equilibrium position $r_0$, we can find that $C \approx Zke^2/r_0^3$ (see Exercise 3.6.2).
By letting the atoms approach each other and letting them interact, the perturbed system's Hamiltonian becomes $H = H_0 + H_1$, where the interaction part of the Hamiltonian is

$$H_1 = \frac{ke^2}{R} + \frac{ke^2}{R + x_1 - x_2} - \frac{ke^2}{R + x_1} - \frac{ke^2}{R - x_2}, \tag{3.2.2}$$

where the $R$ is the equilibrium distance between the atoms as in Figure 3.2.3(c). The first term is due to the ion-ion repulsion, the second term is the negative-negative charge repulsion, the third term is the first ion-second-atom's-negative-charge attraction, and the last term is the the second ion-first-atom's-negative-charge attraction. We see that $H_1$ is complicated and it becomes convenient to simplify it. We therefore expand the terms involving the negative charge coordinates, while taking

$R >> |x_1 - x_2|$, $R >> x_1$, and $R >> x_2$, to 2nd order in a Taylor series. We have

$$\begin{aligned}
\frac{1}{R+x_1-x_2} &= \left(\frac{1}{R}\right)\frac{1}{1+(x_1-x_2)/R} = \frac{1}{R}(1-(x_1-x_2)/R+((x_1-x_2)/R)^2+\cdots) \\
&\approx \frac{1}{R}(1-(x_1-x_2)/R+((x_1-x_2)/R)^2); \\
\frac{-1}{R+x_1} &= -\left(\frac{1}{R}\right)\frac{1}{1+x_1/R} = -\frac{1}{R}(1-x_1/R+(x_1/R)^2+\cdots) \\
&\approx -\frac{1}{R}(1-x_1/R+(x_1/R)^2); \\
\frac{-1}{R-x_2} &= -\left(\frac{1}{R}\right)\frac{1}{1-x_2/R} = -\frac{1}{R}(1+x_2/R+(x_2/R)^2+\cdots) \\
&\approx -\frac{1}{R}(1+x_2/R+(x_2/R)^2).
\end{aligned} \tag{3.2.3}$$

Substituting these approximations into Equation 3.2.2, we get

$$\begin{aligned}
H_1 &\approx \frac{ke^2}{R}\{1+[1-x_1/R+x_2/R+(x_1^2+x_2^2-2x_1x_2)/R^2] \\
&\quad -[1-x_1/R+x_1^2/R^2]-[1+x_2/R+x_2^2/R^2]\} \\
&= -2\frac{ke^2x_1x_2}{R^3}.
\end{aligned} \tag{3.2.4}$$

Combining this expression with $H_0$ we write the approximation for the full Hamiltonian as

$$H = H_0 + H_1 \approx \frac{p_1^2}{2m} + \frac{1}{2}Cx_1^2 + \frac{p_2^2}{2m} + \frac{1}{2}Cx_2^2 - 2\frac{ke^2x_1x_2}{R^3}, \tag{3.2.5}$$

and shows that the presence of the coupling between the atoms, $H_1$, decreases the energy relative to the free atoms Hamiltonian and we can expect some binding. We notice that $H_0$ is diagonal, but $H_1$ mixes the variables $x_1$ and $x_2$, we therefore write $H$ in matrix form as

$$\begin{aligned}
H =& \frac{1}{2m}\left[\begin{pmatrix} p_1 & p_2 \end{pmatrix}\begin{pmatrix} 1 & 0 \\ 0 & 1 \end{pmatrix}\begin{pmatrix} p_1 \\ p_2 \end{pmatrix}\right] + \frac{C}{2}\left[\begin{pmatrix} x_1 & x_2 \end{pmatrix}\begin{pmatrix} 1 & 0 \\ 0 & 1 \end{pmatrix}\begin{pmatrix} x_1 \\ x_2 \end{pmatrix}\right] \\
&- \frac{ke^2}{R^3}\left[\begin{pmatrix} x_1 & x_2 \end{pmatrix}\begin{pmatrix} 0 & 1 \\ 1 & 0 \end{pmatrix}\begin{pmatrix} x_1 \\ x_2 \end{pmatrix}\right],
\end{aligned} \tag{3.2.6}$$

to notice that a transformation is needed in order to diagonalize the non-diagonal $H_1$ matrix but which leaves the diagonal $H_0$ unmodified. This is not without cost. The matrix we seek transforms the original coordinates to a space where $H$ is diagonal. This process is actually very useful and we employ it here. We first define the following

$$x \equiv \begin{pmatrix} x_1 \\ x_2 \end{pmatrix}, \quad p \equiv \begin{pmatrix} p_1 \\ p_2 \end{pmatrix}, \quad X \equiv \begin{pmatrix} X_s \\ X_a \end{pmatrix}, \quad P \equiv \begin{pmatrix} P_s \\ P_a \end{pmatrix}, \tag{3.2.7}$$

and define the $2 \times 2$ matrix $V_E$ such that

$$x = V_E X, \quad p = V_E P, \quad V_E^{-1}V_E = I, \tag{3.2.8}$$

where $I$ is the $2 \times 2$ unit diagonal matrix. With these definitions, the Hamiltonian of Equation 3.2.6 can be rewritten as

$$\begin{aligned}
H =& \frac{1}{2m}\left[\begin{pmatrix} P_s & P_a \end{pmatrix}V_E^{-1}\begin{pmatrix} 1 & 0 \\ 0 & 1 \end{pmatrix}V_E\begin{pmatrix} P_s \\ P_a \end{pmatrix}\right] + \frac{C}{2}\left[\begin{pmatrix} X_s & X_a \end{pmatrix}V_E^{-1}\begin{pmatrix} 1 & 0 \\ 0 & 1 \end{pmatrix}V_E\begin{pmatrix} X_s \\ X_a \end{pmatrix}\right] \\
&- \frac{ke^2}{R^3}\left[\begin{pmatrix} X_s & X_a \end{pmatrix}V_E^{-1}MV_E\begin{pmatrix} X_s \\ X_a \end{pmatrix}\right],
\end{aligned} \tag{3.2.9}$$

where we have also defined the non-diagonal matrix $M \equiv \begin{pmatrix} 0 & 1 \\ 1 & 0 \end{pmatrix}$. The matrix elements of the $V_E$ matrix, or eigenvectors, are to be found such that

$$V_E^{-1} M V_E = D, \tag{3.2.10}$$

where $D$ is a diagonal matrix that contains the eigenvalues of $M$. Notice that $V_E$ does not affect the already diagonal matrices of $H$ since $V_E^{-1} I V_E = V_E^{-1} V_E = I$. Once $V_E$ and $D$ are known, Equation 3.2.9 becomes diagonal and the system's energies become transparent. To find the eigenvalues, the procedure begins by noticing that $MV_E = DV_E$ or $(M-D)V_E = 0$ which means that $|M-D| = 0$, and writing $D = \begin{pmatrix} \lambda & 0 \\ 0 & \lambda \end{pmatrix}$, to solve for the eigenvalues $\lambda$ as follows

$$|M-D| = \left| \begin{pmatrix} 0 & 1 \\ 1 & 0 \end{pmatrix} - \begin{pmatrix} \lambda & 0 \\ 0 & \lambda \end{pmatrix} \right| = \left| \begin{pmatrix} -\lambda & 1 \\ 1 & -\lambda \end{pmatrix} \right| = \lambda^2 - 1 = 0, \tag{3.2.11a}$$

or

$$\lambda = \pm 1 \Rightarrow D = \begin{pmatrix} 1 & 0 \\ 0 & -1 \end{pmatrix}. \tag{3.2.11b}$$

Having the eigenvalues, we can find the eigenvectors by solving and normalizing as follows

$$(M - \lambda_n I) V_{En} = 0, \qquad \sum_i V_{E\,in}^2 = 1, \tag{3.2.11c}$$

where $\lambda_n$ is the $nth$ eigenvalue and $V_{En}$ is the $nth$ column or eigenvector of the $V_E$ matrix. Using the first eigenvalue, $\lambda = 1$, we get

$$\left[ \begin{pmatrix} 0 & 1 \\ 1 & 0 \end{pmatrix} - \begin{pmatrix} 1 & 0 \\ 0 & 1 \end{pmatrix} \right] \begin{pmatrix} V_{E11} \\ V_{E21} \end{pmatrix} = \begin{pmatrix} -1 & 1 \\ 1 & -1 \end{pmatrix} \begin{pmatrix} V_{E11} \\ V_{E21} \end{pmatrix} = \begin{pmatrix} -V_{E11} + V_{E21} = 0 \\ V_{E11} - V_{E21} = 0 \end{pmatrix}. \tag{3.2.11d}$$

Similarly, using $\lambda = -1$

$$\left[ \begin{pmatrix} 0 & 1 \\ 1 & 0 \end{pmatrix} - \begin{pmatrix} -1 & 0 \\ 0 & -1 \end{pmatrix} \right] \begin{pmatrix} V_{E12} \\ V_{E22} \end{pmatrix} = \begin{pmatrix} 1 & 1 \\ 1 & 1 \end{pmatrix} \begin{pmatrix} V_{E12} \\ V_{E22} \end{pmatrix} = \begin{pmatrix} V_{E12} + V_{E22} = 0 \\ V_{E12} + V_{E22} = 0 \end{pmatrix}. \tag{3.2.11e}$$

Equations 3.2.11d and 3.2.11e result in $V_{E11} = V_{E21}$ and $V_{E22} = -V_{E12}$, respectively. Finally, normalizing the eigenvectors, as in Equation 3.2.11c, we have $V_{E11}^2 + V_{E21}^2 = 1 = 2V_{E11}^2$ or $V_{E11} = 1/\sqrt{2} = V_{E21}$. In a similar way we find that $V_{E12} = 1/\sqrt{2} = -V_{E22}$, so that the matrix $V_E$ takes the form

$$V_E = \frac{\begin{pmatrix} 1 & 1 \\ 1 & -1 \end{pmatrix}}{\sqrt{2}}. \tag{3.2.12}$$

It can be verified that $V_E$ satisfies the inverse property of a unitary matrix; i.e., $V_E = V_E^{-1}$ and the triple matrix product $V_E^{-1} M V_E = D$ as it should be, in consistency with Equation 3.2.10. Substituting $D$ for the triple matrix product into Equation 3.2.9 and, again, using $V_E^{-1} I V_E = I$, find

$$\begin{aligned} H = & \frac{1}{2m} \left[ \begin{pmatrix} P_s & P_a \end{pmatrix} \begin{pmatrix} 1 & 0 \\ 0 & 1 \end{pmatrix} \begin{pmatrix} P_s \\ P_a \end{pmatrix} \right] + \frac{C}{2} \left[ \begin{pmatrix} X_s & X_a \end{pmatrix} \begin{pmatrix} 1 & 0 \\ 0 & 1 \end{pmatrix} \begin{pmatrix} X_s \\ X_a \end{pmatrix} \right] \\ & - \frac{ke^2}{R^3} \left[ \begin{pmatrix} X_s & X_a \end{pmatrix} \begin{pmatrix} 1 & 0 \\ 0 & -1 \end{pmatrix} \begin{pmatrix} X_s \\ X_a \end{pmatrix} \right], \end{aligned} \tag{3.2.13}$$

which, after multiplying the matrices through, leads to

$$\begin{aligned} H &= \frac{P_s^2}{2m} + \frac{P_a^2}{2m} + \frac{C}{2}X_s^2 + \frac{C}{2}X_a^2 - \frac{ke^2}{R^3}X_s^2 + \frac{ke^2}{R^3}X_a^2 \\ &= \left(\frac{P_s^2}{2m} + \frac{1}{2}C_sX_s^2\right) + \left(\frac{P_a^2}{2m} + \frac{1}{2}C_aX_a^2\right), \end{aligned} \tag{3.2.14a}$$

where we have made the following definitions for the spring constants associated with the transformed coordinate space

$$C_s \equiv C - \frac{2ke^2}{R^3}, \qquad C_a \equiv C + \frac{2ke^2}{R^3}. \tag{3.2.14b}$$

Once again, we see that the system consists of two springs whose energies are shifted with respect to the unperturbed energies (Equation 3.2.1). Also, the interacting system is characterized by symmetric and antisymmetric displacements and momenta which, according to Equation 3.2.8, are given in terms of the unperturbed system's coordinates by

$$\begin{pmatrix} X_s \\ X_a \end{pmatrix} = V_E^{-1}\begin{pmatrix} x_1 \\ x_2 \end{pmatrix} = \begin{pmatrix} x_1 + x_2 \\ x_1 - x_2 \end{pmatrix}/\sqrt{2}, \tag{3.2.15a}$$

and

$$\begin{pmatrix} P_s \\ P_a \end{pmatrix} = V_E^{-1}\begin{pmatrix} p_1 \\ p_2 \end{pmatrix} = \begin{pmatrix} p_1 + p_2 \\ p_1 - p_2 \end{pmatrix}/\sqrt{2}. \tag{3.2.15b}$$

We can, therefore, associate the transformed coordinates $X_s$, $P_s$ and $X_a$, $P_a$ symmetric and antisymmetric modes of vibration, respectively, with frequencies given by

$$\omega_s = \sqrt{\frac{C_s}{m}} = \sqrt{\frac{C - 2ke^2/R^3}{m}}, \qquad \omega_a = \sqrt{\frac{C_a}{m}} = \sqrt{\frac{C + 2ke^2/R^3}{m}}, \tag{3.2.16a}$$

and associated energies of

$$E_s = \frac{\hbar\omega_s}{2}, \qquad E_a = \frac{\hbar\omega_a}{2}. \tag{3.2.16b}$$

This shows that the symmetric mode is lower in energy than the initial single atomic unperturbed system's energy of $\hbar\omega_0/2$ of Equation 3.2.1b; the antisymmetric mode is higher in energy. If we expand the frequencies to 2nd order in the coupling term (using $\sqrt{1+x} \approx 1 + x/2 - x^2/8)$), we have from Equation 3.2.16a

$$\begin{aligned} \omega_{s,a} &= \sqrt{\frac{C}{m}}\sqrt{1 \mp 2ke^2/CR^3} \\ &\approx \omega_0\left(1 \mp \frac{1}{2}\frac{2ke^2}{CR^3} - \frac{1}{8}\left(\frac{2ke^2}{CR^3}\right)^2\right), \end{aligned} \tag{3.2.17a}$$

Whereas we started with oscillator atoms at infinite separations, each with an energy of $\hbar\omega_0/2$, as they get nearer to each other their energies begin to split. Their coupling increases as they get closer and we end up with the same two oscillators at two different energies $E_s$ and $E_a$. The symmetric state is a bonding state and the antisymmetric state is an antibonding state. The splitting between the energies is such that $E_a - E_0 \approx E_0\left(\frac{1}{2}\frac{2ke^2}{CR^3} - \frac{1}{8}\left(\frac{2ke^2}{CR^3}\right)^2\right)$, and $E_0 - E_s \approx E_0\left(\frac{1}{2}\frac{2ke^2}{CR^3} + \frac{1}{8}\left(\frac{2ke^2}{CR^3}\right)^2\right)$. All this is illustrated in Figure 3.2.4.

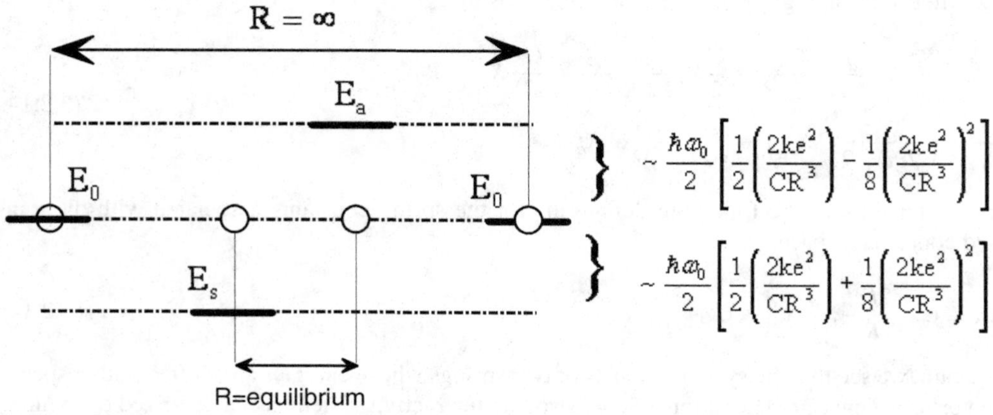


Figure 3.2.4: The energy levels of the two-springs model system are shown when the atoms (circles) are at $R=\infty$ with an energy of $E_0$ each. Also shown are the obtained energy level splittings $E_s$ and $E_a$ when the atoms are brought to an equilibrium distance. The approximate amount of splitting between the levels is also shown as discussed in the text leading to this figure.

**Example 3.2.1.1**

Let's produce a plot of the behavior of the level splitting described in Figure 3.2.4 versus the distance between the atoms. Noticing that from Equation 3.2.17a $\omega_{s,a}=\sqrt{\frac{C}{m}}\sqrt{1\mp 2ke^2/CR^3}$, but $\omega_0=\sqrt{C/m}$ and if we assume a simple hydrogen atom and let $C\sim ke^2/r_0^3$ as mentioned earlier, then we can write $\omega_{s,a}=\omega_0\sqrt{1\mp 2r_0^3/R^3}$. Further, if we define energy units of $\hbar\omega_0$ and distance units of $r_0$, then $\bar{E}_{s,a}\equiv E_{s,a}/(\hbar\omega_0)=\sqrt{1\mp 2/x^3}$, where we have also defined the dimensionless distance $x\equiv R/r_0$. In this way, it suffices to plot the dimensionless $\bar{E}$ versus $x$ to get the plot we seek. Note that as $x\to\infty$, $\bar{E}_{s,a}\to 1$ here as expected. For the equilibrium distance between the atoms, we take a bond length of about 2.5Å and taking $r_0\sim 0.5$Å, then $x_{min}\sim 2.5/0.5=5\equiv d$, but certainly no less than $2^{1/3}$ to keep $\bar{E}_s$ real. The results are shown in Figure 3.2.5.

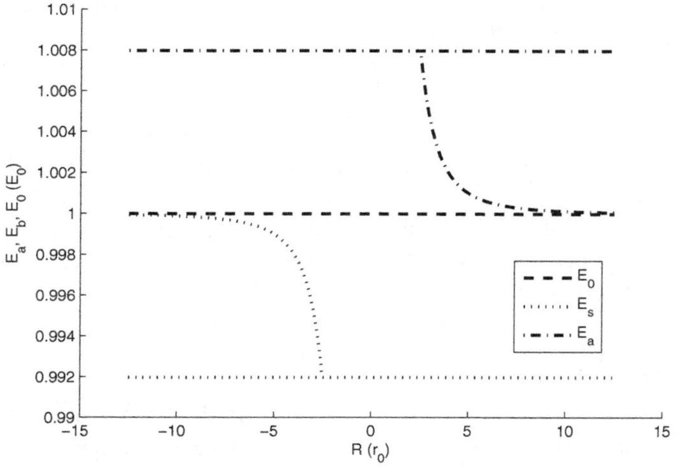


Figure 3.2.5: The energy levels of the two-springs model system are shown as calculated according to Example 3.2.1.1. When the two springs are far apart, each spring has energy $E_0$ associated with it. As the atoms (modeled as springs) are brought closer, their interaction splits the energy level into a symmetric mode of energy $E_s$ and an antisymmetric mode of energy $E_a$. Here energy is in units of $E_0$ and distance $R$ is in units of $r_0$. In the example, the initial atom positions are $\mp 5d/2$ for atoms 1 and 2, respectively, where we take their closest separation as $d \equiv x_{min} = 5$ in units of $r_0$ such that $x = R/r_0$.

We can find the total approximate final energy of the coupled oscillator system as

$$E_{Tf} = \frac{\hbar}{2}(\omega_s + \omega_a) \approx \hbar\omega_0 \left(1 - \frac{1}{8}\left(\frac{2ke^2}{CR^3}\right)^2\right), \tag{3.2.17b}$$

as well as the approximate change in energy between the coupled and the uncoupled systems of oscillators

$$\Delta U = E_{Tf} - E_{Ti} \approx \hbar\omega_0 \left(1 - \frac{1}{8}\left(\frac{2ke^2}{CR^3}\right)^2\right) - \hbar\omega_0 = -\frac{\hbar\omega_0}{8}\left(\frac{2ke^2}{CR^3}\right)^2 \sim -\frac{A}{R^6}, \tag{3.2.17c}$$

where $A \equiv \frac{\hbar\omega_0}{8}\left(\frac{2ke^2}{C}\right)^2$, which vanishes in the limits when $\hbar \to 0$, reflecting its quantum mechanical nature. Equation 3.2.17c is a result of the attractive interaction between the two oscillating atoms and varies as the inverse 6th power of their separation. This is the so-called van der Waals interaction or London interaction or dipole-dipole interaction, which is the principal attractive interaction in inert gas crystals.

A question that might be asked is: what causes atoms to have an equilibrium distance? In the above Example 3.2.1.1, the calculation of the equilibrium separation between the two atomic systems was not carried out. That requires a more sophisticated procedure than the one employed here, which is left for later in the chapter. However, as two atoms get closer to each other, their electron distributions begin to overlap. According to the Pauli exclusion principle, no two electrons can have the same set of quantum numbers. As the electrons' wavefunctions begin to overlap, the Pauli principle prevents the atoms from coming too close. The competition between the attractive interaction and the repulsive interaction establishes an equilibrium distance at which the pair of atoms coexist. We make no effort here to derive such repulsive interaction but a generally used, empirical form, of this interaction, which seems to work well in the inert gasses, is

$$U_{repulsive} \sim \frac{B}{R^{12}}. \tag{3.2.18}$$

This repulsive part, which is proportional to the square of Equation 3.2.17c is generally considered to be an empirical relation.
In Section 3.2.2 we make use of the above interactions; that is, Equations 3.2.17c and 3.2.18, for the attractive and repulsive parts, respectively, of a potential between a pair of atoms originally proposed by Sir John Edward Lennard-Jones.

### 3.2.2 Lennard-Jones Potential

If we combine the attractive interaction between two atomic dipoles, Equation 3.2.17c, with the empirical relation of Equation 3.2.18, we get a potential for a pair of interacting atoms in the form

$$U(R) = \frac{B}{R^{12}} - \frac{A}{R^6} = 4\varepsilon\left[\left(\frac{\sigma}{R}\right)^{12} - \left(\frac{\sigma}{R}\right)^{6}\right], \tag{3.2.19a}$$

with the definitions

$$A \equiv 4\varepsilon\sigma^6, \quad B \equiv 4\varepsilon\sigma^{12}. \tag{3.2.19b}$$

Because of the powers to which the parenthesized terms are raised to, the potential is also known as the Lennard-Jones $6-12$ potential. For large $R$, $\left(\frac{\sigma}{R}\right)^6$ is the dominant term and the atoms attract, but for small $R$, $\left(\frac{\sigma}{R}\right)^{12}$ takes over and the atoms repel. The force between the atoms is given by

$$F = -\frac{\partial U(R)}{\partial R}. \tag{3.2.20}$$

Let's look at the Lennard-Jones (L-J) potential to get an understanding of how it works.

**Example 3.2.2.1**
Here we first find the point at which the minimum of the Lennard-Jones potential occurs. We do this by setting the derivative of the force in Equation 3.2.20 to zero, which when solved for $R$ gives

$$R = 2^{1/6}\sigma \equiv R_{nn} \sim 1.1225\,\sigma, \tag{3.2.21a}$$

where $R_{nn}$ is the nearest neighbor distance or bond length in this model. Evaluating $U(R)$ of Equation 3.2.19a at $R = R_{nn}$, we get

$$U(R = R_{nn}) \equiv U_{min} = -\varepsilon, \tag{3.2.21b}$$

which corresponds to the binding energy of the pair of atoms. Inputting this much energy into the molecule would cause it to dissociate. Finally, let $R_0$ be the distance below $R_{nn}$ at which the potential crosses the $R$ axis. This location can be found as

$$U(R = R_0) = 0 \quad \Rightarrow R_0 = \sigma. \tag{3.2.21c}$$

A plot of this potential for an Argon pair of atoms, for which (using Table 3.2.2) $\varepsilon = 0.0104\,eV$ and $\sigma = 3.40$ Å, is shown in Figure 3.2.6 where the minimum energy and the bond length are identified according to the results of this example.

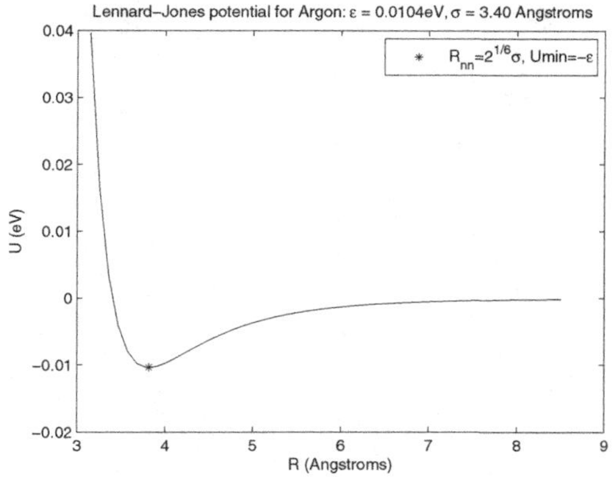


Figure 3.2.6: The Lennard-Jones potential for the solid Argon atom pair potential of Example 3.2.2.1. The parameters used are $\varepsilon = 0.0104\,eV$ and $\sigma = 3.40$ Å from Table 3.2.2. The bond length obtained is $R_{nn} = 2^{1/6}\sigma = 3.82$ Å.

We now develop a Monte-Carlo version of the above example and verify its results by comparing to the analytic result.

**Example 3.2.2.2**

In Example 3.2.2.1 the results for the minimum energy and the bond length were obtained analytically for Argon. We now repeat the same example within a Monte-Carlo framework. Here, the idea is to work with the Lennard-Jones potential; however, we use random numbers to vary the distance in between the two atoms and accept the distance if it leads to a lower minimum of the potential than the previous saved value. When such minimum is not changing significantly within a tolerance value of about $tol = 1 \times^{-3} \varepsilon$ and the number of trials taken is not too large, say, less than 500 iterations, the calculation is stopped. The random numbers, $r_n$, are picked such that they lie in the range $-\delta < r_n < \delta$, where $\delta$ is a desired upper value ($\delta = 1$ is an example). A new random number is chosen for every iteration of the distance until the desired minimum is achieved. For every random number, the atomic distance is varied according to $R_{new} = R_{old}(1 + c_{mod} r_n)$, where $c_{mod}$ is a moderating parameter so that $R$ does not vary too drastically. The simulation that follows uses $c_{mod} = 0.5$. To get started, the simulation uses a guess for the starting distance as shown in the script LJ_1dMCsym.m listed below. The results obtained are shown in Figure 3.2.7. The bond length obtained this way is $R_{nn} = 3.821$Å with $U_{min} = -0.01039eV$ for the value of the minimum potential. Comparing the converged simulation values to the analytic ones, they are in very close agreement, where the analytic values are $2^{1/6}\sigma = 3.816$ Å, and $\varepsilon = 0.0104\,eV$, respectively, as used in Example 3.2.2.1.

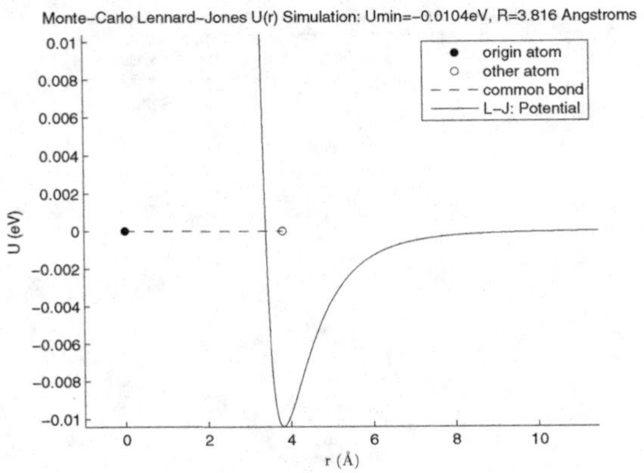


Figure 3.2.7: Result of the Lennard-Jones potential Monte-Carlo simulation for the solid Argon atom pair potential of Example 3.2.2.2. The parameters used are $\varepsilon = 0.0104\,eV$ and $\sigma = 3.40$ Å from Table 3.2.2. The bond length obtained is $R_{nn} = 3.821$Å with $U_{min} = -0.01039eV$ for the value of the minimum potential.

The code LJ_1dMCsym.m used to create Figure 3.2.7 follows.

```
%copyright by J. E Hasbun and T. Datta
%LJ_1dMCsym.m
%Program to do a simulation of the distance between two
%particles in 1 d using the Lennard-Jones potential,
%where U=4*epsilon*((sigma/R)^12-(sigma/R)^6)
clear
%We work with Argon
eps=0.0104;                %eV
sig=3.4;                   %Angstroms
%Define the LJ potential as an anonymous function
U=@(eps,sig,r) 4*eps*((sig./r).^12-(sig./r).^6);
Rguess=2.0*sig;
Rold=Rguess;               %guess starting distance in terms of sigma
Uold=U(eps,sig,Rold); %starting energy
cmod=0.25;                 %moderator for the random number used below
tol=1.e-3*eps;             %convergence tolerance
itermax=500;               %maximum iterations
diffU=10*tol;
plot(0,0,'ko','MarkerSize',5,'MarkerFaceColor','k')
hold on
axis([-1 1.5*Rguess -1 1])
plot(Rold,0,'bo','MarkerSize',5,'MarkerFaceColor','b')
iter=0;
while (diffU > tol & iter < itermax)
  iter=iter+1;
  cla
  h(1)=plot(0,0,'ko','MarkerSize',5,'MarkerFaceColor','k');
  rn=-1.0+2.0*rand(); %random number  -1 < r < 1
  Rnew=Rold*(1+cmod*rn);   %new R based on rand #, with moderation
```

```
  Unew=U(eps,sig,Rnew);     %new energy
  diffU=abs(Unew-Uold)/abs(Unew+Uold);
  if (Unew < Uold)
    Rold=Rnew;
    Uold=Unew;
    diffUold=diffU;
    h(2)=plot(Rold,0,'ko','MarkerSize',5);
    pause(0.05)
  end
end
h(2)=plot(Rold,0,'ko','MarkerSize',5);
h(3)=line([0,Rold],[0,0],'LineStyle','--','Color','k');
fprintf('diffU=%8.3e, Min U=%9.6f (eV), R=%9.6f (A), iter=%6i\n',...
    diffUold,Uold,Rold,iter)
Ns=500;
rl=0.1*Rold;
ru=3.0*Rold;
rs=(ru-rl)/(Ns-1);
r=rl:rs:ru;
Ur=U(eps,sig,r);
Umin=min(Ur);
h(4)=plot(r,Ur,'k-');
axis([-1 1.5*Rguess Umin abs(Umin)])
xlabel('r (\AA)','interpreter','latex')
ylabel('U (eV)')
str=cat(2,'Monte-Carlo Lennard-Jones U(r) Simulation:',' Umin=',...
  num2str(Umin,'%9.4g'),'eV, R=',num2str(Rold,'%9.4g'),' Angstroms');
title(str)
legend(h,'origin atom','other atom','common bond','L-J: Potential')
```

The Lennard-Jones $6-12$ potential of Equation 3.2.19a has found much use in solid state physics and other similar potentials have been proposed that replace the repulsive part with a slightly different form. Due to its mathematical simplicity, an alternate expression that is often used for the repulsive interaction is

$$U_{repulsive}(R) = \lambda \exp(-R/\rho), \tag{3.2.22}$$

where $\lambda$ and $\rho$ are considered parameters. This will find use in a later section.

### 3.2.3 Lennard-Jones Potential for Crystals

We now extend the previous section's Lennard-Jones (L-J) pair potential to the regime where there are many noble gas atoms at equilibrium. It is also assumed that the ion cores are at rest and that the total system's energy is the result of adding all the energies between atomic pairs. Consider a system of $N$ atoms in a crystal, with each pair of atoms interacting through an L-J pair potential, $U(R_{ij}) = 4\varepsilon[(\sigma/R_{ij})^{12} - (\sigma/R_{ij})^6]$ for atoms $i$, $j$ where $i \neq j$. By summing over distinct pairs of atoms, and noting that $U(R_{ij}) = U(R_{ji})$, the total cohesive energy of the system can now be written as

$$\begin{aligned}U_{tot} =& U(R_{12})+U(R_{13})+\ldots+U(R_{1N})+U(R_{23})+U(R_{24})+\ldots+U(R_{2N})\\ &+U(R_{34})+U(R_{35})+\ldots+U(R_{3N})+U(R_{45})+U(R_{46})+\ldots+U(R_{4N})\\ &+\ldots+U(R_{N-1N}) = \sum_{\substack{j,i\\ i>j}}^{N} U(R_{ij}) = \frac{1}{2}\sum_{\substack{j,i\\ i\neq j}}^{N} U(R_{ij}),\end{aligned} \tag{3.2.23}$$

where the factor of $1/2$ prevents double counting. Since there are $N$ pairs of identical atoms, using the last equation, we can sit on the $ith$ atom and carry one of the sums over $j$ then multiply the result by $N$ to get the total energy; that is,

$$U_{tot} = \frac{N}{2}\sum_{\substack{j\\ j\neq i}}^{N} U(R_{ij}) = 4\varepsilon\frac{N}{2}\sum_{\substack{j\\ j\neq i}}^{N}[(\sigma/R_{ij})^{12}-(\sigma/R_{ij})^{6}]. \tag{3.2.24}$$

Letting the nearest neighbor distance be $R$ and since from a given atom all the other atoms are at multiples of this distance, we can write $R_{ij} = p_{ij}R$. Here, $p_{ij}$ is a number which, when multiplied by $R$, gives the separation between the reference atom $i$ and the atom $j$ of the crystal. The total cohesive energy becomes

$$U_{tot} = 4\varepsilon\frac{N}{2}\sum_{\substack{j\\ j\neq i}}^{N}\left[\left(\frac{\sigma}{p_{ij}R}\right)^{12}-\left(\frac{\sigma}{p_{ij}R}\right)^{6}\right] = 4\varepsilon\frac{N}{2}\left[\left(\frac{\sigma}{R}\right)^{12}\sum_j{}'\left(\frac{1}{p_{ij}}\right)^{12}-\left(\frac{\sigma}{R}\right)^{6}\sum_j{}'\left(\frac{1}{p_{ij}}\right)^{6}\right], \tag{3.2.25}$$

where the prime on the summation indicates that $j \neq i$. To see how the sums of the $1/p_{ij}$ term are carried out, we do the following example.

**Example 3.2.3.1**

Calculate the expression $\sum_j{}'\left(\frac{1}{p_{ij}}\right)^{12}$ for the simple cubic system.

**Solution**

From the origin atom, the simple cubic system has nearest neighbors at positions $(100)$, $(010)$, $(001)$, $(\bar{1}00)$, $(0\bar{1}0)$, and $(00\bar{1})$, in units of the lattice constant $a$; that is, six nearest neighbors at $1R$ where $R = a$ so that $p_{01} = 1$. As discussed in Chapter 1, we can make linear combinations of these vectors to obtain farther away neighbors from the origin atom. In this way we find there are twelve 2nd nearest neighbors (one instance is $(110)$ in units of a) at a distance of $\sqrt{2}R$, so that $p_{02} = \sqrt{2}$. Similarly, there are eight 3rd nearest neighbors at a distance $\sqrt{3}R$ and so $p_{03} = \sqrt{3}$. Continuing this way, there are six 4th nearest neighbors with $p_{04} = \sqrt{4}$. If we include only these contributions to the sum, we get the approximate value $6(1/\sqrt{1})^{12} + 12(1/\sqrt{2})^{12} + 8(1/\sqrt{3})^{12} + 6(1/\sqrt{4})^{12} = 6.1999 \approx 6.20$, which is in common use.

Jones [11] showed that the exact value of the lattice sum $\sum_j{}'\left(\frac{1}{p_{ij}}\right)^{12} = 6.2021$. To get close to this value, we would have to include the next 24, 24, 12, and 30 numbers of $5th$, $6th$, $8th$, and $9th$ nearest neighbors, respectively, with diminishing contributions from the farther away neighbors. Notice that there is no $p_{07}$ to speak of that follows the above $p_{0j} = \sqrt{j}$ rule. In the simple cubic, the exact sum can be obtained from the expression

$$\sum_{i=-N}^{N}\sum_{j=-N}^{N}\sum_{k=-N}^{N}\left(\frac{1}{i^2+j^2+k^2}\right)^{s/2} \quad (i\neq 0, j\neq 0, k\neq 0), \tag{3.2.26}$$

which is a triple sum over all integral values of $i$, $j$, $k$ with the exclusion of $i = 0$, $j = 0$, $k = 0$, and where $s$ is the Lennard-Jones exponent of interest, either 12 or 6. Due to the high exponent when $s = 12$, the respective sum converges rapidly; the same cannot be said for the $s = 6$ exponent. From Equation 3.2.26 we can, therefore, compute the exact value $\sum_j' \left(\frac{1}{p_{ij}}\right)^6 = 8.4019$ (see Exercise 3.6.7), but the approximate value of 8.40 is often used.

For the FCC, $\sum_j' \left(\frac{1}{p_{ij}}\right)^{12} \approx 12.13$ and $\sum_j' \left(\frac{1}{p_{ij}}\right)^6 \approx 14.45$; for the BCC $\sum_j' \left(\frac{1}{p_{ij}}\right)^{12} \approx 9.11$ and $\sum_j' \left(\frac{1}{p_{ij}}\right)^6 \approx 12.25$. These sums involve slightly different formulas from that of Equation 3.2.26 and are not discussed here. For the HCP structure, the FCC values are to be used. Considering the inert gas crystals, we substitute the FCC structure values into Equation 3.2.25 to get

$$U_{tot}(R) = 4\varepsilon\frac{N}{2}\left[(12.13)\left(\frac{\sigma}{R}\right)^{12} - (14.45)\left(\frac{\sigma}{R}\right)^6\right]. \tag{3.2.27}$$

Setting the derivative of $U_{tot}$ to zero, $dU_{tot}/dR = -2N\varepsilon\left[12(12.13)\left(\sigma^{12}/R^{13}\right) - 6(14.45)\left(\sigma^6/R^7\right)\right] = 0$, we obtain the FCC bond length at zero Kelvin as

$$\left.\frac{\sigma^6}{R^6}\right|_{R=R_0} = \frac{14.45}{2(12.13)}, \quad \Rightarrow \frac{R_0}{\sigma} = \left(\frac{2(12.13)}{14.45}\right)^{1/6} \approx 1.09, \tag{3.2.28}$$

which is the same for all the elements with FCC or HCP structures according to this approach. Using the values for $R_0 = d_{nn}$ and $\sigma$ from Table 3.2.2 the following values for the $R_0/\sigma$ ratios are obtained: Ne, 1.14; Ar, 1.11; Kr, 1.10; and Xe, 1.09. The theoretical value deviation from this experimental value is greatest for the lightest elements. This deviation is attributed to the quantum mechanical zero point motion; i.e., $<\hat{p}^2>/2m$. Here, we use the quantum mechanical operator for $\hat{p}$ such that $<\hat{p}^2> = (\hbar/i)(\hbar/i)^* <\psi(x)|\partial^2/\partial x^2|\psi(x)>$ and $\psi(x) \approx sin(2\pi x/\lambda)$ so that $<\hat{p}^2>/2m \propto 1/m\lambda^2$, where $\lambda$ is the wavelength associated with the ground state. This implies a smaller correction for the larger mass atoms as confirmed by isotope effect experiments. In the isotope effect observations, for example, a crystal of the isotope $Ne^{20}$ has been observed to have a larger lattice constant than a crystal of $Ne^{22}$.
The cohesive energy that results from Equation 3.2.27 when it is evaluated at the $R_0$ bond length of Equation 3.2.28 is

$$U_{tot}(R_0) = 4\varepsilon\frac{N}{2}\left[(12.13)\left(\frac{1}{1.0902}\right)^{12} - (14.45)\left(\frac{1}{1.0902}\right)^6\right] = -2.1517(4N\varepsilon). \tag{3.2.29}$$

Using the values for $\varepsilon$ in Table 3.2.2 the magnitude of the energies per atom that result based on Equation 3.2.29 are: Ne, 0.027; Ar, 0.090; Kr, 0.121; and Xe, 0.172 all in $eV/atom$. Again, the larger percent errors occur for the lighter elements due to the zero point motion that is unaccounted for.
At this point, using Equation 3.2.27 for the potential in the FCC structure, we can obtain an expression for the bulk modulus, $B = -VdP/dV$ at zero Kelvin. If we let $\alpha_J = 12.13$ and $\beta_J = 14.45$, then

$$U(R) = 4\varepsilon\frac{N}{2}\left[\alpha_J\left(\frac{\sigma}{R}\right)^{12} - \beta_J\left(\frac{\sigma}{R}\right)^6\right]. \tag{3.2.30}$$

Using the thermodynamic expression $dU = -PdV + TdS$ at zero Kelvin (with $dS = 0$) we have for the bulk modulus

$$B = V\left.\frac{d^2U}{dV^2}\right|_{V=V_0}, \tag{3.2.31a}$$

where $V_0$ is the equilibrium volume. Since in the FCC structure the volume for $N$ cells is $V = Na^3/4$, which in terms of the nearest neighbor distance ($R = a/\sqrt{2}$) becomes $V = NR^3/\sqrt{2}$, then with $dV = 3NR^2dR/\sqrt{2}$, we can write for the FCC crystals

$$B = \frac{\sqrt{2}}{9NR}\frac{d^2U}{dR^2}\bigg|_{R=R_0}, \tag{3.2.31b}$$

where from Equation 3.2.28 and the above definitions, the equilibrium nearest neighbor distance $R_0 = (2\alpha_J/\beta_J)^{1/6}\sigma$. Using Equation 3.2.30 we obtain

$$\frac{d^2U}{dR^2} = 2N\varepsilon\left[(12)(13)\alpha_J\frac{\sigma^{12}}{R^{14}} - (6)(7)\beta_J\frac{\sigma^6}{R^8}\right], \tag{3.2.31c}$$

which when substituted into Equation 3.2.31b and simplified, the bulk modulus for the FCC inert gas solids becomes

$$B = \frac{4\varepsilon\beta_J{}^{5/2}}{\sigma^3\alpha_J{}^{3/2}}. \tag{3.2.31d}$$

As a simple example, in the case of the argon solid crystal at zero Kelvin, this expression yields a value of $B = 4(16.7\times10^{-22}J)(14.45^{5/2})/[(3.4\times10^{-10}m)^3(12.13^{3/2})] \approx 3.2\times10^9 N/m^2$.
In the next example, we apply Equation 3.2.23 in order to find the equilibrium distance among three Neon particles in two dimensions.

**Example 3.2.3.2**
Use the Lennard-Jones potential to obtain the equilibrium distance between three Neon particles on a two-dimensional $x-y$ plane.
**Solution**
We use Equation 3.2.23 and write it in the form $U_{tot} = \sum_{j,i \atop i>j}^{3} U(R_{ij}) = 4\varepsilon \sum_{j,i \atop i>j}^{3} [(\sigma/R_{ij})^{12} - (\sigma/R_{ij})^6]$, where we let $R_{ij} = \sqrt{(x_i-x_j)^2+(y_i-y_j)^2}$. We then find the minimum of the potential with respect to the variables $x_2, x_3, y_2$, and $y_3$, where the origin particle is fixed in space at $x_1 = 0$, $y_1 = 0$. This is a numerical problem which we solve with the MATLAB script in the form of the function LJ_2d_min.m that is listed below. The potential is evaluated by the function LJ_funNd(r,Np,Nd) that is part of the entire script, where $r$ holds the positions of the particles, $Np$ is the number of particles, and $Nd$ is the number of dimensions. This function is used by the MATLAB minimizing routine fminsearch with the parameters (@LJ_funNd,r_guess,[],np,nd). Besides making use of the potential function, this minimizing routine needs an initial guess for the positions of the particles, held by the array *r_guess*. Once the function converges using MATLAB's default tolerances (not being varied in this example), the final positions are held by the array *rnew* and the minimum of the potential is held by the variable *Umin*. The rest of the script basically plots the final positions of the particles and connects them with a straight line as shown in Figure 3.2.8 where the final average distance between the particles is $d_{ave} = 3.076$ Å organized in an equilateral triangle and with minimum of the potential at an energy of $Umin = -0.00936\,eV$.

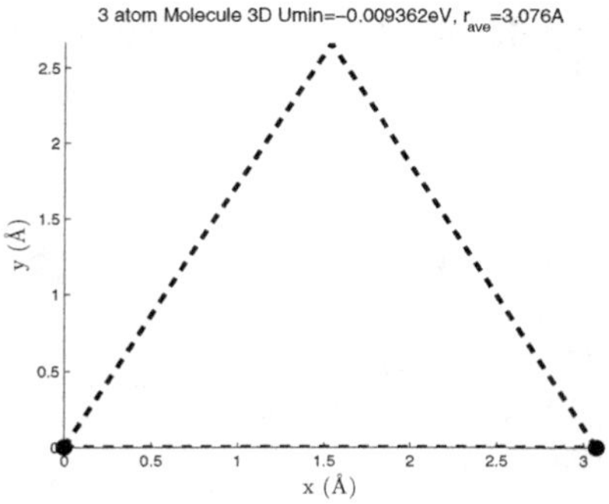


Figure 3.2.8: Final result from the script LJ_2d_min.m of Example 3.2.3.2. The thee Neon particles (black dots) organize themselves on an equilateral triangle with an average side length of $d_{ave} = 3.076\,\text{Å}$ and with a potential energy minimum of $-0.00936\,eV$.

The program listing follows.

```
%copyright by J. E Hasbun and T. Datta
%LJ_2d_min.m
%This program finds the equilibrium distance between three particles
%in 2d. It does so by minimizing the Lennard-Jones potential
%U=4*epsilon* [Sum of i,j > i of ((sigma/Rij)^12-(sigma/Rij)^6)]
function LJ_2d_min
clear
global eps sig
e=1.602176487e-19;        %electronic charge
eps=5.0e-22;              %for Neon in Joules
eps=eps/e;                %for Neon in eV
sig=2.74;                 %in Angstroms
np=3;                     %Number of particles
nd=2;                     %number of dimensions
r_guess(1,1)=0;           %x - origin particle is fixed
r_guess(1,2)=0;           %y
r_guess(2,1)=2;           %x - 2nd particle
r_guess(2,2)=0;           %y
r_guess(3,1)=2;           %x - 3rd particle
r_guess(3,2)=2;           %y
[rnew,Umin]=fminsearch(@LJ_funNd,r_guess,[],np,nd);
%show the final positions of the particles
hold on
axis([-1*max(abs(rnew(:,1))) 2*max(abs(rnew(:,1))) ...
  -2*max(abs(rnew(:,2))) 2*max(abs(rnew(:,2)))])
for i=1:np
  plot(rnew(i,1),rnew(i,2),'ko','MarkerSize',10,'MarkerFaceColor','k')
end
fprintf('Umin=%9.6g (eV)\n',Umin)
%Final particle distances
rave=0.0;
icord=0;
for i=1:np
```

```
  fprintf(' p#%2i x=%9.4f, y=%9.4f (A)\n',i,rnew(i,1),rnew(i,2))
end
for i=1:np-1
  for j=i+1:np
    icord=icord+1;
    rij=((rnew(i,1)-rnew(j,1))^2+(rnew(i,2)-rnew(j,2))^2)^(1/2);
    rave=rij+rave;
    fprintf(' r(%2i,%2i)=%9.4f (A)\n',i,j,rij)
    line([rnew(i,1),rnew(j,1)],[rnew(i,2),rnew(j,2)],'LineStyle','--',...
      'Color','k','LineWidth',3)
  end
end
axis tight
rave=rave/icord;
fprintf(' average r=%9.4f (A)\n',rave)
xlabel('x (\AA)','interpreter','latex')
ylabel('y (\AA)','interpreter','latex')
str=cat(2,'3 atom Molecule 3D',' Umin=',...
     num2str(Umin,'%9.4g'),'eV, r_{ave}=',num2str(rave,'%9.4g'),'A');
title(str)

function U_LJ=LJ_funNd(r,Np,Nd)
%r is a Np x Nd vector
%Np equal number of particles and
%Nd=number of dimensions
global eps sig
%The Lennard-Jones potential evaluated at vector r
U_LJ=0.0;
sig12=sig^12;
sig6=sig^6;
for i=1:Np-1
  for j=i+1:Np
    rij=0.0;
    for k=1:Nd
      rij=(r(i,k)-r(j,k))^2+rij;
    end
    U_LJ=(sig12*(1./rij).^6-sig6*(1./rij).^3)+U_LJ;
  end
end
U_LJ=4*eps*U_LJ;
```

## 3.3 Ionic Crystals

Ideal ionic crystals are those which are formed by the combination of atoms from groups *I* and *VII* of the periodic table. Table 3.3.3 shows the electronic configuration for the elements involved. The alkali metals have a noble gas electron configuration and an extra electron that is loosely bound to the atom.

Table 3.3.3: Most ionic crystals are formed by the combination of atoms from group *I* and group *VII* shown here with their electronic configuration.

| Group *I* | electronic configuration | Group *VII* | electronic configuration |
|---|---|---|---|
| Li | He+2$s$ | F | Ne−2$p$ |
| Na | Ne+2$s$ | Cl | Ar−3$p$ |
| K | Ar+2$s$ | Br | Kr−4$p$ |
| Rb | Kr+2$s$ | I | Xe−5$p$ |
| Cs | Xe+2$s$ | | |

The halide elements are shy of a single electron to be able to complete a noble gas electronic configuration. The halogen's high affinity for electrons and the alkali's disposition to give up its electron easily make the formation of ionic crystals possible. In this way, the halide atoms complete their p-subshell and the alkali atoms retain a filled s-subshell. Under normal conditions, one of the cations (positive ions) Li, Na, K, Rb, or Cs combines with one of the anions (negative ions) F, Cl, Br, or I to crystallize in the NaCl structure, with the exception of CsCl, CsBr, and CsI which are most stable in the CsCl structure.

In this section, we develop a potential that is commonly used to understand ionic systems. The potential minimum is the lattice energy of the crystal system. To see the significance of the lattice energy, consider the ionic NaCl system. The lattice energy (obtained later in Example 3.3.1.2) per NaCl molecule in a crystal is about $-7.9\,eV$. Now, in order to form a single free standing NaCl molecule from the neutral atoms, Na is first ionized with a cost of about $5.14\,eV$, and the electron given to Cl, which, because of its affinity, an energy of about $3.61\,eV$ is given off. The net energy per free molecule is, therefore, about $-7.9+5.4-3.61=-6.11\,eV$. From this, it is seen that the energy minimum per molecule in the crystal lattice is lower than that of a single molecule. This is seen as reason enough for the separated ions to seek to lower their total energy and form the NaCl crystal.

Suppose we use the Coulomb interaction between two already formed ions, one positive and one negative, and calculate their interacting energy. We write $-ke^2/R_0$, and if we take $R_0 \approx 2.81$ Å for NaCl, the result is about $(-9\times10^9 Nm^2/C^2)(1.602\times10^{-19}C)^2/(2.81\times10^{-10}m) = -8.22\times10^{-19}J$ or $-5.13\,eV$, with about 35% error when compared to the above lattice energy of $-7.9\,eV$. The error is due to the lack of accounting for the lattice contribution. The effort to improve on this simple model brings us into the subject of the Madelung energy.

### 3.3.1 Ionic Crystal Potential and Madelung Energy

The Madelung energy helps us to do a better accounting for the interactions between a given ion and the rest of the ions of an ionic crystal. We first consider a pair potential as we did with the Lennard-Jones potential. Here, however, rather than using the $1/r^{12}$ repulsive term, we employ that positive interaction of Equation 3.2.22. Furthermore, referring to a given crystal ion, depending on the sign of the neighboring ions, there will be an ionic term of the form $\pm ke^2/r$. As in the Lennard-Jones potential, there is also a van der Walls $-1/r^6$ attractive interaction, which is neglected due to its smallness compared to the ionic contribution. We thus have the following interaction energy between ions $i$ and $j$ in the crystal

$$U_{ij} = \lambda \exp\left(-\frac{r_{ij}}{\rho}\right) \pm \frac{ke^2}{r_{ij}}, \tag{3.3.32}$$

where the $\pm$ sign is for like charges and unlike charges, respectively. Here $k = 1/4\pi\varepsilon_0$, and $\lambda$, $\rho$ are experimental parameters of the model. If we look at the total interaction energy associated between the *ith* ion and the rest of the crystal ions, we have

$$U_i = \sum{}' U_{ij}, \tag{3.3.33a}$$

which is independent of whether the reference ion $i$ is $\pm$ and where the prime indicates $i \neq j$. Assuming the ion is not near the surface, the total energy for $N$ molecules is

$$U_{total} = NU_i. \tag{3.3.33b}$$

This total energy is also the lattice energy, which is the energy required to separate the crystal into individual ions at an infinite distance apart. We next define $r_{ij} \equiv p_{ij}R$ with $p_{ij}$ is, as before, the number of times $R$ that atom $j$ is away from atom $i$, to obtain for the pair potential

$$U_{ij} = \begin{cases} \lambda \exp\left(-\frac{R}{\rho}\right) - \frac{ke^2}{R} & n.n. \\ \pm \frac{ke^2}{p_{ij}R} & \text{otherwise} \end{cases}, \tag{3.3.34}$$

where we have ignored the repulsive term beyond nearest neighbors (n.n.) and $\lambda$ is the strength of the repulsive term for nearest neighbors. Substituting this into Equations 3.3.33, and carrying out the sum, we get the expression

$$U_{total}(R) = N\left[z\lambda \exp\left(-\frac{R}{\rho}\right) - \frac{ke^2\alpha}{R}\right] \tag{3.3.35a}$$

where $z$ is the number of nearest neighbors, and where we have defined the Madelung constant

$$\alpha \equiv \sum_j{}' \frac{\pm 1}{p_{ij}}, \tag{3.3.35b}$$

where now the $+$ sign is for unlike charges and the $-$ sign is for like charges. We note that for stability reasons it is necessary that $\alpha > 0$. The Madelung constant plays an important role in the theory of ionic crystals. We will come back to its evaluation later. We now use Equation 3.3.35a to obtain the equilibrium distance, $R = R_0$, between the nearest neighbor atoms. The potential is minimized, $dU_{tot}/dR\big|_{R=R_0} = 0$, to obtain the relation for the bond length

$$R_0^2 \exp(-R_0/\rho) = \frac{ke^2\rho\alpha}{z\lambda}, \tag{3.3.36}$$

which, given $\rho$, $\lambda$, and $\alpha$ is to be solved self consistently for $R_0$. Knowing this, we can then find the lattice energy as $U_0 \equiv U_{total}(R = R_0)$ from Equation 3.3.35a or, equivalently, we can substitute $z\lambda \exp(-R_0/\rho) = ke^2\rho\alpha/R_0^2$ from Equation 3.3.36 into $U_{total}(R = R_0)$ to get

$$\begin{aligned} U_0 &= N\left[\frac{ke^2\rho\alpha}{R_0^2} - \frac{ke^2\alpha}{R_0}\right] \\ &= -U_M\left(1 - \frac{\rho}{R_0}\right), \end{aligned} \tag{3.3.37a}$$

where we have defined the Madelung energy

$$U_M \equiv \frac{N\alpha ke^2}{R_0}. \tag{3.3.37b}$$

We now have all the necessary ingredients to perform a crystal system's lattice energy calculation. Before doing so, however, in the following example, we go back and see how we might go about calculating the Madelung constant for the NaCl crystal.

**Example 3.3.1.1**
The Madelung constant for the NaCl contains factors similar to that of a simple cubic (see Example 3.2.3.1). The nearest neighbor distance is $R_0 = a/2$, where $a$ is the lattice constant, so that there are 6 of them (Cl surrounds Na), each with $p_{01} = 1$. Next, there are 12 Na 2nd neighbors at a distance of $\sqrt{2}R_0$, each with $p_{02} = \sqrt{2}$, and so on. From Equation 3.3.35b, including up to the 7th neighbors we have alpha approximated as $\alpha \approx 6 - 12/\sqrt{2} + 8/\sqrt{3} - 6/\sqrt{4} + 24/\sqrt{5} - 24/\sqrt{6} + 12/\sqrt{8} = 4.3113$, which is far from converging and does so very slowly. It turns out that for NaCl, all of the terms can be obtained from the expression

$$\alpha_{NaCl} = \lim_{N\to\infty} \sum_{i=-N}^{N} \sum_{j=-N}^{N} \sum_{k=-N}^{N} \frac{(-1)^{(i+j+k+1)}}{\sqrt{i^2+j^2+k^2}} \quad (\text{except } i=j=k=0) \qquad (3.3.38)$$
$$= 1.747565.$$

Table 3.3.4 lists values of $\alpha$ for three different structures.

Table 3.3.4: Crystal structures Madelung constant values.

| System | $\alpha$ |
|---|---|
| Sodium chloride, NaCl | 1.747565 |
| Cesium chloride, CsCl | 1.762675 |
| Zinc blend, cubic ZnS | 1.6381 |

Table 3.3.5 contains various properties and parameters for the alkali-halide crystals with the NaCl structure. Using parameters from this table, we do the following example.

**Example 3.3.1.2**
Using appropriate parameters from Table 3.3.5 for NaCl, plot the ionic potential, obtain its lattice energy and the equilibrium bond length.
**Solution**
We first solve for the equilibrium bond length by solving Equation 3.3.36 for $R_0$ self consistently. We define $x = R_0/\rho$ and we let $c = ke^2\alpha/(z\lambda\rho) = (8.99 \times 10^9 Nm^2/C^2)(1.602 \times 10^{-19}C)^2(1.747)/(6553.59\,eV \cdot 1.602 \times 10^{-19}J/eV \cdot 0.321 \times 10^{-10}m) \approx 0.0120$. We then solve the dimensionless equation $x^2\exp(-x) = c$ iteratively. This equation has two solutions, we seek the larger of the two. The simple iterative solution is such that $x_{i+1} = -log(c/x_i^2)$ and we iterate this, as $i \to N$ for large $N$, until $x_N \approx x_{N-1}$, assuming a reasonable initial guess of greater than 2, since the larger root occurs beyond where the maximum of $x^2\exp(-x)$ occurs. A simple script for this follows.

```
x=2.3;          %the initial guess
c=0.0120;       %the constant
for i=1:10      %the loop - 10 iterations do fine
x=-log(c/x^2) %the new guess, which will converge
end
```

Alternatively, one can apply the Newton-Raphson method, which is accomplished as
`x=fzero('x^2*exp(-x)-0.012',2.3)`
in MATLAB. In either case, the result is $x \approx 8.76$, to obtain the bond length as $R_0 = x\rho = 8.76 \cdot 0.321\text{Å} \approx 2.81\text{Å}$. Having $R_0$, we can find the Madelung energy per molecule from Equation 3.3.37b as $U_M/N = \alpha ke^2/R_0 = (1.747) \cdot (8.99 \times 10^9 Nm^2/C^2)(1.602 \times 10^{-19}C)^2/2.81 \times 10^{-10}m = 1.4395 \times 10^{-18}J$ or $U_M/N \approx 8.9\,eV/molecule.$ We are ready to obtain the lattice energy per molecule from Equation 3.3.37a as $U_0/N = -(U_M/N)(1 - \rho/R_0) = -(U_M/N)(1 -$

$1/x) = -(8.9\,eV)(1-1/8.76) \approx -7.9\,eV/molecule$. This is the lattice energy per NaCl molecule we referred to in the introductory part of Section 3.3. Since $1eV/\text{molecule} = N_a e/4186J \approx 23.05 kcal/mol$, where $N_a$ is Avogadro's number, we can express the lattice energy as $U_0 \approx -181.7\,kcal/mol$. The bond length, Madelung energy, and lattice energy found thus far are close to the ones calculated more accurately by the script ionic_NaCl.m listed below. The script's theoretical lattice energy result of $-182.5\,kcal/mol$ for NaCl is shown in the last column of Table 3.3.5, which is close to the quoted experimental value. Finally, the plot of the ionic potential for NaCl is shown in Figure 3.3.9.

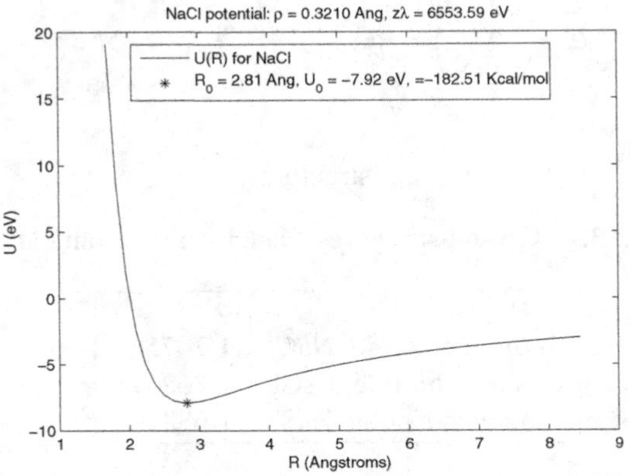


Figure 3.3.9: The ionic NaCl system's potential calculated by the script ionic_NaCl.m using Equations 3.3.35 - 3.3.37. The calculated values for the bond length ($R_0$), the Madelung energy ($U_M/N$), and the lattice energy ($U_0/N$) are 2.8148 Å, $8.9372\,eV/molecule$, and $-7.9180\,eV/molecule$ ($-182.5064\,kcal/mol$), respectively.

The code ionic_NaCl.m used in this example follows.

```
%copyright by J. E Hasbun and T. Datta
%ionic_NaCl.m
%This script performs the calculation of the potential
%U=z*lambda*exp(-R/rho)-UM*R0./R, where UM=alpha*k*e^2/R0 is
%the Madelung energy. The NaCl ionic system is done here.
%It makes use of the Madelung constant for the FCC structure.
%It calculates the minimum energy (lattice energy), U0,
%and the nearest neighbor distance, R0.
clear
Na=6.02214179e23;            %Avogadro's constant (1/mol)
JpK=4.186e3;                 %Joules per Kcal
e=1.602176487e-19;           %electronic charge
eps0=8.854187817e-12;        %Permittivity of free space (C^2/N/m^2)
k=1/4/pi/eps0;               %constant (mks units)
alpha=1.747;                 %NaCl Madelung Constant parameter
rho=0.321e-10;               %NaCl potential decay parameter (m)
Zlamb=6553.59;               %NaCl z*lambda parameter (ev)
const=k*e^2*alpha/rho/(Zlamb*e);          %dimensionless, to get R0
fR0=inline('x.^2.*exp(-x)-c','x','c');    %function=0 to get R0
xg=2-log(2^2*exp(-2))-log(const);         %rough guess for x
```

```
[x,fval] =fzero(fR0,xg,[],const);        %finds x as the zero of fR0
R0=x*rho;                    %R0 in meters
%R0=2.82e-10                 %table value for R0 if desired (m) for NaCl
UM=alpha*k*e^2/R0/e;         %Madelung energy per molecule in eV
U0=-UM*(1-rho/R0);           %U0  in eV
U0_KpM=U0*e*Na/JpK;          %U0(minimum energy)in Kcal per mol
%Define U_of_R per molecule in eV
U_of_R=@(R,Zlamb,rho,UM,R0) Zlamb*exp(-R/rho)-UM*R0./R;
Rl=0.6*R0; Ru=3*R0;
R=Rl:(Ru-Rl)/50:Ru;                %R range
U=U_of_R(R,Zlamb,rho,UM,R0);  %U(R) per molecule in eV
h=plot(R*1e10,U,'k');
hold on
plot(R0*1e10,U0,'k*')
xlabel('R (Angstroms)')
ylabel('U (eV)')
disp('NaCl ionic system')
fprintf('Madelung energy=%8.4f eV, R0=%8.4f Angstroms\n',UM,R0/1e-10)
fprintf('Lattice energy=%9.4f eV =%10.4f Kcal/mol\n',U0,U0_KpM)
str=cat(2,'NaCl potential: \rho = ',num2str(rho/1e-10,'%6.4f'),...
   ' Ang, z\lambda = ',num2str(Zlamb,'%4.2f'),' eV');
title(str)
str2=cat(2,'R_0 = ',num2str(R0/1e-10,'%7.2f'),...
   ' Ang, U_0 = ',num2str(U0,'%7.2f'),' eV, =',...
   num2str(U0_KpM,'%8.2f'),' Kcal/mol');
legend('U(R) for NaCl',str2,0)
```

Minor modifications of this script should yield the theoretical lattice energies for all the systems shown in Table 3.3.5. Based on Equations 3.3.35 - 3.3.37, the only inputs that vary from one system to another are $z\lambda$ and $\rho$.

Finally, we can write an expression for the bulk modulus of ionic crystals at zero Kelvin. We make use of the FCC structure bulk modulus Equation 3.2.31b as well as Equations 3.3.35a, 3.3.36, and 3.3.37b to obtain

$$B = \frac{\sqrt{2}ke^2\alpha}{9R_0^3\rho}\left(1-\frac{2\rho}{R_0}.\right) = \frac{\sqrt{2}}{9R_0^2\rho}\left|\frac{U_M}{N}\right|\left(1-\frac{2\rho}{R_0}\right) = \frac{\sqrt{2}}{9\rho^3x^2}\left|\frac{U_M}{N}\right|\left(1-\frac{2}{x}\right), \tag{3.3.39}$$

where in the last expression, $x = R_0/\rho$, as used in Example 3.3.1.2, and with the rest of the parameters used there, for NaCl; that is, $U_M/N = 8.9\,eV = 1.4259\times10^{-18}J$, $x = 8.76$, $\rho = 0.321\times10^{-10}m$, one gets $B \approx 6.8\times10^{10}N/m^2$. This value is larger by a factor of about 2.8 from the room temperature experimental value shown in Table 3.3.5.

## 3.4 Covalent Bonding

Covalent bonding refers to the bonding that occurs when it is easier for interacting atoms to achieve an outer electron configuration that will lead to a lowering of the total energy without necessarily giving up any electrons. In so doing, each atom gains enough electrons so as to achieve a noble gas atom outer electron configuration. The bonding that takes place leads to the creation of stable molecules and crystals. It happens in such a way that the electrons belonging to the involved atoms

Table 3.3.5: Properties and parameters for the alkali-halide crystals with the NaCl structure (at room temperature and atmospheric pressure [3]) used in the theory. The theoretical results for the lattice energy ($U_0$) were obtained using Equations 3.3.35 - 3.3.37, as in Example 3.3.1.2.

| Crystal | $R_0$ (Å) | Bulk Modulus $(10^{10}N/m^2)$ | $z\lambda$ $(eV)$ | $\rho$ (Å) | $-E_{lattice}$ $(kcal/mol)$ (Experiment) | $-E_{lattice}(-U_0)$ $(kcal/mol)$ (Theory) |
|---|---|---|---|---|---|---|
| LiF | 2.014 | 6.71 | 1847.49 | 0.291 | 242.3 | 245.9 |
| LiCl | 2.570 | 2.98 | 3058.34 | 0.330 | 198.9 | 196.4 |
| LiBr | 2.751 | 2.38 | 3688.73 | 0.340 | 189.8 | 184.7 |
| LiI | 3.000 | 1.71 | 3738.66 | 0.366 | 177.7 | 169.5 |
| NaF | 2.317 | 4.65 | 4000.81 | 0.290 | 214.4 | 219.0 |
| NaCl | 2.820 | 2.40 | 6553.59 | 0.321 | 182.6 | 182.5 |
| NaBr | 2.989 | 1.99 | 8301.21 | 0.328 | 173.6 | 172.9 |
| NaI | 3.237 | 1.51 | 9861.59 | 0.345 | 163.2 | 160.0 |
| KF | 2.674 | 3.05 | 8176.38 | 0.298 | 189.8 | 193.0 |
| KCl | 3.147 | 1.74 | 12795.09 | 0.326 | 165.8 | 165.3 |
| KBr | 3.298 | 1.48 | 14355.47 | 0.336 | 158.5 | 157.7 |
| KI | 3.533 | 1.17 | 17788.30 | 0.348 | 149.9 | 148.1 |
| RbF | 2.815 | 2.62 | 11109.89 | 0.301 | 181.4 | 183.8 |
| RbCl | 3.291 | 1.56 | 19910.42 | 0.323 | 159.3 | 158.9 |
| RbBr | 3.445 | 1.30 | 18911.77 | 0.338 | 152.6 | 152.0 |
| RbI | 3.671 | 1.06 | 24903.62 | 0.348 | 144.9 | 142.9 |

are shared. It takes two electrons to create a covalent bond, one from each atom, and whose spins are antiparallel to each other. An example of covalent bonding is the hydrogen molecule ($H_2$) as well as methane ($CH_4$). In the hydrogen molecule the $s$-shell electrons from each hydrogen atom are shared by the two ions. In methane, the carbon atom shares its four $(s,p)$-shells valence electrons with each of four hydrogen atoms. Semiconductor crystals involve covalent bonding and often include elements in groups II, III, IV, V, and VI of the periodic table. For example, IIB-VIB compounds like ZnS and CdTe form covalent bonding, as do IIIB-VB compounds such as GaAs and InSb. Group IVB crystals of Si and Ge form covalent binding as well with respectable cohesive energies as shown in Table 3.1.1. This is indicative of the covalent bond's strength, especially that of group IVB such as C, itself responsible for the diamond structure, one of the hardest substances known. Its stability is due to the strength of the covalent bond formed among its neighbors. Covalent bonding is most easily understood in terms of a simple quantum mechanical model of a molecule in which electrons are shared among the protons. In this regard, the simplest such molecule that we can think of is that of the single electron molecular hydrogen ion which we study next.

### 3.4.1 The Molecular Hydrogen Ion ($H_2^+$) - An Analytical Calculation

The molecular hydrogen ion consists of two protons and one electron as shown in Figure 3.4.10.

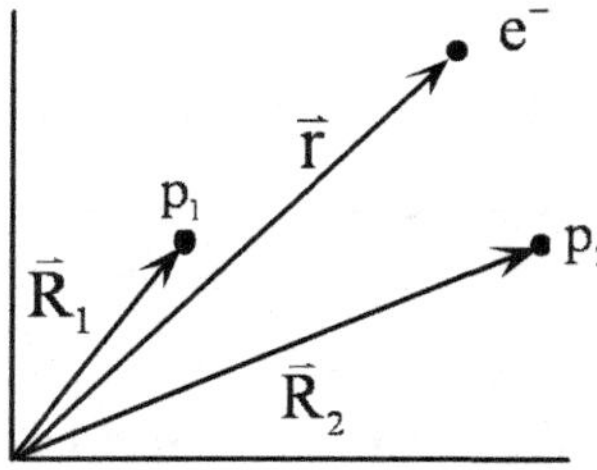


Figure 3.4.10: The $H_2^+$ molecule with protons $p_i$ at positions $\vec{R}_i$ with $i = 1,2$ and a single electron $e^-$ with coordinate $\vec{r}$.

We use the $H_2^+$ system as the simplest example of covalent bonding. It is the simplest model of a molecule in which a single electron is shared between two protons separated by a finite distance. We begin by writing the Hamiltonian of the system as

$$H_{total} = H_{H_2^+} + H_{\text{ions}}, \tag{3.4.40a}$$

where $H_{H_2^+}$ refers to the electron part

$$H_{H_2^+} = -\frac{\hbar^2 \vec{\nabla}_r^2}{2m} - \frac{ke^2}{|\vec{r} - \vec{R}_1|} - \frac{ke^2}{|\vec{r} - \vec{R}_2|}, \tag{3.4.40b}$$

and the $H_{\text{ions}}$ is the proton-proton repulsion

$$H_{\text{ions}} = \frac{ke^2}{|\vec{R}_1 - \vec{R}_2|}. \tag{3.4.40c}$$

The electron part, $H_{H_2^+}$, consists of the electron kinetic energy and its electrical potential energy associated with each of two protons. The electron coordinate is $\vec{r}$, while $\vec{R}_i$ is the coordinate of the ith proton with $i = 1,2$. The protons themselves are considered to be at rest, using the so-called Born-Oppenheimer approximation. In order to treat this problem, it is helpful to recall results from the hydrogen atom. Its normalized ground state electronic wavefunction is $s(r) = \exp(-r/a_0)/[\sqrt{\pi} a_0^{3/2}]$, where $r$ is the electron coordinate from the origin where the proton is assumed to be located and $a_0 = 4\pi\varepsilon_0\hbar^2/me^2 = 0.529\text{Å}$ is the Bohr radius. This wavefunction is peaked around the proton. In a similar fashion, for the case of the ground state of $H_2^+$, the electron distribution is thought to peak around each proton with associated normalized wavefunction

$$s_i(r) = \frac{1}{\sqrt{\pi} a_0^{3/2}} \exp(-|\vec{r} - \vec{R}_i|/a_0), \tag{3.4.41a}$$

for proton $i$ with $i = 1,2$. Exponential wavefunctions of this type are also known as Slater orbitals. Note, though, that the electron is shared by both protons so that the total wavefunction is better represented as a linear combination of the two wavefunctions, or

$$\psi_{s,a}(r) = N[s_1(r) \pm s_2(r)], \tag{3.4.41b}$$

where $s,a$ stand for symmetric (+) and antisymmetric (−) combinations, respectively. The symmetric combination is associated with the bonding or equilibrium state of the molecule, while the antisymmetric state is associated with the antibonding state, as shown in Figure 3.4.11. The factor of $N$ in Equation 3.4.41b is the new wavefunction's normalization constant.

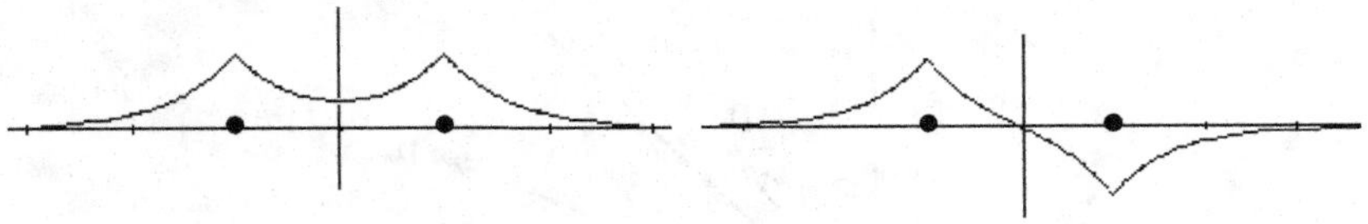


Figure 3.4.11: Symmetric (bonding) and antisymmetric (antibonding) electronic wavefunction combinations described by Equations 3.4.41. The dots represent the equilibrium positions of the two protons.

If in Equation 3.4.40b we refer to the terms $T \equiv -\frac{\hbar^2 \vec{\nabla}_r^2}{2m}$, $V_i \equiv -\frac{ke^2}{|\vec{r}-\vec{R}_i|}$, and using the Dirac notation for the wavefunctions of Equation 3.4.41b, $|s_i> \equiv s_i(r)$, the symmetric state can now be written as $|\psi_s> \equiv \psi_s(r) = N(|s_1> + |s_2>)$. With these definitions and the fact that $<s_i|s_i> = 1$ for $i = 1,2$; that is, the wavefunctions are normalized, we have

$$<s_1|T+V_1|s_1> = <s_2|T+V_2|s_2> = E_{1s}, \tag{3.4.42a}$$

where $E_{1s} = ke^2/2a_0 = 13.606eV$ is the electronic ground state energy of atomic hydrogen, which is an energy unit known as a Rydberg. Another unit of energy that is often used is the Hartree ($27.211\,eV$), which is equivalent to two Rydbergs. We further define

$$\begin{aligned}
&\Delta \equiv <s_1|s_2> = <s_2|s_1>,\\
&A \equiv -<s_2|V_1|s_2> = -<s_1|V_2|s_1>,\\
&\text{and}\\
&B \equiv -<s_1|V_1|s_2> = -<s_2|V_2|s_1>.
\end{aligned} \tag{3.4.42b}$$

Also since $<\psi_s|\psi_s> = 1 = N^2 <s_1+s_2|s_1+s_2> = 2N^2(1+\Delta)$ then

$$N = 1/\sqrt{2(1+\Delta)}. \tag{3.4.42c}$$

The ground or bonding state energy of the $H_2^+$ molecule, associated with the symmetric state, is

$$\begin{aligned}
E_s &= \frac{<\psi_s|H_{H_2^+}|\psi_s>}{<\psi_s|\psi_s>} = N^2(<s_1+s_2|H_{H_2^+}|s_1+s_2>)\\
&= N^2(<s_1|H_{H_2^+}|s_1> + <s_2|H_{H_2^+}|s_2> + 2<s_1|H_{H_2^+}|s_2>).
\end{aligned} \tag{3.4.43}$$

With the use of Equations 3.4.42, $E_s$ can be simplified. We make use of the terms

$$\begin{aligned}
&<s_1|H_{H_2^+}|s_1> = <s_1|T+V_1|s_1> + <s_1|V_2|s_1> = E_{1s} - A,\\
&<s_2|H_{H_2^+}|s_2> = <s_2|T+V_2|s_2> + <s_2|V_1|s_2> = E_{1s} - A,\\
&\text{and}\\
&<s_1|H_{H_2^+}|s_2> = <s_1|V_1|s_2> + <s_1|T+V_2|s_2> = -B + E_{1s} <s_1|s_2> = -B + E_{1s}\Delta,
\end{aligned} \tag{3.4.44}$$

which when substituted back into Equation 3.4.43, we obtain the final expression for the electronic ground state energy of $H_2^+$ in the form

$$E_s = 2N^2(E_{1s}(1+\Delta) - A - B) = E_{1s} - \frac{A+B}{1+\Delta}, \tag{3.4.45}$$

where we have also used Equation 3.4.42c. A similar calculation, using the antisymmetric wavefunction $|\psi_a>\equiv \psi_a(r) = N(|s_1> - |s_2>)$ leads to a normalization constant $N = 1/\sqrt{2(1-\Delta)}$ and an antibonding electronic state energy of

$$E_a = 2N^2(E_{1s}(1-\Delta) - A + B) = E_{1s} - \frac{A-B}{1-\Delta}. \tag{3.4.46}$$

The total energy of the $H_2^+$ molecule is obtained by adding the electronic energy and the ionic energy from Equation 3.4.40 or

$$E_{\text{total}s,a} = E_{s,a} + \frac{ke^2}{R}, \tag{3.4.47}$$

where $R \equiv |\vec{R}_1 - \vec{R}_2|$. To calculate the above energies, the only remaining detail are the expressions for the integrals of Equation 3.4.42b. They are indeed available [12], [13] as follows

$$\Delta \equiv < s_1|s_2 >= \left[1 + \frac{R}{a_0} + \frac{R^2}{3a_0^2}\right] e^{-R/a_0},$$

$$A = - < s_2|V_1|s_2 >= \frac{ke^2}{a_0}\left[\frac{a_0}{R} - \left(\frac{a_0}{R} + 1\right) e^{-2R/a_0}\right], \tag{3.4.48}$$

and

$$B = - < s_1|V_1|s_2 >= \frac{ke^2}{a_0}\left[\frac{R}{a_0} + 1\right] e^{-R/a_0}.$$

In the following example, we perform a calculation based on the analytic formulas of both energy states discussed above for the $H_2^+$ molecule.

**Example 3.4.1.1**

Before starting on the calculation of the bonding and antibonding state energies of Equation 3.4.45, 3.4.46, 3.4.47, and 3.4.48, let's define the units we wish to employ. For distance we will use $a_0$, and for energy we will employ Hartree units ($\varepsilon_b = 2E_{1s} = ke^2/a_0$) mentioned in connection with Equation 3.4.42a. These units seem natural here for the above equations have the same form except that the quantities $k$, $e$, and $a_0$ do not need to appear in the actual calculations. These units are also known as atomic units. In this example, we plot the electronic $H_2^+$ energy from Equation 3.4.45, the total energy from Equation 3.4.47, and also the ion-ion energy (last term of the total energy) versus R ($|\vec{R}_1 - \vec{R}_2|$). From the minimum of the energy, we find the bond length as well as the total bonding energy. For comparison purposes, the energy with the neutral hydrogen atom is superimposed on the plot (-1/2 Hartree). Figure 3.4.12(a) shows that the minimum energy occurs at a bond length of about $R_b = 2.493\,a_0 = 1.319$ Å and a minimum energy value of about $E_{min} = -0.565$ Hartree $= -15.370\,eV$. The magnitude of the difference between the hydrogen atom's ground state, $E_{1s}$, and $E_{min}$ is referred to as the dissociations energy, whose value is $E_{1s} - E_{min} = 0.0650$ Hartree $= 1.77\,eV$ for $H_2^+$. The experimental value of the dissociation energy is known to be about $2.78\,eV$ with a bond length of about 1.06 Å [13] and corrections to the present calculation is beyond our scope.

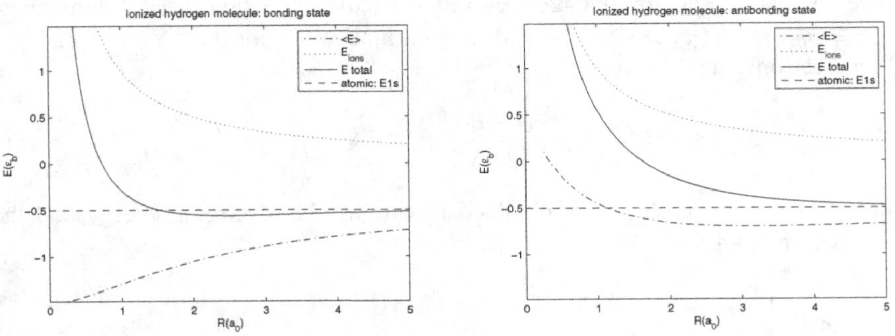


(a) bonding state (b) antibonding state

Figure 3.4.12: (a) The bonding state energy for the $H_2^+$ molecule in Hartree units ($\varepsilon_b$) versus proton-proton separation, $R$, in units of the Bohr radius ($a_0$). The ionic energy is $1/R$, the electronic energy is $< E >$ from Equation 3.4.45, the total energy is from Equation 3.4.47, and the hydrogen atom energy corresponds to $E_{1s} = -1/2$ Hartree $= -13.6\,eV$. The minimum of the total energy occurs at $R_b = 2.493a_0 = 1.319$ Å, with a minimum of $E_{min} = -0.565$ Hartree $= -15.370\,eV$. The molecule's dissociation energy corresponds to $E_{1s} - E_{min} = 0.0650$ Hartree $= 1.77\,eV$. (b) The results are shown for the antibonding state using Equations 3.4.46 and 3.4.47. Its total energy value at the equilibrium position is $-0.289\,Hartree = -7.871\,eV$.

The code employed in the calculation of the bonding state is hydro_mol_ion_sym.m whose listing follows.

```
%copyright by J. E Hasbun and T. Datta
%hydro_mol_ion_sym.m
%This program incorporates the analytic solution to the
%singly ionized hydrogen molecule.
%
clear
e=1.602176487e-19;          %electronic charge
h=6.62606896e-34;           %Planck'constant (J.s)
eps0=8.854187817e-12;       %Permittivity of free space (C^2/N/m^2)
k=1/4./pi/eps0;             %Electrical constant (N.m^2/C^2)
hbar=h/2./pi;               %hbar
me=9.10938215e-31;          %electron mass (kg)
a0=hbar^2/me/e^2/k;         %Bohr radius (m)
Eb=2*hbar^2/(2*me*a0^2); %Hartree energy unit
%Analytic result
R=0.25:0.001:5;                       %plotting range in unit of a0
Ab=1./R-(1./R+1.0).*exp(-2*R);        %A in units of Eb
Bb=(R+1).*exp(-R);                    %B in units of Eb
Del=(1+R+R.^2/3).*exp(-R);            %overlap, R in a0 units
E1s=-1.0/2.0;                         %Bohr energy in Eb units
Eions=1./R;                           %Ion-Ion repulsion energy
Ebo=E1s-(Ab+Bb)./(1.0+Del);           %Bonding energy
Etotbo=Ebo+Eions;                     %Total energy for bonding
[Etmin,index_ymin]=min(Etotbo);  %total energy minimum & its array index
Rf=R(index_ymin);                     %equilibrium bond length at min of energy
Ebomin=Ebo(index_ymin);               %bonding energy at equilibrium
```

```
Eimin=Eions(index_ymin);          %ion energy at equilibrium
%Plots follow
plot(R,Ebo,'k-.')
hold on
plot(R,Eions,'k:')
plot(R,Etotbo,'k-')
line([0 max(R)],[E1s E1s],'Color','k','LineStyle','--') %H atom energy
xlabel('R(a_0)'), ylabel('E(\epsilon_b)')
axis([0 max(R) min(Ebo) -min(Ebo)])
legend('<E>','E_{ions}','E total','atomic: E1s')
title('Ionized hydrogen molecule: bonding state')
hold off
%
disp('H2+ molecule analytic Results - bonding state')
fprintf('R_equilibrium=%9.4f, (a0) or %9.6f (Angs)\n',Rf,Rf*a0/1e-10)
fprintf('<E>=%9.6f (Hartree) or %9.6f (eV)\n',Ebomin,Ebomin*Eb/e)
fprintf('Eions=%9.6f (Hartree) or %9.6f (eV)\n',Eimin,Eimin*Eb/e)
fprintf('Etotal=%9.6f (Hartree) or %9.6f (eV)\n',Etmin,Etmin*Eb/e)
```

A slight modification of this code, using the obtained equilibrium bond length, produces the results of the antibonding state as shown in Figure 3.4.12(b). At the bond length, the total energy of this state is $-0.289 Hartree = -7.871\, eV$ (see Exercise 3.6.15).

The calculation performed in this section has been based on the analytic results for the $H_2^+$ or singly ionized hydrogen molecule. In the Section 3.4.2, we work with the hydrogen molecule and perform a fully numeric calculation of its bonding and antibonding states based on the simplest wavefunction possible.

### 3.4.2 The Hydrogen Molecule ($H_2$) - A Numeric Calculation

Here we seek to extend the idea started in the previous section to describe the sharing of electrons among two protons. The hydrogen molecule, having two electrons, is considerably more complicated than the single electron $H_2^+$ molecule. The reason is that as in $H_2^+$ an electron can be considered to have a wavefunction associated with the probability of being found around each of the two protons, but now, in $H_2$, we have two electrons, so one might consider a total wavefunction that is a product of such wavefunctions similar to those of Equation 3.4.41b around both protons, but one for each electron. For concreteness, let's start with the symmetric state and define the wavefunction

$$\phi_0(r) = \frac{s_1(r) + s_2(r)}{\sqrt{2}}, \tag{3.4.49a}$$

where the $s_i(r)$ are the Slater type orbitals as in Equation 3.4.41a for an electron with coordinate $\vec{r}$ around the ith proton. For the two electrons we, therefore, assume a product of such wavefunctions to form the total wave function

$$\psi_0^H(r_1, r_2) = \phi_0(r_1)\phi_0(r_2), \tag{3.4.49b}$$

for the bonding state. Similarly for the case of the antibonding state we have

$$\phi_1(r) = \frac{s_1(r) - s_2(r)}{\sqrt{2}}, \qquad \text{and} \qquad \psi_1^H(r_1, r_2) = \phi_1(r_1)\phi_1(r_2). \tag{3.4.49c}$$

This is indeed what Hartree proposed [14]. The product of two one-electron probabilities is known as an uncorrelated wavefunction. Thus, the above Hartree wavefunctions $\psi_i^H(r)$ for $i = 1,2$ are uncorrelated. Furthermore, in general, products of one-electron wavefunctions are known as Hartree products. Such products are not in line with the quantum mechanical principle that electrons are indistinguishable under the exchange of coordinates. Also, according to the Pauli exclusion principle, electrons are characterized by antisymmetric wavefunctions in such a way that the probability of finding them at the same position at the same time is zero. For simplicity, and for example, let's consider a totally hypothetical situation in which we build a two-electron wavefunction $\xi_a(r_1,r_2) = \chi_1(r_1)\chi_2(r_2)$. Next, exchanging the electron coordinates leads to $\xi_a(r_2,r_1) = \chi_1(r_2)\chi_2(r_1)$ which is already different from $\xi_a(r_1,r_2)$ because $\chi_1(r) \neq \chi_2(r)$ and $\chi_i(r_1) \neq \chi_i(r_2)$ for $i = 1,2$. However, if we had instead constructed the wavefunction $\xi_b(r_1,r_2) = \chi_1(r_1)\chi_2(r_2) - \chi_1(r_2)\chi_2(r_1)$, we see that if we exchange coordinates we get the wavefunction $\xi_b(r_2,r_1) = \chi_1(r2)\chi_2(r_1) - \chi_1(r_1)\chi_2(r_2) = -\xi_b(r_1,r_2)$ and thus $\xi_b(r_1,r_2)$ is antisymmetric under the exchange of coordinates; furthermore, $|\xi_b(r_1,r_2)|^2 = |\xi_b(r_2,r_1)|^2$, and the so constructed $\xi_b$ maintains the indistinguishability of the electrons while obeying the Pauli principle (also if $r_1 = r_2 = r$, $\xi_b$ vanishes as required). Notice that the way in which $\xi_b(r_1,r_2)$ has been constructed is such that it is not a simple product of one-electron wavefunctions, so that $\xi_b(r_1,r_2)$ is correlated.

In our present discussion, we have not mentioned the intrinsic spin of the electrons which also plays a role in correlations and in determining the kind of wavefunction we use. In the case of the Hartree wavefunction Equation 3.4.49b, $\psi_0^H(r_1,r_2)$ corresponds to the space part in the total wavefunction

$$\psi(x_1,x_2) = \psi_{\text{space}}\psi_{\text{spin}} = \psi_0^H(r_1,r_2)\frac{\alpha_1\beta_2 - \alpha_2\beta_1}{\sqrt{2}}, \tag{3.4.50}$$

where the variables $x_1, x_2$ contain the position and spin coordinates and where $\alpha_i, \beta_j$ stand for the ith electron with spin-up and the jth electron with spin-down, respectively. The spin part being antisymmetric under exchange makes the total wave function antisymmetric so that the space part must be symmetric. Since we do not include spin in our $H_2$ Hamiltonian (below), the spin coordinates do not appear in our discussion any further. For the hydrogen molecule we write the Hamiltonian as

$$H_{total} = H_{\text{ions}} + H_{H_2}(r_1,r_2), \tag{3.4.51a}$$

with

$$\begin{aligned} H_{\text{ions}} &= \frac{ke^2}{|\vec{R}_1 - \vec{R}_2|}, \\ H_{H_2}(r_1,r_2) &= H_{H_2^+}(r_1) + H_{H_2^+}(r_2) + h(r_1,r_2). \end{aligned} \tag{3.4.51b}$$

The ions term is a repulsive interaction and we assume they are fixed as before. The electron part $H_{H_2}(r_1,r_2)$ contains the kinetic term for each electron as well as each electron interacting with each ion; for each electron, there is a term with the same form as the singly ionized hydrogen molecule Hamiltonian. The last term is the electron-electron interaction. These terms are given by

$$H_{H_2^+}(r) = -\frac{\hbar^2}{2m}\vec{\nabla}_r^2 - \frac{ke^2}{|\vec{r} - \vec{R}_1|} - \frac{ke^2}{|\vec{r} - \vec{R}_2|}, \tag{3.4.51c}$$

and

$$h(r_1,r_2) = \frac{ke^2}{|\vec{r}_1 - \vec{r}_2|}. \tag{3.4.51d}$$

As mentioned before, the Hartree wavefunction is not the best assumption, but for the present case, the main disadvantage in using Equations 3.4.49 lies in the description of what happens for large ion separation. In such case, we should end up with two separate hydrogen atoms, yet that is not what

Equations 3.4.49b and 3.4.49c describe. A better wavefunction would be a symmetric wavefunction that has terms like $\psi_L(r_1,r_2) \approx s_1(r_1)s_2(r_2)$ or $s_1(r2)s_2(r_1)$ or a linear combination of them. Such wavefunction is separable, in the sense that in the absence of the electron-electron interaction, for large ion separation, the Hamiltonian $H_{H_2}(r_1,r_2) \to H_L(r_1,r_2) \equiv H_1(r_1)+H_2(r_2)$, since each electron would only interact with a single ion in this limit, and where $H_i(r) = -\frac{\hbar^2}{2m}\vec{\nabla}_r^2 - \frac{ke^2}{|\vec{r}-\vec{R}_i|}$ is the hydrogen atom Hamiltonian associated with the ith ion. In this limit, $H_L(r_1,r_2)|\psi_L> = (E_{1s}+E_{1s})|\psi_L>$, which is what one expects. We will come back to the better wavefunction later. For now, in spite of these disadvantages, we seek to find the ground state energy of the electronic bonding state associated with $H_{H_2}(r_1,r_2)$ in Equation 3.4.51b acting on the wavefunction of Equation 3.4.49b with the definition in Equation 3.4.49a. While we might guess what the wave-function behaves like, we do not know its exact form; for this reason, we will use the variational method. In the variational method, the wavefunction guess depends on one or more unknown parameters. We find the best parameters by minimizing the energy. The idea of the variational principle is that the resulting minimum energy is an upper bound to the exact ground state energy. In this way, we calculate our estimate to the ground state for the $H_2$ molecule by minimizing the quantity for the total energy

$$E_{total} = \frac{ke^2}{|\vec{R}_1 - \vec{R}_2|} + < H_{H_2} >, \tag{3.4.52a}$$

with respect to the variational parameters, where the first term is the ionic repulsive energy, and the second term is the electronic energy given by the expectation value

$$< H_{H_2} > = \frac{< \psi_0^H | H_{H_2}(r_1,r_2) | \psi_0^H >}{< \psi_0^H | \psi_0^H >}, \tag{3.4.52b}$$

involving the variational wavefunction $\psi_0^H(r_1,r_2)$ of Equation 3.4.49b. This wavefunction depends on the $s_i(r)$ basis functions through Equation 3.4.49a and, as will be described later below, these basis functions incorporate the variational parameters of interest here. Before considering what the specific form of our variational wavefunctions $s_i(r)$ look like, let's do a simple analytic example to illustrate the variational process.

**Example 3.4.2.1**

Assuming a variational wavefunction of shape $\psi(r) = ce^{-\alpha r}$, obtain an upper bound to the ground state energy of the hydrogen atom.

**Solution**

Throughout this example, we will use the integral identity $\int_0^\infty r^n e^{-ar} dr = n!/a^{n+1}$.

We first normalize the wavefunction $< \psi|\psi > = \int_0^\pi \sin\theta d\theta \int_0^{2\pi} d\phi \int_0^\infty |\psi(r)|^2 r^2 dr = 1 = 4\pi c^2/(4\alpha^3)$, which gives $c = \sqrt{\alpha^3/\pi}$. Placing this constant back into the original wavefunction, we have $\psi(r) = \sqrt{\alpha^3/\pi} e^{-\alpha r}$. The hydrogen atom Hamiltonian is

$$H_H = -\hbar^2 \vec{\nabla}^2/2m - ke^2/r, \tag{3.4.53}$$

and using spherical symmetry we have that $\vec{\nabla}^2\psi(r) = \frac{1}{r^2}\frac{d}{dr}\left(r^2\frac{d}{dr}\psi(r)\right) = \sqrt{\alpha^3/\pi}\alpha\left[-\frac{2}{r}+\alpha\right]e^{-2\alpha r}$. Having normalized $\psi$, we then have that $\frac{<\psi|H_H|\psi>}{<\psi|\psi>} = <\psi|H_H|\psi> = \frac{\alpha^4}{\pi}\frac{\hbar^2}{2m}\int\left[-\frac{2}{r}+\alpha\right]e^{-2\alpha r}d^3r - \frac{ke^2\alpha^3}{\pi}\int\frac{e^{-2\alpha r}}{r}d^3r$, where $d^3r = r^2 dr \sin\theta d\theta d\phi$. As usual, the $\theta$ and $\phi$ integration gives a factor of $4\pi$, and after performing the integration over $r$ and simplifying, we get $E_0 = <\psi|H_H|\psi> = \frac{\hbar^2\alpha^2}{2m} - ke^2\alpha$. The obtained variational energy depends on the yet unknown variational parameter $\alpha$. We find its value my minimizing $E_0$, or setting $\frac{dE_0}{d\alpha} = \frac{\alpha\hbar^2}{m} - ke^2 = 0$, which gives $\alpha = \frac{kme^2}{\hbar^2}$; however, one notices that the inverse of this is exactly the definition of the Bohr radius. Thus $\alpha = 1/a_0$. Substituting this $\alpha$ back into the obtained ground state, we see that $E_0 = \frac{\hbar^2}{2ma_0^2} - \frac{ke^2}{a_0} = \frac{ke^2}{2a_0} - \frac{ke^2}{a_0}$ or $E_0 = -\frac{ke^2}{2a_0}$. This is in fact the ground state of the hydrogen atom. The assumed form of the variational wavefunction happened to have the correct shape so as to lead to the exact result.

The above example has been carried out analytically and the wavefunction ansatz is a Slater type of orbital; that is, a decaying exponential. This wavefunction happened to be exact for the ground state of the hydrogen atom. In general, it is not a simple process to know what kind of ansatz will work best, and sometimes more than one variational parameter is involved in the ansatz. We will do the hydrogen molecule in a different way. As you probably have surmised, if we were to expand a wavefunction in a larger basis to achieve better accuracy in the solution, more variational parameters come into play. There are also more integrals to be done. The integrals involving exponential bases are difficult and numerical methods have to be used to carry them out, which becomes a time-consuming task, even for computers. While that is one option, it is also possible to employ an ansatz that involves a Gaussian basis. We will use such bases for the hydrogen molecule's wavefunction expansion. One advantage in using a Gaussian bases is that there exist analytic formulas for the integrals. Below are the known integrals between Gaussians [15] which will find use in our hydrogen molecule problem. We first define a Gaussian function in the form

$$g_{1s,\alpha}(r - R_A) \equiv |1s, \alpha, A> = \exp[-\alpha|\vec{r} - \vec{R}_A|^2], \tag{3.4.54a}$$

and the product of two Gaussians as

$$g_{1s,\alpha}(r - R_A)g_{1s,\beta}(r - R_B) = Kg_{1s,\gamma}(r - R_P), \tag{3.4.54b}$$

where $K = \exp[-\alpha\beta|\vec{R}_A - \vec{R}_B|^2/\gamma]$, $\gamma = \alpha + \beta$, and $\vec{R}_P = (\alpha\vec{R}_A + \beta\vec{R}_B)/\gamma$. The overlap integral is given by

$$< 1s, \alpha, A|1s, \beta, B> = \left(\frac{\pi}{\alpha+\beta}\right)^{3/2} \exp\left[-\frac{\alpha\beta|\vec{R}_A - \vec{R}_B|^2}{\alpha+\beta}\right]. \tag{3.4.54c}$$

The kinetic integral is

$$< 1s, \alpha, A| - \vec{\nabla}^2|1s, \beta, B> = \frac{\alpha\beta}{\alpha+\beta}\left[6 - 4\frac{\alpha\beta}{\alpha+\beta}|\vec{R}_A - \vec{R}_B|^2\right]\left(\frac{\pi}{\alpha+\beta}\right)^{3/2} \exp\left[-\frac{\alpha\beta}{\alpha+\beta}|\vec{R}_A - \vec{R}_B|^2\right]. \tag{3.4.54d}$$

With $r_c \equiv |\vec{r} - \vec{R}_C|$ and $Z$ the atomic number, we have for the coulomb integral

$$< 1s, \alpha, A| - Z/r_c|1s, \beta, B> = -\frac{2Z\pi}{\alpha+\beta} \exp\left[-\frac{\alpha\beta}{\alpha+\beta}|\vec{R}_A - \vec{R}_B|^2\right] F_0\left[(\alpha+\beta)|\vec{R}_P - \vec{R}_C|^2\right], \tag{3.4.54e}$$

where $F_0(t) = t^{-1/2}\int_0^{t^{1/2}} e^{-y^2}dy = \sqrt{\pi}\cdot\text{Erf}(\sqrt{t})/(2\sqrt{t})$, and $\text{Erf}(x) = (2/\sqrt{\pi})\int_0^x e^{-x'^2}dx'$. Finally, for the two-electron integral we have

$$\begin{aligned} < 1s, \alpha, A; 1s, \beta, B|\frac{1}{|\vec{r}_1 - \vec{r}_2|}|1s, \gamma, C; 1s, \delta, D> = & \frac{(2\pi)^{5/2}}{(\alpha+\gamma)(\beta+\delta)(\alpha+\beta+\gamma+\delta)^{1/2}} \\ & \cdot\exp\left[-\frac{\alpha\gamma}{\alpha+\gamma}|\vec{R}_A - \vec{R}_C|^2 - \frac{\beta\delta}{\beta+\delta}|\vec{R}_B - \vec{R}_D|^2\right] \\ & \cdot F_0\left[\frac{(\alpha+\gamma)(\beta+\delta)}{(\alpha+\beta+\gamma+\delta)}|\vec{R}_P - \vec{R}_Q|^2\right], \end{aligned} \tag{3.4.54f}$$

where in this expression $\vec{R}_P = (\alpha\vec{R}_A + \gamma\vec{R}_C)/(\alpha+\gamma)$ and $\vec{R}_Q = (\beta\vec{R}_B + \delta\vec{R}_D)/(\beta+\delta)$. To get a good handle on the numerical solution process involving the Gaussian expansion of the ansatz, let's repeat the hydrogen atom Example 3.4.2.1 using this concept. After that, we will be ready to do the hydrogen molecule.

**Example 3.4.2.2**

Assuming a variational wavefunction expanded in terms of Gaussian basis functions in the form $\psi(r) = c_1 e^{-\alpha_1 r^2} + c_2 e^{-\alpha_2 r^2}$, obtain an upper bound to the ground state energy of the hydrogen atom.

**Solution**

We will work with atomic units as before; that is, distance in units of $a_0$ and energy in Hartrees, $\hbar^2/(ma_0^2) = ke^2/a_0$. In these units, the hydrogen atom Hamiltonian of Equation 3.4.53 becomes

$$H_H = -\vec{\nabla}^2/2 - 1/r, \tag{3.4.55}$$

In these units, we write the exact wave function as $\psi_{\text{exact}} = \exp(-r)$, which will be used to see how well the given numeric solution compares to it. Both of the wavefunctions (exact and Gaussian-expanded) will be normalized numerically by solving for $N$ in $N^2 \int |\psi|^2 d^3r = 1$ and then multiplying the unnormalized wavefunction by the resulting $N$. We seek to minimize the variational energy

$$E_0 = \frac{<\psi|H_H|\psi>}{<\psi|\psi>} = \frac{\sum_{i=1}^{2}\sum_{j=1}^{2} c_i^* c_j < G_i|H_H|G_j >}{\sum_{i=1}^{2}\sum_{j=1}^{2} c_i^* c_j < G_i|G_j >}, \tag{3.4.56}$$

where $G_i = e^{-\alpha_i r^2}$ are the Gaussian functions. We also note that we will be working with real coefficients so that we will take $c_i^* = c_i$. The overlap integrals $< G_i|G_j >$ are done through Equation 3.4.54c with $\vec{R}_A = \vec{R}_B = 0$ since we only have one ion and it is located at the origin. The matrix elements of the Hamiltonian are

$$< G_i|H_H|G_j > = \frac{1}{2} < G_i| - \vec{\nabla}^2|G_j > + < G_i| - Z/r_c|G_j >, \tag{3.4.57}$$

which are done through Equations 3.4.54d and 3.4.54e, respectively, with $Z = 1$ for hydrogen and $\vec{R}_A = \vec{R}_B = \vec{R}_c = 0$. Since we now have four variational parameters; i.e., $c_1$, $c_2$, $\alpha$, and $\beta$, we will make use of MATLAB's **fminsearch** function to minimize the numerical $E_0$ of Equation 3.4.56. The function fminsearch requires initial guesses for these parameters. We will use $c_1 = c_2 = 1/2$ and $\alpha_1 = 0.17792$ and $\alpha_2 = 1.9701$ which were obtained by approximately fitting the exact exponential function with two Gaussian functions, similar to the given ansatz. The script, in the form of a function, used for the numerical solution is hydroGausGr0.m. The reason for making it as a function is that, in MATLAB, other functions can be embedded within a function. The various integrals are carried out by various functions as follows: function EneGroundFind2 calculates $E_0$ and is called by fminsearch for minimizing. To calculate $E_0$, EneGroundFind2 calls function OverMatrix, which does the overlap integrals; function DelsMatrix, which does the kinetic term integrals; and function OrintsMatrix, which does the $1/r$ potential term integrals. After convergence, fminsearch returns the final $c's$ and $\alpha's$, which are used by the function WaveG in order to produce the exact analytic and numeric wavefunctions which are then plotted. After running the script, the final values obtained are: $c_1 = 0.47277$, $c_2 = 0.65136$, for the expansion coefficients; $\alpha_1 = 0.20153$, $\alpha_2 = 1.3325$ for the Gaussian exponents; $E_0 = -0.485813$ Hartrees or $-13.219636\,eV$, for the energy. The exact result is $-\frac{1}{2}$ Hartrees $= -13.6057\,eV$. The obtained two-Gaussian wavefunction is compared to the exact ground state exponential wavefunction in Figure 3.4.13.

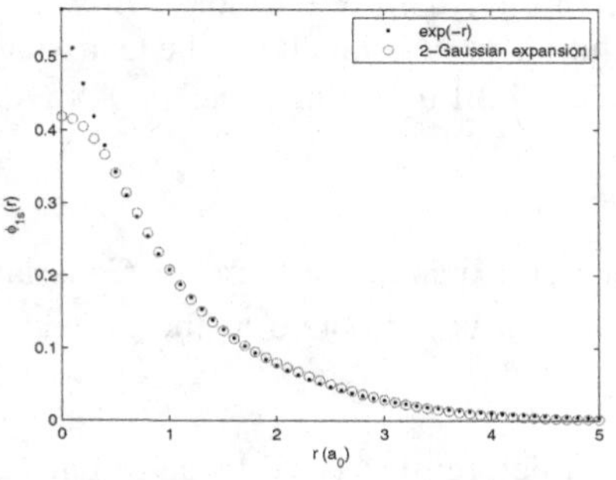


Figure 3.4.13: The two-Gaussian expansion wavefunction is compared to the analytic hydrogen atom ground state wavefunction. The obtained estimate of the error is 0.003694 as calculated by Equation 3.4.58. The error can be made smaller by including more Gaussians (see Exercise 3.6.18).

For the wavefunction and after convergence is achieved, an estimate of the error is made using

$$\text{error} \approx \frac{\sqrt{\sum_{i=1}^{N}\left(s(r_i)-\sum_j c_j G_j(r_i)\right)^2}}{N}, \tag{3.4.58}$$

where $s(r_i) = \exp(-r_i)$ is the exact wavefunction (aside from normalization) and $G_j(r_i) = \exp(-\alpha_j r_i^2)$ the numerical approximation. Thus calculated, the run produces an error estimate of 0.003694. The results can be improved by including more Gaussians (see Exercise 3.6.18). The listing of the code used in this example, hydroGausGr0.m, follows.

```
%copyright by J. E Hasbun and T. Datta
%hydroGausGr0.m
%This script is set up for two gaussians.
%Program to find the ground state if Hydrogen based on expanding
%the ground state wavefunction with Gaussian orbitals
%(psi(r)=sum_i Ci*Gi(r)). The eigenvalue problem is
%[-(hbar^2/(2*m)) del^2 -ke^2/r]psi=E*psi. If we let r=rbar*a0 and
%E=Ebar*Eb, where a0=4pi*epsilon0*hbar^2/me^2 and
%Eb=2*hbar^2/(2m*a0^2)=1 Hartree (atomic units), the SE becomes
%[-del^2/2-1/r]psi=E*psi with E in units of Eb, and r in units of a0.
%The ground state <H>=<psi|H|psi>/<psi|psi> is calculated as
%<H>=sum(all of Ci*Hij*Cj)/sum (all Ci*Sij*Cj) or just multiply matrices
%once all matrix elements and overlaps have been found.
%Expected result is 1/2 a Hartree.
%Matlab's fminsearch minimizes the energy & optimizes the Gaussian
%exponents, as well the C's to have a total of 2*nG parameters to
%optimize, where nG is the number of Gaussians employed.

function hydroGausGr0
clear all
global Cg
e=1.602176487e-19;       %electronic charge
h=6.62606896e-34;        %Planck'constant (J.s)
eps0=8.854187817e-12;    %Permittivity of free space (C^2/N/m^2)
```

```
k=1/4./pi/eps0;          %Electrical constant (N.m^2/C^2)
hbar=h/2./pi;            %hbar
me=9.10938215e-31;       %electron mass (kg)
a0=hbar^2/me/e^2/k;      %Bohr radius (m)
Eb=2*hbar^2/(2*me*a0^2);
%The Gaussian exponents guesses (play with them)
%A way to guess them is to fit the Hydrogen ground state also.
alphaG(1)=1.7792e-001;
alphaG(2)=1.9701e+000;
nG=length(alphaG);
Cg(1:nG)=1/nG;           %starting guesses for the C's as similar weights
%Put the alphas into half of ParsG and the C's in the other half
ParsG=zeros(1,2*nG);   %declare the ParsG array size
ParsG(1:nG)=alphaG(1:nG);
ParsG(nG+1:2*nG)=Cg(1:nG);
%
opts =optimset('TolFun',1e-10,'TolX',1.e-10,'MaxIter',800);
[ParsF,Ene,Eflag,Output]=fminsearch(@EneGroundFind2,ParsG,opts);
disp('Results')
fprintf('Energy=%9.6f (Hartree) or %9.6f (eV)\n',Ene,Ene*Eb/e)
disp('The alphas:')
fprintf('alpha_Guess=['), fprintf('%9.4e, ',alphaG), disp(']') %the guesses
fprintf('alpha_final=['), fprintf('%9.4e, ',ParsF(1:nG)), disp(']')
    %the final ones
disp('The C''s:')
fprintf('C_Guess=['), fprintf('%9.4e, ',Cg), disp(']') %the guesses
fprintf('C_final=['), fprintf('%9.4e, ',ParsF(nG+1:end)), disp(']')
    %the final ones
fprintf('iterations=%4i\n',Output.iterations)
%disp(Output.message)
%
%Plot the wave function expansion for the Ground state
%The ground state is psi(x)=exp(-x)=sum_over_n (C_n*Gaussian_PSI_n(x))
x=0:0.1:5;
[ana,numG,Error]=WaveG(x,ParsF(1:nG),ParsF(nG+1:end));
fprintf('Estimate of the wavefunction error=%12.6f\n',Error)
plot(x,ana,'k.')
hold on
plot(x,numG,'bo')
hold off
xlabel('r (a_0)')
ylabel('\phi_{1s}(r)')
str=cat(2,num2str(nG),'-Gaussian expansion');
legend('exp(-r)',str)
axis([0 max(x) 0 max(max(numG),max(ana))])

function [ana,numG,Error]=WaveG(x,alpha,Coef)
nG=length(alpha);
ana=exp(-x);
for i=1:length(x)
  numG(i)=0;
```

```
  for j=1:nG
    numG(i)=Coef(j)*exp(-alpha(j)*x(i)^2)+numG(i); %Gaussian expansion
  end
end
%Normalization
%Using N^2*4*pi*IntegralOf(r^2 * psi^2 dr), then
%N=1/sqrt(4*pi*IntegralOf(r^2 * psi^2 dr))
dx=(max(x)-min(x))/(length(x)-1);
Cana=1.0/sqrt(dx*trapz(x.^2.*abs(ana).^2))/2/sqrt(pi); %Ground state Norm
   const
ana=ana*Cana;
CNG=1.0/sqrt(dx*trapz(x.^2.*abs(numG).^2))/2/sqrt(pi); %Gaussian Norm const
numG=numG*CNG;
Error=sqrt(sum(abs((ana-numG).^2)))/length(ana);

function EG=EneGroundFind2(pars)
%1st half of pars are the alphas, 2nd half  are the C's
nG=length(pars)/2;
Over=OverMatrix(pars(1:nG));          %Gaussian Overlap S matrix elements
Dels=DelsMatrix(pars(1:nG));          %-Del^2/2 Gaussian matrix elements
Orints=OrintsMatrix(pars(1:nG));      %1/r Gaussian matrix elements
HM=(Dels+Orints);                     %The total hamiltonian matrix
%easiest to do matrix products: (vector)*Hmatrix*(vector)'
hS=pars(nG+1:2*nG)*HM*pars(nG+1:2*nG)';
OS=pars(nG+1:2*nG)*Over*pars(nG+1:2*nG)';
EG=hS/OS;                             %ground state energy

function SM=OverMatrix(par)
%Gaussians overlap integrals
%See J. M. Thijssen' "computational Physics", chapter 3 (H atom case)
nG=length(par);
for i=1:nG
  for j=i:nG
    SM(i,j)=(pi/(par(i)+par(j)))^1.5;
    SM(j,i)=SM(i,j); %Hermitian
  end
end

function DM=DelsMatrix(par)
%Gaussian integrals for the -Del^2/2 term
%%See J. M. Thijssen' "computational Physics", chapter 3 (H atom case)
nG=length(par);
for i=1:nG
  for j=i:nG
    DM(i,j)=3.*par(i)*par(j)*pi^1.5/(par(i)+par(j))^(5./2.);
    DM(j,i)=DM(i,j); %Hermitian
  end
end

function OM=OrintsMatrix(par)
%Performs the Gaussian integrals for the -1/r term
```

```
%%See J. M. Thijssen' "computational Physics", chapter 3 (H atom case)
nG=length(par);
for i=1:nG
  for j=i:nG
    OM(i,j)=-2*pi/(par(i)+par(j));
    OM(j,i)=OM(i,j); %Hermitian
  end
end
```

For the hydrogen molecule of interest, we write the Hamiltonian Equations 3.4.51 in atomic units, as before. This is equivalent to setting $\hbar^2/m = ke^2 = 1$ in the actual equations, with the understanding that distance is in units of $a_0$ and energy in Hartrees. For the ground state Equation 3.4.52b takes the form

$$< H_{H_2} >= 2\frac{< \phi_0(r)|H_{H_2^+}(r)|\phi_0(r) >}{< \phi_0(r)|\phi_0(r) >} + \frac{< \phi_0(r_1)\phi_0(r_2)|h(r_1,r_2)|\phi_0(r_1)\phi_0(r_2) >}{< \phi_0(r)|\phi_0(r) >^2}, \tag{3.4.59}$$

where we have used Equation 3.4.49b and $H_{H_2^+}(r)$ is the singly ionized hydrogen molecule Hamiltonian of Equation 3.4.51c. The first term is twice the energy of the $H_{H_2^+}$ in this model, which is negative, while the second term is the contribution of the electron-electron interaction, which is positive. By substituting the corresponding expressions and studying the form of the terms, we can carry them out by using the integral expressions from Equations 3.4.54, since we assume a Gaussian expansion for each of the $s_i$. As an example, the overlap term in the denominator takes the form

$$< \phi_0(r)|\phi_0(r) >= \sum_{\nu=1}^{2N_G}\sum_{\nu'=1}^{2N_G} c_\nu^* c_{\nu'} < G_\nu(r)|G_{\nu'}(r) >, \tag{3.4.60}$$

where $N_G$ is the number of Gaussians used in the expansion

$$s_n(r_i) = \frac{1}{\sqrt{\pi a_0^3}}\sum_k G_{kn}(r_i), \tag{3.4.61}$$

for $i = 1,2$ electrons, $n = 1,2$ ions, and where $G_{kn}(r_i) = \exp[-\alpha_{kni}|\vec{r}_i - \vec{R}_n|^2]$. In Equation 3.4.60, the first $N_G$ terms in the sum refer to values associated with the Gaussian expansion of $s_1(r)$ and the second $N_G$ terms refer to the terms for the Gaussian expansion of $s_2(r)$; that is, $\phi_0(r) = (s_1(r)+s_2(r))/\sqrt{2} = \sum_{i=1}^{N_G} c_{i1}G_{i1}(r) + \sum_{i=2}^{N_G} c_{i2}G_{i2}(r) = \sum_{\nu=1}^{2N_G} c_\nu G_\nu(r)$. The overlaps have the corresponding analytic values given by Equation 3.4.54c where, in the present case, the definition of Equation 3.4.54a is equivalent to $g_{1s,\alpha_{kn}}(r_i - R_n) = G_{kn}(r_i)$. In a similar way, the various terms in the full expression of Equation 3.4.59 can be calculated. The process can be repeated for the antisymmetric state if we replace $\phi_0(r)$ with $\phi_1(r)$ of Equation 3.4.49c while noting the sign change in the coefficients of $s_2(r)$.

Figure 3.4.14(a) is the result of performing the hydrogen molecule numerical calculation using the Hartree wavefunction discussed for the bonding and antibonding states, respectively, with a two-Gaussian basis. The equilibrium $\alpha$'s are $[0.25532, 1.6428]$ which are the same for both ions with expansion coefficients $[0.39291, 6.7109]$, also the same for both ions. The bonding state total energy (minimum) obtained is $-1.098822\,\text{Hartree} = -29.900469\,eV$ at the equilibrium distance of $1.4027\,a0 = 0.742301\,\text{Å}$. Pauling and Wilsonc [13] quote experimental values of 0.74 Å for the bond length and a total energy of $-31.92\,eV$ for $H_2$. Thus there is room for improvement. The code used for this calculation is hydro_mol_Hartree0.m, which due to its length is not listed but is available on CD or is downloadable. (The code is similar in concept and composition as hydroGausGr0.m for the hydrogen atom of Example 3.4.2.2, except that it incorporates the full integral formulas of

Equations 3.4.54. Be sure to read the comments throughout the code.) The antibonding state total energy obtained at the equilibrium bond length is 0.407476 Hartree $= 11.087982\,eV$. The script also produces the wavefunctions $\phi_{0,1}(r)$ for the bonding and antibonding states, respectively, as shown in Figure 3.4.14(b).

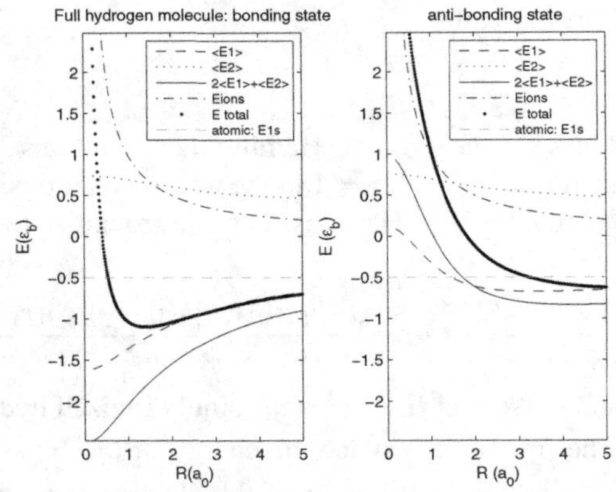


(a) bonding/antibonding energies vs. ion distance

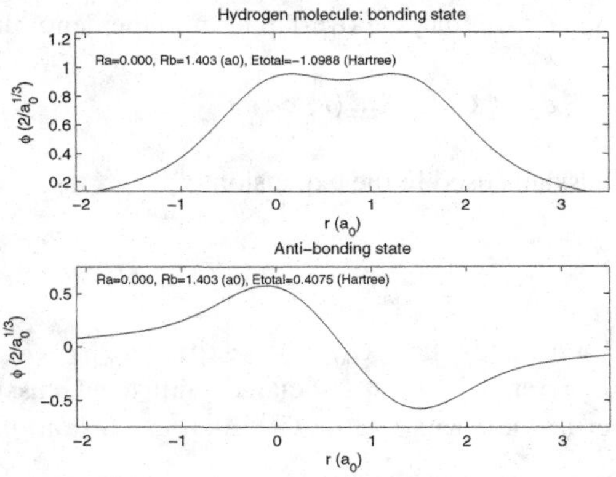


(b) bonding/antibonding equilibrium wavefunctions

Figure 3.4.14: (a) Bonding and antibonding states results for the hydrogen molecule using the Hartree approximation with a two-Gaussian basis. The code used is hydro_mol_Hartree0.m, which, due to its length, is not listed but is available on CD or is downloadable. The various contributions (see Equations 3.4.51, 3.4.52) shown are due to $E_1 = < H_{H_2^+}(r) >$, $E_2 = < h(r_1, r_2) >$, the ions (dash-dot), the total energy (thick dots), and the hydrogen atom ground state (dashed). The bonding state energy minimum is $-1.098822$ Hartree $= -29.900469\,eV$ at the equilibrium bond length of $1.4027\,a0 = 0.742301$ Å. (b) Shown are the equilibrium bonding and antibonding states wavefunctions $\phi_{0,1}(r)$ as calculated by using the Hartree approximation with a two-Gaussian basis with code hydro_mol_Hartree0.m.

We can improve on the equilibrium energy result slightly if we use a four-Gaussian basis (see Exercise 3.6.19). However, ultimately, the Hartree ansatz has to be replaced with a better one. An improvement exists over the Hartree wavefunctions employed here and which correct the flaws discussed at the beginning of this section. The better ground state wavefunction is the so-called

Hartree-Fock wavefunction. It makes use of the Heitler-London approximation for the ground state

$$\psi_0^{HF}(r_1,r_2) = \frac{1}{\sqrt{2}}[s_1(r_1)s_2(r_2)+s_1(r_2)s_2(r_1)], \tag{3.4.62a}$$

and two other possible symmetric choices

$$\psi_1^{HF}(r_1,r_2) = s_1(r_1)s_1(r_2), \quad \text{and} \quad \psi_2^{HF}(r_1,r_2) = s_2(r_1)s_2(r_2), \tag{3.4.62b}$$

to construct the improved ansatz for the total spatial part of the ground state wavefunction in the form

$$\psi_{GS}(r_1,r_2) = \frac{C_1}{\sqrt{2}}\psi_0^{HF}(r_1,r_2) + \frac{C_2}{2}[\psi_1^{HF}(r_1,r_2)+\psi_2^{HF}(r_1,r_2)], \tag{3.4.62c}$$

whose study is beyond our scope.

### 3.4.3 Semiconductors

Earlier, in the introductory part of Section 3.4, it was mentioned that semiconductor crystals engage in covalent bonding and they often involve elements in groups II, III, IV, V, and VI of the periodic table. Compounds made of elements from these groups in the proper combination form various kinds of semiconductors. Well-known semiconductor examples such as ZnS and CdTe are IIB-VIB systems, while GaAs and InSb are IIIB-VB systems, and C, Si and Ge are group IVB systems. The classic example of a diamond lattice, with carbon as the elemental unit, comes about because of carbon's $3s^2 3p^2$ valence electron structure. When the carbon atoms approach themselves to begin crystal formation, the $3s$ and $3p$ energy levels become close in energy and the electronic wavefunctions mix and create the so-called hybrid orbitals. Carbon has two electrons in the $2s$ shell and two electrons in the $2p$ shell, but during hybridization one of the $2s$ electrons is promoted to a $p$ state and the bonding occurs among the $3s, 3p_x, 3p_y$, and $3p_z$ orbitals. The mixing of the orbitals in this way is referred to a $sp^3$ hybridization and the electronic structure of tetrahedral solids may be understood in terms of it. These hybrids are directional and in diamond, for example, four such bonds point from a given carbon atom to its four nearest neighbors, which gives the structure its tetrahedral geometry. Diamond, therefore, has a coordination number of four. The coordination number is the number of nearest neighbors. In tetrahedral structures, such as zinc-blende, diamond (special case of zinc-blende), as well as wurtzite (with different translational symmetry compared to zinc-blende), the electrons' wavefunctions are written in terms of the $sp^3$ hybrids. The "3" indicates that the probability of finding the electron in the $p$-state is three times as much as the probability of finding it in the $s$-state.

In semiconductors, therefore, the bonds are described in terms of a linear combination of atomic orbitals (LCAO). The wavefunction associated with each electron in atom $k$ located at $\vec{R}_k$ is written as

$$\chi_k(\vec{r}-\vec{R}_k) = \sum_n A_{kn}\phi_{kn}(\vec{r}-\vec{R}_k), \tag{3.4.63}$$

where the $A_{kn}$'s are constants, and represents a particular combination of atomic orbitals; i.e., a hybrid. Here the $\phi_{kn}(r)$'s, for $n = 1,2,\ldots$ represent atomic orbitals such as $s_k, p_{kx}, p_{kx}$, and $p_{kz},\ldots$, belonging to atom $k$. For a two-atom system we would construct the total wavefunction

$$\psi(r) = \sum_{k=1}^{2} C_k\chi_k(\vec{r}-\vec{R}_k) = C_1\chi_k(\vec{r}-\vec{R}_1) + C_2\chi_k(\vec{r}-\vec{R}_2), \tag{3.4.64}$$

where the $C_k$'s are constants. A typical hybrid wavefunction involving $s, p$ contributions has lobes which point in a particular direction ($\hat{n} = \alpha\hat{i} + \beta\hat{j} + \gamma\hat{k}$) depending on the values of $\alpha, \beta$, and $\gamma$ in the $p$ character part of the wavefunction. That is, let's assume $s, p$ contributions only, and, from Equation 3.4.63, for a given atom, let's write the hybrid as

$$\chi(\vec{r}) = A_s\phi_s(r) + A_p\phi_p(r), \tag{3.4.65a}$$

where $A_s, A_p$ are constants and where

$$\phi_p(r) \equiv \hat{n}\cdot\vec{r}\frac{f(r)}{r}, \tag{3.4.65b}$$

for some function $f(r)$. Next, using $\vec{r} = x\hat{i} + y\hat{j} + z\hat{k}$ and, for constants $\alpha, \beta, \gamma$ obeying the relation $\alpha^2 + \beta^2 + \gamma^2 = 1$, using $\hat{n} = \alpha\hat{i} + \beta\hat{j} + \gamma\hat{k}$, the lobe's largest contribution is in the direction of $\hat{n}$ (since $\hat{n}\cdot\vec{r} = nr\cos\theta_{n,r}$ with $\theta_{n,r}$ the angle between $\hat{n}$ and $\vec{r}$). Substituting $\vec{r}$ and $\hat{n}$ into Equations 3.4.65 and simplifying, we have the more transparent form

$$\chi(\vec{r}) = A_s\phi_s(r) + A_p\left(\phi_{p_x}(r) + \phi_{p_y}(r) + \phi_{p_z}(r)\right) = A_s\phi_s(\vec{r}) + A_p\left(\alpha x + \beta y + \gamma z\right)\frac{f(r)}{r}, \tag{3.4.66}$$

where we have used $\phi_{p_\nu}(r) = \nu f(r)/r$ with $\nu = x, y$, and $z$. Let's do a simple example of this.

**Example 3.4.3.1**

Let's write the wavefunction for a hybrid lobe that points toward a cube corner. One way to do this is to take $\alpha = \beta = \gamma = 1/\sqrt{3}$. We notice that $\alpha^2 + \beta^2 + \gamma^2 = 1$ is satisfied and the wavefunction is $\chi(\vec{r}) = A_s\phi_s(\vec{r}) + A_p(x+y+z)\frac{f(r)}{\sqrt{3}r}$. This wavefunction points in the direction of $\hat{n} = \hat{i} + \hat{j} + \hat{k}$, which is the cube corner with all positive coordinates. We can obtain values for $A_s$ and $A_p$ by using the orthonormality (orthogonality and normalization) of the hybrid functions. We use the fact that the $p$ orbital wave functions, $\phi_{p_\nu}(r) = \nu f(r)/r$ for $\nu = x, y, z$, as well as $\phi_s(r)$, are also orthonormal. So that we must have $<\chi(r)|\chi(r)>= 1 = A_s^2 + A_p^2 = 1$. Furthermore, since there is only one $s$ orbital that must be shared among all four bond orbitals, we let $4A_s^2 = 1 \rightarrow A_s = 1/2$. Similarly, since three $p$ orbitals are to be shared with four bonds, we also have $4A_p^2 = 3 \rightarrow A_p = \sqrt{3}/2$. The condition $A_s^2 + A_p^2 = 1$ is also satisfied; therefore, making these substitutions, the final wavefunction is $\chi(\vec{r}) = \left(\phi_s(\vec{r}) + (x+y+z)\frac{f(r)}{r}\right)/2$.

Notice that in Example 3.4.3.1 we could just as easily have picked other sign combinations for $\alpha$, $\beta$, and $\gamma$ and in all cases $\alpha^2 + \beta^2 + \gamma^2 = 1$. However, in tetrahedral semiconductors, we are restricted to combinations that lead to four bonds in a tetrahedral geometry. Each atom creates four bonds, each bond has two shared electrons. Whereas, in the tetrahedral crystal systems, each atom alone has only four valence electrons, by hybridization and sharing, each atom ends up with a total of eight electrons, completing its $s$ and $p$ shells, and achieving a noble atom's electronic configuration. This, in turn, accounts for the strength of the tetrahedral bond. In summary, we have the following four hybrid orbitals for the semiconductors with the tetrahedral geometry.

$$\begin{aligned}
\chi_a(\vec{r}) &= \frac{1}{2}\left(\phi_s(r) + (x+y+z)\frac{f(r)}{r}\right), \\
\chi_b(\vec{r}) &= \frac{1}{2}\left(\phi_s(r) + (-x-y+z)\frac{f(r)}{r}\right), \\
\chi_c(\vec{r}) &= \frac{1}{2}\left(\phi_s(r) + (x-y-z)\frac{f(r)}{r}\right), \\
\chi_d(\vec{r}) &= \frac{1}{2}\left(\phi_s(r) + (-x+y-z)\frac{f(r)}{r}\right),
\end{aligned} \tag{3.4.67}$$

and is straightforward to show they are normalized and orthogonal to each other. A more general description of the hybrid $\chi$'s associated with atom $k$ located at $\vec{R}_k$ is

$$\chi_i(\vec{r}-\vec{R}_k)=\frac{1}{2}\left(\phi_s(\vec{r}-\vec{R}_k)+(\vec{r}-\vec{R}_k)\cdot\hat{n}_i\frac{f(|\vec{r}-\vec{R}_k|)}{|\vec{r}-\vec{R}_k|}\right), \tag{3.4.68a}$$

where the index $i$ runs over each of the hybrid lobes $(a,b,c,d)$ and where the $\hat{n}_i$'s are the hybrid lobe directions

$$\hat{n}_a=\hat{\imath}+\hat{\jmath}+\hat{k},\quad \hat{n}_b=-\hat{\imath}-\hat{\jmath}+\hat{k},\quad \hat{n}_c=\hat{\imath}-\hat{\jmath}-\hat{k},\quad \hat{n}_d=-\hat{\imath}+\hat{\jmath}-\hat{k}. \tag{3.4.68b}$$

**Example 3.4.3.2**

Write an $sp^3$ hybrid wavefunction associated with two identical atoms, one located at the origin and the other located at a distance $d$ away in the $[1,1,1]$ direction.

**Solution**

Here we are dealing with one hybrid lobe, the one labeled $\chi_a$ with direction $\hat{n}_a$ given by Equation 3.4.68b. For bookkeeping, we will use the atom's label 1,2 as well. The first atom is located at the origin ($\vec{R}_1=0$), and, from Equation 3.4.68a, the hybrid associated with the first atom is

$$\chi_{1a}(\vec{r}-\vec{R}_1)=\frac{1}{2}\left(\phi_s(r)+\vec{r}\cdot\hat{n}_a\frac{f(r)}{r}\right)=\frac{1}{2}\left(\phi_s(r)+(x+y+z)\frac{f(r)}{r}\right). \tag{3.4.69a}$$

The second atom is located at the corners of a cube. Assume the cube has sides $a$, so that $\vec{R}_2=a\hat{\imath}+a\hat{\jmath}+a\hat{k}$ and $R_2=\sqrt{3}a=d$; that is, $a=d/\sqrt{3}$. Thus the hybrid associated with the second atom is

$$\begin{aligned}\chi_{2a}(\vec{r}-\vec{R}_2)&=\frac{1}{2}\left(\phi_s(|\vec{r}-\vec{R}_2|)+(\vec{r}-\vec{R}_2)\cdot\hat{n}_a\frac{f(|\vec{r}-\vec{R}_2|)}{|\vec{r}-\vec{R}_2|}\right)\\&=\frac{1}{2}\left(\phi_s(|\vec{r}-\vec{R}_2|)+((x-a)\hat{\imath}+(y-a)\hat{\jmath}+(z-a)\hat{k})\cdot\hat{n}_a\frac{f(|\vec{r}-\vec{R}_2|)}{|\vec{r}-\vec{R}_2|}\right)\\&=\frac{1}{2}\left(\phi_s(|\vec{r}-\vec{R}_2|)+(x+y+z-3a)\frac{f(|\vec{r}-\vec{R}_2|)}{|\vec{r}-\vec{R}_2|}\right)\\&=\frac{1}{2}\left(\phi_s(|\vec{r}-\vec{R}_2|)+(x+y+z-\sqrt{3}d)\frac{f(|\vec{r}-\vec{R}_2|)}{|\vec{r}-\vec{R}_2|}\right).\end{aligned} \tag{3.4.69b}$$

With the above understanding, using Equation 3.4.64 to make a linear combination, we can write the total wavefunction for the identical atoms. We take $C_1=C_2=C$ as the normalization constant. The wavefunction we seek is

$$\psi(r)=\frac{C}{2}\left[\left(\phi_s(r)+(x+y+z)\frac{f(r)}{r}\right)+\left(\phi_s(|\vec{r}-\vec{R}_2|)+(x+y+z-\sqrt{3}d)\frac{f(|\vec{r}-\vec{R}_2|)}{|\vec{r}-\vec{R}_2|}\right)\right]. \tag{3.4.70}$$

In the next example, we work to produce a concrete example of the hybrid orbitals described by Equation 3.4.67.

**Example 3.4.3.3**

In order to produce a visual representation of the hybrid bonds associated with an atom in a tetrahedral geometry, let's use hydrogenic $2s$ and $2p$ orbitals for the $\phi_s(r)$ and $\phi_{p_\nu}(r)$'s of Equations 3.4.67. For the normalized $2s$ state we have

$$\phi_{2s}(r)=\left(\frac{Z^3}{32\pi a_0^3}\right)^{1/2}\left(2-\frac{Zr}{a_0}\right)e^{-\frac{Zr}{2a_0}}, \tag{3.4.71a}$$

and for the $2p$ states using

$$\phi_{2pm}(r) = \left(\frac{3}{4\pi}\right)^{1/2} \left(\frac{Z}{2a_0}\right)^{3/2} v_m \frac{Zr}{\sqrt{3}a_0} \frac{e^{-\frac{Zr}{2a_0}}}{r}, \tag{3.4.71b}$$

where $v_m = x, y, z$ for $m = 1, -1, 0$ respectively. These $\phi_{2pm}$ functions are normalized and the last two factors represent $f(r)/r$ of Equation 3.4.65b.
Here, we will take the atomic number $Z = 1$ and for distance we will use units of $a_0$. The wavefunction will be plotted in units of $(1/(32\pi a_0^3))^{1/2}$. With this arrangement, the $2s$ state can be written as

$$\phi_{2s}(r) = (2 - r)e^{-r/2}, \tag{3.4.72a}$$

and each of the $2p$ states is

$$\phi_{2pm}(r) = v_m e^{-r/2}, \tag{3.4.72b}$$

where again $v_m$ takes the coordinate $x, y, z$ for $m = 1, -1, 0$, respectively, but now in units of $a_0$. The actual hybrid bonds to be plotted, are obtained from these equations and Equation 3.4.67 in the present units; that is,

$$\begin{aligned}
\chi_a(\vec{r}) &= \frac{1}{2}(2 - r + x + y + z))e^{-r/2},\\
\chi_b(\vec{r}) &= \frac{1}{2}(2 - r - x - y + z)e^{-r/2},\\
\chi_c(\vec{r}) &= \frac{1}{2}(2 - r + x - y - z)e^{-r/2},\\
\chi_d(\vec{r}) &= \frac{1}{2}(2 - r - x + y - z)e^{-r/2}.
\end{aligned} \tag{3.4.72c}$$

We next use MATLAB's **isosurface** command (plots the constant value surfaces of the Equations 3.4.72c) to visualize the bonds as shown in Figure 3.4.15.

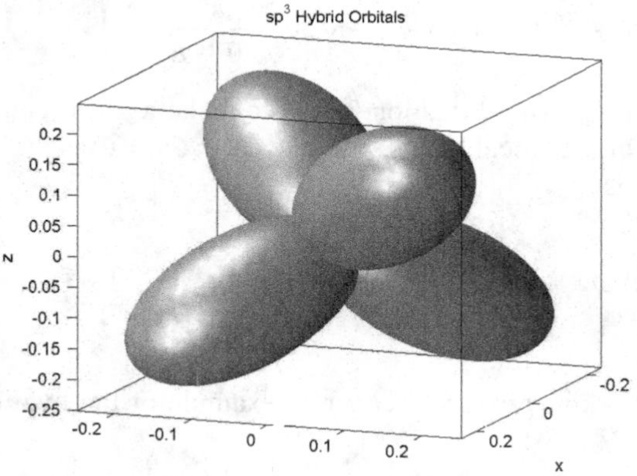


Figure 3.4.15: The hybrid functions from Equations 3.4.72c are shown as constant value surfaces.

The following code (hybrid_one_bond.m) shows how to reproduce one of the bonds shown in Figure 3.4.15 (see Exercise 3.6.22 for the rest of the hybrid bonds).

```
%copyright by J. E Hasbun and T. Datta
%hybrid_one_bond.m
%hybrid wavefunctions plotted in units of sqrt(1/[32*pi*a0^3])
%versus distance in units of a0. We work with the 2s, 2p hybridization
%
clear;
N=128;                       %grid points
ul=1.0; us=2.*ul/N;          %range and step size
[x,y,z]=meshgrid(-ul:us:ul,-ul:us:ul,-ul:us:ul);
r=sqrt(x.^2+y.^2+z.^2);      %distance
fr=0.5*exp(-r/2.0);          %fp(r)/r/2 - needed throughout
% One of the hybrids plot follows
figure
ff1=(2.0-r+x+y+z).*fr;       %one hybrid
s=0.95;                      %isosurface value plotted
isosurface(x,y,z,ff1,s);
colormap(gray)
lighting gouraud             %lighting control type for smooth looks
az=110; el=14;
camlight (az,el)             %camera light from azimuth, elevation
view(az,el)
box on
axis ([-0.25 0.25 -0.25 0.25 -0.25 0.25])
xlabel('x')
ylabel('y')
zlabel('z')
title('sp^3 Hybrid Orbitals')
```

## 3.5 Metals

Metals are characterized by a high electrical conductivity. In these materials, a large number of electrons are free to move about, typically one, two, or more per atom. These electrons are often referred to as conduction electrons. The binding in metals comes from the interaction between the ion cores and the conduction electrons. The attraction between the fixed ions and the uniformly distributed electrons is responsible for the bonding in metals. Metallic materials come in a variety of structures, such as FCC, HCP, or BCC. Alkali (group IA), alkaline (group IIA), and transition metal atoms (Groups IIIA-VIIIA, IB, IIB) of the periodic table form metallic bonds. Here the valence electrons form a *Fermi sea* (an uniform electron background) wherein the ions are embedded.
Individual atoms in a metal contribute one or more valence electrons to the metallic crystal. These electrons interact weakly with all the atoms in the crystal. The electronic wavefunction in metals extends throughout the material with nearly the same amplitude all over. One can think of the wavefunction of the electrons as being nearly plane waves. An example is to write the wavefunction for an electron in a metal as

$$\psi_k(\vec{r}) = e^{i(\vec{k}\cdot\vec{r})}\phi(\vec{r}), \tag{3.5.73}$$

where $\phi(\vec{r})$ is an atomic wavefunction which is periodic over the metal crystal lattice; i.e., $\phi(\vec{r}) = \phi(\vec{r}+\vec{T})$, with $\vec{T}$ a lattice translation vector. Here $\vec{k}$ is a wavevector in the Brillouin zone and is associated with the electron momentum. As we will see later in the text, $\psi_k(\vec{r})$ is called a Bloch function which has certain periodic properties. The wavefunction is a crystal eigenstate with an eigenvalue that depends on $\vec{k}$. This $k$-dependent eigenvalue is more properly called an energy band. In the present case, the energy band is obtained from $-\hbar^2\vec{\nabla}^2\psi_k(\vec{r})/(2m) = E_k\psi_k(\vec{r})$ or $E_k = \hbar^2k^2/(2m)$ within which the electrons are not restricted to gain energy since the band resembles that of a free electron. Thus, electrons can move easily under an applied voltage which is responsible for the low resistance properties of metals. Electrons, which are actually loosely bound to the ion cores, are, therefore, thought of as being nearly free. They behave as if they were particles in an infinite ocean. This is, in fact, the reason for the term "Fermi sea" that is used to describe electrons in metals.

## 3.6 Chapter 3 Exercises

3.6.1. Use Table 3.1.1 to collect the atomic numbers and the boiling points of the elements. (a) Make a plot of the boiling points versus their corresponding atomic numbers. Repeat for the cohesive energies versus the corresponding atomic numbers. Comment on your observations. (b) For every element, let its cohesive energy and its boiling point represent a data pair. Sort these pairs for all the elements in the table according to energy and produce a plot of the boiling point versus cohesive energy. Comment on your observations.

3.6.2. In the unperturbed Hamiltonian of Section 3.2.1, the constant $C$ can be estimated in a hydrogenic model by writing the effective potential seen by the electron in the presence of a positive charge $Ze$ in the form $V_{eff}(r) = -Zke^2/r + L^2/2mr^2$. (a) Show that if we write a Taylor expansion of this potential about the equilibrium position; i.e., $V_{eff}(r) - V_{eff}(r_0) = V'_{eff}(r_0)(r - r_0) + V''_{eff}(r_0)(r - r_0)^2/2! + \ldots$, then, for small displacements from equilibrium, we can make the approximation $V_{eff}(r) - V_{eff}(r_0) \approx C \cdot (r - r_0)^2/2$ to obtain $C \approx Zke^2/r_0^3$. (b) Taking $Z = 1$ give an estimate for the numerical value of this force constant and the associated vibrational frequency of the isolated atom, assuming an energy equivalent to the hydrogen ground state.

3.6.3. Verify that the eigenvector matrix $V_E$ of Equation 3.2.12 satisfies the inverse property of a unitary matrix and that $V_E^{-1} M V_E = D$.

3.6.4. Proceed as described in Example 3.2.1.1 and write a script that reproduces Figure 3.2.5.

3.6.5. Show the necessary steps to obtain the values of (a) $R_{nn}$, (b) $U_{min}$, and (c) $R_0$ described in Example 3.2.2.1. (d) Reproduce the graph of Figure 3.2.6 for $U(R)$ versus $R$.

3.6.6. Run the code LJ_1dMCsym.m listed in Example 3.2.2.2 to reproduce Figure 3.2.7.

3.6.7. After reading Example 3.2.3.1 for the simple cubic system, (a) show that, including contributions up to the 4th nearest neighbors, we get $\sum_j{}' \left(\frac{1}{p_{ij}}\right)^6 \approx 7.89$. (b) Write a simple script that incorporates the calculation of Equation 3.2.26 and obtain the exact values of $\sum_j{}' \left(\frac{1}{p_{ij}}\right)^s = 6.2021$ and 8.4019 for $s = 12$ and 6, respectively.

3.6.8. (a) Use the values of $d_{nn}$ and $\sigma$ from Table 3.2.2 and confirm the obtained ratios of $R_0/\sigma$ discussed in Section 3.2.3. (b) Do the same for the obtained cohesive energies associated with Equation 3.2.29 and find the percent errors of these values compared to the experimental energies of Table 3.2.2. (c) Use a wavefunction of the form $\psi = A\sin(kx)$, where $k = 2\pi/\lambda$ and $A = \sqrt{2/\lambda}$, to calculate the quantity $< \hat{p}^2 > /2m$ over a one-dimensional box of size $\lambda$. Take $\hat{p}$ to be the $x$ component of the quantum mechanical linear momentum.

3.6.9. (a) Starting from Equation 3.2.31a, confirm Equation 3.2.31b. (b) Starting from Equation 3.2.30, confirm Equation 3.2.31c. (c) Finally, starting from Equation 3.2.31b, show the missing steps in obtaining Equation 3.2.31d. (d) What is the value of the bulk modulus for Xe?

3.6.10. (a) Read Example 3.2.3.2 and run the associated code LJ_2d_min.m to reproduce Figure 3.2.8. (b) Modify the code in order to find the Lennard-Jones potential's minimum energy configuration of four Neon particles in three dimensions. What is their minimum energy? What is the average nearest neighbor distance of the particles? Display their final minimum energy configuration on a graph similar to that of the example. What is the structural shape described by the final particle arrangement?

3.6.11. According to Example 3.3.1.1, the Madelung constant for NaCl can be calculated using Equation 3.3.38. Incorporate this formula into a simple program and have it run with $N = 10$, 20, 30, 40, 50, 100, 500, and 1000. Does the value of $\alpha$ seem to be converging to the expected exact value?

3.6.12. Using the method of Example 3.3.1.2, you should be able to modify the code provided in order to carry out calculations for other systems. Obtain the theoretical bond lengths, not shown in Table 3.3.5, as well as reproduce the theoretical energies of all the ionic systems shown in the table.

3.6.13. (a) Show all the missing steps needed in obtaining Equation 3.3.39. (b) Calculate the zero Kelvin bulk modulus for the LiF ionic system and comment on your result.

3.6.14. Use the Hamiltonian of Equation 3.4.40b for the ionized hydrogen molecule along with the antisymmetric wavefunction $|\psi_a> = N(|s_1> - |s_2>)$ to obtain (a) the normalization constant $N = 1/\sqrt{2(1-\Delta)}$ where $\Delta$ is defined in Equation 3.4.42b, and (b) the antibonding electronic state energy of Equation 3.4.46. Hint: the process is similar to that which leads to Equation 3.4.45.

3.6.15. Study code hydro_mol_ion_sym.m from Example 3.4.1.1 and, after running it to reproduce Figure 3.4.12 (a), modify it to reproduce the results shown in Figure 3.4.12(b) for the anti-bonding state of Equations 3.4.46 and 3.4.47.

3.6.16. Using atomic units, obtain a plot of the symmetric and antisymmetric wavefunctions for the $H_2^+$ molecule as indicated by Equations 3.4.41. Hint: set one proton at the origin and the other at the equilibrium bond length $R$ as obtained in Example 3.4.1.1. The total wavefunction may be plotted in units of $\sqrt{1/[\pi a_0^3]}$.

3.6.17. After studying Example 3.4.2.1 and using $\psi(r) = ce^{-\alpha r}$ in addition to the integral identity $\int_0^\infty r^n e^{-ar} dr = n!/a^{n+1}$ verify the following results from the example:
(a) $<\psi|\psi> = 4\pi c^2/(4\alpha^3)$,
(b) $\vec{\nabla}^2\psi(r) = \sqrt{\alpha^3/\pi}\alpha\left[-\frac{2}{r}+\alpha\right]e^{-2\alpha r}$, and
(c) $E_0 = <\psi|H|\psi> = \frac{\hbar^2\alpha^2}{2m} - ke^2\alpha$.

3.6.18. (a) Read Example 3.4.2.2 and run the code hydroGausGr0.m to reproduce the quoted results. (b) By modifying this code and assuming a variational wavefunction expanded in terms of four Gaussian basis functions, obtain an upper bound ground state energy of the hydrogen atom. How do the results compare to that of the example?

3.6.19. (a) Download the code hydro_mol_Hartree0.m or copy it from a CD and reproduce the results of Figure 3.4.14 and associated discussion. (b) Modify the script to make it work with a four-Gaussian basis. Use the following values for the initial guesses of the $\alpha$'s: $[1.7792e-001, 1.9701e+000, 1.9994e+001, 7.0674e-001]$, both of these being identical for each of the ions. For the coefficients of the expansion, start all four of them off with the same value of 0.25. Comment on the results.

3.6.20. If we use the Dirac representation, we can write the first of Equations 3.4.67 as $|\chi_a> = \frac{1}{2}(|s> + |p_x> + |p_y> + |p_z>)$, where $|s>$ represents the orthonormal $s$ orbital $\phi_s(r)$, and where $|p_x> = Cxf(r)/r$, $|p_y> = Cyf(r)/r$, $|p_z> = Czf(r)/r$ are the three $p$ orthonormal orbitals with normalization constant $C$ (Note: for the orthonormal properties of the $s$ and $p$

orbitals, see Exercise 3.6.21). (a) In a similar fashion, with appropriate signs, write the rest of the hybrid wavefunctions. (b) Show that because the $s$ and $p$ orbitals obey orthonormality conditions, the four hybrid orbitals are also orthonormal.

3.6.21. Show that the functions $|s> = \phi_{2s}(r)$ and $|p_m> = \phi_{2pm}(r)$ of Equations 3.4.71 are orthonormal.

3.6.22. Modify the code provided for one hybrid bond (hybrid_one_bond.m) of Example 3.4.3.3 and reproduce Figure 3.4.15.

# 4

# *Lattice Vibrations*

**Contents**



## 4.1 Introduction

Lattice vibrations are collective excitations of a crystal. They come about due to the interaction among the atoms in crystals planes. The atoms in crystals oscillate about their equilibrium positions in a similar way that a mass at the end of a spring oscillates about its equilibrium position. In a crystal, however, its periodicity plays a role in the behavior of the vibration. Whereas the single mass-spring system is characterized by a specific frequency, in a crystal the frequency is a function of the propagation wavevector $\vec{k}$ that describes the crystal wave associated with the vibration. This gives rise to frequency bands. Vibrations in crystals are quantized and are known as phonons. Whereas a photon is a quantum of light, a phonon is a quantized crystal vibration.
Let us recall the simplest quantized vibration that we are familiar with; i.e., the vibration associated with the one-dimensional spring-mass system from introductory quantum mechanics. In that system the classical Hamiltonian is $\hat{H} = \hat{p}^2/2m + m\omega^2x^2/2$ and with the replacement $\hat{p} = -i\hbar d/dx$, the Schrodinger equation of the system, $\hat{H}\psi_n = E_n\psi_n$, is

$$-\frac{\hbar^2}{2m}\frac{d^2\psi_n(x)}{dx^2} + \frac{1}{2}m\omega^2x^2\psi_n(x) = E_n\psi_n(x), \tag{4.1.1}$$

where $m$ is the vibrating mass and $\omega$ is the natural vibration frequency related to the spring stiffness constant $C$; that is, $\omega = \sqrt{C/m}$. If we make the substitutions $x = y\sqrt{\hbar/(m\omega)}$ and $E_n = \hbar\omega\varepsilon_n/2$ into the above equation, we get the dimensionless form

$$-\frac{d^2\psi_n(y)}{dy^2} + y^2\psi_n(y) = \varepsilon_n\psi_n(y), \tag{4.1.2a}$$

where now the harmonic potential is dimensionless; i.e., $V(y) = y^2$. In these units, the wavefunction solutions and eigenvalues are

$$\psi_n(y) = A_n H_n(y) \exp(-\frac{y^2}{2}), \qquad \varepsilon_n = 2n + 1, \tag{4.1.2b}$$

where $A_n = (\frac{1}{n!2^n\sqrt{\pi}})^{1/2}$ for $\psi_n(y)$ is normalized over $y$. Here $H_n(y)$ are the Hermite polynomials, the first two of which are $H_0(y) = 1$ and $H_1(y) = 2y$, the rest can be obtained through the recursion formula $H_{n+1}(y) = 2yH_n(y) - 2nH_{n-1}(y)$. Using the previously mentioned relation between $E_n$ and $\varepsilon_n$, the quantized energy levels for the simple one-dimensional harmonic oscillator are given by

$$E_n = (n + \frac{1}{2})\hbar\omega. \tag{4.1.3}$$

The first four wavefunctions along with their associated energy levels from Equation 4.1.2b are shown in Figure 4.1.1.

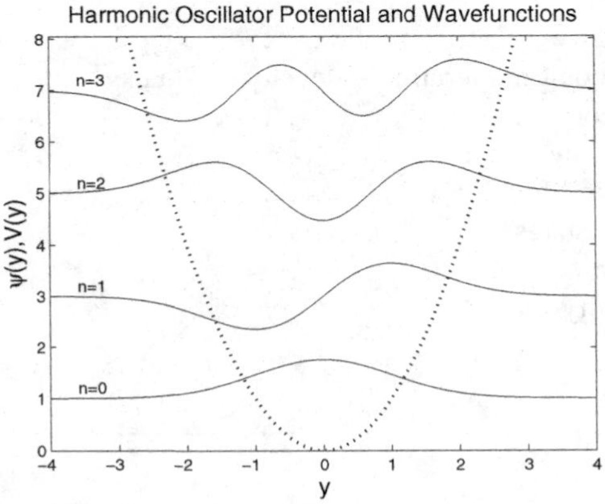


Figure 4.1.1: The harmonic potential (short dashes) is shown in dimensionless units. Also shown are the first four wavefunctions ($n = 0, 1, 2,$ and 3), each displaced by an amount corresponding to its respective eigenvalue $(2n + 1)$ in dimensionless units.

The sample code qholev.m listed below shows how to reproduce part of the figure. As is, it plots the potential and the wavefunction for the third quantum state.

```
%copyright by J. E Hasbun and T. Datta
%qholev.m
%Plots the harmonic oscillator potential and a quantum
%mechanical wavefunction for n=2.
%Dimensionsless unit y is used, such that x=sqrt(hbar/(m*w))y,
%Energy is in units of hbar*omega/2. Omega is the natural frequency.
clear;
dy=2*3/100;                  %step size
y=-3:dy:3;                   %range
H(0+1,:)=ones(1,101);        %Hermite polynomials for n=0,1
H(1+1,:)=2*y;
H(2+1,:)=2*y.*H(1+1,:)-2*1*H(0+1,:); %recursion formula
Expo=exp(-y.^2/2);
```

```
V=@(y) y.^2;                  %potential function in dimensionless units
plot(y,V(y),'k:')
hold on
eps=2*2+1;
A=1/sqrt(2^2*factorial(2)*sqrt(pi));
psi=A*H(2+1,:).*Expo;
plot(y,psi+eps,'k')           %add energy level to psi
xlabel('y')
ylabel('\psi(y),V(y)')
```

In a similar way to Equation 4.1.3 for a simple harmonic oscillator, a crystal phonon has a vibration frequency that is a function of the wavevector $\vec{k}$. If we write it as $\omega_{\vec{k}}$, the energy associated with the phonon spectrum is

$$E_{\vec{k}} = \hbar\omega_{\vec{k}}. \tag{4.1.4}$$

## 4.2 Phonons: One Atom Per Primitive Cell (Linear Chain I)

In this section we consider elastic vibrations in a crystal with one atom per cell. We wish to find the frequency of an elastic wave in terms of the wavevector and the elastic constants of the solid. Mathematically, waves that propagate in the [100], [110], and the [111] cube directions (see Figure 4.2.2) are the simplest to work with. Here, entire planes of atoms move in phase with displacements either parallel or perpendicular to the direction of the wavevector, $\vec{k}$, where we recall that the momentum and the wavevector are related through $\vec{p} = \hbar\vec{k}$.

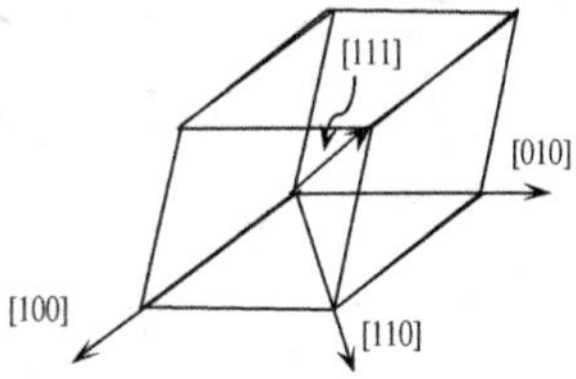


Figure 4.2.2: Shown are some particularly convenient cube directions.

Waves whose amplitude vary parallel or antiparallel to the propagation direction are referred to as longitudinal waves. Those waves whose amplitudes vary perpendicularly to the direction of propagation are called transverse waves. For each wavevector $\vec{k}$, or direction of propagation, there are three modes of vibration, one longitudinal and two transverse as shown in Figure 4.2.3. Notice that the longitudinal wave can be understood in terms of a transverse wave with the same frequency. The amplitude being positive for compressions and negative for rarefactions.

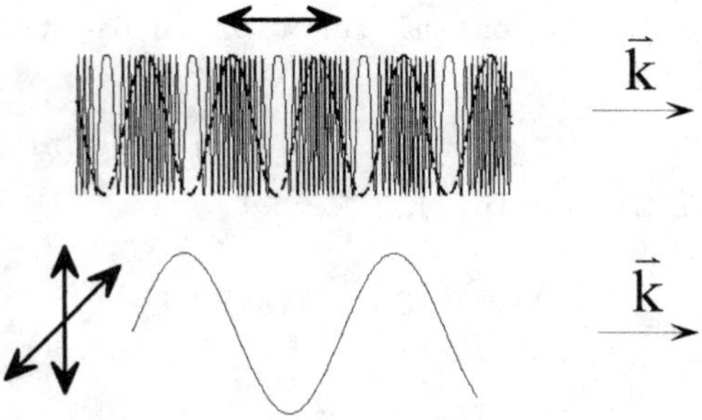


Figure 4.2.3: The top drawing illustrates the longitudinal wave's vibrational mode in the $\pm\vec{k}$ direction. It can be understood in terms of the superimposed transverse wave (dashed). The bottom illustration corresponds to the transverse wave, which consists of two vibrational modes, both perpendicular to the $\vec{k}$ direction of motion. Altogether, the waves consist of a total of three vibrational modes.

Figure 4.2.4, illustrate how it is possible to obtain longitudinal (a) and transverse phonons (b) in crystals. In both figures, the $s+p$ crystal plane's displacement, $u_{s+p}$, away from the equilibrium position is shown, $a$ is the lattice constant, and, again, $\vec{k}$ is the direction of propagation; the index $p = \pm$ integer. Both cases can be understood by the same set of equations treated in terms of a one-dimensional problem. The actual position of the $s+p$ plane is given by $r = (s+p)a$ relative to the origin, but we only make use of the displacements $u_{s+p}$ from equilibrium.

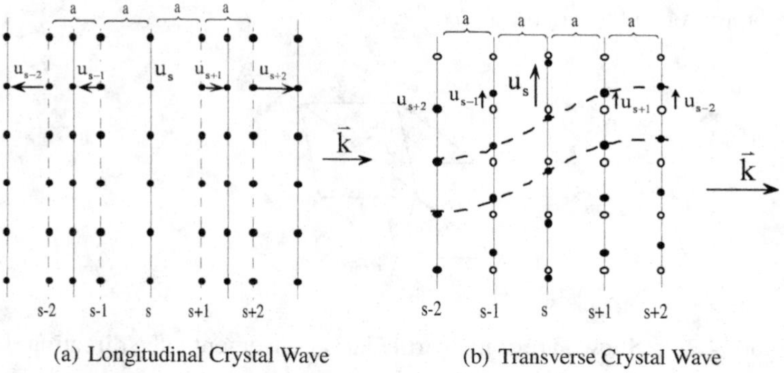


(a) Longitudinal Crystal Wave (b) Transverse Crystal Wave

Figure 4.2.4: (a) Shows an instantaneous longitudinal wave in a crystal with $u_{s+p}$ denoting a parallel displacement, away from the dashed equilibrium positions, of the $s+p$ plane parallel to the direction of motion. Here $p = \pm$ integer. (b) A transverse wave in a crystal with $u_{s+p}$ denoting a transverse displacement, away from the circled equilibrium positions, of the $s+p$ plane perpendicularly to the direction of motion. The dashed curve in (b) illustrates a section of the transverse wave at a particular instant in time. In both cases, the lattice constant $a$ is shown. Both situations can be understood in terms of a one-dimensional crystal wave. The arrows illustrate the magnitude of the displacements in the shown direction.

Here, all the atoms in the crystal are identical and we assume that the force on an atom, of mass $M$, on plane $s$ due to all the atoms in atomic planes $p$, for $1 \le |p| \le \infty$, is proportional to the sum of the differences of the $s+p$ plane's displacement and that of plane $s$; that is,

$$F_s = \sum_p C_p(u_{s+p} - u_s) = M\frac{d^2u_s}{dt^2}, \tag{4.2.5}$$

which is a form of Hooke's Law where $C_p$ represents the stiffness constant associated with plane $p$. Here $u_{s+p} - u_s$ is the displacement difference of the $s+p$ plane and that of the $s$ plane. Equation 4.2.5 represents interactions between plane $s$ and the rest of the planes in the crystal. Notice that if we consider that the main contribution comes from the nearest neighbors, then it suffices to keep the terms involving $p = \pm 1$ only so that Equation 4.2.5 becomes

$$F_s = C(u_{s+1} - u_s) + C(u_{s-1} - u_s) = M\frac{d^2u_s}{dt^2}, \tag{4.2.6}$$

where we have used $C_{-1} = C_1 = C$, for nearest neighbors. For now, we will continue the calculations using Equation 4.2.5 and later specialize to nearest neighbors. We notice that the sum over $p$ in Equation 4.2.5 includes $\pm p$ integers because there are atoms to the left and right of the $s$ plane. Next, recall that a traveling wave of wavevector $\vec{k}$ and frequency $\omega$ in the form

$$E(\vec{r},t) = E_0 \exp\left[i(\vec{k}\cdot\vec{r} - \omega t)\right] \tag{4.2.7}$$

is a solution of the wave equation

$$\frac{1}{v^2}\frac{\partial^2 E}{\partial t^2} = \vec{\nabla}^2 E, \tag{4.2.8}$$

where $\omega = vk$; that is, it is linear in $k$ with $k = \sqrt{k_x^2 + k_y^2 + k_z^2}$. Here for convenience and to be consistent with our picture, we pick $\vec{r} = sa\hat{k}$ where $a$ is the lattice constant and $\hat{k}$ is one of the directions of interest; i.e., $[111], [110]$, or $[100]$. Further, with $\vec{k} = K\hat{k}$, where $K$ is now the magnitude of the wavevector, we have $\vec{k}\cdot\vec{r} = ska$. We thus assume traveling wave solutions of Equation 4.2.5 in the form

$$u_{s+p} = u\exp\left[i((s+p)Ka - \omega t)\right], \tag{4.2.9}$$

with the case of $p = 0$ corresponding to $u_s = u\exp\left[i(sKa - \omega t)\right]$. Substituting $u_s$ and $u_{s+p}$ into Equation 4.2.5 as well as using $d^2u_s/dt^2 = -\omega^2 u_s$, we get

$$M\omega^2 u e^{[i(sKa-\omega t)]} = -\sum_p C_p\left[e^{[i((s+p)Ka-\omega t)]} - e^{[i(sKa-\omega t)]}\right],$$

which simplifies to

$$M\omega^2 = -\sum_p C_p(e^{ipKa} - 1), \tag{4.2.10}$$

with the sum over all planes $p = \pm 1, \pm 2, \ldots$. Separating the positive and negative values of $p$ in this expression we have

$$M\omega^2 = -\sum_{p>0} C_p(e^{ipKa} - 1) - \sum_{p>0} C_{-p}(e^{-ipKa} - 1)$$

but, due to translational symmetry, $C_p = C_{-p}$, to obtain

$$M\omega^2 = -\sum_{p>0} C_p(e^{ipKa} + e^{-ipKa} - 2),$$

which, using $\cos\theta = (\exp(i\theta) + \exp(-i\theta))/2$, gives

$$\omega^2 = \frac{2}{M}\sum_{p>0} C_p[1-\cos(pKa)]. \tag{4.2.11}$$

This equation expresses the general relationship between $\omega$, the wavevectors $K$, and the force constant $C_p$. We are interested in what happens to $\omega$ versus $K$. However, $K$ is a wavevector in $k-$ space, as such, we recall the Brillouin zone (BZ) in one dimension with boundaries at $K = \pm\pi/a$ (associated with $\vec{G}$ vectors at $2\pi/a$). We would like to know what happens to the dispersion relation Equation 4.2.11 at the BZ boundaries. Its derivative at the boundary

$$\frac{d\omega^2}{dK} = \frac{2a}{M}\sum_{p>0} pC_p \sin(pKa)]\Big|_{K=\pm\pi/a} = 0; \tag{4.2.12}$$

that is, it vanishes for all integers $p$. This shows that the $\omega^2$ versus $K$ relation has zero slope at the BZ boundaries. Of special interest is the behavior of the dispersion relation Equation 4.2.11 for the case when $p = 1$; i.e., nearest neighbors. In which case the summation involves the $p = 1$ term alone, to get

$$\omega^2 = \frac{2C}{M}[1-\cos(Ka)], \tag{4.2.13}$$

where we have let $C = C_1$. Since $\sin^2(\theta) = (1 - \cos(2\theta))/2$, we can write this as $\omega^2 = (4C/M)\sin^2(Ka/2)$ or for positive $\omega$

$$\omega = \sqrt{\frac{4C}{M}}\left|\sin\left(\frac{Ka}{2}\right)\right|. \tag{4.2.14}$$

As found in Equation 4.2.12, $\omega$ has a vanishing derivative at the BZ boundaries for which $K = \pm\pi/a$ with a value of $\omega = 2\sqrt{C/M}$ at those points. The zero derivative tells us that the dispersion approaches the BZ boundary perpendicularly. For small $K$; i.e., large $\lambda$, where $K = 2\pi/\lambda$, we can write Equation 4.2.14 as

$$\omega = \lim_{K\to \text{small}} 2\sqrt{\frac{C}{M}}\frac{Ka}{2}\frac{\sin(\frac{Ka}{2})}{\frac{Ka}{2}} \approx \sqrt{\frac{Ca^2}{M}}K. \tag{4.2.15}$$

If in this expression we let $v = \sqrt{Ca^2/M}$ be the wave speed, then the dispersion relation for small $K$ is linear or $\omega \approx vK$, which is the long wavelength limit, as found in connection with Equation 4.2.8. It, of course, vanishes at $K = 0$; these behaviors are evident in Figure 4.2.5 where the plot of the dispersion relation of Equation 4.2.14 is shown.

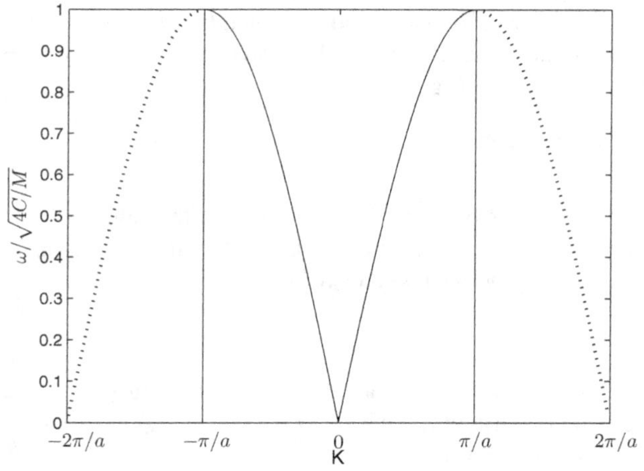


Figure 4.2.5: The dispersion relation ($\omega/\sqrt{\frac{4C}{M}}$) for the one atom per cell of Equation 4.2.14. The region $-\pi/a \leq K \leq \pi/a$ corresponds to the first Brillouin zone (BZ). The dotted lines indicate its periodic behavior beyond the first BZ.

The function of $K$ dispersion relation of Figure 4.2.5 is also known as a frequency band. Its structure is due to the crystal contribution. It is a band in the sense that there exists a continuous frequency spectrum $\omega(K)$ as opposed to one frequency that results from a single spring-mass system. It is thus possible for a phonon in the crystal to acquire a frequency value based on the value of $K$ in the BZ. One of the aspects regarding the plot of Figure 4.2.5 that needs to be asked about is, why do we look at $\omega(K)$ mainly in the first Brillioun zone (BZ)? Let's see why that is. From Equation 4.2.9 we see that the ratio

$$\frac{u_{s+p}}{u_s} = \left.\frac{ue^{[i((s+p)Ka-\omega t)]}}{ue^{[i(sKa-\omega t)]}}\right|_{p=1} = e^{[iKa]}, \tag{4.2.16}$$

and the range of $-\pi \leq Ka \leq \pi$ covers all the independent values of $\exp[iKa]$; therefore, two atoms cannot be out of phase by more than a factor of $\pi$ because $\exp[iKa]$ has a period of $2\pi$, so that all values of the exponential are contained within the first BZ. In one dimension, the extreme values of $K$ are $K_{\max} = \pm\pi/a$. The concept that crystal properties at $K$ values outside of the extreme values can be evaluated within the first BZ is illustrated in the following example.

**Example 4.2.0.1**
Show that the value of the expression $e^{[iKa]}$ at $Ka = 3\pi$, which lies outside of the first BZ in one-dimension, is contained within the first BZ.
**Solution**
We notice that $e^{[iKa]}|_{Ka=3\pi} = e^{[3\pi i]} = e^{[i(2\pi+\pi)]} = e^{[2\pi i]}e^{[i\pi]} = e^{[i\pi]}$, since $e^{[2\pi i]} = 1$. We see the value of $e^{[iKa]}$ at $Ka = 3\pi$ is the same as the value at $Ka = \pi$, which happens to be the first BZ boundary in this case. In this particular example, notice that we could have also written $e^{[iKa]}|_{Ka=3\pi} = e^{[i(4\pi-\pi)]} = e^{[4\pi i]}e^{[-i\pi]} = e^{[-i\pi]}$. The value of $e^{[iKa]}|_{Ka=3\pi}$ is the same at the equivalent points within the first BZ; those points occur at $Ka = \pm\pi$, which happen to be the boundaries.

More generally, the idea of Example 4.2.0.1 is that an exponential $\exp(iKa)$, whose phase, $Ka$, lies outside the first BZ, can be expressed in term of a phase $K'a$ which lies within the first BZ. This is possible because $K$ and $K'$ are related according to the similar concept of a previous chapter; i.e., the

Bragg condition, according to which $\Delta\vec{k} = \Delta\vec{G}$. Here this condition translates to $(K - K') = 2n\pi/a$. If $K$ is outside the first BZ, a translation is carried out, such that $K = K' + 2n\pi/a$, where the factor $2n\pi/a$ is a multiple of the smallest one-dimensional reciprocal lattice vector $G = 2\pi/a$ for an integer $n$. In one dimension, Equation 4.2.16 is summarized as

$$e^{[iKa]} = e^{[2n\pi i]}e^{[iK'a]} = e^{[iK'a]} = \frac{u_{s+p}}{u_s}, \tag{4.2.17}$$

where, as usual, $\exp[2n\pi i] = 1$ has been used, and $K'$ lies in the range $-\pi/a \leq K' \leq \pi/a$ of the first BZ. The important point to remember here is that what enables the translation operation is the general reciprocal lattice vector $2n\pi/a$. In vector form, the above comments are written as

$$\vec{K} = \vec{K}' + \vec{G}, \tag{4.2.18}$$

which is the Bragg condition. An interesting aspect of the traveling wave solution Equation 4.2.9, at the BZ boundaries, can be illustrated if we look at $u_s(K = K_{max}) = u\exp[i(sKa - \omega t)]|_{Ka=\pm\pi} = u\exp[\pm is\pi]\exp[-i\omega t]$. This can be written as

$$u_s(K = K_{max}) = u(\omega)(-1)^s, \tag{4.2.19}$$

where $u(\omega) = \exp[-i\omega t]$ and where we used $\exp[i\pi] = -1$. As can be seen, this equation is not of the general traveling wave form $f(kx - \omega t)$ (see Exercise 4.9.3). However, the wave does oscillate in time. It is a standing wave. It moves neither to the right nor to the left. This is analogous to the Bragg condition; when satisfied, the wave cannot propagate in a lattice except through successive reflections back and forth and a standing wave is set up. The critical value of $K = K_c$ with $K_c = \pm\pi/a$ satisfies the Bragg condition. This can be seen by recalling that $2d\sin(\theta) = n\lambda$ and that the wavevector is related to the wavelength through $K = 2\pi/\lambda$. We now consider a beam incident on a sample, perpendicular to the atomic planes with lattice constant $d = a$; when a standing wave is set up, its nodes occur on each plane. So from Figure 4.2.6, we learn that the standing wave has $\lambda = 2a$. In our case, we can set up the equality $|K_c| = |\pm\pi/a| = 2\pi/\lambda$ which also gives $\lambda = 2a$ and confirms that $K_c$ satisfies the Bragg condition.

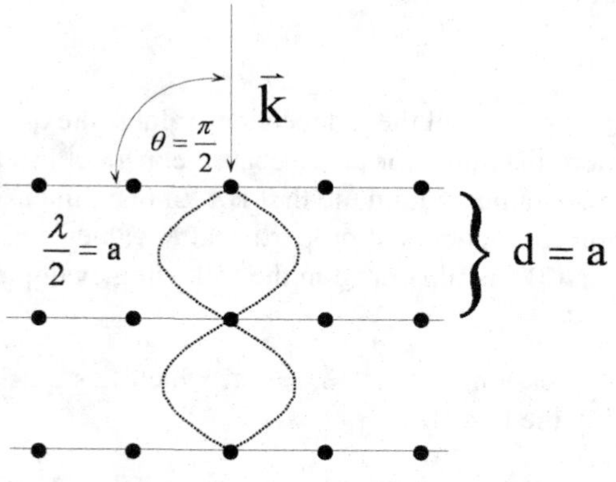


Figure 4.2.6: A standing wave showing the Bragg condition being satisfied at an incident angle of $\theta = \pi/2$.

### 4.2.1 Linear Chain of N+1 Atoms Simulation

Here consider a linear chain of $N + 1$ atoms connected by springs in a similar way as in the above linear chain; however, we restrict ourselves specifically to one dimension. Consider the force on atom $i$ according to Equation 4.2.6, which we write here as

$$f_i = C(y_{i+1} - y_i) + C(y_{i-1} - y_i) = C(y_{i+1} - 2y_i + y_{i-1}) = m_i \frac{dv_i}{dt}, \quad (4.2.20)$$

where now $y_{i\pm1} - y_i$ takes the place of the displacement difference between the $i \pm 1$ atom and that of the $i$th atom whose mass is $m_i$ as shown in Figure 4.2.7.

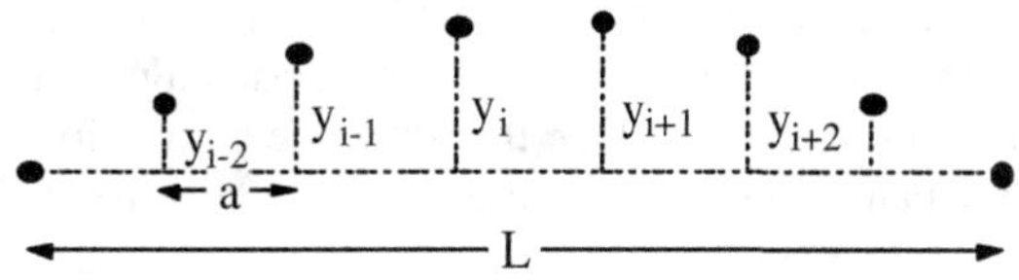


Figure 4.2.7: The one-dimensional linear chain of atoms connected by springs. The equilibrium position is the lattice constant $a$, and the chain length is $L = Na$, where $N$ is the number of cells associated with $N + 1$ atoms in the chain. The atoms' displacements occur in the $y$ direction.

The acceleration of the $i$th particle is written as

$$\frac{dv_i}{dt} = \frac{C}{M}(y_{i+1} - 2y_i + y_{i-1}), \quad (4.2.21a)$$

where, since all the atoms have the same mass, $M = m_i$. For the particle's velocity we write

$$\frac{dy_i}{dt} = v_i = \dot{y}_i. \quad (4.2.21b)$$

The idea is to solve this coupled system of equations (4.2.21) for the displacement, $y_i(t)$, of each atom as a function of time. In order to do so, we will assume initial conditions for $y_i(t = 0)$ and $v_i(t = 0)$. Considering the one-dimensional phonon, a vibrational mode exists for a given wavevector $K$ with frequency $\omega$ and energy $E = \hbar\omega$. The values of allowed $K$ depend on the number of cells of the system. For the $N + 1$ atom system discussed here, there are $N$ cells, each of width $a$, as shown in Figure 4.2.7. If we consider periodic boundary conditions, then from Equation 4.2.9, at a particular time, we have that the vibration of atom $s$ is the same as that of the $s + N$ atom, or

$$e^{isKa} = e^{i(s+N)Ka}, \text{ or } NKa = 2n\pi, \quad \Longrightarrow K = \frac{2n\pi}{L}, \quad (4.2.22)$$

where $L = Na$; i.e., the length of the atomic chain of $N + 1$ atoms. Thus, in the range where $-\pi/a \leq K \leq \pi/a$ (or $0 \leq K \leq 2\pi/a$) there exists $n = N$ values of $K$ or $n = N/2$ values in the range $0 \leq K \leq \pi/a$. In this way, we refer to the $p$th mode of vibration the value of

$$K = K_p = \frac{p\pi}{L} = \frac{p\pi}{Na}, \quad (4.2.23)$$

such that $p = 1$ corresponds to the first vibrational mode, $p = 2$ is the second, etc., and $p = N$ is the $N$th vibrational mode which happens at $K = K_{p=N} = K_{\max} = \pi/a$; i.e., the zone boundary value. Therefore, the simulation employs the initial condition

$$y_{i,p}(t = 0) = A\sin(K_p x_i), \quad (4.2.24)$$

for the $i$th atom located at $x_i = ia$ for $0 \leq i \leq N$ in the $p$th vibrational mode with $K_p$ given by Equation 4.2.23. Here we take $A = 1$Å. In the simulation that follows, we let the lattice spacing be the same as for a copper crystal or $a = 2.56$Å, with the mass of each atom of copper having the value $M = 1.05 \times 10^{-25} kg$. We let the initial conditions be as in Equation 4.2.24 with $p = 1$ and with the particles at rest ($v_i(t = 0) = 0$). The solution to the coupled differential equation system is carried out using the MATLAB 'ode23' solver for time in the range $[t_0, t_f]$, where $t_0 = 0$ and $t_f = 3.75 ps$. Starting with the initial conditions, the solver makes use of the derivatives; i.e., the right-hand sides of Equations 4.2.21 and advances the solution $y_i(t)$ for the position of each atom. The function 'derivs', in the listing below, stores the positions of the $N+1$ particles in the first $N+1$ array locations of the 'der' array, while the last $N+1$ locations contain the speeds of the particles. The 'der' array being used has, therefore, $2(N+1)$ locations. The simulation plots the new position of $i$th particle, $y_i(t)$ for $i = [0, N]$, as a function of time, while pausing for a short period at each time step taken, before the next position is plotted. In this way, the simulation of the particle motion is carried out. The running time can also be seen printed on the screen for which a snapshot is shown in Figure 4.2.8.

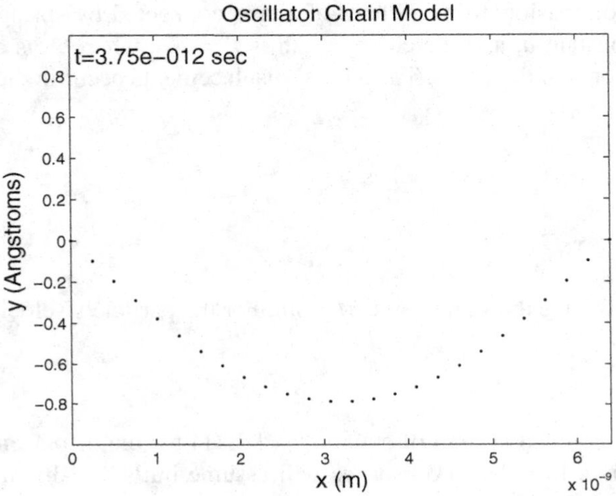


Figure 4.2.8: Snapshot of the linear chain simulation. This simulation makes use of $N+1 = 26$ particles.

The printed time can be used to obtain the atomic chain's oscillation period, which can be confirmed to obey the condition of Equation 4.2.14, here written as

$$\omega_p = 2\sqrt{\frac{C}{M}} \sin\left(\frac{K_p a}{2}\right) = 2\sqrt{\frac{C}{M}} \sin\left(\frac{p\pi a}{2L}\right) = 2\sqrt{\frac{C}{M}} \sin\left(\frac{p\pi}{2N}\right), \tag{4.2.25}$$

for the $p$th vibrational mode (see Exercise 4.9.8) and where $N$ is the number of cells. In the foregoing simulation, we work with $N+1 = 26$ copper atoms with the spring constant obtained from the product of Cu's bulk modulus and its lattice constant, or

$$C = Ba, \tag{4.2.26}$$

where $B \approx 1.42 \times 10^{11} N/m^2$. Below is a listing of the code linear_chain.m which can be used to reproduce Figure 4.2.8 and to carry out Exercises 4.9.8 and 4.9.9.

```
%copyright by J. E Hasbun and T. Datta
%linear_chain.m
%simulation of the time dependent motion of N+1 particle
```

```
%oscillator chain model. We use copper atoms equidistantly displaced
%from each other by the lattice constant a.
%We take initial distribution as yA*sin[j*pi/L], where L=N*a to be the
%initial distribution of particles. Use M=1.05e-25kg for the mass of a
%Copper atom, and run the simulation for a short time enough to
%determine the oscillation period from which w can be determined.

function linear_chain
clear
global yA L C Np1 M;
a=2.56e-10;          %lattice constant in Copper in m
yA=1.0;              %amplitude in Angstroms
N=25;                %number of cells;
Np1=N+1;             %number of particles
L=N*a;               %length over which particles are spread
B=1.42e11;           %Bulk modulus for Copper in N/m^2
C=B*a;               %spring constant in Copper
if(Np1<3), return; end;    %no less than three particles allowed
x0=0:a:L;            %x positions for N particles, evenly space
M=1.05e-25;          %Copper atom mass in kg
nm=1;                %the nm-th mode
t0=0.0;              %set time range
tf=3.75e-12;         %seconds
range=[t0;tf];
y0=profile(x0,nm);          %Np1 particles y initial position (row vector)
ymax=max(y0);
v0=zeros(1,Np1);            %particle not moving initially (row vector)
ic=[y0';v0'];               %the initial conditions, as columns
%opt=odeset('AbsTol',1.e-7,'RelTol',1.e-7); %if needed use this line
%[t,y]=ode23(@derivs,range,ic,opt);         %with this one, but comment the
                                            %next one
[t,y]=ode23(@derivs,range,ic);
%Plot the height of the particles at each x position every time
for i=1:length(t)
    %fprintf('t=%5.2f\n',t(i));
    %fprintf('%6.2f',y(1:N,i));
    %disp(' ');
    clf
    plot(x0,y(i,1:Np1),'k.','MarkerSize',5)
    axis([x0(1) x0(Np1) -ymax ymax])
    str1=cat(2,'t=',num2str(t(i),'%5.2e sec'));
    text((x0(Np1)-x0(1))*0.01,ymax*(1-0.1),str1,'FontSize',14);
    pause(0.05)
end
xlabel('x (m)','FontSize',14)
ylabel('y (Angstroms)','FontSize',14),
title('Oscillator Chain Model','FontSize',14)

function y=profile(x,nm)
global yA L;
%Possible initial height profile of the particles
```

```
y=yA*sin(nm*x*pi/L);      %nth vibrational mode

function [der]=derivs(t,y)
global C Np1 M;
%This function evaluates all the components of the
%derivative vector, putting the results in the array 'der'.
%y(1:Np1): y positions for the N+1 particles
%y(Np1+1:2*Np1): velocities for the N+1 paricles
der=zeros(2*Np1,1);      %initialize der as a column vector
for i=2:Np1-1
    der(i) =y(Np1+i);    %velocities
end
der(1)=0.0;              %end particles are fixed,so
der(Np1)=0.0;            %their y velocities are zero.
for i=Np1+2:2*Np1-1
    der(i)=C*(y(i-Np1-1)+y(i-Np1+1)-2*y(i-Np1))/M; %accelerations
end
der(Np1+1)=0.0;          %end particles don't move, so
der(2*Np1)=0.0;          %their accelerations are zero
```

### 4.2.2 Phase and Group Velocities

Consider two one-dimensional waves of the same amplitude, $A$, traveling in space and time, each described by

$$y_i(x,t) = A\sin(k_i x - \omega_i t), \tag{4.2.27}$$

where $i = 1, 2$ for each wave, respectively. We introduce the small wavevector $\delta k$ such that $k_1 = k + \delta k$ and $k_2 = k - \delta k$; similarly, we introduce the small frequency $\delta\omega$ so that $\omega_1 = \omega + \delta\omega$ and $\omega_2 = \omega - \delta\omega$. Letting $\phi$ be the phase of the wave, the waves' phase velocities can be obtained by the expression

$$v_{ph} = -\frac{(\partial\phi/\partial t)|_x}{(\partial\phi/\partial x)|_t}, \tag{4.2.28}$$

while keeping $x$ constant in the numerator and $t$ constant in the denominator. For the waves of Equation 4.2.27, each wave's phase is $\phi_i = (k_i x - \omega_i t)$, and we see that the corresponding phase velocity is $v_{ph_i} = \omega_i/k_i$. We next add the two waves, and after simplifying we get the expression for the resulting wave in the form

$$y_t(x,t) = 2A\cos\left[\delta k(x - v_G t)\right]\sin(kx - \omega t), \tag{4.2.29}$$

where $\delta k = (k_1 - k_2)/2$, $k = (k_1 + k_2)/2$, $\delta\omega = (\omega_1 - \omega_2)/2$, $\omega = (\omega_1 + \omega_2)/2$, and where we have used $v_G = \delta\omega/\delta k$. This resulting wave, by Equation 4.2.28 and letting $\phi$ be the phase of the sin part, has a phase velocity of $v_{ph} = \omega/k$ and an amplitude $2A\cos\left[\delta k(x - v_G t)\right]$ whose maximum value is twice the amplitude of the initial waves; i.e., $2A$. This amplitude is modulated by the term $\cos\left[\delta k(x - v_G t)\right]$ which explains the behavior of the envelope of the wave, itself traveling with a velocity known as the group velocity, $v_G$ as described above. In one dimension it is usually written as $v_G = d\omega/dk$, consistent with the above definition. This definition can be extended to three dimensions as

$$\vec{v}_G = \vec{\nabla}_k\omega(\vec{k}), \tag{4.2.30}$$

where $\omega(\vec{k})$ is a crystal's dispersion frequency which, more generally, is a function of the wavevector $\vec{k}$. Figure 4.2.9 illustrates what we have described. Two waves, one being the solid line and the other being the dotted line, are added. By the superposition principle, the resulting wave is that of Equation 4.2.29 (circles) and moves with the phase velocity $v_{ph}$. Additionally, the envelope (dashed line) of $y_t(x,t)$ is shown to be moving at $v_G$. The parameters used in the calculation are shown on the figure and in its caption.

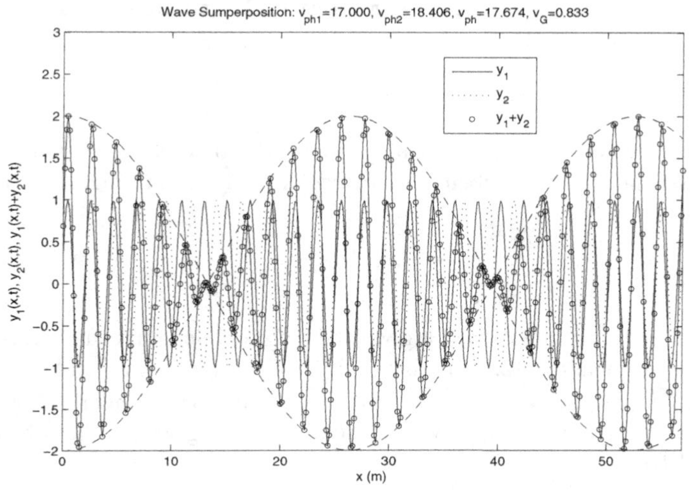


Figure 4.2.9: Two waves $y_1(x,t)$ (solid) and $y_2(x,t)$ (dotted) are superimposed to obtain the resulting wave $y_t(x,t)$ (circled). The parameters used for the waves are $A = 1$, $\omega_1 = 51.0Hz$, $k_1 = 3.0m^{-1}$, $\omega_2 = 50.8Hz$, $k_2 = 2.76m^{-1}$, with phase velocities $v_{ph_1} = \omega_1/k_1 = 17.0m/s$, $v_{ph_2} = \omega_2/k_2 = 18.406m/s$. The resulting wave (Equation 4.2.29) has $\omega = 50.9Hz$, $k = 2.88m^{-1}$, $d\omega = 0.1Hz = \delta\omega$, and $dk = 0.12m^{-1} = \delta k$. The phase velocity is $v_{ph} = \omega/k = 17.674m/s$, while the envelope (dashed line) moves with $v_G = d\omega/dk = 0.833m/s$.

The following starter code group_phase_vel_example.m can be modified to simulate wave motion as well as to calculate the phase and group velocities shown in Figure 4.2.9 (see Exercise 4.9.10). As is, the code animates a traveling wave.

```
%copyright by J. E Hasbun and T. Datta
%group_phase_vel_example.m
%This code simulates the displacent the wave y1=A*sin(k1*x-w1*t)
%as a function of space and time.
clear;
A=1.0;
N=401;                          %x points
w1=10;                          %frequency
k1=0.1;                         %wavevector
tau1=2*pi/w1;                   %period
tmax=tau1;                      %maximum time
lambda1=2*pi/k1;                %wavelength
lmax=2*lambda1;                 %maximum distance
vph1=w1/k1;
fprintf('w1=%5.3f, k1=%5.3f, vph1=w1/k1=%5.3f\n',w1,k1,vph1)
```

```
x=0:lmax/N:lmax;                %position array
t=0:0.01:2*tmax;                %time array
for i=1:length(t)               %loop over the time array
  clf                           %clear figure before replotting
  y1=A*sin(k1*x-w1*t(i));       %calculate the wave over all space
  plot(x,y1,'k');               %replot y1 for each time
  axis([0 lmax -A A])
  pause (0.05)                  %pause momentarily to see figure
end
xlabel('x (m)')
ylabel('y_1(x,t)')
```

For the one atom per primitive cell nearest neighbor model of a phonon, from Equation 4.2.14 we can obtain the associated group velocity. Writing $\omega(K) = \sqrt{4C/M}\sin(Ka/2)$, and using Equation 4.2.30, we have

$$v_G = \sqrt{Ca^2/M}\cos(Ka/2). \tag{4.2.31}$$

For small $K$ this is a constant, and, in fact, it gives the long wavelength result that is obtained from the small $K$ behavior of Equation 4.2.15. In that limit, $\omega(K) \approx \sqrt{Ca^2/M}K$ and $v_G = d\omega(K)/dK = \sqrt{Ca^2/M}$. At large $K$; that is, $K = \pm\pi/a$, $v_G \to 0$, since at the BZ boundaries, the wave, as mentioned earlier, is no longer a traveling wave but rather a standing wave. These behaviors are displayed in Figure 4.2.10.

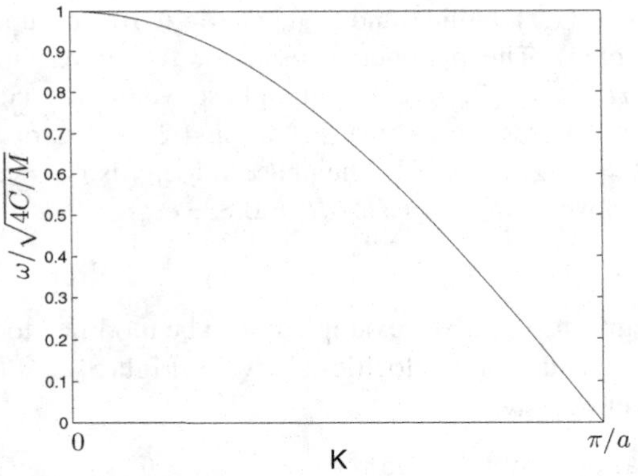


Figure 4.2.10: The group velocity ($v_G/\sqrt{Ca^2/M}$) versus $K$ from Equation 4.2.31 for the one atom per cell phonon dispersion.

## 4.3 Phonons: Two Atoms Per Primitive Cell (Linear Chain II)

In this section we work with a system of two atoms per cell. For each polarization mode in a given direction, the dispersion relation $\omega(K)$ develops two types of branches. These branches are acoustic and optical. Each branch can have longitudinal as well as transverse modes. We can have the following types of branches: one longitudinal acoustic (LA), two transverse acoustic (TA),

one longitudinal optical (LO), and two transverse optical (TO). If there are $p$ atoms per primitive cell and $f$ degrees of freedom per atom, then for a given value of $K$ there are a total of $fp$ branches. Of the $fp$ branches, $f$ are acoustic branches and the remaining $fp-f$ branches are optical. In the monatomic one-dimensional case of the previous section, $p=1$, $f=1$ so that we had a total of $fp=1$ branches with $f=1$ acoustic and $fp-f=1-1=0$ optical. This is the reason we only had the acoustic branch, and it can be recognized because its dispersion relation vanishes as $K \to 0$. The number of degrees of freedom depends on the dimension of the problem, $f=1,2,3$ for one, two, and three dimensions, respectively. As another example, in a three-dimensional system, with two atoms per cell, $f=3$ and $p=2$ so that there are $fp=6$ branches with $f=3$ acoustic (one LA, two TA) and the rest $fp-f=3$ optical (one LO, and two TO). As mentioned at the beginning, below we work with a one-dimensional system with $p=2$ atoms per cell. We will have $f=1$ acoustic (TA or LA) and $pf-f=1$ optical (TO or LO). Since the above counting is for every value of $K$, in the BZ and there are $N$ values of $K$, then the complete total number of branches is $Nfp$ with $Nf$ acoustical and the remaining $Nf(p-1)$ are optical. One might think of $Nfp$ as the total number of degrees of freedom in a crystal; that is, total degrees of freedom=(total number of cells)(atoms/cell)(degrees of freedom/atom).
Below we work with a phonon in a structure with two atoms per cell ($p=2$) and a symmetry direction is chosen so as to convert the problem to one dimension. The system is NaCl with $K$ along the $[111]$ direction as shown in Figure 4.3.11.

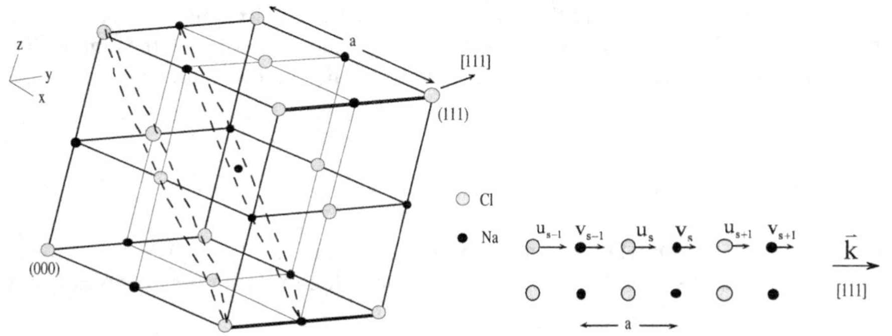


(a) NaCl motion planes due to a phonon

(b) Atomic displacements due to a phonon

Figure 4.3.11: (a) The NaCl crystal vibrational planes (dashed) are shown as the phonon travels with $\vec{K}=[111]$, a high symmetry direction. (b) The Cl and Na atomic planes' displacements are shown as the phonon passes by in the $[111]$ direction.

As the phonon travels along the $\vec{K}=[111]$ direction, it sees the one-dimensional rows of alternating atoms as shown in Figure 4.3.11 (b). Notice there are Cl atoms between the corners $(000)$ and $(111)$ of the cube. The $K$ direction $[111]$ lies along the line connecting these two atoms. Below, we let $m_1$ represent the mass of the Cl atom, with displacement represented by $u$, while $m_2$ is the mass of the Na atom with associated displacement $v$. As before, the force on the atoms is proportional to the sum of the difference of the displacements, and we concentrate on the nearest neighbor contributions only. We have, therefore, from Figure 4.3.11 (b)

$$
\begin{aligned}
F_1 &= m_1\frac{d^2u_s}{dt^2} = C[v_s-u_s]+C[v_{s-1}-u_s] = C[v_s+v_{s-1}-2u_s],\\
F_2 &= m_2\frac{d^2v_s}{dt^2} = C[u_s-v_s]+C[u_{s+1}-v_s] = C[u_s+u_{s+1}-2v_s].
\end{aligned}
\tag{4.3.32}
$$

As done previously (see Equation 4.2.9), we assume traveling wave solutions for each plane in the form

$$u_s = u\exp(i[sKa - \omega t]), \qquad v_s = v\exp(i[sKa - \omega t]), \tag{4.3.33}$$

which when substituted into Equations 4.3.32 give

$$\begin{aligned} -m_1\omega^2 u e^{i(sKa-\omega t)} &= C[ve^{i(sKa-\omega t)} + ve^{i((s-1)Ka-\omega t)} - 2ue^{i(sKa-\omega t)}],\\ -m_2\omega^2 v e^{i(sKa-\omega t)} &= C[ue^{i(sKa-\omega t)} + ue^{i((s+1)Ka-\omega t)} - 2ve^{i(sKa-\omega t)}], \end{aligned} \tag{4.3.34}$$

and which, after canceling the $\exp[i(sKa - \omega t)]$ term and reorganizing, give

$$\begin{aligned} -m_1\omega^2 u &= C[v + ve^{-iKa} - 2u] = Cv[1 + e^{-iKa}] - 2Cu,\\ -m_2\omega^2 v &= C[u + ue^{iKa} - 2v] = Cu[1 + e^{iKa}] - 2Cv, \end{aligned} \tag{4.3.35}$$

where we have assumed the same stiffness constants throughout. We can write these equations in matrix form as follows

$$\begin{pmatrix} -m_1\omega^2 & 0 \\ 0 & -m_2\omega^2 \end{pmatrix}\begin{pmatrix} u \\ v \end{pmatrix} = \begin{pmatrix} -2C & C(1+e^{-iKa}) \\ C(1+e^{iKa}) & -2C \end{pmatrix}\begin{pmatrix} u \\ v \end{pmatrix}, \tag{4.3.36}$$

which is of the form $DX = MX$, where the left- and right-hand side 2x2 matrices are labeled $D$ and $M$, respectively, while $X$ is the single column vector. As in Chapter 3, we seek the eigenvalues; here, these translate to the vibrational frequencies of the system. They can be found from the zeros of the determinant $|D - M|$ or

$$\begin{vmatrix} 2C - m_1\omega^2 & -C(1+e^{-iKa}) \\ -C(1+e^{iKa}) & 2C - m_2\omega^2 \end{vmatrix} = 0. \tag{4.3.37}$$

The determinant is $(2C - m_1\omega^2)(2C - m_2\omega^2) - C^2(1+e^{iKa})(1+e^{-iKa}) = 0$ or $m_1m_2\omega^4 - 2C\omega^2(m_1+m_2) + 2C^2[1 - \cos(Ka)] = 0$, from which solving for $\omega^2$ gives

$$\omega^2 = \frac{C(m_1+m_2)}{m_1m_2} \pm \sqrt{\left(\frac{C(m_1+m_2)}{m_1m_2}\right)^2 - \frac{2C^2}{m_1m_2}[1 - \cos(Ka)]}. \tag{4.3.38}$$

Notice that there are two values of $\omega^2$, but the lower root is the acoustic branch because as $K \to 0$, the branch tends to zero. Looking at the maximum values of $K$, the $\cos(Ka)|_{Ka=\pm\pi} = -1$, so that $\omega^2 = \frac{C(m_1+m_2)}{m_1m_2} \pm \sqrt{\left(\frac{C(m_1+m_2)}{m_1m_2}\right)^2 - \frac{4C^2}{m_1m_2}}$,
which after simplification, becomes

$$\omega^2 = \frac{C(m_1+m_2)}{m_1m_2} \pm \frac{C(m_1-m_2)}{m_1m_2} = \begin{cases} 2C/m_2 = \omega_O^2 \\ 2C/m_1 = \omega_A^2 \end{cases} \tag{4.3.39}$$

From this equation, since Cl's mass is greater than that of Na, or $m_1 > m_2$, then at $Ka = \pm\pi$ we see that $\omega_O^2 = 2C/m_2 > \omega_A^2 = 2C/m_1$, the higher frequency result is the optical branch ($\omega_O^2$); the lower is the acoustic branch ($\omega_A^2$). For small $Ka$, and writing $[1 - \cos(Ka)]) \approx (Ka)^2/2$ and substituting this into Equation 4.3.38 we get

$$\omega^2 = \frac{C(m_1+m_2)}{m_1m_2} \pm \sqrt{\left(\frac{C(m_1+m_2)}{m_1m_2}\right)^2\left[1 - \frac{m_1m_2(Ka)^2}{(m_1+m_2)^2}\right]},$$

or

$$\omega^2 = \frac{C(m_1+m_2)}{m_1m_2} \pm \frac{C(m_1+m_2)}{m_1m_2}\sqrt{1 - \frac{m_1m_2(Ka)^2}{(m_1+m_2)^2}}.$$

Since $Ka$ is small and after using $\sqrt{1-x} \approx 1-x/2$ for small $x$, this (with $x = m_1 m_2 (Ka)^2/(2(m_1+m_2)^2)$) becomes,
$\omega^2 = \frac{C(m_1+m_2)}{m_1 m_2} \pm \frac{C(m_1+m_2)}{m_1 m_2}\left(1 - \frac{m_1 m_2 (Ka)^2}{2(m_1+m_2)^2}\right)$,
after simplifying and taking the limit as $K \to 0$, get

$$\omega^2\Big|_{K\to 0} = \begin{cases} \left(\frac{2C(m_1+m_2)}{m_1 m_2} - \frac{C(Ka)^2}{2(m_1+m_2)}\right)\Big|_{K\to 0} = \frac{2C(m_1+m_2)}{m_1 m_2} = 2C\left(\frac{1}{m_1}+\frac{1}{m_2}\right) = \omega_O^2 \\ \frac{C(Ka)^2}{2(m_1+m_2)}\Big|_{K\to 0} = 0 = \omega_A^2 \end{cases} \quad (4.3.40)$$

The lower frequency dispersion, which vanishes as $K \to 0$ is the acoustic branch ($\omega_A^2$). The higher frequency dispersion is the optical branch ($\omega_O^2$) and it is a constant at $K = 0$. Putting together these results, we see that the acoustic frequency dispersion starts from zero at $Ka = 0$ and reaches the value of $2C/m_1$ at $Ka = \pm\pi$ as seen from Equation 4.3.39; similarly, the optical branch frequency, begins from the value $2C(1/m_1 + 1/m_2)$ at $Ka = 0$ and reaches the value $2C/m_2$ at $Ka = \pm\pi$. Overall, $\omega_O^2 > \omega_A^2$ throughout the BZ. These results are illustrated in Figure 4.3.12, where a plot of

$$\frac{\omega_{O,A}}{\sqrt{C/m_2}} = \sqrt{\frac{(1+m)}{m} \pm \sqrt{\left(\frac{(1+m)}{m}\right)^2 - \frac{2}{m}[1-\cos(Ka)]}} \quad (4.3.41)$$

with $m = m_1/m_2$, is shown and which is obtained from Equation 4.3.38; the positive (negative) root is the optical (acoustic) frequency dispersion.

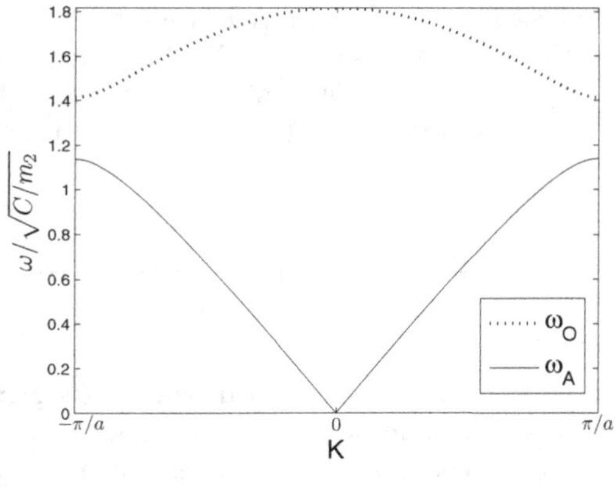


Figure 4.3.12: The optical (short dashed) and acoustic (solid line) frequency dispersions from Equations 4.3.41 are shown in units of $\sqrt{C/m_2}$. The frequency spectrum is that of a NaCl crystal phonon traveling in the [111] direction, where $m_1$ is the Cl atomic mass, and $m_2$ is the atomic mass of Na.

Notice that at the maximum value of $K$ in the BZ, there exists a gap between the acoustical and optical branches. The gap value is obtained from Equation 4.3.39, we have

$$\omega_g = \omega_O|_{Ka=\pi} - \omega_A|_{Ka=\pi} = \sqrt{2C/m_2} - \sqrt{2C/m_1} = \sqrt{2C}(1/\sqrt{m_2} - 1/\sqrt{m_1}) \quad (4.3.42a)$$

or

$$\frac{\omega_g}{\sqrt{C/m_2}} = \sqrt{2}\left(1 - \frac{1}{\sqrt{m}}\right), \quad (4.3.42b)$$

where, again, we have used $m = m_1/m_2$. Notice that if $m_1 = m_2$, then this gap is zero. That corresponds to a system of identical atoms and we need to consider only one branch as in the previous section (see Exercise 4.9.12). This frequency gap is an energy region of quantum mechanically forbidden energies (see Equation 4.1.4) at the BZ boundaries where the waves are damped out. In order for an acoustic phonon to continue to propagate at higher energy than its value at the BZ boundary, it has to gain enough energy to jump the gap. If it does so, through phonon collision or by means of photon absorption, its vibrational mode changes; it becomes an optical mode with higher energy than the acoustic mode.

To obtain a better understanding of the atomic motion for both acoustic and optical modes, we pick a convenient value of $K$ for which we know the frequency. For concreteness, let's work with $K = 0$ for which from Equation 4.3.40 $\omega_O^2 = 2C(1/m_1 + 1/m_2)$, for the optical mode, and $\omega_A^2 = 0$ for the acoustic mode. The idea is to substitute one of these frequency mode values into either of the Equations 4.3.35 and obtain a relation between $u$ and $v$. For example, using the first of the equations with the optical mode frequency at $K = 0$ we get

$$-2Cm_1\left(\frac{1}{m_1}+\frac{1}{m_2}\right)u = 2Cv - 2Cu, \text{ or } m_2u + m_1u = m_2u - m_2v \Rightarrow \frac{u}{v} = -\frac{m_2}{m_1}. \tag{4.3.43}$$

Repeating this process with the acoustic mode frequency gives

$$0 = 2Cv - 2Cu \text{ or } \frac{u}{v} = 1 \Rightarrow u = v. \tag{4.3.44}$$

Equation 4.3.43 says that in the optical mode at $K = 0$, the atoms vibrate proportional to the negative of the ratio of their masses. That is, they move 180° out of phase or against each other with a displacement for which $|u| < |v|$ because $m_2 < m_1$. Since $u$ is associated with the Cl atom here and $v$ is associated with the Na atom, then Cl displaces less than Na. In the acoustic mode, Equation 4.3.44 implies that the atoms vibrate in phase with each other or in the same sense and displace by the same amount. The expressions for general $K$ are complex (see Exercise 4.9.14).

## 4.4 Phonon Momentum

A phonon is a quantized vibration which, as mentioned before, has energy according to Equation 4.1.4. While a phonon does not carry physical momentum due to its coordinates being relative coordinates of the atoms, the phonon acts as if it had a momentum given by

$$\vec{p} = \hbar\vec{K}. \tag{4.4.45}$$

In crystals, there are wavevector selection rules for allowed transitions between quantum states. In a previous chapter we considered the elastic scattering of photons (x-rays) for which we had

$$\vec{k}' = \vec{k} + \vec{G}, \tag{4.4.46}$$

where $\vec{k}'$ is the final photon momentum, $\vec{k}$ is the initial photon momentum, and $\vec{G}$ is a crystal reciprocal lattice vector. This expression states that we can create a photon of momentum $\vec{k}'$ only if there is a $\vec{G}$ (reciprocal lattice vector) which can be added to the initial photon momentum. As one can see, the phonon is not involved in this elastic scattering situation. However, if a photon is inelastically scattered, due to the interaction with a phonon, Equation 4.4.46 is replaced with

$$\vec{k}' \pm \vec{K} = \vec{k} + \vec{G}, \tag{4.4.47}$$

where we have modified Equation 4.4.46 to include a phonon with associated wavevector $\vec{K}$. This equation states that a photon of momentum $\vec{k}'$ can be created with the simultaneous absorption (− sign) or emission (+ sign) of a phonon, given that an initial photon interacted with the crystal through a reciprocal lattice vector $\vec{G}$. This process occurs in such a way that the crystal acts as if it had a momentum $\hbar\vec{G}$.

Inelastic scattering of neutrons by phonons can be used to obtain information about the frequency dispersion relation in crystals. Neutrons interact with a crystal by the scattering from the atomic nuclei and thus carry information about the vibrational properties of the crystal. If we let $\vec{k}$ be the wavevector of the initial neutron, $\vec{k}'$ the wavevector of the scattered neutron, then by momentum conservation

$$\vec{k}+\vec{G}=\vec{k}'\pm\vec{K}, \tag{4.4.48}$$

so that the inelastic interaction of the neutron with the crystal ($\vec{G}$) gives rise to the creation or destruction of a phonon ($\vec{K}$). By energy conservation we expect

$$\hbar\omega_{\vec{K}}=\frac{\hbar^2k^2}{2m_n}-\frac{\hbar^2k'^2}{2m_n}. \tag{4.4.49}$$

Therefore, by detecting the neutron properties, from Equation 4.4.48 one obtains $\pm\vec{K}=\vec{k}+\vec{G}-\vec{k}'$; i.e., the phonon wavevector, and from Equation 4.4.49 the dispersion relation value at the corresponding wavevector is obtained.

## 4.5 Phonon Heat Capacity

The change in the internal energy ($U$) per unit temperature is referred to as the heat capacity. Theoretically, the heat capacity is studied at constant volume,

$$C_V=(\partial U/\partial T)_V, \tag{4.5.50}$$

while experimentally the heat capacity is obtained at constant pressure, $C_P=(\partial U/\partial T)_P$. The two are related through the expression $C_V-C_P=9\alpha^2BVT$, where $\alpha$ is the coefficient of linear expansion, $V$ is the crystal volume, $B$ is the bulk modulus, and $T$ is the temperature. For small enough temperatures, assuming $\alpha$ and $B$ remain constant, the difference between $C_V$ and $C_P$ is negligibly small. Here we work with the heat capacity at constant volume, Equation 4.5.50, which is also referred to as the lattice heat capacity. We will let the energy associated with the temperature be $\tau\equiv k_BT$, with $k_B=1.38065\times10^{23}\,J/K$ the Boltzmann constant, so that at a given temperature the total energy due to the phonons in a crystal is a sum of the energies over all modes, indexed by wavevector $\vec{K}$ and polarization $p$; where, as mentioned before, there are three polarization modes, two transverse and one longitudinal. We also let $U_{K,p}$ be the average energy associated with a phonon of wavevector $\vec{K}$ and polarization mode $p$. We have the total phonon contribution to the internal energy as

$$U=\sum_K\sum_p U_{K,p}=\sum_K\sum_p <n_{K,p}>\hbar\omega_{K,p}, \tag{4.5.51}$$

where $<n_{K,p}>$ is the average thermal equilibrium occupancy for wavevector $K$ and polarization $p$; i.e., the number of phonons at temperature $\tau$. Phonons are similar to photons in that they are both bosons, so that $<n_{K,p}>$ is given by the Planck distribution or

$$<n_{K,p}>=\frac{1}{\exp\left(\frac{\hbar\omega_{K,p}}{\tau}\right)-1}. \tag{4.5.52}$$

This distribution has the following behavior for small and large $\tau$

$$< n_{K,p} > \approx \begin{cases} e^{-\frac{\hbar\omega}{\tau}} & \tau \to 0 \\ \frac{\tau}{\hbar\omega} - \frac{1}{2} & \tau \to \infty \end{cases}, \tag{4.5.53}$$

where we have temporarily used $\omega = \omega_{K,p}$. This behavior of $< n >$ is shown in Figure 4.5.13.

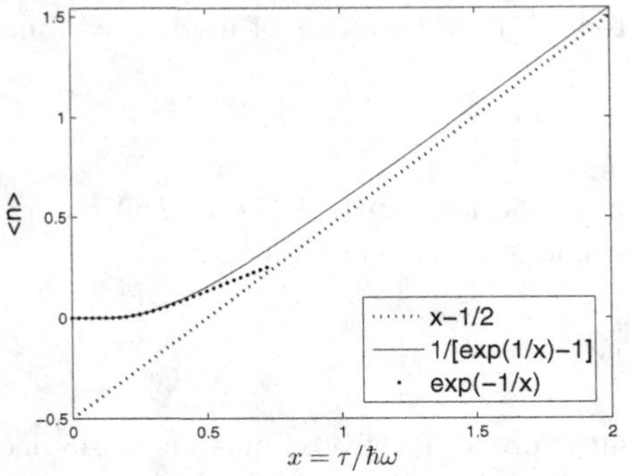


Figure 4.5.13: The Planck distribution, Equation 4.5.52 (solid line), along with its approximations, Equation 4.5.53, for low temperature (dots) as well as for high temperature (dashed). The variable $x$ represents the dimensionless quantity $\hbar\omega/\tau$, where $\tau = k_B T$.

Substituting Equation 4.5.52 into Equation 4.5.51 the internal energy becomes

$$U = \sum_K \sum_p \frac{\hbar\omega_{K,p}}{\exp\left(\frac{\hbar\omega_{K,p}}{\tau}\right) - 1}. \tag{4.5.54}$$

Before we proceed any further with Equation 4.5.54 for the energy and then obtaining an expression for the phonon heat capacity, we need to enumerate the normal modes of vibration in order to convert the sum over $K$ into an integral.

## 4.5.1 Normal Mode Enumeration - Density of States

In order to obtain the number of modes at a given frequency range, we consider phonon modes to be similar to vibrational modes on a one-dimensional string at a particular frequency as shown in Figure 4.5.14.

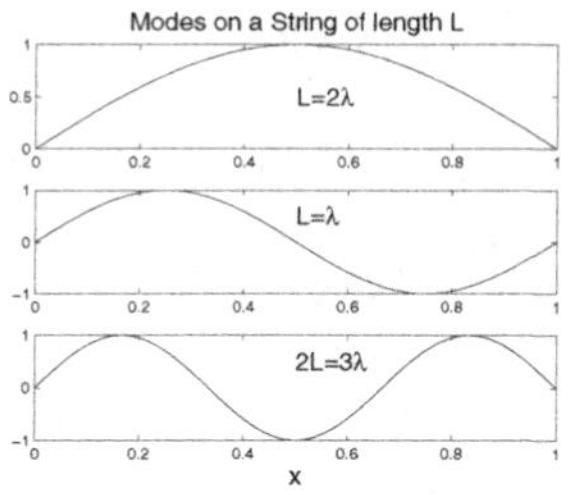


Figure 4.5.14: Examples of wave modes on a one-dimensional string of length $L$ at a particular frequency. The fundamental or first mode has wavelength $\lambda = L/2$, the next second and third modes have $\lambda = L$ and $\lambda = 2L/3$, respectively. In general, the nth mode has $\lambda = 2L/n$.

Waves on a string obey the one-dimensional form of the wave equation with a one-dimensional solution in the form $A\exp\left[i(Kx - \omega t)\right]$ with phase velocity $v = \omega/K$ and group velocity $v_g = d\omega/dK$ (see Exercise 4.9.7). As shown in Figure 4.5.14, the nth mode has a wavelength $\lambda = 2L/n$; however, $K = 2\pi/\lambda$ and we can think of $n$ as the number of modes at a given frequency (since also $K = \omega/v$) or

$$n = \frac{KL}{\pi}, \tag{4.5.55a}$$

and

$$dn = \frac{L}{\pi}dK \equiv D_{1D}(\omega)d\omega, \tag{4.5.55b}$$

where $dn$ here is thought of as the number of modes per unit frequency $\omega$ in one dimension. We have also introduced the quantity $D_{1D}(\omega)d\omega$ which can be rewritten as

$$D_{1D}(\omega)d\omega = \frac{L}{\pi}\frac{d\omega}{d\omega/dK} = \frac{L}{\pi}\frac{d\omega}{v_g}. \tag{4.5.55c}$$

Here $D_{1D}(\omega)$ is defined as the density of modes per unit frequency in one dimension and is also commonly referred to as the density of states. In three dimensions, we think of a sphere with $K_x, K_y, K_z > 0$ which means that the wave vector spans one octant of the sphere, as in Figure 4.5.15.

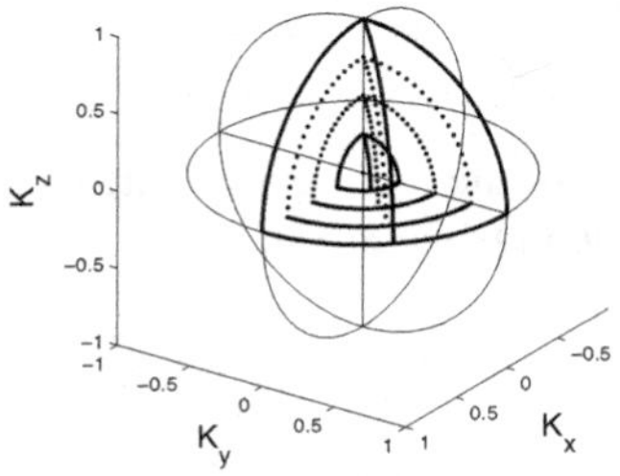


Figure 4.5.15: Three-dimensional modes are shown within one octant of a sphere. The dots are representative examples of vibrational modes; i.e., values of $\vec{K}$, for which $k_x, k_y, k_z > 0$.

We, therefore, write the analog of Equation 4.5.55b as

$$dn_{3d}(K_xK_yK_z) = \tfrac{1}{8}dn_xdn_ydn_z = \tfrac{1}{8}\tfrac{L_x}{\pi}\tfrac{L_y}{\pi}\tfrac{L_z}{\pi}dK_xdK_ydK_z,$$

and if we consider spherical symmetry, we have $L_x = L_y = L_z = L$ and $dK_xdK_ydK_z = K^2dK\sin\theta d\phi$ so that $dn_{3d}(K_xK_yK_z)$ becomes

$$dn_{3d}(\theta,\phi) \equiv \tfrac{1}{8}\left(\tfrac{L}{\pi}\right)^3 K^2dK\sin\theta d\phi.$$

Integrating this over $\theta|_0^{\pi}$ and over $\phi|_0^{2\pi}$ get

$$dn_{3d} = \int_0^{\pi}\int_0^{2\pi} d\theta d\phi dn_{3d}(\theta,\phi) = \frac{4\pi}{8}\left(\frac{L}{\pi}\right)^3 K^2dK = \frac{L^3}{2\pi^2}K^2dK. \tag{4.5.56}$$

In analogy to the one-dimensional case of Equation 4.5.55b, we set

$$dn_{3d} = \frac{L^3}{2\pi^2}K^2dK = \frac{L^3}{2\pi^2}K^2\frac{d\omega}{(d\omega/dK)} = D_{3D}(\omega)d\omega, \tag{4.5.57a}$$

so that we write the three-dimensional density of states

$$D_{3D}(\omega) = \frac{L^3}{2\pi^2}K^2\frac{dK}{d\omega} = \frac{V}{2\pi^2}K^2\frac{dK}{d\omega}, \tag{4.5.57b}$$

where we have used $V = L^3$. The total number of modes is obtained by summing over the possible values of $\vec{K}$ in three dimensions (see Figure 4.5.15); that is, $\sum_K(1)$. However, the development we have elaborated on allows us to integrate Equation 4.5.57a or

$$N = \sum_K(1) \to \int dn_{3d} = \frac{V}{2\pi^2}\frac{K^3}{3} = \left(\frac{L}{2\pi}\right)^3 4\pi\frac{K^3}{3}, \tag{4.5.58}$$

where the "$\to$" indicates the replacement. The idea is that if we have a sum over $K$ in the form $\sum_K f_{\vec{K}}$ where $f_{\vec{K}}$ is any function of $\vec{K}$ then, by the above method, we can carry it out as

$$\sum_K f_{\vec{K}} = \int dn_{3d}f_{\vec{K}} = \int D_{3D}(\omega_K)f_{\omega_K}d\omega_K. \tag{4.5.59}$$

In this fashion, Equations 4.5.51 and 4.5.54 become

$$U = \sum_p\int D(\omega_p)\frac{\hbar\omega_p}{\exp\left(\frac{\hbar\omega_{K,p}}{\tau}\right)-1}d\omega_p = \sum_p\int D(\omega_p) < n_p > \hbar\omega_p d\omega_p = \sum_p U_p, \tag{4.5.60}$$

where we have suppressed the "3D" subscript for $D(\omega)$ and defined

$$U_p \equiv \int D(\omega_p) < n_p > \hbar\omega_p d\omega_p, \tag{4.5.61}$$

with $D(\omega)$ given by Equation 4.5.57b. In the next section, the calculation of Equations 4.5.60 and 4.5.61 are carried out using the so-called Debye model.

### 4.5.2 Debye Theory

In 1912 Peter Debye developed a method of carrying out the integral of Equation 4.5.61 by assuming a frequency spectrum for an elastic medium with a high frequency cut-off known as the Debye frequency $\omega_D$ to be determined below. In this respect, we work with the relation $\omega = vK$ so that $d\omega/dK = v$ or $dK/d\omega = 1/v$ with $v$ the speed of sound in the medium. The associated density of states, using Equation 4.5.57b, is written as

$$D(\omega) = \frac{V}{2\pi^2}\left(\frac{\omega}{v}\right)^2 \frac{1}{v} = \frac{V\omega^2}{2\pi^2 v^3}. \tag{4.5.62}$$

Notice that Equation 4.5.61 does not have a restriction in the integral, so that $0 \leq \omega < \infty$, which is not appropriate because a crystal system can support only so many modes. The total number of modes is actually given by Equation 4.5.58

$$N = \frac{V}{2\pi^2}\frac{K^3}{3} = \frac{V}{2\pi^2}\frac{\omega^3}{3v^3}, \tag{4.5.63}$$

which means that the frequency must lie in the interval $0 \leq \omega \leq \omega_{\max}$. We let $\omega_{\max} = \omega_D$, obtained by solving for it from Equation 4.5.63, or

$$\omega^3\Big|_{\omega=\omega_D} = \omega_D^3 \equiv 6\pi^2 v^3 \frac{N}{V}, \tag{4.5.64}$$

which defines the Debye frequency $\omega_D$ and $N$ is the number of allowed wavectors. This $N$ is also equal to the number of primitive cells in the crystal; i.e., $V/V_c$ with $V_c$ the primitive cell volume and $V$ the crystal volume. Associated with the Debye frequency, since $K = \omega/v$ there is a Debye wavector which, with the use of Equation 4.5.64, is given by

$$K_D \equiv \frac{\omega_D}{v} = \left(\frac{6\pi^2 N}{V}\right)^{(1/3)}. \tag{4.5.65}$$

By incorporating the Debye concepts, we now rewrite Equation 4.5.61 as

$$U_p \equiv \int_0^{\omega_D} d\omega_p D(\omega_p)\frac{\hbar\omega_p}{\exp\left(\frac{\hbar\omega_p}{\tau}\right)-1} = \int_0^{\omega_D} d\omega_p \frac{V\omega_p^2}{2\pi^2 v_p^3}\frac{\hbar\omega_p}{\left[\exp\left(\frac{\hbar\omega_p}{\tau}\right)-1\right]} \tag{4.5.66}$$

where use has been made of the $D(\omega)$ from Equation 4.5.62 and the above understanding for the value of $K$. Since there are three polarization modes possible for each value of $K$ in one dimension (two transverse and one longitudinal and it so remains in three dimensions along certain special directions), we assume that all modes are equivalent; that is, independent of the subscript $p$, to write from Equation 4.5.60 $U = \sum_p U_p \approx 3U_p$ or

$$U = \frac{3V\hbar}{2\pi^2 v^3}\int_0^{\omega_D} d\omega \frac{\omega^3}{\left[\exp\left(\frac{\hbar\omega}{\tau}\right)-1\right]}. \tag{4.5.67}$$

To work with this expression, we let $x = \hbar\omega/\tau$, where we recall that $\tau = k_B T$, and so $dx = (\hbar/\tau)d\omega$ and by the same token $x_D = \hbar\omega_D/\tau \equiv \theta/T$ where we have defined the so-called Debye temperature $\theta \equiv \hbar\omega_D/k_B$. The Debye temperature is a measure of the maximum vibrational energy of a crystal. It can be calculated from

$$\theta = \frac{\hbar\omega_D}{k_B} = \frac{\hbar v}{k_B}\left(6\pi^2\frac{N}{V}\right)^{1/3}, \tag{4.5.68}$$

where we have used Equation 4.5.64. With the above definitions, we can rewrite Equation 4.5.67 as

$$U = \frac{3V\hbar}{2\pi^2 v^3}\left(\frac{\tau}{\hbar}\right)^4 \int_0^{x_D} dx \frac{x^3}{[e^x - 1]}. \tag{4.5.69}$$

Notice that from Equation 4.5.68 $\hbar^3 v^3/k_B^3 = \theta^3/(6\pi^2(N/V))$ so that Equation 4.5.69 can be written in terms of the Debye temperature as

$$U = 9Nk_BT\left(\frac{T}{\theta}\right)^3 \int_0^{x_D} dx \frac{x^3}{[e^x - 1]}. \tag{4.5.70}$$

Also for large $T$, since $x$ is small in this limit, we approximate the integrand's denominator as $e^x - 1 \approx x + x^2/2$ and $1/(x+x^2/2) \approx (1/x)(1-x/2)$; the integrand becomes $x^3(1-x/2)/x = x^2(1-x/2)$ and the integral is $\int_0^{x_D} x^2(1-x/2)dx = x_D^3/3 - x_D^4/8 = (x_D^3/3)(1-3x_D/8) = \theta^3[1-3\theta/(8T)]/(3T^3)$ so that in the high temperature limit we get

$$U\bigg|_{\text{large } T} \approx 9Nk_BT\left(\frac{T}{\theta}\right)^3 \frac{\theta^3}{3T^3}\left(1-\frac{3\theta}{8T}\right) = 3Nk_BT\left(1-\frac{3\theta}{8T}\right). \tag{4.5.71a}$$

For very large T, this expression gives the expected classical result

$$U\bigg|_{\text{large } T} \to U_{large\,T} \sim 3Nk_BT = U_{large\,T}. \tag{4.5.71b}$$

for $N$ vibrators in three dimensions. The lattice heat capacity is obtained from Equations 4.5.50 and 4.5.67 to obtain

$$C_V = \frac{3V\hbar}{2\pi^2 v^3}\frac{\hbar}{k_BT^2}\int_0^{\omega_D}\frac{\omega^4 \exp\left(\frac{\hbar\omega}{\tau}\right)d\omega}{\left[\exp\left(\frac{\hbar\omega}{\tau}\right)-1\right]^2} = \frac{3V\hbar}{2\pi^2 v^3}\frac{1}{k_BT^2}\left(\frac{\tau}{\hbar}\right)^5\int_0^{x_D}\frac{e^x x^4 dx}{(e^x-1)^2} = 9Nk_B\left(\frac{T}{\theta}\right)^3\int_0^{x_D}\frac{e^x x^4 dx}{(e^x-1)^2}, \tag{4.5.72}$$

where we have again used the variable $x = \hbar\omega/\tau$ as well as the definition of Equation 4.5.68. For large $T$ we expect $x$ to be small so that $x^4e^x/(e^x-1)^2 \approx x^4(1+x)/(1+x-1)^2 = x^4(1+x)/x^2 \approx x^2$. Thus for the high $T$ limit we write

$$C_{V\,large\,T} \approx 9Nk_B\left(\frac{T}{\theta}\right)^3\int_0^{x_D} x^2dx = 9Nk_B\left(\frac{T}{\theta}\right)^3 x_D^3/3 = 3Nk_B\left(\frac{T}{\theta}\right)^3\left(\frac{\theta}{T}\right)^3 = 3Nk_B. \tag{4.5.73}$$

This is the classical result for $C_V$ and is, of course, the same that one gets by taking $d/dT$ of Equation 4.5.71b. This limit is known as the Dulong-Petit law. For a mole of a substance, $N$ is replaced with Avogadro's number $N_A = 6.02214 \times 10^{23}\,mol^{-1}$ then with $N_Ak_B = R = 8.31447\,J/mol \cdot K$, the ideal gas constant, the classical result for the molar heat capacity of a solid is $C_{Vm} = 3R = 24.94\,J/mol \cdot K$.

### 4.5.3 Debye $T^3$ Law

It is of interest to obtain analytic expressions for $U$ and $C_V$ for small $T$ ($<< \theta$). In the case of $U$, this is carried out by noting that in Equation 4.5.70, the upper limit is large for low $T$; that is, we make the replacement $x_D = \hbar\omega_D/\tau \to \infty$, and write

$$U_{low\,T} \approx 9Nk_BT\left(\frac{T}{\theta}\right)^3\int_0^{\infty} dx\frac{x^3}{[e^x-1]}. \tag{4.5.74a}$$

By consulting a mathematical handbook [19] we find that $\int_0^\infty dx \frac{x^{n-1}}{[e^x-1]} = n!\sum_s(1/s^n) = n!\pi^4/90$. Substituting these results into $U_{lowT}$, we get for the low temperature limit

$$U_{lowT} \approx 9Nk_BT\left(\frac{T}{\theta}\right)^3 \frac{6\pi^4}{90} = \frac{3\pi^4 Nk_BT^4}{5\theta^3}. \tag{4.5.74b}$$

The low temperature expression for the heat capacity of the solid can be obtained by taking $d/dT$ of $U_{lowT}$, to get

$$C_{V\,lowT} \approx \frac{12\pi^4 Nk_BT^3}{5\theta^3} \sim 233.78Nk_B\left(\frac{T}{\theta}\right)^3, \tag{4.5.75}$$

which is the so-called Debye $T^3$ law. The evaluation of the general form of $U$ from Equation 4.5.70 has to be done numerically and the results are shown in Figure 4.5.16(a) along with its low and high temperature behaviors from Equations 4.5.74b and 4.5.71a, respectively, assuming a Cu crystal. For very large temperatures, the classical limit of Equation 4.5.71b will be reached. Similarly, the numerical results for $C_V$ from Equation 4.5.72 as well as the low and high temperature behaviors, from Equations 4.5.75 and 4.5.73, are shown in Figure 4.5.16 (b), respectively. Notice how well the Debye $T^3$ law describes the low temperature behavior of $C_V$.

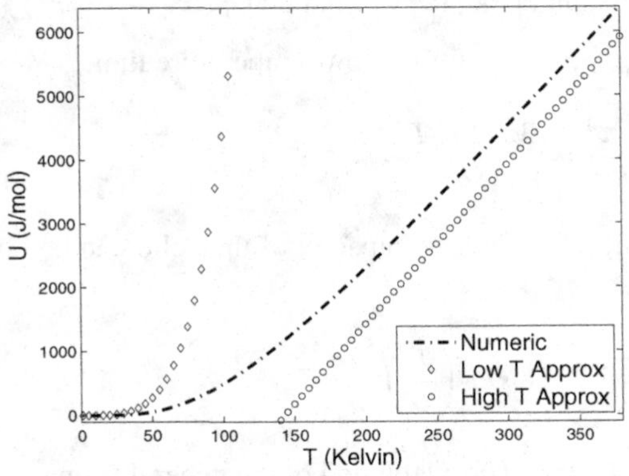


(a) Phonon Internal Energy

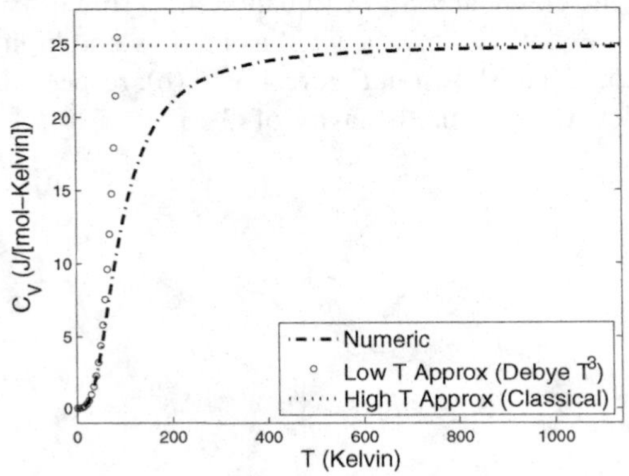


(b) Phonon Heat Capacity

Figure 4.5.16: (a) Shows the numerically calculated $U$ (dash-dot) using Equation 4.5.70, and its low (diamonds, Equation 4.5.74b) and high (open circles, Equation 4.5.71a) temperature behaviors. (b) shows the numerically calculated $C_V$ (dash-dot) using Equation 4.5.72, and its low (open circles, Equation 4.5.75) and high (dots, Equation 4.5.73) temperature behaviors. The low temperature $C_V$ follows the Debye $T^3$ law. The calculation has been carried out for the copper crystal.

The results shown in Figure 4.5.16 (a) for $U$ on a copper crystal has been carried out using the script phonon_U.m. The script finds the Debye frequency using Equation 4.5.64. Here the speed of sound used is obtained by $v = \sqrt{Y/\rho}$ with a Young's modulus of $Y = 76 \times 10^9\, N/m^2$; $N/V = \rho N_A/M_w = \rho/(uM_w)$ with Cu density $\rho = 8890 kg/m^3$; molecular weight $M_w = 63.546 gr/mol$; and atomic mass unit $u \approx 1.6605 \times 10^{-27}\, kg/(gr/mol)$. The respectively calculated values of sound speed, $N/V$, Debye frequency, and Debye temperature are $v = 2923.9\, m/s$, $N/V = 8.43 \times 10^{28}\, 1/m^3$ $\omega_D = 4.99 \times 10^{13}\, s^{-1}$, and $\theta = 381.6 Kelvin$. In the code below the numerical integration is done by the MATLAB function "quad" which uses the function definition "`Uint=@(x) x.^3./(exp(x)-1)`" for the integrand, but there are other ways to define functions also. The listing of the script follows.

```
%copyright by J. E Hasbun and T. Datta
%phonon_U.m
%Calculates the phonon internal energy for a Copper crystal
%based on the Debye Model.
```

```
clear;
h=6.62606896e-34;       %Planck'constant (J.s)
hbar=h/2./pi;           %hbar
kB=1.3806504e-23;       %Boltzmann Contant  (J/K)
u=1.660538782e-27;      %atomic mass unit
Na=(1/u)*1.e-3;         %Avogadro's number
Mw=63.546;              %Cu gr/mol. Note 1 gr/mol=1u
Y=0.76e11;              %Cu (cast) Young's mod. in Pa (M. Marder's text)
rho=8890;               %Cu density kg/m^3
vs=sqrt(Y/rho);         %speed of sound
nv=rho/u/Mw;            %N/V
omD=vs*(6*pi^2*nv)^(1/3);  %Debye frequency
thD=hbar*omD/kB;        %Debye temperature
fprintf('N/V=%6.3e 1/m^3, vs=%6.3f m/s\n',nv,vs)
fprintf('Debye freq.=%6.3e 1/m^3, Debye Temp=%6.3f m/s\n',omD,thD)
Uint=@(x) x.^3./(exp(x)-1);                 %U integrand
Tol=1.e-3;              %small number for lowest T
T=0:5:thD;              %T variable of U
T(1)=1000*Tol;          %use this value instead of T=0 low limit;
for i=1:length(T)
  xD=thD/T(i);
  %integrate; use small number for lower limit; upper limit is xD
  U(i)=9*Na*kB*T(i)*(T(i)/thD)^3*quad(Uint,Tol,xD);
  UlT(i)=3*pi^4*Na*kB*T(i)^4/thD^3;  %Low T approx
  UhT(i)=3*Na*kB*T(i)-9*Na*kB*thD/8; %High T approx
end
plot(T,U,'k-.','LineWidth',2)
hold on
plot(T,UlT,'kd','MarkerSize',4)
plot(T,UhT,'ko','MarkerSize',4)
legend('Numeric','Low T Approx','High T Approx',4);
axis([0 max(T) -100 max(U)])
xlabel('T (Kelvin)')
ylabel('U (J/mol)')
```

Notice that the above theoretically obtained Debye temperature is off when compared to the experimental value for Cu shown in Table 4.5.1 and gives an error of about 10%. The theoretical result can be improved by doing a better calculation of the sound velocity, see reference [21].

Table 4.5.1: Experimental Debye temperatures (in Kelvin) for the elements

| Element | Θ | Element | Θ | Element | Θ | Element | Θ |
|---|---|---|---|---|---|---|---|
| Am | 121 | Eu | 118 | Na | 157 | Sm | 169 |
| Ar | 92 | Fe | 477 | Nb | 276 | Sn | 199 |
| Ag | 227 | Ga | 325 | Nd | 163 | Sr | 147 |
| Al | 433 | Ge | 373 | Ne | 74.6 | Ta | 245 |
| As | 282 | Gd | 182 | Ni | 477 | Tb | 176 |
| Au | 162 | H | 122 | Np | 259 | Te | 152 |
| Ba | 111 | He | 34-108 | Os | 467 | Th | 160 |
| Be | 1481 | Hf | 252 | Pa | 185 | Ti | 420 |
| Bi | 120 | Hg | 72 | Pb | 105 | Tl | 78.5 |
| B | 1480 | Ho | 190 | Pd | 271 | Tm | 200 |
| C(graphite) | 412 | I | 109 | Pr | 152 | U | 248 |
| C(diamond) | 2250 | In | 112 | Pt | 237 | V | 399 |
| Ca | 229 | Ir | 420 | Pu | 206 | W | 383 |
| Cd | 210 | K | 91.1 | Rb | 56.5 | Xe | 64.0 |
| Ce | 179 | Kr | 71.9 | Re | 416 | Y | 248 |
| Co | 460 | La | 150 | Rh | 512 | Yb | 118 |
| Cr | 606 | Li | 344 | Ru | 555 | Zn | 329 |
| Cs | 40.5 | Lu | 183 | Sb | 220 | Zr | 290 |
| Cu | 347 | Mg | 403 | Sc | 346 | | |
| Dy | 183 | Mn | 409 | Se | 153 | | |
| Er | 188 | Mo | 423 | Si | 645 | | |
| Source:[20] | | | | | | | |

In the following example, we will use some material data obtained at low temperature and extract the associated Debye temperature of the material.

**Example 4.5.3.1**
It is possible to obtain low temperature $C_V$ data for solid argon from reference [3]. Some of the data has been digitized and copied here in the form of MATLAB lines that can be incorporated into a script. The lines are

```
%TData is the array of temperature values cubed
TData=[0.13,0.33,0.60,0.89,1.14,1.45,1.65,2.03,2.30,...
  2.56,2.92,3.41,3.57,3.99,4.12,4.59,5.04,5.26,5.39,...
  6.13,6.29,7.07,7.09,7.20,7.47,7.87];
%CvData is the array of specific heat values in mJ/mol/K
CvData=[0.49,1.00,1.58,2.31,2.96,3.76,4.19,5.21,5.87,...
  6.45,7.39,8.70,9.21,10.37,10.59,11.68,12.84,13.57,...
  13.86,15.82,15.96,17.85,18.21,18.36,19.09,20.10];
```

Since the low temperature behavior of $C_V$ is given by Equation 4.5.75, plotting the experimental data versus the temperature raised to the third power has a straight line shape. Applying linear regression to the data pairs $x_i$ and $y_i$, where $i$ runs through the $N$ data points. Here note that $x_i$ and $y_i$ are stand-ins for the experimental value pairs of $T^3$ and $C_V$, respectively. With this understanding, the best line fit, $y = mx + b$, has coefficients given by

$$m = \frac{N\sum x_i y_i - \sum x_i \sum y_i}{N\sum x_i^2 - (\sum x_i)^2}, \tag{4.5.76a}$$

for the slope, and

$$b = \frac{\sum y_i - m\sum x_i}{N}, \tag{4.5.76b}$$

for the intercept. However, since $C_V \to 0$ as $T \to 0$, we need to force the line to have an intercept that coincides with the origin and set $b$ in Equation 4.5.76b to zero. In which case we have that

$\sum y_i = m \sum x_i$. By substituting this expression into the 2nd term of the numerator of Equation 4.5.76a and solving for $m$, we find that the slope of the best line fit is given by

$$m = \frac{\sum x_i y_i}{\sum x_i^2}, \tag{4.5.77a}$$

with units of $J/K^4$. This is the slope of the line that is based on the experimental data. It enables us to write

$$C_{V\,low T} = mT^3, \tag{4.5.77b}$$

and which, when compared to Equation 4.5.75, allows us to extract the experimental value of a material's Debye temperature by setting $m = 233.78Nk_B/\theta^3$ so that for Avogadro's number of particles $\theta = (233.78R/m)^{1/3}$, where once again $R = N_A k_B = 8.31447 J/(mol \cdot K)$. Figure 4.5.17 shows the above data and a superimposed plot of the fit according to Equation 4.5.77 in the low temperature range of the data. The extracted Debye temperature is $\theta = 91.3K$. Comparing this with the value of $92K$ for solid argon (Ar) from Table 4.5.1, we get an absolute value of the percent error of about 0.76% and is a testament of how well the Debye $T^3$ law describes the low temperature behavior of the experimental specific heat data.

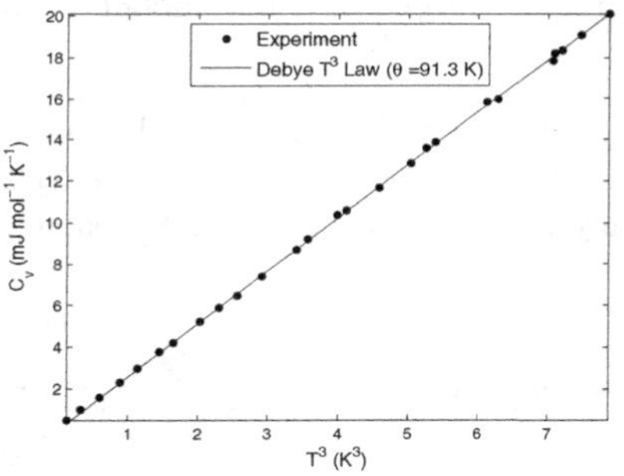


Figure 4.5.17: Shown is a plot of the solid Ar data (dots) from Example 4.5.3.1 with $C_V$ in units in $mJ/mol/K$. The line is the best fit whose slope $m = 2.5568 mJ/K^4$ and with $R = 8.3145 \times 10^3 mJ/(mol \cdot K)$, the extracted Debye temperature is $\theta = (233.78 \cdot 8.3145 \times 10^3/2.5568)^{1/3} \approx 91.3K$.

### 4.5.4 The Einstein Model

In 1907 Albert Einstein had carried out a simpler derivation of the above internal energy ($U$) and heat capacity ($C_V$). He did it by assuming a simple form of the density of states. Regarding Equation 4.5.60, he reasoned that if all oscillator modes are assumed to have the same frequency, $\omega_E$ (the Einstein characteristic frequency), one can thus write the density of states as a delta function, or

$$D_p(\omega_p) = N\delta_p(\omega_p - \omega_E), \tag{4.5.78}$$

for mode $p$ and $N$ atoms in one dimension. Here $\omega_E$ is supposedly the single frequency value at which the delta function is most representative of the density of states (see Figure 4.6.20). This assumption is in stark contrast to Equation 4.5.62. Following Equation 4.5.61, we thus write the Einstein $U$ for the one mode as

$$U_E = N\int \delta_p(\omega_p - \omega_E)\frac{\hbar\omega_p}{e^{\hbar\omega_p/\tau} - 1}d\omega_p = N\frac{\hbar\omega_E}{e^{\hbar\omega_E/\tau} - 1}. \tag{4.5.79}$$

The heat capacity that follows from this expression is

$$C_{VE} = (\partial U_E/\partial T)_V = Nk_B \left(\frac{\hbar\omega_E}{\tau}\right)^2 \frac{\hbar\omega_E}{\left(e^{\hbar\omega_E/\tau} - 1\right)^2}. \tag{4.5.80}$$

In three dimensions, $N$ is to be replaced by $3N$ since there are three modes per oscillator in that case. If we let $x = \hbar\omega_E/\tau$, for high $T$, $\exp(x) \approx 1 + x$ so that in this limit $\exp(x)/(\exp(x) - 1)^2 \approx (1+x)/x^2 \sim 1/x^2$ so that $C_{VE} \sim Nk_B[\hbar\omega_E/\tau]^2[\tau/(\hbar\omega_E)]^2 = Nk_B$ in one dimension and $3Nk_B$ in three dimensions, which is the Dulong and Petit value. At low temperature $x$ is large so that $\exp(x)/(\exp(x) - 1)^2 \approx \exp(x)/(\exp(x))^2 = \exp(-x)$, to obtain the expression for the Einstein heat capacity at low temperature as

$$C_{VE\,lowT} = Nk_B \left(\frac{\hbar\omega_E}{\tau}\right)^2 e^{-\hbar\omega_E/\tau} = Nk_B \left(\frac{\theta_E}{T}\right)^2 e^{-\theta_E/T}, \tag{4.5.81}$$

where $\theta_E \equiv \hbar\omega_E/k_B$, is the Einstein characteristic temperature. Again, in three dimensions, we replace $N$ with $3N$. As is evident, this expression goes to zero exponentially and does not follow the Debye $T^3$ behavior that so well describes the experimental data. Both the Einstein and the Debye models produce the expected behavior at height $T$. The following example illustrates the differences between the Debye model and the much simpler Einstein model for the heat capacity.

**Example 4.5.4.1**

We now use some $C_V$ data for diamond from Touloukian and Buyco (1970) [22] as digitized from reference [20]. The digitized data is listed below in the form of MATLAB lines that can be incorporated into a script. (The temperature is in Kelvin and the heat capacity was converted to $J/(mol \cdot K)$. The lines are

```
%Temperature in K
T_exp=[15.94,16.52,17.33,17.75,18.84,19.76,20.49,22.54,23.93,...
  25.71,27.29,29.66,31.88,33.44,36.35,39.98,41.94,43.48,46.70,...
  48.41,51.38,55.20,59.30,63.71,68.45,72.67,77.16,80.94,85.93,...
  89.10,91.27,98.04,100.43,106.63,107.94,113.24,123.10,126.13,...
  133.88,135.51,138.81,147.35,152.75,162.15,178.42,182.80,201.05,...
  213.46,221.31,232.11,252.30,264.70,284.38,335.99,369.58,401.74,...
  468.92,497.76,547.43,623.94,686.32,782.17,829.99,912.82,1003.74];
%Constant volume specific heat in J/mol/K
Cv_exp=[0.001696,0.001872,0.002136,0.002436,0.002779,0.003169,...
  0.003615,0.004551,0.005365,0.006324,0.007961,0.008503,0.01143,...
  0.01262,0.01439,0.01696,0.01935,0.02207,0.02518,0.02872,0.03067,...
  0.03736,0.04404,0.05544,0.06754,0.07961,0.1002,0.1181,0.1393,...
  0.1812,0.2067,0.2281,0.2689,0.3275,0.3736,0.4551,0.5191,0.6535,...
  0.7454,0.7961,0.9081,1.036,1.221,1.393,1.812,2.207,2.436,2.968,...
  3.615,3.861,4.404,5.365,6.535,8.227,9.384,10.7,12.21,13.93,14.87,...
  16.42,18.72,20,20,21.36,21.36];
```

In the upper graph of Figure 4.5.18, the calculation of the Debye heat capacity from Equation 4.5.72 (dashed) is compared with the above diamond data (dots). The solid line is the Einstein heat capacity from Equation 4.5.80.

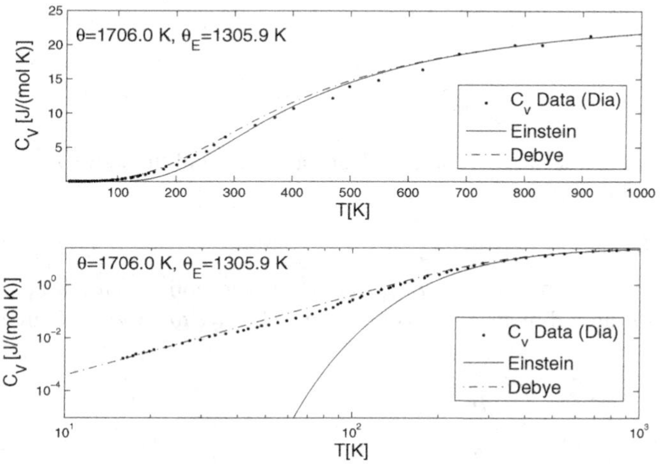


Figure 4.5.18: The Debye formula for the heat capacity is shown as a dashed curve (Equation 4.5.72). The Einstein formula is the solid curve (Equation 4.5.80), and the diamond data of Touloukian and Buyco (1970), as digitized from reference [20], are the solid dots. The lower graph is a log-log plot (based 10) version of the upper graph and illustrates the deviation of the Einstein model in the low temperature regime. The Debye ($\theta$) and Einstein ($\theta_E$) temperature values used are $1706.0\,K$ and $1305.943\,K$, respectively, with corresponding frequencies $\omega_D = 2.234e+014\,s^{-1}$ and $\omega_E = 1.710e+014\,s^{-1}$. Here $\omega_E$ has been taken to be proportional to $\omega_D$ such that it is slightly adjusted to improve the comparison with the data.

Notice that the Einstein heat capacity does very well through most of the entire range of temperature. The largest discrepancy occurs at the very low temperature regime. This is best seen in the lower part of the figure where the same plots are presented on a log-log scale (based 10) using MATLAB's loglogplot command. Here the linear behavior of the data is most evident and the resulting graph is representative of the definitive Debye $T^3$ law.

## 4.6 General Density of States

The success of the Debye model over the Einstein model is due to the use of a better model of the density of states (DOS). It therefore becomes of interest to have a general expression for the number of states or modes available per frequency $\omega$, given a frequency dispersion $\omega_K$; i.e., $D(\omega)$. The number of allowed $\vec{K}$ values in a phonon frequency range between $\omega$ and $\omega + d\omega$ is written as

$$D(\omega)d\omega = \left(\frac{L}{2\pi}\right)^3 \int_{\text{shell}} d^3K, \tag{4.6.82}$$

which is an integral over the volume of a "shell" in $K$ space bounded by two surfaces on which the phonon frequency is constant as depicted in Figure 4.6.19.

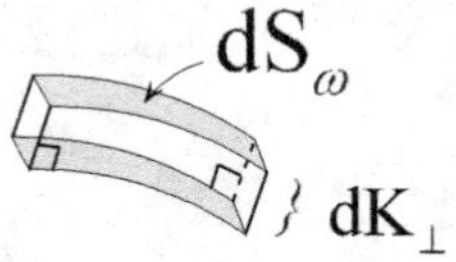


Figure 4.6.19: Element of the shell volume with surface $dS_\omega$ and thickness $dK_\perp$ discussed in connection with Equation 4.6.82.

This volume element is written as $d^3K = dS_\omega dK_\perp$. However, notice that in one dimension $dK_\perp = d\omega/(d\omega/dK_\perp)$ and in three dimensions this becomes $dK_\perp = d\omega/|\vec{\nabla}_K\omega|$, so that Equation 4.6.82 becomes

$$D(\omega)d\omega = \left(\frac{L}{2\pi}\right)^3 \int_{\text{shell}} dS_\omega \frac{d\omega}{|\vec{\nabla}_K\omega|}, \qquad \text{or} \qquad D(\omega) = \frac{V}{(2\pi)^3}\int_{\text{shell}} \frac{dS_\omega}{v_G}, \tag{4.6.83}$$

where we have used the definition for the group velocity $\vec{v}_G$ from Equation 4.2.30 and made the substitution $V = L^3$ for the volume. In the following example, we show how this equation for the DOS yields the Debye model's DOS discussed before.

**Example 4.6.0.1**

To obtain the Debye DOS from Equation 4.6.83, we assume the free particle energy dispersion $E = \hbar^2k^2/(2m) = \hbar\omega$ or $\omega = \hbar k^2/(2m)$ so that $d\omega = (\hbar k/m)dk$. We also have that $v_G = d\omega/dk = \hbar k/m$ and for the surface element we write $dS_\omega = k^2 \sin\theta d\theta d\phi$ to obtain from Equation 4.6.83

$$D(\omega)d\omega = \tfrac{V}{(2\pi)^3}\int_{\theta,\phi} k^2 \sin\theta d\theta d\phi \tfrac{\hbar(k/m)dk}{\hbar(k/m)} = 4\pi\tfrac{V}{(2\pi)^3}k^2dk,$$

and leads to $D(\omega) = \frac{V}{2\pi^2}k^2\frac{dk}{d\omega}$. Once again, as we did in the Debye model of Section 4.5.2, we take the sound velocity in the medium as $\omega = vk$ or $k = \omega/v$ and $dk/d\omega = 1/v$, which when substituted into the above expression we get $D(\omega) = \frac{V}{2\pi^2}\frac{\omega^2}{v^3}$, which is the Debye model result of Equation 4.5.62. This DOS has a quadratic shape and $\omega$ is restricted to the range of $0 < \omega < \omega_D$ as shown in Figure 4.6.20 (solid line). In contrast, the Einstein model's DOS corresponds to a single peak at the frequency, $\omega = \omega_E$ (dashed line) as in Equation 4.5.78. A DOS similar to the short-dashed curve is characteristic of a real crystal's DOS. The discontinuities it contains are due to singular points associated with a crystal's band structure.

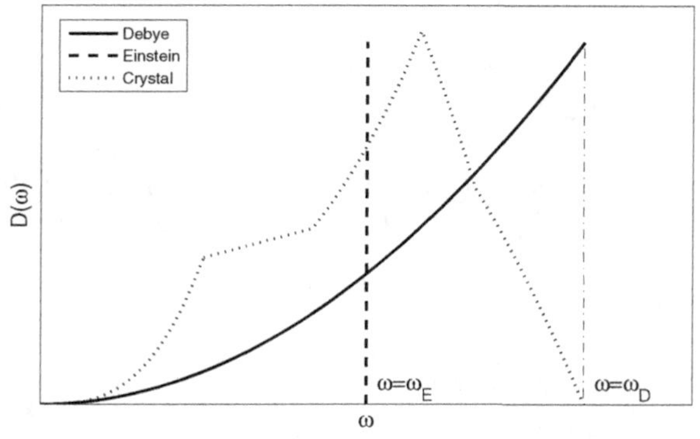


Figure 4.6.20: Density of states for the Debye model (solid), the Einstein model (dashed), and a possible crystal (short dashed).

## 4.7 Thermal Expansion

The general tendency of materials to expand when heated is referred to as thermal expansion. A deep understanding of this property of materials can be gained by means of the crystal potential. In an earlier chapter, the Lennard-Jones (LJ) potential was used in connection with the cohesive energy of noble gas crystals. Here, further below, we will apply it to obtain a grasp for thermal expansion. Although the potential is not as realistic as we wish, it affords us the opportunity to experiment with it in trying to predict thermal expansion. Recall that, for small displacements from equilibrium, it is possible to expand a potential in a power series. Here we write an approximation to the potential that includes up to fourth order in $x$; that is,

$$U(x) = cx^2 - gx^3 + fx^4, \tag{4.7.84}$$

where $x$ is measured from the equilibrium bond length. Given a potential, a crystal's average atomic displacement versus temperature can be estimated classically through a Boltzmann distribution to write for the average atomic distance

$$< x >= \frac{\int_{-\infty}^{\infty} x e^{-\beta U(x)} dx}{\int_{-\infty}^{\infty} e^{-\beta U(x)} dx}, \tag{4.7.85}$$

where $\beta \equiv 1/(k_B T)$. If the last two terms of Equation 4.7.84 are ignored and the resulting harmonic potential is used into Equation 4.7.85, the result is $< x >= 0$ because $\int_{-\infty}^{\infty} x e^{-\beta cx^2} dx = 0$ since the integrand is odd. We see, therefore, that a harmonic potential is incapable of describing thermal expansion. The symmetry of the potential about the equilibrium position leads to a zero average position at any temperature. Thus, in order to explain thermal expansion, terms beyond the harmonic approximation are needed. The extra terms in Equation 4.7.84, beyond the quadratic, are anharmonic terms and produce the appropriate potential antisymmetry needed to obtain a nonzero average displacement versus temperature. To see this, for small displacements, we write the integrand in the numerator of Equation 4.7.85 as $e^{-\beta U(x)} = e^{-\beta cx^2} e^{-\beta(-gx^3+fx^4)} \approx e^{-\beta cx^2}(1 - \beta(-gx^3 + fx^4))$. For the denominator, we only keep the harmonic term and let the integrand take the form $e^{-\beta U(x)} \approx e^{-\beta cx^2}$.

With these approximations Equation 4.7.85 becomes

$$<x>=\frac{\int_{-\infty}^{\infty} e^{-\beta cx^2}(x+\beta gx^4-\beta fx^5)dx}{\int_{-\infty}^{\infty} e^{-\beta cx^2}dx}. \tag{4.7.86}$$

The integrals can be carried out using $\int_0^\infty x^m e^{-ax^2}dx = \Gamma[(m+1)/2]/(2a^{(m+1)/2})$, where $\Gamma(m+1/2) = 1\cdot 3\cdot 5\cdot \ldots \cdot (2m-1)\sqrt{\pi}/2^m$ given that $\Gamma(1/2) = \sqrt{\pi}$. We have

$$\begin{array}{ll} \int_{-\infty}^{\infty} e^{-\beta cx^2}dx = \frac{\sqrt{\pi}}{\sqrt{\beta c}}, & \int_{-\infty}^{\infty} xe^{-\beta cx^2}dx = 0 \\ \beta g\int_{-\infty}^{\infty} x^4 e^{-\beta cx^2}dx = \frac{3g\sqrt{\pi}}{4c^{5/2}\beta^{3/2}}, & f\beta\int_{-\infty}^{\infty} x^5 e^{-\beta cx^2}dx = 0 \end{array}. \tag{4.7.87}$$

Substituting these results into Equation 4.7.86, we find

$$<x>=\frac{3gk_BT}{4c^2}, \tag{4.7.88}$$

indicating a linear dependence of the lattice displacement with temperature. Albeit different, this is reminiscent of the simple linear relation $\Delta L = \alpha L_0 \Delta T$ commonly used to describe changes in length, $\Delta L$, with changes in temperature, $\Delta T$, where $\alpha$ is a coefficient of linear expansion, $L_0$ is the initial length. In our present approach, we can identify the coefficients in the fourth order expansion of Equation 4.7.84 if we make use of the Lennard-Jones (L-J) potential, with parameters $\varepsilon$ and $\sigma$, for an FCC crystal (see Chapter 3), which we write in the form

$$U_{LJ}(r) \equiv \frac{U_{tot}(r)}{N} = 2\varepsilon\left[(12.13)\left(\frac{\sigma}{r}\right)^{12} - (14.45)\left(\frac{\sigma}{r}\right)^{6}\right], \tag{4.7.89}$$

for the total energy per atom. Expanding this to fourth order about $R_0$ we have

$$\begin{aligned} U_{LJ}(r) \approx{}& U_{LJ}(R_0) + U'_{LJ}(R_0)(r-R_0) + \frac{1}{2!}U''_{LJ}(R_0)(r-R_0)^2 + \frac{1}{3!}U'''_{LJ}(R_0)(r-R_0)^3 \\ &+ \frac{1}{4!}U^{iv}_{LJ}(R_0)(r-R_0)^4, \end{aligned} \tag{4.7.90a}$$

where, as in Chapter 3, since $U'_{LJ}(R_0) = 0$, the equilibrium position is $R_0 = \alpha\sigma$ with $\alpha = (2(12.13)/14.45)^{1/6} \approx 1.09$. By the same token, recall that when $R_0$ is substituted into Equation 4.7.89, $U_{LJ}(R_0) = -\frac{14.45^2}{2(12.13)}\varepsilon \approx -8.61\varepsilon$. The rest of the derivatives are

$$\begin{aligned} U''_{LJ}(R_0) &= 2\varepsilon\left[(12.13)(12)(13)\frac{\sigma^{12}}{R_0^{14}} - (14.25)(6)(7)\frac{\sigma^6}{R_0^8}\right] \approx \frac{2\varepsilon(260.699)}{\sigma^2}, \\ U'''_{LJ}(R_0) &= 2\varepsilon\left[-(12.13)(12)(13)(14)\frac{\sigma^{12}}{R_0^{15}} + (14.25)(6)(7)(8)\frac{\sigma^6}{R_0^9}\right] \approx -\frac{2\varepsilon(5021.752)}{\sigma^3}, \\ U^{iv}_{LJ}(R_0) &= 2\varepsilon\left[(12.13)(12)(13)(14)(15)\frac{\sigma^{12}}{R_0^{16}} - (14.25)(6)(7)(8)(9)\frac{\sigma^6}{R_0^{10}}\right] \approx \frac{2\varepsilon(81,377.812)}{\sigma^4}. \end{aligned} \tag{4.7.90b}$$

If we let $x = r - R_0$ and compare the effective potential; that is, $U_{eff}(r) \equiv U_{LJ}(r) - U_{LJ}(R_0)$ from Equation 4.7.90a to that of Equation 4.7.84, we can identify the coefficients as follows

$$c = \frac{U''_{LJ}(R_0)}{2!} \approx \frac{2\varepsilon(260.699)}{2!\sigma^2},\; g = \frac{-U'''_{LJ}(R_0)}{3!} \approx \frac{2\varepsilon(5021.752)}{3!\sigma^3},\; f = \frac{U^{iv}_{LJ}(R_0)}{4!} \approx \frac{2\varepsilon(81,377.812)}{4!\sigma^4}. \tag{4.7.91}$$

In Figure 4.7.21, a comparison is made between the Lennard-Jones (LJ) potential of Equation 4.7.89, its fourth order expansion, according to the above discussion - Equations 4.7.84 and 4.7.91, where we have included the potential minimum $U_{LJ}(R_0) \approx -8.61\varepsilon$ - and the harmonic approximation for energy in units of $\varepsilon$ and distance in units of $\sigma$. As mentioned above in connection with Equations 4.7.90, the minimum of the potential occurs at $R_0 = \alpha\sigma$, so that in units of $\sigma$ it happens at $r(\sigma) = \alpha \approx 1.09$.

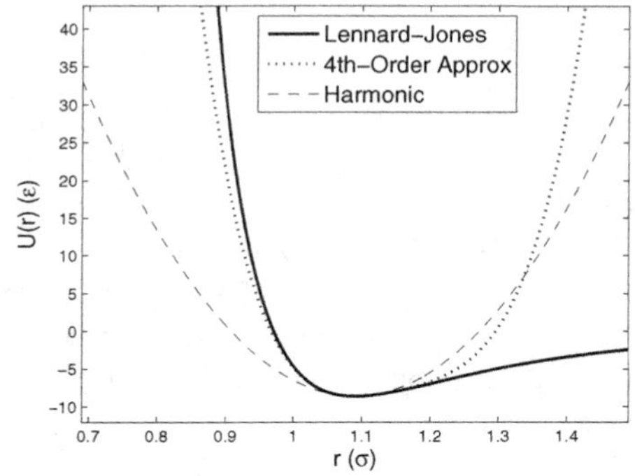


Figure 4.7.21: The Lennard-Jones (LJ) potential (Equation 4.7.89) is shown compared with its fourth order expansion (Equations 4.7.84 and 4.7.91) as well as the harmonic approximation. Here energy is in units of $\varepsilon$ and distance in units of $\sigma$.

In particular, notice the asymmetry caused by the anharmonic terms as one compares with the quadratic form of the potential. The listing of the script, thermal_potential.m, that reproduces the figure follows. Notice that we defined three "functions" within the MATLAB script to calculate the potentials.

```
%copyright by J. E Hasbun and T. Datta
%thermal_potential.m
%The script plots the Lennard-Jones (LJ) potential and it's expanded
%approximation to 4rth order in (r-R). We use energy/atom in
%units of epsilon, and distance in units of sigma.
%It also calculates the average

function thermal_potential
clear
global alpha U0
alpha=(2*12.13/14.45)^(1/6);   %LJ bond length in Sigma units
U0=14.45^2/2/12.13;            %LJ min energy magnitude in epsilon units
rl=0.1; rs=0.001; ru=15;
r=rl:rs:ru;
plot(r,ULJ(r),'k','LineWidth',2)
hold on
plot(r,ULJexp(r,0),'k:','LineWidth',2)
plot(r,ULJexp(r,1),'k--')
h=legend('Lennard-Jones','4th-Order Approx','Harmonic','Location','North');
set(h,'FontSize',14)
axis ([alpha-0.4 alpha+0.4 -U0*(1+0.4) 5*U0])
```

```
xlabel('r (\sigma)','FontSize',14)
ylabel('U(r) (\epsilon)','FontSize',14)

function Ut=ULJ(x)
%Lennard-Jones potential function.
%Energy in units of epsilon, distance in units of sigma.
Ut=2*(12.13./x.^12-14.45./x.^6);

function Ua=ULJexp(x,k)
global alpha U0
%Fourth order Lennard-Jones potential expanded function.
%Energy in units of epsilon, distance in units of sigma.
Cpp=12.13*12*13/alpha^14-14.45*6*7/alpha^8;
Cppp=-12.13*12*13*14/alpha^15+14.45*6*7*8/alpha^9;
Cpppp=12.13*12*13*14*15/alpha^16-14.45*6*7*8*9/alpha^10;
if (k==0)                          %up to 4rth order in x-alpha
  Ua=-U0+2*Cpp*(x-alpha).^2/2+2*Cppp*(x-alpha).^3/6+...
    2*Cpppp*(x-alpha).^4/24;
else
  Ua=-U0+2*Cpp*(x-alpha).^2/2; %the quadratic case only
end
```

Essentially, as the temperature increases and the atoms vibrate away from equilibrium, the average distance increases with temperature. The fourth order expansion of Equation 4.7.88 predicts a linear behavior. Since the constants for this approximation are given by Equations 4.7.91, we can actually plot it and compare with real data. To do so, we will plot the lattice constant for argon. The relation is as before, if we have the average nearest neighbor distance, $<r>$, the lattice constant for the FCC structure, from Chapter 1, is $<a>=\sqrt{2}<r>$. In the case of the simple analytic fourth order approximation result, Equation 4.7.88, since $<x>$ is measured from the equilibrium value at $T=0$, we have $<r>=R_0+<x>$. We can also perform the full one-dimensional calculation numerically for $<r>$ using the actual potential through Equation 4.7.85 rather than the approximate expression, Equation 4.7.86, with $x$ replaced by $r$; that is,

$$<r>=\frac{\int_{-\infty}^{\infty} re^{-\beta U(r)}dr}{\int_{-\infty}^{\infty} e^{-\beta U(r)}dr}, \tag{4.7.92}$$

where, for $U(r)$, we can use the fourth order expansion, Equations 4.7.90, as well as the full LJ potential of Equation 4.7.89. We will be comparing the three results against experimental data [23] for solid Ar. The data is also used to extract the value of $\sigma$ needed in the LJ potential; that is, at $T=0$ we have $a_{T=0}/\sqrt{2}=R_0=\alpha\sigma$ which yields $\sigma\approx 3.438$ Å. The Debye temperature of $T_D=92\,K$ for Ar is used to get $\varepsilon$. We do this by setting $\varepsilon=k_BT_D$, so that $\varepsilon\approx 0.00793\,eV$. This is essentially the energy units in which the potentials are calculated. Figure 4.7.22 shows the experimental data (circles) for Ar's lattice constant versus temperature, $<r>$ from Equation 4.7.92 with $U(r)$ from Equation 4.7.89 (solid) as well as from Equations 4.7.90 (long dashes), and the result from $<r>=R_0+<x>$ (short dashes), where $<x>$ is from the analytic Equations 4.7.88 and 4.7.91.

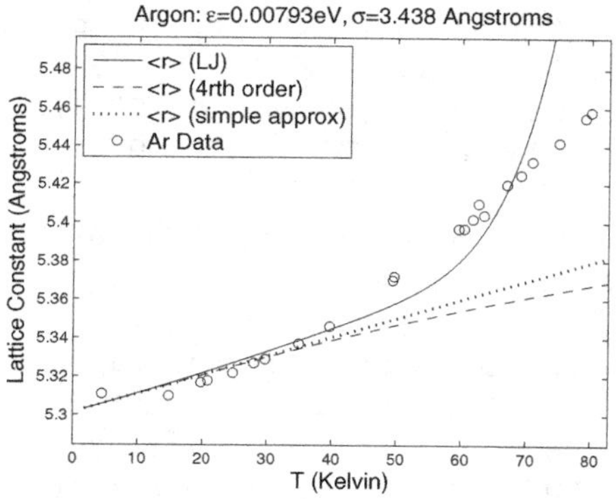


Figure 4.7.22: The lattice constant versus temperature for solid Ar. The open circles are the experimental data [23]. The calculated lattice constants are from $\sqrt{2} < r >$, where $< r >$ is calculated using Equation 4.7.92 through the LJ potential of Equation 4.7.89 (solid), as well as the fourth order approximation Equations 4.7.90 (long dashes). The short-dashed curve is due to the simpler approximation for which $< r >= R_0 + < x >$ where $< x >$ is from Equation 4.7.88.

Except for very low temperatures, the experimental data shows a non-linear behavior with temperature. The approximations seem to break down at around $40K$. Although the full LJ potential tends to explain the curving effect with temperature, it also needs improvement since the data does not bow as rapidly.

The script thermal_latt_const.m, listed below, contains the main ingredients needed to obtain the results shown in Figure 4.7.22, except for the incorporation of the LJ potential, which is left as an exercise. We note that the actual calculation is done in dimensionless units. So that energy is in units of $\varepsilon$, and length is in units of $\sigma$, but in the end distance is converted to Å by multiplying by $\sigma$. The extrapolation to obtain $\sigma$, as mentioned above, is carried out in the script by first using MATLAB's 'polyfit' function, which finds the polynomial of order 1 coefficient, based on eight data pairs near $T = 0$. Then MATLAB's function 'polyval' is used to extrapolate the value of the lattice constant at $T = 0$ from which $\sigma$ is obtained. Also, note that the quantity $< x >$ from Equation 4.7.88 can be expressed in units of $\sigma$ where we write $k_B T = k_B T_D(T/T_D) = (T/T_D)\varepsilon$. This factor of $\varepsilon$ cancels similar factors in the constants of Equations 4.7.91. Finally, notice that in this script, the integration is numerically carried out by MATLAB's simple "trapz" function. Feel free to try other methods as well as to vary the range of integration to improve the comparison.

```
%copyright by J. E Hasbun and T. Datta
%thermal_latt_const.m
%This script calculates the average lattice constant for Argon
%and compares with the experimental values.

function thermal_latt_const
clear
global alpha U0         %variables to be recognized by functions
e=1.602176487e-19;      %electronic charge
kB=1.3806504e-23;       %Boltzmann Contant  (J/K)
%Data for solid Argon digitized from
%"Measurements of X-Ray Lattice Constant,
%Thermal Expansivity, and Isothermal Compressibility
%of Argon Crystals"  O. G. Peterson, T D. N. Batchelder,
%& R. O. Simmons, Phys. Rev. vol. 150, No. 2 (1966)
```

```
%T_e is in Kelvin, a_e is Argon's lattice constant converted to angs.
T_e=[4.516,14.837,19.802,20.881,24.762,27.999,29.727,...
  34.910,39.665,49.405,49.627,59.586,60.446,61.750,...
  62.631,63.476,66.961,69.123,70.862,74.976,79.097,...
  79.966,83.006];
a_e=[5.311,5.310,5.317,5.318,5.322,5.327,5.329,5.337,...
  5.346,5.370,5.372,5.397,5.397,5.402,5.410,5.404,...
  5.420,5.425,5.432,5.442,5.455,5.458,5.469];
alpha=(2*12.13/14.45)^(1/6); %LJ bond length in Sigma units
U0=14.45^2/2/12.13;          %LJ min energy magnitude in epsilon units
%Let the extrapolated value at T=0 be the a0 for Argon, then
%R0=a0/sqrt(2) and since R0/sigma=alpha, we find that sigma=R0/alpha
p=polyfit(T_e(1:8),a_e(1:8),1); % fit low T data with a quadratic
a0=polyval(p,0);  %extrapolate to get the T=0 lattice constant;
R0=a0/sqrt(2);    %T=0 bond length for Argon (fcc nearest neighbor)
sigma=R0/alpha;   %our sigma - Lennard-Jones distance parameter
TD=92;            %argon Debye temperature
eps=kB*TD/e;      %energy unit (eV) - Lennard-Jones energy parameter
fprintf('[a0,R0]=[%6.3f, %6.3f]Angs, sigma=%6.3f Angs\n',a0,R0,sigma)
fprintf('Debye-T=%6.2f K, epsilon(eV)=%6.3g eV\n',TD,eps)
%Integrals are evaluated for energy units of epsilon, distance in
%units of sigma and temperature in units of argon's Debye temperature.
%After integration the conversion is done for temperature and distance
rl=0.1; rs=0.001; ru=15; %integration range in units of sigma
r=rl:rs:ru;
T=0.02:0.01:0.9;
for i=1:length(T)
  %integral based on the fourth order expansion
  yLJexp1=rs*trapz(r,r.*Boltz(ULJexp(r),T(i)));
  yLJexp2=rs*trapz(r,Boltz(ULJexp(r),T(i)));
  raveLJexp(i)=yLJexp1/yLJexp2;      %4rth order approx average bond length
end
%Simple analytic approx
Cpp=12.13*12*13/alpha^14-14.45*6*7/alpha^8;
Cppp=-12.13*12*13*14/alpha^15+14.45*6*7*8/alpha^9;
const=abs(Cppp/Cpp^2)/4;
%for the simple approx add alpha: the equilibrium point in sigma units
r_simple=const*T+alpha;
%For the FCC: lattice constant=sqrt(nearest neighbor distance)
raveLJexp=raveLJexp*sqrt(2);
r_simple=r_simple*sqrt(2);
%Convert temperature to Kelvin and distance to angstroms
T=T*TD;
raveLJexp=raveLJexp*sigma;
r_simple=r_simple*sigma;
hold on
plot(T,raveLJexp,'k--') %Full fourth order expansion case
plot(T,r_simple,'k:')   %Simple analytic result
plot(T_e,a_e,'ko')      %Ar data
legend('<r> (4rth order)','<r> (simple approx)','Ar Data',2)
axis ([0 T(end) a_e(1)*(1-0.005) a_e(end)*(1+0.005)])
```

```
xlabel('T (Kelvin)','FontSize',14)
ylabel('Lattice Constant (Angstroms)','FontSize',14)
str=cat(2,'Argon: \epsilon=',num2str(eps,'%6.3g'),...
  'eV, \sigma=',num2str(sigma,'%6.3f'),' Angstroms');
title(str,'FontSize',14);


function y=Boltz(x,T)
%The Boltzmann factor
y=exp(-x/T);

function Ua=ULJexp(x)
global alpha U0
%Fourth order Lennard-Jones potential expanded function.
%Energy in units of epsilon, distance in units of sigma.
Cpp=12.13*12*13/alpha^14-14.45*6*7/alpha^8;
Cppp=-12.13*12*13*14/alpha^15+14.45*6*7*8/alpha^9;
Cpppp=12.13*12*13*14*15/alpha^16-14.45*6*7*8*9/alpha^10;
%up to 4rth order in x-alpha
Ua=-U0+2*Cpp*(x-alpha).^2/2+2*Cppp*(x-alpha).^3/6+...
  2*Cpppp*(x-alpha).^4/24;
```

## 4.8 Thermal Conductivity

Thermal conductivity refers to the ability of a substance to permit thermal energy flow. Thermal energy flow has contributions from phonons and electrons. In this section, we discuss the flow of thermal energy due to phonons. In steady state, the flow of heat within a substance or thermal energy flux per unit volume is proportional to the temperature gradient within the substance. It can be written as

$$j_{ph} = -K\frac{dT}{dx}, \tag{4.8.93}$$

where $K$ is the thermal conductivity coefficient in Watts/(meter·K). The minus sign in this expression indicates that flow of heat happens in the direction of decreasing temperature - from hot to cold. It is possible to obtain an approximation for the thermal conductivity coefficient if we consider the temperature difference, $\Delta T$, between two points in a crystal, with cross-sectional area $dA$, at different temperatures, so that $\Delta T = \frac{dT}{dx}\Delta x$. We also let $\Delta x$ be the average distance of travel between collisions; i.e., the mean free path a particle travels with a speed of sound in a time $\tau$ between collisions. Here $\tau$ is an average time, and we write $\Delta x \approx v_x\tau$, where $v_x$ is the speed in the $x$-direction. Thus we can write

$$\Delta T = \frac{dT}{dx}v_x\tau. \tag{4.8.94}$$

The net flux of particles through $\Delta x$ is $nv_x$, where $n$ is the number of particles per volume passing area $dA$ per second. We thus have that the unaveraged energy flux is given by

$$\left(\frac{\text{number of particles}}{\text{area sec}}\right)(\text{energy per particle}) = nv_x\Delta Q = j \equiv \text{energy flux}.$$

However, $\Delta Q = c\Delta T$, with $c$ the heat capacity, so that with Equation 4.8.94

$$j = -nv_x c\Delta T = -nv_x^2 c\tau \tfrac{dT}{dx},$$

where we have put in a minus sign to account for the direction of heat flow; i.e., from high to low $T$. Defining the average energy flux $j_{ph} \equiv < j > = -n < v_x^2 > c\tau dT/dx$ and using the relation that $v_x^2 + v_y^2 + v_z^2 = v^2$, with $v_x^2 \approx v_y^2 \approx v_z^2$, then $< v_x^2 > \approx < v^2 > /3$, and we get the result that

$$j_{ph} = -\frac{nc\tau < v^2 >}{3}\frac{dT}{dx}. \tag{4.8.95}$$

If the phonon velocity $v$ is approximately constant, then $< v^2 > \approx v^2$. We also let $C = nc$ be the heat capacity per volume in units of $J/(m^3K)$ and let $\ell = v\tau$ be the mean free path. With these definitions Equation 4.8.95 becomes

$$j_{ph} = -\frac{1}{3}Cv\ell\frac{dT}{dx}. \tag{4.8.96}$$

Finally, comparing this expression with that of Equation 4.8.93, we see that the thermal conductivity coefficient is given by

$$K = \frac{1}{3}Cv\ell, \tag{4.8.97}$$

in units of $J/(m \cdot K \cdot sec)$. Peter Debye was the first person to use this formula and obtain crystal values of $K$. For example, NaCl has $C = (1.00, 1.88) \times 10^6 J/(m^3 \cdot K)$, $K = (27, 7)W/(m \cdot K)$, and $\ell = (100, 23)$ Å at temperatures of $(83, 273)$ Kelvin, respectively.

## 4.9 Chapter 4 Exercises

4.9.1. (a) Substitute the quantum harmonic oscillator wavefunction of Equation 4.1.2b for $n = 0$ into Equation 4.1.2a and show that the resulting energy state agrees with the expected value.
(b) Repeat for the $n = 1$ state.
(c) Using the fact that the wave function is normalized $\int |\psi(y)|^2 dy = 1 = \int |\psi(x)|^2 dx$ write the wavefunction of Equation 4.1.2b in terms of $x$, where $x$ is in standard units of distance. Hint: recall that $x$ and $y$ are related by a certain factor; in addition, consult a quantum mechanics text to be sure the obtained wavefunction has the correct form.

4.9.2. Reproduce Figure 4.1.1 by modifying code qholev.m.

4.9.3. (a) Show that $E(\vec{r},t) = E_0 \exp\left[i(\vec{k}\cdot\vec{r} - \omega t)\right]$ is a solution of the wave Equation 4.2.8. (b) Repeat for the general expression $E(\vec{r},t) = f(\vec{k}\cdot\vec{r} - \omega t)$.

4.9.4. Write a program that reproduces Figure 4.2.5.

4.9.5. The value of the expression $e^{[iKa]}$ at $Ka = 1.6\pi$ is equivalent to that which is evaluated at what value of $Ka$ in the range $-\pi \le Ka \le \pi$? Hint: refer to Example 4.2.0.1.

4.9.6. Show that when the two waves, each represented by Equation 4.2.27 for $i = 1, 2$, are added, the resulting wave is given by Equation 4.2.29.

4.9.7. Show that in one dimension Equation 4.2.8 becomes $(1/v^2)d^2y/dt^2 = d^2y/dx^2$ with solution $y(x,t) = y_0 \exp\left[i(kx - \omega t)\right]$. Show that, in the present case, Equation 4.2.30 for the group velocity takes on the one-dimensional expression $v_g = d\omega/dk$ and equals $v$. Also show that Equation 4.2.28 for the phase velocity yields the same result.

4.9.8. Run the code linear_chain.m in order to reproduce Figure 4.2.8 and confirm that the $p = 1$ vibrational mode has a frequency that obeys Equation 4.2.25.

4.9.9. Modify the code linear_chain.m of Subsection 4.2.1 in order to perform a linear chain vibration for the 3rd mode with 57 Cu atoms and confirm that its vibrational frequency is consistent with that predicted by Equation 4.2.25.

4.9.10. Write a script that reproduces Figure 4.2.9. Hint: use starter example code group_phase_vel_example.m provided in Subsection 4.2.2.

4.9.11. Give the small $K$ limit of Equation 4.2.31 to second order in $K$. What is the limit of the resulting expression as $K \to 0$?

4.9.12. Show that in the limit of $m_1 = m_2 = m$, Equation 4.3.38 gives the one atom per cell result of Equation 4.2.13. Beware of the fact that the lattice constant in the two atoms per cell system is twice that of the one atom per cell system.

4.9.13. Write the code that reproduces Figure 4.3.12 and numerically confirm the graphical magnitudes of the optical and acoustic dispersion frequencies for both values of the wavevector at $K = 0, \pi$. What is the expected numerical value of the frequency gap?

4.9.14. In Equations 4.3.43 and 4.3.44, the traveling wave atomic amplitudes were obtained at $K = 0$. Use Equations 4.3.35 in order to obtain a general expression for $u/v$ versus $K$. After checking the $K = 0$ limits for consistency, obtain the limiting values of $u/v$ for the optical and acoustic mode frequencies at $K = \pi/a$. Obtain a plot of $u/v$ in the range of $0 < K < \pi$. Explain your results. Hint: you will need to incorporate Equation 4.3.41 in your calculations. It helps if you have already worked out Exercise 4.9.13.

4.9.15. (a) Starting with Equation 4.5.52, using the proper limits show how the expressions in Equation 4.5.53 result. (b) Reproduce the plot of Figure 4.5.13.

4.9.16. In one dimension, the classical result for the internal energy associated with an oscillator can be obtained by using the classical expression for the average of $E$

$$U = < E > = \frac{\int_0^\infty dE E \exp(-\frac{E}{k_B T})}{\int_0^\infty dE \exp(-\frac{E}{k_B T})}.$$

By performing the indicated integrations and extending the result to $N$ oscillators in three-dimensions, show that the classically expected Equation 4.5.71b is obtained.

4.9.17. Show all the steps involved in obtaining the three expressions for $C_V$ of Equation 4.5.72.

4.9.18. Study the script phonon_U.m associated with Figure 4.5.16 (a) and modify it in order to reproduce the results for $C_V$ shown in Figure 4.5.16 (b).

4.9.19. After reading Example 4.5.3.1, incorporate the provided lines of data into a script to do the fitting procedure described there and reproduce Figure 4.5.17 regarding solid Ar's Debye temperature.

4.9.20. After reading Example 4.5.4.1, incorporate the provided lines of data into a script in order to reproduce Figure 4.5.18 for the Debye and Einstein's heat capacities and their comparison with the experimental data.

4.9.21. Run the script thermal_potential.m in Section 4.7 to reproduce Figure 4.7.21.

4.9.22. Study script thermal_latt_const.m in Section 4.7. Modify it in order to incorporate the calculation of Equation 4.7.92, using the Lennard-Jones potential of Equation 4.7.89, and reproduce Figure 4.7.22.

# 5

# *Free Electron Gas*

**Contents**



## 5.1 Introduction

A free electron gas refers to the electrons in a crystal whose behavior is treated as if they were free from the binding forces that keep them confined to the crystal. These are the same as the valence electrons associated with the atoms in a crystal. They are also the same as the conduction electrons that move about freely within the crystal volume. The "free electron gas" model is used to understand the physical properties of metals. The simplest of which are the alkali metals such as lithium, sodium, potassium, cesium, and rubidium. In the case of lithium, for example, the valence electron is in the $2s$ state, which becomes a conduction electron in the associated energy band of the crystal. A crystal of lithium is monovalent. It contains $N$ electrons and $N$ positive lithium cores. Each lithium core contains 2 electrons occupying the $1s$ shell. The extra electron in the $2s$ state becomes part of the so-called Fermi sea or free electron gas. The electrons in the Fermi sea are described quantum mechanically and obey the Pauli exclusion principle. Below, we start off by describing a free one-dimensional electron gas quantum mechanically.

## 5.2 Free One-Dimensional Electron Gas

Following the introduction, consider an electron of mass $m$, in a box of length $L$. The box is described by the potential

$$V(x) = \begin{cases} 0, & 0 \le x \le L, \\ \infty, & x < 0 \text{ and } x > L \end{cases}, \tag{5.2.1}$$

where the electron is confined to be free only within the boundaries of the one-dimensional box. The Schrodinger equation in one dimension is

$$H\psi_n(x) = -\frac{\hbar^2}{2m}\frac{d^2\psi_n(x)}{dx^2} = \varepsilon_n\psi_n(x), \tag{5.2.2}$$

to be solved in the region of existence of $\psi$; i.e., $0 \le x \le L$. The wave function does not exist outside the box because of the infinitely hard walls that are impossible to penetrate, even quantum mechanically. The index $n$ is a level index associated with the $nth$ electron orbital energy, $\varepsilon_n$, and wavefunction $\psi_n(x)$. Consistent with the above picture, we have the boundary conditions (BCs) $\psi_n(0) = 0 = \psi_n(L)$, so that a particular solution that satisfies Equation 5.2.2 and the stated BCs is

$$\psi_n(x) = A\sin(k_n x), \tag{5.2.3a}$$

where $A$ is the normalization constant and

$$k_n = \frac{n\pi}{L}. \tag{5.2.3b}$$

One also notes that if we set $k_n = (n\pi/L) = 2\pi/\lambda_n$, then $L = n\lambda_n/2$; i.e., the length of the box is equal to a multiple of half wavelengths associated with the $nth$ state and we can also write

$$\psi_n(x) = A\sin\left(\frac{2\pi}{\lambda_n}x\right) = A\sin\left(\frac{n\pi}{L}x\right). \tag{5.2.3c}$$

Since the particle exists within the box, the wavefunction obeys the normalization condition

$$\int_0^L |\psi_n(x)|^2 dx = 1, \tag{5.2.4a}$$

which with the use of Equation 5.2.3c yields the normalization constant value of

$$A = \sqrt{\frac{2}{L}}. \tag{5.2.4b}$$

From Equation 5.2.3a, it is seen that $-(\hbar^2/2m)d^2\psi_n(x)/dx^2 = -(\hbar^2/2m)k_n^2\psi_n$, which when compared to Equation 5.2.2, gives the eigenvalues

$$\varepsilon_n = \frac{\hbar^2}{2m}k_n^2 = \frac{\hbar^2}{2m}\frac{4\pi^2}{\lambda_n^2} = \frac{\hbar^2}{2m}\frac{n^2\pi^2}{L^2}, \tag{5.2.5}$$

for the one-dimensional particle in a box of size L. Figure 5.2.1 shows four such levels with their associated wavefunctions and wavelengths.

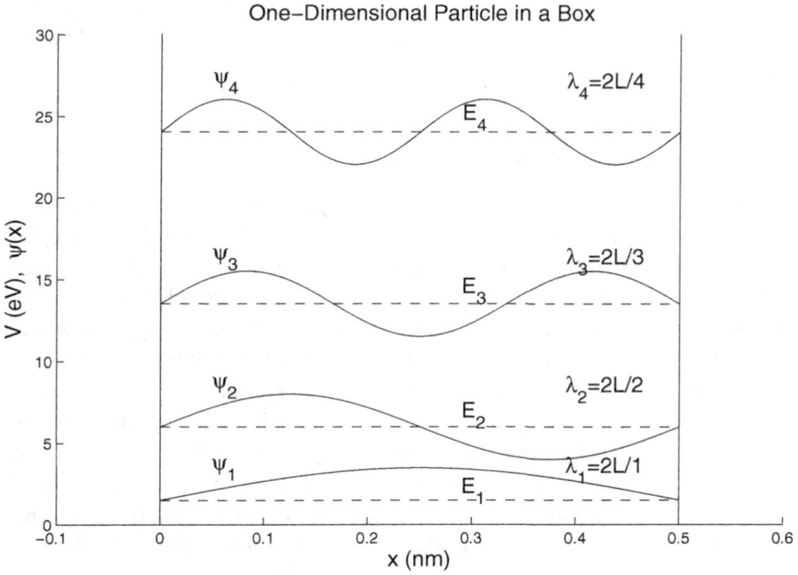


Figure 5.2.1: The first four levels, wavefunctions, and wavelengths for a one-dimensional particle in a box. The box width is $L = 0.5nm$, the $x$ coordinate is in units of $nm$, and the energy unit is in electron volts ($eV$). The wavefunctions, as plotted, have been displaced by their corresponding energy eigenvalue for visualization purposes.

A starter code that does one level and wavefunction for the one-dimensional particle in the box and which can be modified to reproduce Figure 5.2.1 is one_d_particle_in_box_one_level.m, whose listing follows.

```
%copyright by J. E Hasbun and T. Datta
%one_d_particle_in_box_one_level.m
%One-dimensional particle in a box wavefunction solution
%and energy level.
clear
L=0.5;                     %well width in nm
h=6.62606896e-34;          %Planck'constant (J.s)
hbar=h/2./pi;              %hbar
e=1.602176487e-19;         %electronic charge
hbar_eV=hbar/e;            %hbar in eV.s
c=299792458;               %speed of light (m/s)
cnm=c*1e9;                 %speed of light (nm/s)
me=0.511003e6/cnm^2;       %electron mass eV.s^2/nm^2
N=301;                     %number of x points to plot
dx=L/(N-1);                %step size for the x variable
%Wave function and energy level follow
hold on                    %overlays plots
line([0 0],[0 10],'Color','k')          %left potential wall
line([L L],[0 10],'Color','k')          %right potential wall
E1=pi^2*hbar_eV^2/(2*me*L^2);           %1st energy level in eV
x=0:dx:L;                               %x as an array
```

```
psi=sqrt(2/L)*sin(pi*x/L);                %wavefunction versus x array
plot(x,psi+E1,'k')                        %wavefunction plot displaced by En
axis([-0.1 L+0.1 0 10])                   %window plot view
xlabel('x (nm)')
ylabel('V (eV),  \psi(x)')
title('One-Dimensional Particle in a Box')
```

Equation 5.2.5 refers to the ground state electronic energy levels in the one-dimensional system. It is of interest to find out how $N$ electrons would be organized in the electronic configuration. Since electrons obey the Pauli exclusion principle, which states that no two electrons can have identical quantum numbers, then the way the electronic energy levels are occupied with electrons depends on their quantum numbers. In the present system, the electrons have two quantum numbers. These are the orbital quantum number, $n$, and the intrinsic spin quantum number $m_s = \pm 1/2$, depending on the orientation, up or down. Thus for a given index $n$, we can place two electrons in each level, one chosen with spin up and the other with spin down while still obeying the Pauli exclusion principle. Since the energy levels are independent of the spin quantum number, the pair of electrons are degenerate. Degeneracy is the property associated with orbitals of the same energy. If, for example, we had 5 electrons in the system, in the ground state we would place the first two electrons in the $n = 1$ state with opposite spins; similarly, the second two electrons would go in the $n = 2$ state, and the last electron would go in the $n = 3$ state with either spin up or down. From Equation 5.2.5, for a given orbital energy, we can find the $nth$ orbital number with energy $\varepsilon_n$ as

$$n = \sqrt{\frac{2m\varepsilon_n L^2}{\pi^2 \hbar^2}}. \tag{5.2.6}$$

Knowing this, we can find the number of electrons that occupy energies up to this orbital quantum number. Thus in the ground state, at $T = 0K$, the number of electrons occupying states up to the $nth$ orbital is $2n$, since there are two $m_s$ values for each $n$ due to spin. Therefore, if we consider a system with $N$ electrons in it, the total number of occupied orbitals in the ground state is the Fermi orbital number; i.e.,

$$n \to \nu_F = N/2, \tag{5.2.7}$$

which is to be understood as the orbital quantum number associated with the maximum filled orbital that is possible. Again, following Equation 5.2.5, associated with the Fermi orbital is the Fermi energy

$$\varepsilon_F = \frac{\hbar^2}{2m}\left(\frac{N\pi}{2L}\right)^2, \tag{5.2.8}$$

which is the highest filled energy level of the $N$ electron system in the ground state.

**Example 5.2.0.1**
Use the density of states concept discussed in a previous chapter to obtain the Fermi energy for a free one-dimensional electron gas.
**Solution**
We start from the three-dimensional formula of the previous chapter

$$D(\omega)d\omega = \left(\frac{L}{2\pi}\right)^3 \int_{\text{shell}} d^3k, \tag{5.2.9}$$

and notice that there is factor of $L/2\pi$ for each dimension. In one dimension, the integral over the shell is replaced with a single k-space point so that in the one-dimensional case we have

$$D(\omega)d\omega = 2\left(\frac{L}{2\pi}\right) dk, \tag{5.2.10}$$

where the factor of 2 has been inserted to take electron spin into account. Performing the integration, we have

$$\int_0^{\omega_F} D(\omega)d\omega = 2\left(\frac{L}{2\pi}\right)\int_0^{k_F} dk = \frac{L}{\pi}k_F = \nu_F,$$

to obtain the Fermi wavevector $k_F = \nu_F\pi/L$. The Fermi energy is then $\varepsilon_F = \hbar^2k_F^2/2m = (\hbar^2/2m)(\nu_F\pi/L)^2$. For $N$ electrons, the number of occupied states at $T = 0K$ is given by $\nu_F = N/2$, and the final result agrees with Equation 5.2.8. Finally, the Fermi frequency, $\omega_F$, alluded to in the above integral, while not used, is given by $\omega_F = \varepsilon_F/\hbar$.

## 5.3 The Fermi-Dirac Distribution

Above $T = 0K$, all the electrons that would normally lie below the Fermi energy or Fermi edge experience thermal excitation to higher levels by an energy that is on the order of $k_BT$. Electrons within $k_BT$ below the Fermi edge can jump to states that are about $k_BT$ above the Fermi edge. The distribution of electrons for $T > 0$, therefore, changes as a function of temperature. The occupation probability is given by the Fermi-Dirac distribution function (or Fermi function for short)

$$f(\varepsilon) = \frac{1}{\exp[(\varepsilon-\mu)/k_BT]+1}, \tag{5.3.11}$$

which is a function of temperature. At $T = 0K$, $\mu$ is equal to the Fermi energy $\varepsilon_F$. This distribution applies to spin 1/2 particles referred to as Fermions, to which category the electron belongs, and takes into consideration their quantum nature. A few of the limits of $f(\varepsilon)$ follow. At $T = 0K$

$$f(\varepsilon) = \begin{cases} \frac{1}{0+1} = 1, & \text{if } \varepsilon < \mu, \\ \frac{1}{\infty+1} = 0, & \text{if } \varepsilon > \mu \end{cases}, \tag{5.3.12a}$$

which means that all electrons lie below the Fermi level and none above it at $T = 0K$. The occupation probability has a value between 0 and 1. It takes the value of 1/2 whenever $\varepsilon = \mu$, at any temperature. It means that the probability of finding an orbital occupied at the chemical potential is always 1/2, which is the average of the probability of being occupied and being unoccupied. This information can be used to find the value of $\mu$. For large energy; i.e., $\varepsilon - \mu >> k_BT$, $\exp[(\varepsilon-\mu)/k_BT] >> 1$ so that

$$f(\varepsilon) \approx \exp[-(\varepsilon-\mu)/k_BT] \approx \exp[-\varepsilon/k_BT] = f_B(\varepsilon), \tag{5.3.12b}$$

where $\mu$ can be ignored in this limit. This limiting case is the Boltzmann distribution function or the classical limit. All these features can be seen in Figure 5.3.2 where we plot the Fermi-Dirac distribution versus energy in units of $\varepsilon_b = 1.602\times10^{-19}J$; i.e., $1eV$. We let $\mu$ take on the value of $2.5\varepsilon_b$ and the temperature is in units of $T_b = \varepsilon_b/k_B \approx 1.16\times10^4\,K$.

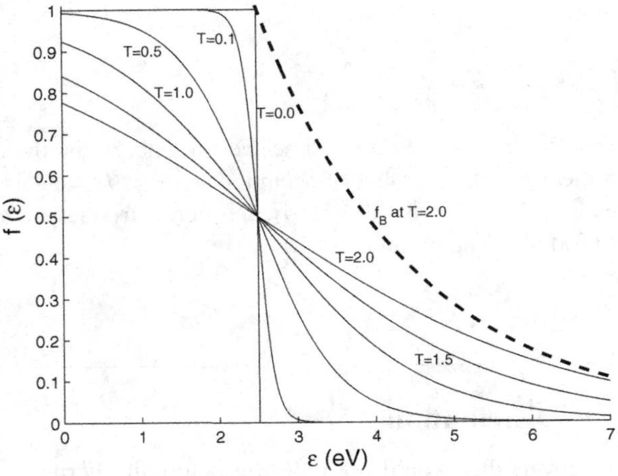


Figure 5.3.2: The Fermi-Dirac distribution versus energy in units of $\varepsilon_b = 1eV$ for various temperatures in units of $T_b = 1.16 \times 10^4\,K$. The chemical potential, $\mu = 2.5\,eV$, corresponds to the energy at which $f(\varepsilon) = 0.5$. The dashed curve corresponds to the Boltzmann expression of Equation 5.3.12b keeping the same value of $\mu$.

## 5.4 Free Three-Dimensional Electron Gas

The three-dimensional electron gas description follows that of the one-dimensional analogue we saw in Section 5.2. We write the Schrodinger equation in the vector form for a three-dimensional cubical box as

$$-\frac{\hbar^2}{2m}\left(\frac{\partial^2}{\partial x^2}+\frac{\partial^2}{\partial x^2}+\frac{\partial^2}{\partial x^2}\right)\psi_{\vec{k}}(\vec{r}) = \varepsilon_{\vec{k}}\psi_{\vec{k}}(\vec{r}), \tag{5.4.13}$$

where we have used $\vec{k}$ which is a three-dimensional vector that involves indices for each dimension as we will see below. The solution to Equation 5.4.13 is similar to that of the one-dimensional case and we write it as

$$\psi_{\vec{k}}(\vec{r}) = A\sin\left(\frac{n_x\pi}{L}x\right)\sin\left(\frac{n_y\pi}{L}y\right)\sin\left(\frac{n_z\pi}{L}z\right), \tag{5.4.14}$$

where $n_x, n_y, n_z$ are positive integers and $A = (2/L)^{3/2}$. We think of the electrons in a crystal lattice being described by this wavefunction. Furthermore, this wavefunction obeys the periodicity of the lattice, which can be shown if, for example, we look at one of the coordinates, $\psi(x+L,y,z) = \psi(x,y,z)$. In which case, looking at the only affected part of the wavefunction, $\sin(n_x\pi x/L + n_x\pi) = \sin(n_x\pi x/L)\cos(n_x\pi) + \cos(n_x\pi x/L)\sin(n_x\pi) = \sin(n_x\pi x/L)$, if $n_x$ is an even integer in this example. A more convenient way to guarantee periodicity is to assume plane wave solutions in the form

$$\psi_{\vec{k}}(\vec{r}) = \exp(i\vec{k}\cdot\vec{r}), \tag{5.4.15}$$

with $\vec{k} = k_x\hat{i} + k_y\hat{j} + k_z\hat{k}$ and $\vec{r} = x\hat{i} + y\hat{j} + z\hat{k}$. This plane wave already satisfies Equation 5.4.13 since $-(\hbar^2/2m)\vec{\nabla}^2\psi_{\vec{k}}(\vec{r}) = (\hbar^2/2m)k^2\psi_{\vec{k}}(\vec{r}) = \varepsilon_{\vec{k}}\psi_{\vec{k}}(\vec{r})$, where $\varepsilon_{\vec{k}} = (\hbar^2/2m)k^2$ with $k^2 = k_x^2 + k_y^2 + k_z^2$. In order to have periodicity with the plane wave solution, we can do a simple example displacement along the $x$ direction to write

$$\psi_{\vec{k}}(\vec{r}+\hat{i}L) = e^{i\vec{k}\cdot\vec{r}+ik_xL} = e^{i\vec{k}\cdot\vec{r}}e^{ik_xL} = e^{i\vec{k}\cdot\vec{r}}, \tag{5.4.16}$$

where we have taken $e^{ik_xL} = 1$ so that $k_xL = 2n_x\pi$ or $k_x = 2n_x\pi/L$. Similarly, we can find that $k_y = 2n_y\pi/L$ and $k_z = 2n_z\pi/L$ with $n_x$, $n_y$, and $n_z$ integers. With these definitions we rewrite

$$\varepsilon_{\vec{k}} = \frac{\hbar^2}{2m}k^2 = \frac{\hbar^2}{2m}\left(k_x^2+k_y^2+k_z^2\right) = \frac{\hbar^2}{2m}\left(\frac{4\pi^2}{L^2}\right)\left(n_x^2+n_y^2+n_z^2\right), \tag{5.4.17}$$

which is the energy of an electron with wavevector $\vec{k}$. The quantum mechanical momentum of an electron is obtained by using the momentum operator

$$\vec{p} = \frac{\hbar}{i}\vec{\nabla} = \frac{\hbar}{i}\left(\hat{i}\frac{\partial}{\partial x}+\hat{j}\frac{\partial}{\partial y}+\hat{k}\frac{\partial}{\partial z}\right), \tag{5.4.18}$$

and obtaining its expectation value with the wavefunction of Equation 5.4.15. Using $\vec{\nabla}\psi_{\vec{k}} = i\left(k_x\hat{i}+k_y\hat{j}+k_z\hat{k}\right)\psi_{\vec{k}} = i\vec{k}\psi_{\vec{k}}$, we see that

$$\vec{p}\psi_{\vec{k}}(\vec{r}) = \hbar\vec{k}\psi_{\vec{k}}(\vec{r}), \tag{5.4.19}$$

and the expectation value of the momentum is $<\vec{p}>=<\psi_{\vec{k}}|\vec{p}|\psi_{\vec{k}}>=\hbar\vec{k}<\psi_{\vec{k}}|\psi_{\vec{k}}>=\hbar\vec{k}$, since $<\psi|\psi>=1$. The particle velocity is given by $\vec{v}=<\vec{p}>/m$ or

$$\vec{v} = \hbar\vec{k}/m. \tag{5.4.20}$$

We can also obtain the Fermi energy using

$$\varepsilon_F = \frac{\hbar^2k_F^2}{2m}, \tag{5.4.21}$$

where $k_F$ is the magnitude of a three-dimensional wavevector that is characteristic of the number of electrons up to the highest occupied orbital at $T = 0\,K$, as discussed in Section 5.2. Given that there are $N$ electrons up to the Fermi level ($\varepsilon_F = \hbar\omega_F$), we work with the three-dimensional density of states expression

$$\int_0^{\omega_F} D(\omega)d\omega = \int_0^{\varepsilon_F} D(\varepsilon)d\varepsilon = N, \tag{5.4.22a}$$

which determines the Fermi level and where we have defined $D(\varepsilon) = D(\omega)/\hbar$. From Equation 5.2.9 we have

$$D(\omega)d\omega = 4\pi\frac{2V}{(2\pi)^3}k^2dk, \tag{5.4.22b}$$

where the factor of $4\pi$ is due to the shell integral, the factor of 2 is due to the electron spin, and $V$ is the volume. For specificity, let $N_F$ be the number of electrons up to the Fermi level at $T = 0\,K$ with wavevector $k_F$, so that from Equations 5.4.22 we have

$$4\pi\frac{2V}{(2\pi)^3}\int_0^{k_F}k^2dk = N_F = \frac{Vk_F^3}{3\pi^2}. \tag{5.4.23}$$

From this we can solve for the Fermi wavevector

$$k_F = \left(\frac{3\pi^2N_F}{V}\right)^{1/3}, \tag{5.4.24}$$

which when substituted into Equation 5.4.21 we get the Fermi level energy

$$\varepsilon_F = \frac{\hbar^2}{2m}\left(\frac{3\pi^2 N_F}{V}\right)^{2/3} = \frac{\hbar^2}{2m}\left(3\pi^2 n_F\right)^{2/3}, \tag{5.4.25}$$

where $n_F \equiv N_F/V$ is the material's free electron density. The associated Fermi velocity is

$$v_F = \frac{\hbar k_F}{m} = \frac{\hbar}{m}\left(3\pi^2 n_F\right)^{1/3}. \tag{5.4.26}$$

**Example 5.4.0.1**

Estimate the Fermi energy, the Fermi velocity, and the Fermi temperature for copper.

**Solution**

Copper is a transition metal with one $4s$ valence electron per Cu atom available for conduction. Write the density of electrons as $n = N/V = \rho_{Cu}/m_{Cu}$ where $m_{Cu} = 63.546u$, $1u \sim 1.66\times 10^{-27}kg$, and $\rho_{Cu} = 8.93\times 10^3 kg/m^3$ to get $n = 8.47\times 10^{28}\,m^{-3}$. The Fermi energy is $\hbar^2/(2m)(3\pi^2 n)^{2/3} \approx (1.055\times 10^{-34})^2/(2\cdot 9.11\times 10^{-31})(3\pi^2\cdot 8.47\times 10^{28})^{2/3} = 1.128\times 10^{-18}J \approx 7.03\,eV$. The Fermi velocity is $v_F = (\hbar/m)(3\pi^2 n)^{1/3} \approx (1.055\times 10^{-34}/9.11\times 10^{-31})(3\pi^2\cdot 8.47\times 10^{28})^{1/3} = 1.57\times 10^6\,m/s \approx 0.005c$, where $c$ is the speed of light. The Fermi temperature is $T_F = \varepsilon_F/k_B \approx 1.128\times 10^{-18}/1.38\times 10^{-23} = 8.16\times 10^4\,K$. In general, the calculation of these quantities for other metals is similar; however, the number of electrons available for conduction is equal to the number of atoms per volume $(N/V)$ multiplied by the number of valence electrons per atom. For the copper metal done here, the number of valence electrons is unity. For other metals, as in the case of aluminum, it is 3. See Table 5.4.1 for other example elements.

Table 5.4.1: Free electron calculated values of the Fermi energy, Fermi velocity, and Fermi temperature for example elements. The density, atomic mass, and valence electrons (val e's) used in the calculations are shown.

| Element | $\rho(10^3 kg/m^3)$ | $m(u)$ | val e's | $\varepsilon_F(eV)$ | $v_F(10^6 m/s)$ | $T_F(10^4 K)$ |
|---|---|---|---|---|---|---|
| Cu | 8.93 | 3.55 | 1.00 | 7.03 | 1.57 | 8.16 |
| Al | 2.70 | 6.98 | 3.00 | 11.66 | 2.03 | 13.5 |
| Li | 0.53 | 6.94 | 1.00 | 4.70 | 1.29 | 5.45 |
| Na | 0.97 | 2.99 | 1.00 | 3.14 | 1.05 | 3.65 |
| K | 0.86 | 9.10 | 1.00 | 2.04 | 0.848 | 2.37 |
| Ag | 10.50 | 107.87 | 1.00 | 5.50 | 1.39 | 6.39 |
| Au | 19.28 | 196.97 | 1.00 | 5.52 | 1.39 | 6.41 |
| Zn | 7.14 | 65.39 | 2.00 | 9.43 | 1.82 | 10.9 |

Note that given the Fermi level, from Equation 5.4.25 we can also solve for the number of electrons that fill the states up to the Fermi level; that is, $N_F = (V/3\pi^2)(2m\varepsilon_F/\hbar^2)^{3/2}$ or for $\varepsilon \leq \varepsilon_F$ and $N \leq N_F$, we can write

$$N = \frac{V}{3\pi^2}\left(\frac{2m\varepsilon}{\hbar^2}\right)^{3/2}. \tag{5.4.27}$$

From Equations 5.4.22b and 5.4.24 we have $D(\omega) = 4\pi(2V/(2\pi)^3)k^2 dk/d\omega$ and $k = \left(3\pi^2 N/V\right)^{1/3}$ so that $dk/d\omega = (3\pi^2/V)^{1/3}(N^{-2/3}/3)dN/d\omega$. By substituting these quantities into $D(\omega)$ we get

$$D(\omega) = 4\pi\frac{2V}{(2\pi)^3}\left(\frac{3\pi^2 N}{V}\right)^{2/3}\left(\frac{3\pi^2}{V}\right)^{1/3}\left(\frac{1}{3}N^{-2/3}\right)\frac{dN}{d\omega} = \frac{dN}{d\omega}. \tag{5.4.28}$$

As in Equations 5.4.22, if we now define $D(\varepsilon) = D(\omega)/\hbar = (1/\hbar)dN/d\omega \equiv dN/d\varepsilon$, with the use of Equation 5.4.27 we can get the density of states per energy $\varepsilon$ or density of orbitials

$$D(\varepsilon) = \frac{dN}{d\varepsilon} = \frac{V}{2\pi^2}\left(\frac{2m}{\hbar^2}\right)^{3/2}\varepsilon^{1/2}. \tag{5.4.29}$$

Multiplying the density of states ($D(\varepsilon)$) by the probability of finding an electron at this energy ($f(\varepsilon)$), at a particular temperature, gives the number of occupied states or orbitals per energy; that is,

$$\text{density of occupied orbitals} = D(\varepsilon)f(\varepsilon). \tag{5.4.30}$$

At $T = 0K$, $f(\varepsilon) = 1$ so that the density of occupied orbitals per energy equals the density of states, which are all filled up to the Fermi level.

**Example 5.4.0.2**

To illustrate the behavior of the density of occupied orbitals versus energy and temperature, we will plot Equation 5.4.30 in dimensionless units. First though, in Equation 5.4.29, we write $V = L^3$ and let $\varepsilon$ be in units of $E_b \equiv \hbar^2/2mL^2 = 1.602 \times 10^{-19} J$ or $1\,eV$ so that $L = \sqrt{\hbar^2/2mE_b} = 1.95 \times 10^{-10} m$ or 1.95Å. We then write $\varepsilon = \bar{\varepsilon}E_b$ and simplify the density of states of Equation 5.4.29 as $D(\varepsilon) = \bar{D}(\bar{\varepsilon})/E_b$, with the dimensionless density of states in the form $\bar{D}(\bar{\varepsilon}) \equiv \sqrt{\bar{\varepsilon}}/2\pi^2$. The Fermi function can also be written in dimensionless form if, in addition to writing $\varepsilon$ in $E_b$ units, we also let $\mu = \bar{\mu}E_b$ and $T = \bar{T}T_B$ where $k_BT_B \equiv E_b$ so that $T_B = E_b/k_B = 1.16 \times 10^4 K$ is the temperature unit. Thus all the barred quantities are dimensionless. In these units, Equation 5.3.11 for the Fermi function becomes $f(\varepsilon) = \{\exp[(\bar{\varepsilon} - \bar{\mu})/\bar{T}] + 1\}^{-1} = f(\bar{\varepsilon})$ and the density of orbitals of Equation 5.4.30 is $f(\bar{\varepsilon})\sqrt{\bar{\varepsilon}}/2\pi^2$. Using these units, Figure 5.4.3 shows plots of the Fermi function, the three-dimensional density of states, and the density of orbitals for two different temperatures. At very low temperature (left side of the figure), the density of orbitals and the density of states are equal up to the Fermi level ($\varepsilon_F = \mu$ at $T = 0K$) due to the vanishing of the Fermi function above $\varepsilon_F$. For a higher temperature (right side of the figure), the density of orbitals spreads out past $\mu$ due to thermal activation on the order of $k_BT$ and electrons occupy higher states with a probability dictated by the Fermi function.

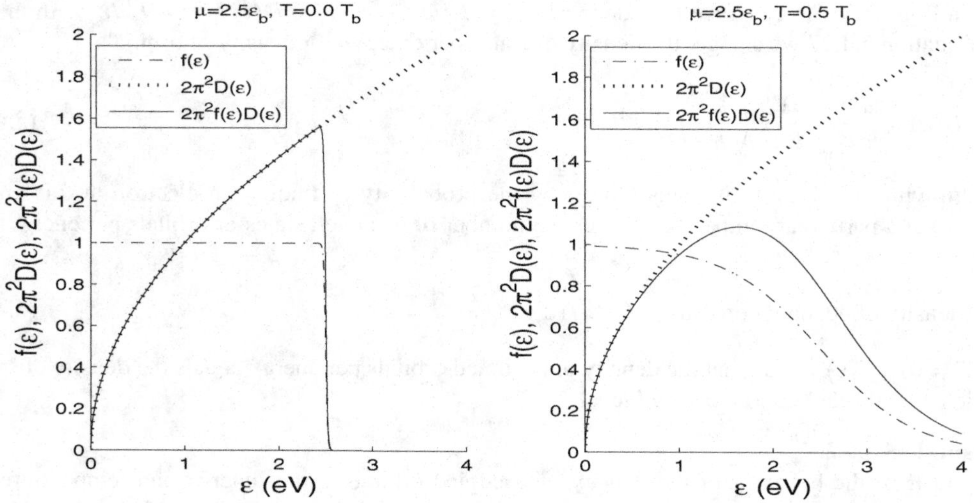


Figure 5.4.3: A plot of the Fermi function Equation 5.3.11, the density of states Equation 5.4.29, and the density of orbitals Equation 5.4.30 versus energy in $E_b$ units, where $E_b = 1\,eV$ as described in Example 5.4.0.2. The left figure is for a temperature near zero and the right figure is for a temperature of $0.5T_b$ where $T_b = E_b/k_B = 1.16 \times 10^4 K$ is the unit of temperature. The chemical potential used is $\mu = 2.5E_b$. The density of states and the density of orbitals were enhanced by the factor of $2\pi^2$ for visualization purposes.

The MATLAB script orbitals_vs_energy.m listed below can be used to reproduce Figure 5.4.3's plots and shows how the units were incorporated for future purposes.

```
%copyright by J. E Hasbun and T. Datta
%orbitals_vs_energy.m
%The number of orbitals per energy versus energy
%as the product of the 3-Dim dos times the Fermi-Dirac
%distribution function is plotted versus energy.
clear
h=6.62606896e-34;             %Planck'constant (J.s)
hbar=h/2./pi;                 %hbar
me=9.10938215e-31;            %electron mass (kg)
e=1.602176487e-19;            %electronic charge
Eb=e;                         %energy in Joules (=1eV)
kB=1.3806504e-23;             %Boltzmann Constant (J/K)
Tb=Eb/kB;                     %temperature unit
Lb=(hbar^2/2/me/Eb)^(1/2);    %length unit
fprintf('Energy unit: %3.2f eV,\n',Eb/e)
fprintf('Temperature unit: %4.3g K\n',Tb)
fprintf('Length unit: %4.3g m\n',Lb)
%temperature variable in units of Tb
Tv=[0.01,0.5];
Emax=4;
Emin=0;
Es=(Emax-Emin)/500;
mu=2.5;
```

```
eps=0:Es:Emax;
fe= @(eps,mu,T) 1./(exp((eps-mu)./T)+1);  %FD distribution definition
De= @(eps) eps.^(1/2)/2/pi^2;             %Density of states 3D definition
for j=1:length(Tv)
  T=Tv(j);
  subplot(1,2,j)
  hold on
  for i=1:length(eps)
    fd(i)=fe(eps(i),mu,T);
    dos(i)=De(eps(i));
    ne(i)=fd(i)*dos(i);                   %orbitals/energy
  end
  plot(eps,fd,'k-.')
  plot(eps,2*pi^2*dos,'k:','LineWidth',2)
  plot(eps,2*pi^2*ne,'k-')
  legend('f(\epsilon)','2\pi^2D(\epsilon)',...
    '2\pi^2f(\epsilon)D(\epsilon)',0)
  xlabel('\epsilon (eV)','Fontsize',14)
  str1='f(\epsilon), 2\pi^2D(\epsilon), 2\pi^2f(\epsilon)D(\epsilon)';
  ylabel(str1,'Fontsize',14)
  str2=cat(2,'\mu=',num2str(mu,'%3.1f'),'\epsilon_b, T=',...
    num2str(T,'%3.1f'),' T_b');
  title(str2)
end
```

As we have seen, the density of occupied orbitals depends on energy and temperature and this is related to the number of electrons. Integrating the density of occupied orbitals over the energy gives the number of electrons. Therefore, it becomes natural to replace Equation 5.4.22a with

$$\int_0^\infty D(\varepsilon)f(\varepsilon)d\varepsilon = N, \tag{5.4.31}$$

where the upper limit is $\infty$ due to non-zero contributions from the Fermi function with energy. Evidently, this equation says that at temperatures higher than zero, the chemical potential changes in such a way that this expression is consistent with the conservation of electron number. If the number of electrons is known, it determines the position of the chemical potential at a given temperature. At $T = 0K$, $\mu \to \varepsilon_F$, $f(\varepsilon < \varepsilon_F) \to 1$ else it vanishes, and Equation 5.4.31, therefore, becomes equal to Equation 5.4.22a. We illustrate this concept with the following example.

**Example 5.4.0.3**

In dimensionless units Equation 5.4.31 has the same form $N = \int_0^\infty D(\bar{\varepsilon})f(\bar{\varepsilon})d\bar{\varepsilon}$. If we know the Fermi level, at $T = 0K$ the number of electrons is $N = \int_0^{\varepsilon_F} D(\varepsilon)d\varepsilon = 1/(3\pi^2)\left(2mV^{2/3}/\hbar^2\right)^{3/2}\varepsilon_F^{3/2}$, which, as expected, is consistent with Equation 5.4.27. With $V = L^3$ and using the same units developed in Example 5.4.0.2, this becomes $N = \bar{E}_F^{3/2}/(3\pi^2)$, which with a value of $\bar{E}_F = 2.5$, it gives $N \approx 0.134$ and electron density of $N/V = N/L^3 = 1.795 \times 10^{28} m^{-3}$ where $L = 1.95$Å. We next let the value of $N$ be constant and inquire about what the value of the chemical potential is at a temperature other than zero. To find $\mu$ we write Equation 5.4.31 (in dimensionless units for convenience) as $F_{\bar{\mu}} = N - \int_0^\infty \bar{D}(\bar{\varepsilon})f(\bar{\varepsilon})d\bar{\varepsilon}$. For a given temperature and the correct $\bar{\mu}$, the result of the integral equals $N$, in which case $F_{\bar{\mu}} = 0$. There exists a well-known scheme to obtain the argument of a function whose value causes the vanishing of the function that depends on the said argument. The method is referred to as the Newton-Raphson method. The idea is that for a function of $x$, say $F(x)$, the value of $x$ that causes $F(x) = 0$ is approximated by the iteration $x_{i+1} = x_i - F(x_i)/F'(x_i)$

where $x_0$ is a close enough starting guess to the root of the function, $x_{i\to\infty}$. If the derivative of the function, $F'(x)$, is not known, the approximate expression $F'(x) \approx (F(x+\Delta) - F(x))/\Delta$, for small $\Delta$ can be used. In this manner, for the above-mentioned electron density of $N/V = 1.795 \times 10^{28} m^{-3}$ (here $F_{\bar{\mu}} = 0$ for the correct $\bar{\mu}$), at a temperature of $0.5T_b$, we get $\bar{\mu} \approx 2.4$ or $\mu \approx 2.4E_b$ which is about $0.1E_b$ lower than the corresponding value of $\varepsilon_F$ at $T = 0K$. The following function listing of chem_pot_at_T.m performs this calculation.

```
%copyright by J. E Hasbun and T. Datta
%chem_pot_at_T.m
%Chemical potential at a given temperature. The integral
%of the density of orbitals is integrated over energy.
%The Newton-Raphson method is used to get the chemical potential
%self-consistently. The integration is done by the trapezoidal rule.
function chem_pot_at_T
clear
global fe De
h=6.62606896e-34;              %Planck'constant (J.s)
hbar=h/2./pi;                  %hbar
me=9.10938215e-31;             %electron mass (kg)
e=1.602176487e-19;             %electronic charge
Eb=e;                          %energy in Joules (=1eV)
kB=1.3806504e-23;              %Boltzmann Constant (J/K)
Tb=Eb/kB;                      %temperature unit
Lb=(hbar^2/2/me/Eb)^(1/2);     %length unit
fprintf('Energy unit: %3.2f eV,\n',Eb/e)
fprintf('Temperature unit: %4.3g K\n',Tb)
fprintf('Length unit: %4.3g m\n',Lb)
Ef=2.5;              %Fermi level at T=0 (in Eb units)
N=Ef^(3/2)/3/pi^2; %Electron number corresponding to T=0
N_over_V=(2*me/hbar^2)^(3/2)*(Ef*Eb)^(3/2)/3/pi^2; %actual N/V
fe= @(eps,mu,T) 1./(exp((eps-mu)./T)+1);  %FD distribution definition
De= @(eps) eps.^(1/2)/2/pi^2;             %Density of states 3D definition
fprintf('T=0 case: Ef=%3.2g, N=%4.3f\n',Ef,N)
N_iter=0;            %iteration counter
N_iterMax=5;         %maximum iteration number
mu_old=Ef;           %initial guess is the Fermi level at T=0
mu_new=mu_old;
del=1.e-3;
mu_corr=10*del;      %use the correction for tolerance
mu_max=8.0*Ef;       %use a large number for the integral's upper limit
Es=(mu_max-0)/500; %even divisions=> odd variables
eps=0:Es:mu_max;
%Let's find how the Fermi level changes with temperature next
%temperature variable in units of Tb
T=0.5;               %temperature variable
while (abs(mu_corr) > del & N_iter < N_iterMax)
  mu1=mu_new;
  mu2=mu1+del;                 %vary mu by small ammount
  y1=norb(eps,mu1,T);         %calculate with mu
  y2=norb(eps,mu2,T);         %calculate with mu+del
  %For integration use the trapezoidal
```

```
  N1=trapz(eps,y1);          %integral=> occupied states with mu
  N2=trapz(eps,y2);          %integral=> occupied states with mu+del
  ff=N-N1;                   %function whose zero we seek at mu
  ffd=N-N2;                  %function at mu+del
  %Newton Raphson: x(i+1)=x(i)-f(xi)/f'(xi), where f'~(f(x+del)-f(x))/del
  %or x(i+1)=x(i)-del/(f(x+del)/f(x)-1)
  N_iter=N_iter+1;
  mu_corr=del/(ffd/ff-1.0);
  mu_new=mu_old-mu_corr;     %Newton-Raphson step
  mu_old=mu_new;
end
fprintf('T (Tb)=%4.2f, N_iter=%3i\n',T,N_iter)
fprintf('ff=%3.2g, mu_corr=%3.2g, result_N=%4.3f\n',ff,mu_corr,N1)
fprintf('N/V (1/m^3)=%5.3e, final mu (eV)=%5.3f\n',N_over_V,mu_new)
```

```
function y=norb(eps,mu,T)
global fe De
for i=1:length(eps)
  y(i)=fe(eps(i),mu,T)*De(eps(i));
end
```

In this function, we have made use of the "global" statement so that the defined Fermi distribution and the density of states can be recognized by other functions that invoke the same global statement without the need to repeat the definitions. Also this function can be suitably modified to obtain the behavior of the chemical potential versus temperature as shown in Figure 5.4.4. Here, $\mu$ drops as the temperature increases due to the antisymmetry of the Fermi-Dirac distribution about $\varepsilon = \mu$. As the temperature increases, $\mu$ must decrease to keep the number of electrons constant. In addition to the numeric chemical potential, the figure also shows the behavior of a low temperature approximation developed further below (see Equation 5.4.39b).

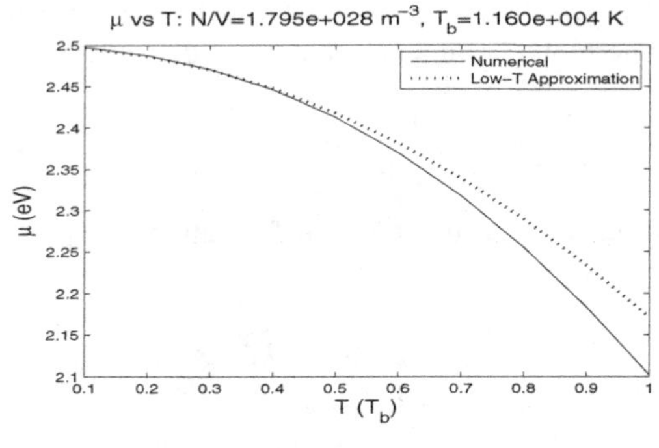


Figure 5.4.4: The numerically obtained (solid curve) chemical potential versus temperature (from Equation 5.4.31) for an electron density of $N/V = 1.795 \times 10^{28} m^{-3}$ as in Example 5.4.0.3. The dashed curve is the low temperature approximation for the chemical potential of Equation 5.4.39b.

One aspect to notice from the relation expressed in Equation 5.4.27 is that one can write $\ln N = (3/2)\ln\varepsilon + \text{const}$, where $\text{const} = \ln\left(V(2m/\hbar^2)^{3/2}\right)$. We can, therefore, write $\int (dN/N) =$

$(3/2)\int(d\varepsilon/\varepsilon)$ which, by lifting the integrals, becomes $dN/N = (3/2)d\varepsilon/\varepsilon$ or

$$\frac{dN}{d\varepsilon} = \frac{3N}{2\varepsilon} = D(\varepsilon), \tag{5.4.32}$$

which is also consistent with Equations 5.4.27 and 5.4.29.

It is possible to obtain an analytic approximation for $\mu$ at low temperature. To this end, we define

$$h(\varepsilon) = \int_0^{\varepsilon} \frac{\partial D(\varepsilon)}{\partial \varepsilon} d\varepsilon, \tag{5.4.33}$$

such that as $\varepsilon \to 0$, $h(\varepsilon) \to 0$, also we see that $h'(\varepsilon) = D(\varepsilon)$. By substituting $h'(\varepsilon)$ into Equation 5.4.31 and integrating by parts, we get

$$N = f(\varepsilon)h(\varepsilon)\big|_0^\infty - \int_0^\infty h(\varepsilon)\frac{\partial f(\varepsilon)}{\partial \varepsilon} d\varepsilon = -\int_0^\infty h(\varepsilon)\frac{\partial f(\varepsilon)}{\partial \varepsilon} d\varepsilon, \tag{5.4.34}$$

where we used the fact that $f(\infty) = 0$ and $h(0) = 0$. Since $f(\varepsilon)$ is nearly constant everywhere except at $\varepsilon = \mu$ and, therefore, $\partial f(\varepsilon)/\partial\varepsilon$ is close to zero everywhere except near $\varepsilon = \mu$, we can expand $h(\varepsilon)$ in a Taylor series about $\mu$. We do so to second order; i.e.,

$$h(\varepsilon) \approx h(\mu) + (\varepsilon - \mu)h'(\mu) + \frac{1}{2}(\varepsilon - \mu)^2 h''(\mu). \tag{5.4.35}$$

Substituting this approximation for $h(\varepsilon)$ into Equation 5.4.34, we find

$$N = -h(\mu)I_1 - h'(\mu)I_2 - \frac{1}{2}h''(\mu)I_3, \tag{5.4.36a}$$

where

$$I_1 = \int_0^\infty \frac{\partial f(\varepsilon)}{\partial \varepsilon} d\varepsilon = f(\varepsilon)\big|_0^\infty = (0-1) = -1. \tag{5.4.36b}$$

For integrals $I_2$ and $I_3$, since $f'(\varepsilon)$ is mostly non-zero near $\varepsilon = \mu$ we can replace the integration limits; that is, $\int_0^\infty \approx \int_{-\infty}^\infty$. Furthermore, the integrand in $I_2$ is odd so that

$$I_2 = \int_0^\infty (\varepsilon - \mu)\frac{\partial f(\varepsilon)}{\partial \varepsilon} d\varepsilon \approx \int_{-\infty}^\infty (\varepsilon - \mu)\frac{\partial f(\varepsilon)}{\partial \varepsilon} d\varepsilon = 0. \tag{5.4.36c}$$

Similarly for $I_3$ and, in addition, we will use the transformation $x = (\varepsilon - \mu)/k_B T$ for integration purposes to write

$$\begin{aligned} I_3 &= \int_0^\infty (\varepsilon - \mu)^2 \frac{\partial f(\varepsilon)}{\partial \varepsilon} d\varepsilon = -\frac{1}{k_B T}\int_{-\infty}^\infty \frac{(\varepsilon - \mu)^2 e^{\frac{(\varepsilon-\mu)}{k_B T}}}{\left(1 + e^{\frac{(\varepsilon-\mu)}{k_B T}}\right)^2} d\varepsilon \\ &= -(k_B T)^2 \int_{-\infty}^\infty \frac{x^2 e^x}{(1+e^x)^2} dx = -\frac{\pi^2}{3}(k_B T)^2, \end{aligned} \tag{5.4.36d}$$

where the derivative of the Fermi function was also carried out. The integration was done by parts from which one gets $4\int_0^\infty x/(1+e^x) = \pi^2/3$ using standard mathematical tables [25]. Equation 5.4.36a simplifies to

$$N = h(\mu) + \frac{\pi^2}{6}h''(\mu)(k_B T)^2. \tag{5.4.37}$$

Recalling the definition of $h(\varepsilon)$ in Equation 5.4.33, $h'(\varepsilon) = D(\varepsilon)$ and $h''(\varepsilon) = D'(\varepsilon)$; furthermore, we also know that at $T = 0K$, $N = \int_0^{\varepsilon_F} D(\varepsilon)d\varepsilon$, then Equation 5.4.37 gives

$$N = \int_0^{\mu} D(\varepsilon)d\varepsilon + D'(\mu)\frac{\pi^2}{6}(k_BT)^2 = \int_0^{\varepsilon_F} D(\varepsilon)d\varepsilon. \tag{5.4.38a}$$

Letting $D(\varepsilon)$ take on a roughly constant value; i.e., $D(\varepsilon) \approx D(\varepsilon_F)$, in addition to $D(\mu) \approx D(\varepsilon_F)$ and similarly $D'(\mu) \approx D'(\varepsilon_F)$ then

$$N \approx D(\varepsilon_F)\int_0^{\mu} d\varepsilon + D'(\varepsilon_F)\frac{\pi^2}{6}(k_BT)^2 = D(\varepsilon_F)\int_0^{\varepsilon_F} d\varepsilon \tag{5.4.38b}$$

or

$$D(\varepsilon_F)\mu + D'(\varepsilon_F)\frac{\pi^2}{6}(k_BT)^2 = D(\varepsilon_F)\varepsilon_F. \tag{5.4.38c}$$

The result of Equation 5.4.38c can be used to obtain a low temperature approximation for the chemical potential if we solve for $\mu$ to get

$$\mu \approx \varepsilon_F - \frac{\pi^2}{6}(k_BT)^2\frac{D'(\varepsilon_F)}{D(\varepsilon_F)}. \tag{5.4.39a}$$

Finally, using the free electron gas density of states of Equation 5.4.29 we see that $D'(\varepsilon) = D(\varepsilon)/(2\varepsilon)$, so that

$$\mu \approx \varepsilon_F - \frac{\pi^2}{12}\frac{(k_BT)^2}{\varepsilon_F}. \tag{5.4.39b}$$

The comparison between this low temperature approximation and the full numerical calculation (Equation 5.4.31) for the chemical potential is shown in Figure 5.4.4. For the parameters used in the calculation, the low temperature approximation seems to do very well in the range between $0K$ and about $0.4T_b$ (approximately $4600K$).

## 5.5 Electron Gas Heat Capacity

The heat capacity of a metal has a contribution due to the phonons, which we studied in a previous chapter, as well as the contribution due to the electrons studied here. As it turns out, only those electrons in orbitals with an energy range $k_BT$ of the Fermi energy are thermally excited and these are the ones that contribute to the electronic heat capacity. To see this, we consider the average energy associated with the electron gas defined as

$$U_{el} = \int_0^{\infty} \varepsilon f(\varepsilon)D(\varepsilon)d\varepsilon, \tag{5.5.40a}$$

where $f(\varepsilon)$ is the Fermi-Dirac distribution Equation 5.3.11, and $D(\varepsilon)$ is the density of states of Equation 5.4.29. The electronic heat capacity follows from this expression as

$$C_{el} = \frac{dU_{el}}{dT}. \tag{5.5.40b}$$

The analytical scheme we follow below parallels the development of the previous section, in particular Equations 5.4.33-5.4.36. Again we define the function $h(\varepsilon)$

$$h(\varepsilon) = \int_0^{\varepsilon} \varepsilon D(\varepsilon) d\varepsilon, \tag{5.5.41}$$

so that $h'(\varepsilon) \equiv \varepsilon D(\varepsilon)$, where as $\varepsilon \to 0$, $h(\varepsilon) = \to 0$. With this understanding, we can write Equation 5.5.40a as

$$U_{el} = \int_0^{\infty} f(\varepsilon) h'(\varepsilon) d\varepsilon. \tag{5.5.42}$$

This expression is integrated by parts to obtain

$$U_{el} = f(\varepsilon)h(\varepsilon)|_0^{\infty} - \int_0^{\infty} h(\varepsilon)\frac{\partial f(\varepsilon)}{\partial \varepsilon} d\varepsilon = -\int_0^{\infty} h(\varepsilon)\frac{\partial f(\varepsilon)}{\partial \varepsilon} d\varepsilon, \tag{5.5.43}$$

Even though in Equation 5.5.41 we have defined $h(\varepsilon)$ in a different way from that defined in Equation 5.4.33, the operations in carrying out the integrals shown in the last of Equation 5.5.43 are identical to those of Equations 5.4.34-5.4.36d. Thus, in analogy to Equation 5.4.37, we, therefore, have the result

$$U_{el} \approx h(\mu) + \frac{\pi^2}{6} h''(\mu)(k_B T)^2. \tag{5.5.44}$$

From Equation 5.5.41 we see that we can write

$$h(\mu) = \int_0^{\varepsilon_F} \varepsilon D(\varepsilon) d\varepsilon + \int_{\varepsilon_F}^{\mu} \varepsilon D(\varepsilon) d\varepsilon, \tag{5.5.45a}$$

in which if we approximate $\varepsilon D(\varepsilon) \approx \varepsilon_F D(\varepsilon_F)$ in the range of $\varepsilon_F$ and $\mu$ for low temperature, we get

$$h(\mu) \approx h(\varepsilon_F) + \varepsilon_F D(\varepsilon_F)(\mu - \varepsilon_F), \tag{5.5.45b}$$

where we have used the definition of Equation 5.5.41 for $h(\varepsilon)$ and $\int_{\varepsilon_F}^{\mu} d\varepsilon = \mu - \varepsilon_F$. Substituting the result from Equation 5.4.39a for $(\mu - \varepsilon_F)$ into $h(\mu)$ we get

$$h(\mu) \approx h(\varepsilon_F) - \frac{\pi^2}{6}(k_B T)^2 D'(\varepsilon_F)\varepsilon_F. \tag{5.5.45c}$$

From Equation 5.5.41, $h'(\varepsilon) = \varepsilon D(\varepsilon)$ and $h''(\varepsilon) = D(\varepsilon) + \varepsilon D'(\varepsilon)$, so that by substituting $h''(\mu) \approx D(\varepsilon_F) + \varepsilon_F D'(\varepsilon_F)$, in addition to substituting Equation 5.5.45c, into Equation 5.5.44 for the internal energy, we get

$$\begin{aligned} U_{el} &\approx h(\varepsilon_F) - \frac{\pi^2}{6}(k_B T)^2 D'(\varepsilon_F)\varepsilon_F + \frac{\pi^2}{6}(D(\varepsilon_F) + \varepsilon_F D'(\varepsilon_F))(k_B T)^2 \\ &= h(\varepsilon_F) + \frac{\pi^2}{6}(k_B T)^2 D(\varepsilon_F). \end{aligned} \tag{5.5.46a}$$

From this result, Equation 5.5.40b, for the heat capacity, yields

$$C_{el} = \frac{\pi^2}{3} D(\varepsilon_F) k_B^2 T. \tag{5.5.46b}$$

Finally, from Equation 5.4.32 we also have $D(\varepsilon_F) = 3N/2\varepsilon_F$, to write

$$C_{el} \sim \frac{\pi^2}{3}\left(\frac{3N}{2\varepsilon_F}\right)k_B^2T = \frac{\pi^2}{3}\left(\frac{3N}{2k_BT_F}\right)k_B^2T = \frac{\pi^2}{2}\left(\frac{T}{T_F}\right)Nk_B, \qquad (5.5.46c)$$

where we have used $\varepsilon_F = k_BT_F$. This result shows a linear temperature dependence. It also shows that only a small fraction $(T/T_F)$ of the total electrons are thermally excited and which contribute to the electronic heat capacity; i.e., those electrons in an energy range $k_BT$ of the Fermi energy. In contrast, classically and incorrectly, one would have expected an energy of $U_{cl} = Nk_BT$ so that $C_{cl} = Nk_B$.

It is possible to calculate $U_{el}$ numerically from Equation 5.5.40a, which we do in the dimensionless energy units of the previous section; that is, $U_{el} = \bar{U}_{el}E_b$ where $\bar{U}_{el} = \int_0^\infty \bar{\varepsilon}f(\bar{\varepsilon})D(\bar{\varepsilon})d\bar{\varepsilon}$. The heat capacity becomes $C_{el} = dU_{el}/dT = (E_b/T_b)d\bar{U}_{el}/d\bar{T} = \bar{C}_{el}k_B$ where $\bar{C}_{el} = d\bar{U}_{el}/d\bar{T}$. The derivative can be approximately calculated numerically using the formula $df(x)/dx \sim (f(x+\Delta) - f(x))/\Delta$. In doing this exercise, we bear in mind that it is important to consider whether the temperature dependence of the chemical potential makes a difference in the calculations. In this respect, Figure 5.5.5 shows the results of two sets of calculations for both $U_{el}$ and $C_{el}$. In one set, the constant Fermi energy is used with a value of $\varepsilon_F = 2.5\,eV$. In the other set, the approximate temperature dependent $\mu$ of Equation 5.4.39b has been employed with the same $\varepsilon_F$. Furthermore, in dimensionless units, the analytic expression works out to $\bar{C}_{el} = \pi^2\bar{D}(\bar{E}_F)\bar{T}/3 = \pi^2(\bar{T}/\bar{T}_F)N/2$ also in units of $k_B$.

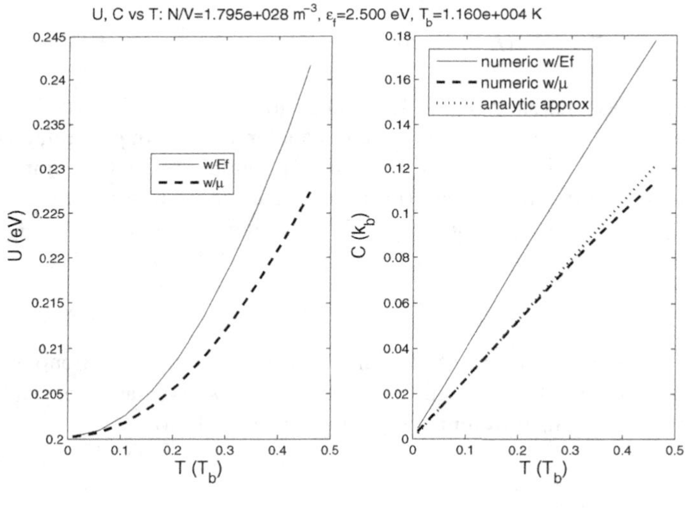


Figure 5.5.5: The left panel shows the average electronic energy (Equation 5.5.40a) versus temperature for an electron density of $N/V = 1.795 \times 10^{28} m^{-3}$, a constant $\varepsilon_F = 2.5\,eV$ (solid line). The dashed line corresponds to a similar calculation but making use of $\mu(T)$ from Equation 5.4.39b rather than $\varepsilon_F$. Both calculations use the dimensionless units of Section 5.4. The right panel is the comparison of the numerical and the analytic approximation for the electronic heat capacity, also in dimensionless units (see text). The calculation employing the constant $\varepsilon_F$ corresponds to the solid line. The dashed line is the calculation that makes use of $\mu(T)$ (Equation 5.4.39b) instead. The low temperature analytic approximation (Equation 5.5.46c) is best at very low temperature but the linear temperature behavior is evident (dotted line). Refer to Examples 5.4.0.2 and 5.4.0.3 for further details regarding the units employed.

In Figure 5.5.5 we see that the effect of incorporating the temperature dependence of the chemical potential resulted in an improved comparison between the heat capacities at low temperature. The

calculated heat capacity that makes use of $\mu$ (dashed curve) is in better agreement (in magnitude and in shape) with the behavior of the analytic low temperature $C_{el}$ than the case for which the constant $\varepsilon_F$ is used (solid). At higher temperature, the numerical heat capacity (dashed curve) begins to deviate from the low temperature approximation, as should be expected. In summary, this exercise has been very instructive in conveying the significance of including the temperature dependence of the chemical potential in calculations carried out over a large temperature range.
To conclude this section, altogether, a material's heat capacity naturally involves the contribution due to phonons, as discussed in a previous chapter, as well as the contribution due to electrons, given above. In a metal at low temperature, therefore, the experimental heat capacity is often fitted with an expression of the form $C = \gamma T + AT^3$, where $\gamma$ and $A$ are fit coefficients. The term linear with $T$ is due to the electronic contribution and the cubic term in $T$ is the phonon's Debye's $T^3$ law contribution.

## 5.6 Electrical Conductivity (Drude Model)

Electrons are subject to external forces due to electric and magnetic fields. The total force on a charged particle in the presence of these fields is given by the so-called Lorentz force

$$\vec{F} = m\frac{d\vec{v}}{dt} = q(\vec{E} + \vec{v} \times \vec{B}) = -e(\vec{E} + \vec{v} \times \vec{B}), \tag{5.6.47}$$

where for electrons we have taken $q = -e$. In what follows, we will ignore the presence of the magnetic field whose contribution will be discussed in a later section. In general, electrons in solids experience scattering due to collisions with impurities, phonons, and lattice imperfections. In its simplest form, under an electric field alone (in a specific direction), we could describe the force experienced by an electron in a solid with the one-dimensional expression

$$m\frac{dv}{dt} = -eE - m\frac{v}{\tau}, \tag{5.6.48}$$

where one thinks of $1/\tau$ as the scattering collision rate (due to impurities, phonons, etc.). The term $-v/\tau$ acts as a resistive force to the motion. Furthermore, if $\tau$ is taken as a constant, and assuming the electrons begin from rest, this differential equation has the solution

$$v(t) = -\frac{eE\tau}{m}\left(1 - e^{-t/\tau}\right), \tag{5.6.49a}$$

which after a long time has passed gives the steady state value of the electronic speed or drift speed

$$v(t \to \infty) \equiv v_\infty = -\frac{eE\tau}{m}. \tag{5.6.49b}$$

Thus, according to this simple model, the maximum velocity achieved by electrons is $v_\infty$ which is limited by $\tau$; i.e., the collision time or time between collisions. For $n$ electrons per unit volume, the steady state current density is

$$j = nqv_\infty = -nev_\infty = \frac{ne^2E\tau}{m}. \tag{5.6.50}$$

According to Ohm's law the current density is proportional to the electric field or

$$j = \sigma E, \tag{5.6.51}$$

where $\sigma$ is the proportionality constant known as the conductivity. Equating these last two expressions, we find the electric conductivity as

$$\sigma = \frac{ne^2\tau}{m}. \tag{5.6.52}$$

Other quantities of interest are the electrical resistivity defined as

$$\rho \equiv \frac{1}{\sigma} = \frac{m}{ne^2}\frac{1}{\tau}, \tag{5.6.53a}$$

and the electronic mobility $\mu$ defined as the constant of proportionality in the expression $v_\infty \propto E$; that is, $v_\infty = \mu E$. Comparing this with Equation 5.6.49b we see that

$$\mu = -\frac{e\tau}{m}. \tag{5.6.53b}$$

The idea is that electrons travel for a time $\tau$ between collisions and therefore have associated with them a mean free path given by $\ell = v_\infty \tau$. For electrons with an energy on the order of the Fermi energy, with speeds on the order of the Fermi speed $v_\infty = v_F$ the mean free path is obtained from $\ell = v_F\tau$. For pure copper at low temperature $\tau \sim 2\times10^{-9}\,s$ and from Table 5.4.1 we have that $v_F = 1.57\times10^6\,m/s$ so that $\ell \sim 3.14\times10^{-3}m/s = 3.14mm$. This value is smaller at room temperature because, in general, $\tau$ is a function of temperature and tends to decrease as the temperature rises. Again, from Example 5.4.0.1, for copper $n = 8.47\times10^{28}\,m^{-3}$ so that, using the electronic charge and mass, we get $\sigma = 4.77\times10^{12}\,(\Omega\cdot m)^{-1}$, $\rho = 2.09\times10^{-13}\,\Omega\cdot m$ and $|\mu| = 352m^2/(Volt\cdot sec)$. (See Exercise 5.10.12 for a room temperature estimate.)

Experimentally, the analysis of metal resistivity data involves obtaining the scattering rate; i.e., $1/\tau$, which involves the contributions from the various mechanisms. If we restrict ourselves to impurities and phonons, we would write the scattering rate as a sum of the individual scattering rates or

$$\frac{1}{\tau} = \frac{1}{\tau_i} + \frac{1}{\tau_{ph}}, \tag{5.6.54a}$$

where $\tau_i$ and $\tau_{ph}$ are the collision times associated with impurities and phonons, respectively. The resistivity becomes

$$\rho = \rho_i + \rho_{ph} = \frac{m}{ne^2}\left(\frac{1}{\tau_i} + \frac{1}{\tau_{ph}}\right), \tag{5.6.54b}$$

which is known as Matthiessen's rule. Since at $T = 0$ the phonon contribution $\rho_{ph} \to 0$ then $\rho|_{T\to 0} \to \rho_i(T = 0)$ is the extrapolated value of the resistivity at $T = 0K$ and the resistivity ratio $\rho(T)/\rho_i(0)|_{T=T_{RT}}$ is a measure of the sample purity, where $T_{RT}$ is the room temperature. In solids, phonons can experience scattering from other phonons as well, and this leads to the so-called thermal resistivity of a material. One could have a three-phonon process in which momentum is conserved, such as $\vec{K}_1 + \vec{K}_2 = \vec{K}_3$; that is two phonons interact with the creation of a third phonon. Another mechanism is the so-called umklapp process in which the three-phonon scattering occurs with the additional creation of a reciprocal lattice vector; that is, $\vec{K}_1 + \vec{K}_2 = \vec{K}_3 + \vec{G}$. In both of these processes, energy is conserved; however, the most important contribution to thermal resistivity is the umklapp process.

## 5.7 Electronic Motion in Magnetic Fields and the Classical Hall Effect

Within solids, electrons are made to flow by subjecting them to electric fields to which they readily respond. Electrons also respond to magnetic fields and it seems natural to investigate their motion in the presence of both of these fields. Notwithstanding such responsivity to electromagnetic fields, the electronic motion is damped by scattering mechanisms inherent within the materials through which they move. Here we consider the motion of electrons within a solid in the presence of an electric and a magnetic field with the additional contribution due to scattering. In this regard, we consider Equation 5.6.47 and add the simple resistive force similar to that of Equation 5.6.48; i.e.,

$$m\frac{d\vec{v}}{dt} = -e(\vec{E} + \vec{v} \times \vec{B}) - m\frac{\vec{v}}{\tau}, \tag{5.7.55}$$

where we now perform the analysis in three dimensions. We let the electric field take the form $\vec{E} = (E_x, E_y, E_z)$ and, for simplicity, we let the magnetic field lie along the $z-$direction or $\vec{B} = (0, 0, B_z)$ so that with $\vec{v} = (v_x, v_y, v_z)$, $\vec{v} \times \vec{B} = (v_y B_z, -v_x B_z, 0)$ and Equation 5.7.55 becomes

$$m\left(\frac{d}{dt} + \frac{1}{\tau}\right)(v_x, v_y, v_z) = -e(E_x + v_y B_z, E_y - v_x B_z, E_z). \tag{5.7.56}$$

As in the previous section, our interest lies in the steady state behavior of the electrons. To this end, we set $d\vec{v}/dt = 0$ on the left side of these equations to obtain

$$\begin{aligned} m\tfrac{v_x}{\tau} &= -e(E_x + v_y B_z) \\ m\tfrac{v_y}{\tau} &= -e(E_y - v_x B_z) \\ m\tfrac{v_z}{\tau} &= -eE_z. \end{aligned} \tag{5.7.57}$$

Thus, in the steady state, the $z$ component of the velocity is determined by the $E_z$ component of the electric field. However, the $x$ and $y$ components of the velocity are coupled. For example, if from the second of these we solve for $v_y = -e(E_y - v_x B_z)\tau/m$ and substituting this back into the first, we get the equation $-eE_x - eB_z(-eE_y + ev_x B_z)\tau/m = mv_x/\tau$ or solving for $v_x$ get,

$$v_x = -\frac{e\tau}{m}\left(\frac{E_x - \omega_c \tau E_y}{1 + (\omega_c \tau)^2}\right), \tag{5.7.58a}$$

where $\omega_c = eB_z/m$ is the cyclotron frequency. From this expression, one notices that the presence of $B_z$ decreases the terminal speed in the $x$-direction. A similar process for the $y$ component of the velocity gives

$$v_y = -\frac{e\tau}{m}\left(\frac{E_y + \omega_c \tau E_x}{1 + (\omega_c \tau)^2}\right). \tag{5.7.58b}$$

One of the applications of the above result is the classical Hall effect. Referring to Figure 5.7.6, we consider a Hall bar; that is, a sample material with current along the $x$ direction and whose width and height are along the $y$, $z$ directions.

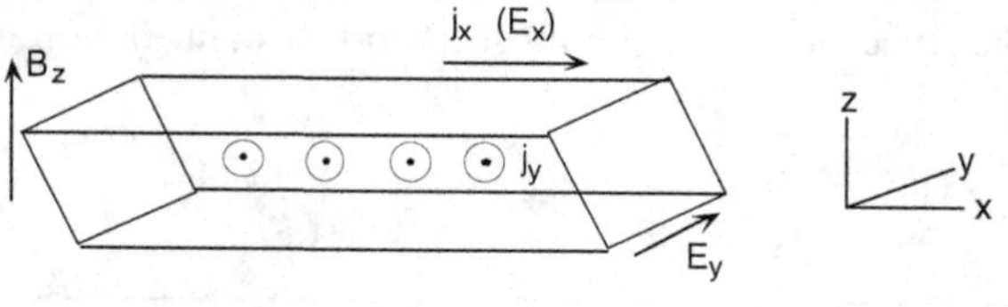


Figure 5.7.6: Hall Bar illustration of a sample used in a typical Hall effect experiment. The electric field component ($E_x$) is due to an applied voltage. The magnetic field ($B_z$) is perpendicular to the applied electric field. The shown $E_y$ component of the electric field is the so-called Hall field.

When electric ($E_x$) and magnetic ($B_z$) fields are present as shown in the figure, the $x$ direction flowing current, $j_x$, due to $E_x$, interacts with the magnetic field, $B_z$ (through $v_x \times B_z$). The magnetic interaction is responsible for inducing a current in the $y$ direction and thereby a transverse electric field $E_y$ with associated current $j_y$. Since the Hall bar is finite in the $y$ direction, the current $j_y$ flows until the setup $E_y$ field, due to charge pile-up, equalizes the magnetic force responsible for $j_y$. At this point, the transverse current stops and $v_y = 0$. From Equation 5.7.58b, we see that this happens when

$$E_y = -\frac{e\tau}{m} E_x B_z, \tag{5.7.59a}$$

which is referred to as the Hall field, named after Edwin Hall who discovered the effect in 1879. Also note that when $v_y = 0$ Equation 5.7.57 gives $v_x = -eE_x\tau/m$ so that $j_x = -nev_x = ne^2E_x\tau/m$. From Equation 5.7.59a, we see that $E_y/B_z = -e\tau E_x/m$ and one can define the ratio

$$R_H \equiv \frac{E_y/B_z}{j_x} = \frac{-e\tau E_x/m}{ne^2E_x\tau/m} = -\frac{1}{ne}, \tag{5.7.59b}$$

known as the Hall coefficient. In an experiment, since $j_x$ and $B_z$ are both known, the idea is to measure $E_y$ to obtain the Hall coefficient. If $R_H < 0$ as in Equation 5.7.59b, the charge carriers responsible for $j_x$ are electrons, else if $R_H > 0$, the charge carriers are holes, which are charge carriers, similar to electrons, but with positive charge. Such charge carriers are found in semiconductor material systems and are created when electrons are excited from the valence band to the conduction band. The empty space left after the electron is excited is the hole, which, in the presence of a field, moves in the opposite direction to the motion of the electron. From Equation 5.6.51, since $\rho = 1/\sigma$, Ohm's law can be expressed as $E_x = j_x\rho_L$; that is

$$\rho_L = E_x/j_x, \tag{5.7.60a}$$

where we have used the notation $\rho_L$ to denote the longitudinal resistivity here. Since it does not depend on the magnetic field, the classical longitudinal resistivity is therefore constant. In the case of the Hall effect experiment, there is a similar definition for the transverse resistivity or Hall resistivity, which using the above definitions, is written as

$$\rho_H \equiv \frac{E_y}{j_x} = \frac{-e\tau E_x B_z/m}{ne^2E_x\tau/m} = -\frac{B_z}{ne} = R_H B_z. \tag{5.7.60b}$$

This expression shows that the classical Hall resistivity of a Hall bar is linear with the magnetic field and that the slope is positive ($R_H > 0$) for holes and negative ($R_H < 0$) for electrons as shown in Figure 5.7.7.

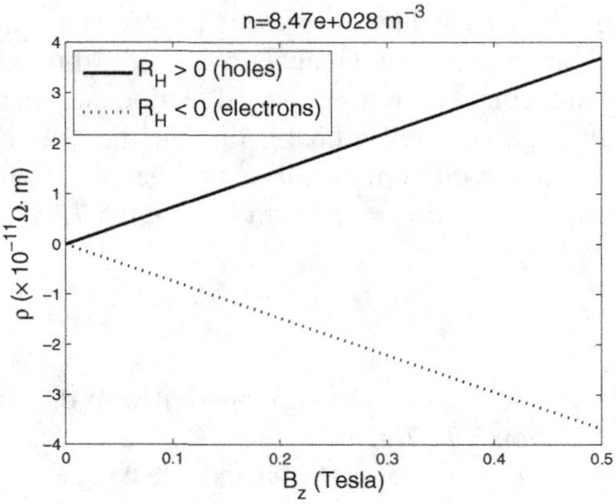


Figure 5.7.7: The classical Hall resistivity versus the applied magnetic field for electrons and holes for which the particle density is assumed to be the same as in Cu ($8.47 \times 10^{28}\, m^{-3}$).

## 5.8 The Quantum Hall Effect

The quantum Hall effect pertains to Klaus von Klitzing's discovery [26] that for very clean samples and large magnetic fields, the transverse or Hall resistivity exhibits plateaus at specific field values. The 1985 Nobel prize in physics was awarded to him for the achievement. These plateaus represent deviations from the expected classically linear behavior of the Hall resistivity with the magnetic field. The plateaus in the Hall resistivity are also followed by the concomitant disappearance of the longitudinal resistivity at the same field values. This is illustrated in Figure 5.8.8.

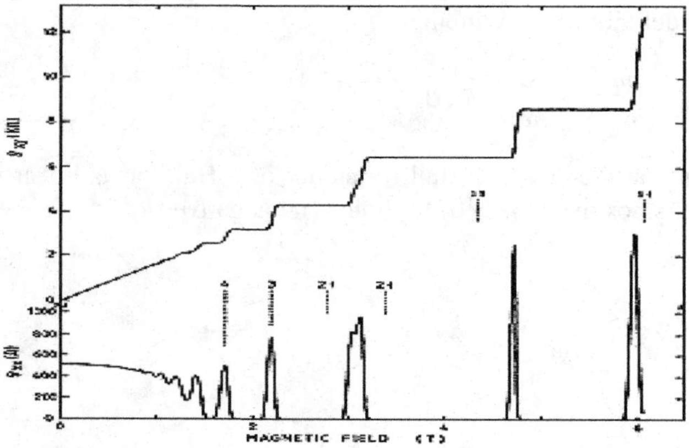


Figure 5.8.8: Experimental curves for the Hall resistivity $\rho_H$ (upper curve) and the longitudinal resistivity $\rho_L$ (lower curve) of a heterostructure as a function of the magnetic field at a fixed carrier density corresponding to zero gate voltage. The temperature of the experiment is about $8\,mK$ [26]. This figure is that of Figure 14 from von Klitzing's work, which has been reprinted with permission [26].

In this experiment, a very thin Hall bar sample is prepared in which the electrons within it are thought of as a two-dimensional electron gas as shown in Figure 5.8.9.

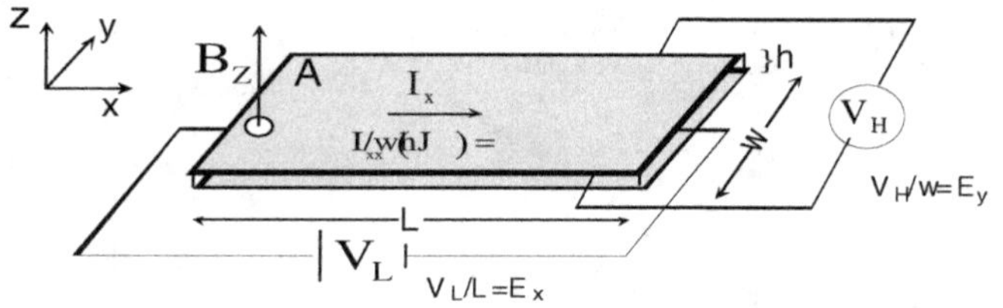


Figure 5.8.9: An illustration of a two-dimensional sample setup (heterostructure) for measuring the transverse or Hall voltage ($V_H$) due to $E_y$ under an applied voltage ($V_L$) due to $E_x$ and a magnetic field $B_z$. In this case, the Hall bar is very thin ($h$ is very small). Here the sample area is $A = wL$.

The magnetic field through the sample ($B_z$) is expressed as multiples of the electron number times a minimum flux value or quantized flux unit, $\phi_0 = h/e$, in units of webers. Thus, for $N$ electrons on a sample of area $A$, the ratio of the number of floxoids ($N\phi_0$) to the total flux through the sample ($B_zA$), where we have ignored the electron spin, is written as

$$\nu = \frac{N\phi_0}{B_zA} = \frac{N}{B_zA}\frac{h}{e}. \tag{5.8.61}$$

Here $\nu$ is known as the Landau level filling factor. It is a number that is large at small $B_z$ and small at large $B_z$. For example, from Equation 5.8.61 we see that $B_zA = Nh/(e\nu)$ and from Equation 5.7.60b, for positive charges $q = |e|$, and replacing the volume density ($n$) with $N/A$, for two dimensions, we have that

$$\rho_H = \frac{B_zA}{Nq} = \frac{Nh}{Ne^2\nu} = \frac{1}{\nu}\frac{h}{e^2}, \tag{5.8.62}$$

which becomes quantized in units of $h/e^2$ whenever $\nu$ takes on integer values with a simultaneous disappearance of $\rho_L$. By the way, notice that in two-dimensions the unit of resistivity is now in Ohms. The experimental positions of the plateaus are accurate enough to enable the refinement of the fundamental ratio of $h/e^2$.

To illustrate the concept, below is a listing of the simple-minded script hall_example.m for which we have picked a sample area $A = 1 \times 10^{-4} m^2$ with $N = 3 \times 10^{10}$ electrons. We vary the magnetic field in the range between 0 and 10 tesla. Whenever the value of $\nu$ is an integer, we place a square. This indicates the plateau position for the transverse quantum mechanical resistivity as shown in Figure 5.8.10.

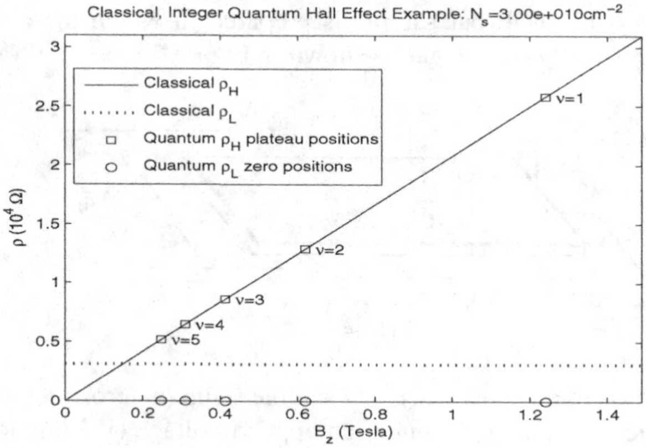


Figure 5.8.10: The results of the simple script hall_example.m to illustrate the quantum mechanical behavior of the Hall resistivity plateaus (squares) compared to the classical behavior (solid line). The zeros of the longitudinal resistivity at the integer values of the Landau level filling factor ($\nu$) are indicated by the open circles. The classical value of the longitudinal resistivity is chosen to be an arbitrary constant.

Simultaneously, at the position of every plateau, a circle is placed at the same field value to indicate the vanishing of the quantum mechanical longitudinal resistivity. The classical transverse resistivity is the expected straight line behavior with the magnetic field and, finally, some constant is chosen for the classical longitudinal resistivity. Here, it should be understood that for small magnetic fields, the value of $\nu$ becomes high and the integer values become closer to each other to indicate the classical regime in this simple-minded example. This can be seen by running the script with an increasing value of electrons. However, a more realistic and quantitative description of the entire process is beyond the scope of this text. Below follows the script listing.

```
%copyright by J. E Hasbun and T. Datta
%hall_example.m
%Simple illustration of the classical transverse and
%logitudinal resistivities along with their quantum mechanical
%versions.
clear
e=1.602176487e-19;          %electronic charge
h=6.62606896e-34;           %Planck'constant (J.s)
fxQ=h/e;                    %quantized unit of flux (webers)
rhoQ=fxQ/e;                 %quantized resistivity (ohm) (in 2D)
A=1.e-4;                    %sample area m^2
N=3.e10;                    %number of electrons
Bmin=0; Bs=0.25; Bmax=10; %range of B
B=Bmin:Bs:Bmax;
nB=length(B);
for i=1:nB
  rhoH(i)=B(i)*A/N/e;  %classical rho_H in Ohm (classical)
end
rhoL=0.015*rhoH(nB);   %class. rho_L in Ohm - let it be some constant
%Quantized Hall resistance
%Need nu (the Landau level fill factor), use N=nu*e*B*A/h=nu*B*A/fxQ
```

```
nu_min=N*fxQ/Bmax/A;       %minimum value of nu
nu_max=N*fxQ/Bs/A;         %maximum value of nu
nu_low=ceil(nu_min);       %round the value
nu_high=ceil(nu_max);      %highest nu for rho to reach rhoH(nB)
nu=nu_low:1:nu_high;       %vary nu in integer values
nNu=length(nu);
for j=1:nNu
  rhoHQ(j)=rhoQ/nu(j);  %quantum rho_H=(h/e^2)/nu
  %Find the B field values of rho_L=0, use def: BA=N*(h/e)/nu
  BQ(j)=N*fxQ/A/nu(j);
end
plot(B,rhoH/1e4,'k')       %classical rho_H
hold on
%classical rho_L
line([B(1) B(nB)],[rhoL/1e4 rhoL/1e4],'Color','k','LineStyle',':',...
  'LineWidth',2)
xlabel('B_z (Tesla)')
ylabel('\rho (10^4 \Omega)')
plot(BQ,rhoHQ/1e4,'ksq') %quantum rho_H
plot(BQ,0.0,'ko')           %quantum rho_L zero positions
for j=1:nNu
  str=cat(2,'  \nu=',num2str(nu(j),'%i'));
  text(BQ(j),rhoHQ(j)/1e4,str)
end
axis([0 6*BQ(nNu) 0 6*rhoHQ(nNu)/1e4])
legend('Classical \rho_H','Classical \rho_L','Quantum \rho_H
    plateau positions',...  'Quantum \rho_L zero positions',0)
str=cat(2,'Classical, Integer Quantum Hall Effect Example: ',...
    'N_s=',num2str(N/(A*1e4),'%6.2e'),'cm^{-2}');
title(str)
```

The story of the quantum Hall effect does not end with the quantized integer values of $\nu$. In fact, the 1998 Nobel prize in physics was awarded to Robert B. Laughlin, Horst L. Störmer, and Daniel C. Tsui for their work on the fractional quantum Hall effect. The description of the prize reads "The three researchers are being awarded the Nobel Prize for discovering that electrons acting together in strong magnetic fields can form new types of 'particles', with charges that are fractions of electron charges." That is, the workers were awarded the prize for their discovery of a new form of a quantum fluid with fractionally charged excitations. In the fractional quantum Hall effect, the Landau level filling factor, $\nu$, takes on fractional values such that $\nu = p/q$ where $p$ and $q$ are integers with no common factors. Some examples are $\nu = 1/3, 2/5, 3/7$, etc.

## 5.9 Electronic Thermal Conductivity of Metals

From the thermal conductivity Section 4.8 in Chapter 4, we had that the coefficient of thermal conductivity for particles with velocity $v$, heat capacity $C$, and mean free path $\ell$ is

$$K = \frac{1}{3} C v \ell, \tag{5.9.63}$$

where $C$ is the electronic heat capacity per volume. For pure metals, the greatest contribution to the thermal conductivity is due to the electrons; therefore, for our purpose, we consider the electronic contribution here as given by Equation 5.5.46c. We let $C = C_{el}/V$, $v = v_F$, $k_B T_F = m v_F^2/2$, and $\ell = v_F \tau$, to obtain for the coefficient of thermal conductivity

$$K = \frac{1}{3} C v \ell = \frac{1}{3V} \frac{\pi^2}{2} \left( \frac{N}{k_B T_F} \right) k_B^2 T v_F \ell = \frac{\pi^2}{3 \cdot 2V} \frac{N k_B^2 T}{m v_F^2/2} v_F^2 \tau = n \frac{\pi^2}{3} \frac{k_B^2 T \tau}{m} \tag{5.9.64}$$

where $n = N/V$. A quantity of interest is the ratio of the electronic thermal conductivity to that of the electrical conductivity; i.e.,

$$\frac{K}{\sigma} = \frac{\pi^2 k_B^2 n T \tau}{3m(ne^2 \tau/m)} = \frac{\pi^2}{3} \left( \frac{k_B}{e} \right)^2 T. \tag{5.9.65}$$

This linear behavior of the ratio $K/\sigma$with $T$ is known as the Wiedeman-Franz law. Furthermore, the ratio

$$L \equiv \frac{K}{\sigma T} = \frac{\pi^2}{3} \left( \frac{k_B}{e} \right)^2 = 2.44 \times 10^{-8} \frac{W\Omega}{K^2}, \tag{5.9.66}$$

is known as the Lorentz number and whose experimental values helped confirm the free electron theory of metals. For example, Cu has experimental values of $L$ equal to $2.23 \times 10^{-8} W\Omega/K^2$ and $2.33 \times 10^{-8} W\Omega/K^2$ at $273K$ and $373K$, respectively, which are in good agreement with Equation 5.9.66.

## 5.10 Chapter 5 Exercises

5.10.1. Apply the normalization condition Equation 5.2.4a to the expression of Equation 5.2.3c and show that the normalization constant is given by Equation 5.2.4b.

5.10.2. Modify the starter script one_d_particle_in_box_one_level.m of Subsection 5.2 to reproduce Figure 5.2.1.

5.10.3. Write a script that reproduces the plots of the Fermi-Dirac distribution function as shown in Figure 5.3.2.

5.10.4. Proceed as in Equation 5.4.16, but make a three-dimensional displacement instead and show the periodicity of the plane wave Equation 5.4.15.

5.10.5. Use Equation 5.2.9 appropriately for a three-dimensional electron gas to obtain the expression of Equation 5.4.22b.

5.10.6. After reading Example 5.4.0.1, calculate the Fermi level, the Fermi velocity, and the Fermi temperature for aluminum.

5.10.7. Write a script to reproduce the calculations of the Fermi level, the Fermi velocity, and the Fermi temperature shown in Table 5.4.1.

5.10.8. Show that by integrating Equation 5.4.29 over $\varepsilon$ as shown in Equation 5.4.22a, there results a Fermi energy consistent with that of Equation 5.4.25.

5.10.9. Modify the function chem_pot_at_T.m of Example 5.4.0.3 in order to reproduce the results of $\mu$ versus $T$ shown in Figure 5.4.4.

5.10.10. Write a script whose purpose is to reproduce the plots shown in Figure 5.5.5 for the numerical average energy and heat capacity of an electron gas versus temperature shown in Equations 5.5.40a and 5.5.40b, respectively. One set of calculations involves obtaining $U_{el}$ and $C_{el}$ using $\varepsilon_F$, the other set employs $\mu$ as given by the approximation in Equation 5.4.39b. Be sure to compare the numerical heat capacities with the analytic approximation of Equation 5.5.46c.

5.10.11. Letting $v(t=0) \equiv v_0$, show the steps that lead to the solution of the Equation 5.6.48 and show how it gives Equation 5.6.49a under the appropriate conditions.

5.10.12. Read Subsection 5.6 and use copper's room temperature collision time value of $\tau \sim 2.46 \times 10^{-14}\,s$ to obtain copper's mean free path, conductivity, resistivity, and mobility. Repeat the calculation for aluminum whose room temperature $\tau \sim 7.16 \times 10^{-15}\,s$. Feel free to use the information provided in Table 5.4.1, Example 5.4.0.1, and also see Exercise 5.10.6. Finally, assume Fermi energy values for the materials' drift speeds.

5.10.13. Show the necessary steps to obtain Equation 5.7.58b.

5.10.14. Show that, in two-dimensions, the classical unit of the Hall resistivity is Ohms and that this is also the case for the quantum unit of resistivity $h/e^2$.

5.10.15. Run the code example hall_example.m to reproduce Figure 5.8.10 of Subsection 5.8. Rerun the code for a higher number of electrons and explain your observations.

# 6

# *Introduction to Electronic Energy Bands*

**Contents**



## 6.1 Introduction

Picturing the electrons in crystals as free electrons, to a certain extent, is quite useful, but the picture fails to account for the difference between different materials such as metals, semimetals, semiconductors, and insulators; and various properties such as hall coefficients, the relationships between conduction electrons in metals and valence electrons in free atoms, etc. Energy bands is a term used to describe the electronic energies of electrons in crystalline materials. The energy bands in crystals are the analogue of electronic energy levels in atoms. Atoms have electronic level structures, whereas crystals have electronic band structures. Associated with the electronic energy bands in crystals are the energy gaps. The energy bands and the energy gaps are similar to the frequency dispersion and frequency gaps discussed in Chapter 4 regarding phonons in one and two atoms per primitive cell. The energy gaps in crystals correspond to energy regions in the band structures where electrons are forbidden to propagate. These gaps are the equivalent of electronic level separation in atoms. Thus, solutions to the crystal form of the Schrodinger equation exist only within the allowed energy regions for which propagation occurs; i.e., within the energy bands. The atomic orbitals in atoms can either be empty or occupied. The atomic behavior depends on the occupational state of the orbitals. In a similar manner, the amount of band filling determines the crystalline behavior as well. Figure 6.1.1 shows a simplified energy band model of an insulator, a metal, and a semiconductor.

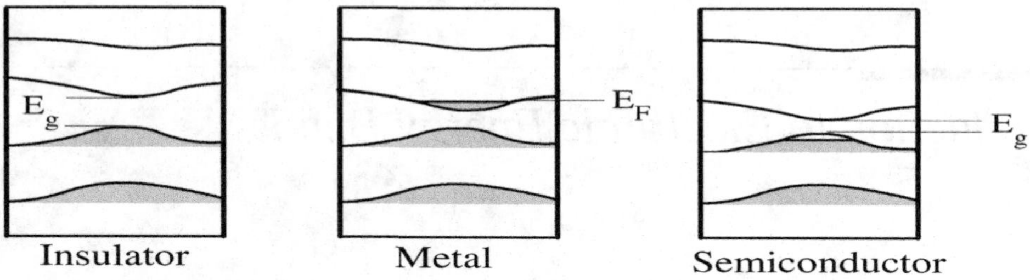


Figure 6.1.1: Simplified energy band model of an insulator, a metal, and a semiconductor. The insulator has filled (shaded region) valence and a large energy gap ($E_g$) exists to the nearest energy band. The metal has a band that is not completely filled and electrons can gain energy easily and are good conductors. Metals are also characterized by their Fermi energy, $E_F$. Semiconductors have a nearly full valence band and a small energy gap exists to the next nearest band and are poor insulators.

The insulator has filled valence and a large energy gap exists to the nearest empty conduction energy band. The metal has a band that is not completely filled. The electrons occupy energies up to the Fermi level ($E_F$) and can gain energy easily, within the limits allowable by the band, under an applied voltage. These are generally good conductors. Semiconductors have a nearly full valence band (shaded region) and a small energy gap exists to the nearest empty conduction band. These are poor insulators and they can conduct electricity if the voltage applied is large enough to excite electrons across the energy gap to the nearest conduction band. To account for the above-mentioned possible behaviors of the electronic properties of materials, the free electron model needs to be extended to include the crystal lattice periodicity. This involves the interaction of the electrons with all the crystal atoms.

## 6.2 Nearly Free Electron Model - Gaps at the Brillouin Zone Boundaries

Figure 6.2.2 contains a one-dimensional illustration of a crystal potential with bound energy levels and nearly free electron energy levels. The deeply bound energy levels are called core levels and the electrons found in those states are referred to as core electrons. As the energies of the electrons increase, the electronic wavefunctions begin to delocalize. Thus the upper bound levels sample more of the rest of the crystal's atomic potential than the core electrons. Finally, the nearly free electrons sample the collective crystal's potential due to all the ions. The electronic wavefunctions are delocalized and are more like plane waves throughout the crystal.

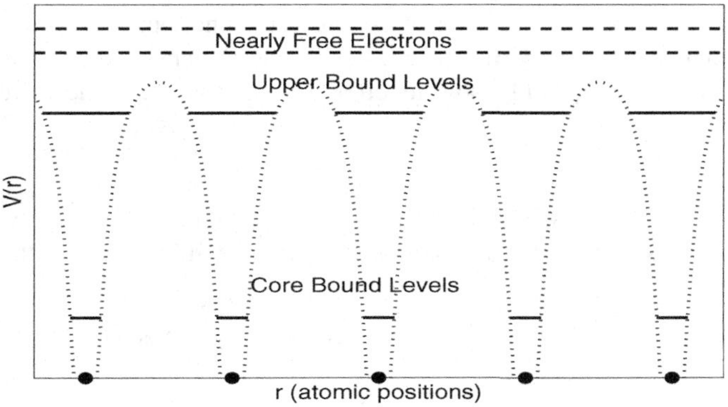


Figure 6.2.2: The crystal potential (dotted) created by the ions (black dots) is shown to be responsible for having electrons in bound levels as well as nearly free electrons at higher energies near the potential maxima. The bound levels are localized in the vicinity of their parent atoms. The nearly free electrons are bound to the crystal but are delocalized, sample the entire crystal's collective potential due to the atoms, and their wavefunctions are plane waves.

In the free electron model of a crystal, the allowed energy values are given by

$$\varepsilon_k = \frac{\hbar^2}{2m}|\vec{k}|^2 = \frac{\hbar^2}{2m}\left(k_x^2 + k_y^2 + k_z^2\right), \tag{6.2.1}$$

with associated wavefunction $\psi_k(r) \approx e^{i\vec{k}\cdot\vec{r}}$ and momentum $\vec{p} = \hbar\vec{k}$. For periodic boundary conditions on a cube of sides $L$, we have $\psi_k(r+L) = e^{i\vec{k}\cdot\vec{r}}e^{i\vec{k}\cdot\vec{L}} = \psi_k(r)$ so that $e^{i\vec{k}\cdot\vec{L}} = e^{ik_xL}e^{ik_yL}e^{ik_zL} = 1$ or $k_xL = k_yL = k_zL = 2n\pi$ with integer $n$. Thus $k_x, k_y, k_z$ take on the values $\{0, \pm 2\pi, \pm 4\pi, etc.\}/L$. The allowed energies are in this way illustrated in one dimension by the dots in Figure 6.2.3.

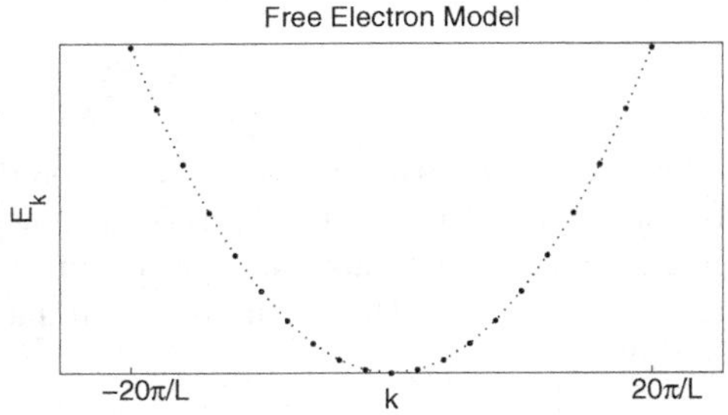


Figure 6.2.3: One-dimensional free electron energy eigenvalues for $-2n\pi/L < k < 2n\pi/L$ for $n = 1, 2..., 10$. The black dots correspond to the allowed values of the wavevector $k$. The short-dashed curve shows the quadratic behavior of $\varepsilon_k$ versus $k$.

In the nearly free electron model, the Bragg condition (see Chapter 2) is incorporated; that is,

$$k^2 = (\vec{k}+\vec{G})\cdot(\vec{k}+\vec{G}) = |\vec{k}+\vec{G}|^2, \tag{6.2.2}$$

which, as seen before, leads to Brillouin zone (BZ) boundaries at $k = \pm G/2$. Also as seen in Chapter 2, for a cubic solid of lattice constant $L = a$, the wavevectors $k$ work out to be $\pm\pi/a$ so that the reciprocal lattice vectors are $G = 2n\pi/a$ for integer $n$. X-ray reflections occur at the BZ boundaries and, at these special values of the wavevectors $k$, the wavefunctions are not traveling waves, instead they are standing waves. We, therefore, consider a one-dimensional case and look at the zone boundaries. Let $k = k_x = \pm\pi/a$, and let's write left and right traveling plane waves

$$\psi_L = e^{-i\pi x/a}, \qquad \psi_R = e^{i\pi x/a}, \tag{6.2.3}$$

respectively, at the BZ boundaries. Standing waves are set up within the BZ as waves get reflected back and forth. We form symmetric and antisymmetric linear combinations of these waves to get

$$\psi_+ \equiv A\left(\psi_R + \psi_L\right) = A\left(e^{i\pi x/a} + e^{-i\pi x/a}\right) = 2A\cos\frac{\pi x}{a} = \sqrt{\frac{2}{a}}\cos\frac{\pi x}{a},$$

and (6.2.4)

$$\psi_- \equiv A\left(\psi_R - \psi_L\right) = A\left(e^{i\pi x/a} - e^{-i\pi x/a}\right) = 2Ai\sin\frac{\pi x}{a} = i\sqrt{\frac{2}{a}}\sin\frac{\pi x}{a},$$

respectively, where we have used $A = 1/\sqrt{2a}$ as a normalization constant such that $\int_0^a |\psi_\pm(x)|^2 dx = 1$. The charge pile up associated with each of these wavefunctions is proportional to their probability densities $\rho_\pm$. We define these densities as

$$\rho_+(x) \equiv |\psi_+(x)|^2 = \frac{2}{a}\cos^2\frac{\pi x}{a} \quad \text{and} \quad \rho_-(x) \equiv |\psi_-(x)|^2 = \frac{2}{a}\sin^2\frac{\pi x}{a}. \tag{6.2.5}$$

The energy corresponding to each probability density is approximated by averaging the crystal potential, $u(x)$, over a unit cell or

$$U_\pm = \int_0^a u(x)\rho_\pm(x)dx. \tag{6.2.6}$$

Since the ions are located at periodic positions of period $a$, we can approximate the crystal potential with a function that has the lattice periodicity and whose maxima occur at the ionic positions; that is,

$$u(x) = u_0\cos\frac{2\pi x}{a}, \tag{6.2.7}$$

where $u_0$ is a constant. The probability densities and the potential are plotted in Figure 6.2.4. Notice that the symmetric probability, $\rho_+$, peaks at the ion positions indicating charge pile up associated with bonding. In contrast, the antisymmetric probability, $\rho_-$, has minima at the ion positions and correspond to charge depletion or antibonding. The crystal potential $u(x)$ in plotted in units of $u_0$ versus distance in units of $a$ as well.

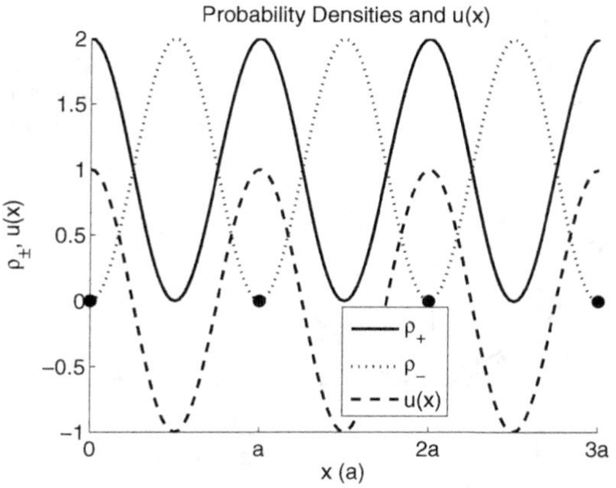


Figure 6.2.4: The probability densities, $\rho_+$ (solid) and $\rho_-$ (short dashed), of Equation 6.2.5 for a lattice constant $a$ of unit length. The ion positions correspond to the black dots. The crystal potential of Equation 6.2.7 is the dashed curve with maxima at the ion positions and minima in between.

Using $\int_0^a \cos^2(\pi x/a)\cos(2\pi x/a)dx = a/4$ and $\int_0^a \sin^2(\pi x/a)\cos(2\pi x/a)dx = -a/4$, we see that Equations 6.2.6 along with 6.2.5 and 6.2.7 yield

$$U_\pm = \pm\frac{u_0}{2}, \qquad \text{and } U_+ - U_- \equiv E_g = \frac{u_0}{2} - \left(-\frac{u_0}{2}\right) = u_0, \tag{6.2.8}$$

which indicates that there exists an energy gap between the symmetric and antisymmetric states at the BZ boundaries, as shown in Figure 6.2.5; that is, at $k = \pm\pi/a = G/2$.

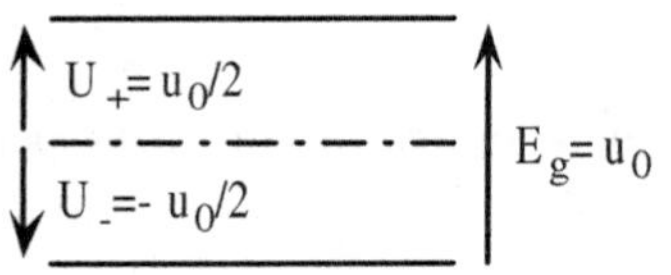


Figure 6.2.5: The symmetric and antisymmetric states' energies and resulting gap at the Brillouin zone boundaries $k = \pm\pi/2$.

The above treatment has been fruitful because of the use of standing waves at the zone boundaries. In this way, the nearly free electron model modifies the free electron model by incorporating the Bragg condition, which leads to gaps at the zone boundaries. This is shown in Figure 6.2.6.

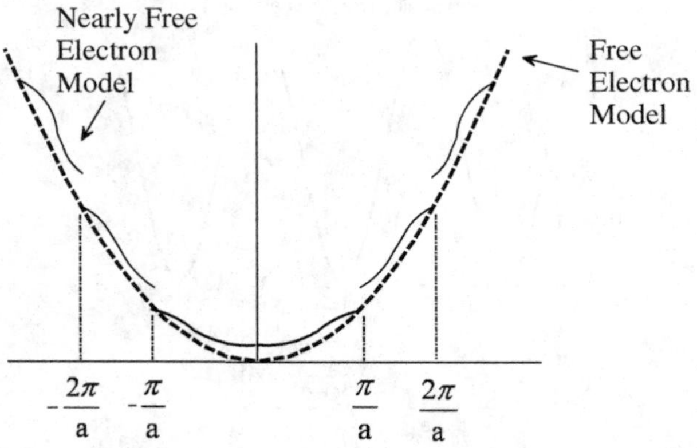


Figure 6.2.6: Comparison between the free and nearly free electron models. In the nearly free electron model, energy gaps open up at the zone boundaries with the incorporation of the Bragg condition.

The nearly free electron model also assumes the potential seen by the electrons is weak compared to that seen by the more deeply bound electrons. However, in general, we need to find a treatment that would lead us to understand the behavior of the energies versus the wavevector $k$. We will come back to this later below.

## 6.3 Bloch Functions

In 1928 Felix Bloch discovered the theorem that for a realistic periodic potential the Schrodinger equation must have solutions of the form

$$\psi_{\vec{k}}(\vec{r}) = u_{\vec{k}}(\vec{r})e^{i\vec{k}\cdot\vec{r}}, \tag{6.3.9a}$$

which is referred to as a Bloch function. Here the atomic function, $u_{\vec{k}}$, obeys the lattice periodicity condition

$$u_{\vec{k}}(\vec{r}) = u_{\vec{k}}(\vec{r}+\vec{T}), \tag{6.3.9b}$$

where $\vec{T}$ is a lattice translation vector. The idea is that the eigenfunctions of the wave equation for a periodic potential are the product of a plane wave $e^{i\vec{k}\cdot\vec{r}}$ multiplied by a function $u_{\vec{k}}(\vec{r})$ that has the periodicity of the crystal lattice. A simple illustration of Bloch's theorem for a simple non-degenerate case (no two wavefunctions with the same energy exist) can be carried out for a one-dimensional crystal. Consider a periodic chain of $N$ atoms as shown in Figure 6.3.7 where the $n = 0$ atom is equivalent to the $Nth$ atom.

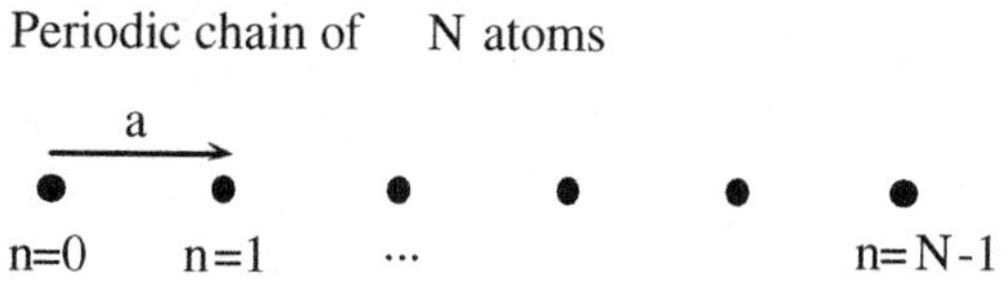


Figure 6.3.7: Periodic chain of $N$ atoms with period $a$. The $n = 0$ atom is equivalent to the $Nth$ atom.

The chain has period $a$ and a possible lattice translation vector is $T = ma$, where $m$ is an integer in the range $[0, N-1]$. Let's consider a translation $T = a$ and write the translated wavefunction as $\psi_k(x+a) = u_k(x+a)e^{ik(x+a)} = u_k(x)e^{ikx}e^{ika}$, where we have used Equation 6.3.9b. Similarly, for a translation of $T = 2a$, we have that $\psi_k(x+2a) = u_k(x+2a)e^{ik(x+2a)} = u_k(x)e^{ikx}e^{i2ka}$. Furthermore, continuing this way and, in fact, for a full translation around the chain (recall that the $Nth$ atom is equivalent to the zeroth atom) we have $\psi_k(x+Na) = \psi_k(x)$. This shows, therefore, that for a translation of $T = Na$, $\psi_k(x+Na) = u_k(x+Na)e^{ik(x+Na)} = u_k(x)e^{ikx}e^{iNka} = \psi_k(x)$. This also implies that $e^{iNka} = 1$ or, for integer $n$, $Nka = 2n\pi$ which results into $k = 2n\pi/(Na)$ in order to guarantee periodicity. With $k$ known, our wavefunction solution for the one-dimensional crystal becomes $\psi_k(x) = u_k(x)e^{i2n\pi x/(Na)}$ which is of the Bloch form as in Equation 6.3.9a. We can see that if we go back and repeat the translation $T = a$, for example, we get $\psi_k(x+a) = u_k(x)e^{ikx}e^{ika} = \psi_k(x)e^{ika} = \psi_k(x)e^{i2n\pi/N}$; similarly, the translation of $T = Na$ results in $\psi_k(x+Na) = u_k(x)e^{ikx}e^{iNka} = \psi_k(x)e^{iN2n\pi a/(Na)} = \psi_k(x)$, since $e^{i2n\pi} = 1$. In this way we know the wavefunction throughout the one-dimensional crystal. Note that because of the Bloch theorem, when describing the properties of the crystal, we restrict ourselves to the range $0 < x < a$ since one does not need a larger range. One unit cell is enough to describe the crystal wavefunction. We next move on to a more realistic but still one-dimensional model of a crystal and study its energy structure.

## 6.4 The Kronig-Penney Model

In 1930 Ralph Kronig and William Penney developed a one-dimensional model of a crystal in the form of an array of square quantum wells. The centers of the wells represent the ionic positions and the barriers occur in between the ions. The model is seen in Figure 6.4.8 and it is known as the Kronig-Penney model. We will work with this model in this section.

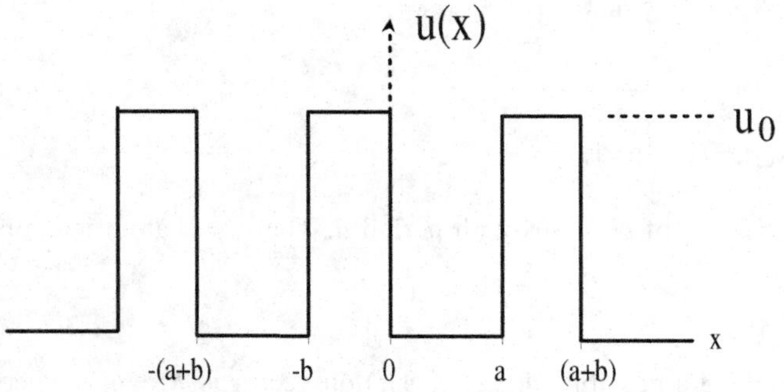


Figure 6.4.8: The array of square quantum wells that represents the one-dimensional Kronig-Penney model of a crystal. The wells are of width $a$ and the barriers are of height $U_0$ and thickness $b$.

We first write the one-dimensional Schrodinger equation for this model as

$$-\frac{\hbar^2}{2m}\frac{d^2\psi(x)}{dx^2} + u(x)\psi(x) = \varepsilon\psi(x), \tag{6.4.10}$$

where $u(x)$ is the periodic potential energy and $\varepsilon$ is the energy eigenvalue. In the region $0 < x < a$ we have that $u(x) = 0$ and the wavefunction is a linear combination of plane waves traveling to the right and to the left or

$$\psi_I(x) = Ae^{ikx} + Be^{-ikx}, \tag{6.4.11}$$

where the wavevector $k$ is related to the energy as $k = \sqrt{2m\varepsilon/\hbar^2}$. In the region $-b < x < 0$ where $u = u_0$, we have exponential solutions

$$\psi_{II}(x) = Ce^{qx} + De^{-qx}, \tag{6.4.12}$$

with the quantity $q$ bearing information about the potential height as $q = \sqrt{2m(u_0 - \varepsilon)/\hbar^2}$. Because crystals are periodic, the solution in the region $a < x < a+b$ is related to the solution in the $-b < x < 0$ region by Bloch's theorem; that is,

$$\psi_{III}(x) = \psi(a < x < a+b) = \psi(-b < x < 0)e^{ik'(a+b)} = \psi_{II}(x)e^{ik'(a+b)}, \tag{6.4.13}$$

where $a + b = T$ is a one-dimensional lattice translation vector. Here $k'$ will be determined later. At this point, all the ingredients to solve the problem have been stated. The constants $A, B, C$, and $D$ are to be chosen such that $\psi(x)$ and $\psi'(x)$ are continuous at $x = 0$ and $x = a$. The $x = 0$ boundary condition leads to

$$\begin{aligned} \psi_I(0) = \psi_{II}(0) &\Longrightarrow A + B = C + D \\ \psi_I'(0) = \psi_{II}'(0) &\Longrightarrow ikA - ikB = qC - qD \end{aligned} \tag{6.4.14a}$$

and the $x = a$ boundary condition leads to

$$\begin{aligned} \psi_I(a) = \psi_{III}(a) &\Longrightarrow Ae^{ika} + Be^{-ika} = (Ce^{-qb} + De^{qb})e^{ik'(a+b)} \\ \psi_I'(a) = \psi_{III}'(a) &\Longrightarrow ik(Ae^{ika} - Be^{-ika}) = q(Ce^{-qb} - De^{qb})e^{ik'(a+b)} \end{aligned} \tag{6.4.14b}$$

where we have used $\psi_{III}(x=a) = \psi_{II}(x=-b)\exp[ik'(a+b)]$ and similarly for the derivatives. The above Equations 6.4.14 can be organized in matrix form as follows

$$\begin{pmatrix} 1 & 1 & -1 & -1 \\ ik & -ik & -q & q \\ e^{ika} & e^{-ika} & -e^{-qb}e^{ik'(a+b)} & -e^{qb}e^{ik'(a+b)} \\ ike^{ika} & -ike^{-ika} & -qe^{-qb}e^{ik'(a+b)} & qe^{qb}e^{ik'(a+b)} \end{pmatrix} \begin{pmatrix} A \\ B \\ C \\ D \end{pmatrix} = \begin{pmatrix} 0 \\ 0 \\ 0 \\ 0 \end{pmatrix}. \qquad (6.4.15)$$

A solution of this matrix equation exists if the determinant of the coefficients matrix on the left also vanishes. The vanishing of the determinant provides a condition for the allowed values of $k$ which in turn are related to the allowed values of $\varepsilon$. Separating the real and imaginary parts of the determinant and setting each to zero leads to the eigenvalue equation

$$\left(\frac{q^2-k^2}{2qk}\right)\sin ka \sinh qb + \cos ka \cosh qb = \cos k'(a+b), \qquad (6.4.16)$$

and, in the following example, we investigate its possible solutions.

**Example 6.4.0.1**

In order to obtain the energy solutions contained in Equation 6.4.16, referring to the left panel of Figure 6.4.9, notice that they must exist within a region of space where the right-hand side (RHS) of the equation is real. That happens between the extreme values of the cosine function; i.e., $\pm 1$ (dashed lines). The idea then is to solve the equation self-consistently for all the values of $k$ for which the left-hand side (LHS) of the equation is between $\pm 1$ as $k'$ is varied. The quantity $q$ contains information about the potential height and, for visualization purposes, given a value of $q$ we can actually plot the LHS versus $k$ (solid line) and show the energy solutions that lie within the expected region (dots). These solutions are responsible for the band structure in the Kronig-Penney model. We can obtain accurate energy solutions versus $k'$ and plot them on a separate graph. Here $k'$ happens to be a Brillouin zone (BZ) wavector variable and its range is $-n\pi/(a+b) < k' < n\pi/(a+b)$, where $n$ is an integer. Whenever $(a+b)k'/\pi$ takes on an integer value, a new band is formed. Thus the band structure for the Kronig-Penney model is obtained as shown on the right panel of Figure 6.4.9. In particular, notice the gaps that occur at the BZ boundaries. The energies have been obtained for positive values of $k'$ and symmetry was used to plot them for $-k'$. The calculations have been carried out using dimensionless units. That is, we employ distance units of $a_b = 1$Å and let the wavevectors have units of $k_b = 1/a_b$ so that $k = \bar{k}k_b$. Here, the dimensionless units are the 'barred' quantities and $a = \bar{a}a_b$, $b = \bar{b}a_b$. The specific values of $\bar{a}$ and $\bar{b}$ are properties of the potential well and barrier widths, respectively, in Angstroms. We also let the energy take on dimensionless units such that $\varepsilon = \hbar^2\bar{k}^2/(2ma_b^2) = \bar{k}^2\varepsilon_b = \bar{\varepsilon}\varepsilon_b$. We see that, in these units, $\bar{\varepsilon} = \bar{k}^2$ or $\bar{k} = \sqrt{\varepsilon}$. With the above value of $a_b$, the energy unit is $\varepsilon_b = \hbar^2/(2ma_b^2) = 3.81\,eV$ shown on the figure's title. In the calculation, $u_0 = \bar{u}_0\varepsilon_b$ and the very important quantity $q = \sqrt{2m(u_0-\varepsilon)/\hbar^2} = \sqrt{2m\varepsilon_b/\hbar^2}\sqrt{\bar{u}_0-\bar{\varepsilon}} = k_b\bar{q}$; so that, $\bar{q} = \sqrt{\bar{u}_0-\bar{\varepsilon}}$. With these definitions the numerical form of Equation 6.4.16 remains the same except that the variables are all dimensionless as explained above. In this way, the results of the calculation shown in Figure 6.4.9 are for a potential height of $u_0 = 150\varepsilon_b = 571.5\,eV$, with well and barrier widths of 2.0 and 0.025 Å, respectively.

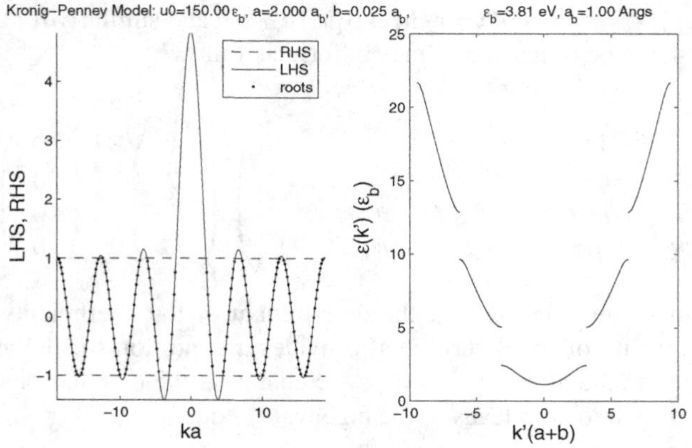


Figure 6.4.9: The left panel shows the possible energy solutions within the allowed region; that is, within a maximum and minimum range of the RHS of Equation 6.4.16 of $\pm 1$, respectively (dashed line). In the plot of the LHS of Equation 6.4.16, the solid dots are the source of the energy solutions that lead to the band structure. The smooth solid line is the full plot of the LHS. The right panel shows the actual band structure associated with Kronig-Penney model, for the shown range, for the potential parameters of $u_0 = 150\varepsilon_b = 571.5\,eV$, $a = 2$ Å and $b = 0.025$ Å.

The code employed in obtaining Figure 6.4.9 is Kronig_Penney_model_numeric whose listing follows below. The program is written as a MATLAB function; i.e., it employs other defined functions. In particular, notice that there is a 'searchguess' function. Its purpose is to seek new bands whenever $(a+b)k'/\pi$ takes on an integer value. The search's energy step depends on the potential and, like many programs, it may need tweaking as needed. The roots of Equation 6.4.16 are done by MATLAB's built-in 'fzero' function as shown. The LHS and RHS of the equation are conveniently defined as 'fL' and 'fR', respectively, for use throughout the program.

```
%copyright by J. E Hasbun and T. Datta
%Kronig_Penney_model_numeric.m
%Solves for the energy eigenvalues versus k' in the kronig-Penney
%model of a one dimensional crystal

function Kronig_Penney_model_numeric
clear
global fL fR u0 a b
h=6.62606896e-34;              %Planck'constant (J.s)
hbar=h/2./pi;                  %hbar (J.s)
me=9.10938215e-31;             %electron mass (kg)
e=1.602176487e-19;             %electronic charge
ab=1e-10;                      %1 angstom unit of distance
kb=1/ab;                       %wevevector unit
Eb=hbar^2*kb^2/2/me;           %energy unit in joules
Eb_eV=Eb/e;                    %energy unit in eV
u0=150;                        %potential height in Eb units
a=2; b=0.025;                  %well, barrier widths in ab units
kp=(0:0.005:3.0)*pi/(a+b);     %vary k'
k=(-6:0.05:6)*pi/a;            %vary k (associated with energy)
%right hand side of energy equation (x=a, y=b)
```

```
fR=@(kv,x,y) cos(kv*(x+y));
%left hand side of energy equation (x=a, y=b)
fL=@(xE,V,x,y) ((V-2*xE)./(2*sqrt(u0-xE).*sqrt(xE))).*...
  sinh(sqrt(V-xE)*y).*sin(sqrt(xE)*x)+cosh(sqrt(V-xE)*y).* ...
  cos(sqrt(xE)*x);
%Evaluation and plotting (yR=RHS, yL=LHS)
yR=fR(kp,a,b);
yL=fL(k.^2,u0,a,b);  %Note: in dimensionless units energy=k^2
subplot(1,2,1)       %LHS and min(RHS), max(RHS) plotted versus k
line([min(k*a) max(k*a)],[max(yR), max(yR)],'Color','k',...
  'LineStyle','--')  %min(RHS)
hold on
plot(k*a,yL,'k')     %LHS
for i=1:length(k)
  if ((yL(i) <=  max(yR)) & (yL(i) >=  min(yR)))
    plot(k(i)*a,yL(i),'k.','MarkerSize',5)
  end
end
legend('RHS','LHS','roots',1)
line([min(k*a) max(k*a)],[min(yR), min(yR)],'Color','k',...
  'LineStyle','--')  %max(RHS)
xlabel('ka','FontSize',14)
ylabel('LHS, RHS','FontSize',14)
str1=cat(2,'Kronig-Penney Model: u0=',num2str(u0,'%5.2f'),...
  ' \epsilon_b',', a=',num2str(a,'%5.3f'),' a_b',', b=',...
  num2str(b,'%5.3f'),' a_b, ');
title(['          ',str1])
axis tight
hold off
%
subplot(1,2,2)       %Energy eigenvalues versus k_prime
%find the lowest energy guess for the first kp point
eps_guess=searchguess(1.e-3,kp(1));
nr=0;                %root counter
test_old=0.0;        %variable to check when kp*pi/(a+b)=integer
for i=1:length(kp)
  yRi=fR(kp(i),a,b);
  if (yRi <= 1 & yRi >= -1)          %if solutions exist
    nr=nr+1;                         %root counter
    kr(nr)=kp(i);                    %store the related k'
    %find the energies for which FofE=fL-fR=0, for each k'
    eps(nr) = fzero(@(xE) FofE(xE,kp(i)),eps_guess);
    eps_guess=eps(nr);               %use this as next guess
    test_new=mod(kp(i)*(a+b)/pi,1); %to check if k'=integer
    %There is an energy gap when kp*pi/(a+b)=integer, so we need
    %to search for a higher energy guess at those points
    if (test_new < test_old)
      eps_guess=searchguess(eps_guess,kp(i));
      %fprintf(' ********** xguess=%8.3f\n',xguess);
    end
    test_old=test_new;
```

```
  end
end
%
%use the calculated values and apply symmetry to get the whole
%BZ plot for the energy bands
kp_a=kr*(a+b);          %k' variable where a root was found
ep_a=eps;               %corresponding found energy root
kp_b=kp_a-kp_a(end);    %get the reflection of k'
ep_b=ep_a(end:-1:1);    %use reflection on energy also
plot(kp_a,ep_a,'k.','MarkerSize',1)
hold on
plot(kp_b,ep_b,'k.','MarkerSize',1)
xlabel('k{''}(a+b)','FontSize',14)
ylabel('\epsilon(k'') (\epsilon_b)','FontSize',14)
str2=cat(2,'\epsilon_b=',num2str(Eb_eV,'%5.2f'),...
  ' eV',', a_b=',num2str(ab/1e-10,'%3.2f'),...
  ' Angs');
title(['          ',str2])

function [y]=searchguess(x,kk)
%This function does a simple search for a root by tracking
%a sign change in the FofE function
global u0 a
del=13.2e-3*u0/a+0.2*x;    %step size to search for a root
                           %notice it depends on u0, a - may tweek
xi=x;
xf=xi;
for i=1:50*del
  xf=xf+del;
  if(FofE(xi,kk)*FofE(xf,kk) <= 0.0)
    y=xf;          %root exits, so return as guess
    return
  end
  xi=xf;
end

function [y]=FofE(eps,kk)
global fL fR u0 a b
%This function is the difference of left and right functions
y=fL(eps,u0,a,b)-fR(kk,a,b);
```

## 6.5 Electron in a General Periodic Potential - the Central Equation

In this section, we consider a general but unspecified one-dimensional periodic potential $u(x)$ for a crystal lattice of constant $a$. It builds on the concept of the nearly free electron model of Section 6.2, but it is not limited to the BZ boundaries. We first recall that, for a periodic potential, $u(x)$ is translationally invariant; i.e., $u(x+a) = u(x)$. Such potential may be expanded as a Fourier series in the reciprocal lattice vectors $G$. That is,

$$u(x) = \sum_G u_G e^{iGx}, \tag{6.5.17}$$

for $-\infty < G < \infty$ and where $u_G$ is the Fourier transform of $u(x)$. To point out a property of the potential, we note that if we let $u_G = u_{-G}$ and we neglect $u_{G=0}$, then the sum can be written as

$$\sum_G u_G e^{iGx} = \sum_{G<0} u_G e^{iGx} + \sum_{G>0} u_G e^{iGx} = \sum_{G>0} u_{-G} e^{-iGx} + \sum_{G>0} u_G e^{iGx} = \sum_{G>0} u_G \left(e^{-iGx} + e^{iGx}\right),$$

or

$$u(x) = 2\sum_{G>0} u_G \cos Gx, \tag{6.5.18}$$

which also ensures $u(x)$ is real and symmetric about $x = 0$. We will come back to this form of the potential later but in what follows we continue to use the form given by Equation 6.5.17. We are ready to set up the Schrodinger equation for the periodic potential $u(x)$ in the case of one electron per atom. We write

$$\left(\frac{p^2}{2m} + u(x)\right)\psi(x) = \varepsilon\psi(x). \tag{6.5.19}$$

We will let $k$ be the allowed wavevectors, according to the boundary conditions, and expand $\psi(x)$ as a Fourier series in those wavevectors or

$$\psi(x) = \sum_k C_k e^{ikx}. \tag{6.5.20}$$

With the above definitions, we have $\left(\frac{p^2}{2m}\right)\psi(x) = -\frac{\hbar^2}{2m}\frac{d^2}{dx^2}\psi(x) = \frac{\hbar^2}{2m}\sum_k k^2 C_k e^{ikx}$, and $u(x)\psi(x) = \sum_G\sum_k u_G e^{iGx} C_k e^{ikx}$, so that Equation 6.5.19 becomes

$$\sum_k \frac{\hbar^2 k^2}{2m} C_k e^{ikx} + \sum_G\sum_k u_G C_k e^{i(k+G)x} = \varepsilon\sum_k C_k e^{ikx}. \tag{6.5.21}$$

We will now use the integral property

$$\int_0^a e^{i(k-k')x}dx = a\delta_{k,k'}, \tag{6.5.22a}$$

where

$$\delta_{k,k'} \equiv \begin{cases} 1 \text{ if } k = k' \\ 0 \text{ if } k \neq k' \end{cases}, \tag{6.5.22b}$$

is the Kronecker delta function. Multiplying Equation 6.5.21 by $e^{-ik'x}$, integrating over $x$ on $[0, a]$, and using the Kronecker delta function property we get

$$a\frac{\hbar^2 k'^2}{2m} C_{k'} + a\sum_G u_G C_{k'-G} = a\varepsilon C_{k'}. \tag{6.5.23a}$$

Dividing this by $a$, replacing the symbol $k'$ back to $k$, and using $\lambda_k \equiv \hbar^2 k^2/2m$, we find the equation

$$(\lambda_k - \varepsilon)C_k + \sum_G u_G C_{k-G} = 0, \tag{6.5.23b}$$

for each value of $k$. This is the central equation. The idea is to find the $C_{k-G}$'s, and then, by Equation 6.5.20, the wavefunction $\psi(x)$. In order for Equation 6.5.23b to be satisfied, the determinant of the coefficients is set to zero. In turn, this gives the energy bands versus the wavevector k. Before considering this equation in greater detail, we return to the wavefunction $\psi(x)$ of Equation 6.5.20 and verify that it obeys Bloch's theorem. First we rewrite it as a sum over the $G$'s as

$$\psi_k(x) = \sum_G C_{k-G} e^{i(k-G)x} = \sum_G C_{k-G} e^{-iGx} e^{ikx} \tag{6.5.24a}$$

and define $\Psi_k(x) \equiv \sum_G C_{k-G} e^{-iGx}$ to write $\psi_k(x) = \Psi_k(x) e^{ikx}$. Displacing $x$ by a translation $T$ we have that

$$\psi_k(x+T) = \Psi_k(x+T) e^{ik(x+T)}; \tag{6.5.24b}$$

however, $\Psi_k(x+T) = \sum_G C_{k-G} e^{-iG(x+T)} = \sum_G C_{k-G} e^{-iGx} e^{-iGT} = \Psi_k(x) e^{-iGT}$. But, from Chapter 2, we know that $e^{-iGT} = 1$ due to $GT = 2n\pi$, and Equation 6.5.24b becomes

$$\psi_k(x+T) = \Psi_k(x) e^{ikx} e^{ikT} = \psi_k(x) e^{ikT}, \tag{6.5.24c}$$

which is of the Bloch form as in Equation 6.3.9. Here, $e^{ikT}$ is the phase factor by which the Bloch function is multiplied when a crystal lattice translation $T$ is made, as discussed in Section 6.3.

**Example 6.5.0.1**

In this example, we show how Equation 6.5.23b is used to get the free electron result. In such case we know that the potential of Equation 6.5.17 is zero and we set $u_G = 0$ then Equation 6.5.23b gives $(\lambda_k - \varepsilon) C_k = 0$ and all $C_{k-G} = 0$ except for $C_k$. Thus from Equation 6.5.20 we have $\psi(x) = C_k e^{ikx}$; i.e., a single plane wave, where we can take $C_k$ as a constant. Also, here we see that $\varepsilon = \lambda_k = \hbar^2 k^2 / 2m$, which is what we expect for the free electron case.

We will come back to this section later, but before that we discuss what is referred to as the reduced Brillouin zone (BZ) scheme. We introduce this concept next.

## 6.6 Empty Lattice Approximation

Whenever we ignore the crystal potential in Equation 6.5.23b, the energy band structure is essentially that of a free electron as in Example 6.5.0.1. This is what is termed as the empty lattice approximation. The general expression for the free electron energy in three dimensions is

$$\varepsilon_k = \frac{\hbar^2 \vec{k}^2}{2m}, \qquad -\infty < k < \infty. \tag{6.6.25}$$

This expression is referred to as the extended BZ scheme and represents a parabolic behavior for $\varepsilon_k$ versus $\vec{k}$ for all $\vec{k}$. However, band structures are for the most part conveniently plotted in terms of $\varepsilon_k$ versus $\vec{k}$ in the first BZ. If the wavevectors fall outside the first BZ, they are carried back into the first BZ by subtracting a suitable reciprocal lattice vector. In the expression of Equation 6.6.25, $\vec{k}$ can take on any value, except that we would like to make the replacement

$$\vec{k} \to \vec{k} + \vec{G}, \tag{6.6.26}$$

and restrict the new $\vec{k}$ to the first BZ. With this replacement, the energy becomes

$$\varepsilon_k = \frac{\hbar^2}{2m} \left( \vec{k} + \vec{G} \right)^2 \qquad \vec{k} \text{ on 1st BZ}, \tag{6.6.27a}$$

with $\vec{G}$ taking on appropriate reciprocal lattice points so as to keep $\vec{k}$ in the first BZ. This method of plotting the energy is referred to as the reduced BZ scheme. More specifically, knowing the specific $\vec{G}$'s for a particular structure and restricting the components of $\vec{k}$ to the first BZ we work with

$$\varepsilon_k = \frac{\hbar^2}{2m}\left((k_x+G_x)^2+(k_y+G_y)^2+(k_z+G_z)^2\right). \tag{6.6.27b}$$

**Example 6.6.0.1**

In this example we obtain a plot of the free electron energy bands for the simple cubic system in the reduced zone scheme in the $[100]$ direction; that is, we take $[k_x,k_y,k_z]=[k_x,0,0]$ in Equation 6.6.27b. The energy becomes $\varepsilon_k=(\hbar^2/2m)\left((k_x+G_x)^2+G_y^2+G_z^2\right)$. For the simple cubic the reciprocal lattice vectors are $\vec{b}_1=\frac{2\pi}{a}\hat{\imath}$, $\vec{b}_2=\frac{2\pi}{a}\hat{\jmath}$, and $\vec{b}_3=\frac{2\pi}{a}\hat{k}$. The first BZ has boundaries at $k_x=\pm\vec{b}_1/2$, $k_y=\pm\vec{b}_2/2$, and $k_z=\pm\vec{b}_3/2$, so that the range of $k_x$ is $[-\pi/a,\pi/a]$. The reciprocal lattice points lie at the reciprocal lattice vectors given by $\vec{G}=\nu_1\vec{b}_1+\nu_2\vec{b}_2+\nu_3\vec{b}_3=(2\pi/a)(\hat{\imath}+\hat{\jmath}+\hat{k})$, where $\nu_1$, $\nu_2$, and $\nu_3$ are $\pm$ integers, including zero. There is a band of energy for each set of $\nu_1,\nu_2$, and,$\nu_3$ values as shown in Table 6.6.1.

Table 6.6.1: A few examples of the different energy bands associated with the empty lattice approximation for the simple cubic structure. Bands for which $\varepsilon_k(k_x)$ is identical are degenerate bands. For example, bands 2, 3 are degenerate, and similarly are bands 4, 5, 6, and 7, etc.

| $\nu_1$ | $\nu_2$ | $\nu_3$ | $2m\varepsilon_k(k_x)/\hbar^2$ | $2m\varepsilon_k(0)/\hbar^2$ | Band |
|---|---|---|---|---|---|
| 0 | 0 | 0 | $k_x^2$ | 0 | 1 |
| 1 | 0 | 0 | $(k_x+2\pi/a)^2$ | $4(\pi/a)^2$ | 2 |
| $\bar{1}$ | 0 | 0 | $(k_x-2\pi/a)^2$ | $4(\pi/a)^2$ | 3 |
| 0 | 1 | 0 | $k_x^2+(2\pi/a)^2$ | $4(\pi/a)^2$ | 4 |
| 0 | $\bar{1}$ | 0 | $k_x^2+(2\pi/a)^2$ | $4(\pi/a)^2$ | 5 |
| 0 | 0 | 1 | $k_x^2+(2\pi/a)^2$ | $4(\pi/a)^2$ | 6 |
| 0 | 0 | $\bar{1}$ | $k_x^2+(2\pi/a)^2$ | $4(\pi/a)^2$ | 7 |
| 1 | 1 | 0 | $(k_x+2\pi/a)^2+(2\pi/a)^2$ | $8(\pi/a)^2$ | 8 |
| 1 | 0 | 1 | $(k_x+2\pi/a)^2+(2\pi/a)^2$ | $8(\pi/a)^2$ | 9 |
| 1 | $\bar{1}$ | 0 | $(k_x+2\pi/a)^2+(2\pi/a)^2$ | $8(\pi/a)^2$ | 10 |
| 1 | 0 | $\bar{1}$ | $(k_x+2\pi/a)^2+(2\pi/a)^2$ | $8(\pi/a)^2$ | 11 |
| $\bar{1}$ | 1 | 0 | $(k_x-2\pi/a)^2+(2\pi/a)^2$ | $8(\pi/a)^2$ | 12 |
| $\bar{1}$ | 0 | 1 | $(k_x-2\pi/a)^2+(2\pi/a)^2$ | $8(\pi/a)^2$ | 13 |
| ... | ... | ... | ... | ... | ... |

Degenerate bands have the same energy.

We will use our previous units for which $a=\bar{a}a_b$ with $a_b=1$Å and energy $\varepsilon=\bar{\varepsilon}_kE_b$ with $E_b=\hbar^2/(2ma_b^2)=3.81\,eV$ we can write $\bar{\varepsilon}_k=(\bar{k}_x+\bar{G}_x)^2+\bar{G}_y^2+\bar{G}_z^2$. The results are shown in Figure 6.6.10.

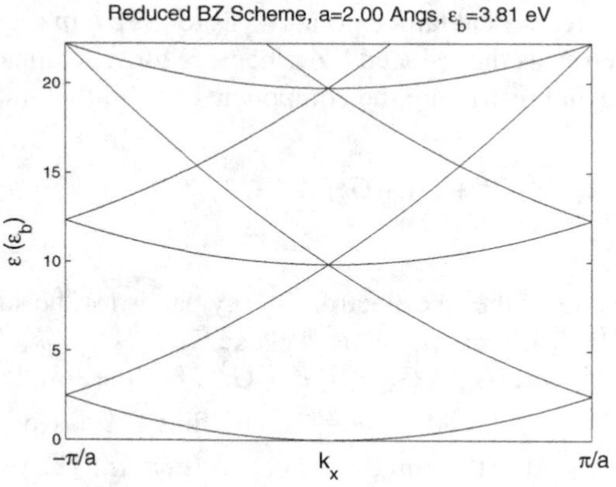


Figure 6.6.10: The free electron energy bands along the [100] direction for the simple cubic system in the reduced zone scheme as discussed in Example 6.6.0.1 and for which some energy band examples are listed in Table 6.6.1.

In the present case we have that $\bar{G}_x = v_1 2\pi/\bar{a}$, $\bar{G}_y = v_2 2\pi/\bar{a}$ and $\bar{G}_z = v_3 2\pi/\bar{a}$. For the calculations we let $\bar{k}$ lie in the range $\pm\pi/\bar{a}$. We have let $\bar{a} = 2$ and carried out the calculations outlined above using the script empty_lattice_SC.m whose listing follows below. Notice that the variable "vmax", currently set to 1, determines the bands calculated; the variable "dir" is currently set to perform the calculations in the [100] direction.

```
%copyright by J. E Hasbun and T. Datta
%empty_lattice_SC.m
%The empty lattice approximation is used for the
%simple cubic system in the reduced zone scheme
clear
%SC
h=6.62606896e-34;               %Planck'constant (J.s)
hbar=h/2./pi;                   %hbar (J.s)
me=9.10938215e-31;              %electron mass (kg)
e=1.602176487e-19;              %electronic charge
ab=1e-10;                       %1 angstom unit of distance
kb=1/ab;                        %wavevector unit
Eb=hbar^2*kb^2/2/me;            %energy unit in joules
Eb_eV=Eb/e;                     %energy unit in eV
%fprintf('Energy unit Eb=%5.3g J, or %5.3f eV\n',Eb,Eb_eV)
a = 2;             %Lattice constant in ab units,
a1=a*[1,0,0];      %using the cubic cell vectors
a2=a*[0,1,0];
a3=a*[0,0,1];
%Reciprocal lattice vectors follow
Vt=dot(a1,cross(a2,a3));  %system's unit cell volume
b1=2*pi*cross(a2,a3)/Vt;
b2=2*pi*cross(a3,a1)/Vt;
b3=2*pi*cross(a1,a2)/Vt;
vmax=1;                          %max v integer (determines bands)
dir=[1,0,0];                     %BZ direction chosen
```

```
%max k's on the first BZ - Simple Cubic
Gs=b1+b2+b3;                     %make a small G
kx_max=Gs(1)/2;                  %max kx from Gs bisector
ky_max=Gs(2)/2;                  %max ky  "     "
kz_max=Gs(3)/2;                  %max kz  "     "
kmax=[kx_max,ky_max,kz_max];
%chosen k direction
dirk=[dir(1)*kmax(1),dir(2)*kmax(2),dir(3)*kmax(3)];
x=-1:0.1:1;                      %direction step size
hold on
for v1=-vmax:vmax
  for v2=-vmax:vmax
    for v3=-vmax:vmax
      G=[v1*b1+v2*b2+v3*b3]; %reciprocal lattice vector
      for i=1:length(x)
        kdir=x(i)*dirk;         %step in the chosen direction
        eps(i)=((kdir(1)+G(1))^2+(kdir(2)+G(2))^2+(kdir(3)+G(3))^2);
      end
      plot(x,eps,'k')
    end
  end
end
epm=(Gs(1)^2+Gs(2)^2+Gs(3)^2); %for plotting purposes
axis([-1 1 0 0.75*epm])
xlabel('k_x','FontSize',14)
ylabel('\epsilon (\epsilon_b)','FontSize',14)
str=cat(2,'Reduced BZ Scheme, a=',num2str(a,'%3.2f'),' Angs, ',...
  '\epsilon_b=',num2str(Eb_eV,'%3.2f'),' eV');
title(str,'FontSize',13)
set(gca,'XTick',[-1,1])
set(gca,'XTickLabel','')
set(gca,'XTickLabel',{'-pi/a','pi/a'})
```

## 6.7 Solution of the Central Equation

In this section, we go back to study Equation 6.5.23b and consider a possible solution. This equation actually represents a set of equations for the $C_k$'s for all reciprocal lattice vectors $G$. It has a solution if the determinant of the $C_k$ coefficients vanishes. We now recall the special form of the potential $u(x)$ as expressed in Equation 6.5.18. Suppose that we approximate it with

$$u(x) = 2u_G \cos gx; \qquad (6.7.28)$$

that is, only one component of $G$ is kept, $G = g$, with $g$ the shortest possible value of $G$. We can enumerate the $G$'s as $G_1 = g$, $G_2 = 2g$, $G_3 = 3g$, and so on. Since only one value of $G$ is retained in the $u$'s, Equation 6.5.23b becomes

$$(\lambda_k - \varepsilon)C_k + u_g C_{k-g} + u_g C_{k+g} = 0, \qquad (6.7.29)$$

where, as before, we have used $u_g = u_{-g}$. This equation is to be solved for all $k$, and there is only one of these equations for each value of $k$. If in this equation we make the replacement $k \to k - g$

we get

$$(\lambda_{k-g}-\varepsilon)\,C_{k-g}+u_gC_{k-2g}+u_gC_k=0; \tag{6.7.30a}$$

replacing $k \to k+g$ we have,

$$(\lambda_{k+g}-\varepsilon)\,C_{k+g}+u_gC_k+u_gC_{k+2g}=0; \tag{6.7.30b}$$

continuing this way and replacing $k \to k+2g$ we have

$$(\lambda_{k+2g}-\varepsilon)\,C_{k+2g}+u_gC_{k+g}+u_gC_{k+3g}=0; \tag{6.7.30c}$$

and so on. There is an infinite set of these equations and, starting with Equation 6.7.29, we can organize them in matrix form as follows

$$\begin{pmatrix} \cdots & \vdots & \vdots & \vdots & \vdots & \vdots & \cdots \\ \cdots & (\lambda_{k-g}-\varepsilon) & u_g & 0 & 0 & 0 & \cdots \\ \cdots & u_g & (\lambda_k-\varepsilon) & u_g & 0 & 0 & \cdots \\ \cdots & 0 & u_g & (\lambda_{k+g}-\varepsilon) & u_g & 0 & \cdots \\ \cdots & 0 & 0 & u_g & (\lambda_{k+2g}-\varepsilon) & u_g & \cdots \\ \cdots & \vdots & \vdots & \vdots & \vdots & \vdots & \cdots \end{pmatrix} \begin{pmatrix} \vdots \\ C_{k-g} \\ C_k \\ C_{k+g} \\ C_{k+2g} \\ C_{k+3g} \\ \vdots \end{pmatrix} = \begin{pmatrix} \vdots \\ 0 \\ 0 \\ 0 \\ 0 \\ 0 \\ \vdots \end{pmatrix}. \tag{6.7.30d}$$

The vanishing of the determinant of this matrix yields the eigenvalues and, while the matrix is infinitely long, only a small block, whose determinant is to vanish, is needed to be considered here. The eigenvalues versus $k$ produce the energy bands. Each band is characterized by an index $n$; therefore, $\varepsilon_{nk}$ is the $n$th energy band versus $k$. The bands result because each determinant has various roots for every $k$ point and each root falls on a particular band. The bands are usually sorted from lowest to highest values. The order of the coefficients' matrix determines the number of eigenvalues per $k$ point. Bands with the same energy are said to be degenerate.

**Example 6.7.0.1**

In this example we consider a smaller block from the matrix of Equation 6.7.30d. We limit the coefficients' coupling to $C_{k-g}$ and $C_k$ to obtain the $2\times 2$ block

$$\begin{pmatrix} (\lambda_{k-g}-\varepsilon) & u_g \\ u_g & (\lambda_k-\varepsilon) \end{pmatrix} \begin{pmatrix} C_{k-g} \\ C_k \end{pmatrix} = 0, \tag{6.7.31a}$$

where the determinant of the coefficients is set to zero to obtain

$$\begin{vmatrix} (\lambda_{k-g}-\varepsilon) & u_g \\ u_g & (\lambda_k-\varepsilon) \end{vmatrix} = 0 = (\lambda_{k-g}-\varepsilon)(\lambda_k-\varepsilon)-u_g^2. \tag{6.7.31b}$$

Solving this equation for $\varepsilon$ results in the two energy bands as a function of $k$

$$\varepsilon_k = \frac{\lambda_k+\lambda_{k-g}}{2} \pm \sqrt{\left(\frac{\lambda_k-\lambda_{k-g}}{2}\right)^2 + u_g^2}. \tag{6.7.31c}$$

In this expression the value of $u_g$ determines the gap splitting that occurs at the BZ boundary. To see this, in the calculations that follow, we let $g = 2\pi/a$ and let $k$ lie parallel to $g$. We then consider the BZ boundary at $k = \pm g/2$, to get $\lambda_k = \hbar^2k^2/2m$, $\lambda_{k-g} = \hbar^2|(\vec{k}-\vec{g})|^2/2m = \hbar^2(k^2+$

$g^2 - 2|k||g|\cos(0))/2m$. At the zone boundary $k = \pm g/2$ so that $\lambda_k = \lambda_{k-g} = \hbar^2 g^2/8m$ and the energies of Equation 6.7.31c become $\varepsilon_{k=\pm g/2} = \lambda_k \pm u_g$. The gap value at the zone boundary is thus $\varepsilon_{gap} = \varepsilon_+ - \varepsilon_- = 2u_g$. If we continue with these zone boundary energy values, we can go back to Equation 6.7.31a and see that, for example, at $k = g/2$, $k - g = -g/2$ and $(\lambda_{-g/2} - \varepsilon_{g/2})C_{-g/2} + u_g C_{g/2} = \pm u_g C_{-g/2} + u_g C_{g/2} = 0$, where we have used $\lambda_k - \varepsilon_{k=\pm g/2} = \pm u_g$ as found above. We thus have that $C_{g/2} = \pm C_{-g/2}$. If we next let $C_{-g/2} = A$ then $C_{g/2} = \pm A$ and from Equation 6.5.20, since there are only two allowed wavevectors, $\psi(x) = C_{g/2}e^{igx/2} \pm C_{g/2}e^{igx/2}$, so that with $g = 2\pi/a$ the wavefunctions correspond to the symmetric and antisymmetric standing wave combinations of Equations 6.2.4 of Section 6.2 where $A = \sqrt{2/a}$. If we compare Equation 6.2.7 with Equation 6.7.28 we also see that $u_0 = 2u_G$. Finally, as found above, near the zone boundary $\lambda_k \sim \lambda_{k-g}$ and we can approximate the bands' behavior near $g/2$ with

$$\varepsilon_k \sim \frac{\lambda_k + \lambda_{k-g}}{2} \pm u_g = \frac{\hbar^2(k^2 + k^2 + g^2 - 2kg)}{4m} \pm u_g, \tag{6.7.32a}$$

so that we find that in the limit as $k \to g/2$

$$\lim_{k\to g/2} \frac{d\varepsilon_k}{dk} \sim \lim_{k\to g/2} \frac{\hbar^2(4k - 2g)}{4m} \to 0. \tag{6.7.32b}$$

We see, therefore, that the bands approach the zone boundaries with zero slope. This will be evident later when we perform the calculations of the bands as the next example demonstrates.

**Example 6.7.0.2**

Example 6.7.0.1 obtained the analytic energy bands for the case when a $2 \times 2$ block of Equation 6.7.30d is considered. In this example, we consider the same case except that it will be carried out numerically. This has the advantage that it can be modified to carry out the bands of an $N \times N$ block. The idea is that rather than obtaining the eigenvalues analytically through the zero of the determinant (as in Equation 6.7.31b), we solve for the eigenvalue problem. That is, we rewrite Equation 6.7.31a as

$$\begin{pmatrix} \lambda_{k-g} & u_g \\ u_g & \lambda_k \end{pmatrix} \begin{pmatrix} C_{k-g} \\ C_k \end{pmatrix} = \begin{pmatrix} \varepsilon_k & 0 \\ 0 & \varepsilon_k \end{pmatrix} \begin{pmatrix} C_{k-g} \\ C_k \end{pmatrix}, \tag{6.7.33}$$

and search for the eigenvalue of the coefficients' matrix on the left for every value of $k$, sorted from lowest to highest. These eigenvalues are the energy bands $\varepsilon_{nk}$ for each band $n$. Plotting these values versus $k$ for the $2 \times 2$ system should yield the same results as in Example 6.7.0.1. As we have done before, we make use of distance units of $a_b = 1$Å and let the wavevector have units of $k_b = 1/a_b$ so that $k = \bar{k}k_b$. Similarly, $g = 2\pi/a = \bar{g}k_b$, so that $\bar{g} = 2\pi/\bar{a}$. The energy unit $\varepsilon_b = \hbar^2/(2ma_b^2) = 3.81\,eV$. In this way we write $\lambda_k = \bar{k}^2\varepsilon_b$, $u_g = \bar{u}_g\varepsilon_b$. The calculations have been carried out with the script central_eq_bands.m using $\bar{a} = 1$ and $\bar{u}_g = 2$ as shown in Figure 6.7.11. As in the previous example, we let $k$ lie parallel to $g$ and vary $k$ in the range $-g/2 \le k \le g/2$. Also, recall that the gap at the zone boundary is $2u_g$. Further, notice that the bands approach the zone boundaries with zero slope, as expected. The condition of $u_g = 0$ is gapless and corresponds to the empty lattice approximation. The results shown here are identical to the analytic bands found in Example 6.7.0.1 (see Exercise 6.9.7).

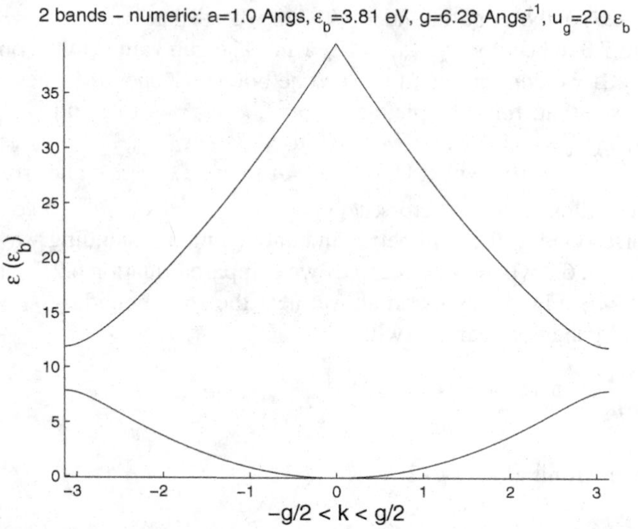


Figure 6.7.11: The numerical results for the bands corresponding to the $2 \times 2$ system of Example 6.7.0.2, obtained by finding the numerical eigenvalues of Equation 6.7.33 versus $k$. The energy unit is $\varepsilon_b = \hbar^2/(2ma_b^2) = 3.81\,eV$, the wavevector unit is $k_b = 1\text{Å}^{-1}$, and the value of potential used is $u_g = 2\varepsilon_b$.

The code listing, central_eq_bands.m, used in obtaining Figure 6.7.11 follows. It can be modified to do a larger block of equations which would yield more bands. Notice that the eigenvalues are obtained through the use of MATLAB's internal eigenvalue solver, "`eig()`".

```
%copyright by J. E Hasbun and T. Datta
%central_eq_bands.m
%Plots the numeric solutions for the central equation
%for an Nb band system. The eigenvalues are obtained here
%after setting up the matrix of the coefficients.
clear;
clear all;
h=6.62606896e-34;                %Planck's constant (J.s)
hbar=h/2./pi;                    %hbar (J.s)
me=9.10938215e-31;               %electron mass (kg)
e=1.602176487e-19;               %electronic charge
ab=1e-10;                        %1 angstom unit of distance
kb=1/ab;                         %wavevector unit
Eb=hbar^2*kb^2/2/me;             %energy unit in joules
Eb_eV=Eb/e;                      %energy unit in eV
%fprintf('Energy unit Eb=%5.3g J, or %5.3f eV\n',Eb,Eb_eV)
a=1.0;              %lattice constant in ab units
g=2*pi/a;
ak=-g/2:0.01:g/2; %k will be chosen parallel to g
ug=2;               %potential magnitude in Eb units
Nb=2;               %number of bands to do
%mg is the value of m in (k-(i-m)g) in the matrix elements
%The mg value chosen seems to work well for symmetry reasons.
```

```
if(mod(Nb,2)==0), mg=Nb/2; else mg=(Nb+1)/2; end
aM=zeros(Nb,Nb);  %define the matrix of coefficients
for ik=1:length(ak)
  for i=1:Nb
    %diagonal terms
    if(abs(i-mg) <= 1.e-3)
      aM(i,i)=ak(ik)^2;
    else
      thkg=acos(dot(ak(ik),g)/abs(ak(ik)*g));  %angle between k and g
      aM(i,i)=ak(ik)^2+((i-mg)*g)^2-2*ak(ik)*((i-mg)*g)*cos(thkg);
    end
  end
  for i=1:Nb-1
    aM(i,i+1)=ug;  %complete the tridiagonal matrix
    aM(i+1,i)=ug;
  end
  %get eigenvalues from smallest to highest for each k. Apparently,
  %there is no sorting needed here.
  Eiv(:,ik)=eig(aM);
end
hold on
for i=1:Nb
  plot(ak,Eiv(i,:),'k')
end
xlabel('-g/2 < k < g/2','FontSize',14)
ylabel('\epsilon (\epsilon_b)','FontSize',14)
str=cat(2,num2str(Nb,'%2.0f'),' bands - numeric: a=',...
  num2str(a,'%2.1f'),' Angs, \epsilon_b=',...
  num2str(Eb_eV,'%3.2f'),' eV, g=',num2str(g,'%3.2f'),...
' Angs^{-1}, u_g=',num2str(ug,'%3.1f'),' \epsilon_b');
title(str,'FontSize',12)
axis tight
```

We conclude this section by summarizing our findings. In the absence of a crystal potential, the energy bands are gapless and we have the empty lattice approximation. With the inclusion of a crystal potential, the resulting energy bands experience gaps that depend on the value of the potential. Finally, the energy bands approach the BZ boundaries with zero slope.

## 6.8 Counting Band Orbitals

For a one-dimensional lattice of lattice constant $a$, there are $N$ primitive cells so the values of the electron wavevector, $k = 0, \pm 2\pi/L, \pm 4\pi/L, \ldots, N\pi/L$, are all allowed because there are $N$ cells in a lattice of length $Na$. This is so because the zone boundary lies at $N\pi/L = N\pi/Na = \pi/a$. Since there are $N$ cells, each cell contributes one independent value of $k$ to each band. Taking into account the two possible orientations of the electron spin, then each band can sustain $2N$ orbitals. For example, if a given crystal has 1 atom per cell, each band would be half-filled, as is the case of a metal. This is the reason why metals are commonly modeled in terms of one atom per cell. If there were two electrons per cell, then for $N$ cells, the associated band would be full, as is the case for an insulator.

The same would be the case for a crystal with two atoms per cell with each atom contributing one valence electron to the band. Crystals with even number of valence electrons can be insulators. However, one has to consider if the bands overlap. If they overlap, the crystal is a metal because in that case the bands are not full. For diamond, silicon, and germanium, each has 2 atoms per cell with four valence electrons. This means that there are eight electrons per primitive cell. There will be four valence bands which at $T = 0K$ will be occupied by the eight electrons, because each band is occupied by two electrons (up and down spin). These systems will thus be insulators at $T = 0K$.

## 6.9 Chapter 6 Exercises

6.9.1. Show that the Bragg condition Equation 6.2.2 leads to the special k-space values $k = \pm G/2$ associated with Brillouin zone boundaries.

6.9.2. (a) Run code Kronig_Penney_model_numeric of Example 6.4.0.1 to reproduce Figure 6.4.9. (b) Modify the program in order to calculate the Kronig-Penney model band structure for a potential with $u_0 = 80\varepsilon_b$, $a = 4$ Å, and $b = 0.025$ Å.

6.9.3. Work out the details of the assertion of Equation 6.5.22.

6.9.4. Show that $u_G$ is the Fourier transform of $u(x)$, where $u(x)$ is the potential of Equation 6.5.17. Also, show that $u_{G=0}$ corresponds to the crystal potential average over a unit cell $(0 < x < a)$.

6.9.5. After reading Subsection 6.5, start with Equation 6.5.21 and show all the steps necessary to achieve Equation 6.5.23b.

6.9.6. After reading Example 6.6.0.1, reproduce Figure 6.6.10 for the [100] direction. By modifying the provided script, repeat the calculations along the [111] direction in the simple cubic system.

6.9.7. Read Examples 6.7.0.1, 6.7.0.2 and run the script central_eq_bands.m to reproduce Figure 6.7.11. Using the same parameters of Example 6.7.0.2, write a script that will perform the calculations of the analytic energy bands of Equation 6.7.31c. Investigate what happens as you vary the value of the potential $u_g$. Comment on your observations.

6.9.8. After reproducing Figure 6.7.11 using the script central_eq_bands.m of Example 6.7.0.2, modify it to enable it to do a $4 \times 4$ block system of equations. Run your new script for a value of $\bar{u}_g = 5$. Comment on your observations, especially the splitting at $k = 0$ as well as at the zone boundaries.

# 7

# *Semiconductor Crystal Properties*

**Contents**



## 7.1 Introduction

In solid state physics there are generally four categories of crystalline materials: metals, semimetals, semiconductors, and insulators. They are so categorized based on their electrical properties such as the conductivity ($\sigma$) or resistivity ($\rho = 1/\sigma$) and the electron concentration. Metallic crystals are characterized by their high electron concentration ($n$) and high $\sigma$ (low $\rho$); insulators have associated low $n$ and low $\sigma$ (high $\rho$). Semimetals and semiconductors lie in between the metals and the insulators. For the most part, metallic materials have $\sigma > 10^5\,(\Omega \cdot m)^{-1}$ with high electron or charge carrier concentrations as, for example, Cu and Na. Semimetals have slightly smaller conductivities and lower electron concentrations as, for example, As and Sb. Semiconductors are characterized by $10^{-6} < \sigma < 10^5\,(\Omega \cdot m)^{-1}$ with carrier concentrations that depend on temperature, as for example, elemental solids of Si and Ge as well as compound solids of InAs, and GaAs. Insulators have $\sigma < 10^{-6}\,(\Omega \cdot m)^{-1}$. The carriers in metals are the electrons, but in semiconductors the carriers can be electrons or holes depending on the type of impurity or dopant the semiconductor is doped with. When a semiconductor is pure; i.e., it contains no dopants, the semiconductor is referred to as intrinsic, else it is an extrinsic semiconductor. We thus see that the material conductivities are closely tied with the carrier concentrations as shown in Figure 7.1.1.

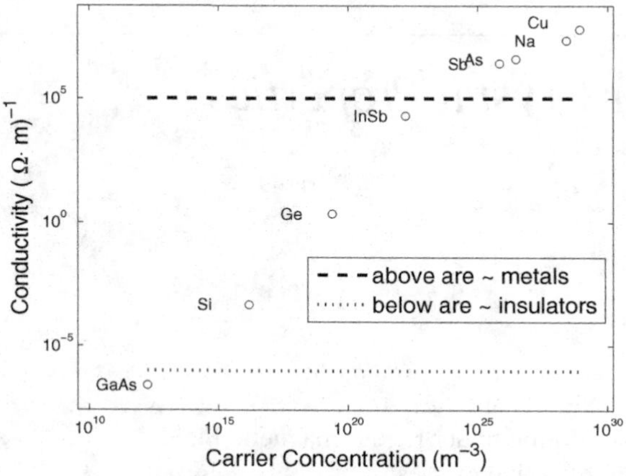


Figure 7.1.1: A few illustrative room temperature conductivity values (circles) for metals and semiconductors versus their corresponding carrier concentration, assuming no impurities. The metals and semimetals are found above the dashed line, while semiconductors lie approximately between the dotted and the dashed lines. The region below the dotted line is presumably the insulator region.

The following code listing, mat_sigmaVs_conc.m, can be used to reproduce most of Figure 7.1.1.

```
%copyright by J. E Hasbun and T. Datta
%mat_sigmaVs_conc.m
%plots the data of some materials' conductivities vs their carrier
%concentrations
clear
%Using typical conductivity values and intrinsic carrier concentrations
mat={'GaAs' 'Si' 'Ge' 'InSb' 'Sb' 'As' 'Na' 'Cu'};
sig=[2.55e-07 4.59e-04 2.23 2.00e+04...
   2.56e6 3.85e6 2.11e+07 5.88e+07];
nv=[1.79e+12 1.45e+16 2.4e+19 1.60674e+22...
  7e25 3e26 2.60e+28 8.45e+28];
%
loglog(nv,sig,'ko','MarkerSize',5)
hold on
nm=length(mat);
text(nv*(1-0.99),sig*2,mat,'FontSize',10)
line([nv(1) nv(end)],[1e5 1e5],'Color','k',...
  'LineStyle','--','LineWidth',2);    %metals above this line
line([nv(1) nv(end)],[1e-6 1e-6],'Color','k',...
  'LineStyle',':','LineWidth',2);     %insulators below this line
xlabel('Carrier Concentration (m^{-3})','FontSize',14)
ylabel('Conductivity ( \Omega\cdot m)^{-1}','FontSize',14)
```

Semiconductor materials contain elements from groups IV, such as Si and Ge; groups III-V, such as GaAs and GaSb; and groups II-VI, such as ZnSe and CdTe. At $T = 0K$, semiconductors are characterized by an electronic band structure scheme as shown in Figure 7.1.2.

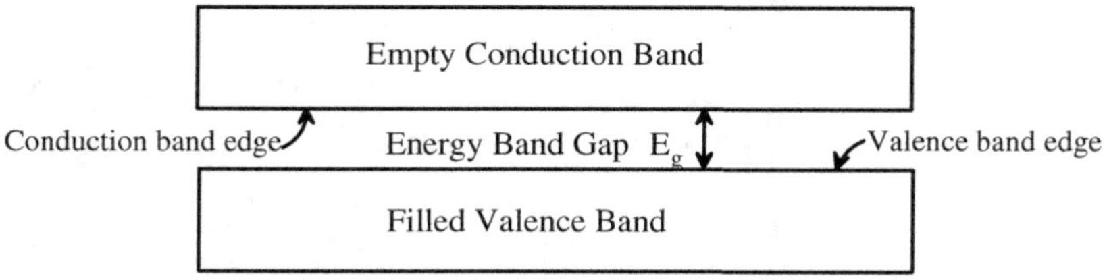


Figure 7.1.2: A typical $T = 0K$ valence and conduction band structure scheme for semiconductors, including the band gap, $E_g$.

The conduction band edge ($E_c$) refers to the bottom of the conduction band. Similarly, the valence band edge ($E_v$) is associated with the top of the valence band. The band gap, $E_g = E_c - E_v$, represents an energy barrier between the valence and conduction bands. It is the difference between the conduction and valence band edges. At $T = 0K$, the valence bands are filled with electrons. There are two electrons per band. Since semiconductors have four valence electrons per atom and two atoms per cell, it takes eight electrons to fill the four valence bands. As the temperature increases from $T = 0K$, the electrons are thermally excited from the valence to the conduction band; in the process, the empty orbitals left by the excited electrons become holes. Both electrons and holes contribute to the electrical conductivity. Holes are similar to electrons, but they have a positive charge. We will discuss their mass later below.

### 7.1.1 Band Gap Significance

The importance of the band gap stems from the observation that in intrinsic semiconductors, as we will see later, the charge carrier concentration $n \propto \exp(-E_g/2k_BT)$ and that the conductivity $\sigma \propto n$ so that $\sigma \propto \exp(-E_g/2k_BT)$ or $\ln(\sigma/\sigma_0) = -E_g/2k_BT$, where $\sigma_0$ is taken to be the conductivity value when $E_g = 0$. Thus the intrinsic semiconductor carrier concentration and thereby the conductivity is largely controlled by the ratio $E_g/2k_BT$. If the band gap is large, the conductivity is small, whereas if the band gap is small, the conductivity is large. Experimentally, band gaps can be found by studying the temperature dependence of the conductivity (see later below) as well as by means of optical absorption. Semiconductors can have direct or indirect absorption processes. One refers to direct or indirect band gap materials whenever the shortest energy difference between the valence and conduction bands occurs at BZ wavevector values of $k = 0$ or $k \neq 0$, respectively, as shown in Figure 7.1.3. Recall from a previous chapter that a hole is the absence of an electron, so that the process of exciting an electron from the valence band to the conduction band simultaneously creates a hole in the valence band. Thus, in (a) the absorbed photon produces an electron hole pair, with opposite momenta, and energy equal to the direct semiconductor band gap $E_g = \hbar\omega_g$. In (b) the absorbed photon produces an electron hole pair with opposite momenta as well as a phonon with momentum $k_\Omega = -k_c$, due to momentum conservation, where $k_c$ is the value of the wavevector where the indirect conduction band minimum occurs. In both cases, the electron ends up in the conduction band (CB), while the hole ends up in the valence band (VB).

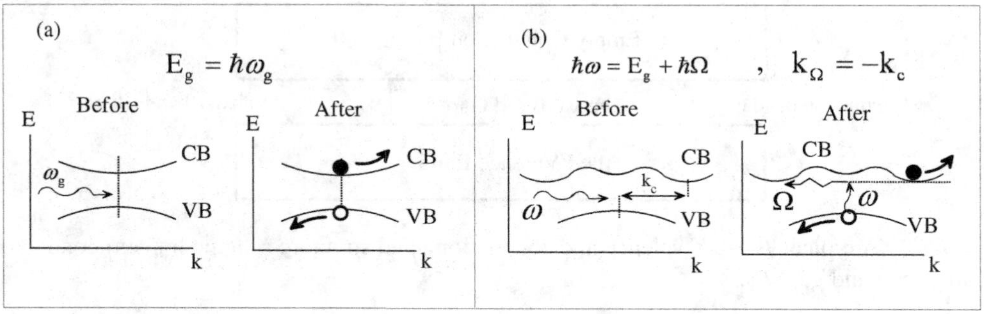


Figure 7.1.3: (a) Direct band gap photon absorption process with the formation of an electron-hole pair with opposite momenta. Here, the photon energy equals the band gap energy $E_g = \hbar\omega_g$. (b) Indirect band gap photon absorption with the creation of a phonon of energy $\hbar\Omega$ and an electron hole pair with opposite momenta. The photon energy is now $\hbar\omega = E_g + \hbar\Omega$ and the phonon momentum is $k_\Omega = -k_c$.

For direct band gap semiconductors electrons and holes can recombine, which leads to photon emission. Direct electron and hole recombination does not occur for indirect materials. Band gap information can be obtained through optical absorption. The onset of photon absorption in direct band gap materials occurs sharply and this can be used to identify the band gap. In indirect semiconductor materials, the transition is not as sharp since the onset of photon absorption is accompanied with phonon emission or absorption as shown in Figure 7.1.4.

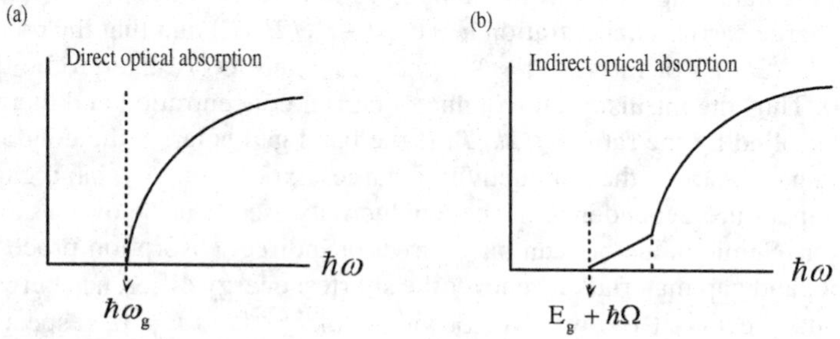


Figure 7.1.4: (a) Illustration of the optical absorption process. The onset of optical absorption occurs at $\hbar\omega_g$ for the direct gap system. The region before optical absorption is transparent. (b) The onset of optical absorption, $\hbar\omega = E_g + \hbar\Omega$, is illustrated for an indirect gap material.

Values of some semiconducting material band gaps are listed in Table 7.1.1. The band gap can also be deduced from the temperature dependence of the conductivity or the intrinsic carrier concentration through measurements of the Hall voltage versus temperature.

Table 7.1.1: Examples of energy band gaps (i=indirect, d=direct) for a few semiconducting crystals (Source: [3]).

| Crystal | Type | $E_g$ ($eV$ at $0K$) | $E_g$ ($eV$ at $300K$) |
|---|---|---|---|
| Diamond | i | 5.4 | |
| Si | i | 1.17 | 1.11 |
| Ge | i | 0.744 | 0.66 |
| InSb | d | 0.24 | 0.17 |
| InAs | d | 0.43 | 0.36 |
| InP | d | 1.42 | 1.27 |
| GaP | i | 2.32 | 2.25 |
| GaAs | d | 1.52 | 1.43 |
| GaSb | d | 0.81 | 0.68 |
| AlSb | i | 1.65 | 1.6 |
| CdS | d | 2.582 | 2.42 |
| CdSe | d | 1.840 | 1.74 |
| CdTe | d | 1.607 | 1.44 |

## 7.2 Electron and Hole Motion under Electromagnetic Fields

In semiconductors, a hole is the absence of an electron; therefore, its energy is the negative of the electron energy or $\varepsilon_h(k_h) = -\varepsilon_e(k_e)$, where the momentum wavectors are $k_e$ and $k_h$ for the electron and hole, respectively. By momentum conservation $k_h = -k_e$ and we assume that by symmetry $\varepsilon_{e,h}(-k_{e,h}) = \varepsilon_{e,h}(k_{e,h})$. Using the previous definition for group velocity, from Chapter 4, $\vec{v}_G = \vec{\nabla}_k \omega(\vec{k})$, and with $\omega(\vec{k}) = \varepsilon(\vec{k})/\hbar$, we have for both electrons and holes the group velocity expression

$$\vec{v}_{h,e}(\vec{k}) = \frac{1}{\hbar}\vec{\nabla}_k \varepsilon_{h,e}(\vec{k}). \tag{7.2.1}$$

Note, however, that for an electron, viewing it as a wave in one dimension for example, $v_e = (1/\hbar)d\varepsilon_e/dk_e$ and for the hole $v_h = (1/\hbar)d\varepsilon_h/dk_h = (1/\hbar)d(-\varepsilon_e)/d(-k_e) = (1/\hbar)d\varepsilon_e/dk_e = v_e$ and thus, in three dimensions, we can safely conclude that $\vec{\nabla}_k \varepsilon_h(\vec{k_h}) = \vec{\nabla}_k \varepsilon_e(\vec{k_e}) = \vec{\nabla}_k \varepsilon(\vec{k})$ or more generally $\vec{v}_h(\vec{k_h}) = \vec{v}_e(\vec{k_e}) = \vec{v}(\vec{k})$. The Lorentz force for a charge $q$ is $\vec{F} = q(\vec{E} + \vec{v} \times \vec{B})$; if we let the electron charge be $-e$, the hole charge be $e$, the force be $\vec{F} = d\vec{p}/dt = \hbar d\vec{k}/dt$, and using the above relations, we can write

$$\frac{d\vec{k}}{dt} = -\frac{e}{\hbar}(\vec{E} + \vec{v}_e(\vec{k_e}) \times \vec{B}) = -\frac{e}{\hbar}(\vec{E} + \frac{1}{\hbar}\vec{\nabla}_k \varepsilon(\vec{k}) \times \vec{B}), \tag{7.2.2a}$$

for the electron, and

$$\frac{d\vec{k}}{dt} = \frac{e}{\hbar}(\vec{E} + \vec{v}_h(\vec{k_h}) \times \vec{B}) = \frac{e}{\hbar}(\vec{E} + \frac{1}{\hbar}\vec{\nabla}_k \varepsilon(\vec{k}) \times \vec{B}), \tag{7.2.2b}$$

for the hole. We then see that each of these equations is identical to that seen by a charge $q$ with $q = -e$ for the electron and $q = e$ for the hole. An interesting aspect of Equation 7.2.2 can be seen if, for example, we set $\vec{E} = 0$, in which case the term $\vec{\nabla}_k \varepsilon(\vec{k}) \times \vec{B}$ is perpendicular to both $\vec{\nabla}_k \varepsilon(\vec{k})$ and $\vec{B}$, which implies that the charges move in a direction that is perpendicular to both the gradient of $\varepsilon(\vec{k})$ and the magnetic field, and this motion describes a surface of constant energy. This is illustrated in the following example.

**Example 7.2.0.1**
We next consider the hole motion from Equation 7.2.2b in the absence of an electric field and assuming that the magnetic field is given by $\vec{B} = B_z\hat{k}$. Let's also assume that $\varepsilon(\vec{k}) = (\hbar^2/2m)(k_x^2 + k_y^2)$. This leads to $\vec{\nabla}_k\varepsilon(\vec{k}) = (\hbar^2/m)(k_x\hat{i} + k_y\hat{j})$ and $\vec{\nabla}_k\varepsilon(\vec{k}) \times \vec{B} = (\hbar^2/m)B_z(k_y\hat{i} - k_x\hat{j})$ so that, ignoring the $z$ component in Equation 7.2.2b, we get the equations of motion as
$\frac{dk_x}{dt}\hat{i} + \frac{dk_y}{dt}\hat{j} = \frac{e}{m}B_z\left(k_y\hat{i} - k_x\hat{j}\right)$, or $\frac{dk_x}{dt} = \frac{e}{m}B_z k_y$ and $\frac{dk_y}{dt} = -\frac{e}{m}B_z k_x$,
which are two coupled sets of equations. We can solve these if we take the derivative of the first and substitute the second into the result; that is,
$\frac{d^2k_x}{dt^2} = \frac{e}{m}B_z\frac{dk_y}{dt} = -\frac{e^2}{m^2}B_z^2 k_x$,
so that, with $\omega_c \equiv eB_z/m$, we can write the solution as $k_x(t) = A\sin(\omega_c t)$, where $A$ is a constant. Repeating this process for $k_y$, we can similarly write $k_y(t) = A\cos(\omega_c t)$. These results are consistent with the above expressions for $dk_x/dt$ and $dk_y/dt$. Furthermore, we notice that $k_x^2(t) + k_y^2(t) = A^2$; that is, the motion is a circle of constant radius, which implies a constant energy surface.

## 7.3 Electron and Hole Effective Masses

In semiconductors the concept of effective mass is tied to the curvature of the valence band (for holes) or the conduction band (for electrons) near a band edge. To see how this works, we recall the example of the analytic solutions to the $2 \times 2$ block of the central equation in Chapter 6. We write those solutions for $\lambda_k - \lambda_{k-g} << u_g$ as

$$\begin{aligned}\varepsilon_k &= \frac{\lambda_k + \lambda_{k-g}}{2} \pm u_g\sqrt{1 + \left(\frac{\lambda_k - \lambda_{k-g}}{2}\right)^2 \frac{1}{u_g^2}} \\ &\sim \frac{\lambda_k + \lambda_{k-g}}{2} \pm u_g\left(1 + \frac{1}{2}\left(\frac{\lambda_k - \lambda_{k-g}}{2}\right)^2 \frac{1}{u_g^2}\right),\end{aligned} \tag{7.3.3}$$

where, as before, $\lambda_k = \hbar^2k^2/2m$ and $\lambda_{k-g} = \hbar^2(k-g)^2/2m$. In this example, the band edge of interest occurs when $k \approx g/2$; i.e., near the zone boundary. To that end, we write

$$\lambda_k = \frac{\hbar^2}{2m}(k - g/2 + g/2)^2 = \frac{\hbar^2}{2m}(K + g/2)^2 = \frac{\hbar^2}{2m}(K^2 + Kg + (g/2)^2), \tag{7.3.4a}$$

where we have introduced the definition $K \equiv k - g/2$. In a similar way, we have

$$\lambda_{k-g} = \frac{\hbar^2}{2m}(K - g/2)^2 = \frac{\hbar^2}{2m}(K^2 - Kg + (g/2)^2), \tag{7.3.4b}$$

so that

$$\frac{\lambda_k + \lambda_{k-g}}{2} = \frac{\hbar^2}{2m}(K^2 + (g/2)^2) \quad \text{and} \quad \frac{\lambda_k - \lambda_{k-g}}{2} = \frac{1}{2}\frac{\hbar^2}{2m}(2Kg). \tag{7.3.4c}$$

If we let $\lambda \equiv \hbar^2(g/2)^2/2m$ and substitute Equations 7.3.4c into Equation 7.3.3, we get

$$\begin{aligned}\varepsilon_k &\approx \frac{\hbar^2}{2m}(K^2 + (g/2)^2) \pm u_g \pm \frac{\hbar^2}{2m}K^2\left(\frac{2\lambda}{u_g}\right) = \frac{\hbar^2}{2m}(g/2)^2 \pm u_g + \frac{\hbar^2}{2m}K^2\left(1 \pm \left(\frac{2\lambda}{u_g}\right)\right) \\ &= \varepsilon_\pm + \frac{\hbar^2}{2m}K^2\left(1 \pm \left(\frac{2\lambda}{u_g}\right)\right) \equiv \varepsilon_{K\pm},\end{aligned} \tag{7.3.5}$$

where we have defined $\varepsilon_{\pm} \equiv \hbar^2(g/2)^2/2m \pm u_g$; i.e., the upper (+) and lower (−) band values at the BZ boundary when $k = g/2$ or $K = 0$. The quantity $\varepsilon_k$ thus obtained represents the behavior of the bands edges versus $k$ near the BZ boundary value of $g/2$, where $g = 2\pi/a$. We can incorporate these approximations into the script central_eq_bands.m of Chapter 6 to obtain the plot shown in Figure 7.3.5 in the same units previously defined there.

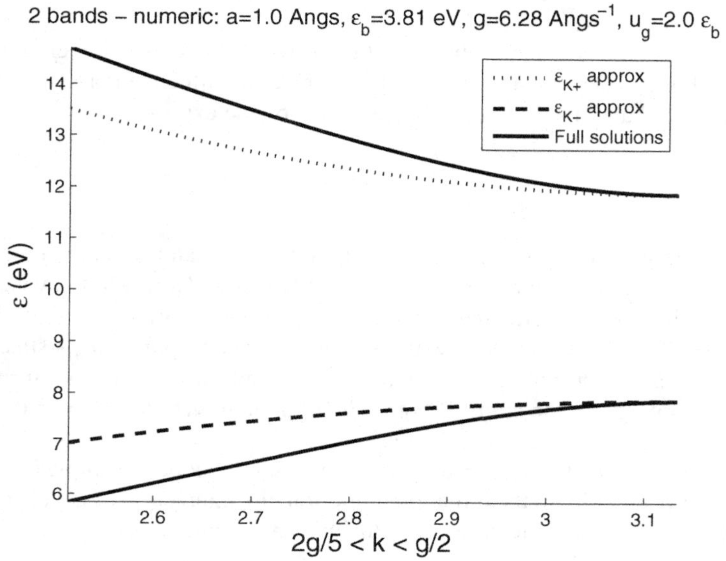


Figure 7.3.5: The approximations from Equations 7.3.5 for the upper (dotted) and lower (dashed) bands are shown compared to the numerical results from Chapter 6 near the BZ boundary of $k = g/2$.

An interesting aspect regarding Equation 7.3.5 can be observed if we take the second derivative of $\varepsilon_k$ with respect to $k$. Since $d^n\varepsilon_k/dk^n = d^n\varepsilon_k/dK^n$ we can, in fact, see that

$$\frac{1}{\hbar^2}\frac{d^2\varepsilon_k}{dk^2} = \frac{\left(1 \pm \left(\frac{2\lambda}{u_g}\right)\right)}{m} \equiv \frac{1}{m^*_{e,h}}; \tag{7.3.6}$$

that is, the curvature of the upper or lower band near the band edge is associated with an electron or hole effective mass, respectively, and labeled as $m^*$ with corresponding subscripts. Later below we will use a slightly more convenient notation for the carrier effective mass. The electron has a positive effective mass which is consistent with the curvature of the conduction band; however, the valence band curvature is negative near the band edge, so that a negative value is usually introduced in order to have a positive hole mass. While the effective mass formula obtained above is an observation that results from looking at the simple two-band case near the band edges, it can be made more general as shown in Exercise 7.7.4. In semiconductors or materials with similar band properties, in the presence of a force acting on them, the charge carriers move as if their mass were equal to the effective mass. The above Equation 7.3.6 can be generalized to the three-dimensional case when the energy surface is anisotropic. The expression is written as elements of a tensor matrix (see Exercise 7.7.5); that is,

$$\left(\frac{1}{m^*}\right)_{\mu\nu} = \frac{1}{\hbar^2}\frac{\partial^2}{\partial k_\mu \partial k_\nu}\varepsilon_{\vec{k}}, \tag{7.3.7}$$

where $\mu$ and $\nu$ span the $xyz$ coordinates.

### 7.3.1 Effective Masses for Various Semiconductors

For semiconductors with isotropic energy surfaces with direct band gaps, an electron energy near the conduction band edge can be described by

$$\varepsilon_k = E_g + \frac{\hbar^2 k^2}{2m_c}; \tag{7.3.8a}$$

that is, an electron near the conduction band minimum will behave as if it were a free electron in vacuum, albeit with the electron mass replaced by the electron effective mass, $m^* = m_c$. Similarly, a hole near the valence band maximum can be described by the expression

$$\varepsilon_k = -\frac{\hbar^2 k^2}{2m_h}, \tag{7.3.8b}$$

where the zero of energy is taken to be at the top of the valence band edge, the hole effective mass is $m^* = m_h$, and the minus sign reflects the curvature of the band. However, the valence band edges in semiconductors involve hole effective masses such as $m_{hh}$ (heavy hole), $m_{lh}$ (light hole), and $m_{soh}$ (split-off hole). These are associated with two bands derived from the $p_{3/2}$ states, of the parent atoms, for the heavy and light holes, respectively, and the splitting that occurs for the $p_{1/2}$ states, due to spin-orbit coupling, for the split-off hole. The light hole mass tends to be the smaller of the three due to its related band's greater curvature.

The carrier effective mass can be measured by exposing a semiconductor sample to a magnetic field. Assuming a field perpendicular to the direction of motion, the interaction of the charge carriers with the field; i.e., $F_B = qvB = m^* v^2/r$ leads to $v = qBr/m^* = r\omega_c$, where

$$\omega_c = \frac{qB}{m^*} \tag{7.3.9}$$

is the cyclotron frequency. If the sample is further exposed to an oscillating field, the carriers will exhibit resonant absorption at the cyclotron frequency. Knowing the magnitude of $B$ and the absorption frequency enables the measuring of the effective carrier mass. Table 7.3.2 shows typical effective mass values for some direct-gap semiconductor materials.

Table 7.3.2: Values of carrier effective masses ($m_c$, $m_{hh}$, $m_{lh}$, and $m_{soh}$ for the electron, the heavy hole, the light hole, and the split-off hole, respectively, in units of the electron mass $m$) for some direct-gap semiconductors. (Source: [3]).

| Crystal | $m_c/m$ | $m_{hh}/m$ | $m_{lh}/m$ | $m_{soh}/m$ |
|---|---|---|---|---|
| InSb | 0.015 | 0.39 | 0.021 | 0.11 |
| InAs | 0.026 | 0.41 | 0.025 | 0.08 |
| InP | 0.073 | 0.4 | 0.078 | 0.15 |
| GaSb | 0.047 | 0.3 | 0.06 | 0.14 |
| GaAs | 0.066 | 0.5 | 0.082 | 0.17 |

**Example 7.3.1.1**

Here, let's obtain the value of the cyclotron frequency of electrons in InSb in a magnetic field of $1\,T$. Since $m = 0.5 \times 10^6\, eV/c^2$ and using the fact that $1\,T = 1\,V \cdot s/m^2$, we have that $\omega_c = eB/m_c = 1\,(eV \cdot s/m^2)c^2/(0.015 \cdot 0.5 \times 10^6\, eV)$ or $\omega_c = 1.2 \times 10^{13} rad/s$ or $1.9 \times 10^{12}\, Hz$.

## 7.4 Intrinsic Carrier Concentration

In the absence of any dopants (impurities that donate electrons or holes), the number of electrons available for conduction depends on the temperature through the Fermi function

$$f_e(\varepsilon) = \frac{1}{e^{(\varepsilon-\mu)/\tau} + 1}, \tag{7.4.10a}$$

which gives the probability that a conduction electron orbital is occupied and where $\tau \equiv k_B T$. However, for the temperatures of interest here we take $\varepsilon - \mu >> \tau$ so that $\exp[(\varepsilon - \mu)/\tau] >> 1$ and we can approximate the Fermi function with

$$f_e(\varepsilon) \approx e^{(\mu-\varepsilon)/\tau}. \tag{7.4.10b}$$

Assuming a simple parabolic behavior, and measuring from the bottom of the conduction band ($E_c$) we write the conduction electron energy as

$$\varepsilon = E_c + \frac{\hbar^2 k^2}{2m_c}. \tag{7.4.11}$$

where $m_c$ is the electron effective mass. In order to obtain the number of electrons versus energy, we use the density of states from Chapter 5 as regards to the three-dimensional electron gas. Also, measuring from $E_c$, we have the expression (for $\varepsilon \geq E_c$)

$$D_e(\varepsilon) = \frac{V}{2\pi^2}\left(\frac{2m_c}{\hbar^2}\right)^{3/2} (\varepsilon - E_c)^{1/2}. \tag{7.4.12}$$

As discussed before, we find the number of electrons in the conduction band by integrating the product of $D_e(\varepsilon) f_e(\varepsilon)$ over $\varepsilon$. Here, we look for the electron concentration $n$ = number of electrons per volume or

$$n = \frac{1}{V}\int_{E_c}^{\infty} D_e(\varepsilon) f_e(\varepsilon) d\varepsilon \approx \frac{1}{2\pi^2}\left(\frac{2m_c}{\hbar^2}\right)^{3/2} e^{\mu/\tau} \int_{E_c}^{\infty} (\varepsilon - E_c)^{1/2} e^{-\varepsilon/\tau} d\varepsilon. \tag{7.4.13}$$

If we let $x = (\varepsilon - E_c)/\tau$ the integral term in this expression takes the form

$$\int_{E_c}^{\infty} (\varepsilon - E_c)^{1/2} e^{-\varepsilon/\tau} d\varepsilon = e^{-E_c/\tau} \tau^{3/2} \int_0^{\infty} x^{1/2} e^{-x} dx = e^{-E_c/\tau} \tau^{3/2} \sqrt{\pi}/2,$$

where we have used the integral result $\int_0^\infty x^n e^{-ax} dx = \Gamma(n+1)/a^{n+1}$ with $\Gamma(3/2) = \sqrt{\pi}/2$ for $n = 1/2$. Substituting this back into Equation 7.4.13 gives

$$n = n_0 e^{(\mu - E_c)/\tau}, \tag{7.4.14a}$$

where we have defined

$$n_0 = 2\left(\frac{m_c k_B T}{2\pi\hbar^2}\right)^{3/2}, \tag{7.4.14b}$$

a quantity that will be used later. In these expressions, the electron concentration is determined once $\mu$ is known. We can perform a similar calculation for holes in the valence band. The analogue of Equation 7.4.11 for the holes, measuring from the top of the valence band ($E_v$), is

$$\varepsilon = E_v - \frac{\hbar^2 k^2}{2m_h}. \tag{7.4.15}$$

Since a hole is the absence of an electron, the hole occupation probability is

$$f_h(\varepsilon) = 1 - f_e(\varepsilon) = 1 - \frac{1}{e^{(\varepsilon-\mu)/\tau}+1} = \frac{e^{(\varepsilon-\mu)/\tau}}{e^{(\varepsilon-\mu)/\tau}+1} = \frac{1}{1+e^{(\mu-\varepsilon)/\tau}} \approx e^{(\varepsilon-\mu)/\tau}, \tag{7.4.16}$$

assuming $\mu - \varepsilon >> \tau$ for holes, similar to what we did in Equation 7.4.10b for electrons. The density of states for the holes becomes (for $\varepsilon \leq E_v$)

$$D_h(\varepsilon) = \frac{V}{2\pi^2}\left(\frac{2m_h}{\hbar^2}\right)^{3/2}(E_v - \varepsilon)^{1/2}. \tag{7.4.17}$$

The hole concentration ($p$ = number of holes per volume) in the valence band is obtained by integrating the product of $D_h(\varepsilon)f_h(\varepsilon)$ over $\varepsilon$; that is,

$$p = \frac{1}{V}\int_{-\infty}^{E_v} D_h(\varepsilon)f_h(\varepsilon)d\varepsilon \approx \frac{1}{2\pi^2}\left(\frac{2m_h}{\hbar^2}\right)^{3/2} e^{-\mu/\tau}\int_{-\infty}^{E_v}(E_v-\varepsilon)^{1/2}e^{\varepsilon/\tau}d\varepsilon. \tag{7.4.18}$$

The integral in this expression is similar to that of Equation 7.4.13 and is carried out in a similar way to obtain

$$p = p_0 e^{(E_v-\mu)/\tau}, \tag{7.4.19a}$$

with the definition

$$p_0 = 2\left(\frac{m_h k_B T}{2\pi\hbar^2}\right)^{3/2}. \tag{7.4.19b}$$

As for electrons, once $\mu$ is known, the hole concentration can be determined. Multiplying Equations 7.4.14 and 7.4.19 together, one finds that their product

$$np = n_0 p_0 e^{(E_v-E_c)/k_BT} = 4\left(\frac{k_BT}{2\pi\hbar^2}\right)^3 (m_c m_h)^{3/2} e^{-E_g/k_BT} \tag{7.4.20}$$

is independent of the chemical potential $\mu$ and, as before, $E_g = E_c - E_v$. This result is significant because it is an expression of the law of mass action at a given temperature. It says that if the number of electrons (holes) increases the number of holes (electrons) must decrease in such a way as to keep the product constant, which, in turn, points to mass conservation. This expression also applies to the case when impurities are present in extrinsic semiconductors. The only assumption employed was that $\mu$ is far from the band edges. Because the product $np = \text{constant}(T)$, independent of impurity concentration, introducing impurities to increase $n$ ($p$) must simultaneously decrease $p$ ($n$). In this way, impurities are used to affect the electronic properties of semiconductors. For an intrinsic semiconductor where no impurities are present, the number of electrons ($n = n_i$) equals the number of holes ($p = p_i$) so that from Equation 7.4.20 we can write

$$n_i = p_i = \sqrt{n_i p_i} = \sqrt{n_i^2} = 2\left(\frac{\tau}{2\pi\hbar^2}\right)^{3/2}(m_c m_h)^{3/4} e^{-E_g/2\tau}; \tag{7.4.21}$$

that is, the intrinsic carrier concentration depends exponentially on $-E_g/2\tau$. Additionally, by having $n = p$, in this case, from Equations 7.4.14 and 7.4.19 we can also write $n_0 \exp[(\mu - E_c)/\tau] = p_0 \exp[(E_v - \mu)/\tau]$ or $\exp(2\mu/\tau) = (p_0/n_0)\exp[(E_v + E_c)/\tau]$, which yields the chemical potential as

$$\mu = \frac{E_v + E_c}{2} + \frac{\tau}{2}\ln\left(\frac{p_0}{n_0}\right) = E_v + \frac{E_g}{2} + \frac{3\tau}{4}\ln\left(\frac{m_h}{m_c}\right), \tag{7.4.22}$$

where we have used $E_c = E_v + E_g$. From this expression, recalling that $\tau = k_B T$, we see that at $T = 0$, the chemical potential reaches the value of $E_v + E_g/2 = E_F$; i.e., in semiconductors, the Fermi level lies in the middle of the gap. This is illustrated in Figure 7.4.6. The chemical potential $\mu = E_F$ at $T = 0K$ and varies with temperature, as long as the electron and hole effective masses are not equal.

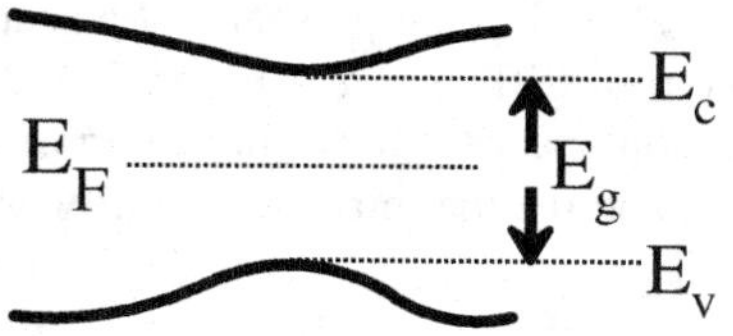


Figure 7.4.6: Illustration of the Fermi level, $E_F$, position in the middle of the semiconductor gap. Also shown is the conduction band edge, $E_c$; the valence band edge, $E_v$; and the band gap, $E_g = (E_c - E_v)$.

Figure 7.4.7 contains a plot of the electron concentration, Equation 7.4.14, with the help of Equation 7.4.22, versus temperature (see Exercise 7.7.8).

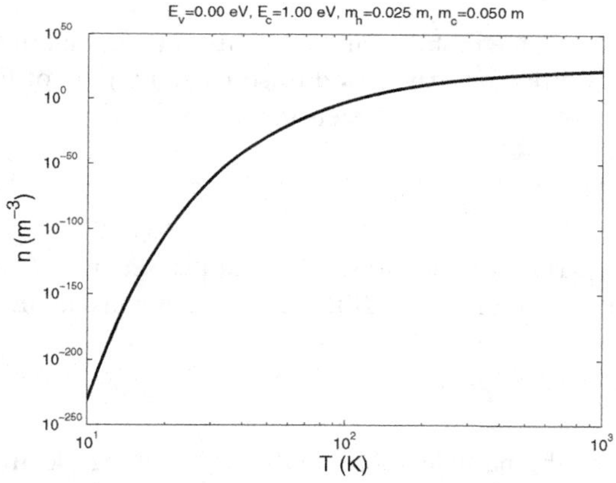


Figure 7.4.7: Electron concentration $(1/m^3)$ for a system with $E_v = 0\,eV$, $E_c = 1\,eV$, $m_c = 0.05m$, and $m_h = 0.025m$, where $m$ is the electron mass (see Exercise 7.7.8). Both, vertical and horizontal, scales are logarithmic.

### 7.4.1 Intrinsic Carrier Mobility

We have discussed the electronic mobility and the electronic conductivity before in Chapter 5. Recall that for a simple model of electron transport, the electronic conductivity is written as $\sigma = ne^2\tau_{coll}/m$, where $n$ is the electron concentration, $\tau_{coll}$ the time between collisions ($1/\tau_{coll}$ is the scattering rate), and $m$ is the electron mass. Here we can generalize this expression a bit further by including the hole motion; that is, the total intrinsic carrier conductivity is a collective contribution from electrons and holes; that is,

$$\sigma_i = \frac{n_i e^2 \tau_e}{m_c} + \frac{p_i e^2 \tau_h}{m_h}, \tag{7.4.23}$$

where we have assigned different collision times for each carrier type ($\tau_e$ for electrons and $\tau_h$ for holes) and included their appropriate effective masses. We also notice from Equation 7.4.21 that because the intrinsic carrier concentrations are dependent on temperature, the conductivity is similarly affected by temperature. We will come back to this later. Also in Chapter 5, we had written the electronic mobility as $\mu_E = v_E/E$, where $v_E$ is the terminal speed reached by an electron in a material under the action of an electric field $E$, and in the presence of a scattering force that is proportional to the speed. For this simple electron drift model, we had found that $v_E = -eE\tau_{coll}/m$ so that $\mu_E = -e\tau_{coll}/m$. However, as mentioned above, for a single carrier type, $\sigma = ne^2\tau_{coll}/m = -ne\mu_E$, which is positive. Thus, it suffices to define the intrinsic mobility as

$$\mu_i = |v|/E, \tag{7.4.24a}$$

which, for the simple electron drift model, becomes, for each semiconductor carrier type,

$$\mu_e = \frac{e\tau_e}{m_c} \quad \text{(electrons)} \qquad \text{and} \qquad \mu_h = \frac{e\tau_h}{m_h} \quad \text{(holes)}, \tag{7.4.24b}$$

with all quantities: the electron charge, the collision times, and the effective masses taking on positive values. Combining these expressions with Equation 7.4.23, we define the intrinsic conductivity as

$$\sigma = n_i e\mu_e + p_i e\mu_h. \tag{7.4.24c}$$

There exists a connection between the mobility and the diffusion constant in a material. To be clear, according to Fick's 1st law of diffusion, particles diffuse from a region of high concentration to a region of low concentration; that is, for a one-dimensional case,

$$J = -D\frac{dC}{dx}, \tag{7.4.25a}$$

where $J$ is the particle flux (particles that diffuse per area per second), $C$ is the concentration of particles (number of particles per volume) and $D$ is the diffusion constant here expressed as

$$D = \frac{\bar{v}^2\tau_{coll}}{3}, \tag{7.4.25b}$$

with $\bar{v}$ the average velocity of the particles. Letting the particles be electrons, for example, and, using the equipartition theorem, $\bar{v}^2 = 3k_BT/2m_c$, we see that

$$D = k_BT\frac{\tau_{coll}}{m_c} = \frac{\mu_e k_BT}{e}, \tag{7.4.25c}$$

and describes a relationship between mobility and diffusivity. This expression is a simple version of what is known as Einstein's drift-diffusion relation. As mentioned before, the intrinsic carrier concentrations are functions of temperature (Equation 7.4.21); we thus notice that according to Equation 7.4.24c, one could make the approximation

$$\sigma_i \approx Ae^{-E_g/2k_BT}, \tag{7.4.26}$$

so that a knowledge of the experimental intrinsic $\sigma_i$ versus temperature contains information about the energy gap $E_g$ of a material.

**Example 7.4.1.1**

Figure 7.4.8 shows the natural logarithm of the experimental electrical conductivity of intrinsic germanium versus the inverse of the temperature (dots). We can take the natural logarithm of Equation 7.4.26 and obtain $ln(\sigma) = ln(A) - E_g/2k_BT$ which is of the form $y = mx + b$; i.e., a straight line with $y \equiv ln(\sigma)$, $x \equiv 1/T$, intercept $b \equiv ln(A)$, and slope $m = -E_g/2k_B$. This indicates that the natural logarithm of the experimental $\sigma$ plotted versus $1/T$ behaves as a straight line whose slope yields the band gap or $E_g = -2mk_B$. The data used in the figure are provided in the code snippet listing below. The straight line fit was carried out using the MATLAB internal commands 'polyfit' and 'polyval', and used here in a similar way as we did in Chapter 4. The former obtains the fit coefficients, and the latter calculates the theoretical line fit shown in the figure.

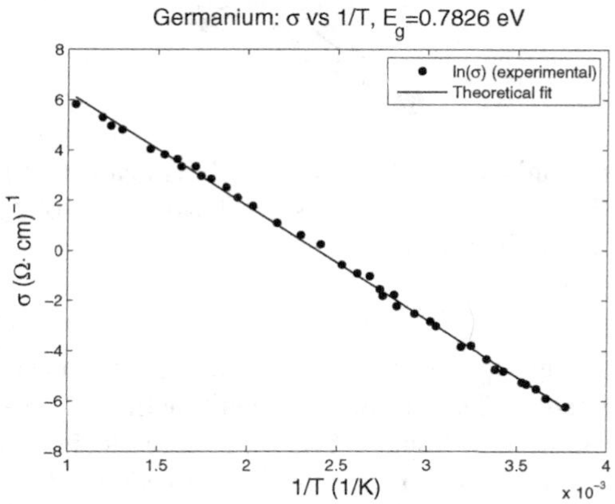


Figure 7.4.8: The logarithm of the experimental $\sigma$ for germanium versus $1/T$ (dots). The straight line fit analysis described in Example 7.4.1.1 yields a band gap of $E_g \sim 0.78\,eV$ (Data from [27]).

From the graph, a slope of $-4.54 \times 10^3\,K$ is obtained. This results into a band gap for semiconducting Ge of $\sim 0.78\,eV$, which is close to the value of $\sim 0.74\,eV$ from Table 7.1.1 at low temperature. The data used in the analysis described in this example follows (see Exercise 7.7.9).

```
%x0 in 1/T  (inverse Kelvin)
x0=[0.001041,0.001190,0.001237,0.001300,0.001457,0.001535,...
  0.001606,0.001629,0.001708,0.001739,0.001794,0.001880,...
  0.002029,0.001943,0.002163,0.002296,0.002406,0.002524,...
  0.002610,0.002680,0.002735,0.002751,0.002814,0.002829,...
  0.002931,0.003018,0.003049,0.003245,0.003190,0.003331,...
  0.003425,0.003378,0.003527,0.003551,0.003606,0.003661,0.003770];
%y0 in 1/(ohm-centimeter)) - conductivity
y0=[342.405943,203.074316,145.019614,124.955666,57.005514,...
  45.581777,37.834939,28.039634,28.078411,19.307971,17.270054,...
  12.341431,5.844582,8.178657,2.982614,1.836023,1.264278,...
  0.555157,0.396723,0.354948,0.210164,0.161683,0.168048,...
  0.107177,0.079539,0.059012,0.048949,0.022346,0.021503,...
  0.013238,0.008144,0.008771,0.005202,0.004828,0.004006,...
  0.002756,0.001970];
```

## 7.5 Impurities in Semiconductors

The act of adding impurities in semiconductors is referred to as doping and the impurities doing the doping are called dopants. As an example, let's consider a silicon atom, a group IV element of the periodic table with an outer valence electron configuration of $3s^2 3p^2$. In a Si crystal each silicon atom is tetrahedrally bonded to each of its four nearest neighbors through the sharing of electrons as shown in Figure 7.5.9.

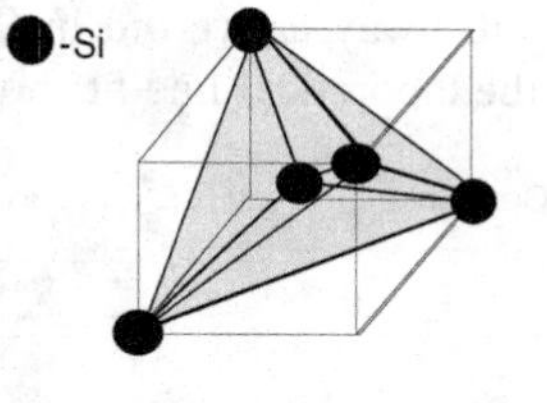


Figure 7.5.9: Illustration of silicon's tetrahedral bonds with its nearest neighbors. Silicon's valence electron configuration is $3s^2 3p^2$. Each electron is shared with each of its neighbors and thus Si forms four $sp^3$ hybridized or covalent bonds.

Now imagine that an atom from group V of the periodic table, with outer electron configuration of $3s^2 3p^3$, replaces the center Si atom. The group V element, acting as an impurity and having one more electron, will make the same four bonds with its Si nearest neighbors, but the extra electron will be free to contribute to the electronic properties of the semiconductor. Impurities with extra electrons are known as electron donors or simply donor impurities. Should the impurity atom be one that belongs to group III of the periodic table, with a valence electron configuration of $3s^2 3p$ and lacking an electron compared to Si, will act as an impurity that will accept an electron or donate a hole. Such impurities are known as electron acceptors or simply acceptor impurities.

### 7.5.1 Impurity States and Conductivity

Impurities are atoms that are present in crystals, naturally or deliberately, and are different from the atoms that make up a pure crystal. Impurities tend to be present as a minority in number compared to the host atoms. It has been argued [28] that an estimate of what is a high impurity concentration can be obtained by considering a semiconductor cube with a volume whose sides are related to an electron's de Broglie wavelength or about 100 Å. Considering further that a statistically meaningful high number of impurity atoms is about 1000 within this quantum box. The impurity concentration here is about $1000/(100\text{Å})^3$ or $\sim 1 \times 10^{21}/cm^3 = 1 \times 10^{27}/m^3$.
By virtue of losing an electron, donor impurities are slightly positive and a donated electron tends to interact with the ionized donor, albeit within a crystal. There exists, therefore, a Coulomb interaction between the electron and the donor. This interaction is written as

$$V(r) = -\frac{ke^2}{r}, \qquad k = \frac{1}{4\pi\varepsilon}, \qquad \varepsilon = K\varepsilon_0, \tag{7.5.27}$$

where, as is evident, the interaction is a modified version of the standard Coulomb interaction. There is an extra factor in the denominator; i.e., the constant $K$ that accounts for the crystal's contribution to the interaction and is the material's dielectric constant. Some dielectric constant values are given in Table 7.5.3.

Table 7.5.3: Dielectric constant values for a few semiconducting crystals (Sources: when available [3], else [16]).

| Crystal | $K$ |
|---|---|
| Diamond | 5.5 |
| Si | 11.7 |
| Ge | 15.8 |
| InSb | 17.88 |
| InAs | 14.55 |
| InP | 12.37 |
| GaP | 9.1 |
| GaAs | 13.13 |
| GaSb | 15.69 |
| AlSb | 10.3 |
| CdS | 5.2 |
| CdSe | 5.8 |
| CdTe | 7.2 |

The problem of the binding energy associated with the donated electron bound to the donor through the potential of Equation 7.5.27 is similar to the hydrogen atom binding energy modified by the dielectric constant as well as by the electronic effective mass within the crystal. We thus write the impurity binding energy as

$$E_D = \frac{m_c(ke^2)^2}{2K^2\hbar^2} = \frac{m_c}{mK^2}E_0, \tag{7.5.28a}$$

where in the hydrogen Bohr ground state ($E_0 = 13.6\,eV$) we have replaced the electron mass ($m$) with the effective mass, $m_c$, as well as replaced $k$ with $k/K$. Similarly, we have for the donor's Bohr radius

$$a_D = \frac{K\hbar^2}{m_c ke^2} = \frac{mK}{m_c}a_0, \tag{7.5.28b}$$

where $a_0 = 0.529$ Å is the hydrogen Bohr radius. The donor orbital is depicted in Figure 7.5.10(a). Notice that it may encompass many host atoms depending on the dielectric constant and the effective mass. The energy $E_D$ is the energy it takes to ionize the donor impurity. Once ionized, the impurity electron becomes available for conduction. This is the reason why the donor ionization energy is located at $E_D$ below the conduction band as shown in Figure 7.5.10(b). The case of acceptor states is similar in that the negative charged impurity interacts with the hole. The acceptor ionization energy is obtained as

$$E_A = \frac{m_h(ke^2)^2}{2K^2\hbar^2} = \frac{m_h}{mK^2}E_0. \tag{7.5.29a}$$

When ionized, the acceptor hole becomes available for conduction in the valence band. The associated acceptor Bohr radius is

$$a_A = \frac{K\hbar^2}{m_h ke^2} = \frac{mK}{m_h}a_0. \tag{7.5.29b}$$

Both, the donor and acceptor levels are depicted in Figure 7.5.10(b).

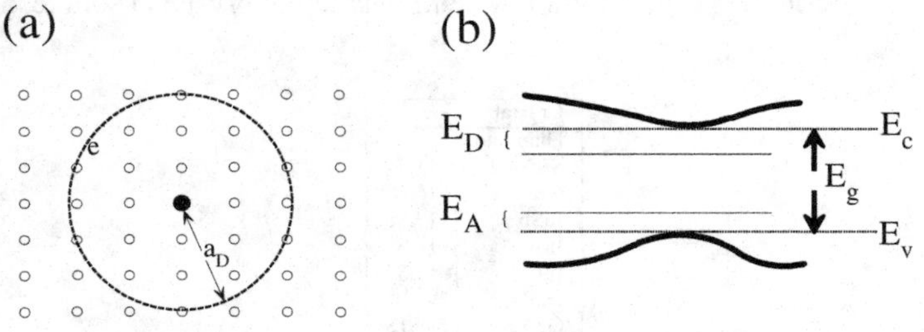


Figure 7.5.10: (a) A depiction of a donor impurity (dark circle) and the orbit of the donated electron bound to it (dashed). The orbit may encompass many host atoms (open circles). (b) The position of the donor level ($E_D$) with respect to the conduction band to where the bound electron goes if ionized is shown. Similarly, the acceptor level ($E_A$) is shown with respect to its associated valence band to where the bound hole goes if ionized.

**Example 7.5.1.1**
If we assume an electron effective mass of $m_c \approx 0.2m$ in Si, with the use of $K = 11.7$ from Table 7.5.3, Equations 7.5.28 give for this system $E_D \approx 20\,meV$ and $a_D = 31$ Å.

If donors are present in considerably higher amounts than acceptors, the material will contain an excess of electrons and it is referred to as an n-type material. However, if the acceptors are present to higher amounts than donors, the material will contain an excess of holes and it is referred to as a p-type material. The sign of the Hall voltage is a possible way to test whether the material is n-type or p-type. Another possible test is the sign of the thermoelectric potential (discussed later below). The impurity levels can be shallow or deep depending on their location within the semiconductor gap. Shallow donors are those whose levels lie within about 10% of $E_g$ from the conduction band edge. Similarly, shallow acceptors lie within about 10% of $E_g$ from the valence band edge. From Table 7.1.1, in the case of Si for which the indirect band gap is about 1.2eV, the shallow states lie within about $100\,meV$ of the respective band edge. Deep levels lie closer to the middle of the gap. One aspect about the effect of impurities in semiconductors is how they can affect the electronic conductivity of a semiconductor. This can be seen by recalling the law of mass action (Equation 7.4.20), the product $np$ is a constant at a given temperature and is independent of impurity concentration, we can thus write $np = p_i n_i = n_i^2$, where we have also used Equation 7.4.21. If we then assume that in the presence of a concentration $N_D$ of donor impurities, with $N_D \gg n_i$, then the concentration of electrons is $n \approx N_D$ to obtain that the concentration of holes is $p = n_i^2/N_D$, which will be much less than $n_i$. This can be seen by letting $N_D = C \cdot n_i$ with $C >> 1$ so that $p = n_i^2/N_D = n_i^2/Cn_i = n_i/C$, which is much smaller than $n_i$, which is itself much smaller than $N_D$. With these results, the conductivity of Equation 7.4.24c in the presence of a high concentration of donor impurities can be written as

$$\sigma = ne\mu_e + pe\mu_h = N_D e\mu_e + \left(\frac{n_i^2}{N_D}\right) e\mu_h \approx N_D e\mu_e, \tag{7.5.30}$$

which indicates that the conductivity is mostly due to the donated electrons. If in contrast to the above excess of donors we have a large concentration ($N_A$) of acceptors, with $N_A \gg n_i$; that is, at a temperature when all the acceptors have been ionized, then the concentration of holes is $p \approx N_A$. The electron concentration is obtained from $np = n_i^2$ or $n \approx n_i^2/N_A$ which is much smaller than $n_i$, which is itself much smaller than $N_A$, similar to the above high concentration of donor arguments. In this way for a high concentration of acceptor impurities we have

$$\sigma = ne\mu_e + pe\mu_h = \left(\frac{n_i^2}{N_A}\right) e\mu_e + N_A e\mu_h \approx N_A e\mu_h, \tag{7.5.31}$$

which is mostly due to the excess of holes.

## 7.6 Extrinsic Carrier Concentration

In this section we discuss the behavior of extrinsic semiconductors; that is, when there are impurities present, it is possible for them to become ionized by thermal means so that the number of carriers (electrons or holes) become much higher than the intrinsic carrier concentration. When acceptors and donors are present, the chemical potential $\mu$ is found by the requirement that the entire crystal remain neutral. The neutrality condition is based on ensuring that the total concentration of electrons equals the total concentration of holes, or

$$n+N_A^- = p+N_D^+, \tag{7.6.32}$$

where $N_A^-$ is the concentration of ionized acceptors which gives rise to extra holes, and $N_D^+$ the concentration of ionized donors which gives rise to extra electrons. We next seek the electron occupation probability of a donor. This is done through the thermodynamical statistical expression for the average electron number [10]

$$<n>= \frac{\sum_j N_j \exp[-(\varepsilon_j - \mu N_j)/\tau]}{\sum_j \exp[-(\varepsilon_j - \mu N_j)/\tau]}, \tag{7.6.33}$$

where a donor impurity could be empty, or contain one electron of either spin (assuming the same energy). The two-electron case is prohibited due to a high coulomb electron interaction. The one-electron case is taken to be $\varepsilon_d \equiv E_c - E_D$ (see Figure 7.5.10) higher than the empty case. The electron donor occupation probability becomes

$$\begin{aligned} f_{0D}(\varepsilon_d) \equiv <n_{\text{donor}}> &= \frac{0e^{[-(0-0\mu)/\tau]} + 1e^{[-(\varepsilon_d-1\mu)/\tau]} + 1e^{[-(\varepsilon_d-1\mu)/\tau]}}{e^{[-(0-0\mu)/\tau]} + e^{[-(\varepsilon_d-1\mu)/\tau]} + e^{[-(\varepsilon_d-1\mu)/\tau]}} \\ &= \frac{2e^{[-(\varepsilon_d-\mu)/\tau]}}{1+2e^{[-(\varepsilon_d-\mu)/\tau]}} = \frac{1}{1+\frac{1}{2}e^{[(\varepsilon_d-\mu)/\tau]}}. \end{aligned} \tag{7.6.34}$$

Here note that $\varepsilon_d$ is the donor level energy, which is $E_D$ lower than the conduction band edge. For the electron acceptor occupation probability, we can have two electrons (zero holes), one electron (one hole) of either spin, but not zero electrons (two holes for the acceptor is the analog of two electrons for a donor) due to high coulomb hole repulsion. Here the two-electron state is $\varepsilon_a \equiv E_v + E_A$ (see Figure 7.5.10) higher than the one-electron state. We have for the electron acceptor occupation probability

$$\begin{aligned} <n_{\text{acceptor}}> &= \frac{1e^{[-(0-1\mu)/\tau]} + 1e^{[-(0-1\mu)/\tau]} + 2e^{[-(\varepsilon_a-2\mu)/\tau]}}{e^{[-(0-1\mu)/\tau]} + e^{[-(0-1\mu)/\tau]} + e^{[-(\varepsilon_a-2\mu)/\tau]}} \\ &= \frac{2e^{[\mu/\tau]} + 2e^{[-(\varepsilon_a-2\mu)/\tau]}}{2e^{[\mu/\tau]} + e^{[-(\varepsilon_a-2\mu)/\tau]}} = \frac{1+e^{[-(\varepsilon_a-\mu)/\tau]}}{1+\frac{1}{2}e^{[-(\varepsilon_a-\mu)/\tau]}}. \end{aligned} \tag{7.6.35a}$$

However, the number of holes is the maximum number of electrons (2) minus the mean electron acceptor number ($<n_{\text{acceptor}}>$) or

$$f_{0A}(\varepsilon_a) \equiv 2- <n_{\text{acceptor}}>= 2 - \frac{1+e^{[-(\varepsilon_a-\mu)/\tau]}}{1+\frac{1}{2}e^{[-(\varepsilon_a-\mu)/\tau]}} = \frac{1}{1+\frac{1}{2}e^{[-(\varepsilon_a-\mu)/\tau]}}. \tag{7.6.35b}$$

Again, we note that $\varepsilon_a$ is the acceptor level energy, which is $E_A$ higher than the valence band edge. The neutrality condition, Equation 7.6.32 requires the concentration of ionized impurities. To this end, the concentration of ionized donors is the concentration of donors ($N_D$) minus the concentration of occupied donors ($N_D f_{0D}(\varepsilon_d)$) or

$$N_D^+ = N_D(1 - f_{0D}(\varepsilon_d)) = N_D\left(1 - \frac{1}{1+\frac{1}{2}e^{(\varepsilon_d-\mu)/\tau}}\right) = \frac{N_D}{1+2e^{(\mu-\varepsilon_d)/\tau}}. \tag{7.6.36a}$$

Similarly, for the ionized acceptors we have

$$N_A^- = N_A(1 - f_{0A}(\varepsilon_a)) = N_A\left(1 - \frac{1}{1+\frac{1}{2}e^{[-(\varepsilon_a-\mu)/\tau]}}\right) = \frac{N_A}{1+2e^{(\varepsilon_a-\mu)/\tau}}. \tag{7.6.36b}$$

It is worth mentioning that for Ge, Si, and GaAs the factor of 2 in the denominator of Equation 7.6.36b is replaced by a factor of 4 because each acceptor impurity level in these systems, in addition to possibly containing no holes or accepting one hole of either spin, is doubly degenerate as a result of two degenerate valence bands at $\vec{k} = 0$ [29]. Here this complication is ignored. Because the chemical potential's position in semiconductors changes based on the concentration of impurities. Its position can be found through the neutrality condition Equation 7.6.32 with the use of Equations 7.4.14, 7.4.19, and 7.6.36; that is,

$$2\left(\frac{m_c\tau}{2\pi\hbar^2}\right)^{3/2} e^{(\mu-E_c)/\tau} + \frac{N_A}{1+2e^{(\varepsilon_a-\mu)/\tau}} = 2\left(\frac{m_h\tau}{2\pi\hbar^2}\right)^{3/2} e^{(E_v-\mu)/\tau} + \frac{N_D}{1+2e^{(\mu-\varepsilon_d)/\tau}}. \tag{7.6.37}$$

The idea is that, for a given system (that is $E_c$ and $E_v$ are known), given the concentration of acceptors ($N_A$ and $\varepsilon_a$) and donors ($N_D$ and $\varepsilon_d$) the chemical potential is found from Equation 7.6.37 self-consistently for a given temperature ($\tau = k_B T$). Once $\mu$ is obtained, we can go back to Equations 7.4.14 and 7.4.19 to obtain the concentrations of electrons ($n$) and holes ($p$). Incidentally, notice that if in Equation 7.6.37 we take $N_A = N_D = 0$, the result for $\mu$ is identical to that of Equation 7.4.22.

**Example 7.6.0.1**
Assuming that $N_A = 0$ in Equation 7.6.37, what is the approximate concentration of electrons, also assuming that $\mu - \varepsilon_d \gg \tau$?
**Solution**
From Equation 7.6.37, we first notice that, for $\mu - \varepsilon_d \gg \tau$, $\frac{1}{1+2e^{(\mu-\varepsilon_d)/\tau}} \sim (1/2)e^{(\varepsilon_d-\mu)/\tau}$, so that $N_D^+ \approx (N_D/2)e^{(\varepsilon_d-\mu)/\tau}$. We can also ignore the hole concentration or $p \sim 0$ and write $n \sim N_D^+$. This also implies that $nN_D^+ = n^2$ or $n^2 = n_0 e^{(\mu-E_c)/\tau}(N_D/2)e^{(\varepsilon_d-\mu)/\tau}$, where we have also used Equation 7.4.14. Taking the square root of $n^2$, get

$$n \approx \sqrt{\frac{n_0 N_D}{2}} e^{(\varepsilon_d-E_c)/2\tau} = \sqrt{\frac{n_0 N_D}{2}} e^{-E_D/2\tau}, \tag{7.6.38a}$$

where we recall that $E_D$ is the donor impurity binding energy. Furthermore, the chemical potential for this case is given by (see Exercise 7.7.11)

$$\mu = E_c - \frac{E_D}{2} + \frac{\tau}{2}\ln\left(\frac{N_D}{2n_0}\right). \tag{7.6.38b}$$

To obtain the full solution for $\mu$ from Equation 7.6.37 we need to do it in a self-consistent way as mentioned above. That is, vary $\mu$ until the left- and right-hand sides are equal for a given donor and acceptor concentrations. The left panel of Figure 7.6.11 contains the obtained chemical potential $\mu$ versus $N_D$ ($N_A$) for small $N_A$ ($N_D$). Using the converged value of $\mu$ the electron (hole) concentration,

$n$ ($p$), versus $N_D$ ($N_A$) for small $N_A$ ($N_D$) from Equations 7.4.14 and 7.4.19 have been calculated. These are shown on the right panel of the figure. We have also added the approximation results from Equation 7.6.38 for $\mu$ versus $N_D$ and $n$ versus $N_D$ when $N_A = 0$ on the left and right panels, respectively, for illustration purposes.

Referring to the left panel of Figure 7.6.11, for low impurity concentrations, notice that $\mu$ starts off with the value of $\mu_0$ (the intrinsic value) which lies near the middle of the gap and, as the donor concentration increases (for a fixed low value of $N_A$), it travels upward toward the conduction band. Similarly (for a fixed low value of $N_D$) $\mu$ moves toward the valence band as the acceptor concentration is increased. The right panel shows that $n$ increases as $N_D$ increases (for a fixed low value of $N_A$), but $p$ decreases. In a similar fashion, $n$ decreases as $N_A$ is increased (for a fixed low value of $N_D$) while $p$ increases. The approximations of $\mu$ (left panel) and $n$ (right panel) that result from Equations 7.6.38 versus $N_D$ become closer to the full results when $N_D$ is quite high, which is consistent with the conditions of the approximation.

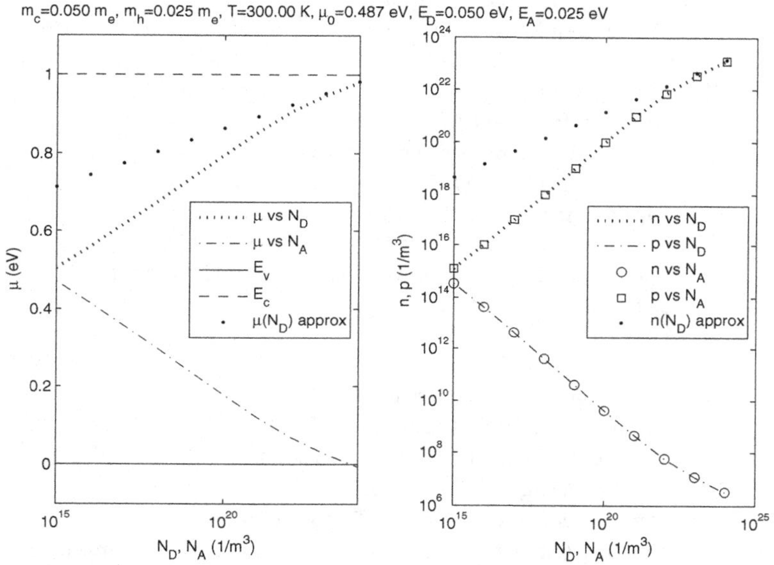


Figure 7.6.11: The left panel shows the chemical potential $\mu$ versus $N_D$ ($N_A$) with $N_A = 1 \times 10^{14}\,m^{-3}$ ($N_D = 1 \times 10^{14}\,m^{-3}$). The flat lines correspond to the valence ($0\,eV$) and conduction ($1\,eV$) band values used. The approximation for $\mu$ from Equation 7.6.38b, when $N_A$ is assumed zero is also shown as indicated by the legend. The right panel shows $n$ ($p$), versus $N_D$ ($N_A$) for when $N_A = 1 \times 10^{14}\,m^{-3}$ ($N_D = 1 \times 10^{14}\,m^{-3}$) from Equations 7.4.14, using the converged values of $\mu$ shown on the left panel. The effective masses used are $m_c = 0.05\,m_e$, $m_h = 0.025\,m_e$, the temperature is $T = 300\,K$, the donor binding energy is $E_D = 0.05\,eV$, and the acceptor binding energy is $E_A = 0.025\,eV$. In the figure, the value $\mu_0 = 0.487\,eV$ corresponds to the intrinsic value of the chemical potential which was used as an initial guess for low impurity concentrations.

The details of the process of performing the $\mu$, $n$, and $p$ caculations discussed in Figure 7.6.11 are encompassed by a simplified version of the plotting code. The following code charge_neutral.m is a simple example of the calculations. The code finds the value of $\mu$ self-consistently using the MATLAB function "fzero" with an initial guess of $\mu_0$ for values of $N_D$ and $N_A$. The "fzero" routine varies $\mu$ as it searches for a value of zero in the function "FN", which corresponds to the difference of the left- and right-hand sides of Equation 7.6.37. When the condition is satisfied, a converged value of $\mu$ has been found. Using the converged value of $\mu$ the code proceeds to obtain $n$. The energy units used in the actual calculations are in *eV* and the distance units are in *nm*. The concentrations in units of $m^{-3}$ are obtained by conversion from *nm* to *m* as explained in the code's comments. After the calculations, the results are printed. The listing of the code follows.

```
%copyright by J. E Hasbun and T. Datta
%charge_neutral.m
%It uses the charge neutrality condition to obtain the chemical
%potential (mu) self-consistently based on the donor (ND) and
%acceptor (NA) concentration. Once mu is obtained, the electron
%concentration (n) and the hole (p) concentration can be calculated
%at a given temperature.
clear;
hbarC=197;          %hbar*c in units of eV*nm
kB=8.6174e-5;       %Boltzmann constant in units of eV/K
Ev=0.0;             %valence band edge in eV
Ec=1.0;             %conduction band edge in eV
Eg=Ec-Ev;           %band gap
ED=0.050;           %donor binding energy in eV
EA=0.025;           %acceptor binding energy in eV
epsd=Ec-ED;         %donor impurity level
epsa=Ev+EA;         %acceptor impurity level
mh=0.025;           %hole effective mass value in units of me
mc=0.05;            %electron effective mass value in units of me
meCs=0.5e6;         %electron mass in eV
T=300;              %Kelvin
tau=kB*T;
ND=1e24;            %donor concentration in 1/m^3
NA=1e14;            %acceptor concentration in 1/m^3
%Actual calculations use units of distance in nm
NDnm=ND*(1e-9)^3; %converting 1/m^3 to 1/nm^3
NAnm=NA*(1e-9)^3; %as used here in the FN formula
mu0=Ev+Eg/2+(3/4)*tau*log(mh/mc); %intrinsic mu - use as guess
%Based on the neutrality condition, the function FN must be
%zero for the correct chemical potential mu (variable x).
n0=2*(mc*meCs*tau/hbarC^2/2/pi)^(3/2);
p0=2*(mh*meCs*tau/hbarC^2/2/pi)^(3/2);
FN =@(x) n0*exp((x-Ec)/tau)+NAnm/(1+2*exp((epsa-x)/tau))- ...
    p0*exp((Ev-x)/tau)-NDnm/(1+2*exp((x-epsd)/tau));
%
mu=fzero(FN,mu0); %solve for mu using mu_0 as guess
%use the new mu to get the electron and hole concentrations
nnm=n0*exp((mu-Ec)/tau); %electron conc.
pnm=p0*exp((Ev-mu)/tau); %hole conc.
n=nnm/(1e-9)^3; %converting to 1/m^3
%The approximations for when NA=0
n_D_app=sqrt(n0*NDnm/2)*exp(-ED/tau/2); %elect. conc.
n_D_approx=n_D_app/(1e-9)^3; %converting to 1/m^3
mu_approx=Ec-ED/2+(tau/2)*log(NDnm/2/n0);%chem. pot.
fprintf('mh=%6.3f me, mc=%6.3f me, T=%6.2f K\n',mh,mc,T)
fprintf('Ev=%6.3f eV, Ec=%6.3f eV, mu0=%6.3f eV\n',Ev,Ec,mu0)
```

```
fprintf('NA=%6.3g 1/m^3, ND=%6.3g 1/m^3, mu=%6.3f eV\n',NA,ND,mu)
fprintf('n=%6.3g 1/m^3\n',n)
disp('Next are the approximate results assuming NA=0')
fprintf('n_approx=%6.3g 1/m^3, mu_approx=%6.3f eV\n',n_D_approx,mu_approx)
```

The above script charge_neutral.m can be modified in order to reproduce the results shown in Figure 7.6.11 (see Exercise 7.7.12).

### 7.6.1 Ohmic Contacts

An ohmic contact is a kind of a junction between a metal and a semiconductor whose resistance is low enough as not to limit the current flow. The semiconductor's resistance is what limits the flow of current instead. The ohmic contact term itself does not necessarily involve the $I-V$ relation normally associated with a device's ohmic behavior. An example of an ohmic contact's band structure is shown in Figure 7.6.12 where the left panel shows the metal's work function, $\phi_m$, and Fermi level, $E_{FM}$, as well as an n-type semiconductor's work function, $\phi_n$, with its Fermi level, $E_{Fn}$. The right panel shows what happens after contact is made. The band bending that occurs is due to electron transfer from the metal to the semiconductor side.

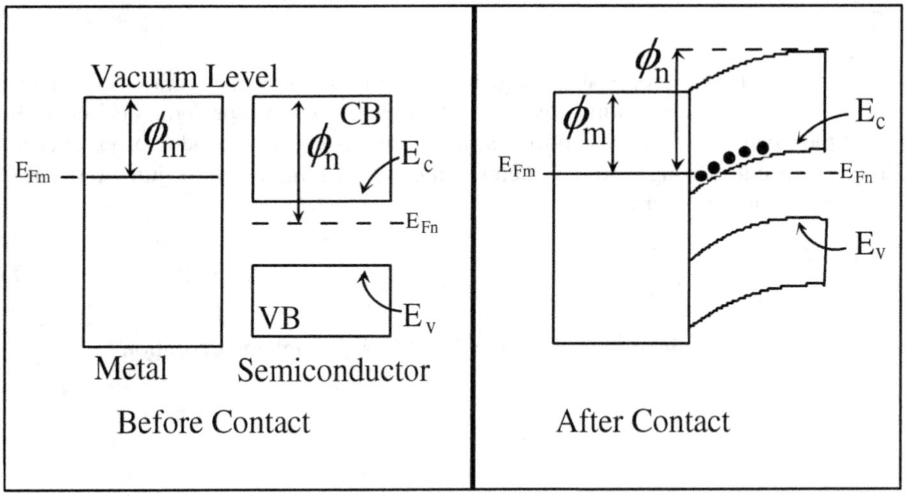


Figure 7.6.12: Metal-n-type semiconductor before contact (left panel) and after contact (right panel). The metal and semiconductor Fermi levels are $E_{Fm}$ and $E_{Fn}$, respectively. Similarly, $\phi_m$ and $\phi_n$ are the corresponding materials' work functions. The band bending (right panel) occurs because the Fermi level becomes uniform throughout the system.

The work function, $\phi$, of a material is the energy difference between the vacuum level and the Fermi level. The vacuum level is the energy position where the electron, having no kinetic energy left, is free from the solid. In the ohmic contact between a metal and an n-type semiconductor, the metal's work function is less than the semiconductor's work function. Because $\phi_m < \phi_n$, there are more energetic electrons in the metal than in the semiconductor's conduction band (CB). As the electrons from the metal's side at $E_{Fm}$ tunnel to the semiconductor's CB, they pile up near the junction as shown by the filled circles until equilibrium is reached in such a way that the accumulated electrons prevent any more electrons from tunneling from the metal to the semiconductor; in other words, the Fermi level becomes uniform across the whole metal semiconductor junction system. We next suppose that there is a DC current flowing from the semiconductor to the metal; that is, electrons flow from the metal to the semiconductor. In that case the electrons' energy will change from $E_{Fm}$ to $E_C + 3k_BT/2$, where $3k_BT/2$ is an average kinetic energy gained. The energy difference corresponds to $\Delta E = E_c + 3k_BT/2 - E_{Fm}$ per electron in the contact region, which is an energy gain. This gain in energy means that the electron must absorb heat from its environment (lattice vibrations). The energy flux associated with the electron charge flux is

$$j_{ue} = n\left(E_c - E_{Fm} + \frac{3}{2}k_BT\right)v, \tag{7.6.39a}$$

where for electrons $v = -\mu_E E$. This thermoelectric process that occurs at a junction between two dissimilar materials is known as the Peltier effect. The Peltier coefficient $\Pi$ is defined by the relation

$$j_{ue} = \Pi_e j_{qe}, \tag{7.6.39b}$$

where for, electrons, $j_{qe} = -nev$. With this quantity, we can find the Peltier coefficient for the electrons as

$$\Pi_e = \frac{j_{ue}}{j_{qe}} = -\frac{E_c - E_{Fm} + \frac{3}{2}k_BT}{e}, \tag{7.6.39c}$$

which is negative because $j_{ue}$ is opposite in sign to $j_{qe}$. Here, the current direction is from the n-type semiconductor to the metal (electrons move from the metal to the n-type semiconductor), but heat is absorbed at the junction. In the case of holes (metal and p-type semiconductor junction), heat is absorbed by the current flowing from the metal to the p-type semiconductor junction. The current density is $j_{qh} = pev$ and the energy flux are

$$j_{uh} = n\left(E_{Fm} - E_v + \frac{3}{2}k_BT\right)v, \tag{7.6.40a}$$

The Peltier coefficient is obtained as in Equation 7.6.39b or through the definition

$$j_{uh} = \Pi_h j_{qh}, \tag{7.6.40b}$$

to obtain

$$\Pi_h = \frac{j_{uh}}{j_{qh}} = \frac{E_{Fm} - E_v + \frac{3}{2}k_BT}{e}, \tag{7.6.40c}$$

which is positive because $j_{uh}$ has the same sign as $j_{qh}$. The hole current is from the metal to the p-type semiconductor (the same as the direction of the current) and heat is absorbed at the junction. Measuring the voltage associated with the Peltier effect, in which one end of the junction is heated, is one way to determine what type of semiconductor material is present in the sample. This is because the sign of the voltage will be different depending on whether the sample is n-type or p-type. It is also possible to combine the above effects in such a way as to create practical thermoelectric cooling devices. Finally, it should be mentioned that the Peltier effect is different from the Joule heating ($I^2R$) that is associated with the electrical resistance of a material and which is due to the collisions suffered by the moving particles through that material. Such collisions also give rise to temperature changes in the sample.

## 7.7 Chapter 7 Exercises

7.7.1. If during an optical absorption experiment in a Si semiconductor at low temperature a photon of $\omega = 1.8 \times 10^{15}\, rad/sec$ emits an acoustic phonon, what is the phonon's energy and radial frequency and wavevector, given that the phonon phase velocity is $\approx 4 \times 10^3\, m/s$? Hint: assume a linear dispersion.

7.7.2. In Example 7.2.0.1, (a) show that the solutions for $k_x(t)$ and $k_y(t)$ are consistent with the initial equations of motion for the wavevectors. (b) What is the value of the constant $A$ in the solutions for $k_x(t)$ and $k_y(t)$ and how is it related to the constant energy surface referred to in the example?

7.7.3. Modify the code of script central_eq_bands.m from Chapter 6 to incorporate the band edge approximations of Equation 7.3.5 in order to reproduce Figure 7.3.5 keeping the same units of distance and energy as previously employed.

7.7.4. Treating a particle as a wave packet moving with speed $v_g$ in one dimension and considering a force acting on it, show that an expression for the mass of the particle in terms of its energy has the form of Equation 7.3.6.

7.7.5. Show that Equation 7.3.7 results by repeating Exercise 7.7.4 and considering a three-dimensional version of the problem including a more general energy surface, $\varepsilon_{\vec{k}}$.

7.7.6. Obtain the resonant frequency of band edge electrons in a sample of GaAs in a magnetic field of $940\,G$.

7.7.7. Beginning with Equation 7.4.18 obtain Equation 7.4.19.

7.7.8. Reproduce Figure 7.4.7. You might find it useful to employ the following constants in appropriate units: $\hbar c = 197\, eV \cdot nm$, $k_B = 8.6174 \times 10^{-5}\, eV/K$, and the mass of the electron $mc^2 = 0.5 \times 10^6\, eV$.

7.7.9. Write the code necessary to reproduce the plots and the analysis of Figure 7.4.8 as described in Example 7.4.1.1. Feel free to use the code starter provided in the example.

7.7.10. Calculate the energy state, the radius, and the de Broglie wavelength associated with a donor state in Ge, assuming an electron effective mass of $m_c \approx 0.1m$.

7.7.11. After reading Example 7.6.0.1, show the steps to obtain Equation 7.6.38b for $\mu$ in the presence of a donor concentration $N_D$.

7.7.12. Run the code charge_neutral.m provided in Section 7.6 to be sure it runs correctly. The printed calculations should match the following results:

```
mh= 0.025 me, mc= 0.050 me, T=300.00 K
Ev= 0.000 eV, Ec= 1.000 eV, mu0= 0.487 eV
NA=1e+014 1/m^3, ND=1e+024 1/m^3, mu= 0.981 eV
n=1.31e+023 1/m^3
Next are the approximate results assuming NA=0
n_approx=1.4e+023 1/m^3, mu_approx= 0.983 eV
```

Modify the code in order to reproduce the results shown in Figure 7.6.11.

# 8

# *Simple Band Structure Calculations*

**Contents**



## 8.1 Introduction

One of the general goals of solid state theory is to be able to obtain the electronic energy levels and wave functions in crystal systems. Such knowledge, in turn, enables the calculation and, therefore, the prediction of solid state crystal properties. This is not a simple task to accomplish, for it involves solving the Schrodinger equation for many electrons interacting with themselves as well as all the ions in a crystal. One of the stumbling blocks in tackling this problem exactly is often a computational limitation; that is, if a sample contains $N_1$ atoms and each atom contains $N_2$ electrons, then there are $N_1N_2$ particles. Since each particle has its own coordinates and each of the $N_1$ atoms interacts with each of the $N_2$ electrons, the actual problem involves $N_1^{N_2}$ variables. For example, if there are two atoms and each contains four electrons, there are 16 variables associated with their interactions due to each atom's interaction with each of the total of eight electrons available; this counting does not include the ions' interactions among themselves nor the electrons interacting among themselves for that matter.

Approximate band structure calculation methods, therefore, have often been developed in order to minimize the computational demand and the computational time required in meeting such demands. One of the universal approximations made in tackling this problem is to apply the so-called Born-Oppenheimer approximation, according to which the nuclei are assumed to be static. In this way, one solves the electronic problem without worrying about the nuclei. As far as the nuclei are concerned, the idea is that since the electrons are very light compared to the nuclei and move more rapidly than the nuclei, they can follow the slower moving nuclei quite well. In this manner, the electron distribution determines the potential in which the nuclei move.

Furthermore, all band structure calculations operate within the single-electron approximation; that is, the electronic behavior of solids is considered to be that of a single electron moving in the presence of a crystal potential. Finding the potential itself is one of the tasks of the approximation. By doing so, the number of variables of the problem is shortened from $N_1^{N_2}$ to $N_1N_2$.

Popular sophisticated approaches to obtain the electronic energies and wavefunctions are among the Hartree-Fock and density functional methods. In the Hartree-Fock theory, the main computational obstacle in performing the electronic calculations lies in the coulomb interaction between the electrons; i.e., the so-called electron-electron (e-e) interaction. Thus, in this approximation the e-e interaction is replaced by an effective potential in such a way that each electron moves in a field produced by the sum over all the other electrons. While obeying the Pauli exclusion principle, this method is self-consistent in that iterations are performed so as to vary the wave functions while minimizing the total electronic system energy. A simplified version of this method, the Hartree approximation, was employed in Chapter 3 for the hydrogen molecule.

In the density functional theory (DFT), rather than obtaining the wavefunction explicitly, the method seeks to find the electron density so that, once found, one can deduce the potential in which the electrons move. Knowing the potential allows for the solution of the Schrodinger equation to the many body problem. This is also a self-consistent approach, since the electron density is varied while seeking a minimum of the total system energy.

Approximations may make use of pseudo-potentials to approximate the atomic potentials. They may also use wavefunction approximations such as the linear combination of atomic orbitals method, the plane wave method, and the linear augmented plane wave approach.

The pseudo-potential method may play a role in the above approximations in that rather than trying to find the exact potential associated with each atom in the crystal, the electrons' contribution to the potential is separated into two parts. One part is for the core electrons and another is for the valence electrons. Ultimately, the core electrons are considered to be rigid and part of the core, while the outer valence electrons are dealt with explicitly. The resulting wavefunctions or basis functions are considered pseudo-wavefunctions, associated with the pseudo-potential, in the sense that they are approximations to the real wavefunctions of the crystal. The basis functions thus obtained describe the electrons in a given atom. These basis functions may be used to obtain crystal wavefunctions.

The concept of the linear combination of atomic orbitals (LCAO) method is based on the idea that once the atomic pseudo-potentials are known, a wavefunction for a particular state of the crystal is a quantum mechanical superposition of component atomic wavefunctions, each centered on a particular atom and weighted by an appropriate coefficient. This method was actually used in obtaining the molecular hydrogen ground state wavefunction in Chapter 3 for the hydrogen molecule, albeit using the Hartree approximation.

The plane wave method is based on the idea that the crystal is a periodic array of atoms, so that one can create crystal wavefunctions that involve plane wave expansions due to the fact that such approach automatically obeys Bloch's theorem. However, plane wave expansions do not converge readily in the interior of an atomic cell. This disadvantage leads to the improved linear augmented plane wave (LAPW) method where the basis set (used in the expansion) is made to behave like a

plane wave outside the atomic cell but obeys a spherically symmetric Schrodinger equation inside the cell. The atomic cell is modeled using the so-called muffin tin potential.
In contrast to the above-mentioned methods, there exists an approach that is simple to apply and has been used in semiconductors to various degrees of success. This is the so-called tight binding method. The tight binding method is based on the idea of a collection of isolated atoms which are slowly brought closer together in order to form a crystal. The wave function is a linear combination of atomic orbitals, but the only orbitals considered are those of the valence electrons. In this method, the crystal potential is written as a sum of identical atomic potentials, so that the atomic orbitals of the individual atoms are also identical. The entire crystal wavefunction obeys Bloch's theorem but matrix elements used in the solution of the energy eigenvalues have been tabulated a priori or are obtained by variational methods that improve the comparison with experiment. Due to its simplicity, here we concentrate on calculations based on the tight binding model.

## 8.2 The Single Band Tight Binding Model

In this section, we study the tight binding model for crystal systems with one electron per atom. The electron is assumed to be in the "s" state moving in the presence of a crystal potential $U(r)$ of an isolated atom and whose wavefunction is $\phi(r)$. This assumes that the other crystal atoms' influence is small. The crystal wavefunction is written as

$$\psi_{\vec{k}}(\vec{r}) = \sum_j C_{\vec{k}j}\phi(\vec{r}-\vec{r}_j), \tag{8.2.1}$$

where the sum is over all $N$ lattice points $\vec{r}_j$. If the coefficients of the expansion, $C_{\vec{k}j}$ are taken to have the plane waveform,

$$C_{\vec{k}j} = \frac{1}{\sqrt{N}}e^{i\vec{k}\cdot\vec{r}_j}, \tag{8.2.2}$$

then Equation 8.2.1 is of the Bloch form. This can be seen if in Equation 8.2.1 we make the displacement $\vec{r} \to \vec{r}+\vec{T}$ to get

$$\begin{aligned} \psi_{\vec{k}}(\vec{r}+\vec{T}) &= \sum_j C_{\vec{k}j}\phi(\vec{r}+\vec{T}-\vec{r}_j) \\ &= e^{i\vec{k}\cdot\vec{T}}\frac{1}{\sqrt{N}}\sum_j e^{i\vec{k}\cdot(\vec{r}_j-\vec{T})}\phi(\vec{r}-(\vec{r}_j-\vec{T})) \\ &= e^{i\vec{k}\cdot\vec{T}}\frac{1}{\sqrt{N}}\sum_i e^{i\vec{k}\cdot\vec{r}_i}\phi(\vec{r}-\vec{r}_i) = \sum_i C_{\vec{k}i}\phi(\vec{r}-\vec{r}_i) \\ &= e^{i\vec{k}\cdot\vec{T}}\psi_{\vec{k}}(\vec{r}), \end{aligned} \tag{8.2.3}$$

where we used $\vec{r}_i = \vec{r}_j - \vec{T}$; i.e., another equivalent lattice point with identical wavefunction. The last equation is the Bloch condition discussed in Chapter 3. We next use the Dirac notation and write the wavefunction of Equation 8.2.1 as

$$|\vec{k}> = \frac{1}{\sqrt{N}}\sum_j e^{i\vec{k}\cdot\vec{r}_j}|j), \tag{8.2.4}$$

where $|\vec{k}>$ represents the normalized crystal wavefunction $\psi_{\vec{k}}(\vec{r})$ and $|j) = |\phi_j)$ represents the basis function $\phi(\vec{r}-\vec{r}_j)$. Notice that the functions $|\vec{k}>$ are automatically normalized because the basis functions $|j)$ are orthonormal (see Exercise 8.8.1). In the tight binding approximation one calculates the first order energy of the crystal by finding the diagonal matrix elements of the Hamiltonian as

$$<\vec{k}|\hat{H}|\vec{k}>=\frac{1}{N}\sum_{j}\sum_{m}e^{i\vec{k}\cdot(\vec{r}_j-\vec{r}_m)}(m|\hat{H}|j), \tag{8.2.5a}$$

where the $(m|\hat{H}|j)\equiv\int dV\phi(\vec{r}-\vec{r}_m)\hat{H}\phi(\vec{r}-\vec{r}_j)$; i.e., a Hamiltonian volume integral between two basis functions, one on site $m$ and the other on site $j$. If we let $\vec{\rho}_m\equiv\vec{r}_m-\vec{r}_j$ so that $\phi(\vec{r}-\vec{r}_m)=\phi(\vec{r}-\vec{\rho}_m-\vec{r}_j)$. The energy becomes

$$<\vec{k}|\hat{H}|\vec{k}>=\sum_{m}e^{-i\vec{k}\cdot\vec{\rho}_m}\frac{1}{N}\sum_{j}\int dV\phi(\vec{r}-\vec{\rho}_m-\vec{r}_j)\hat{H}\phi(\vec{r}-\vec{\rho}_j)=\sum_{m}e^{-i\vec{k}\cdot\vec{\rho}_m}\int dV\phi(\vec{r}-\vec{\rho}_m)\hat{H}\phi(\vec{r}), \tag{8.2.5b}$$

where we have taken $\sum_j\int dV\phi(\vec{r}-\vec{\rho}_m-\vec{r}_j)\hat{H}\phi(\vec{r}-\vec{\rho}_j)=N\int dV\phi(\vec{r}-\vec{\rho}_m)\hat{H}\phi(\vec{r})$, since all sites are identical. If we next neglect all integrals except those that involve the same atom ($\rho_{m=0}=\rho_0$) and those between nearest neighbors (nn) connected by $\rho\equiv\rho_m$, we have

$$\begin{aligned}<\vec{k}|\hat{H}|\vec{k}>&=\sum_{m=0,nn}e^{-i\vec{k}\cdot\vec{\rho}_m}\int dV\phi(\vec{r}-\vec{\rho}_m)\hat{H}\phi(\vec{r})\\&=e^{-i\vec{k}\cdot\vec{\rho}_0}\int dV\phi(\vec{r}-\vec{\rho}_0)\hat{H}\phi(\vec{r})+\sum_{m=nn}e^{-i\vec{k}\cdot\vec{\rho}_m}\int dV\phi(\vec{r}-\vec{\rho})\hat{H}\phi(\vec{r})\\&=-\alpha-\gamma\sum_{m=nn}e^{-i\vec{k}\cdot\vec{\rho}_m}\equiv\varepsilon_k,\end{aligned} \tag{8.2.5c}$$

where we have defined $\alpha\equiv-\int dV\phi(\vec{r}-\vec{\rho}_0)\hat{H}\phi(\vec{r})$, $\gamma\equiv-\int dV\phi(\vec{r}-\vec{\rho})\hat{H}\phi(\vec{r})$, and we have let $\rho_0=0$. Here $\varepsilon_k$ corresponds to a single energy band. It can be specialized to particular systems once the crystal structure is identified. Later, we will consider the simple cubic structures. In the above equations, $\alpha$ is the diagonal energy associated with each atomic electron, also referred to as the on-site energy; $\gamma$ is the overlap or off-diagonal energy, which also goes by the name of hopping energy because it connects one site to another. While often $\alpha$ and $\gamma$ are treated as parameters (variables whose values are changed to fit a given experiment), for the case of hydrogen in the $1s$ state, the diagonal energy is $\alpha=(me^4k^2/2\hbar^2)=ke^2/(2a_0)=13.606\,eV$ or one Rydberg (Ry). Similarly, the off-diagonal energy is that between two hydrogen atoms in the $1s$ state, $\rho$ apart, and is given by $\gamma=(me^4k^2/\hbar^2)(1+\rho/a_0)\exp(-\rho/a_0)$, where the Bohr radius is $a_0=\hbar^2/(me^2k)$. The quantity $me^4k^2/\hbar^2$ is the same as $2Ry=27.212\,eV$ or one Hartree.
Often the single band tight binding Hamiltonian for a crystal is written as

$$\hat{H}=\sum_{p}|p)(-\alpha)(p|+\sum_{pq}{}'|p)(-\gamma)(q|=\hat{H}_D+\hat{H}_{OD}, \tag{8.2.6a}$$

where the prime indicates that $p\neq q$, and $\hat{H}_D$, $\hat{H}_{OD}$ are the diagonal and off-diagonal parts of $\hat{H}$, respectively. Using Equation 8.2.4, we see that the diagonal energy is

$$\begin{aligned}<\vec{k}|\hat{H}_D|\vec{k}>&=\frac{1}{N}\sum_{p}\sum_{j'}\sum_{j}e^{-i\vec{k}\cdot\vec{r}_{j'}}e^{i\vec{k}\cdot\vec{r}_j}(j'|p)(p|j)(-\alpha)\\&=\frac{1}{N}\sum_{p}\sum_{j',j}e^{-i\vec{k}\cdot(\vec{r}_j-\vec{r}_{j'})}\delta_{j',p}\delta_{p,j}(-\alpha)\\&=\frac{1}{N}\sum_{p}(-\alpha)=\frac{1}{N}N(-\alpha)=-\alpha,\end{aligned} \tag{8.2.6b}$$

where we have used the orthonormality of the basis; i.e., $(p|j) = \delta_{p,j}$, which equals unity for $j = p$, else it vanishes. Similarly, repeating the process for the off-diagonal term of $\hat{H}$ (see Exercise 8.8.2), we get $<\vec{k}|\hat{H}_{OD}|\vec{k}> = -\gamma \sum_{m=nn} e^{-i\vec{k}\cdot\vec{\rho}_m}$. So by adding these two results one gets the expected energy

$$\varepsilon_k = <\vec{k}|\hat{H}_D + \hat{H}_{OD}|\vec{k}> = -\alpha - \gamma \sum_{m=nn} e^{-i\vec{k}\cdot\vec{\rho}_m}, \tag{8.2.6c}$$

as in Equation 8.2.5c.

### 8.2.1 The Simple Cubic (SC)

Here we obtain the simple cubic (SC) system's tight binding band based on Equation 8.2.6c. The SC structure has six nearest neighbors associated with it. They are located at $\vec{\rho}_m = (\pm a, 0, 0), (0, \pm a, 0), (0, 0, \pm a)$; so that, with $\vec{k} = k_x\hat{i} + k_y\hat{j} + k_z\hat{k}$,

$$\sum_{m=nn} e^{-i\vec{k}\cdot\vec{\rho}_m} = e^{-ik_xa} + e^{ik_xa} + e^{-ik_ya} + e^{ik_ya} + e^{-ik_za} + e^{ik_za} = 2\cos(k_xa) + 2\cos(k_ya) + 2\cos(k_za). \tag{8.2.7a}$$

The simple cubic band is, therefore,

$$\varepsilon_{\vec{k},SC} = -\alpha - 2\gamma(\cos(k_xa) + \cos(k_ya) + \cos(k_za)). \tag{8.2.7b}$$

In Figure 8.2.1, the simple cubic band $E_{\vec{k}} = (\varepsilon_{\vec{k},SC} + \alpha)/(2\gamma)$ is plotted versus $k_x$ and $k_y$ for a value of $k_z = 0.0$ for a lattice constant of $a = 1$ in arbitrary units.

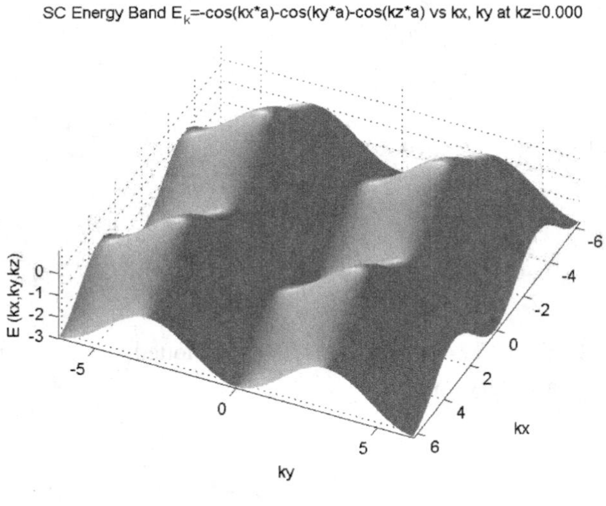


Figure 8.2.1: The simple cubic tight binding band $E_{\vec{k}} = (\varepsilon_{\vec{k},SC} + \alpha)/(2\gamma)$ for $k_z = 0.0$. The range of $k_x$ and $k_y$ is between $\pm 2\pi/a$ with $a = 1$ in arbitrary units.

The script used to create Figure 8.2.1 is TBbandSC.m whose listing follows.

```
%copyright by J. E Hasbun and T. Datta
%TBbandSC.m
%Tight binding model band for the simple cubic system
```

```
clear, clc;
a=1.0;
q=2*pi/a;       %range of k
kz=0.0;         %Keep kz fixed
[kx,ky]=meshgrid(-q:0.1:q,-q:0.1:q);  %variables for the 2D plot
E=-cos(kx*a)-cos(ky*a)-cos(kz*a);     %SC energy band
figure;
surfl(kx,ky,E);                       %2D plot
colormap(gray)
shading interp
xlabel('kx'), ylabel('ky'), zlabel('E (kx,ky,kz)')
str=cat(2,'SC Energy Band E_k=-cos(kx*a)-cos(ky*a)-cos(kz*a) vs kx, ky at
   kz=',... num2str(kz,'%4.3f'));
title(str)
axis([-q q -q q min(min(E)) max(max(E))])
view(115,70)                          %viewpoint azimuth, elevation
```

### 8.2.2 The Body Centered Cubic

The body centered cubic (BCC) structure has eight nearest neighbors associated with it. They are located at $\vec{\rho}_m = (\pm a/2, \pm a/2, \pm a/2) = (a/2)[(1,-1,1),(1,1,1),(1,-1,-1),(1,1,-1),(-1,-1,1),$ $(-1,1,1),(-1,-1,-1),(-1,1,-1)]$. We have

$$\begin{aligned}\sum_{m=nn} e^{-i\vec{k}\cdot\vec{\rho}_m} =& e^{ia(-k_x+k_y-k_z)/2} + e^{ia(-k_x-k_y-k_z)/2}\\ &+ e^{ia(-k_x+k_y+k_z)/2} + e^{ia(-k_x-k_y+k_z)/2}\\ &+ e^{ia(k_x+k_y-k_z)/2} + e^{ia(k_x-k_y-k_z)/2}\\ &+ e^{ia(k_x+k_y+k_z)/2} + e^{ia(k_x-k_y+k_z)/2}\\ =& 8\cos\left(\frac{k_x a}{2}\right)\cos\left(\frac{k_y a}{2}\right)\cos\left(\frac{k_z a}{2}\right).\end{aligned} \tag{8.2.8a}$$

Substituting this into Equation 8.2.6c, the BCC band becomes

$$\varepsilon_{\vec{k},BCC} = -\alpha - 8\gamma\cos\left(\frac{k_x a}{2}\right)\cos\left(\frac{k_y a}{2}\right)\cos\left(\frac{k_z a}{2}\right). \tag{8.2.8b}$$

In Figure 8.2.2, the BCC band $E_{\vec{k}} = (\varepsilon_{\vec{k},BCC} + \alpha)/(8\gamma)$ is plotted versus $k_x$ and $k_y$ for a value of $k_z = \pi/(2a)$ with $a = 1$ in arbitrary units.

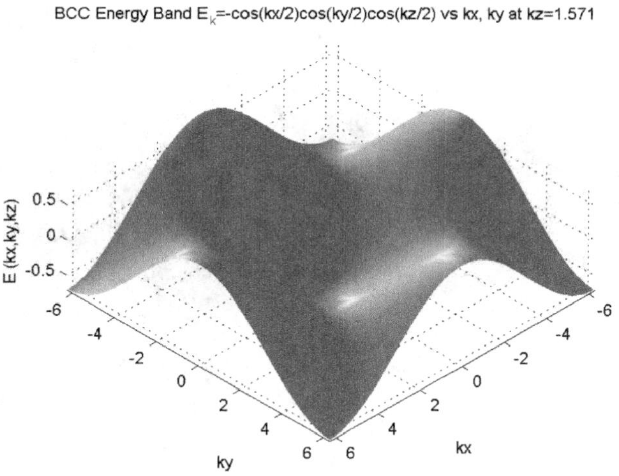


Figure 8.2.2: The body centered cubic tight binding band $E_{\vec{k}} = (\varepsilon_{\vec{k},BCC} + \alpha)/(8\gamma)$ for $k_z = \pi/(2a)$. The range of $k_x$ and $k_y$ is between $\pm 2\pi/a$ with $a = 1$ in arbitrary units.

### 8.2.3 The Face Centered Cubic

The face centered cubic (FCC) structure has twelve nearest neighbors associated with it. They are located at $\vec{\rho}_m = (a/2)[(1,-1,0),(1,1,0),(-1,-1,0),(-1,1,0),(0,-1,1),(0,1,1),(0,-1,-1),$ $(0,1,-1),(1,0,1),(-1,0,1),(1,0,-1),(-1,0,-1)]$. After performing $\sum\limits_{m=nn} e^{-i\vec{k}\cdot\vec{\rho}_m}$ and using Equation 8.2.6c, the FCC resulting band is

$$\varepsilon_{\vec{k},FCC} = -\alpha - 4\gamma\left(\cos\left(\frac{k_x a}{2}\right)\cos\left(\frac{k_y a}{2}\right) + \cos\left(\frac{k_y a}{2}\right)\cos\left(\frac{k_z a}{2}\right) + \cos\left(\frac{k_x a}{2}\right)\cos\left(\frac{k_z a}{2}\right)\right). \quad (8.2.9)$$

In Figure 8.2.3, the FCC band $E_{\vec{k}} = (\varepsilon_{\vec{k},FCC} + \alpha)/(4\gamma)$ is plotted versus $k_x$ and $k_y$ for a value of $k_z = \pi/(2a)$ with $a = 1$ in arbitrary units.

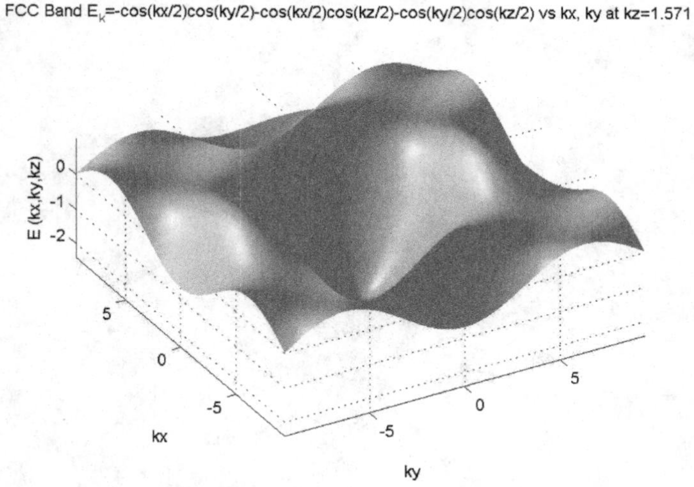


Figure 8.2.3: The face centered cubic tight binding band $E_{\vec{k}} = (\varepsilon_{\vec{k},FCC} + \alpha)/(4\gamma)$ for $k_z = \pi/(2a)$. The range of $k_x$ and $k_y$ is between $\pm 3\pi/a$ with $a = 1$ in arbitrary units.

## 8.3 The Density of States and the Fermi Surface

As discussed in Chapter 5, given $N$ electrons, the Fermi energy is the energy of the highest filled orbital at $T = 0$. For a spherical electron energy dispersion, the Fermi energy is

$$\varepsilon_F = \frac{\hbar^2 k_F^2}{2m}, \tag{8.3.10a}$$

where $k_F$ is the magnitude of a three-dimensional wavevector that is characteristic of the number of electrons up to the highest occupied orbital at $T = 0K$. For a more general energy dispersion, the Fermi energy is obtained through the expression (See Chapter 5)

$$N = \int_0^{\varepsilon_F} D(\varepsilon) d\varepsilon. \tag{8.3.10b}$$

Where $D(\varepsilon)$ is the density of states. As regards to the density of states, we will now define the way in which we will use it. For a function of $\vec{k}$, each wavevector $\vec{k}$ is associated with a Brillouin zone volume of

$$V_{BZ} \equiv (2\pi/L)^3 = (2\pi)^3/V_c, \tag{8.3.11a}$$

for a system with unit cell volume $V_c = L^3$. Thus, similar to Chapter 5 (see the normal mode enumeration section), a sum of a function $f_{\vec{k}}$ over the wavevectors $\vec{k}$ is expressed in integral form as

$$\sum_{\vec{k}} f_{\vec{k}} = \frac{1}{V_{BZ}} \int_{\text{all k space}} d\vec{k} f_{\vec{k}} = \frac{N}{V_{BZ}} \int_{V_{BZ}} d\vec{k} f_{\vec{k}}, \tag{8.3.11b}$$

where we have used the fact that, due to periodicity, for a crystal, the total $k$-space volume is $N$ times the Brillouin zone volume (see Exercise 8.8.8) or

$$V_{k-\text{space}} = \int\limits_{\text{all k space}} d\vec{k} = N \int\limits_{V_{BZ}} d\vec{k} = NV_{BZ}. \tag{8.3.11c}$$

Ignoring the spin of the electron, for the density of states definition, a function of energy $\varepsilon_k$ can be written as

$$\frac{1}{N}\sum_{\vec{k}} f(\varepsilon_{\vec{k}}) = \frac{1}{V_{BZ}}\int\limits_{V_{BZ}} d\vec{k} f(\varepsilon_{\vec{k}}) = \frac{1}{V_{BZ}}\int d\varepsilon \int\limits_{V_{BZ}} d\vec{k}\delta(\varepsilon-\varepsilon_{\vec{k}})f(\varepsilon) = \int d\varepsilon f(\varepsilon)D(\varepsilon), \tag{8.3.12a}$$

where we have defined the density of electron states as

$$D(\varepsilon) \equiv \frac{1}{V_{BZ}}\int\limits_{V_{BZ}} \delta(\varepsilon-\varepsilon_{\vec{k}})\,d\vec{k}. \tag{8.3.12b}$$

Similarly, if the electron spin is included (signified by $\sigma$), we have

$$\frac{1}{N}\sum_{\vec{k},\sigma} f(\varepsilon_{\vec{k}}) = \frac{2}{V_{BZ}}\int\limits_{V_{BZ}} d\vec{k} f(\varepsilon_{\vec{k}}) = \int d\varepsilon f(\varepsilon)D_{\Sigma\sigma}(\varepsilon)$$

with (8.3.12c)

$$D_{\Sigma\sigma}(\varepsilon) \equiv \frac{2}{V_{BZ}}\int\limits_{V_{BZ}} \delta(\varepsilon-\varepsilon_{\vec{k}})\,d\vec{k},$$

where we have defined $D_{\Sigma\sigma}(\varepsilon) \equiv 2D(\varepsilon)$ to include the factor of 2 due to the electron spin. This density of states expression can be used in conjunction with Equation 8.3.10b in order to find the Fermi level $E_F$. Incidentally, Equations 8.3.12b and 8.3.12c are sometimes referred to as differential density of states. The total density of states (or integrated density of states) is a version of Equation 8.3.10b but the integral is a function of energy; that is,

$$N(\varepsilon) = \int_{-\infty}^{\varepsilon} D(\varepsilon')d\varepsilon', \tag{8.3.13}$$

and it gives the total number of states as a function of $\varepsilon$. The lower limit is chosen to be $-\infty$ because the energy band has $\varepsilon < 0$ contributions. If the electron spin is included, the proper form of $D(\varepsilon)$ is to be used to multiply by a factor of 2.

**Example 8.3.0.1**

Let's use Equation 8.3.12c to obtain the three-dimensional free electron gas density of states. Due to spherical symmetry, we first write $\int d\vec{k} = \int k^2 dk \int \sin\theta d\theta \int d\phi = 4\pi \int k^2 dk$, so that from Equation 8.3.12c we get

$D_{free}(\varepsilon) = 4\pi\frac{2}{V_{BZ}}\int k^2\delta(\varepsilon-\varepsilon_{\vec{k}})dk$.

For a free electron gas, the energy band is $\varepsilon_k = \hbar^2k^2/2m$ or $2kdk = 2md\varepsilon_k/\hbar^2$ and $k^2dk = (m/\hbar^2)\sqrt{2m\varepsilon_k/\hbar^2}\,d\varepsilon_k$

The density of states becomes

$D_{free}(\varepsilon) = \frac{8\pi}{V_{BZ}}\int(\frac{m}{\hbar^2})\sqrt{\frac{2m\varepsilon_k}{\hbar^2}}\delta(\varepsilon-\varepsilon_{\vec{k}})d\varepsilon_k = \frac{8\pi}{V_{BZ}}\frac{m}{\hbar^3}\sqrt{2m\varepsilon} = \frac{8\pi V}{(2\pi)^3}\frac{m}{\hbar^3}\sqrt{2m\varepsilon} = \frac{V}{2\pi^2}\left(\frac{2m}{\hbar^2}\right)^{3/2}\sqrt{\varepsilon}$,

which is the density of states of a three-dimensional electron gas obtained in Chapter 5.

The Fermi surface for a three dimensional electron gas is a sphere. This can be seen from Equation 8.3.10a by writing $2m\varepsilon_F/\hbar^2 = k_x^2 + k_y^2 + k_z^2$. Thus, for a given value of $\varepsilon_F$, the wavevector values that solve this equation describe a sphere of radius $2m\varepsilon_F/\hbar^2$. For a more general energy band, the shape of the Fermi surface can be solved numerically by finding all wavevectors $\vec{k}$ that satisfy the equation $\varepsilon_{\vec{k}} = E_F$. The number of electrons $N$ involved in Equation 8.3.10b determines the value of $E_F$ associated with the Fermi surface. In the case of the single band cubic systems considered in Section 8.2, only two electrons can be placed in each band, one for each spin (up and down). A band with a single electron is referred to as a half-filled band. Below we consider these systems with a Fermi energy value such that the band has a single electron.

### 8.3.1 The Simple Cubic Fermi Surface

From Equation 8.2.7b, the simple cubic band we plotted in Figure 8.2.1 is written as

$$E_{\vec{k},SC} \equiv (\varepsilon_{\vec{k},SC} + \alpha)/(2\gamma) = -\cos(k_x a) - \cos(k_y a) - \cos(k_z a) \qquad (8.3.14)$$

This band is symmetric with a range of $-3 \leq \varepsilon_{\vec{k},SC} \leq 3$, so that the middle of the band lies at $\varepsilon_{\vec{k},SC} = 0 = E_F$. At this value of the Fermi level, the band is half-filled. Figure 8.3.4 shows the simple cubic's Fermi surface thus obtained.

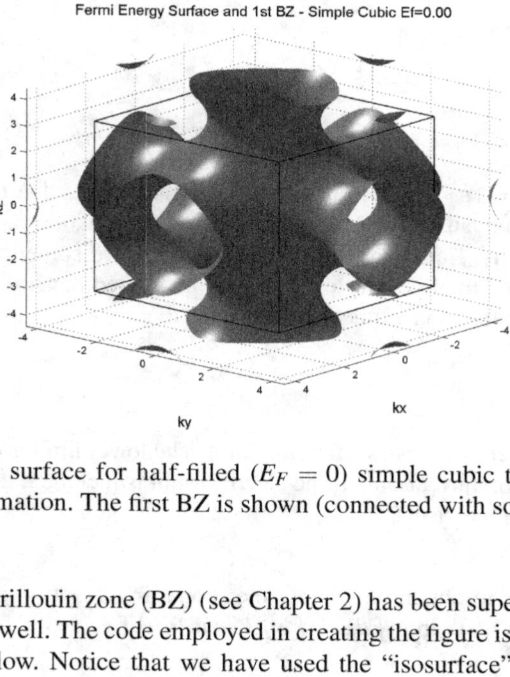


Figure 8.3.4: The Fermi surface for half-filled ($E_F = 0$) simple cubic tight binding band in the nearest-neighbor approximation. The first BZ is shown (connected with solid lines).

The simple cubic's first Brillouin zone (BZ) (see Chapter 2) has been superimposed onto the Fermi surface in Figure 8.3.4 as well. The code employed in creating the figure is TBfermiIsoSurfBzSc.m, whose listing follows below. Notice that we have used the "isosurface" MATLAB command in conjunction with "meshgrid" to create the constant energy surface.

```
%copyright by J. E Hasbun and T. Datta
%TBfermiIsoSurfBzSc.m, Simple Cubic
%To find the full Fermi surface, we use MATLAB's isosurface
%command for a given Fermi energy value.
%The obtained Fermi Surface is superimposed with the Brillouin Zone.
%Given kx, ky, and kz, surfaces of contant energy E=Ef are sought.
function TBfermiIsoSurfBzSc
```

```
clear,clc
a=1.0;
q=pi/a;
qq=(7./5.)*q;
Ef=0.0;                                  %half filled band
[kx,ky,kz]=meshgrid(-qq:0.1:qq,-qq:0.1:qq,-qq:0.1:qq);
E=-cos(kx*a)-cos(ky*a)-cos(kz*a);    %SC energy band
hf=figure;
%isosurface connects points that have the specified value
%much the way contour lines connect points of equal elevation.
isosurface(kx,ky,kz,E,Ef);    %the isosurface to get the Fermi surface
colormap(gray)
xlabel('kx','FontSize',14), ylabel('ky','FontSize',14)
str=cat(2,'Fermi Energy Surface and 1st BZ - Simple Cubic Ef=',...
    num2str(Ef,'%4.2f'));
zlabel('kz','FontSize',14), title(str,'FontSize',14)

%Let's draw the first BZ
%SC BZ - with volume: (2*pi/a)^3
set(hf,'Position',[90 78 560 420])
view(130,20)                   %viewpoint azimuth, elevation
grid on
axis([-qq qq -qq qq -qq qq])

%corners
c1=[1/2,1/2,1/2]; c2=[-1/2,1/2,1/2]; c3=[-1/2,-1/2,1/2]; c4=[1/2,-1/2,1/2];
c5=[1/2,1/2,-1/2]; c6=[-1/2,1/2,-1/2]; c7=[-1/2,-1/2,-1/2];
   c8=[1/2,-1/2,-1/2];
c1=c1*2*q; c2=c2*2*q; c3=c3*2*q; c4=c4*2*q; c5=c5*2*q; c6=c6*2*q;
c7=c7*2*q; c8=c8*2*q;
%c1 top corner connectors
liner(c1,c2,'-','k',1.0)
liner(c1,c4,'-','k',1.0)
liner(c1,c5,'-','k',1.0)
%c2 top corner connectors
liner(c2,c6,'-','k',1.0)
liner(c2,c3,'-','k',1.0)
%c3 top corner connectors
liner(c3,c4,'-','k',1.0)
liner(c3,c7,'-','k',1.0)
%c5 bottom corner connectors
liner(c5,c6,'-','k',1.0)
liner(c5,c8,'-','k',1.0)
%c7 bottom corner connectors
liner(c7,c6,'-','k',1.0)
liner(c7,c8,'-','k',1.0)
%c8 bottom corner connectors
liner(c8,c4,'-','k',1.0);

function liner(v1,v2,lin_style_txt,lin_color_txt,lin_width_num)
%Draws a line given initial vector v1 and final vector v2
```

```
%lin_style_txt: line style text format
%lin_color_txt: line color text format
%lin_width_num: line width number format
line([v1(1),v2(1)],[v1(2),v2(2)],[v1(3),v2(3)],...
    'LineStyle',lin_style_txt,'color',...
    lin_color_txt,'linewidth',lin_width_num)
```

### 8.3.2 The Body Centered Cubic Fermi Surface

The body centered cubic's band plotted in Figure 8.2.2 is

$$E_{\vec{k},BCC} \equiv (\varepsilon_{\vec{k},BCC} + \alpha)/(8\gamma) = -\cos(\frac{k_x a}{2})\cos(\frac{k_y a}{2})\cos(\frac{k_z a}{2}), \tag{8.3.15}$$

which is symmetric with a range of $-1 \leq \varepsilon_{\vec{k},BCC} \leq 1$. Thus, the Fermi energy of $E_F = 0 = \varepsilon_{\vec{k},BCC}$ yields a Fermi surface for a half filled band as well. This along its first BZ (see Chapter 2) are shown in Figure 8.3.5.

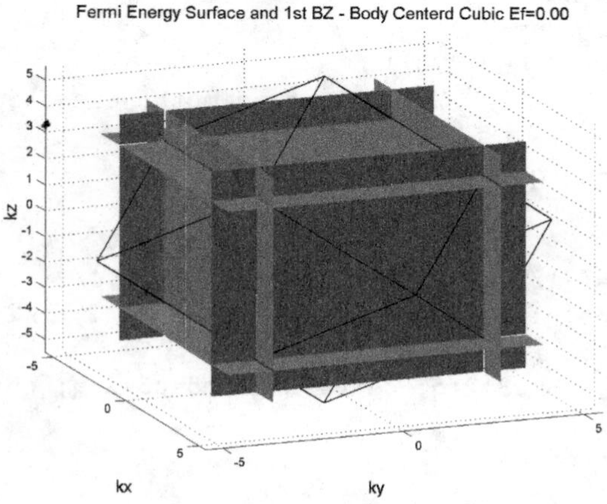


Figure 8.3.5: The Fermi surface for half-filled ($E_F = 0$) body centered cubic tight binding band in the nearest-neighbor approximation. The first BZ is shown (connected with solid lines).

### 8.3.3 The Face Centered Cubic Fermi Surface

The face centered cubic's band plotted in Figure 8.2.3 is

$$E_{\vec{k},FCC} \equiv (\varepsilon_{\vec{k},FCC} + \alpha)/(4\gamma) = -\left(\cos(\frac{k_x a}{2})\cos(\frac{k_y a}{2}) + \cos(\frac{k_y a}{2})\cos(\frac{k_z a}{2}) + \cos(\frac{k_x a}{2})\cos(\frac{k_z a}{2})\right), \tag{8.3.16}$$

and is symmetric with a range of $-3 \leq \varepsilon_{\vec{k},FCC} \leq 3$. Again, the Fermi energy of $E_F = 0 = \varepsilon_{\vec{k},FCC}$ yields a Fermi surface for a half-filled band. This and its first BZ (see Chapter 2) are shown in Figure 8.3.6.

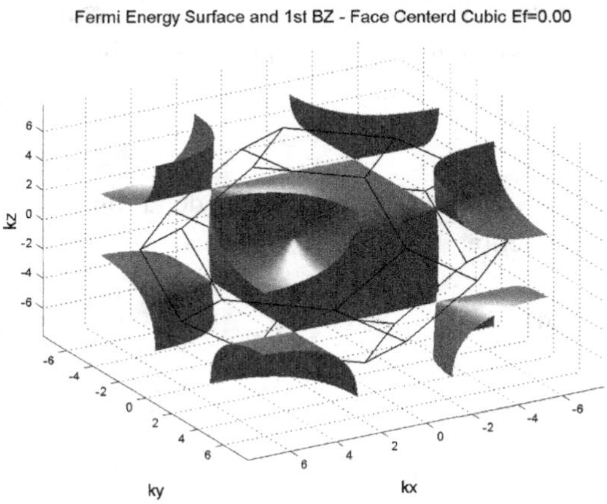


Figure 8.3.6: The Fermi surface for half-filled ($E_F = 0$) face centered cubic tight binding band in the nearest-neighbor approximation. The first BZ is shown (connected with solid lines).

## 8.4 Green's Function and the Density of States

A convenient way to introduce the concept of Green's function is to consider a particle that is located at site $j$ described by the basis function $|j)$ at time $t = 0$. If we now wish to write the probability amplitude of finding the particle at site $j$ associated with basis function $|i)$ at time $t$, one writes [20]

$$(i|\hat{G}(t)|j) = (i|e^{-i\hat{H}t/\hbar}|j), \tag{8.4.17a}$$

where $\hat{G}(t) \equiv e^{-i\hat{H}t/\hbar}$ is the time dependent Green's function operator, $\hat{H}$ is the energy or Hamiltonian operator with eigenfunctions $|\varepsilon_n>$ and eigenvalues $\varepsilon_n$; i.e., $\hat{H}|\varepsilon_n>= \varepsilon_n|\varepsilon_n>$. Here, we work with the Fourier transform of $\hat{G}(t)$, defined as

$$\hat{G}(\varepsilon) = \frac{1}{i\hbar}\int_0^\infty dt e^{i\varepsilon t/\hbar}\hat{G}(t). \tag{8.4.17b}$$

Assuming that $\varepsilon$ has a small positive imaginary part, after substituting for $\hat{G}(t)$, this expression (see Exercise 8.8.9) gives

$$\hat{G}(\varepsilon) = (\varepsilon - \hat{H})^{-1}, \tag{8.4.17c}$$

which is the energy dependent Green's function operator. If Green's function operator of Equation 8.4.17c acts on the $|\vec{k}>$ state of Equation 8.2.4, we have

$$\hat{G}(\varepsilon)|\vec{k}>= (\varepsilon - \hat{H})^{-1}|\vec{k}>= (\varepsilon - \varepsilon_{\vec{k}})^{-1}|\vec{k}>= G_{\vec{k}}(\varepsilon)|\vec{k}>, \tag{8.4.18a}$$

where we have used Equation 8.2.5c along with the expansion $(1-x)^{-1} = \sum_{n=0}^{\infty} x^n$ so that $(\varepsilon - \hat{H})^{-1} = (1/\varepsilon)\sum_{n=0}^{\infty}(\hat{H}/\varepsilon)^n$ (see Exercise 8.8.10). Since $<\vec{k}|\vec{k}>= 1$, we thus obtain the single band $\vec{k}$-space Green's function

$$G_{\vec{k}}(\varepsilon) =<\vec{k}|\hat{G}(\varepsilon)|\vec{k}>= (\varepsilon - \varepsilon_{\vec{k}})^{-1}, \tag{8.4.18b}$$

where, once again, $\varepsilon$ contains a small positive imaginary part. This is often incorporated by replacing $\varepsilon$ with $\varepsilon + i\Delta$ where $\Delta$ is a small real number. At this point, we will make use of the identity

$$\lim_{\Delta \to 0} \frac{1}{x + i\Delta} = P\left(\frac{1}{x}\right) - i\pi\delta(x), \tag{8.4.19}$$

where $P$ stands for the principle value and $\delta(x)$ is a Dirac delta function. With this expression, we can write the $\vec{k}$-space Green's function of Equation 8.4.18b as

$$G_{\vec{k}}(\varepsilon) = P\left(\frac{1}{\varepsilon - \varepsilon_{\vec{k}}}\right) - i\pi\delta(\varepsilon - \varepsilon_{\vec{k}}), \tag{8.4.20a}$$

or

$$\delta(\varepsilon - \varepsilon_{\vec{k}}) = -\frac{1}{\pi}\,\mathrm{Im}\left(G_{\vec{k}}(\varepsilon)\right), \tag{8.4.20b}$$

where "Im($F$)" stands for the "imaginary part of $F$". Comparing Equation 8.4.20b with Equations 8.3.12b, we see that the $\vec{k}$-space Green's function can be used to obtain the density of states as

$$D(\varepsilon) = -\frac{1}{\pi V_{BZ}}\,\mathrm{Im}\left(\int_{V_{BZ}} G_{\vec{k}}(\varepsilon)d\vec{k}\right) = -\frac{1}{\pi}\,\mathrm{Im}\left(\frac{1}{N}\sum_{\vec{k}} G_{\vec{k}}\right). \tag{8.4.21a}$$

If we include the electron spin, the density of states is

$$D_{\Sigma\sigma}(\varepsilon) = 2D(\varepsilon) = -\frac{2}{\pi V_{BZ}}\,\mathrm{Im}\left(\int_{V_{BZ}} G_{\vec{k}}(\varepsilon)d\vec{k}\right) = -\frac{2}{\pi}\,\mathrm{Im}\left(\frac{1}{N}\sum_{\vec{k},\sigma} G_{\vec{k}}\right). \tag{8.4.21b}$$

## 8.5 The Site Green's Function

In this section we obtain the site Green's function, show its relationship with the $k$-space Green's function, and also relate it to the density of states. To this end, we start with the eigenfunctions in the Dirac notation. In $\vec{k}$-space we have from Equation 8.2.4

$$|\vec{k}> = \frac{1}{\sqrt{N}}\sum_i e^{i\vec{k}\cdot\vec{r}_i}|i), \tag{8.5.22a}$$

and its Fourier transform

$$|i) = \frac{1}{\sqrt{N}}\sum_{\vec{k}} e^{-i\vec{k}\cdot\vec{r}_i}|\vec{k}>. \tag{8.5.22b}$$

We note that

$$(j|\vec{k}> = \frac{1}{\sqrt{N}}\sum_i e^{i\vec{k}\cdot\vec{r}_i}(j|i) = \frac{1}{\sqrt{N}}e^{i\vec{k}\cdot\vec{r}_j}, \tag{8.5.22c}$$

since by orthogonality $(j|i) = \delta_{ji}$. Similarly $<\vec{k}|j) = (j|\vec{k}>^{\dagger}$. Furthermore, the states $|\vec{k}>$ are normalized; that is $<\vec{k}|\vec{k}> = 1$ and obey the completeness condition

$$\sum_{\vec{k}} |\vec{k}><\vec{k}| = 1. \tag{8.5.23}$$

We next work with the Green's function operator, Equation 8.4.17c, and act on the completeness condition, to write

$$\begin{aligned}\hat{G}(\varepsilon) &= (\varepsilon-\hat{H})^{-1}\sum_{\vec{k}}|\vec{k}><\vec{k}| = \sum_{\vec{k}}(\varepsilon-\hat{H})^{-1}|\vec{k}><\vec{k}| = \sum_{\vec{k}}|\vec{k}>\frac{1}{\varepsilon-\varepsilon_{\vec{k}}}<\vec{k}| \\ &= \sum_{\vec{k}}|\vec{k}>G_{\vec{k}}(\varepsilon)<\vec{k}|,\end{aligned} \tag{8.5.24}$$

where we have used the $\vec{k}$-space Green's function of Equation 8.4.18b. The site Green's function is now calculated from this equation as

$$G_{ij} \equiv (i|\hat{G}|j) = \sum_{\vec{k}}(i|\vec{k}>G_{\vec{k}}<\vec{k}|j) = \frac{1}{N}\sum_{\vec{k}}e^{i\vec{k}\cdot(\vec{r}_i-\vec{r}_j)}G_{\vec{k}}, \tag{8.5.25}$$

where Equation 8.5.22c has been used. Equation 8.5.25 expresses the relationship between the real-space and the $\vec{k}$-space Green's functions. The site Green's function $G_{ij}$ is sometimes referred to as the propagator between initial site $i$ and final site $j$. If the initial and final real space states are equal, say $|0)$, the diagonal site Green's function is

$$G_{00} = (0|\hat{G}|0) = \frac{1}{N}\sum_{\vec{k}}G_{\vec{k}} = \frac{1}{V_{BZ}}\left(\int_{V_{BZ}} G_{\vec{k}}\,d\vec{k}\right). \tag{8.5.26a}$$

By comparing this expression with Equation 8.4.21a, we see that the density of states is related to the site diagonal Green's function as

$$D(\varepsilon) = -\frac{1}{\pi}\mathrm{Im}\,(G_{00}) = -\frac{1}{\pi}\mathrm{Im}\left(\frac{1}{N}\sum_{\vec{k}}G_{\vec{k}}\right) = -\frac{1}{\pi V_{BZ}}\mathrm{Im}\left(\int_{V_{BZ}} G_{\vec{k}}(\varepsilon)d\vec{k}\right), \tag{8.5.26b}$$

which is also known as the local density of states. If we include the electron spin, this expression is to be multiplied by a factor of 2, as in Equation 8.4.21b. Finally, recall that once $D(\varepsilon)$ has been obtained, the integrated or total density of states can be calculated using Equation 8.3.13.

**Example 8.5.0.1**

If in the simple cubic band Equation 8.2.7b the parameter $\alpha = 0$ and considering only one dimension, the band energy for a one-dimensional tight binding model of a one-dimensional crystal in the nearest neighbor approximation is obtained. The resulting band is

$$\varepsilon_k = -2\gamma\cos(ka), \tag{8.5.27}$$

where only a one-dimensional momentum variable is needed; i.e., $k$. We wish to calculate $G_{00}(\varepsilon)$ here, to write

$$G_{00} = \frac{1}{N}\sum_k G_k = \frac{L}{2\pi}\int_0^{2\pi/a} dk\frac{1}{\varepsilon+2\gamma\cos(ka)}, \tag{8.5.28}$$

where the one-dimensional version of Equation 8.5.26a has been used with the Brillouin zone volume being replaced as $V_{BZ} \to 2\pi/L$ for one dimension and where $0 < k < 2\pi/a$ (also $L = a$ below). Because the Green function integrand above has singularities whenever $\varepsilon = -2\gamma\cos(ka)$, we will more conveniently transform the integral to a contour integral about the unit circle. To this end, let

$z = \exp(ika)$ and we see that $|z| = 1$, $dz = ia\exp(ika)dk$ or $dk = dz/(iaz)$, and $\cos(ka) = (z + 1/z)/2$ so that

$$G_{00} = \frac{a}{2a\pi i}\oint \frac{dz}{z(\varepsilon + 2\gamma(z+1/z)/2)} = \frac{1}{2\pi i\gamma}I, \tag{8.5.29a}$$

where

$$I = \oint \frac{dz}{z^2 + z\varepsilon/\gamma + 1}. \tag{8.5.29b}$$

We first consider the region where $\varepsilon < 0$ or $\varepsilon = -|\varepsilon|$ in which case

$$I = \oint \frac{dz}{(z - z_+)(z - z_-)}, \tag{8.5.30}$$

where $z_+ = |\varepsilon|/(2\gamma) + \sqrt{(|\varepsilon|/(2\gamma))^2 - 1}$ and $z_- = |\varepsilon|/(2\gamma) - \sqrt{(|\varepsilon|/(2\gamma))^2 - 1}$. Since $z_+z_- = 1 = |z|$, and $|z_-| < |z_+|$, we see that $|z_-|$ must lie within the unit circle over which we are performing the contour integration, and $|z_+|$ lies outside. Thus we only consider the pole at $z = |z_-|$ and by the residue theorem we thus obtain

$$I = 2\pi i \lim_{z\to z_-} \frac{(z - z_-)}{(z - z_+)(z - z_-)} = \frac{2\pi i}{(z_- - z_+)} = -\frac{\pi i}{\sqrt{(|\varepsilon|/(2\gamma))^2 - 1}}. \tag{8.5.31a}$$

Substituting this back into Equation 8.5.29a, we get for $\varepsilon < 0$

$$G_{00,\varepsilon<0} = -\frac{1}{2\pi i\gamma}\frac{\pi i}{\sqrt{(|\varepsilon|/(2\gamma))^2 - 1}} = -\frac{1}{\sqrt{\varepsilon^2 - 4\gamma^2}}. \tag{8.5.31b}$$

Repeating this process for the case of $\varepsilon > 0$ or $\varepsilon = |\varepsilon|$, we find

$$I = \frac{\pi i}{\sqrt{(|\varepsilon|/(2\gamma))^2 - 1}}, \tag{8.5.32a}$$

which when substituted into Equation 8.5.29a gives

$$G_{00,\varepsilon>0} = \frac{1}{2\pi i\gamma}\frac{\pi i}{\sqrt{(|\varepsilon|/(2\gamma))^2 - 1}} = \frac{1}{\sqrt{\varepsilon^2 - 4\gamma^2}}. \tag{8.5.32b}$$

The results of Equations 8.5.31b and 8.5.32b can be combined into a single result for both $\pm\varepsilon$ in the form

$$G_{00} = \frac{\varepsilon}{|\varepsilon|}\frac{1}{\sqrt{(\varepsilon + i\Delta)^2 - 4\gamma^2}}, \tag{8.5.33a}$$

and where we have added the small positive imaginary part, $\Delta$, to $\varepsilon$ to enable the density of states calculation, as mentioned before. The density of states could simply be obtained using Equation 8.5.26b; that is,

$$D(\varepsilon) = -\frac{1}{\pi}\mathrm{Im}\,(G_{00}). \tag{8.5.33b}$$

Assuming that $\gamma$ is on the order of Hartrees ($Ha$), we use such energy units and let$\gamma = 1/2$, we can obtain a plot of the real and imaginary parts of $G_{00}$ as shown in Figure 8.5.7.

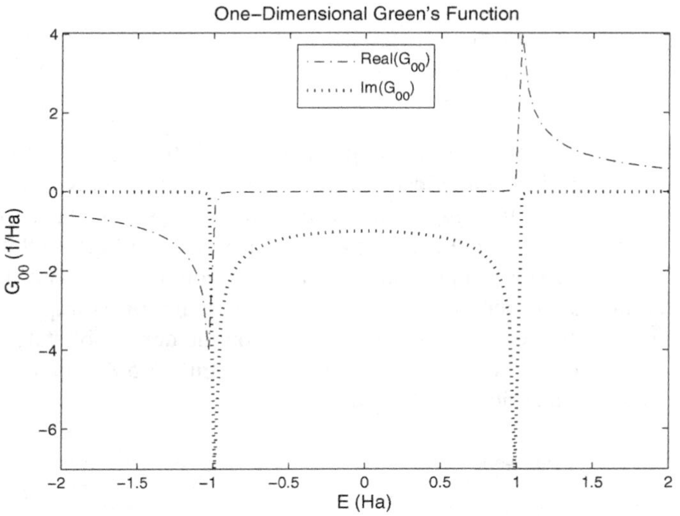


Figure 8.5.7: Shown are the real (dashed) and imaginary (dotted) parts of the Green function $(1/Ha)$ of Equation 8.5.33a versus energy $(Ha)$ for the one-dimensional tight binding band of Equation 8.5.27 of Example 8.5.0.1.

The density of states will have a similar shape as the imaginary part of $G_{00}$, except that it will be opposite in sign and decreased in magnitude by a factor of $\pi$ as mentioned above. The density of states band has a width of $2(2\gamma)$, which is consistent with Equation 8.5.27 since the *cosine* function has a total width of 2 units. With $\gamma = 1/2$, the total width is, therefore, $2Ha$ in our units. The code used in obtaining Figure 8.5.7 is Green_1D.m, which is listed below.

```
%copyright by J. E Hasbun and T. Datta
%Green_1D.m
%It calculates the real and imaginary parts of a one-dimensional
%Green's function for a nearest neighbor tight binding model
%where the band energy Ek=-2*gamma*cos(ka)
clear
el=-2;
eu=2;
Ne=100;
es=(eu-el)/(Ne-1);
gam=1/2;                         %band energy parameter (in Ha)
zim=complex(0.,1.0);
delta=1.e-3;                     %for the small imaginary part
for i=1:Ne
  e0(i)=el+(i-1)*es;
  g00(i)=e0(i)/(sqrt((e0(i)+zim*delta)^2-4*gam^2)*abs(e0(i)));
end
plot(e0,real(g00),'k-.')
hold on
plot(e0,imag(g00),'k:','LineWidth',2)
axis tight
legend('Real(G_{00})','Im(G_{00})',0)
```

```
xlabel('E (Ha)','FontSize',12)
ylabel('G_{00} (1/Ha)','FontSize',12)
title('One-Dimensional Green''s Function','FontSize',12)
```

**Example 8.5.0.2**

It is often the case when the integration of the $\vec{k}$ dependent Green's function over $\vec{k}$ cannot be carried out analytically. In such situations a numerical method is applied that enables its calculation. Since, in general, the Green function contains singularities (due to poles at the eigenvalues of the problem), standard methods of carrying out its integration are not appropriate. In Section 8.8 of this chapter, we present the method we use to perform the numerical integration. Here we show how the numeric method is incorporated in order to reproduce the analytic calculation of Example 8.5.0.1; however, this time, we concentrate on the real part of $G_{00}$ as well as on the density of states $D(\varepsilon)$. So doing provides the testing ground of the numerical method used. Figure 8.5.8 contains the comparison between the analytic and the numerical results.

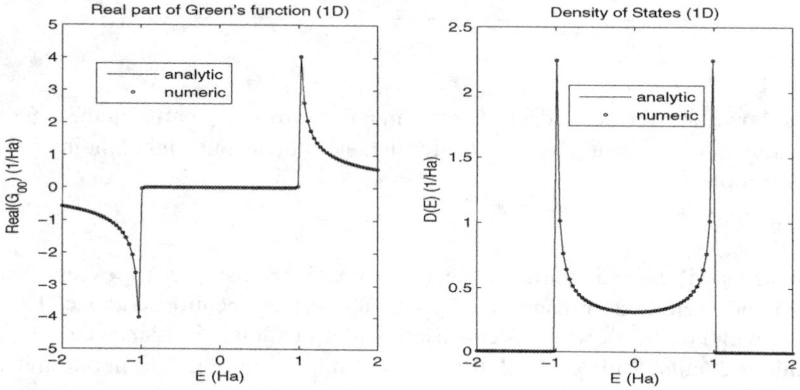


Figure 8.5.8: Shown is a comparison of the analytic (solid line) and numerical (dotted line) results for the real part (left panel) of the Green function ($1/Ha$) and the density of states (right panel) from Equation 8.5.33 versus energy ($Ha$) for the one-dimensional tight binding band of Equation 8.5.27 of Example 8.5.0.1. Example 8.5.0.2 details the code used (Green_1DdosCalc.m) in performing these calculation.

As mentioned in Example 8.5.0.1, the density of states, expectedly, resembles the imaginary part of $G_{00}$, except that it is opposite in sign and decreased in magnitude by a factor of $\pi$. The listing of the code, Green_1DdosCalc.m, employed in this example follows below. Notice how the function singInt.m of Section 8.8 is called upon to carry out the integration for every energy value.

```
%copyright by J. E Hasbun and T. Datta
%Green_1DdosCalc.m
%In addition to calculating the one-dimensional Green's function in
%the nearest neighbor tight binding model analytically, a comparison
%is made with the purely numerical calculation. The numerical
%calculation uses the Roth's integration scheme for singular
%functions; i.e., singint.m (separate listing in the Appendix).
function Green_1DdosCalc
clear
el=-2;                              %upper energy value
eu=2;                               %lower energy value
```

```
Ne=100;                          %number of energy points
es=(eu-el)/(Ne-1);               %energy step
gam=1/2;                         %band energy parameter (in Ha)
zim=complex(0.,1.0);             %imaginary number
delta=1.e-4;                     %small part for plotting, integrating
for i=1:Ne
  e0(i)=el+(i-1)*es;
  g00(i)=e0(i)/(sqrt((e0(i)+zim*delta)^2-4*gam^2)*abs(e0(i)));
  dosa(i)=-imag(g00(i))/pi;      %analytic density of states
end
%The plots are done below
%Numerical calculation follows
a=1;
kl=0;                            %upper limit
ku=2*pi/a;                       %lower limit
Nk=301;                          %number of k-points
dk=(ku-kl)/(Nk-1);
factor=a/2.0/pi;                 %factor for the numerical integral
for j=1:Nk
  k(j)=kl+(j-1)*dk;
  Ek(j)=-2*gam*cos(k(j)*a);      %the band energy defined only once
  top(j)=1.0;                    %the numerator for integration
end
%Integration over k for each energy value
for i=1:Ne
  for j=1:Nk
    deno(j)=e0(i)-Ek(j)+zim*delta;     %the denominator
  end
  ge(i)=factor*singInt(top,deno,dk);   %ge is the numerical integral
  dosn(i)=-imag(ge(i))/pi;             %numerical density of states
end
%
%Plotting
subplot(1,2,1)
plot(e0,real(g00),'k');                %Real part of g00 - analytic
hold on
plot(e0,real(ge),'ko','MarkerSize',2) %numeric
legend ('analytic','numeric',0)
xlabel('E (Ha)')
ylabel('Real(G_{00}) (1/Ha)')
title('Real part of Green''s function (1D)')
%
subplot(1,2,2)
plot(e0,dosa,'k')                      %density of states - analytic
hold on
plot(e0,dosn,'ko','MarkerSize',2)      %numeric
xlabel('E (Ha)')
ylabel('D(E) (1/Ha)')
title('Density of States (1D)')
legend ('analytic','numeric',0)
```

Finally, an example related to the total density of states (see Equation 8.3.13) follows.

**Example 8.5.0.3**
In this example, we perform a total density of states calculation for the one-dimensional density of states of the Example 8.5.0.2 using Equation 8.3.13. Thus, here we will use $D(\varepsilon)$ and integrate it over the energy to get the total density of states. Taking spin into account, we expect that a single band contains 2 electrons, so that we multiply by a factor of 2 in order to show the total electron count for the one-dimensional band. The results obtained are shown in Figure 8.5.9.

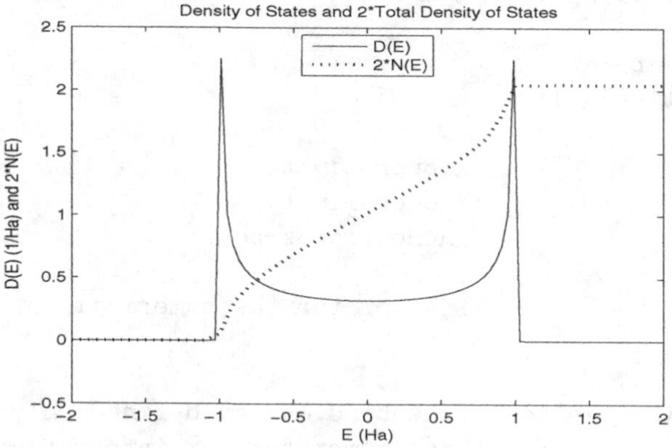


Figure 8.5.9: The total density of states (dotted line) and the density of states (solid line) are shown. The total or integrated density of states is obtained using Equation 8.3.13. It has been multiplied by a factor of 2 to include electron spin.

In the total density of states $(2 * N(E))$, notice that by the end of the energy covering the width of the $D(E)$ band, the number of electrons is 2, which, as expected, shows the number of electrons per band. The code used in obtaining Figure 8.5.9 is Green_1Totdos.m which is listed below. The code makes use of the Romberg numerical integration approach because it tends to do better than other methods, such as Simpson's rule, for example. This is indicated by the line using the function "rombergInt". In order to use the Romberg integration scheme, however, an interpolator has been employed in order to interpolate the density of states, that is used as input. The array "e0" corresponds to the energy values at which the density of states (array "dosa") has been calculated. The interpolator used is indicated by the handle "@fForRomb" in the call to the Romberg method; i.e., "rombergInt(el,eT(nt),@fForRomb)". Here "el" and "eT(nt)" are the lower and upper energy integration limits, respectively. The function "fForRomb" is part of the code listed below, and the interpolating function is interpFunc.m which is listed in Section 8.8 of this chapter. The Romberg integration function, rombergInt.m, is also listed in this appendix. They are listed separately for later use in other programs.

```
%copyright by J. E Hasbun and T. Datta
%Green_1Totdos.m
%Uses the density of states from the analytic Green's function
%for the one-dimensional tight binding model and integrates it
%to obtain the total density of states N(E). The Romberg method
%is used to perform the integration over the density of states D(E).
%Romberg uses the interpolated D(E) to produce N(E).
function Green_1Totdos
clear
```

```
global e0 dosa                  %needed for interpolation
el=-2;                          %upper energy value
eu=2;                           %lower energy value
Ne=100;                         %number of energy points
es=(eu-el)/(Ne-1);              %energy step
gam=1/2;                        %band energy parameter (in Ha)
zim=complex(0.,1.0);            %imaginary number
delta=1.e-4;                    %small part for plotting, integrating
for i=1:Ne
  e0(i)=el+(i-1)*es;
  g00=e0(i)/(sqrt((e0(i)+zim*delta)^2-4*gam^2)*abs(e0(i)));
  dosa(i)=-imag(g00)/pi;        %analytic density of states
end
%Total density of states and plots. Integrate with Romberg's method.
x=0.1;            %energy step
eT=el:x:eu;       %energy range (using previous e1, eu limits)
for nt=1:length(eT)
  intdos(nt)=rombergInt(el,eT(nt),@fForRomb);  %integrate on [el,eT]
  fprintf('E0=%9.4f, integrated dos=%14.6e\n',eT(nt),intdos(nt));
end
%density of states (dosa)
plot(e0,dosa,'k')
hold on
%total dos multiplied by 2 to include electron spin
plot(eT,2*intdos,'k:','LineWidth',2)
legend ('D(E)','2*N(E)','Location','North')
xlabel('E (Ha)')
ylabel('D(E) (1/Ha) and 2*N(E)')
title('Density of States and 2*Total Density of States')

function y=fForRomb(p)
%Function used by rombergInt integration and which interpolates
%the density of states versus e0.
global e0 dosa              %variables passed from the main program
y=interpFunc(p,e0,dosa);    %the Langrange interpolator for Romberg
```

In the Example 8.5.0.3, we used the analytic density of states in order to carry out the integrated density of states as a simplified computation. Often, the density of states is also calculated numerically, as in Example 8.5.0.2; in such cases, more computer time is consumed when both $D(\varepsilon)$ and $N(\varepsilon)$ are computed numerically.

## 8.6 Density of States for the Cubics

As we have seen, the process of obtaining the density of states of a crystal system requires knowledge of its band structure; it also helps to have the Green function, if the methods employed here are used. In Example 8.5.0.2, a one-dimensional integration over $k$ is all that was needed to obtain the $D(\varepsilon)$. In three dimensions, in the absence of analytical results, an integration over $\vec{k}$ is ultimately needed. In this section we discuss the method used to carry out such calculations with minimal numerical effort. The density of states for the cubic systems are fortunately known analytically and they provide a way to assess whatever numerical methods we employ. It is important to realize that obtaining the density of states of a crystal system enables the study of other properties associated with the crystal's electronic properties. The reason is that the density of states provides knowledge as to the distribution of electron states per unit energy in a particular system. Below we will delve deeper into the cubic systems and, in doing so, we gain knowledge that will later be employed in obtaining semiconducting systems' band structures for which analytical results are not known.

### 8.6.1 The Simple Cubic Density of States

We will work with the simple cubic (SC) band discussed in Section 8.2.1 and refer to the energy band in the form $E_{\vec{k}} = (\varepsilon_{\vec{k},SC} + \alpha)/(2\gamma)$ or since $\gamma$ is usually on the order of Hartree ($Ha$) units, we write

$$E_{\vec{k},SC} = -\left(\cos(k_x a) + \cos(k_y a) + \cos(k_z a)\right), \tag{8.6.34a}$$

with associated $\vec{k}$-space Green function

$$G_{\vec{k},SC}(E) = \frac{1}{E - E_{\vec{k},SC} + i\Delta}, \tag{8.6.34b}$$

and with $\Delta$ taking a small positive value. The density of states involves an integral over the SC's Brillouin zone (BZ); that is,

$$D(E) = -\frac{1}{\pi V_{SC-BZ}} \mathrm{Im}\left(\int_{V_{SC-BZ}} G_{\vec{k},SC}(E)d\vec{k}\right), \tag{8.6.35}$$

where, from Chapter 2, $V_{SC-BZ} = (2\pi/a)^3$, with $a$ the lattice constant. The simple cubic's Fermi surface and BZ have been described in Section 8.3.1. Here, we see that the integral of Equation 8.6.35 can be simplified if symmetry considerations are taken into account. The simple cubic BZ has cubic shape and contains four special symmetry points, discussed previously in Chapter 2, and which describe the four corners of a tetrahedron. The tetrahedron so depicted is known as an irreducible tetrahedron. These symmetry points are $\Gamma$, $X$, $R$, and $M$. The tetrahedron has volume $V_{SC,IBZ} = |\vec{X}\cdot(\vec{R}\times\vec{M})|/6 = V_{SC-BZ}/48$, so that 48 times the simple cubic's irreducible BZ tetrahedron, or irreducible BZ (IBZ), volume equals the simple cubic's first BZ volume. By performing symmetry operations of the IBZ, the entire SC first BZ can be reconstructed. Similarly, one performs calculations on the IBZ to minimize the computational effort. In this manner, for the simple cubic, multiplying the density of states calculated within the IBZ by 48 is equivalent to performing the density of states calculation on the entire BZ. Still, carrying out integrations of density of states over three-dimensional $\vec{k}$-space is a time-consuming process. The reason is that to get accurate results, one has to employ many points in momentum space for each dimension. Fortunately, methods have been developed to minimize this effort as well. Here we employ the ray method [33] which is detailed in Section 8.8. Thus using Equations 8.6.34, 8.6.35 and including the ray method details of Section 8.8, the density of states is written as

$$D(E) = -\frac{1}{\pi V_{BZ}} \mathrm{Im} \int_{V_{BZ}} G_{\vec{k}}(E)d\vec{k} \approx -\frac{3V_T}{\pi V_{BZ}} N_{SC-IBZ} \sum_i \delta_i \mathrm{Im}\left\{\int_0^1 d\alpha\, \alpha^2 G[E, \alpha(\vec{q}_1 + \vec{v}_{ai})]\right\}, \tag{8.6.36a}$$

where $N_{SC-IBZ} = 48$; i.e., the number of irreducible tetrahedrons in the simple cubic's BZ and where the specific form of the Green function is

$$G[E, \alpha(\vec{q}_1 + \vec{v}_{ai})] = \lim_{\Delta\to 0} \frac{1}{E - E(\alpha(\vec{q}_1 + \vec{v}_{ai})) + i\Delta}, \tag{8.6.36b}$$

since the wavevector became replaced with $\vec{k} = \alpha(\vec{q}_1 + \vec{v}_a)$ in the ray method. Essentially, the entire $\vec{k}$-space integral (using the energy band of Equation 8.6.34a) has been transformed to a sum over one-dimensional integrals. These integrals are the ones that involve singularities and they will be handled by the methods of Section 8.8 (see Example 8.5.0.2). The quantities $V_T$, $\delta_i = \Delta s_i/S$, $S$ and the vectors $\vec{q}_1$ as well as $\vec{v}_a$ are detailed in Section 8.8. The total density of states is carried out as in Example 8.5.0.3, though we do not worry about the electron spin factor of 2 at this time. For completeness, we will also carry out a comparison between the numeric and the exact results for the simple cubic density of states [34]. In this way, the density of states (numeric and exact) as well as the total density of states are shown in Figure 8.6.10.

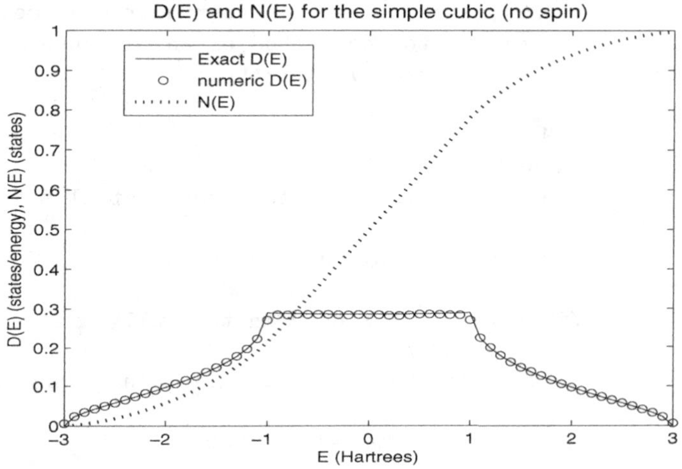


Figure 8.6.10: The exact (solid) [34], numeric (circles) density of states, $D(E)$, for the simple cubic. The integrated density of states (dots), $N(E)$, is shown without including electron spin into account.

The numeric density of states, $D(E)$, appears to do very well compared to the exact results. This is fortunate because our numeric method (discussed above) is the one that will be used for the semiconductor band structure calculations later in the chapter. The total density of states, $N(E)$, shows that the band is filled with one state when the energy reaches the band width. If one includes the factor of 2 for spin, the result is that two electrons occupy the entire band. The entire calculational approach explained in this section has been incorporated into the script sc_dos.m that is listed below and which reproduces Figure 8.6.10.

```
%copyright by J. E Hasbun and T. Datta
%sc_dos.m
%Density of states for the single band, simple cubic system
%according to the ray approach for the tetrahedron method of
%An-Ban Chen, Phys Rev. B V16, 3291 (1977). We also use the
%singInt integration method for the Green's function.
%For the ray method, the 2D vectors on the faces of the thin
%tetrahedrons are obtained from the TTareas function. These vectors are
%average vectors.
function sc_dos
clear, clc;
global e0 dos
delta=1.5e-2;
```

```
im=complex(0.0,1.0);
x=0.1;              %energy step
e2=3.0;
e1=-e2;
e0=e1:x:e2;         %energy range
ntmax=length(e0);
a=1.0;              %lattice constant
tpa=2*pi/a;
VBZ=tpa^3;          %total SC BZ volume
%SC case tetrahedron (only one needed) in units of 2*pi/a;
X=[1/2,0,0]; R=[1/2,1/2,1/2]; M=[1/2,1/2,0]; %tetrahedron symmetry points
%Tetrahedron q vectors according to the method of An-Ban Chen & B. I. Reser.
q(1,:)=R*tpa; q(2,:)=(X-R)*tpa ; q(3,:)=(M-X)*tpa;
%total tetrahedron volume
VSC_t=abs(dot(X,cross(R,M))/6)*VBZ;
lim=9; lom=lim-1; %lim must be odd
%number of divisions along q2, and q3 => total number of TT's is n^2
n=25;
%VSC_t is the standard tetrahedron volume, and there are 48 of them in
%the SC cube's total BZ
factor=VSC_t*48;        %initital factor used in the full integral result
%TT=thin tetrahedron, T=tetrahedron
DD=1/n^2;   %ratio of TT area to T area (since all thin triangles are the
   same)
factor=3.0*DD*factor; %factor is modified further by 3*DD
par1=0.0;
par2=1.0;
dk=(par2-par1)/lom;
al=par1:dk:par2;    %alpha range to integrate (main TT axis)
dv=1/n;             %TT division size
be=0:dv:1;          %beta
ga=be;              %gamma
%The function TTareas produces the average vectors "va" on the faces of
%the TT's. Notation: va(ith TT vector,coordinate(x,y,z))
%Ntt=number of TT's, areas(ith TT area)
[Ntt,areas,va]=TTareas(q(2,:),q(3,:),n,be,ga);
terr=0.0;           %total error
dos=zeros(1,ntmax);
for nt=1:ntmax
  dos(nt)=0.0;
  for it=1:Ntt          %Ntt TT's
    for ko=1:lim    %loop over alpha
      for v=1:3  %build the k vector, kx=k(1), ky=k(2), kz=k(3)
        %The va vectors are average vectors on the TT faces
        k(v)=al(ko)*(q(1,v)+va(it,v)); %va vectors used here
      end
      top(ko)=al(ko)*al(ko);
      deno(ko)=e0(nt)+cos(k(1))+cos(k(2))+cos(k(3))+im*delta;
    end
    gy=singInt(top,deno,dk);
    dos(nt)=-imag(gy)/pi+dos(nt);  %single TT dos, add all n^2 TT's
```

```
  end
  dos(nt)=factor*dos(nt)/VBZ;      %numeric dos
  ge(nt)=jelittoScDos(e0(nt))/VBZ; %exact dos
  err=abs(dos(nt)-ge(nt));         %the error per energy
  terr=terr+err;                   %cumulative error
  fprintf('E0=%9.4f, dos=%9.4f, ge=%9.4f, err=%14.6e\n',...
      e0(nt),dos(nt),ge(nt),err)
end
terr=terr/ntmax;
fprintf('Total error=%14.6e\n',terr)
%Total integrated density of states, and plot
fprintf('Integrated Density of States for the simple cubic')
intdos=zeros(1,ntmax);
for nt=1:ntmax
  intdos(nt)=rombergInt(e1,e0(nt),@fForRomb);  %integrate on [e1,e0]
  fprintf('E0=%9.4f, integrated dos=%14.6e\n',e0(nt),intdos(nt));
end
plot(e0,ge,'k'), hold on         %exact dos
plot(e0,dos,'ko','MarkerSize',5) %numeric dos
%Next we do the total density of states without the factor of 2 for spin.
plot(e0,intdos,'k:','LineWidth',2);
xlabel('E (Hartrees)'), ylabel('D(E) (states/energy), N(E) (states)')
str=cat(2,'D(E) and N(E) for the simple cubic (no spin)');
title(str, 'Fontsize',12)
legend('Exact D(E)','numeric D(E)','N(E)',0)

function ge=jelittoScDos(ee)
%Jelitto's DOS for the simple cubic (exact)
eaa=abs(ee);
if (eaa <= 3.) & (eaa >=  1.)
  a1=3.-eaa;
  a2=a1^2;
  a=sqrt(a1);
  b=80.3702-16.3846*a1;
  d=0.78978*(a2);
  f=-44.2639+3.66394*a1;
  h=-0.17248*(a2);
  ge=a*((b+d)+(f+h)*sqrt(eaa-1.));
else
  if(eaa < 1.)
    ge=70.7801+1.0053*(ee^2);
  else
    ge=0.;
  end
end

function y=fForRomb(p)
%function used by romberg integration and which interpolates
%tdos versus e0
global e0 dos
y=interpFunc(p,e0,dos);
```

### 8.6.2 The Body Centered Cubic Density of States

We now work with the body centered cubic (BCC) band discussed in Section 8.2.2 and refer to the energy band in the form $E_{\vec{k}} = (\varepsilon_{\vec{k},BCC} + \alpha)/(8\gamma)$. In Hartree ($Ha$) units we write

$$E_{\vec{k},BCC} = -\cos\left(\frac{k_x a}{2}\right)\cos\left(\frac{k_y a}{2}\right)\cos\left(\frac{k_z a}{2}\right), \tag{8.6.37a}$$

with associated $\vec{k}$-space Green function

$$G_{\vec{k},BCC}(E) = \frac{1}{E - E_{\vec{k},BCC} + i\Delta}, \tag{8.6.37b}$$

with, as before, $\Delta$ small and positive. The density of states integral over the Brillouin zone (BZ) is

$$D(E) = -\frac{1}{\pi V_{BCC-BZ}} \mathrm{Im}\left(\int_{V_{BCC-BZ}} G_{\vec{k},BCC}(E) d\vec{k}\right), \tag{8.6.38}$$

where, from Chapter 2, $V_{BCC-BZ} = 2(2\pi/a)^3$, with $a$ the lattice constant. The body centered cubic's Fermi surface and BZ have been described in Section 8.3.2. The integral of Equation 8.6.38 can be simplified with symmetry considerations. The body centered cubic's BZ is a rhombic dodecahedron and, similar to the simple cubic, contains four special symmetry points that describe the four corners of an irreducible tetrahedron. The symmetry points are $\Gamma$, $P$, $H$, and $N$, which were discussed in Chapter 2. This tetrahedron has volume $V_{BCC,IBZ} = |\vec{P}\cdot(\vec{H}\times\vec{N})|/6 = V_{BCC-BZ}/24$, so that 24 times the body centered cubic's irreducible BZ tetrahedron volume equals the BCC's first BZ volume. Multiplying the density of states calculated within the IBZ by 24 is equivalent to performing the density of states calculation on the entire BZ. Using the same approach as in Section 8.6.1, the density of states is obtained using Equations 8.6.36 with the band of Equation 8.6.37a. We again compare the numeric and the exact results for the BCC density of states [34]. The density of states (numeric and exact) as well as the total density of states are shown in Figure 8.6.11.

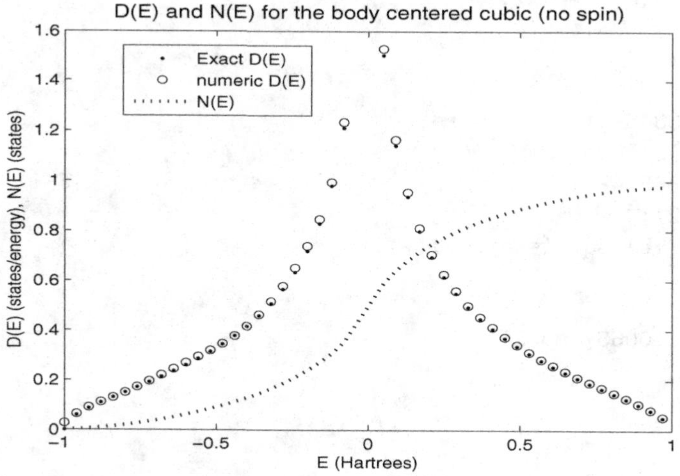


Figure 8.6.11: The exact (dots) [34] and numeric (circles) density of states, $D(E)$, for the body centered cubic are shown. The integrated density of states (dotted line), $N(E)$, is shown without taking electron spin into account.

Again, as in the simple cubic, the numeric density of states, $D(E)$, appears to do very well compared to the exact results. The total density of states, $N(E)$, shows that the band is filled with one state when the energy reaches the band width and if we include spin, two electrons occupy the entire band. The script used in this calculation is bcc_dos.m, listed below, and which reproduces Figure 8.6.11.

```
%copyright by J. E Hasbun and T. Datta
%bcc_dos.m
%Density of states for the single band, body centered cubic system
%according to the ray approach for the tetrahedron method of
%An-Ban Chen, Phys Rev. B V16, 3291 (1977). We also use the
%singInt integration method for the Green's function.
%For the ray method, the 2D vectors on the faces of the thin
%tetrahedrons are obtained from the TTareas function. These vectors are
%average vectors.
function bcc_dos
clear, clc;
global e0 dos
delta=1.5e-2;
im=complex(0.0,1.0);
x=0.04;                              %energy step
e2=1.0;
e1=-e2;
eminl=-0.05; eminr=-eminl;
e0a=e1:x:eminl; e0b=eminr:x:e2; %not to do in between eminl and eminr
e0=[e0a,e0b];                        %energy range, but skip small e range
ntmax=length(e0);
a=1.0;          %lattice constant
tpa=2*pi/a;
VBZ=2*tpa^3;    %total BCC BZ volume
%BCC case tetrahedron (only one needed) in units of 2*pi/a;
P=[1/2,1/2,1/2]; H=[1,0,0]; N=[1/2,1/2,0]; %tetrahedron symmetry points
%Tetrahedron q vectors according to the method of An-Ban Chen & B. I. Reser.
q(1,:)=P*tpa; q(2,:)=(H-P)*tpa ; q(3,:)=(N-H)*tpa;
%total tetrahedron volume
VBCC_t=abs(dot(P,cross(H,N))/6)*VBZ;
lim=9; lom=lim-1;                   %lim must be odd
%number of divisions along q2, and q3 => total number of TT's is n^2
n=25;
%VBCC_t is the standard tetrahedron volume part, and there are 24 of them in
%the body center cube's total BZ
factor=VBCC_t*24;     %initital factor used in the full integral result
%TT=thin tetrahedron, T=tetrahedron
DD=1/n^2;    %ratio of TT area to T area (since all thin triangle are the
   same)
factor=3.0*DD*factor; %factor is modified further by 3*DD
par1=0.0;
par2=1.0;
dk=(par2-par1)/lom;
al=par1:dk:par2;    %alpha range to integrate (main TT axis)
dv=1/n;             %TT division size
be=0:dv:1;          %beta
```

```
ga=be;              %gamma
%The function TTareas produces the average vectors "va" on the faces of
  the TT's
%Notation: va(ith TT vector,coordinate(x,y,z))
%Ntt=number of TT's, areas(ith TT area)
[Ntt,areas,va]=TTareas(q(2,:),q(3,:),n,be,ga);
terr=0.0;                                    %total error
dos=zeros(1,ntmax);
for nt=1:ntmax
  dos(nt)=0.0;
  for it=1:Ntt         %Ntt TT's
    for ko=1:lim   %loop over alpha
      for v=1:3  %build the k vector, kx=k(1), ky=k(2), kz=k(3)
        %The va vectors are average vectors on the TT faces
        k(v)=al(ko)*(q(1,v)+va(it,v)); %va vectors used here
      end
      top(ko)=al(ko)*al(ko);
      deno(ko)=e0(nt)+cos(k(1)/2)*cos(k(2)/2)*cos(k(3)/2)+im*delta;
    end
    gy=singInt(top,deno,dk);
    dos(nt)=-imag(gy)/pi+dos(nt);  %single TT dos, add all n^2 TT's
  end
  dos(nt)=factor*dos(nt)/VBZ;
  ge(nt)=jelittoBccDos(e0(nt))/VBZ;
  err=abs(dos(nt)-ge(nt));
  terr=terr+err;
  fprintf('E0=%9.4f, dos=%9.4f, ge=%9.4f, err=%14.6e\n',...
    e0(nt),dos(nt),ge(nt),err)
end
terr=terr/ntmax;
fprintf('Total error=%14.6e\n',terr)
%Total integrated density of states, and plot
fprintf('Integrated Density of States for the body centered cubic')
intdos=zeros(1,ntmax);
for nt=1:ntmax
  intdos(nt)=rombergInt(e1,e0(nt),@fForRomb);  %integrate on [e1,e0]
  fprintf('E0=%9.4f, integrated dos=%14.6e\n',e0(nt),intdos(nt));
end
plot(e0,ge,'k.'), hold on             %exact dos
plot(e0,dos,'ko','MarkerSize',5) %numeric dos
%Next we do the total density of states without the factor of 2 for spin.
plot(e0,intdos,'k:','LineWidth',2);
xlabel('E (Hartrees)'), ylabel('D(E) (states/energy), N(E) (states)')
str=cat(2,'D(E) and N(E) for the body centered cubic (no spin)');
title(str, 'Fontsize',12)
legend('Exact D(E)','numeric D(E)','N(E)',0)

function ge=jelittoBccDos(ee)
%Jelitto's DOS for the body centered cubic (exact)
eaa=abs(ee);
if (eaa < 1.)
```

```
    ge=2.0*sqrt(1.-eaa)*(log(5.845/eaa))^2;
    ge=ge*(16.6791+3.6364*eaa+2.4880*eaa^2);
else
    ge=0.0;
end

function y=fForRomb(p)
%function used by romberg integration and which interpolates
%tdos versus e0
global e0 dos
y=interpFunc(p,e0,dos);
```

### 8.6.3 The Face Centered Cubic Density of States

We now work with the face centered cubic (FCC) band discussed in Section 8.2.3 and refer to the energy band in the form $E_{\vec{k}} = (\varepsilon_{\vec{k},FCC} + \alpha)/(4\gamma)$. In Hartree ($Ha$) units we write

$$E_{\vec{k},FCC} = -\left(\cos\left(\frac{k_x a}{2}\right)\cos\left(\frac{k_y a}{2}\right) + \cos\left(\frac{k_y a}{2}\right)\cos\left(\frac{k_z a}{2}\right) + \cos\left(\frac{k_x a}{2}\right)\cos\left(\frac{k_z a}{2}\right)\right), \tag{8.6.39a}$$

with associated $\vec{k}$-space Green function

$$G_{\vec{k},FCC}(E) = \frac{1}{E - E_{\vec{k},FCC} + i\Delta}, \tag{8.6.39b}$$

with, as before, $\Delta$ small and positive. The density of states integral over the Brillouin zone (BZ) is

$$D(E) = -\frac{1}{\pi V_{FCC-BZ}} \mathrm{Im}\left(\int_{V_{FCC-BZ}} G_{\vec{k},FCC}(E) d\vec{k}\right), \tag{8.6.40}$$

where, from Chapter 2, $V_{FCC-BZ} = 4(2\pi/a)^3$, with $a$ the lattice constant. The face centered cubic's Fermi surface and BZ have been described in Section 8.3.3. The integral of Equation 8.6.40 can be simplified with symmetry considerations. The face centered cubic's BZ is a fourteen-sided polyhedron and, similar to the other two cubics, contains symmetry points that describe three irreducible tetrahedrons. The symmetry points are $\Gamma$, $K$, $U$, $L$, $W$, and $X$ which were also discussed in Chapter 2. The tetrahedrons' vertices are $\Gamma, L, K, W$; $\Gamma, L, U, W$; and $\Gamma, X, U, W$. They have respective volumes $V_{T1} = L \cdot (K \times W)/6 = V_{FCC-BZ}/32$, $V_{T2} = L \cdot (U \times W)/6 = V_{FCC-BZ}/32$, and $V_{T3} = X \cdot (U \times W)/6 = V_{FCC-BZ}/48$, so that the tetrahedrons have a collective volume of $V_{FCC,IBZ} = V_{FCC-BZ}/12$. Thus 12 times the face centered cubic's irreducible BZ tetrahedrons' volume equals the FCC's first BZ volume. In a similar way to what we have done with the other cubics, multiplying the density of states calculated within the IBZs by 12 is equivalent to performing the density of states calculation on the entire BZ. Using the same approach as in Section 8.6.1, the density of states is obtained using Equation 8.6.36 with the band of Equation 8.6.39a. We again compare the numeric and the exact results for the FCC density of states [34]. The density of states (numeric and exact) as well as the total density of states are shown in Figure 8.6.12.

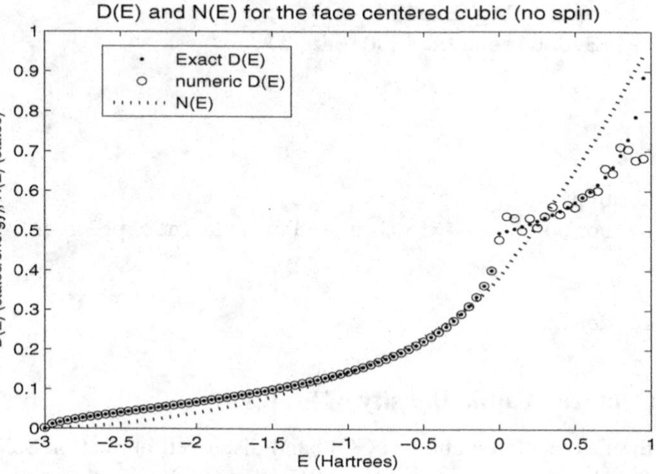


Figure 8.6.12: Shown are the exact (dots) [34] and numeric (circles) density of states, $D(E)$, for the face centered cubic. The integrated density of states (dotted line), $N(E)$, is shown without taking electron spin into account.

The numeric density of states, $D(E)$, for the FCC is a bit more challenging and, to improve it, it is necessary to increase the number of thin tetrahedrons used (variable $n$) and the number of $k$-points in the integration (variable $lim$) within the program code described below. This would demand computational time, of course, and the user is welcome to try. Overall, the agreement between the numerical and the exact results is acceptable. The total density of states, $N(E)$, shows that the band is filled with one state when the energy reaches the band width and if we include spin, two electrons occupy the entire band. The script used in this calculation is fcc_dos.m, listed below, and it reproduces Figure 8.6.12.

```
%copyright by J. E Hasbun and T. Datta
%fcc_dos.m
%Density of states for the single band, face centered cubic system
%according to the ray approach for the tetrahedron method of
%An-Ban Chen, Phys Rev. B V16, 3291 (1977). We also use the
%singInt integration method for the Green's function.
%For the ray method, the 2D vectors on the faces of the thin
%tetrahedrons are obtained from the TTareas function. These vectors are
%average vectors.

function fcc_dos
clear, clc;
global e0 dos
delta=1.5e-3;
im=complex(0.0,1.0);
x=0.05;               %energy step
e2=-3.0;
e1=1.0-delta;         %not to go too close to 1.0
e0=e2:x:e1;           %energy range
ntmax=length(e0);
a=1.0;                %lattice constant
```

```
tpa=2*pi/a;
VBZ=4*tpa^3;          %total FCC BZ volume
%FCC case tetrahedrons (three needed) in units of 2*pi/a;
%points used in units of 2*pi/a
L=[1/2,1/2,1/2]; K=[3/4,3/4,0]; U=[1,1/4,1/4]; W=[1,1/2,0]; X=[1,0,0];
%1st tetrahedron (Vectors in units of 2*pi/a)
A(1,:)=L; B(1,:)=K; C(1,:)=W;  %L, K, W points
%tetrahedron volume = 1/32 of the total BZ vol, so use in corresp. integral
VFCC_t1=abs(dot(A(1,:),cross(B(1,:),C(1,:)))/6);
factor(1)=VFCC_t1;
%2nd tetrahedron (Vectors in units of 2*pi/a)
A(2,:)=L; B(2,:)=U; C(2,:)=W;  %L, U, W points
%tetrahedron volume = 1/32 of the total BZ vol, so use in corresp. integral
VFCC_t2=abs(dot(A(2,:),cross(B(2,:),C(2,:)))/6);
factor(2)=VFCC_t2;
%3rd tetrahedron (Vectors in units of 2*pi/a)
A(3,:)=X; B(3,:)=U; C(3,:)=W;  %X, U, W points
%tetrahedron volume = 1/48 of the total BZ vol, so use in corresp. integral
VFCC_t3=abs(dot(A(3,:),cross(B(3,:),C(3,:)))/6);
factor(3)=VFCC_t3;
lim=21; lom=lim-1; %lim must be odd
%number of divisions along q2, and q3 => total number of TT's is n^2
n=10;
%VFCC_t=(VFCC_t1+VFCC_t2+VFCC_t3) %total volume=sum of 3 tetrahedrons
%VFCC_t is the standard tetrahedron volume part, and there are 12 VFCC_t's
%in the face center cube's total BZ (1/32+1/32+1/48=1/12)
factor_t=12*VBZ;          %to be used in the full integral result
%TT=thin tetrahedron, T=tetrahedron
DD=1/n^2;  %ratio of TT area to T area (since all thin triangles are the
   same)
factor_t=3.0*DD*factor_t; %factor is modified further by 3*DD
par1=0.0;
par2=1.0;
dk=(par2-par1)/lom;
al=par1:dk:par2;   %alpha range to integrate (main TT axis)
dv=1/n;            %TT division size
be=0:dv:1;         %beta
ga=be;             %gamma
%Tetrahedron q vectors according to the method of An-Ban Chen & B. I. Reser.
%Notation: q(tetrahedron(1,2,3),coordinate(x,y,z),vector(1,2,3))
q(:,:,1)=A*tpa; q(:,:,2)=(B-A)*tpa; q(:,:,3)=(C-B)*tpa;
%The function TTareas produces the average vectors "va" on the faces of the
   TT's
%Notation: va(ith TT vector,coordinate(x,y,z),tetrahedron(1,2,3))
%Ntt=number of TT's, areas(tetrahedron(1,2,3),ith TT area)
for tet=1:3      %the fcc has three tetrahedrons
  [Ntt(tet),areas(tet,:),va(:,:,tet)]=TTareas(q(tet,:,2),q(tet,:,3),
    n,be,ga);
end
terr=0.0;                %total error
dos=zeros(1,ntmax);
```

```
for nt=1:ntmax
  dos(nt)=0.0;
  for tet=1:3              %the fcc has three tetrahedrons
    dos_tet(tet)=0.0;
    for it=1:Ntt(tet)  %Ntt TT's for the tet tetrahedron
      for ko=1:lim   %loop over gamma
        for v=1:3  %build the k vector, kx=k(1), ky=k(2), kz=k(3)
          %The va vectors are average vectors on the TT faces
          k(v)=al(ko)*(q(tet,v,1)+va(it,v,tet)); %va vectors used here
        end
        top(ko)=al(ko)*al(ko);
        deno(ko)=e0(nt)+cos(k(1)/2)*cos(k(2)/2)...
          +cos(k(1)/2)*cos(k(3)/2)...
          +cos(k(2)/2)*cos(k(3)/2)+im*delta;
      end
      gy=singInt(top,deno,dk);
      dos_tet(tet)=-imag(gy)/pi+dos_tet(tet);  %single TT dos, add all n^2
        TT's
    end
    %weigh the corresponding tetrahedron contribution by its factor
    dos(nt)=factor(tet)*dos_tet(tet)+dos(nt);  %dos=sum over 3 tetrahedrons
  end
  %finally, multiply by the total three-tetrahedon contribution factor
  dos(nt)=factor_t*dos(nt)/VBZ;
  ge(nt)=jelittoFccDos(e0(nt))/VBZ;
  err=abs(dos(nt)-ge(nt));
  terr=terr+err;
  fprintf('E0=%9.4f, dos=%9.4f, ge=%9.4f, err=%14.6e\n',...
    e0(nt),dos(nt),ge(nt),err)
end
terr=terr/ntmax;
fprintf('Total error=%14.6e\n',terr)
%Total integrated density of states, and plot
fprintf('Integrated Density of States for the face centered cubic')
intdos=zeros(1,ntmax);
for nt=1:ntmax
  intdos(nt)=rombergInt(e2,e0(nt),@fForRomb);  %integrate on [eL,e0]
  fprintf('E0=%9.4f, integrated dos=%14.6e\n',e0(nt),intdos(nt));
end
plot(e0,ge,'k.'), hold on            %exact dos
plot(e0,dos,'ko','MarkerSize',5) %numeric dos
%Next we do the total density of states without the factor of 2 for spin.
plot(e0,intdos,'k:','LineWidth',2);
xlabel('E (Hartrees)'), ylabel('D(E) (states/energy), N(E) (states)')
str=cat(2,'D(E) and N(E) for the face centered cubic (no spin)');
title(str, 'Fontsize',12)
legend('Exact D(E)','numeric D(E)','N(E)',0)

function ge=jelittoFccDos(ee)
%Jelitto's DOS for the face centered cubic (exact)
if((ee > -3.0)) & (ee < 0.0)
```

```
    eaa=abs(ee);
    a1=3.0+ee;
    a2=a1^2;
    a=sqrt(a1);
    b=-85.9325+101.103*a1;
    d=-16.2885*(a2);
    f=56.8683-47.1215*a1;
    h=2.9045*(a2);
    ge=4.*a*((b+d)+(f+h)*sqrt(eaa));
else
    if((ee >= 0.0) & (ee <= 1.0))
        b11=122.595-19.4100*ee+1.76011*ee^2;
        b22=(-44.8100+7.18628*ee)*log(1.-ee);
        ge=4.0*(b11+b22);
    else
        ge=0.;
    end
end

function y=fForRomb(p)
%function used by romberg integration and which interpolates
%tdos versus e0
global e0 dos
y=interpFunc(p,e0,dos);
```

## 8.7 Simple Tight Binding Semiconductor Multiband Structures

Here we extend the tight binding method employed in the single band Section 8.2 to semiconducting systems. In particular, we work with a multiband system which arises from the interactions between the valence electrons in the *s* and *p* orbitals. Semiconducting systems characterized by these interactions include crystals involving group IV elements, such as C, Si, Ge, and Sn. Additionally, compounds made of groups III and V elements have band structures that arise from the interaction of electrons in the *s* and *p* orbitals; some of these compounds are GaAs, InSb, AlP, etc. The purpose of the approach we follow below is meant to be instructive rather than accurate. For example, it does not produce accurate band gaps, nor does it produce inderect band gaps, as is the case for Si and Ge. However, the method illustrates the important features associated in obtaining bands structures, namely, the interaction between the valence electrons that ultimately leads to solutions of the eigenvalue problem versus wavevector. These types of semiconductors have the zinc-blende structure in common, but in order to treat compounds, we work with two species, which we refer to as cations and anions as shown in Figure 8.7.13.

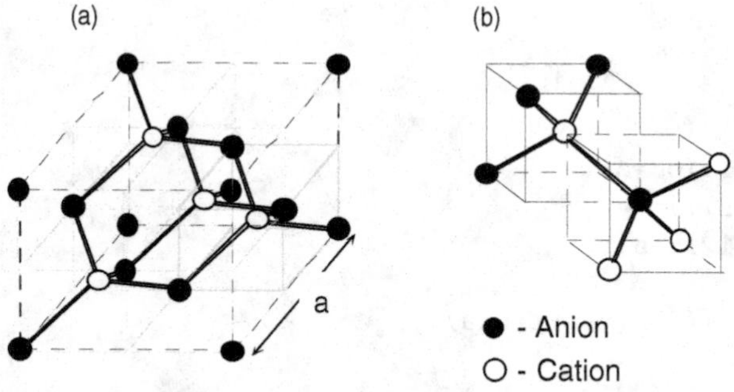


Figure 8.7.13: (a) The standard zinc-blende unit cell. (b) Each cation is surrounded by four nearest neighbor anions and, similarly, each anion is surrounded by four nearest neighbor cations. The standard diamond unit cell structure is obtained by letting the atoms be identical.

The cation is the positive species (the metalic atom which holds electrons more loosely) and the anion is the negative species (the non-metalic atom which has high electron affinity). For III-V compounds, the cation is the group III species, while the group V species is the anion. In the case of group IV elements, there is only one species. As mentioned in Chapter 1, the zinc-blende structure is considered to be two interlaced FCC structures. It has eight atoms per cell, four cations and four anions. The anions occupy one sublattice and the cations occupy the other as illustrated in Figure 8.7.13. We next develop the tight binding Hamiltonian that will be used to obtain the band structures. We will later develop Green's function which will allow us to obtain the density of states.

## 8.7.1 The Band Structure

Let's begin by defining the symbols used throughout, then the method applied in obtaining the band structures is described. We first let the letter $\mu$ represent an atomic orbital, such as $s$ or $p_{x,y,z}$, and let $\nu$ represent the sublattice (anion or cation). We then write the Bloch function of orbital $\mu$ for sublattice $\nu$ as a linear combination of atomic orbitals (LCAO) on each sublattice site as

$$|\mu\nu\vec{k}>= \frac{1}{\sqrt{N}}\sum_i e^{i[\vec{k}\cdot(\vec{R}_i+\vec{\tau}_\nu)]}|\mu\nu\vec{R}_i). \tag{8.7.41}$$

Here $|\mu\nu\vec{k}>$ is a Bloch function with the periodicity of the lattice and the $|\mu\nu\vec{R}_i)$ are atomic functions with orbital $\mu$ ($s$, $p_x$, $p_y$, $p_z$), lattice vector $\vec{R}_i$, sublattice vector $\vec{\tau}_\nu$ (for the cation, say, $\vec{\tau}_c = (0,0,0)a$, and for the anion $\vec{\tau}_a = (1,1,1)a/4$, with $a$ the lattice constant), and $N$ is the number of wavevectors in the BZ. ($N$ is also the number of cells in the crystal.) The position of the cation is taken as $\vec{R}_i = (0,0,0)a = \vec{\tau}_c$. In this way, the cation is surrounded by four anions at positions $\vec{d}_1 = (1,1,1)a/4$, $\vec{d}_2 = (1,\bar{1},\bar{1})a/4$, $\vec{d}_3 = (\bar{1},1,\bar{1})a/4$, and $\vec{d}_4 = (\bar{1},\bar{1},1)a/4$. The eigenvalues are obtained from the equation

$$\hat{H}|\vec{k}n>= E_{\vec{k}n}|\vec{k}n>, \tag{8.7.42}$$

where $\hat{H}$ is the crystal Hamiltonian, $n$ is a band index, and $E_{\vec{k}n}$ are the eigenvalues of the crystal wavefunction $|\vec{k}n>$. A plot of the energy eigenvalues $E_{\vec{k}n}$ versus $\vec{k}$ produces the crystal energy bands. To obtain these energies, the crystal wave function is expanded as a linear combination of Bloch functions, $|\mu\nu\vec{k}>$, by writing

$$|\vec{k}n>= \sum_{\mu\nu}|\mu\nu\vec{k}>< \mu\nu\vec{k}|\vec{k}n> \tag{8.7.43a}$$

where we have used the completeness condition

$$\sum_{\mu\nu} |\mu\nu\vec{k}><\mu\nu\vec{k}| = 1. \tag{8.7.43b}$$

In a similar fashion, the Bloch functions of band index $n$ are thought to obey $\sum_n |\vec{k}n><\vec{k}n| = 1$, so that one can write $|\mu\nu\vec{k}>=\sum_n |\vec{k}n><\vec{k}n|\mu\nu\vec{k}>$. In Equation 8.7.43a, the quantities $<\mu\nu\vec{k}|\vec{k}n>$ are the eigenvectors. So with the help of Equation 8.7.43b, muliplying Equation 8.7.42 by $<\mu\prime\nu\prime\vec{k}|$ from the left, we obtain

$$\sum_{\mu\nu} <\mu\prime\nu\prime\vec{k}|\hat{H}|\mu\nu\vec{k}><\mu\nu\vec{k}|\vec{k}n>= E_{\vec{k}n}\sum_{\mu\nu} <\mu\prime\nu\prime\vec{k}|\mu\nu\vec{k}><\mu\nu\vec{k}|\vec{k}n>. \tag{8.7.44}$$

This expression can be reorganized by making use of the orthogonality of the Bloch functions

$$<\mu\prime\nu\prime\vec{k}|\mu\nu\vec{k}>= \delta_{\mu\mu\prime}\delta_{\nu\nu\prime}, \tag{8.7.45}$$

to yield the secular equation

$$\sum_{\mu\nu} \left({}^{\mu\mu\prime}H_{\vec{k}}^{\nu\nu\prime} - E_{\vec{k}n}\delta_{\mu\mu\prime}\delta_{\nu\nu\prime}\right) <\mu\nu\vec{k}|\vec{k}n>= 0, \tag{8.7.46}$$

which we employ to obtain the energy bands and where we have defined ${}^{\mu\mu\prime}H_{\vec{k}}^{\nu\nu\prime} \equiv <\mu\prime\nu\prime\vec{k}|\hat{H}|\mu\nu\vec{k}>$. Equation 8.7.46 says that, given the matrix elements of the Hamiltonian between two atomic functions; i.e., ${}^{\mu\mu\prime}H_{\vec{k}}^{\nu\nu\prime}$, then by performing a diagonalization procedure we can obtain the energy bands $E_{\vec{k}n}$ as a function of $\vec{k}$ as well as the eigenvectors. Here, we do not intend to solve for the matrix elements of $\hat{H}$: however, instead we think of ${}^{\mu\mu\prime}H_{\vec{k}}^{\nu\nu\prime}$ as parameters in a similar way that Harrison proposed [16]. To this end, the total Hamiltonian that we use has diagonal matrix elements between $s$ and $p$ orbitals for the anion and the cation as well as off-diagonal matrix elements. The Hamiltonian is, therefore, an $8\times 8$ matrix with a $2\times 2$ block notation in the form

$$H_{\vec{k}} = \begin{pmatrix} H^{cc} & H^{ca} \\ H^{ac} & H^{aa} \end{pmatrix}_{\vec{k}} \tag{8.7.47}$$

where each $H^{\nu\nu\prime}$ block is a $2\times 2$ matrix containing $s, p_x, p_y$, and $p_z$ interatomic matrix elements and $\nu$ is the sublattice index ($a =$ anion, $c =$ cation). The full $8\times 8$ Hamiltonian matrix of Equation 8.7.47 is shown in Table 8.7.1.

Table 8.7.1: The full nearest neighbor tight binding $8\times 8$ matrix Hamiltonian model for semiconductors with zinc-blende structure, which is referred to hereafter as the Harrison model (Source: [16]).

| | $s^c$ | $s^a$ | $p_x^c$ | $p_y^c$ | $p_z^c$ | $p_x^a$ | $p_y^a$ | $p_z^a$ |
|---|---|---|---|---|---|---|---|---|
| $s^c$ | $\varepsilon_s^c$ | $E_{ss}g_0$ | 0 | 0 | 0 | $E_{sp}g_1$ | $E_{sp}g_2$ | $E_{sp}g_3$ |
| $s^a$ | $E_{ss}g_0^*$ | $\varepsilon_s^a$ | $-E_{sp}g_1^*$ | $-E_{sp}g_2^*$ | $-E_{sp}g_3^*$ | 0 | 0 | 0 |
| $p_x^c$ | 0 | $-E_{sp}g_1$ | $\varepsilon_p^c$ | 0 | 0 | $E_{xx}g_0$ | $E_{xy}g_3$ | $E_{xy}g_2$ |
| $p_y^c$ | 0 | $-E_{sp}g_2$ | 0 | $\varepsilon_p^c$ | 0 | $E_{xy}g_3$ | $E_{xx}g_0$ | $E_{xy}g_1$ |
| $p_z^c$ | 0 | $-E_{sp}g_3$ | 0 | 0 | $\varepsilon_p^c$ | $E_{xy}g_2$ | $E_{xy}g_1$ | $E_{xx}g_0$ |
| $p_x^a$ | $E_{sp}g_1^*$ | 0 | $E_{xx}g_0^*$ | $E_{xy}g_3^*$ | $E_{xy}g_2^*$ | $\varepsilon_p^a$ | 0 | 0 |
| $p_y^a$ | $E_{sp}g_2^*$ | 0 | $E_{xy}g_3^*$ | $E_{xx}g_0^*$ | $E_{xy}g_1^*$ | 0 | $\varepsilon_p^a$ | 0 |
| $p_z^a$ | $E_{sp}g_3^*$ | 0 | $E_{xy}g_2^*$ | $E_{xy}g_1^*$ | $Exxg_0^*$ | 0 | 0 | $\varepsilon_p^a$ |

The Hamiltonian matrix elements shown in the table include nearest neighbor interactions only and the model assumes that the diagonal energies of the $p$ orbitals are equal. According to the approach, the energy parameters shown are given by the following expressions.

$$E_{ss} = V_{ss\sigma}, \quad E_{sp} = \frac{V_{sp\sigma}}{\sqrt{3}}, \quad E_{xx} = \frac{V_{pp\sigma}+2V_{pp\pi}}{3}, \quad E_{xy} = \frac{V_{pp\sigma}-V_{pp\pi}}{3}, \tag{8.7.48a}$$

where the $V$'s are given by ([16])

$$V_{\ell\ell' g} = n_{\ell\ell' g}\frac{\hbar^2}{md^2}, \tag{8.7.48b}$$

with the orbital index taking on values of $\ell = s, p$, the bond index is $g = \sigma, \pi$, and $d$ is the bond length. Depending on these indices, the $n_{ll'g}$ parameter takes on the values

$$n_{ss\sigma} = -1.40, \quad n_{sp\sigma} = 1.84, \quad n_{pp\sigma} = 3.24, \quad n_{pp\pi} = -0.81. \tag{8.7.48c}$$

As mentioned before and associated with Equation 8.7.41, the nearest neighbor anion positions, to the cation at the center, are the vectors $\vec{d_1} = (1,1,1)a/4$, $\vec{d_2} = (1,\bar{1},\bar{1})a/4$, $\vec{d_3} = (\bar{1},1,\bar{1})a/4$, so that the $\vec{k}$ dependence of the Hamiltonian comes in through the sum over the nearest neighbor sites; that is,

$$\begin{aligned} g_0(\vec{k}) &= e^{i\vec{k}\cdot\vec{d_1}} + e^{i\vec{k}\cdot\vec{d_2}} + e^{i\vec{k}\cdot\vec{d_3}} + e^{i\vec{k}\cdot\vec{d_4}}, \quad g_1(\vec{k}) = e^{i\vec{k}\cdot\vec{d_1}} + e^{i\vec{k}\cdot\vec{d_2}} - e^{i\vec{k}\cdot\vec{d_3}} - e^{i\vec{k}\cdot\vec{d_4}}, \\ g_2(\vec{k}) &= e^{i\vec{k}\cdot\vec{d_1}} - e^{i\vec{k}\cdot\vec{d_2}} + e^{i\vec{k}\cdot\vec{d_3}} - e^{i\vec{k}\cdot\vec{d_4}}, \quad g_3(\vec{k}) = e^{i\vec{k}\cdot\vec{d_1}} - e^{i\vec{k}\cdot\vec{d_2}} - e^{i\vec{k}\cdot\vec{d_3}} - e^{i\vec{k}\cdot\vec{d_4}}. \end{aligned} \tag{8.7.49}$$

At this point all parameters of the model have been specified and the energy bands can be obtained versus $\vec{k}$ by diagonalizing the matrix shown in Table 8.7.1. The model consists of four diagonal energies $(\varepsilon_s^c, \varepsilon_p^c, \varepsilon_s^a, \varepsilon_p^a)$, four off-diagonal energies $(E_{ss}, E_{sp}, E_{xx}, E_{xy})$, and the bond length $(d)$, which make this a nine-parameter band model for semiconductor band structures. While the entire band structure can be obtained versus $\vec{k}$, usually along symmetry directions. For example, the $\Delta$ direction is from the $\Gamma$ point ($\vec{k} = (0,0,0)2\pi/a$) to the $X$ symmetry point ($(1,0,0)2\pi/a$); the $\Lambda$ direction is from the $\Gamma$ point to the $L$ symmetry point ($(1/2,1/2,1/2)2\pi/a$); and the $\Sigma$ direction is from the $\Gamma$ point to the $K$ symmetry point ($(3/4,3/4,0)2\pi/a$). The symmetry points are shown in Figure 8.7.14.

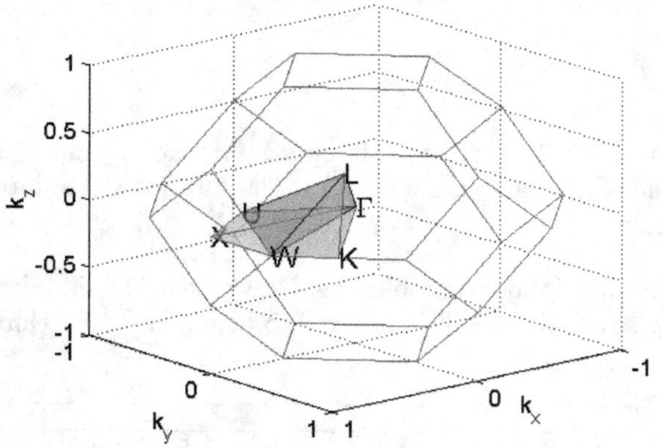


Figure 8.7.14: The zinc-blende system's BZ is the same as the FCC with volume $4(2\pi/a)^3$. Here the irreducible part of the FCC's BZ is shown with the three irreducible tetrahedrons, 12 of which (together) make up the 14-sided polyhedron. The coordinates of each tetrahedron are high symmetry points in the FCC BZ. These high symmetry points, in units $2\pi/a$, are $\Gamma = [0,0,0]$, $L = [1/2,1/2,1/2]$, $K = [3/4,3/4,0]$, $W = [1,1/2,0]$, $U = [1,1/4,1/4]$, and $X = [1,0,0]$ and are shown along with the tetrahedrons whose positions they share.

For calculation purposes, below in Table 8.7.2 we provide some energy parameter values and bond lengths for various elements in groups III, IV, and V of the periodic table.

Table 8.7.2: Some diagonal energies and bond lengths for groups II, IV, and V elements of the periodic table. Note that in the zinc-blende system the lattice length, $a$, is obtained from the nearest neighbor distance, $d$ (or bond length), relation $d = \sqrt{3}a/4$. (Source: [16]).

| atom | $-\varepsilon_s(eV)$ | $-\varepsilon_p(eV)$ | System | d(Å) |
|---|---|---|---|---|
| C | 17.52 | 8.97 | C | 1.54 |
| Al | 10.11 | 4.86 | Si | 2.35 |
| Si | 13.55 | 6.52 | AlP | 2.36 |
| P | 17.10 | 8.33 | Ge | 2.44 |
| S | 20.80 | 10.27 | GaAs | 2.45 |
| Ga | 11.37 | 4.90 | InSb | 2.81 |
| Ge | 14.38 | 6.36 | | |
| As | 17.33 | 7.91 | | |
| In | 10.12 | 4.69 | | |
| Sb | 14.80 | 7.24 | | |

With the above parameters, we are ready to obtain the energy bands. From Equation 8.7.46, we diagonalize the Hamiltonian matrix which corresponds to setting the determinant of the coefficients to zero and solving for the eigenvalues $E$ versus $\vec{k}$; that is,

$$\left| \begin{pmatrix} H_{11} & H_{12} & \cdots & H_{18} \\ H_{21} & H_{22} & \cdots & H_{28} \\ \vdots & \vdots & \vdots & H_{18} \\ H_{81} & H_{82} & \cdots & H_{88} \end{pmatrix} - \begin{pmatrix} E & 0 & \cdots & 0 \\ 0 & E & \cdots & 0 \\ \vdots & \vdots & \ddots & 0 \\ 0 & 0 & \cdots & E \end{pmatrix} \right| = 0, \qquad \Rightarrow E_{\vec{k},n=1\ldots8}. \tag{8.7.50}$$

We do this along previously mentioned symmetry directions and obtain the energy eigenvalues, which, when sorted from the lowest to highest, yield the energy bands $E_{\vec{k},n=1\ldots8}$. Before proceeding with the outline above, let's consider diagonalizing the matrix at $\vec{k} = 0$; that is, at the $\Gamma$ point. In this case, the $g_i$'s of Equation 8.7.49 take on the values of $g_1 = g_2 = g_3 = 0$, and $g_0 = 4$. In this case, all off-diagonal matrix elements in Table 8.7.1 vanish except those that couple $s^c$ with $s^a$, as well as those that couple $p_i^c$ with $p_i^a$ for $i = x, y, z$. This means that at the $\Gamma$ point, the system collapses to two $2 \times 2$ matrices with determinants in the form

$$\left| \begin{pmatrix} \varepsilon_s^c & 4E_{ss} \\ 4E_{ss} & \varepsilon_s^a \end{pmatrix} - \begin{pmatrix} E & 0 \\ 0 & E \end{pmatrix} \right| = 0, \tag{8.7.51a}$$

for the $s$ states and

$$\left| \begin{pmatrix} \varepsilon_p^c & 4E_{xx} \\ 4E_{xx} & \varepsilon_p^a \end{pmatrix} - \begin{pmatrix} E & 0 \\ 0 & E \end{pmatrix} \right| = 0, \tag{8.7.51b}$$

for the $p$ states, and since this is identical for all $p_i$ for $i = x, y, z$, the $p$ states are triply degenerate at the $\Gamma$ point. Equation 8.7.51a results in

$$E^2 - E(\varepsilon_s^c + \varepsilon_s^a) + (\varepsilon_s^c \varepsilon_s^a - 16E_{ss}^2) = 0 \Rightarrow E = \frac{\varepsilon_s^c + \varepsilon_s^a}{2} \pm \sqrt{\left(\frac{\varepsilon_s^c - \varepsilon_s^a}{2}\right)^2 + (4E_{ss})^2} \tag{8.7.52a}$$

for the $s$ states and Equation 8.7.51b yields

$$E^2 - E(\varepsilon_p^c + \varepsilon_p^a) + (\varepsilon_p^c \varepsilon_p^a - 16E_{xx}^2) = 0 \Rightarrow E = \frac{\varepsilon_p^c + \varepsilon_p^a}{2} \pm \sqrt{\left(\frac{\varepsilon_p^c - \varepsilon_p^a}{2}\right)^2 + (4E_{xx})^2}, \tag{8.7.52b}$$

for the $p$ states. Based on this model, we can use these energy solutions in order to obtain the band gap of a semiconductor, which we do in the next example.

**Example 8.7.1.1**
Using parameters suitable for Si, from Table 8.7.2, since in pure Si the anion and the cation are the same species, we have (in eV) $\varepsilon_s^a = -13.55$, $\varepsilon_p^a = -6.52$, $\varepsilon_s^c = \varepsilon_s^a$, $\varepsilon_p^c = \varepsilon_p^a$, and $d = 2.35$Å. With Equation 8.7.48, we also get (in eV) $E_{ss} - 1.9317$, $E_{sp} = 1.4658$, and $E_{xx} = 0.7451$. Using these parameters, the bottom of the conduction band is obtained from the upper root of Equation 8.7.52a, or $E_c = -5.8231\, eV$. The top of the valence band is given by the lower root of Equation 8.7.52b, to obtain $E_v = -9.5004\, eV$. The resulting band gap is $E_g = E_c - E_v = -5.82 - (-9.50) = 3.68\, eV$. In this model, Si comes out as a direct band gap material, whereas in reality it is an indirect gap solid. Furthermore, actually, Si's band gap is about $1\, eV$, so that the Harrison model is meant to be qualitative rather than quantitative. However, from this example, we learn that the conduction band has $s$-like character, while the valence band is of $p$-like character, which is quite illuminating!

In Figure 8.7.15, the above Harrison model's band obtained band structure for Si is compared to the more sophisticated pseudo-potential method's results from reference [35].

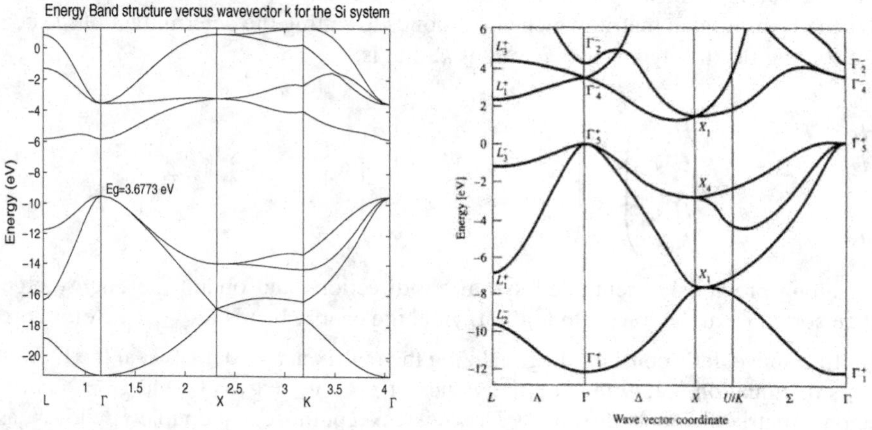


(a) Harrison Model Si Band Structure

(b) Empirical Pseudo-Potential Method Si Band Structure [35] (reprinted with permission)

Figure 8.7.15: (a) The Harrison model obtains a large direct band gap with a value of about $3.68\, eV$ (see Example 8.7.1.1). The parameters of the model are from Table 8.7.2. (b) Reference [35]'s band structure uses the top of the valence band as the zero of energy. The indirect band gap is between the $\Gamma$ and the $X$ points with a value of about $1\, eV$.

All the bands are split at the $K$ point, but three valence bands become degenerate at the $\Gamma$ point as mentioned in Example 8.7.1.1. Also, the system's valence band structure, having four valence bands, with each band being occupied by two electrons (spin up and down), can hold eight electrons. This is consistent with the structure's eight atoms per cell. Each atom has four valence electrons, but each atom shares them among four neighbors, thus each atom contributes a total of one electron to each band, and in this way we have eight electrons per cell. This can also be viewed from the bonding perspective; i.e., the tetrahedral nature of each atom's neighbors. Since each atom makes a bond with each of its four neighbors and each bond takes two electrons, a total of eight electrons are involved in the tetrahedral bonds.
The code used to produce the Si band structure shown in Figure 8.7.15(a) is band_structure_Si.m and is listed below. The functions it makes use of; that is, initialize_Si.m, which provides the system's parameters; diagHamil.m, for setting the diagonal energies; offDiagHamil.m for setting the

off diagonal matrix elements; and sorter.m, which sorts the eigenvalues and eigenvectors; are listed in Section 8.8 of this chapter.

The script band_structure_Si.m that follows calculates the eigenvalues of the $8 \times 8$ Harrison matrix versus $\vec{k}$ along the $\Lambda, \Delta, \Sigma$, and X to K symmetry points directions. After building the Hamiltonian matrix, the eigenvalues are found by the MATLAB command 'eig', then sorted and placed into the 'bandEnergy' array suitable for plotting along the various directions.

```
%copyright by J. E Hasbun and T. Datta
%band_structure_Si.m
%This program uses Harrison's parametrized scaling approach.
%Off diagonal elements scale 1/(bond length)^2
%This is a nearest neighbor tight binding method.
%It's an s-p parametrized approach for 3-5 semiconductors.
%
function band_structure_Si
global H zim NB
%*********** constants ***********
NB=8;          %number of bands (hamiltonian dimension =NBxNB also)
Ns=12;         %number of k steps to do in a given direction
zim=1i;        %the complex number i by itself
H=zeros(NB,NB);%initilize the hamiltonian
%Initialize energies, compound: c=cation, a=anion, es=s-energy, ep=p-energy
%The compound is "system"
[a,esc,esa,epc,epa,ess,esp,exx,exy,system]=initialize_Si();
diagHamil(esc,esa,epc,epa) %diagonal elements of H
%Note: symmetry directions (points): Lambda (L),delta (X), Sigma (K)
sympt=['L';'X';'K'];                       %symmetry points
ba0=2*pi/a;                                %reciprocal lattice vector magnitude
bkpt=ba0*[1/2,1,3./4.];                    %corresponding maximum k point
% ****************************************************
% *************** four different directions --loop *******
% ****************************************************
direction=['Lamb';'Delt';'Sigm';'XtoK']; %chosen directions for calculations
for idir=1:4
  %reset maximum k value when the direction is changed
  if strcmp(direction(idir,:),'Delt')       %Delta direction
    a1=1.;
    b1=0.;
    c1=0.;
    ba=ba0;                                 %X point
  else
    if strcmp(direction(idir,:),'Sigm')     %Sigma direction
      a1=1.;
      b1=1.;
      c1=0.;
      ba=ba0*(3./4.);                       %K point
    else
      if strcmp(direction(idir,:),'Lamb')   %Lambda direction
        a1=1.;
        b1=1.;
        c1=1.;
        ba=ba0/2.;                          %L point
```

```
      else
        %the vector direction from X to K=[(1,1,0)(3/4)-(1,0,0)](2*pi/a)
        %=[-1/4,3/4,0](2pi/a) we will need to start at X, and move in
        %this direction to end up at the U point
        if strcmp(direction(idir,:),'XtoK') %direction toward K point
          a1=-1./4.;
          b1=3/4;
          c1=0;
          ba=ba0;
        else
          break
        end
      end
    end
  end
  st=ba/Ns;
  fprintf('system: %s\n',system)
  fprintf('k-direction a1=%6.3f, b1=%6.3f, c1=%6.3f, (%s)\n',a1,b1,c1,...
    direction(idir,:))
  fprintf('upper limit ba=%5.2f, step size st =%5.2f\n',ba,st)
  % ******************************************************
  % *************** k--loop ****************
  % ******************************************************
  %the first direction set of bands are done from high k value to zero
  %the second direction set of bands are done from zero to high k value
  if strcmp(direction(idir,:),'Lamb')
    bkL=ba;
    bkS=-st;
    bkU=0.0;
  else
    bkL=0.0;
    bkS=st;
    bkU=ba;
  end
  kc=0;                      %reset counter for each direction
  for bk=bkL:bkS:bkU
    kc=kc+1;
    %Notice bx,by,bz includes a factor of 2*pi/a here (see definition of
       ba above)
    if strcmp(direction(idir,:),'XtoK') %in the XtoK direction, start at
       X(2pi/a)[1,0,0]
      bx=ba0+bk*a1;%and continue in the XtoK direction (2pi/a)[-1/4,3/4,0]
        in steps of bk
      by=0+bk*b1;
      bz=0+bk*c1;
    else
      bx=bk*a1;
      by=bk*b1;
      bz=bk*c1;
    end
```

```
    %Tight binding interations - nearest neighbors
    offDiagHamil(bx,by,bz,a,ess,esp,exx,exy) %off-diagonal elements of H
    [z,w]=eig(H); %z=eigenvectors, w=eigenvalues
    %let ww contain the sorted eigenvalues and zz contain the sorted
        eigenvectors
    [ww,zz]=sorter(diag(w),z);
    kval(kc,idir)=bk;
    bandEnergy(:,kc,idir)=ww;
    fprintf('eig.val. at %6.2f %6.2f %6.2f with direction: %s\n',bx,by,bz,
       direction(idir,:))
    fprintf('%8.3f %8.3f %8.3f %8.3f %8.3f %8.3f %8.3f %8.3f\n',ww(1:NB))
  end
end
[tvb,kiv]=max(bandEnergy(NB/2,:,2));     %top of the valence band
[bcb,kic]=min(bandEnergy(NB/2+1,:,2));   %bottom of the conduction band
Eg=bcb-tvb;                              %band gap
fprintf('Top Val. Band tvb =%8.4f %s, and occurs at k=%i\n',tvb,'eV',
   kval(kiv,2))
fprintf('Bot. Cond. Band bcb =%8.4f %s, and occurs at k=%i\n',bcb,'eV',
   kval(kic,2))
fprintf('Band gap Eg=bcb-tvb =%8.4f %s\n',Eg,'eV')
%Next, check with Harrison's formulas for the top of the VB and bottom of
   the CB
%Most of the time, Ev=Ev1 and Ec=Ec1 for 3-5 semiconductors.
%To make Carbon work, Ev2 can be lower than Ec1 so we fixed that here as
%follows:
Ev1=(epc+epa)/2-sqrt(((epc-epa)/2)^2+(4*exx)^2); %Top of VB
Ev2=(epc+epa)/2+sqrt(((epc-epa)/2)^2+(4*exx)^2); %Top of VB
Ec1=(esc+esa)/2+sqrt(((esc-esa)/2)^2+(4*ess)^2); %Bottom of CB
Ec2=(esc+esa)/2-sqrt(((esc-esa)/2)^2+(4*ess)^2); %Bottom of CB
fprintf('Ev1,Ev2,Ec1,Ec2=%8.4f %8.4f %8.4f %8.4f\n',Ev1,Ev2,Ec1,Ec2)
Ev=max(Ev1,Ec2);  %highest of these two roots
Ec=min(Ec1,Ev2);  %lowest of these two roots
EgH=Ec-Ev; %based on Harrison's formulas at the gamma (k=0) point
fprintf('Harrison formulas of the top of the VB and Bottom of the CB, and
   gap:\n')
fprintf('Ev=%8.4f %s, at Gamma\n',Ev,'eV')
fprintf('Ec%8.4f %s,  at Gamma\n',Ec,'eV')
fprintf('Eg=Ec-Ev =%8.4f %s\n',EgH,'eV')
%Below we shift k by reciprocal lattice vector mag (ba0) to plot correctly
plot(ba0-kval(:,1),bandEnergy(:,:,1),'k')             %Lamda direction
hold on
plot(ba0+kval(:,2),bandEnergy(:,:,2),'k')             %Delta direction
plot(3.5*ba0-kval(:,3),bandEnergy(:,:,3),'k')         %Sigma direction
%max k value at the K point is (3/4)*(2pi/a)+, and shift by recip. lat.
%to plot correctly
plot(2*ba0+(3./4.)*kval(:,4),bandEnergy(:,:,4),'k') %XtoK direction
str1=cat(2,'Energy Band structure versus wavevector k for the ',system,'
   system');
title(str1,'Fontsize',12)
ylabel('Energy (eV)','Fontsize',12)
```

```
%xlabel('(k-space)','Fontsize',8)
emin1=min(bandEnergy(1,:,1));
emin2=min(bandEnergy(1,:,2));
emin3=min(bandEnergy(1,:,3));
emin4=min(bandEnergy(1,:,4));
emax1=max(bandEnergy(NB,:,1));
emax2=max(bandEnergy(NB,:,2));
emax3=max(bandEnergy(NB,:,3));
emax4=max(bandEnergy(NB,:,4));
xmin1=min(ba0-kval(:,1));
xmin2=min(ba0+kval(:,2));
xmin3=min(3.5*ba0-kval(:,3));
xmin4=min(2*ba0+(3./4.)*kval(:,4));
xmax1=max(ba0-kval(:,1));
xmax2=max(ba0+kval(:,2));
xmax3=max(3.5*ba0-kval(:,3));
xmax4=max(2*ba0+(3./4.)*kval(:,4));
emin=min([emin1,emin2,emin3,emin4]); emax=max([emax1,emax2,emax3,emax4]);
xmin=min([xmin1,xmin2,xmin3,xmin4]); xmax=max([xmax1,xmax2,xmax3,xmax4]);
axis ([xmin xmax emin emax])
str2=cat(2,'Eg=',num2str(Eg),' eV');
text(ba0+0.05,tvb*(1-0.05),str2)
text(ba0,min(emin1,emin2)*(1+0.075),'\Gamma')
text(ba0-bkpt(1),emin*(1+0.075),sympt(1,:))                    %L point
text(ba0+bkpt(2),emin*(1+0.075),sympt(2,:))                    %X point
text(3.5*ba0-bkpt(3),emin*(1+0.075),sympt(3,:))                %K point
text(3.5*ba0,emin*(1+0.075),'\Gamma')
line ([ba0 ba0],[emin emax],'LineStyle','-','Color','k')  %Gamma line
line ([ba0+bkpt(2) ba0+bkpt(2)],[emin emax],...                %X line
  'LineStyle','-','Color','k')
line ([3.5*ba0-bkpt(3) 3.5*ba0-bkpt(3)],[emin emax],...    %K line
  'LineStyle','-','Color','k')
```

### 8.7.2 The Density of States

It should be mentioned here that while we continue working with Harrison's eight-band semiconductor model which obtains a large gap, improvements to it exist (see the ten-band $sp^3s^*$ model of Vogl. et al. [36] as regards to the gap - direct and indirect), but are beyond our present scope. Having obtained the energy bands $E_{\vec{k}n}$ of a crystal system, the Green function for the crystal is obtained when we let the Green function operator of Equation 8.4.17c act on the crystal state of Equation 8.7.43a for band $n$, to write

$$\hat{G}(\varepsilon)|\vec{k}n>=(\varepsilon-\hat{H})^{-1}|\vec{k}n>=(\varepsilon-E_{\vec{k}n})^{-1}|\vec{k}n>=G_{\vec{k}n}(\varepsilon)|\vec{k}n>, \tag{8.7.53a}$$

and in a similar way to what was done in Equation 8.4.18b, since $<\vec{k}n|\vec{k}n>=1$ we have

$$G_{\vec{k}n}(\varepsilon)=<\vec{k}n|\hat{G}(\varepsilon)|\vec{k}n>=(\varepsilon-E_{\vec{k}n})^{-1}, \tag{8.7.53b}$$

where, as before, $\varepsilon$ contains a small positive imaginary part. For the density of states, as in Equation 8.4.21a, adding the contribution from all the bands, the differential density of states is written as

$$D(\varepsilon)=-\frac{1}{\pi V_{BZ}}\,\mathrm{Im}\left(\sum_n\int_{V_{BZ}}G_{\vec{k}n}(\varepsilon)d\vec{k}\right), \tag{8.7.54}$$

where the $V_{BZ}$ is the same as the BZ volume of the FCC crystal previously discussed. The factor of 2 for spin has not been included. For our Harrison model, there are eight bands, so that Equation 8.7.54 involves eight integrals over the BZ, one for each band. Fortunately, the machinery to carry out the integration already exists, as it was detailed in Section 8.6.
The total density of states is written as in Equation 8.3.13 or

$$N(\varepsilon) = \int_{-\infty}^{\varepsilon} D(\varepsilon')d\varepsilon'. \tag{8.7.55}$$

Figure 8.7.16 contains the results of the calculation for the density of states. The left panel is based on the Harrison model, while the right panel is due to Chelikowsky et al. [35] and is included for comparison purposes.

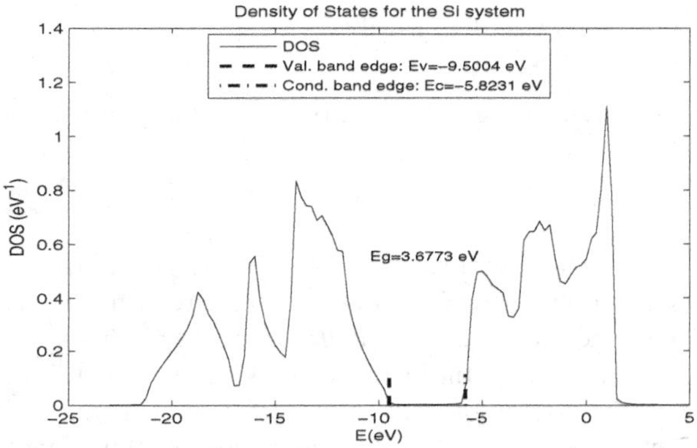


(a) Harrison Model Density of States for Si

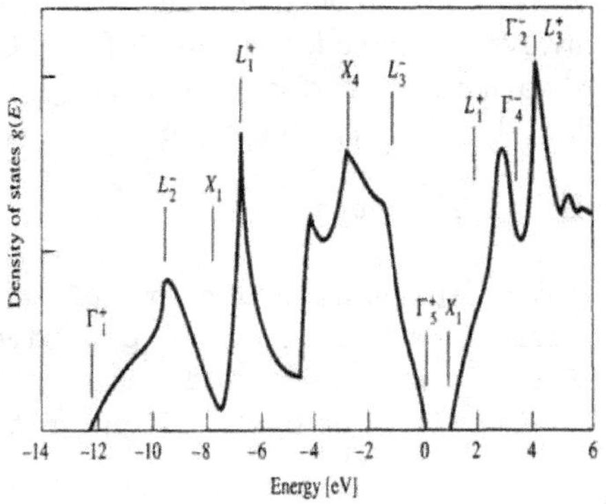


(b) Empirical Pseudo-Potential Method Density of States for Si [35] (reprinted with permission)

Figure 8.7.16: (a) The Harrison model obtained density of states for the Si system. The parameters are as in Figure 8.7.15(a). (b) Reference [35]'s obtained density of states for Si.

Harrison's model density of states again shows the large band gap. Additionally, the top of the valence band in the right panel of Figure 8.7.16 starts at the zero of energy. Aside from the expected differences, the shapes of the curves do bear resemblance, which makes the eight-band model instructively interesting. The resulting total density of states is shown in Figure 8.7.17.

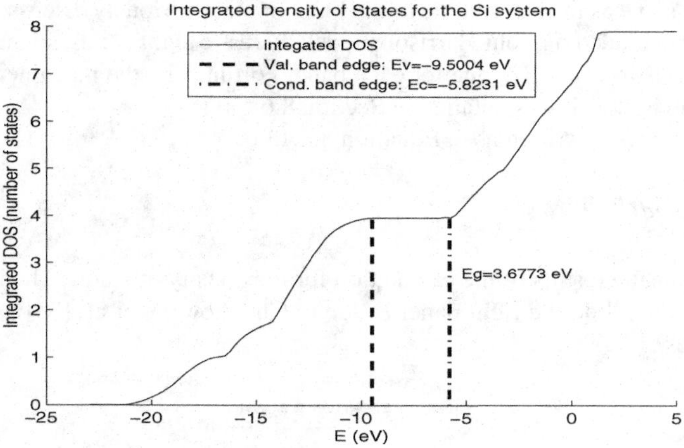


Figure 8.7.17: The total density of states obtained by integrating the Harrison model density of states of Figure 8.7.16(a).

We notice that the number of states reached when $\varepsilon = E_v + E_g/2$ is 4, so that since each state can hold two electrons (spin up and down), a total of 8 electrons fill the valence bands, as expected. The Fermi level lies, therefore, in the middle of the gap. This is in line with what was pointed out in Chapter 7's, Section 7.4, Intrinsic Carrier Concentration for semiconductors. The code used in obtaining both, the density of states of Figure 8.7.16(a) and the integrated density of states of Figure 8.7.17 is band_total_dos_Si.tex which makes use of all the functions (initialize_Si.m, diagHamil.m, offDiagHamil.m, sorter.m, TTareas.m, singInt.m, interpFunc.m, and rombergInt.m) listed in the appendices for this chapter. The listing of the script follows. It first finds the band structure as in band_structure_Si.m, and proceeds to obtain the density of states by integrating over each band's Green function (which involves the three tetrahedrons of the FCC BZ). After the density of states from each band is summed up, the script performs a final integration over the density of states ($D(\varepsilon)$) to obtain the total integrated density of states ($N(\varepsilon)$).

```
%copyright by J. E Hasbun and T. Datta
%band_total_dos_Si.m
%Here, we also find the total integrated density of states
%versus energy. This program uses Harrison's s-p parametrized
%scaling approach for 3-5 semiconductors. Off diagonal elements
%scale 1/(bond length)^2. This is a nearest neighbor tight binding
%method.
function band_total_dos_Si
global H zim NB
global e0 tdos
%*********** constants ***********
NB=8;          %number of bands (hamiltonian dimension =NBxNB also)
H=zeros(NB,NB);%initialize the hamiltonian
delta=1.5e-2;
zim=complex(0.0,1.0);
%a=lattice constant
%Initialize energies, compound: c=cation, a=anion, es=s-energy, ep=p-energy
%The compound is "system"
[a,esc,esa,epc,epa,ess,esp,exx,exy,system]=initialize_Si();
```

```
%x=0.25;             %energy step
%eL=-23.0; eU=5.0;  %lowest and highest energy from the band structure
disp('Return to accept the default values within brackets')
disp('Once the calculation begins, wait for the results')
eL =input('  low energy limit [-23] = ');
if(isempty(eL)), eL=-23.0; end
eU =input('  high energy limit [5] = ');
if(isempty(eU)), eU=5.0; end
x =input('  energy step [0.25] = ');
if(isempty(x)), x=0.25; end
e0=eL:x:eU;          %energy range
ntmax=length(e0);
diagHamil(esc,esa,epc,epa) %diagonal elements of H
tpa=2*pi/a;
VBZ=4*tpa^3;     %total FCC BZ volume
%FCC case tetrahedrons (three needed) in units of 2*pi/a;
%points used in units of 2*pi/a
L=[1/2,1/2,1/2]; K=[3/4,3/4,0]; U=[1,1/4,1/4]; W=[1,1/2,0]; X=[1,0,0];
%1st tetrahedron (Vectors in units of 2*pi/a)
A(1,:)=L; B(1,:)=K; C(1,:)=W;  %L, K, W points
%tetrahedron volume = 1/32 of the total BZ vol, so use in corresp. integral
VFCC_t1=abs(dot(A(1,:),cross(B(1,:),C(1,:)))/6);
factor(1)=VFCC_t1;
%2nd tetrahedron (Vectors in units of 2*pi/a)
A(2,:)=L; B(2,:)=U; C(2,:)=W;  %L, U, W points
%tetrahedron volume = 1/32 of the total BZ vol, so use in corresp. integral
VFCC_t2=abs(dot(A(2,:),cross(B(2,:),C(2,:)))/6);
factor(2)=VFCC_t2;
%3rd tetrahedron (Vectors in units of 2*pi/a)
A(3,:)=X; B(3,:)=U; C(3,:)=W;  %X, U, W points
%tetrahedron volume = 1/48 of the total BZ vol, so use in corresp. integral
VFCC_t3=abs(dot(A(3,:),cross(B(3,:),C(3,:)))/6);
factor(3)=VFCC_t3;
lim=21; lom=lim-1; %lim must be odd
%number of divisions along q2, and q3 => total number of TT's is n^2
n=10;
%VFCC_t=(VFCC_t1+VFCC_t2+VFCC_t3) %total volume=sum of 3 tetrahedrons
%VFCC_t is the standard tetrahedron volume part, and there are 12 VFCC_t's
%in the face center cube's total BZ (1/32+1/32+1/48=1/12)
factor_t=12*VBZ;           %to be used in the full integral result
%TT=thin tetrahedron, T=tetrahedron
DD=1/n^2;   %ratio of TT area to T area (since all thin triangles are the
   same)
factor_t=3.0*DD*factor_t; %factor is modified further by 3*DD
par1=0.0;
par2=1.0;
dk=(par2-par1)/lom;
al=par1:dk:par2;   %alpha range to integrate (main TT axis)
dv=1/n;            %TT division size
be=0:dv:1;         %beta
ga=be;             %gamma
```

```
%Tetrahedron q vectors according to the method of An-Ban Chen & B. I. Reser.
%Notation: q(tetrahedron(1,2,3),coordinate(x,y,z),vector(1,2,3))
q(:,:,1)=A*tpa; q(:,:,2)=(B-A)*tpa; q(:,:,3)=(C-B)*tpa;
%The function TTareas produces the average vectors "va" on the faces of
   the TT's
%Notation: va(ith TT vector,coordinate(x,y,z),tetrahedron(1,2,3))
%Ntt=number of TT's, areas(tetrahedron(1,2,3),ith TT area)
for tet=1:3      %the fcc has three tetrahedrons
  [Ntt(tet),areas(tet,:),va(:,:,tet)]=TTareas(q(tet,:,2),q(tet,:,3),
     n,be,ga);
end
%Initialize arrays
dos=zeros(NB,ntmax);
dos_tet=zeros(NB,3);
top=zeros(1,lim);
deno=zeros(NB,lim);
gy=zeros(1,NB);
tdos=zeros(1,ntmax);
%Main Loop
fprintf('system: %s\n',system)
for nt=1:ntmax
  for nB=1:NB               %initialize each band dos at each energy
    dos(nB,nt)=0.0;
  end
  for tet=1:3               %the fcc has three tetrahedrons
    for nB=1:NB          %initialize each tetrahedron dos for each band
      dos_tet(nB,tet)=0.0;
    end
    for it=1:Ntt(tet)  %Ntt TT's for the tet tetrahedron
      for ko=1:lim   %loop over gamma
        for v=1:3  %build the k vector, kx=k(1), ky=k(2), kz=k(3)
          %The va vectors are average vectors on the TT faces
          k(v)=al(ko)*(q(tet,v,1)+va(it,v,tet)); %va vectors used here
        end
        top(ko)=al(ko)*al(ko);
        %Tight binding interations - nearest neighbors
        offDiagHamil(k(1),k(2),k(3),a,ess,esp,exx,exy) %off-diagonal
            els of H
        [z,w]=eig(H); %z=eigenvectors, w=eigenvalues
        %ww contains the sorted eigenvalues and zz  the sorted eigenvectors
        [ww,zz]=sorter(diag(w),z);
        deno(:,ko)=e0(nt)-ww(:)+zim*delta; %w-Ek for the NB energy bands
      end
      for nB=1:NB
        gy(nB)=singInt(top,deno(nB,:),dk);     %integrate each band over k
        dos_tet(nB,tet)=-imag(gy(nB))/pi+dos_tet(nB,tet);%single TT dos,
            TT's
      end
    end
    %for each band weigh the corresponding tetrahedron contribution by its
        factor
```

```
    for nB=1:NB
      dos(nB,nt)=factor(tet)*dos_tet(nB,tet)+dos(nB,nt);  %dos=sum over
         tetraheds
    end
  end
  %finally, for each band multiply by the three-tetrahedon contribution
     factor
  for nB=1:NB
    dos(nB,nt)=factor_t*dos(nB,nt);
    %fprintf('E0=%9.4f, dos=%9.4f\n',e0(nt),dos(nB,nt)
  end
end
%
%total dos sum over all the bands
fprintf('Density of States for the system: %s\n',system)
for nt=1:ntmax
  tdos(nt)=0.0;
  for nB=1:NB
    tdos(nt)=tdos(nt)+dos(nB,nt);
  end
  %Total dos=integral of d^3k over the BZ / BZ volume
  tdos(nt)=tdos(nt)/VBZ;
  fprintf('E0=%9.4f, tdos=%14.6e\n',e0(nt),tdos(nt));
end
%
%For the Harrison model, we know the band edges, and gap
Ev1=(epc+epa)/2-sqrt(((epc-epa)/2)^2+(4*exx)^2); %Top of VB
Ev2=(epc+epa)/2+sqrt(((epc-epa)/2)^2+(4*exx)^2); %Top of VB
Ec1=(esc+esa)/2+sqrt(((esc-esa)/2)^2+(4*ess)^2); %Bottom of CB
Ec2=(esc+esa)/2-sqrt(((esc-esa)/2)^2+(4*ess)^2); %Bottom of CB
disp('Energies in eV ')
fprintf('Ev1,Ev2,Ec1,Ec2=%8.4f %8.4f %8.4f %8.4f\n',Ev1,Ev2,Ec1,Ec2)
Ev=max(Ev1,Ec2);  %highest of these two roots
Ec=min(Ec1,Ev2);  %lowest of these two roots
EgH=Ec-Ev; %gap - based on Harrison's formulas at the gamma (k=0) point
fprintf('Ev=%9.4f, Ec=%9.4f, Eg=%9.4f\n',Ev,Ec,EgH);
indEv=1+floor((Ev-eL)/x);
indEc=1+floor((Ec-eL)/x);
strv=cat(2,'Val. band edge: Ev=',num2str(Ev,'%9.4f'),' eV');
strc=cat(2,'Cond. band edge: Ec=',num2str(Ec,'%9.4f'),' eV');
strg=cat(2,'Eg=',num2str(EgH,'%9.4f'),' eV');
mxd=max(tdos);
%
%plot the total density of states versus energy
figure(1)
p0(1)=plot(e0,tdos,'k-');
hold on

%
% %Back to figure 1 and add the new information
% figure(1)
```

```
%place a line at Ev and Ec
p0(2)=line([Ev Ev],[0 mxd*0.1],'LineStyle','--','color','k','LineWidth',2);
p0(3)=line([Ec Ec],[0 mxd*0.1],'LineStyle','-.','color','k','LineWidth',2);
pL=legend(p0,'DOS',strv,strc,0);
set(pL,'FontSize',9);
text(Ev-abs(Ev)*0.1,mxd*0.5,strg,'FontSize',9)
xlabel('E(eV)'), ylabel('DOS (eV^{-1})')
str=cat(2,'Density of States for the ',system,' system');
title(str)
hold off
pause(1)
%
%Total integrated density of states, and plot
fprintf('Integrated Density of States for the system: %s\n',system)
intdos=zeros(1,ntmax);
for nt=1:ntmax
  intdos(nt)=rombergInt(eL,e0(nt),@fForRomb);  %integrate on [eL,e0]
  fprintf('E0=%9.4f, integrated dos=%14.6e\n',e0(nt),intdos(nt));
end
%
figure(2)
hold on
p1(1)=plot(e0,intdos,'k-');
%place a line at Ev and Ec of respective heights tdos(indEv) & tdos(indEc)
p1(2)=line([Ev Ev],[0 intdos(indEv)],'LineStyle','--','color','k',
    'LineWidth',2);
p1(3)=line([Ec Ec],[0 intdos(indEc)],'LineStyle','-.','color','k',
    'LineWidth',2);
pL=legend(p1,'integated DOS',strv,strc,0);
set(pL,'FontSize',9);
text(Ec+0.1*abs(Ec),intdos(indEc)*(1-0.3),strg,'FontSize',9)
xlabel('E (eV)'), ylabel('Integrated DOS (number of states)')
str=cat(2,'Integrated Density of States for the ',system,' system');
title(str)
hold off

function y=fForRomb(p)
%function used by romberg integration and which interpolates
%tdos versus e0
global e0 tdos
y=interpFunc(p,e0,tdos);
```

## 8.8 Chapter 8 Exercises

8.8.1. Show that the states $|\vec{k}>$ of Equation 8.2.4 are normalized.

8.8.2. Using the tight binding Hamiltonian described in Equation 8.2.6a, follow the procedure described in obtaining Equation 8.2.6b and show that $<\vec{k}|H_{OD}|\vec{k}>= -\gamma \sum_{m=nn} e^{-i\vec{k}\cdot\vec{\rho}_m}$, so that $\varepsilon_k$ of Equation 8.2.6c results.

8.8.3. Show all the missing details in obtaining the energy bands of the (a) BCC, Equation 8.2.8b, and (b) that of the FCC, Equation 8.2.9.

8.8.4. Run the script TBbandSC.m of Section 8.2.1 to reproduce Figure 8.2.1 for the SC energy band. Modify the script in order to reproduce Figure 8.2.2 for the BCC energy band.

8.8.5. Run the script TBbandSC.m of Section 8.2.1 to reproduce Figure 8.2.1 for the SC energy band. Modify the script in order to reproduce Figure 8.2.3 for the FCC energy band.

8.8.6. Run the script TBfermiIsoSurfBzSc.m of Section 8.3.1 to reproduce Figure 8.3.4 for the SC Fermi surface. Modify the script in order to reproduce Figure 8.3.5 for the BCC Fermi surface for a half-filled band as discussed in the text. For the Brillouin zone details, refer to Chapter 2.

8.8.7. Run the script TBfermiIsoSurfBzSc.m of Section 8.3.1 to reproduce Figure 8.3.4 for the SC Fermi surface. Modify the script in order to reproduce Figure 8.3.6 for the FCC Fermi surface for a half-filled band as discussed in the text. For the Brillouin zone details, refer to Chapter 2.

8.8.8. In one dimension, the $k$-space variable takes on $N$ values; i.e., $k = 2n\pi/(NL)$, for integer $n = 0 \cdots N-1$, and $L$ is the unit cell dimension. Show that $\frac{1}{N}\sum_k f(k) = \frac{L}{2\pi}\int_0^{2\pi/L} f(k)dk$, where periodic boundary conditions for $f(k)$ are used. What do you get if this result is extended to three dimensions?

8.8.9. By assuming that $\varepsilon$ has a small positive imaginary part in Equation 8.4.17b, show that the result is Equation 8.4.17c.

8.8.10. Show that when the Green function operator of Equation 8.4.17c acts on the single band state $|\vec{k}>$ of Equation 8.2.4, the result is the $\vec{k}$-space single band Green's function of Equation 8.4.18b.

8.8.11. Prove the identity of Equation 8.4.19.

8.8.12. Referring to Section 8.5, (a) use Equation 8.5.22 to show that the $|\vec{k}>$ states are normalized; that is, $<\vec{k}|\vec{k}>= 1$. (b) By using the orthogonality of the states $|j)$, show the consistency of the expression for the completeness condition, Equation 8.5.23.

8.8.13. Read Example 8.5.0.1 and after working through the steps in obtaining Equation 8.5.31 for $\varepsilon < 0$, repeat the process and obtain the results shown in Equation 8.5.32 for $\varepsilon > 0$.

8.8.14. After reading Examples 8.5.0.1 and 8.5.0.2, run the codes Green_1D.m and Green_1DdosCalc.m (which uses singInt.m of 8.8) to reproduce Figures 8.5.7 and 8.5.8.

8.8.15. After reading Example 8.5.0.3, run the code Green_1Totdos.m for the total density of states of the one-dimensional tight binding band and reproduce Figure 8.5.9.

8.8.16. After reading Section 8.6.1, run the script sc_dos.m and reproduce Figure 8.6.10.

8.8.17. Read Example 8.7.1.1 and write suitable code that incorporates Equations 8.7.48 and 8.7.52 as well as parameters from Table 8.7.2 in order to reproduce the quoted results for Si.

8.8.18. (a) Run the script band_structure_Si.m associated with silicon's band structure; that is, Figure 8.7.15(a) and reproduce it. (b) Modify the script and obtain the band structure for the GaAs system. What is the GaAs band gap obtained by the model and how does it compare with the actual one? Comment on the results.

8.8.19. (a) Run the script band_total_dos_Si.tex associated with silicon's density of states; that is, Figures 8.7.16(a), 8.7.17 to reproduce them. (b) Modify the script and obtain the corresponding plots for the GaAs system. Comment on the results.

# *Appendix A: Singular Function Integration Using SingInt.m*

In this appendix, we detailed the scheme used in order to carry out the numerical integration of functions that contain singularities (see Example 8.5.0.2). Such is the case that arises in the integrations of Green's functions. Roth's approach [32] for obtaining numerical values of nearly singular integrals uses a method similar to the Simpson's integration rule for the integral $\int_a^b f(x)dx$, except that, rather than a single function, the integral of interest involves two functions; i.e., a numerator and a denominator, each approximated by quadratic functions. The idea is that after identifying the numerator and the denominator of the integrand, one separates them and treats them as detailed below in order to obtain a numerical value for the integral of interest. When the integration region is divided into $N$ subintervals, we have

$$I = \int_a^b \frac{f(x)}{g(x)} dx = \sum_{n=1,3,5,\ldots}^{N-2} I_n \approx \sum_{n=1,3,5,\ldots}^{N-2} S_n. \tag{8.1}$$

where $I_n$ is the value of the integral in the subinterval $[x_n, x_{n+2}]$ with $x_{n+2} = x_n + 2h$, $h = (b - a)/(N-1)$, and $S_n$ the numerical approximation. An integrand may become singular due to the denominator $g(x)$ passing through a zero. Thus, the idea is to approximate $I_n$ as the ratio of two quadratic functions in $x$, written in a normalized fashion, as

$$I_n \approx S_n \equiv h \int_{-h}^{h} \frac{d_n x^2 + e_n x + f_{n+1}}{a_n x^2 + b_n x + g_{n+1}} dx; \tag{8.2}$$

that is, on each subinterval, both functions, the numerator $f(x)$ and the denominator $g(x)$, have been approximated by respective quadratic functions, where $f_n = f(x_n)$ and similarly $g_n = g(x_n)$. In the above expression
$a_n = (g_{n+2} + g_n - 2g_{n+1})/2$, $b_n = (g_{n+2} - g_n)/2$,
and
$d_n = (f_{n+2} + f_n - 2f_{n+1})/2$, $e_n = (f_{n+2} - f_n)/2$.
If for any reason an integrand were to become singular, by separating it into a ratio of two functions, the denominator would pass through a small value and it would be natural that the analytic result of Equation 8.2 would approximate the integral closely. Performing the integration of the ratio of the quadratics analytically, the result for each $n$ subinterval is

$$S_n = \frac{h}{3}\left[\frac{3}{2} f_n (I_{3n} - I_{2n}) + \frac{3}{2} f_{n+2}(I_{3n} + I_{2n}) + 3 f_{n+1}(I_{1n} - I_{3n})\right], \tag{8.3}$$

where
$I_{1n} = \frac{1}{a_n(x_{n+} - x_{n-})} \ln\left\{\frac{(1-x_{n+})(1+x_{n-})}{(1-x_{n-})(1+x_{n+})}\right\}$,
$I_{2n} = \frac{1}{2a_n} \ln\left\{\frac{a_n + b_n + g_n}{a_n - b_n + g_n}\right\} - \frac{b_n}{2a_n} I_{1n}$,
and

$I_{3n} = \frac{2}{a_n} - \frac{b_n}{a_n} I_{2n} - \frac{g_n}{a_n} I_{1n}$,
with
$x_{n\pm} \equiv \frac{-b_n \pm \sqrt{b_n^2 - 4a_n g_n}}{2a_n}$.
This completes the numerical approximation to the integral involving an integrand that is expected to contain singularities; i.e., Equation 8.1. In the event that the denominator is not expected to pass through a zero, the function $g(x)$ is replaced by unity and the above method is equivalent to Simpson's rule. Below is a listing of the code singInt.m that incorporates the above Roth's algorithm. Example 8.5.0.2 illustrates the use of this method in carrying out the numerical calculation analog of the Green function and density of states of Example 8.5.0.1.

```
%copyright by J. E Hasbun and T. Datta
%singInt.m
function y=singInt(cnum,cden,h)
%Estimates the integral of f(x)/g(x) dx on [-h,h]
%where h=(b-a)/(n-1) and n=number of function evaluations.
%The program sums all the contributions on interval [a,b]
%This is Roth's method of integrating singular functions
%[L. M. Roth Phys. Rev. B Vol. 7, p4321-4337 (1973)]
%as implemented by J. E. Hasbun.
%
n=length(cnum); %cden must be the same length as well
y=complex(0.0,0.0);
for k=1:2:n-2
    f0=cnum(k); f1=cnum(k+1); f2=cnum(k+2);
    g0=cden(k); g1=cden(k+1); g2=cden(k+2);
    a=abs(g0+g2-2.0*g1);
    %The amount added to g1 in the next line is slightly adjustable
    if(a < 1.e-7), g1=g1+1i*1.e-7; end
    a=(g0+g2-2.0*g1)/2.0;
    b=(g2-g0)/2.0;
    c=g1;
    q=sqrt(b*b-4.0*a*c);
    xp=(-b+q)/a/2.0;
    xm=(-b-q)/a/2.0;
    xpmxm=xp-xm;
    i1=log((1-xp)*(1+xm)/(1-xm)/(1+xp))/xpmxm/a;
    i2=(log(g2/g0)-b*i1)/a/2.0; %note: a+b+c=g2, a-b+c=g0
    i3=(2.0-b*i2-c*i1)/a;
    y=y+3.0*(f0*(i3-i2)+f2*(i3+i2))/2.0+3.0*f1*(i1-i3);
end
y=y*h/3.0;
```

## Supporting Functions for Semiconductor Band Structures

The program initialize_Si.m discussed in Section 8.7.1 in relation to Figure 8.7.15 makes use of the functions initialize_Si.m, which provides the system's parameters; diagHamil.m, for setting the diagonal energies; offDiagHamil.m for setting the off diagonal matrix elements; and sorter.m, which sorts the eigenvalues and eigenvectors. Their listings follow.

```
%copyright by J. E Hasbun and T. Datta
%initialize_Si.m
function [a,esc,esa,epc,epa,ess,esp,exx,exy,system]=initialize_Si()
system='Si';
%The off diagonal coefficients
esssig=-1.40;
espsig=1.84;
eppsig=3.24;
epppi=-0.81;
h=6.626075e-34;   %Planck's constant in J-sec
hbar=h/2/pi;
me=9.1093897e-31; %electron mass (kg)
e=1.60217733e-19; %electron charge (C). Also recall 1 Joule=(1/e)eV
const=(hbar^2/me)*(1/e)*(1e10)^2; %hbar^2/me in eV-Angtrom^2=7.62eV-Angs^2
%============== semiconductor parameters ===========================
%Paramaters from Harrison's Electronic Structures and the Properties of solids
%Elements: cation=anion (eV), bond length (angstroms) inputs
%c=cation, a=anion, es=s-energy, ep=p-energy
esa=-13.55; epa=-6.52; esc=esa; epc=epa; d=2.35; %Silicon
a=4*d/sqrt(3);     %Lattice constant
%zinc-blende ion near neighbor positions are located at
%{[1,1,1]a/4, [1,-1,-1]a/4, [-1,1,-1]a/4, [-1,-1,1]a/4, so let scale=a/4
scale=a/4;          %near neighbor position vector magnitudes
ro=scale*sqrt(3.); %internuclear or bond length=sqrt(1^2+1^2+1^2)*a/4
r1=ro^2;
vsssig=const*esssig/r1;
vspsig=const*espsig/r1;
vppsig=const*eppsig/r1;
vpppi=const*epppi/r1;
ess=vsssig;
esp=vspsig/sqrt(3);
exx=vppsig/3.+(2./3.)*vpppi;
exy=(vppsig-vpppi)/3.;
```

```
%copyright by J. E Hasbun and T. Datta
%diagHamil.m
function diagHamil(esc,esa,epc,epa)
%c=cation, a=anion, es=s-energy, ep=p-energy
%builds the diagonal part of the 8x8 Harrison hamiltonian
global H
H(1,1)=esc;
H(2,2)=esa;
H(3,3)=epc;
H(4,4)=epc;
H(5,5)=epc;
H(6,6)=epa;
H(7,7)=epa;
H(8,8)=epa;
```

```
%copyright by J. E Hasbun and T. Datta
%offDiagHamil.m
function offDiagHamil(bx,by,bz,dasb,ess,esp,exx,exy)
%c=cation, a=anion, es=s-energy, ep=p-energy
%builds the off-diagonal part of the 8x8 Harrison hamiltonian
%bx,by,bz are the wavevector directions
global H zim NB
do=dasb/4.;   %scale for anion positions is lattice contant/4
cx=cos(bx*do);
cy=cos(by*do);
cz=cos(bz*do);
sx=sin(bx*do);
sy=sin(by*do);
sz=sin(bz*do);
go=4.*(cx*cy*cz-zim*(sx*sy*sz));
g1=4.*(-cx*sy*sz+zim*(sx*cy*cz));
g2=4.*(-sx*cy*sz+zim*(cx*sy*cz));
g3=4.*(-sx*sy*cz+zim*(cx*cy*sz));
H(1,2)=ess*go;
H(1,6)=esp*g1;
H(1,7)=esp*g2;
H(1,8)=esp*g3;
H(2,3)=-esp*conj(g1);
H(2,4)=-esp*conj(g2);
H(2,5)=-esp*conj(g3);
H(3,6)=exx*go;
H(3,7)=exy*g3;
H(3,8)=exy*g2;
H(4,6)=exy*g3;
H(4,7)=exx*go;
H(4,8)=exy*g1;
H(5,6)=exy*g2;
H(5,7)=exy*g1;
H(5,8)=exx*go;
for i=1:NB-1
  for j=i+1:NB
    H(j,i)=conj(H(i,j)); %hermitian matrix
  end
end
```

```
%copyright by J. E Hasbun and T. Datta
%sorter.m
function [ww,zz]=sorter(ww,zz)
%sorts eigenvalues and eigenvectors
%On input ww=unsorted eigenvalues, zz=unsorted eigenvectors
%On output ww=sorted eigenvalues, zz=sorted eigenvectors
NB=length(ww);
for io=1:NB
```

```
  for jo=io:NB-1
    if  ww(jo+1) < ww(io)
      wo=ww(jo+1);
      ww(jo+1)=ww(io);
      ww(io)=wo;
      z0=zz(:,jo+1);
      zz(:,jo+1)=zz(:,io);
      zz(:,io)=z0;
    end
  end
end
```

# Appendix B: Romberg Integration (rombergInt.m) and the Interpolating Function (interpFunc.m)

The code for the Romberg integration scheme, rombergInt.m, as well as the interpolating function, interpFunc.m (see Example 8.5.0.3) are listed below. To learn more details about Romberg integration as well as interpolation methods see DeVries and Hasbun [17], where much of the code concepts were obtained from.

---

```
%copyright by J. E Hasbun and T. Datta
%rombergInt.m
function [result]=rombergInt(a,b,funcInt)
T=zeros(15,5);              %Initial T
error=9999;                 %start with a large error
N = 1;
h  = (b-a)/N;
m = 1;
T(m,1) = 0.5*h*(funcInt(a)+funcInt(b));
%fprintf('m=%2.0f, T=%15.10f\n',m,T(m,1))
while ( (m < 15 & abs(error) > 1.e-8) | (m <= 3)) %require a min # of steps
    m = m + 1;
    N = 2*N;
    h  =(b-a)/N;
    extra = 0.0;
    for i=1:2:N-1
        extra = extra + funcInt(a+i*h);
    end
    T(m,1) = 0.5*T(m-1,1)+h*extra;
    T(m,2) = (4.0*T(m,1) - T(m-1,1) )/ 3.0;
    if(m>=3), T(m,3) = (16.0*T(m,2) - T(m-1,2)) / 15.0; end
    if(m>=4), T(m,4) = (64.0*T(m,3) - T(m-1,3)) / 63.0; end
    if(m>=5), T(m,5) =(256.0*T(m,4) - T(m-1,4)) /255.0; end
    %fprintf('m=%2.0f, T=%15.10f,%15.10f,%15.10f,%15.10f,%15.10f\n',...
    %     m,T(m,1),T(m,2),T(m,3),T(m,4),T(m,5))
    %Use error to check convergence...
    %add a small part ot the denominator to avoid division by zero
    error =(T(m,min(m,5))-T(m,min(m,5)-1))/sqrt(T(m,min(m,5))^2+1.e-12^2);
end
result = T(m,min(m,5));
```

---

```
%copyright by J. E Hasbun and T. Datta
%interpFunc.m
```

```
function y=interpFunc(rho,x,fx)
%Interpolates the function fx using Lagrange interpolation
%at point rho given the function values fx at x in array form
%search the index of x where rho lies
no=3;     %interpolator type, 1=linear, 2=quadratic, 3=cubic, etc.
xlen=length(x);
xmin=min(x); xmax=max(x);
dx=(xmax-xmin)/xlen;
i0=1+floor((rho-xmin)/dx);
if(i0 >= xlen-no),i0=xlen-no; end
y = 0.;
for j = i0:i0+no
  %Evaluate the j-th coefficient
  Lj = 1.0;
  for k =i0:i0+no
    if(j ~= k)
      Lj = Lj * (rho-x(k) )/( x(j)-x(k) );
    end
  end
  %Add contribution of j-th term to the polynomial
  y = y + Lj * fx(j);
end
```

# Appendix C: The Ray Method For K-Space Density of States Integration (TTareas.m)

The ray integration method [33], discussed in Section 8.6, for performing $\vec{k}$-space integrations over an irreducible tetrahedron (T) is based on subdividing it into thin tetrahedrons (TT) as shown in the figure below.

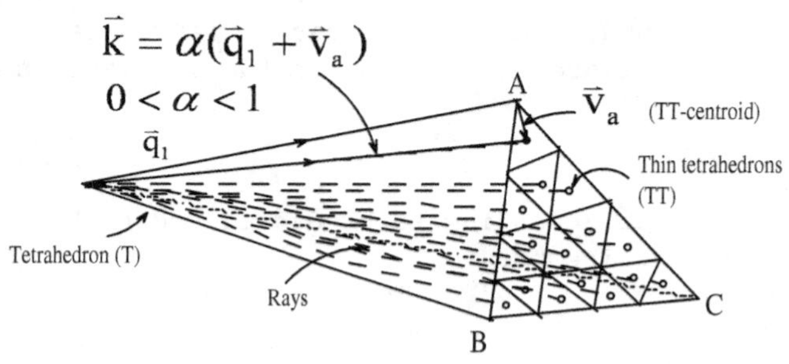


This figure shows how the larger tetrahedron (T) is subdivided into the thin tetrahedrons (TT) shown. Each ray is drawn from the origin to the center of a TT face; that is, its centroid.

In this manner, the integral $J(E) = \int f_{\vec{k}} d\vec{k}$ can approximately be carried out over each TT and then summing over all the TTs to get the total value of $J(E)$. We thus have the integral for the entire (T) as

$$J(E) = \int_{V_T} f_{\vec{k}} d\vec{k} \approx \sum_i I_i, \tag{8.1a}$$

where $I_i$ is the integral over each TT, given by

$$I_i = 3V_T \delta_i \int_0^1 d\alpha\, \alpha^2 f[\alpha(\vec{q}_1 + \vec{v}_{ai})]. \tag{8.1b}$$

In these expressions $V_T$ is the T volume, $\delta_i = \Delta s_i / S$, where $\Delta s_i$ is the TT face area, and $S$ is the T face area. The vectors $\vec{q}_1$ and $\vec{v}_a$ are shown in the figure. The function TTareas.m (listed below) produces the average vectors $\vec{v}_a$ on the faces of the TT's associated with the ray method of integration and which are used to carry out the main integral in the calling program.

```
%copyright by J. E Hasbun and T. Datta
%TTareas.m
function [Ntt,areas,va]=TTareas(q2,q3,n,be,ga)
%Trangles based on the ray method of An-Ban Chen, Phys Rev. B V16, 3291
    (1977)
%Divides a large triangular area into smaller triangles and finds their
%twodimensional vector directions and the smaller triangles areas
```

```
%finds the areas of the thin tetrahedron triangles according to how it's
%described in the Chen paper
%Here is an example of how the vertices 1,2,3,4,5,6 of a triangle are
%constructed. Example, we divide a large triangle in 2^2 smaller triangles
%1=(beta1*q2,gamma1*q3): (1,1)          /1\
%2=(beta2*q2,gamma1*q3): (2,1)         /   \
%3=(beta3*q2,gamma1*q3): (3,1)        /2---4\
%4=(beta2*q2,gamma2*q3): (2,2)       / \ * / \
%5=(beta2*q2,gamma2*q3): (2,3)      /   \ /   \
%6=(beta3*q2,gamma3*q3): (3,3)     3-----5-----6
%On the first loop pass, v1=1, v2=2, v3=4. On the next pass, v1=2
%v2=3, v3=5, v4=4 (extra triangle). On the last pass, we only have
%one triangle v1=4, v2=5, and v3=6.
%The triangle with the * is referred here to as an extra triangle.
%Once the vertices are found, the areas of each triangle can be found
%The numbers in parenthesis are the indices of the beta and gamma
%coefficients
%Inputs
%3D vectors q2, q3,
%arrays beta and gamma made according to be=0:dv:1 and ga=be
%number of subtriangles desired n.
%Output
%Nt: number of triangles
%areas: their areas
%va: average vector positions of these triangles based on the
%    gamma and beta as in Chen's paper
Ntt=0;
for io=1:n              %gamma loop
  for jo=io:n           %beta loop
    Ntt=Ntt+1;
    v1=be(jo)*q2+ga(io)*q3;       %small triangle vertices 1,2,3
    v2=be(jo+1)*q2+ga(io)*q3;
    v3=be(jo+1)*q2+ga(io+1)*q3;
    va(Ntt,:)=(v1+v2+v3)/3;       %average vertix vector (triangle centroid)
    vv1=v2-v1; vv2=v3-v2;         %small triangle vectors
    areas(Ntt)=norm(cross(vv1,vv2)/2); %area of little triangle
    if(jo > io)                   %extra triangle, with indices of v1, v3, v4
      %[jo,io]
      %[jo+1,io+1]
      %[jo,io+1]
      Ntt=Ntt+1;
      v4=be(jo)*q2+ga(io+1)*q3;
      va(Ntt,:)=(v1+v3+v4)/3;     %average vertix vector (triangle centroid)
      vv1=v3-v1; vv2=v4-v3;       %its associated vectors
      areas(Ntt)=norm(cross(vv1,vv2))/2; %area of little triangle
    end
  end
end
```

# *Appendix D: Supporting Functions for Semiconductor Band Structures*

The program initialize_Si.m discussed in Section 8.7.1 in relation to Figure 8.7.15 makes use of the functions initialize_Si.m, which provides the system's parameters; diagHamil.m, for setting the diagonal energies; offDiagHamil.m for setting the off diagonal matrix elements; and sorter.m, which sorts the eigenvalues and eigenvectors. Their listings follow.

---

```
%copyright by J. E Hasbun and T. Datta
%initialize_Si.m
function [a,esc,esa,epc,epa,ess,esp,exx,exy,system]=initialize_Si()
system='Si';
%The off diagonal coefficients
esssig=-1.40;
espsig=1.84;
eppsig=3.24;
epppi=-0.81;
h=6.626075e-34;   %Planck's constant in J-sec
hbar=h/2/pi;
me=9.1093897e-31; %electron mass (kg)
e=1.60217733e-19; %electron charge (C). Also recall 1 Joule=(1/e)eV
const=(hbar^2/me)*(1/e)*(1e10)^2; %hbar^2/me in eV-Angtrom^2=7.62eV-Angs^2
%=============== semiconductor parameters ============================
%Paramaters from Harrison's Electronic Structures and the Properties of solids
%Elements: cation=anion (eV), bond length (angstroms) inputs
%c=cation, a=anion, es=s-energy, ep=p-energy
esa=-13.55; epa=-6.52; esc=esa; epc=epa; d=2.35; %Silicon
a=4*d/sqrt(3);     %Lattice constant
%zinc-blende ion near neighbor positions are located at
%{[1,1,1]a/4, [1,-1,-1]a/4, [-1,1,-1]a/4, [-1,-1,1]a/4, so let scale=a/4
scale=a/4;         %near neighbor position vector magnitudes
ro=scale*sqrt(3.); %internuclear or bond length=sqrt(1^2+1^2+1^2)*a/4
r1=ro^2;
vsssig=const*esssig/r1;
vspsig=const*espsig/r1;
vppsig=const*eppsig/r1;
vpppi=const*epppi/r1;
ess=vsssig;
esp=vspsig/sqrt(3);
exx=vppsig/3.+(2./3.)*vpppi;
exy=(vppsig-vpppi)/3.;
```

---

```
%copyright by J. E Hasbun and T. Datta
%diagHamil.m
function diagHamil(esc,esa,epc,epa)
%c=cation, a=anion, es=s-energy, ep=p-energy
%builds the diagonal part of the 8x8 Harrison hamiltonian
global H
H(1,1)=esc;
H(2,2)=esa;
H(3,3)=epc;
H(4,4)=epc;
H(5,5)=epc;
H(6,6)=epa;
H(7,7)=epa;
H(8,8)=epa;
```

---

```
%copyright by J. E Hasbun and T. Datta
%offDiagHamil.m
function offDiagHamil(bx,by,bz,dasb,ess,esp,exx,exy)
%c=cation, a=anion, es=s-energy, ep=p-energy
%builds the off-diagonal part of the 8x8 Harrison hamiltonian
%bx,by,bz are the wavevector directions
global H zim NB
do=dasb/4.;   %scale for anion positions is lattice contant/4
cx=cos(bx*do);
cy=cos(by*do);
cz=cos(bz*do);
sx=sin(bx*do);
sy=sin(by*do);
sz=sin(bz*do);
go=4.*(cx*cy*cz-zim*(sx*sy*sz));
g1=4.*(-cx*sy*sz+zim*(sx*cy*cz));
g2=4.*(-sx*cy*sz+zim*(cx*sy*cz));
g3=4.*(-sx*sy*cz+zim*(cx*cy*sz));
H(1,2)=ess*go;
H(1,6)=esp*g1;
H(1,7)=esp*g2;
H(1,8)=esp*g3;
H(2,3)=-esp*conj(g1);
H(2,4)=-esp*conj(g2);
H(2,5)=-esp*conj(g3);
H(3,6)=exx*go;
H(3,7)=exy*g3;
H(3,8)=exy*g2;
H(4,6)=exy*g3;
H(4,7)=exx*go;
H(4,8)=exy*g1;
H(5,6)=exy*g2;
H(5,7)=exy*g1;
H(5,8)=exx*go;
for i=1:NB-1
```

```
  for j=i+1:NB
    H(j,i)=conj(H(i,j)); %hermitian matrix
  end
end
```

---

```
%copyright by J. E Hasbun and T. Datta
%sorter.m
function [ww,zz]=sorter(ww,zz)
%sorts eigenvalues and eigenvectors
%On input ww=unsorted eigenvalues, zz=unsorted eigenvectors
%On output ww=sorted eigenvalues, zz=sorted eigenvectors
NB=length(ww);
for io=1:NB
  for jo=io:NB-1
    if  ww(jo+1) < ww(io)
      wo=ww(jo+1);
      ww(jo+1)=ww(io);
      ww(io)=wo;
      z0=zz(:,jo+1);
      zz(:,jo+1)=zz(:,io);
      zz(:,io)=z0;
    end
  end
end
```

# 9

## *Impurities and Disordered Systems*

**Contents**



### 9.1 Introduction

From what we have seen in the previous chapters, studies of solid state crystalline materials have been made possible by their inherent periodicity. Indeed, Bloch's theorem has enabled us to carry out calculations on infinite or bulk systems that would otherwise be an impossibility. However, pure systems are ideal systems and, unless there exist machines that are capable of producing 100% pure crystals of any size, we need to understand real systems. Real systems often contain atoms that are different from the host atomic species. Depending on the concentration of the foreign species (say less than about 1 for every $100,000$ host atoms) we refer to them as impurities; for large concentrations of foreign atoms, a system becomes so heavily doped that they could be considered disordered systems. In some cases, that which we would call impurity becomes a major component and we refer to such a system as an alloy. Impurities play a noticeable role in a system's electrical properties, such as the conductivity, and thus are of interest in solid state physics. A similar situation exists for disordered systems. Random alloys in which at least two atomic species take random positions in a crystal's lattice sites fall within the realm of disordered systems.
Real crystals contain imperfections or defects associated with a crystal's geometry and affect its electrical properties and mechanical strength. Defects can be of several types. One type is the kind which is localized within a crystal region of atomic dimensions such as point defects. An impurity atom is of this kind. Impurity atoms can be substitutional or interstitial. A substitutional impurity replaces a host atom at the normal lattice site, while an impurity at an interstitial position occupies the space between normal lattice site positions. Both situations are depicted in Figure 9.1.1. The vicinity around the substitutional or the interstitial impurities is accompanied by a lattice distortion.

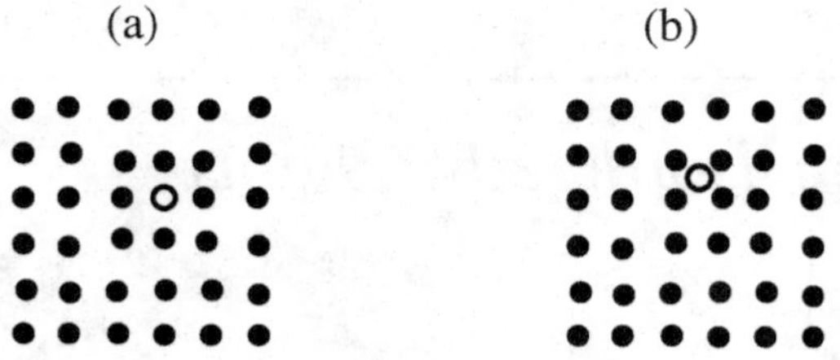

Figure 9.1.1: (a) A crystal of atoms (filled circles) with a substitutional impurity (open circle). (b) The crystal as in (a) with an interstitial impurity (open circle).

A point defect may involve the absence of a host atom due to its motion to another substitutional position leaving behind what is termed as a vacancy. This is known as a Schottky defect; however, a Frenkel defect is a vacancy that exists due to a host atom moving from a lattice site to an interstitial position. Some point defects are associated with impurities with magnetic moments that could interact with a passing electron's spin.

Another type of defect is a dislocation which is a line defect of displaced atoms. Color centers are defects that result from the irradiation of crystals with $x$-rays, $\gamma$-rays, neutrons, or electrons. Such radiation causes the crystal to change color. Stacking faults are defects due to the out-of-sequence organization of atomic layers of what would otherwise be a perfect crystal. For example, if a cubic close packed (CCP) crystal (see Chapter 1) with atomic layers ABCABC . . . stacking were to deviate from that order as in ABABCA..., it would lead to a stacking fault. Grain boundaries are another kind of defect due to the randomly oriented crystallites found at the junctions of most solids that are not single crystals.

Compound systems such as, for example, zinc sulfide are not considered disordered systems because the zinc atom is immediately surrounded by four sulfur atoms and each sulfur atom is similarly surrounded by four zinc atoms in a zinc-sulfide or zinc-blende structure (see also Chapter 1). This is, therefore, a compound crystal system and is not considered a disordered system. However, if we had a system such as $Si_{1-x}Ge_x$ in which $x$ is the concentration of the Ge atomic species, the system is referred to as a binary alloy and it is considered a disordered system. In this case, if $x = 0$, the system is a silicon crystal; however, if $x = 1$, the system is a germanium crystal. Additionally, if $x$ is other than zero or one, the Si and Ge atoms take lattice sites at random and this is the reason why we refer to $Si_{1-x}Ge_x$ as a random alloy as shown in Figure 9.1.2.

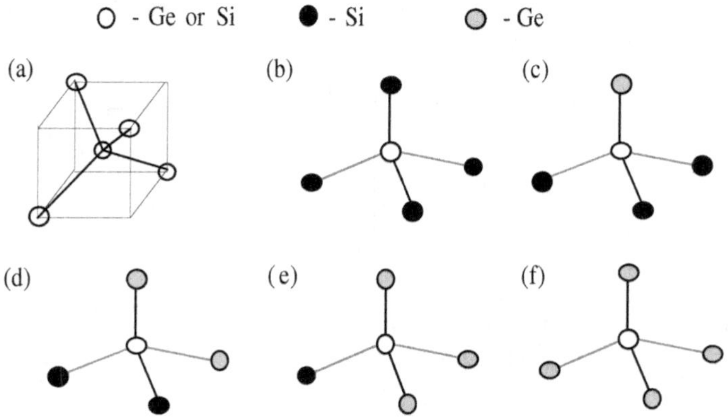


Figure 9.1.2: (a) The basic block of the zinc-blende structure (see Chapter 1) in which each atom is surrounded by four nearest neighbors in a tetrahedral geometry. (b)-(f) The five possible tetrahedral configurations (at random) that Ge and Si can take in the structure of a random $Si_{1-x}Ge_x$ alloy and where the center atom can be either Si or Ge with $x$ the concentration of Ge.

If Ge is a substitutional impurity in a Si crystal, it so happens to be an isoelectronic or compensated impurity because it belongs to the same group in the periodic table. Impurities such as donors and acceptors (discussed in Chapter 7) are uncompensated impurities. In this chapter, we consider a single isoelectronic substitutional impurity in what would otherwise be a perfect crystal and discuss the theory for obtaining the associated impurity level. We also consider a popular theory for studying the electronic properties of binary alloys. The tight binding method and Green's function approach, introduced in previous chapters, will be employed.

## 9.2 The Single Impurity Level

In this section, we study a simple theory that can be applied in order to obtain the energy level associated with a substitutional isoelectronic impurity in a crystal system. The model used is very useful in investigations of scattering. It is a classic example of what is generally known as a single impurity scattering. We start with a single band tight binding crystal Hamiltonian (considered before in Chapter 8) with a single substitutional impurity placed at site $\ell$,

$$\hat{H} = \sum_{n} |n)\varepsilon_0(n| + \sum_{nm}{}' |n)V_{nm}(m| + |\ell)\varepsilon_\mu(\ell|, \tag{9.1}$$

where $\varepsilon_\mu = \varepsilon_I - \varepsilon_0$; that is, if $\varepsilon_\mu$ is set to zero, we have the pure crystal, else a host site is replaced with the impurity at the $\ell$th site. The prime in Equation 9.1 indicates that $n \neq m$ and we assume that the off-diagonal matrix elements $V_{nm}$ remain unmodified and that the only effect of adding the impurity is to change the diagonal energy for the $\ell$th site from $\varepsilon_0$ to $\varepsilon_I$. To illustrate the idea of this concept, imagine a one-dimensional crystal modeled as quantum barriers whose heights ($\varepsilon_0$) are all the same, except for one, where the impurity is located with a height of $\varepsilon_I$. The off-diagonal or hopping term corresponds to the separation width between the barriers, which are taken to be all identical. This is what is shown in Figure 9.2.3.

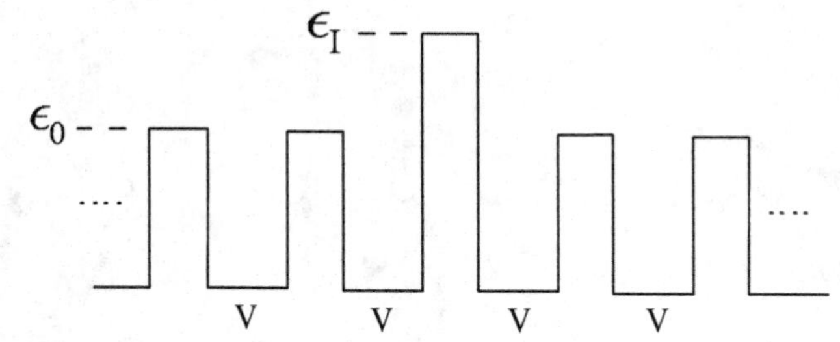


Figure 9.2.3: A single impurity in a crystal is similar to a barrier potential whose height is different from the rest of the barriers that compose the crystal.

As in the previous chapter, we will be working with a nearest neighbor model and write the total Hamiltonian of Equation 9.1 as

$$\hat{H} = \hat{H}_0 + \hat{H}_I \tag{9.2a}$$

where the unperturbed Hamiltonian is that of the crystal; that is,

$$\hat{H}_0 = \sum_n |n)\varepsilon_0(n| + \sum_{nm}{}' |n)V_{nm}(m|, \tag{9.2b}$$

while the impurity Hamiltonian is considered to be the perturbation

$$\hat{H}_I = |\ell)\varepsilon_\mu(\ell|. \tag{9.2c}$$

As we did in Chapter 8, we make use of the crystal wavefunction

$$|\vec{k}> = \frac{1}{\sqrt{N}} \sum_j e^{i\vec{k}\cdot\vec{r}_j} |j), \tag{9.3}$$

and, for specificity, we will work with the simple cubic system in the nearest neighbor approximation for which, from Chapter 8, we already know the matrix element of the unperturbed Hamiltonian (with $\varepsilon_0 = -\alpha$ and $V_{nm} = -\gamma$)

$$\varepsilon_{\vec{k}} = <\vec{k}|\hat{H}_0|\vec{k}> = -\alpha - 2\gamma(\cos(k_x a) + \cos(k_y a) + \cos(k_z a)). \tag{9.4}$$

The Green function associated with the above unperturbed system Hamiltonian is similar to the previous chapter result; that is,

$$\hat{G}_0 = (E - \hat{H}_0)^{-1} \tag{9.5a}$$

and whose diagonal matrix element is

$$G_{0\vec{k}}(E) \equiv <\vec{k}|\hat{G}_0|\vec{k}> = \frac{1}{E - E_{\vec{k}}} \tag{9.5b}$$

where, as before, $E$ is assumed to contain a small positive imaginary part and where we have made use of the usual definition $E_{\vec{k}} = (\varepsilon_{\vec{k}} + \alpha)/(2\gamma) = -(\cos(k_x a) + \cos(k_y a) + \cos(k_z a))$, so that since $\gamma$ is on the order of Hartrees ($H_a$), the energy is in these units. This corresponds to taking a diagonal site energy of $\varepsilon_0 = -\alpha = 0$ and $V_{nm} = -\gamma = -1/2$ in the present units. We next consider Green's function with the inclusion of the perturbing Hamiltonian, $\hat{H}_I$. We have

$$\hat{G} = (E - \hat{H})^{-1} = (E - \hat{H}_0 - \hat{H}_I)^{-1} = (\hat{G}_0^{-1} - \hat{H}_I)^{-1}, \tag{9.6a}$$

or

$$(\hat{G}_0^{-1} - \hat{H}_I)\hat{G} = 1 \quad \Rightarrow \quad \hat{G} = \hat{G}_0 + \hat{G}_0\hat{H}_I\hat{G}, \tag{9.6b}$$

as the Green function associated with the perturbation. We seek the matrix elements of this perturbed $\hat{G}$ and write

$$\begin{aligned} <\vec{k}|\hat{G}|\vec{k}'> &= <\vec{k}|\hat{G}_0|\vec{k}'> + <\vec{k}|\hat{G}_0\hat{H}_I\hat{G}|\vec{k}'> \\ &= <\vec{k}|\hat{G}_0|\vec{k}'> + <\vec{k}|\hat{G}_0\sum_{\vec{k}''}|\vec{k}''><\vec{k}''|\hat{H}_I\sum_{\vec{k}'''}|\vec{k}'''><\vec{k}'''|\hat{G}|\vec{k}'>, \end{aligned} \tag{9.7}$$

where we have used the state $|\vec{k}>$'s completeness condition. With the use of Equations 9.2c and 9.3, notice that

$$<\vec{k}''|\hat{H}_I|\vec{k}'''> = \frac{1}{\sqrt{N}}\sum_n e^{-i\vec{k}''\cdot\vec{r}_n}(n|\ell)\varepsilon_\mu\frac{1}{\sqrt{N}}\sum_{n'} e^{i\vec{k}'''\cdot\vec{r}_{n'}}(\ell|n') = \frac{\varepsilon_\mu}{N}e^{i(\vec{k}'''-\vec{k}'')\cdot\vec{r}_\ell} = \frac{\varepsilon_\mu}{N}, \tag{9.8}$$

where we have used the orthogonality condition $(\ell|n) = \delta_{\ell n}$ and taken the impurity to be located at the origin; i.e., $\vec{r}_\ell = 0$, in the last step. Substituting Equation 9.8 back into Equation 9.7, get

$$G_{\vec{k}\vec{k}'} = G_{0\vec{k}\vec{k}'} + \frac{\varepsilon_\mu}{N}\sum_{\vec{k}''}G_{0\vec{k}\vec{k}''}\sum_{\vec{k}'''}G_{\vec{k}'''\vec{k}'} \tag{9.9a}$$

where we have defined $G_{\vec{k}\vec{k}'} \equiv <\vec{k}|\hat{G}|\vec{k}'>$ and similarly $G_{0\vec{k}\vec{k}'} \equiv <\vec{k}|\hat{G}_0|\vec{k}'>$. With the further definitions

$$\sum_{\vec{k}'}G_{0\vec{k}\vec{k}'} \equiv G_{0\vec{k}}, \qquad \sum_{\vec{k}}G_{\vec{k}\vec{k}'} \equiv G_{\vec{k}'}, \tag{9.9b}$$

we can write Equation 9.9a as

$$G_{\vec{k}\vec{k}'} = G_{0\vec{k}\vec{k}'} + \frac{\varepsilon_\mu}{N}G_{0\vec{k}}G_{\vec{k}'}. \tag{9.9c}$$

Summing this expression over $\vec{k}$, while making use of the definition of Equation 9.9b, we get

$$G_{\vec{k}'} = G_{0\vec{k}'} + \varepsilon_\mu\frac{1}{N}\sum_{\vec{k}}G_{0\vec{k}}G_{\vec{k}'} = G_{0\vec{k}'} + \varepsilon_\mu F_0 G_{\vec{k}'}. \tag{9.9d}$$

where

$$F_0 \equiv \frac{1}{N}\sum_{\vec{k}}G_{0\vec{k}} = <0|\hat{G}_0|0> = \frac{1}{V_{BZ}}\left(\int_{V_{BZ}}\frac{1}{E-E_{\vec{k}}}d\vec{k}\right), \tag{9.9e}$$

is the site diagonal matrix element of the unperturbed crystal's Green function of Equation 9.5b with the integration as discussed in Chapter 8. Finally, replacing $\vec{k}' \to \vec{k}$ in Equation 9.9d, one obtains the perturbed Green function in momentum space as

$$G_{\vec{k}}(E) = G_{0\vec{k}}(E) + \varepsilon_\mu F_0(E)G_{\vec{k}}(E), \tag{9.9f}$$

where we show its explicit dependence on energy $E$ in a similar way that $G_{0\vec{k}}(E)$ does, which is evident from Equation 9.5b. Equation 9.9f can be solved for $G_{\vec{k}}$ to obtain

$$G_{\vec{k}}(E) = \frac{G_{0\vec{k}}(E)}{1-\varepsilon_\mu F_0(E)}, \tag{9.10a}$$

which is an important result. It shows the perturbed system contains a pole whenever the energy has a value that satisfies the relationship

$$\varepsilon_\mu = \frac{1}{F_0(E)}\bigg|_{E=E_{bo}}, \tag{9.10b}$$

for the real part of $F_0(E)$, and which allows us to identify the impurity bound state energy $E_{bo}$. Thus, the impurity acts as an electron trap with this energy level. The density of states associated with the impurity is

$$D_I(E) = -\frac{1}{\pi}\text{Im}\left(\frac{1}{N}\sum_{\vec{k}} G_{\vec{k}}(E)\right) = -\frac{1}{\pi}\text{Im}\left(\frac{F_0(E)}{1-\varepsilon_\mu F_0(E)}\right), \tag{9.10c}$$

where Equation 9.9e has been used. Notice that if the impurity perturbation were to be taken as $\varepsilon_\mu = 0$, Equation 9.10c reduces to

$$D(E) = -\frac{1}{\pi}\text{Im}\left(F_0(E)\right), \tag{9.10d}$$

which is the unperturbed crystal's density of states. Before we apply the above concepts regarding the single impurity, in order to obtain the impurity bound state $E_{bo}$, we need to perform a search for the energy $E$ in such a way that when Equation 9.10b is satisfied $E = E_{bo}$. The search involves varying $E$ and, in doing so, the function $F_0(E)$ from Equation 9.9e may have to be evaluated several times until the solution is found. The integration over the 3-dimensional $\vec{k}$-space takes much more effort. It is, therefore, worthwhile to find an efficient way to carry out such integrations. This is the subject of the following section.

## 9.3 Repeated Integrations Over 3-dimensional $\vec{k}$-space - An Efficient Way

In this section, we develop the approach that will be used throughout the rest of the chapter in order to carry out repeated integrations of functions such as that of Equation 9.9e for different energies $E$. Notice that the integrations we are referring to are over 3-dimensional $\vec{k}$-space. For simplicity, we work with a single band and recall that the density of states in such a crystal system is given by

$$\begin{aligned} D(E) &= \frac{1}{N}\sum_{\vec{k}} \delta(E-E_{\vec{k}}) = \frac{1}{V_{BZ}}\int_{V_{BZ}} \delta(E-E_{\vec{k}})\,d\vec{k} \\ &= -\frac{1}{\pi V_{BZ}}\text{Im}\left(\int_{V_{BZ}} G_{0\vec{k}}(E)\,d\vec{k}\right) = -\frac{1}{\pi V_{BZ}}\text{Im}\left(\int_{V_{BZ}} \frac{1}{E-E_{\vec{k}}}\,d\vec{k}\right), \end{aligned} \tag{9.11}$$

where $G_{0\vec{k}}(E)$ is the unperturbed Green function from Equation 9.5b. The above result is made possible by the relationship between Green's function and the density of states discussed in Chapter 8. In this vein, notice that the function $F_0$ of Equation 9.9e can be written as

$$\begin{aligned} F_0 &= \frac{1}{V_{BZ}}\int_{V_{BZ}} \frac{1}{E-E_{\vec{k}}}\,d\vec{k} = \frac{1}{N}\sum_{\vec{k}}\frac{1}{E-E_{\vec{k}}} = \frac{1}{N}\sum_{\vec{k}}\int dE'\frac{\delta(E'-E_{\vec{k}})}{E-E'} \\ &= \int \frac{dE'}{E-E'}\frac{1}{N}\sum_{\vec{k}}\delta(E'-E_{\vec{k}}) = \int \frac{D(E')dE'}{E-E'}, \end{aligned} \tag{9.12}$$

where the density of states definition of Equation 9.11 has been used. This expression effectively shows how to convert an integration over 3-dimensional $\vec{k}$-space to a one-dimensional energy integral. The integral that remains to be done is one that involves singularities, however, as is usually the case insofar as Green functions are involved. This of course is made possible if the density of states of the crystal system of interest is known a priori. In the following example, we do a calculation of $F_0(E)$ versus $E$.

**Example 9.3.0.1**

Let's calculate $F_0(E)$ versus $E$ for the simple cubic system and show its real and imaginary parts. From Equation 9.12, we will be carrying out the integral, $F_0 = \int \frac{D(E')dE'}{E-E'+i\Delta}$, for small $\Delta$. For the density of states, $D(E)$, we will use the exact analytic result from [34], as discussed in Chapter 8. The integration is to be carried out using our function singInt.m for singular integrals developed in the previous chapter. The results from the above integration are shown in Figure 9.3.4. The simple cubic's density of states is also included in the plot.

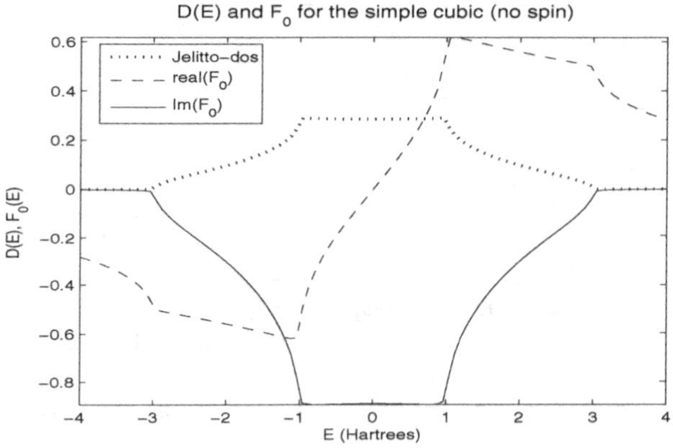


Figure 9.3.4: The real (dashed) and imaginary (solid) parts of the function $F_0(E)$ from Example 9.3.0.1. The simple cubic's density of states (dos) used in the calculation is also shown (dotted).

Notice that the imaginary part of $F_0(E)$ resembles the density of states. This is no accident as shown in Exercise 9.8.1. The code used to do the present calculation is F0_simple_cubic.m listed below. Notice that the function singInt.m (from Chapter 8) has to be available in the path to complete the calculation. Additionally, we have created a separate function to do the analytical density of states; i.e., jelittoScDosAnal.m and must be available in the path as well. It is also listed below.

```
%copyright by J. E Hasbun and T. Datta
%F0_simple_cubic.m
%f0=(1/N)sum_over_k (1/(E0-Ek)) where
%Ek=-cos(kx*a)-cos(ky*a)-cos(kz*a). If we use the density of states
%corresponding to the simple cubic, then
%f0(E)=int(g(E') dE'/(E-E'+im*delta))

function F0_simple_cubic
clear, clc;
delta=1.5e-2;
im=complex(0.0,1.0);
e2=4.0;                 %range for energy E
e1=-e2;
```

```
ntmax=101;
es=(e2-e1)/(ntmax-1); %energy step
a=1.0;                %lattice constant
tpa=2*pi/a;
VBZ=tpa^3;            %total SC BZ volume
for nt=1:ntmax
  e0(nt)=e1+(nt-1)*es;                      %energy E
  ge(nt)=jelittoScDosAnal(e0(nt))/VBZ;      %exact SC dos
end
plot(e0,ge,'k:','LineWidth',2), hold on %exact dos plot
xlabel('E (Hartrees)'), ylabel('D(E), F_0(E)')
str=cat(2,'D(E) and F_0 for the simple cubic (no spin)');
title(str, 'Fontsize',12)
%Repeat the e loop
e2p=3.0;                                    %range for E'
e1p=-e2p;
ntpmax=201;
esp=(e2p-e1p)/(ntpmax-1);                   %E' step
for nt=1:ntmax
  %Integrate over the e' loop
  for ntp=1:ntpmax
    if(nt==1)                          %calculate this part only once
      e0p(ntp)=e1p+(ntp-1)*esp;        %energy E'
      top(ntp)=jelittoScDosAnal(e0p(ntp))/VBZ; %use SC density of states
                                               %as the numerator
    end
    deno(ntp)=e0(nt)-e0p(ntp)+im*delta;        %the denominator
  end;
  f0(nt)=singInt(top,deno,esp);        %F0 at the energy E integration
end
plot(e0,real(f0),'k--')                %real part of F0 vesus E
hold on
plot(e0,imag(f0),'k-')                 %imaginary part of f0 versus E
axis tight
legend('Jelitto-dos','real(F_0)','Im(F_0)',0)
```

---

```
%copyright by J. E Hasbun and T. Datta
%jelittoScDosAnal.m
function ge=jelittoScDosAnal(ee)
%Jelitto's DOS for the simple cubic (exact)
eaa=abs(ee);
if (eaa <= 3.) & (eaa >=  1.)
  a1=3.-eaa;
  a2=a1^2;
  a=sqrt(a1);
  b=80.3702-16.3846*a1;
  d=0.78978*(a2);
  f=-44.2639+3.66394*a1;
  h=-0.17248*(a2);
  ge=a*((b+d)+(f+h)*sqrt(eaa-1.));
```

```
else
  if(eaa < 1.)
    ge=70.7801+1.0053*(ee^2);
  else
    ge=0.;
  end
end
```

## 9.4 The Single Impurity Level Calculation

We are ready to carry out the calculation of the impurity level using Equation 9.10b. The general idea is to find the energy $E$ at which the real part of $F_0(E) = 1/\varepsilon_\mu$; that is,

$$\varepsilon_\mu = \left.\frac{1}{\mathrm{Re}(F_0(E))}\right|_{E=E_{bo}}, \tag{9.13}$$

where $\varepsilon_\mu = \varepsilon_I - \varepsilon_0$ with $\varepsilon_I$ the diagonal energy associated with the foreign atom and $\varepsilon_0$ the diagonal energy of the host site, which in the present case it is $-\alpha = 0$. Referring to Figure 9.3.4 and looking at the real part of $F_0$, we notice there could be various solutions at which $F_0(E) = 1/\varepsilon_\mu$. Depending on the sign of $\varepsilon_\mu$, there could be two solutions for $E < 0$ and two solutions for $E > 0$. For each pair of solutions, there could be one inside the band and one outside. Impurity solutions inside the band are referred to as resonances, and solutions outside the band are generally known as deep levels, depending on how far they lie from the band. Generally, if an impurity level lies a few $meV$'s to tens of $meV$'s away from the band, they are referred to as shallow impurities (donors or acceptors follow this rule also). The deep levels lie far from a band and perhaps are a few hundred $meV$'s away from a single band or they are between bands; i.e., near the middle of a band gap. Provided the approximations made are applicable, the method developed here could be suitable for both shallow and deep levels. This is reflected in the way we search for the solution to Equation 9.13. We normally begin with a guess to the solution and then we iterate the equation until we reach convergence. For deep levels of interest here, we start with a guess that lies outside the band and whose sign depends on the sign of $\varepsilon_\mu$. We will employ the Newton-Raphson technique to locate the solution. We used this method before, in Chapter 5, when we sought the chemical potential versus temperature. Here, though, the energy is to be varied so we write

$$E_{i+1} = E_i - \frac{F(E_i)}{F'(E_i)}, \tag{9.14}$$

where $F(E) \equiv 1/\mathrm{Re}(F_0(E)) - \varepsilon_\mu$; i.e., the function whose zero we seek, and $F'(E)$ is its derivative. Because $F_0(E)$ involves an integration, every time $E$ is varied, a new integration has to be made and the process repeated to convergence. For this reason, it is important to choose an efficient method of seeking the solution. The Newton-Raphson method happens to be one of the best methods in this regard. The derivative can be evaluated numerically using the approximation $F'(E) \sim (F(E+\Delta) - F(E))/\Delta$ for small enough $\Delta$, which involves an extra evaluation of the integral, but it is a worthwhile effort here. The following example considers a certain impurity in a simple cubic system.

**Example 9.4.0.1**

Here we consider an impurity with diagonal energy $\varepsilon_I = 3.25 H_a$ (where in our units $2\gamma = 1 H_a$). Since we have taken the host diagonal energy $\varepsilon_0 = 0 H_a$ then $\varepsilon_\mu = 3.25 H_a$. We seek the bound state associated with the perturbation. This calculation has to be carried out numerically but can also be illustrated graphically. First, by running the script impurity_simple_cubic.m which carries out the procedure described above, in this section, prior to this example, the bound state is found according to Equation 9.13 at $E_{bo} = 3.75H_a$. The listing of the script follows.

```
%copyright by J. E Hasbun and T. Datta
%impurity_simple_cubic.m
%Here we obtain f0=(1/N)sum_over_k (1/(E0-Ek))
%from the density of states. That is,
%f0(E)=int(g(E') dE'/(E-E'+im*delta)) then calculate the
%impurity level from 1/Real(f0)-Emu=0

function impurity_simple_cubic
clear;
delta=1.e-3;
im=complex(0.0,1.0);
a=1.0;                %lattice constant
tpa=2*pi/a;
VBZ=tpa^3;            %total SC BZ volume
e2p=3.0;
e1p=-e2p;
ntpmax=251;
esp=(e2p-e1p)/(ntpmax-1);
%Get the simple cubic density of states to be reused
for ntp=1:ntpmax
  e0p(ntp)=e1p+(ntp-1)*esp;                        %E prime
  top(ntp)=jelittoScDosAnal(e0p(ntp))/VBZ; %integration numerator
end
%Search for the impurity bound state
%Use Newton-Raphson x(i+1)=x(i)-dE/(F(E+dE)/F(E)-1), where
%F(E)=1/f0(E)-E_mu is the function whose zero we seek
tol=1.e-3;                   %convergence tolerance
Ei=3.25; E_0=0;              %impurity and host diagonal energies
Emu=Ei-E_0;                  %The perturbation Emu=E_i-E_0
Eguess=1.2*Emu;              %initial guess (for a deep level)
iFg=10*tol;                  %convergence check variable
iter=0;
maxiter=30;                  %maximum iterations
%Iteration loop
while(abs(iFg) >= tol & iter < maxiter)
  iter=iter+1;                        %iteration counter
  de=max(0.5,abs(Eguess));            %energy variation
  Eguess_de=Eguess+de;                %vary the energy guess
  for ntp=1:ntpmax
    denog(ntp)=Eguess-e0p(ntp)+im*delta;         %1st denominator
    denog_de(ntp)=Eguess_de-e0p(ntp)+im*delta;   %2nd denominator
  end;
  f0g=real(singInt(top,denog,esp));              %integration
  f0g_de=real(singInt(top,denog_de,esp));        %integration
  iFg=1/f0g-Emu;                     %The function whose zero is seek
  iFg_de=1/f0g_de;                   %the varied function
  Ecorr=de/(iFg_de/iFg-1.0);         %Newton-Raphson correction
```

```
  Eguess=Eguess-Ecorr;            %new guess
end
Ebo=Eguess;                       %bound state due to the impurity
disp('Results: note that Ecorr is small at convergence')
fprintf('iter=%4i, Ecorr=%6.4g, 1/f0g=%6.4g, iFg=%6.4g\n',iter,...
  Ecorr,1/f0g,iFg)
fprintf('Peturbation: Emu=%6.4f, Bound state found: Ebo=%6.4f\n',Emu,Ebo)
```

The way the solution has been found can be explained graphically as shown in Figure 9.4.5. The inverse of $\varepsilon_\mu$ intersects the real part of $F_0(E)$ at an energy of $E = E_{bo}$ on the horizontal axis. The intersection is marked by a circle and corresponds to a graphical solution of the problem.

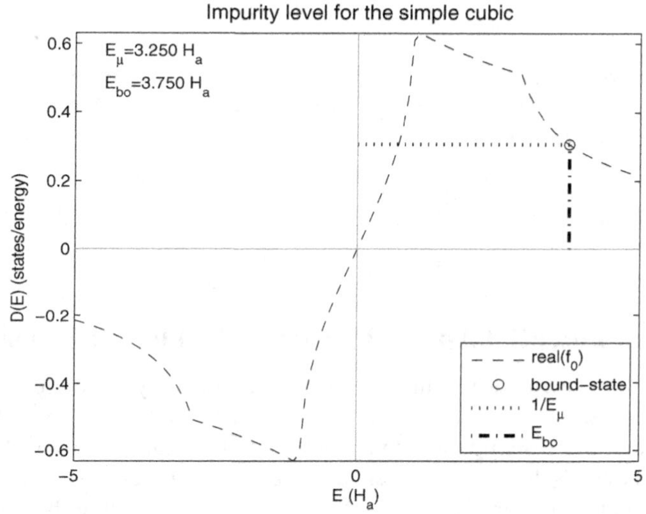


Figure 9.4.5: The impurity bound state $E_{bo} = 3.75\,H_a$ associated with $\varepsilon_\mu = 3.25\,H_a$ and which leads to the solution of Equation 9.13 is shown by the circle, which marks the intersection between the real part of $F_0(E_{bo})$ and $1/\varepsilon_\mu$. The real part of $F_0(E)$ is as in Figure 9.3.4.

The perturbation affects the density of states, and the impurity contribution to it is shown by a large peak at the impurity bound state. If we ignore spin, integrating the areas under the density of states, the result ought to be unity, since in each case the obtained band holds one electron. The plot in Figure 9.4.6 shows the DOSs for the unperturbed and the perturbed cases. The area under each curve is closed to unity (see Exercise 9.8.4)

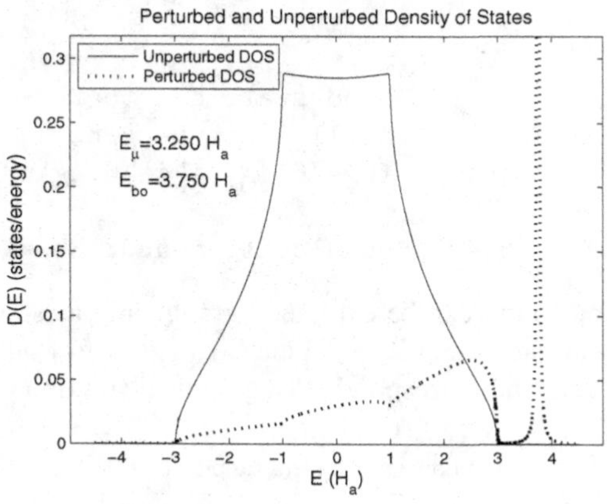


Figure 9.4.6: The density of states of the unperturbed system (solid, see Equation 9.10d) is compared to the perturbed density of states (dotted, see Equation 9.10c). The area under each curve amounts to a single electron (ignoring spin).

## 9.5 The Coherent Potential Approximation (CPA) for Disordered Systems

The previous sections have made use of the Green function to study the effects of an impurity in a crystal system. In this section, we delve into the regime of systems for which the number of impurities is so high that the system is not a pure crystal but is rather an alloy. An alloy falls into the class of systems known as disordered systems. The disorder comes from the presence of atoms in the material whose particular arrangement is not necessarily repeatable throughout the system. Thus, the important property of periodicity is lost in such systems and the standard idea of Bloch's theorem does not apply. We say the system does not have long range order. The study of disordered systems may sound like a difficult task, and it is; however, there are techniques that can be applied and still be able to use the methods we have learned from studying periodic systems. The Green function technique we have developed thus far goes by the name of multiple scattering. We will continue here with that approach and adopt the operator symbol $\hat{v}$, which is more standard, to refer to the perturbing part of the Hamiltonian. In this way, Equation 9.2a is rewritten as

$$\hat{H} = \hat{H}_0 + \hat{v}. \tag{9.15a}$$

We still let the unperturbed system be described by the Green function of Equation 9.5. The perturbed system, however, is similar to Equation 9.6b; that is,

$$\hat{G} = \hat{G}_0 + \hat{G}_0\hat{v}\hat{G} \quad \Rightarrow \quad \hat{G} = [1 - \hat{G}_0\hat{v}]^{-1}\hat{G}_0. \tag{9.15b}$$

This quantity $\hat{G}$ can be related to another quantity $\hat{T}$ known as the scattering matrix. The relationship is such that we can write

$$\hat{G} = \hat{G}_0 + \hat{G}_0\hat{T}\hat{G}_0, \tag{9.16a}$$

which when compared with the first of Equation 9.15b, we see that

$$\hat{T}\hat{G}_0 = \hat{v}\hat{G}, \tag{9.16b}$$

and, therefore, we can write an equation for the $T$-matrix as

$$\hat{T} = \hat{v} + \hat{v}\hat{G}_0\hat{T} \quad \Rightarrow \quad \hat{T} = [1 - \hat{v}\hat{G}_0]^{-1}\hat{v}, \tag{9.16c}$$

where Equation 9.16a has been used. Equations 9.16(a-c) are very convenient for describing properties of random system. In particular, the coherent potential approximation (CPA) [38] for random alloys was originally formulated using the above-mentioned multiple scattering approach in conjunction with a tight binding model. Here we work with the simplest of alloys, namely those which consist of two species of atoms, A and B, substitutionally located in random site positions with respective concentrations $x$ and $y = 1 - x$. We can represent such alloys in the form $A_xB_y$. To capture the concept, in what follows we will refer to Figure 9.5.7. In panel (a) an alloy is illustrated composed of atomic species A and B located at random in substitutional sites.

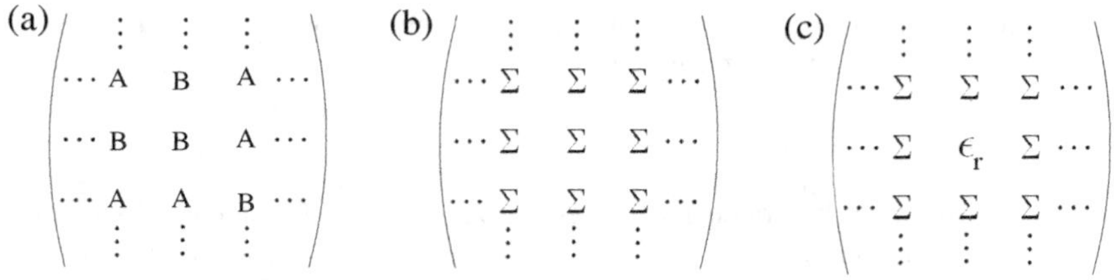


Figure 9.5.7: (a) A random substitutional alloy of A and B species; (b) a crystal of coherent potentials; (c) replacing site 0 with a random site $\varepsilon_r$ that can take on values of either $\varepsilon_A$ or $\varepsilon_B$.

The idea of the CPA is to replace all these alloy sites by coherent sites as shown in panel (b). The tight binding Hamiltonian associated with such a crystal is

$$\hat{H}_0 = \sum_n |n)\Sigma(n| + \sum_{nm}{}' |n)V_{nm}(m|; \tag{9.17}$$

that is, the disorder is assumed to be diagonal, and where, as before, the prime indicates $n \neq m$. Here, the actual diagonal energies that normally take on values of $\varepsilon_A$ and $\varepsilon_B$ have been replaced by the, as yet, unknown coherent potentials $\Sigma$. The off-diagonal terms are taken to be independent of the alloy species. Notice that by doing this, the crystal so described remains translationally invariant. The Green function corresponding to the Hamiltonian of Equation 9.17 is

$$\hat{G}_0 = (E - \hat{H}_0)^{-1}. \tag{9.18}$$

Using the Bloch function

$$|\vec{k}> = \frac{1}{\sqrt{N}} \sum_m e^{i\vec{k}\cdot\vec{r}_m} |m), \tag{9.19}$$

in the nearest neighbor approximation, the matrix elements of $\hat{H}_0$ and, thereby, the alloy system's $\hat{G}_0$ are

$$< k|\hat{H}_0|k > = \Sigma + V_{\vec{k}} \tag{9.20a}$$

and

$$< k|\hat{G}_0|k > = (E - \Sigma - V_{\vec{k}})^{-1} \equiv G_{0\vec{k}} \tag{9.20b}$$

with

$$V_{\vec{k}} = \sum_{m\neq 0}^{nn} V_{0m} \exp(i\vec{k}\cdot\vec{R}_m), \quad \text{and} \quad F_{\Sigma 0}(E) \equiv (0|\hat{G}_0|0) = \frac{1}{N}\sum_{\vec{k}} \frac{1}{E-\Sigma-V_{\vec{k}}}, \tag{9.20c}$$

where in the sum's upper limit *nn* stands for nearest neighbors. Also, as before, the density of states for the alloy according to the CPA is

$$D_{\Sigma}(E) = -\frac{1}{\pi}\mathrm{Im}[F_{\Sigma 0}(E)]. \tag{9.20d}$$

As mentioned above, the $\Sigma$ is unknown and we need a way to determine it. The CPA procedure is, first, to take out one of the coherent potentials of Figure 9.5.7(b), and in its place put in an actual atom whose diagonal energy is $\varepsilon_r$ as shown in panel (c) of the figure, where $r$ can be either A or B. This is tantamount to considering a single impurity in the alloy crystal. We thus think of this as a perturbation (similar to Section 9.2) of the form

$$\hat{H}_r = |0)(\varepsilon_r - \Sigma)(0|, \tag{9.21}$$

where the supposed impurity has been located at the zeroth site. The idea is that $\varepsilon_r$ can take on diagonal energy values of the original alloy atoms; i.e., $\varepsilon_A$ or $\varepsilon_B$ with respective concentrations $x$ and $y = 1-x$. The total system's Hamiltonian is, therefore, that of Equation 9.15a where we make the identification

$$\hat{v} = \hat{H}_r. \tag{9.22}$$

This means that the perturbed Green function of the system is similar to that of Equation 9.15b; that is, for the atom $r$

$$\hat{G}_r \equiv [1-\hat{G}_0\hat{H}_r]^{-1}\hat{G}_0. \tag{9.23a}$$

Taking the site diagonal elements of this, while using Equations 9.20 and 9.21 as reference, we have

$$F_r \equiv (0|\hat{G}_r|0) = [1-F_{\Sigma 0}(E)(\varepsilon_r-\Sigma)]^{-1}F_{\Sigma 0}(E), \tag{9.23b}$$

where use has also been made of the site wavefunctions' completeness condition, $\sum_n |n)(n| = 1$, as well as their orthonormality property, $(n|m) = \delta_{nm}$ (see Exercise 9.8.6). The first step in the process of obtaining $\Sigma$ has been achieved. The second and final step entails the averaging of the function $F_r$ of Equation 9.23b over the atomic species A and B, which are thought to be present with concentrations $x$ and $y$, respectively. The CPA recipe sets this average equal to the function $F_{\Sigma 0}(E)$ of Equation 9.20c; that is,

$$< F_r >_{Ave} = \frac{xF_{\Sigma 0}(E)}{1-F_{\Sigma 0}(E)(\varepsilon_A-\Sigma)} + \frac{yF_{\Sigma 0}(E)}{1-F_{\Sigma 0}(E)(\varepsilon_B-\Sigma)} = F_{\Sigma 0}(E), \tag{9.24}$$

and represents an equation to obtain $\Sigma$ self-consistently. This averaging can be expressed graphically and is shown in Figure 9.5.8.

$$\left\langle \begin{Vmatrix} & \vdots & \vdots & \vdots & \\ \cdots & \Sigma & \Sigma & \Sigma & \cdots \\ \cdots & \Sigma & \epsilon_A/\epsilon_B & \Sigma & \cdots \\ \cdots & \Sigma & \Sigma & \Sigma & \cdots \\ & \vdots & \vdots & \vdots & \end{Vmatrix} \right\rangle_{Ave} = \begin{pmatrix} & \vdots & \vdots & \vdots & \\ \cdots & \Sigma & \Sigma & \Sigma & \cdots \\ \cdots & \Sigma & \Sigma & \Sigma & \cdots \\ \cdots & \Sigma & \Sigma & \Sigma & \cdots \\ & \vdots & \vdots & \vdots & \end{pmatrix}$$

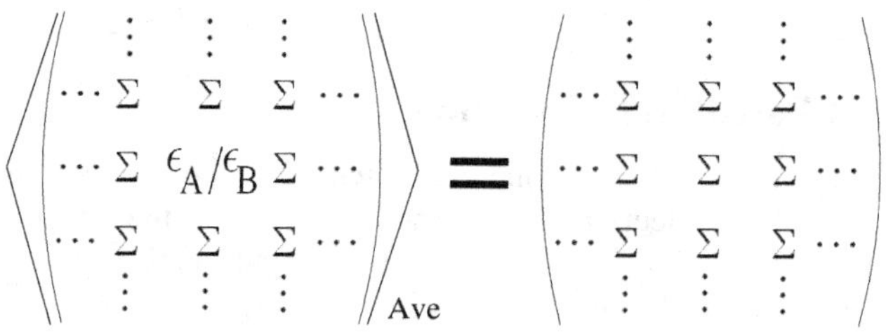

Figure 9.5.8: The CPA's approach to obtain the coherent potential $\Sigma$ self-consistently entails the averaging over the zeroth site in such a way as to be consistent with a crystal of coherent potentials.

If we define the quantity

$$\varepsilon = x\varepsilon_A + y\varepsilon_B, \tag{9.25a}$$

we can rearrange Equation 9.24 for $\Sigma$ as

$$\Sigma = \varepsilon - (\varepsilon_A - \Sigma)F_{\Sigma 0}(E)(\varepsilon_B - \Sigma) \tag{9.25b}$$

where we restate the $\Sigma$ dependent $F_{\Sigma 0}$ and the density of states from Equation 9.20

$$F_{\Sigma 0}(E) = \frac{1}{N}\sum_{\vec{k}} \frac{1}{E - \Sigma - V_{\vec{k}}}, \qquad D_{\Sigma}(E) = -\frac{1}{\pi}\text{Im}[F_{\Sigma 0}(E)]; \tag{9.25c}$$

that is, once $\Sigma$, which is a function of energy $E$, converges from Equation 9.25b, a final evaluation of $F_{\Sigma 0}(E)$ is made in order to obtain the final density of states, $D_{\Sigma}(E)$, in the alloy system. One immediate approximation for $\Sigma$ is the so-called virtual crystal approximation (VCA), which is obtained if the second term of Equation 9.25b is ignored; in which case, $\Sigma \approx \varepsilon$. In that case, the system's diagonal energy is simply a mixture of the original alloy components' diagonal energies weighted by their respective concentrations. As a rule of thumb, the VCA seems to be a reasonable approximation if the alloy diagonal energies $\varepsilon_A$ and $\varepsilon_B$ are close enough. In contrast, if the diagonal energies differ markedly from each other, the CPA is the better approximation to employ.

It is interesting to observe that the self-consistent expression for $\Sigma$ expressed in Equation 9.25b can also be obtained using the $T$-matrix of Equation 9.16c. The steps are as follows. If as in Equation 9.22, $\hat{v} = \hat{H}_r$, we have for the corresponding operator

$$\hat{T}_r = [1 - \hat{H}_r\hat{G}_0]^{-1}\hat{H}_r. \tag{9.26a}$$

Taking the site-diagonal matrix element of this, in a similar way to how we arrived at Equation 9.23b (see Exercise 9.8.7), we get for the $T$-matrix associated with site $r$

$$T_r \equiv \langle 0|\hat{T}_r|0\rangle = \frac{\varepsilon_r - \Sigma}{1 - F_{\Sigma 0}(E)(\varepsilon_r - \Sigma)}. \tag{9.26b}$$

The final step is to average (denoted by $< \cdots >$) this $T$-matrix over the atomic species A and B according to their concentrations then to set the result to zero. This gives

$$< T_r >= 0 = \frac{x(\varepsilon_A - \Sigma)}{1 - F_{\Sigma 0}(E)(\varepsilon_A - \Sigma)} + \frac{y(\varepsilon_B - \Sigma)}{1 - F_{\Sigma 0}(E)(\varepsilon_B - \Sigma)}, \tag{9.26c}$$

and this also gives the self-consistent condition for $\Sigma$ expressed in Equation 9.25b (see Exercise 9.8.8). Based on the above, the physical interpretation of the CPA is that one replaces the exact system by an average effective system; that is, each site in the alloy sees an effective potential or coherent potential. This potential is determined by the requirement that when one replaces it by an actual alloy site, either $\varepsilon_A$ or $\varepsilon_B$, there is no further scattering on the average.

## 9.6 The Coherent Potential Calculation

In this section, the calculation of the coherent potential, $\Sigma$, discussed in the previous section will be carried out. As in Section 9.2, we will work with the simple cubic host system band in the tight binding approximation. This is given by Equation 9.4; that is, $\varepsilon_{\vec{k}} = -\alpha - 2\gamma(\cos(k_x a) + \cos(k_y a) + \cos(k_z a))$. However, in $F_{\Sigma 0}(E)$ of Equation 9.20c, $\Sigma$ replaces $-\alpha$ and we let $V_{\vec{k}} = -(\cos(k_x a) + \cos(k_y a) + \cos(k_z a))$ so that, as before, we are working in units of $2\gamma \approx 1Ha$. Furthermore, since the density of states for the simple cubic is known, the function $F_{\Sigma 0}(E)$ from Equation 9.25c is calculated by the method of Section 9.3; that is,

$$F_{\Sigma 0}(E) = \frac{1}{N}\sum_{\vec{k}} \frac{1}{E - \Sigma - V_{\vec{k}}} = \frac{1}{N}\sum_{\vec{k}} \int dE' \frac{\delta(E' - V_{\vec{k}})}{E - E' - \Sigma} = \int \frac{D(E')dE'}{E - E' - \Sigma}, \tag{9.27a}$$

where the energy $E$, once again, is assumed to contain a small positive imaginary part and where we have used the definition of the density of states as we have calculated it for the cubics in the previous chapter; i.e.,

$$D(E) = \frac{1}{N}\sum_{\vec{k}} \delta(E - V_{\vec{k}}). \tag{9.27b}$$

The final aspect about the calculation we are about to perform is that $\Sigma$ in Equation 9.25b is a self-consistent expression, meaning that it must be solved for numerically in an iterative manner until the left and right sides of the equation are equal; i.e., consistent with each other. As it stands, starting from an initial guess, the equation does not yield to a rapid and efficient convergence. We, therefore, rearrange it so as to improve the iteration process. By multiplying through and rearranging terms, one obtains

$$\Sigma = \frac{\varepsilon - (\varepsilon_A \varepsilon_B - \varepsilon\Sigma)F_{\Sigma 0}(E)}{1 - (x\varepsilon_B + y\varepsilon_A - \Sigma)F_{\Sigma 0}(E)} \qquad \text{(Full Alloy)}, \tag{9.28}$$

and is the form we employ in this section. This expression is useful because in the event that $\varepsilon_A \sim \varepsilon_B \approx \varepsilon$ then

$$\Sigma \approx \varepsilon \qquad \text{(VCA Limit)}, \tag{9.29}$$

which is the energy result that the VCA, mentioned earlier in connection with Equation 9.25, would give. This is seen to work well for the case when the alloy species are nearly the same and when it corresponds to the crystal limit when $\varepsilon_A = \varepsilon_B$. There is an interesting limiting form of $\Sigma$ that one can obtain directly from Equation 9.28; i.e., when the species are very different. This is called the split-band limit. The limit is obtained by assuming, for example, that $\varepsilon_A \to \infty$ and $\varepsilon_B \to 0$, so that $\varepsilon = x\varepsilon_A$, then the above expression gives

$$\Sigma \approx -\frac{x\varepsilon_A(1 + \Sigma F_{\Sigma 0})}{(y\varepsilon_A - \Sigma)F_{\Sigma 0}}, \tag{9.30a}$$

where since $\varepsilon_A$ is assumed large, the 1 has been ignored in the denominator of Equation 9.28. Furthermore, since in this limit $\varepsilon_A >> \Sigma$, we can ignore the $\Sigma$ in the denominator as well, and multiplying through by $(y\varepsilon_A)F_{\Sigma 0}$, remembering that $x + y = 1$, while finally solving for $\Sigma$, we get

$$\Sigma = -\frac{x}{F_{\Sigma 0}(E)} = -\frac{(1-y)}{F_{\Sigma 0}(E)} \qquad \text{(Split Band Limit)}, \tag{9.30b}$$

which represents a self-consistent equation for $\Sigma$ in the split-band limit and corresponds to a strong scattering regime. One final limiting form of $\Sigma$ is the so-called dilute alloy in which one of the alloy components is considered very dilute. For example, let's assume that $x$, the concentration of the A species is very small, then we expect that $\Sigma \sim \varepsilon_B$, so that we can replace $F_{\Sigma 0}$ of Equation 9.27a with $F_{\varepsilon_B 0}$, where $F_{\varepsilon_B 0}(E) \equiv \frac{1}{N}\sum_{\vec{k}} \frac{1}{E-\varepsilon_B-V_{\vec{k}}}$. In this case, Equation 9.25b becomes $\Sigma \approx x\varepsilon_A + (1-x)\varepsilon_B + (\varepsilon_A - \Sigma)F_{\varepsilon_B 0}(\Sigma - \varepsilon_B) = x(\varepsilon_A - \varepsilon_B) + \varepsilon_B + (\varepsilon_A - \Sigma)F_{\varepsilon_B 0}(\Sigma - \varepsilon_B)$, or $(\Sigma - \varepsilon_B)[1-(\varepsilon_A - \Sigma)F_{\varepsilon_B 0}] = x(\varepsilon_A - \varepsilon_B)$, which results in

$$\Sigma \sim \varepsilon_B + \frac{x\delta}{1-\delta F_{\varepsilon_B 0}(E)} \qquad \text{(Dilute Alloy Limit)}, \tag{9.31}$$

where $\delta \equiv \varepsilon_A - \varepsilon_B$. This expression for $\Sigma$ is not self-consistent, but it remains a function of $E$ and the density of states is still given by Equation 9.27.

**Example 9.6.0.1**

In this example, we consider a hypothetical alloy consisting of two species A and B whose structure is that of a simple cubic with a single band in the tight binding approximation. We wish to obtain the density of states (DOS) of the alloy system $A_xB_{1-x}$ and take the diagonal energies associated with the atomic species as $\varepsilon_A = -2.5$ and $\varepsilon_B = 2.5$ in units of $Ha$. When calculating the density of states, we will use the expressions from Equations 9.27 and 9.28 for concentrations of $x = 1.0$, 0.9, and 0.5. The case for $x = 1.0$ corresponds to the pure crystal composed of the A species and the associated density of states is that of a simple cubic band centered at $\varepsilon_A = -2.5$ as can be seen in Figure 9.6.9(a). The pure crystal DOS contains sharp edges or the so-called Van Hove singularities associated with pure crystal band structures, whereas in the alloy density of states such singularities tend to be smoothed out. This is evident in the results shown in panels (b) and (c).

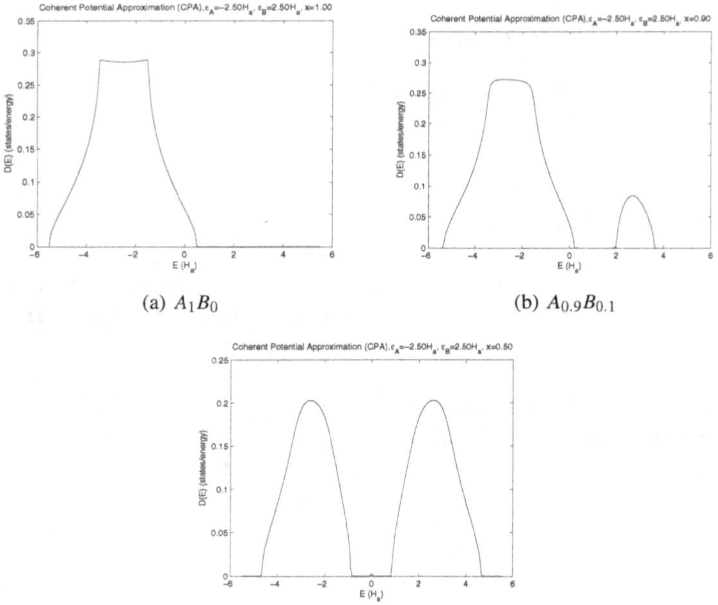


(a) $A_1B_0$ (b) $A_{0.9}B_{0.1}$

(c) $A_{0.5}B_{0.5}$

Figure 9.6.9: The alloy $A_xB_{1-x}$ density of states as obtained using the coherent potential approximation (CPA) for (a) the A crystal ($x = 1.0$), (b) the case with concentration $x = 0.1$, and (c) the case with concentration $x = 0.5$. Here the diagonal energies are $\varepsilon_A = -2.5Ha$, $\varepsilon_B = 2.5Ha$, and the energy $E$ is in units of Hartrees ($Ha$).

We notice that as the concentration $x$ of the A species decreases, the density of states develops the band associated with the B species with concentration $y = 1 - x$. This band is centered about $\varepsilon_B = 2.5Ha$. The density of states associated with the B species increases as $y$ increases. What do you expect will happen in the limit as $x \to 0$? The code employed in obtaining Figure 9.6.9 is CPA.m whose listing follows (recall that some of the functions used in the code have been discussed previously).

```
%copyright by J. E Hasbun and T. Datta
%CPA.m
%We calculate the coherent potential approximation (CPA) sigma.
%The virtual crystal approximation (VCA) is used as guess
%The resulting sigma is used to obtain the density of states
%for the alloy.
function CPA
clear; clc;
delta=1.e-3;
im=complex(0.0,1.0);
Ea=-2.5;                %alloy system species 1 diagonal energy
Eb=2.5;                 %alloy system species 2 diagonal energy
scbw=3.0;               %simple cubic known band width
x=1;
y=1-x;                  %y=concentration of Eb species
e2=Ea-scbw;             %energy range to work with
e1=-e2;
ntmax=201;
es=(e2-e1)/(ntmax-1);
a=1.0;                  %lattice constant
tpa=2*pi/a;
VBZ=tpa^3;              %total SC BZ volume
e2p=3.0;
e1p=-e2p;
ntpmax=251;
esp=(e2p-e1p)/(ntpmax-1);
Evca=x*Ea+y*Eb;                         %VCA energy
sig_guess=Evca;                         %initial sigma guess - VCA energy
cv=0.85;                                %new guess helper parameter
ncmax=25;                               %maximum iterations
tol=1.e-4;
str1=cat(2,'nt=%4i, e0=%4.2f, nc=%2i, sig=%5.4g, drs=%5.4g');
str2=cat(2,', dis=%5.4g, dos=%5.4f\n');
str=cat(2,str1,str2);
for nt=1:ntmax                          %Main energy loop
  nc=0;
  converge=0;
  drs=10*tol;
  dis=10*tol;
  e0(nt)=e1+(nt-1)*es;                  %the energy E
  sig(nt)=sig_guess;                    %use former sigma guess
  while (converge==0 & nc < ncmax)
    nc=nc+1;
    if(drs <  tol & dis < tol)
```

```
      converge=1;            %use the converged sigma as guess
      sig_guess=sig(nt);     %for next energy
    end
    %Integrate over the E' loop
    for ntp=1:ntpmax
      if(nt==1)                  %calculate this part only once
        e0p(ntp)=e1p+(ntp-1)*esp;
        top(ntp)=jelittoScDosAnal(e0p(ntp))/VBZ;
      end
      deno(ntp)=e0(nt)-e0p(ntp)-sig(nt)+im*delta;
    end;
    f0(nt)=singInt(top,deno,esp);
    %CPA rule for sigma's next guess. The 2nd form converges better.
    %signew=Evca-(Ea-sig(nt))*f0(nt)*(Eb-sig(nt));
    signew=(Evca-(Ea*Eb-Evca*sig(nt))*f0(nt))/...
      (1-(y*Ea+x*Eb-sig(nt))*f0(nt));
    %The actual guess we make is a mixture between new and old
    sig(nt)=cv*signew+(1-cv)*sig(nt);
    drs=abs(real(sig(nt)-signew));    %use for convergence criteria
    dis=abs(imag(sig(nt)-signew));
  end
  dos(nt)=-imag(f0(nt))/pi;
  fprintf(str,nt,e0(nt),nc,sig(nt),drs,dis,dos(nt));
end
str=cat(2,'Coherent Potential Approximation (CPA), \epsilon_A=',...
  num2str(Ea,'%5.2f'),'H_a, \epsilon_B=',num2str(Eb,'%5.2f'),...
  'H_a, x=',num2str(x,'%4.2f'));
title(str)
plot(e0,dos,'k')
xlabel('E (H_a)'), ylabel('D(E) (states/energy)')
title(str)
```

It is interesting to see how the self-consistent energy $\Sigma$ behaves versus $E$, and this can most simply be illustrated with the split-band limit of Equation 9.30b. The result is shown on the left panel of Figure 9.6.10 for a concentration of $x = 0.5$. The right panel of the figure contains the density of states for the single band associated with the B species, shown to be centered at zero. The band associated with the A species is presumably centered at $\infty$.

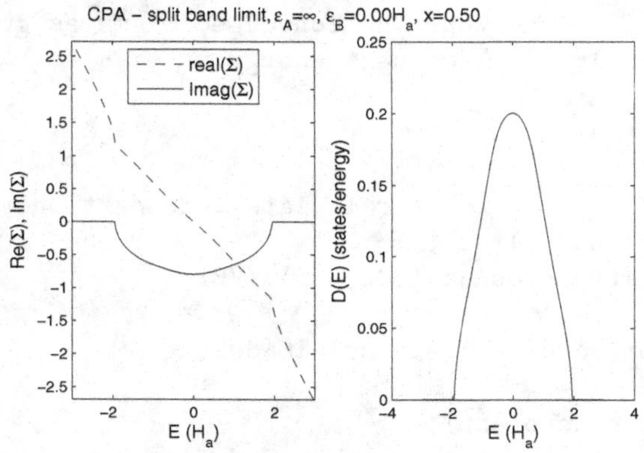


Figure 9.6.10: The left panel shows the $A_xB_{1-x}$ split-band limit results for the real and imaginary parts of $\Sigma$ from Equation 9.30b with $\varepsilon_B = 0$ and $x = 0.5$. The right panel shows the corresponding density of states.

The dilute alloy case can be computed using the approximate self-energy $\Sigma$ of Equations 9.27 and 9.31. The main difference is that we do not have to worry about iterating $\Sigma$ in this case. The results for the density of states are shown in Figure 9.6.11. Here we take $\varepsilon_A = -2.5Ha$, $\varepsilon_A = 2.5Ha$, and $x = 0.05$, since the approximation applies for small $x$.

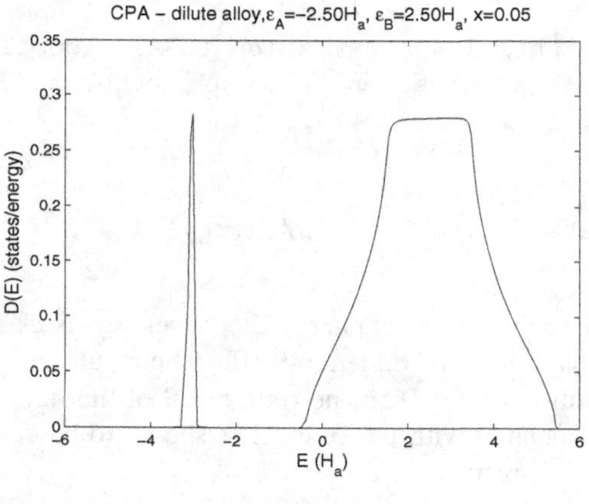


Figure 9.6.11: The $A_xB_{1-x}$ dilute alloy density of states with $\varepsilon_A = -2.5Ha$, $\varepsilon_A = 2.5Ha$, and $x = 0.05$.

Notice the sharp peak to the left of the main B band. The peak is associated with the dilute concentration of the A atoms and resembles an impurity band. Finally, the CPA has been found to be very successful in calculations involving real multisubband systems as illustrated in Figure 9.6.12 for a $Si_{1-x}Ge_x$ alloy.

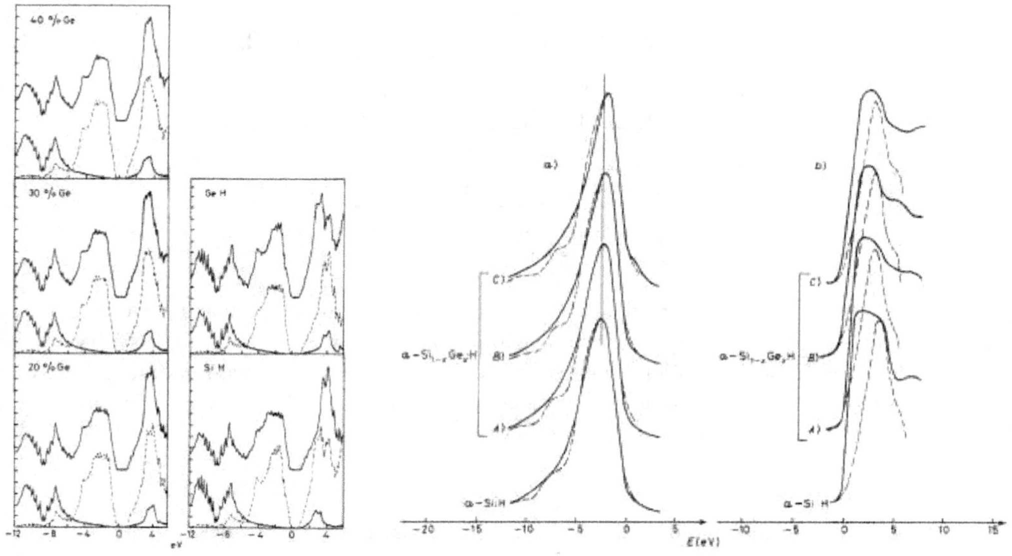


(a) $Si_{1-x}Ge_x$: Theory - CPA calculation [39]

(b) $Si_{1-x}Ge_x$: Comparison between theory and experiment [39]

Figure 9.6.12: (a) The density of states (solid lines) for $Si_{1-x}Ge_x$ alloy for concentrations of $x = 0.4, 0.3, 0.2$ and including the pure Ge and Si results [39]. (b) Comparison between theory (dashed) and experiment (solid) for concentrations of $x = 0.2, 0.3, 0.4$ as indicated by the labels A,B,C respectively [39]. Reproduced with permission.

## 9.7 An Insight into an Effective Medium

The coherent potential approximation described above falls into a class of approaches to study disordered systems. A more general term is an effective medium theory. It is interesting that if we apply the same concept to a random mixture of metal wires and treat them as if they were resistors, as is shown below, the exact result is obtained. Let's work in one dimension and consider a wire of length $\ell$ and cross-sectional area $A$ made of a random mixture of different metal wires, where each individual metal wire has a respective conductivity $g_i$ with $i$ denoting metals 1 and 2. We write the Ohm's voltage across the metal mixture as

$$V = \frac{I\ell}{g_{eff}A}, \tag{9.32a}$$

where $g_{eff}$ is the effective conductivity of the random wire mixture, which is to be estimated through the CPA. The effective wire is illustrated in Figure 9.7.13(a). We next remove a section of the effective wire of length $a$ and replace it by a segment of wire 1 with conductivity $g_1$ as in Figure 9.7.13(b). The conductivity of the resulting wire is given by

$$V_1 = \frac{Ia}{g_1A} + \frac{I(\ell - a)}{g_{eff}A}. \tag{9.32b}$$

The process is repeated as in Figure 9.7.13(c) with wire 2 to obtain the voltage

$$V_2 = \frac{Ia}{g_2A} + \frac{I(\ell - a)}{g_{eff}A}. \tag{9.32c}$$

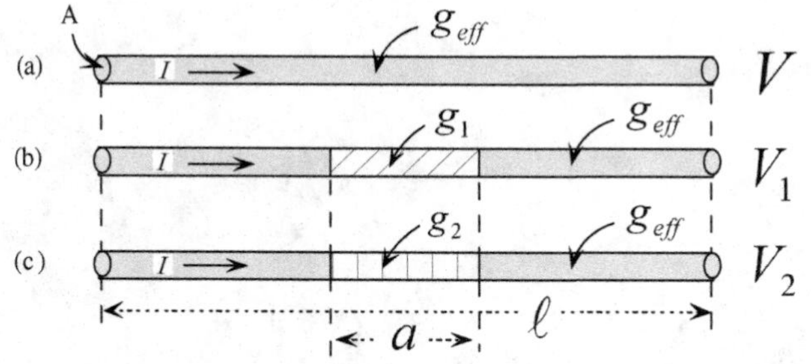


Figure 9.7.13: (a) An effective wire of length $\ell$, cross-sectional area $A$, current $I$, voltage $V$ and effective conductivity $g_{eff}$. (b) A wire of cross-sectional area $A$, current $I$, voltage $V_1$ with effective conductivity $g_{eff}$ for the section of length $\ell - a$ and conductivity $g_1$ for the section of length $a$. (c) A wire of cross-sectional area $A$, current $I$, voltage $V_2$ with effective conductivity $g_{eff}$ for the section of length $\ell - a$ and conductivity $g_2$ for the section of length $a$.

An approximate value of the conductivity of the effective wire is $g_{eff} \approx \bar{g}$ where

$$\bar{g} = xg_1 + yg_2, \tag{9.33}$$

with $x$ and $y$ the fractional compositions of metal 1 and 2, respectively, and where $x + y = 1$. This is the virtual crystal approximation version (VCA) for the present system. To go beyond the VCA, it has been pointed out that the CPA, for the configuration considered here, is equivalent to setting the average voltage fluctuation equal to zero [40]. To this end we have

$$< V_i - V >= xV_1 + yV_2 - V = 0. \tag{9.34a}$$

By substituting the corresponding Equations 9.32 into this, straightforward algebra leads to

$$\frac{1}{g_{eff}} = \frac{x}{g_1} + \frac{y}{g_2}, \tag{9.34b}$$

which is the CPA result for the system and it is also the expected exact result since resistances (which are inversely proportional to conductivities) in series add arithmetically. If we make the definition $\delta \equiv (g_2 - g_1)$ we can express the effective conductivity in the form

$$g_{eff} = \bar{g} - \frac{xy\delta^2}{g_1(1 - x\frac{\delta}{g1})}, \tag{9.35}$$

so that for small $\delta$, we can ignore the $\delta^2$ term and we see that the CPA gives $g_{eff} \sim \bar{g}$, which is the VCA limit result of Equation 9.33. Glancing at the VCA, one notices that the expression acts as a linear interpolation between the pure metal values $g_1$ and $g_2$ versus concentration $x$. The effective medium result goes beyond the linear interpolation. The best way to assess how well this effective medium describes actual observation is to make a comparison with real data. This is accomplished with the use of the resistivity as recommended from Ho et al. [41] based on experimental data. The resistivity can be inverted to obtain the conductivity and Equation 9.34b can be used to obtain the $g$ values versus $x$. We will work with the $Cu_xNi_{1-x}$ alloy for this comparison. We will let $g_1$ be the conductivity of Cu when $x = 1$ and $g_2$ be the conductivity of Ni when $x = 0$. According to the data $g_1 = 0.596 \times 10^8 (\Omega \cdot m)^{-1}$ and $g_2 = 0.144 \times 10^8 (\Omega \cdot m)^{-1}$. Figure 9.7.14 shows the recommended room temperature data (circles) along with the calculations from the VCA (dashed) of Equation 9.33 and the effective medium theory (solid) of Equation 9.35 versus the Cu concentration

$x$. Notice the linear behavior of the VCA as mentioned earlier. The effective medium theory does show more bowing and is an improvement but it still lacks information for a better comparison. A more sophisticated version of the theory appears to be needed in this case; Ho et al. [41] use the Bloch-Grüneisen formula, which is beyond the scope of the present text.

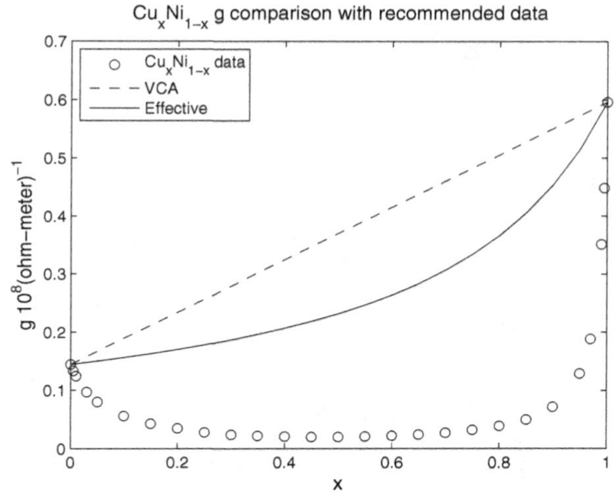


Figure 9.7.14: The $Cu_xNi_{1-x}$ alloy recommended room temperature data [41] (circles) along with the calculated conductivity using the VCA (dashed line), of Equation 9.33, and the effective medium theory (solid line), of Equation 9.35, versus $x$.

The partial code alloy_CuNi_cond.m listed below will reproduce the data as well as the VCA conductivity of Figure 9.7.14 and can be modified to include the effective medium calculation to reproduce the entire figure (see Exercise 9.8.14).

```
%copyright by J. E Hasbun and T. Datta
%alloy_CuNi_cond.m
%Case of 293 K
%To use the effective medium conductivity formula to obtain
%the conductivity versus concentration.
%We work with the Cu(x)Ni(1-x) alloy.
%Cu(x)Ni(1-x) - g in units of 10^8/(ohm-meter)
clear, clc
%recommended data from C. Y. Ho et al.,
%J. Phys. Chem. Ref. Data V12 (2) 183 (1983)
x1d=[0.000,0.005,0.010,0.030,0.050,0.100,0.150,0.200,...
  0.250,0.300,0.350,0.400,0.450,0.500,0.550,0.600,0.650,...
  0.700,0.750,0.800,0.850,0.900,0.950,0.970,0.990,0.995,1.000];
g1d=[0.144,0.133,0.124,0.097,0.080,0.056,0.043,0.035,0.028,...
  0.024,0.022,0.021,0.020,0.020,0.021,0.022,0.024,0.027,...
  0.032,0.039,0.050,0.072,0.129,0.188,0.351,0.448,0.596];
Nd=length(x1d);
plot(x1d,g1d,'ko')
hold on
g1=g1d(Nd);     %pure copper g
g2=g1d(1);      %pure nickel g
```

```
x=0:0.05:1;
Nx=length(x);
for i=1:Nx
  y=1-x(i);
  gbar(i)=x(i)*g1+y*g2; %VCA
end
plot(x,gbar,'k--') %VCA Cu-Ni
```

## 9.8 Chapter 9 Exercises

9.8.1. After reading Example 9.3.0.1, running the script F0_simple_cubic.m, and reproducing Figure 9.3.4, modify the code in order to show that the imaginary part of $F_0$ is related to the density of states $D(E)$. Why?

9.8.2. Use the code impurity_simple_cubic.m of Example 9.4.0.1 and after confirming the result for the stated value of $E_{bo}$ given a perturbation energy of $\varepsilon_\mu = 3.25H_a$, modify the code to seek the bound level for a value of $\varepsilon_\mu = -3.25H_a$. What is the result?

9.8.3. Modify the code impurity_simple_cubic.m of Example 9.4.0.1 and obtain the impurity levels ($E_{bo}$) corresponding to perturbation energies $\varepsilon_\mu$ in the range $[3.25, 5.25]$ in steps 0.1. Produce a plot of $E_{bo}$ versus $\varepsilon_\mu$ and describe your observations.

9.8.4. Write a script to calculate the density of states of the unperturbed simple cubic system from Equation 9.10d and that of the perturbed system of Equation 9.10c for the impurity of Example 9.4.0.1. In your code, include a calculation of the total density of states in order to show the conservation of electron number; i.e., the area should be unity (ignoring spin) for each case. You will need a large energy array due to the impurity contribution. For this exercise, you may use the Romberg integration function rombergInt.m along with its associated interpolator, interpFunc.m, both introduced in Chapter 8. This exercise should effectively reproduce Figure 9.4.6.

9.8.5. Make use of the definitions of Equations 9.16a and 9.16b to obtain the $T$-matrix relation of Equation 9.16c.

9.8.6. Show the intermediate steps that lead from Equation 9.23a to Equation 9.23b.

9.8.7. Starting from Equation 9.26a, show the steps that lead to the expression for $T_r$ shown in Equation 9.26b.

9.8.8. Starting from Equation 9.26c, show the steps that lead to the self-consistent expression for $\Sigma$ shown in Equation 9.25b.

9.8.9. Give the steps that lead from Equation 9.25b to Equation 9.28, thus showing that they are equivalent.

9.8.10. Read Example 9.6.0.1 and reproduce the results shown in Figure 9.6.9. Run the code CPA.m for the case of when $x = 0$ and explain your observations.

9.8.11. Modify the code CPA.m of Example 9.6.0.1 in order to reproduce the results shown in Figure 9.6.10 for the split-band limit of the Alloy $A_xB_{1-x}$ density of states.

9.8.12. Modify the code CPA.m of Example 9.6.0.1 in order to reproduce the results shown in Figure 9.6.11 for the dilute Alloy $A_xB_{1-x}$ density of states.

9.8.13. Starting from Equation 9.34b, show the steps leading to Equation 9.35.

9.8.14. Modify the code alloy_CuNi_cond provided in Section 9.7 in order to include the effective medium calculation for the conductivity to reproduce Figure 9.7.14 in its entirety.

# 10

# *Magnetism I*

**Contents**



## 10.1 Introduction

As a young child, Albert Einstein was fascinated by the magnetic compass. The instrument mesmerized him so much that many years later he wrote, "That experience made a deep and lasting impression on me. Something deeper had to be hidden behind things." A lodestone, a naturally magnetized piece of magnetite with the chemical composition $Fe_3O_4$, was used as an early form of magnetic compass (see Figure 10.1.1). Shen Kuo, the polymathic Chinese scientist, was the first person to explicitly document the use of the compass as a tool for navigation purposes. Later, the compass played an instrumental role in the discovery of North America by Christopher Columbus. While magnetism as an effect was much valued in the ancient times for its utilitarian purposes, the reason behind its origin was not well understood. Michael Faraday's formulation of electrodynamics provided an explanation for the origin of electromagnetic radiation, but could not offer an explanation for the existence of spontaneous magnetism in nature.

At present, magnetic materials and their properties form the backbone of modern technological applications ranging from motors, generators, electromagnets, transformers, and loud speakers to magnetic resonance imaging (MRI) machines, sensors, giant magnetoresistance (GMR) read heads, and data storage in the form of magnetoresistive random-access memory (MRAM) to name a few.

In this chapter you will learn the basics of magnetism. We will assume the existence of magnetic moments and provide explanations for the various phenomena displayed by *dia-*, *para-*, and *ferro-*magnetism. You will learn about hysteresis, domain formation, the magnetocaloric effect utilized for refrigeration purposes, mean field theories of paramagnetic-ferromagnetic phase transition, and the Monte Carlo simulation method to study the magnetic phase transition in a ferromagnetic Ising model. Finally, you will learn that magnetism is purely quantum mechanical. The intrinsic spin angular momentum of an electron, the Coulomb electrostatic repulsion, and the Pauli exclusion principle are the key ingredients to unraveling the origins of magnetism. More will be said on the quantum origins of magnetism in Chapter 11.

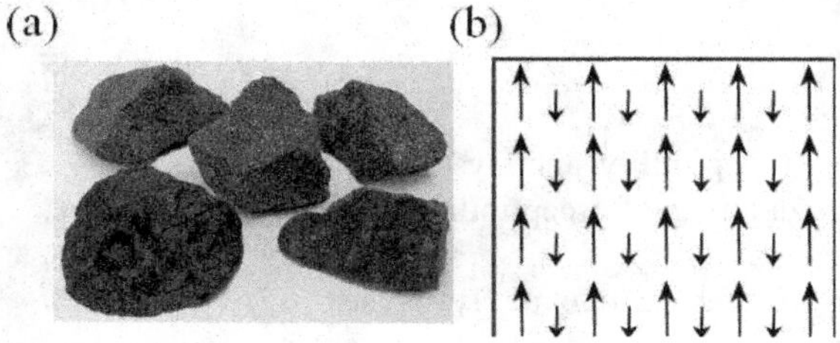


Figure 10.1.1: (a) Lodestone; (b) the naturally occurring magnetite lodestone is ferrimagnetic in nature. The ordering pattern is as shown. The unequal magnetic moments in the two sublattices give rise to a net residual magnetization.

## 10.2 Bohr Magneton

The basic building block of a magnet is its magnetic dipole moment, $\vec{\mu}$. In classical electromagnetism, this can be visualized as a current carrying loop where an electron revolves around the nucleus, Figure 10.2.2. The orbital motion generates an atomic magnetic dipole moment with a north and a south pole. Since the electron undergoes circular motion in an orbit, the generated magnetic moment can be related to the orbital angular momentum, $\vec{L}$. This fact can be utilized to estimate the size of $\vec{\mu}$. The basic definition of $\vec{\mu}$ is

$$\vec{\mu} = IA\hat{n}, \tag{10.1}$$

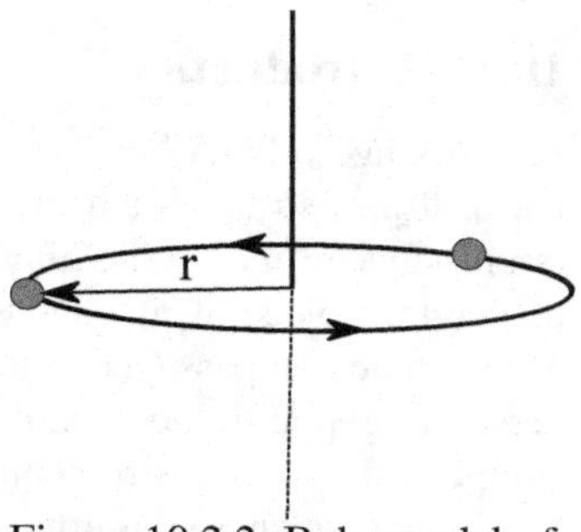


Figure 10.2.2: Bohr model of H atom.

where I is the current in Amperes, A is the area enclosed within the loop, and $\hat{n}$ is the unit normal vector to the loop according to the right-hand rule. For a circular orbit of radius r we have

$$\mu = \pi r^2 I. \tag{10.2}$$

The current from the circulating electron is given by

$$I = \frac{\text{charge}}{\text{time}} = \frac{-e}{(2\pi r/v)}, \tag{10.3}$$

where $v$ is its tangential velocity. Using Equations (10.1) and (10.3), we have

$$\mu = \pi r^2 I = \frac{(-e)vr}{2}. \tag{10.4}$$

From a classical physics perspective, all possible values of angular momentum, in principle, should be allowed. However, we know from the atomic Bohr model that this is not the case and only certain quantized energy levels are allowed. Moving forward, we assume the angular momentum of the electron to be $\hbar$. This gives

$$L = \hbar = m_e vr. \tag{10.5}$$

Utilizing a mixture of classical and quantum mechanics arguments to solve a physics problem is known as a semi-classical approach. Based upon this we have

$$\frac{\hbar}{m_e} = vr, \tag{10.6}$$

which substituted into Equation (10.4) gives for the dipole moment

$$\mu = \frac{(-e)\hbar}{2m_e} \equiv -\mu_B. \tag{10.7}$$

In the above equation we introduced the Bohr magneton

$$\mu_B = \frac{e\hbar}{2m_e} = 9.274 \times 10^{-24} A \cdot m^2, \tag{10.8}$$

as a convenient and natural unit for describing atomic sized magnetic moments. We can estimate the z−component of the magnetic dipole moment arising from the orbital motion. We have using $\vec{L}$ instead of $\hbar$

$$\vec{\mu}_L = -\frac{e}{2m_e}\vec{L}, \tag{10.9}$$

where the z−component is given by

$$\vec{\mu}_{L_z} = -\frac{e}{2m_e}L_z == -\frac{e}{2m_e}m_l\hbar = -m_l\mu_B. \tag{10.10}$$

In the above, we have used the relation $L_z = m_l\hbar$ where $m_l$ is the angular momentum quantum number.

**Example 10.2.0.1**
Swedish chemist Georg Brandt is credited with the discovery of cobalt, which is a silver-grey transition metal with ferromagnetic properties at room temperature. Minerals containing cobalt were used by ancient Egyptian and Mesopotamian civilizations to give glass its deep blue color. The ground state electronic configuration of cobalt is [Ar]$3d^7 4s^2$ as shown in Figure 10.2.3. *Assuming* that each unpaired electron contributes one $\mu_B$, what is the net magnetic moment of an isolated Co atom?

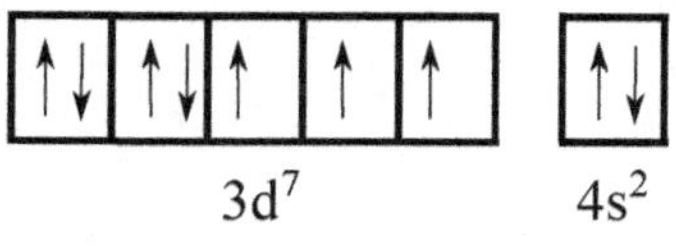


Figure 10.2.3: Electronic configuration of cobalt.

**Solution**

From the ground state electronic configuration of cobalt, we see that there are three unparied electrons. Since each electron contributes one Bohr magneton, based on our assumption, the net magnetic moment will be $3\mu_B$. However, in reality the true value of the magnetic moment per atom of cobalt is $1.72\mu_B$. This discrepancy can be attributed to the assumption of a localized-moment picture, which does not hold for the ferromagnetic metals. To explain the observed experimental value, we need to use concepts of band theory introduced earlier in Chapter 6. A full treatment of the itinerant magnetic behavior is beyond the scope of this introductory textbook.

## 10.3 Magnetization, Susceptibility, and Hysteresis

### 10.3.1 Magnetization and Susceptibility

Table 10.3.1: The various common magnetic ordering types along with examples and magnetic ordering pattern. Range of dimensionless magnetic susceptibility values $\chi$ for the different types of magnets are also listed. Figure 10.8.22 lists the magnetic state of elements in the periodic table.

| **Type of Magnet** | **Examples** | **Magnetic Ordering Pattern** | **Susceptibility** ($\chi_{mol}$, $m^3kg^{-1}$) |
|---|---|---|---|
| **(a) Diamagnet** | Cu, Au, Pb, $Ti^{4+}$, $Sc^{3+}$, $Al_2O_3$ | *No ordering pattern* | $10^{-5} - 10^{-8}$ (Negative) |
| **(b) Paramagnet** | Al, Gd, Pt, $U^{4+}$ , $TiO_2$, $(NH_4)_2Mn(SO_4)_2.2H_2O$ | 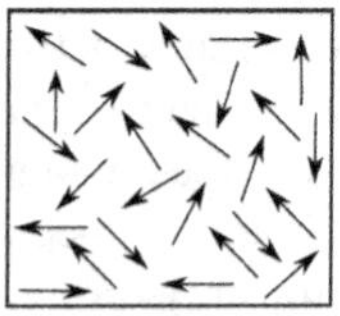 | $10^{-2} - 10^{-5}$ (Positive) |
| **(c) Ferromagnet** | Fe, Ni, Co, $Fe_2O_3$, $ZrZn_2$, MnNiSb | 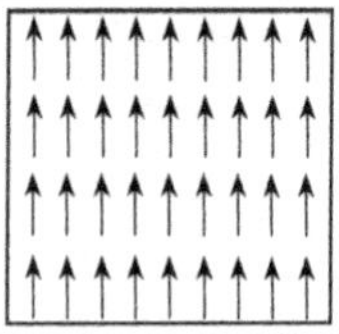 | $10^{3} - 10^{5}$ (Positive) |
| **(d) Ferrimagnet** | $Fe_3O_4$, $TbFe_2$, $Y_3Fe_5O_{12}$(YIG), $BaFe_{12}O_{19}$, | 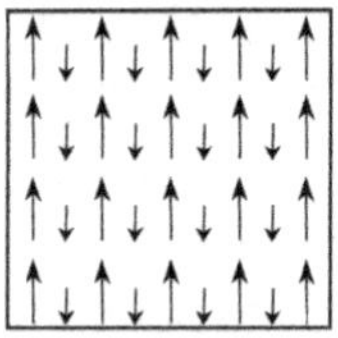 | $10 - 10^{3}$ (Positive) |
| **(e) Antiferromagnet** | Cr, $FeCl_2$, $MnF_2$, MnO, NiO, CoO | 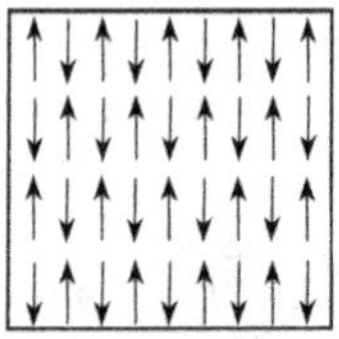 | $10^{-2} - 10^{-5}$ (Positive) |

A magnetic solid consists of a large number of atoms, each with net magnetic dipole moment. In general, magnetic moments in a solid are randomly oriented and the net dipole moment sum is equal to zero. However, in the presence of an externally applied magnetic field, the dipoles may align themselves in the direction of the field. In such a case, the solid acquires a net magnetic moment and becomes magnetized. The sum total of these moments, averaged over the sample volume, can be used to define a quantity called

$$\text{Magnetization}(\vec{M}) = \frac{\text{Net magnetic moment}(\vec{\mathscr{M}})}{\text{Volume}(V)}. \tag{10.11}$$

The net magnetic moment $\vec{\mathcal{M}}$ is equal to the sum of all the dipole moments $\vec{\mu}$ in the magnet.
The magnetic field vector $\vec{B}$ and the auxilliary field vector $\vec{H}$ provide a means to analyze magnetostatic behavior of magnetic materials. The units of $\vec{B}$ are Tesla (T). The units of $\vec{H}$ are $\text{Am}^{-1}$. In free space (vacuum), where there is no magnetization, $\vec{B}$ and $\vec{H}$ are linearly related by

$$\vec{B} = \mu_o \vec{H}, \tag{10.12}$$

where $\mu_0 = 4\pi \times 10^{-7}$ N/A$^2$ is the permeability of vacuum. The response of a magnetic system to an external magnetic field, $\vec{B}$, can be classified into three categories. They are

1. Magnetic dipole moments align opposite to the field (**diamagnetism**)
2. Magnetic dipole moments align in the direction of the field (**paramagnetism**)
3. Magnetic dipole moments align with the field and retain their orientation even after the external magnetic
4. field has been removed (**ferromagnetism**).

The classification of *magnet*-isms does not just end with the above three. There is, in addition to the above: ferrimagnetism, antiferromagnetism, helimagnetism, asperomagnetism, sperimagnetism, speromagnetism, and superparamagnetism. Table 10.3.1 lists examples of material compounds and elements for some of these magnetic types. A periodic table of magnetic elements at room temperature is provided at the end of the chapter in Figure 10.8.22.
In a magnetic solid, the relation between $\vec{B}$ and $\vec{H}$ is modified due to the magnetization, $\vec{M}$, of the medium. In the case of a linear, isotropic, and homogeneous magnetic media, Equation (10.12) is modified to

$$\vec{B} = \mu_o(\vec{H} + \vec{M}). \tag{10.13}$$

The magnetization is given by

$$\vec{M} = \chi \vec{H}, \tag{10.14}$$

where $\chi$ is the dimensionless magnetic susceptibility. The unit for $\vec{M}$ is $\text{Am}^{-1}$. From the above, we have

$$\vec{B} = \mu_o(1+\chi)\vec{H} = \mu_o \mu_r \vec{H} \tag{10.15}$$

where $\mu_r = 1 + \chi$ is the relative permeability. In Equation (10.14) the magnetization response is stated in relation to $\vec{H}$ and not $\vec{B}$. The reason for this choice is rooted in the fact that $\vec{B}$ is sensitive to the total current density (conduction and amperian magnetization) of the medium, whereas $\vec{H}$ is related to the conduction current density only. Since there is no viable experimental method to measure the bound circulating atomic currents giving rise to the amperian current density, it is practical to express the magnetization response of a system as a function of $\vec{H}$, which is controlled by the current in the circuit. The relationship between $\vec{M}$ and $\vec{H}$ will be explored further in Section 10.3.2 by analyzing hysteresis curves in magnetic media.
Bulk magnetic susceptibility can be measured using a variety of experimental methods. Classical susceptibility measurement techniques involving forces exerted on a magnetized specimen are utilized in the Faraday's balance and Gouy's balance setup. The main difference between the two methods is in the use of an inhomogeneous (Faraday) and homogeneous (Gouy) magnetic field. In Exercise 10.9.4 you will obtain a relationship between the magnetic susceptibility and the force experienced by a magnetic specimen in a Gouy balance method. These methods are suitable for finely divided solids and liquids. Presently, *S*uperconducting *QU*uantum *I*nterference *D*evice (SQUID) magnetometers, sensitive to the detection of a magnetic flux quantum, allow for an accurate measurement of the magnetic moment of a sample, from which the magnetization and magnetic susceptibility can be obtained. SQUID magnetometers can perform both DC and AC magnetic moment

measurement and is especially useful for low temperature studies. Magnetic resonance methods involving Nuclear Magnetic Resonance (NMR) can also be used to measure the paramagnetic susceptibility of solutions (see Chapter 11).

The susceptibility curves for a dia-, ferro-, and antiferromagnet are displayed in Figure 10.5.13. Table 10.3.1 lists the range of dimensionless magnetic susceptibility values along with their magnetism classification scheme. In general magnetic susceptibility of crystals is not isotropic, and can depend on the spatial direction. In that case we define a susceptibility tensor, $\chi_{ij}$, which takes into account the anisotropy such that the magnetization response can occur in directions other than the direction of the applied field. Magnetic susceptibility data for materials may be stated as a dimensionless number $\chi$, as a molar susceptibility $\chi_{mol}$ m$^3$ mol$^{-1}$ value

$$\chi_{\text{mol}} = \chi V_{\text{mol}}, \tag{10.16}$$

where $V_{\text{mol}}$ is the molar volume, or as a mass susceptibility $\chi_g$ m$^3$ kg$^{-1}$ value

$$\chi_g = \frac{\chi}{\rho}, \tag{10.17}$$

where $\rho$ is the density. Since mass is easier to measure, mass susceptibility measurements are typically reported. However, for materials which are in a gaseous state, molar susceptibility is a more convenient choice of unit. Therefore, for practical purposes, it is important to know how to convert between the various forms of susceptibility values.

**Example 10.3.1.1**

Spanish general Antonio de Ulloa is credited with the discovery of platinum (Pt) which is a highly unreactive precious silvery white metal with excellent resistance to corrosion. The metal is used in catalytic converters, electrodes, platinum resistance thermometers, and even in jewelry. It has a dimensionless magnetic susceptibility of $\chi$ = 2.61 × 10$^{-4}$. What is the mass susceptibility of Pt? The density of platinum is $\rho$ =21450 kg m$^{-3}$ and its relative atomic mass is 0.19509 kg mo$l^{-1}$.

**Solution**

Mass susceptibility is defined as

$$\chi_g = \frac{\chi}{\rho}. \tag{10.18}$$

in units of m$^3$ kg$^{-1}$. We then have for Pt

$$\chi_g = \frac{2.61 \times 10^{-4}}{21450} = 1.22 \times 10^{-8} \text{m}^3 \ \text{kg}^{-1}. \tag{10.19}$$

Mass susceptibilities are generally three to four orders of magnitude less than the dimensionless susceptibility.

### 10.3.2 Hysteresis

Magnetic moments in a ferromagnetic solid, such as Fe, are all aligned in the same direction only within a small region of space called a magnetic domain (Figure 10.3.4). A typical domain is 100$\mu$m in size, separated by a domain wall of width $\approx$ 0.1$\mu$m. In general, formation of domains and their eventual size depends on a delicate magnetic (free) energy minimization process. Once formed, a set of randomly oriented domains give rise to a net zero magnetization. However, if the ferromagnet is subjected to an applied external field, $\vec{H}_a$, the domains can grow and reorganize to align as best as possible with the external field to give rise to an overall net magnetization.

Figure 10.3.4 shows a typical hysteresis loop plotted as function of $\vec{M}$ vs. $\vec{H}$. Initially, the magnetization response of the material follows a non-linear curve when the field is increased from a zero

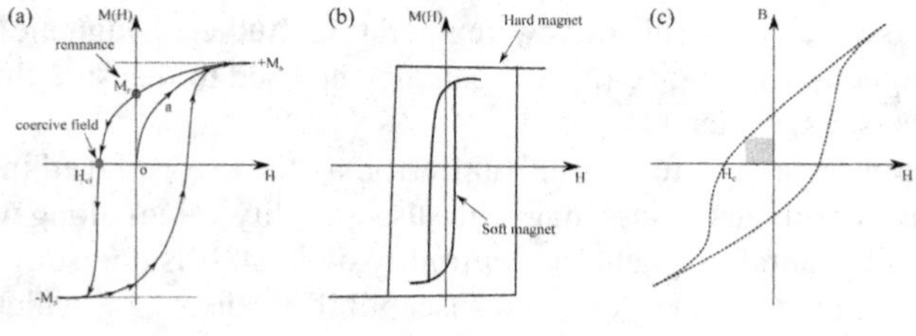


Figure 10.3.4: (a) Hysteresis loop. Remnant magnetization is denoted by the letter $M_r$. The intrinsic coercivity of the sample is given by $H_{ci}$. (b) Hysteresis loop of a hard and soft magnet. (c) The area within a hysteresis loop is a direct measure of the energy loss per unit volume of the material per hysteresis cycle. B-H curve used in the computation of the energy product.

field value, path o $\rightarrow$ a. Upon increasing the field further, those domains which are in the direction of field grow, while others shrink. At higher values of the field, domain rotation and alignment with the crystallographic easy axis occurs leading to magnetic saturation $+M_s$. If the driving field is now decreased and eventually removed, the ferromagnetic material retains a considerable amount of its magnetization. This is known as remnant magnetization or **remnance** which is typically of the same order of magnitude as the spontaneous magnetization. It is indicated by the letter $M_r$ on the $M(H)$-axis. The applied field can be reversed even further to drive the value of the residual magnetization to zero. The field at which this happens is known as the *intrinsic* **coercive field**, $H_{ci}$, and is labelled on the negative $H$-axis. Coercive fields can range from less than 1 Am$^{-1}$ in soft magnets to $10^6$ Am$^{-1}$ in hard magnets. As the field is further increased in the opposite direction, the material is fully saturated in the opposite direction, $-M_s$. The entire process is known as **hysteresis**. In general, hysteresis[1] means to lag behind. In the context of magnetic materials, the resulting magnetization tends to lag behind the external applied field giving rise to a hysteresis loop as the external field is repeatedly cycled back and forth. Hysteresis curves are displayed by ferromagnets. Paramagnets do not show any hysteresis plots.

**Example 10.3.2.1**

The maximum possible magnetization of a ferromagnetic material is called **saturation magnetization** $M_s$ – equal to the product of the net magnetic moment for each atom and the number of atoms present. With this definition in mind, calculate (a) the saturation magnetization and (b) the corresponding saturation magnetic field $B_s$ for Co. The net magnetic moment per atom of cobalt is $1.72\mu_B$. The room temperature density is $\rho$=8.9 g/cm$^3$ and atomic weight $A_{Co}$ is 58.92 g/mol.

**Solution**

(a) The saturation magnetization is obtained from

$$M_s = 1.72\mu_B N, \tag{10.20}$$

where N is the number of atoms per m$^3$ calculated as

$$\begin{aligned} N &= \frac{\rho N_A}{A_{Co}}, \\ &= \frac{(8.9\times10^6\ \mathrm{g/m^3})(6.023\times10^{23}\ \mathrm{atoms/mol})}{58.93\ \mathrm{g/mol}}, \\ &= 9.10\times10^{28}\mathrm{atoms\ m^{-3}}. \end{aligned} \tag{10.21}$$

[1]The word hysteresis originates from the Greek word *husterein*.

Using N, the value of the saturation magnetization is

$$\begin{aligned} M_s &= (1.72)(9.27\times10^{-24}\ \text{Am}^2)(9.10\times10^{28}), \\ &= 1.45\times10^{6}\ \text{Am}^{-1}. \end{aligned} \tag{10.22}$$

(b) The corresponding saturation magnetic field is given by

$$\begin{aligned} B_s &= \mu_o M_s, \\ &= (4\pi\times10^{-7})(1.45\times10^{6}) = 1.82\ T. \end{aligned} \tag{10.23}$$

The key to understanding hysteresis behavior is to study the growth and motion of magnetic domains. As domains grow, they push across the domain walls. Metallurgical properties of the magnet dictate the level of strain and impurity that can pin the motion of the domain walls. Pinning increases the coercivity of the material which is a desirable trait in hard magnets. Permanent magnets, used in generators and motors, are made from **hard magnets**. The thick loop in Fig. 10.3.4(b) corresponds to a hard magnetic material. Rare earth magnets form high quality permanent magnets. In contrast, materials with high purity have few dislocations or dopants. They offer less resistance to the motion of domain walls and are easily magnetized and demagnetized. These materials are good candidates for **soft magnets** which are used in electromagnets and transformer cores, where they are able to reverse the direction of their magnetization rapidly. The thin hysteresis loop in Fig. 10.3.4 corresponds to a soft magnetic material. Table 10.3.2 lists examples and applications of hard and soft magnets used in modern-day technology.

Table 10.3.2: Examples of hard and soft magnetic materials and their technological applications.

| Type | Materials | Applications |
|---|---|---|
| Hard | Alnico<br>$SrFe_{12}O_{19}$<br>$SmCo_5$<br>$Nd_2Fe_{14}B$ | Electric motors, loudspeakers<br>Magnetic recording<br>Sensors, computer hard drives<br>Magnetic resonance imaging (MRI)<br>Voice coil motors (VCMs) |
| Soft | Fe-Co (permendur)<br>Si steel<br>Fe-Si-Al powder (sendust)<br>Yttrium iron garnet (YIG) | Electromagnets, relays<br>Transformers, motors, generators<br>Inductors<br>Microwave isolaters, phase shifters |

An important parameter estimating the strength of a hard magnet is given by the **energy product** $(\vec{B}\cdot\vec{H})_{\text{max}}$ in units of kJ/m$^3$ related to the B(H) loop. Consider the hysteresis loop shown in Figure 10.3.4(c). The energy product corresponds to the area of the largest B−H rectangle that is constructed within the second quadrant of the hysteresis curve. The energy product value is representative of the energy required to demagnetize a permanent magnet and gives a measure of the energy made available by the magnet to perform work in outer space. The larger this value is, the harder is the magnet. The theoretical upper bound for the energy product is given by $\mu_o M_s^2/4$ where $M_s$ is the saturation magnetization. See Exercise 10.9.6.

Domain formation is an energy balancing act between the demagnetizing field energy (which helps to minimize the magnetostatic energy) and the cost of creating a domain wall. The size of the domain wall itself is a competition between exchange and anisotropy energy. We will not dwell on all the details of domain theory; rather, we focus on highlighting the important energy terms which occur in the magnetic free energy, $\mathscr{F}$. The main constitutents are

$$\mathscr{F} = \mathscr{E}_{ex} + \mathscr{E}_{anis} + \mathscr{E}_d + \mathscr{E}_{m-s} + \mathscr{E}_z + \mathscr{E}_{stress}, \tag{10.24}$$

where the symbols stand for

- $\mathscr{E}_{ex}$: Exchange energy,
- $\mathscr{E}_{anis}$: Magnetocrystalline anisotropy energy,
- $\mathscr{E}_{d}$: Demagnetizing field energy,
- $\mathscr{E}_{m-s}$: Magnetostriction energy,
- $\mathscr{E}_{z}$: Zeeman energy,
- $\mathscr{E}_{stress}$: Applied stress energy.

In the next chapter, you will learn about the exchange energy. In this chapter we will focus on the demagnetizing field energy, the Zeeman energy term, magnetocrystalline anisotropy energy, and magnetostriction term. You will learn about the applied stress energy term in Exercise 10.9.13. Figure 10.3.5 highlights the processes through which domains arise. A uniformly magnetized sample possesses substantial magnetostatic (demagnetizing) field energy (Section 10.3.2). To minimize this contribution, Figure 10.3.5(a), the sample splits into two domains, Figure 10.3.5(b) at the cost of creating a domain wall, but reducing the overall magnetostatic energy. This process continues further in step (c), but field lines still exist outside the magnetic sample. To remove this, in step (d) we see the formation of a set of closure domains where the energy cost due to the demagnetizing field is zero, the magnetization component at each domain wall boundary is continous, and the energy is minimized. Iron forms domain patterns such as this.

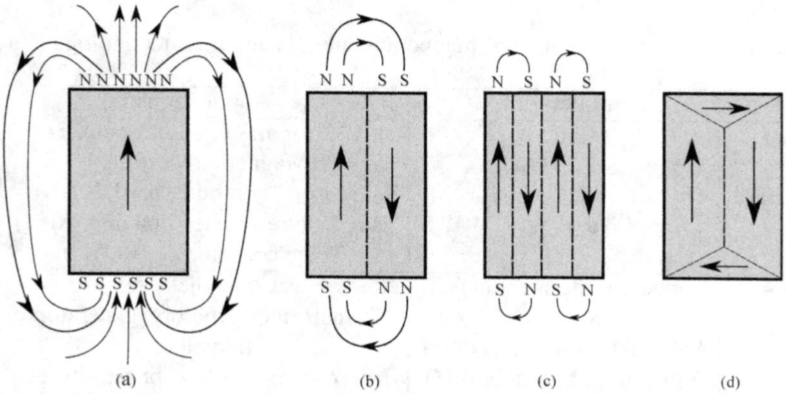


Figure 10.3.5: Origin of domains via minimization of magnetic free energy. Dashed lines represent domain walls. Arrows represent the direction of magnetization with a domain.

Experimentally, magnetic domains are visualized by photomicrographs of domain boundaries obtained either by the technique of magnetic powder patterns or by optical studies using Faraday rotation. The powder pattern technique was developed by F. Bitter and consists of placing a drop of colloidal suspension of finely divided ferromagnetic material such as magnetite on the surface of the ferromagnetic crystal. The colloidal particles strongly concentrate about the boundaries between domains where strong local magnetic fields exist which attract magnetic particles. In the next few sections, you will learn the details of the individual energy constituents which help create a domain.

*Demagnetizing field*

In a ferromagnetic body, the magnetization vector abruptly terminates at the surface resulting in a non-zero divergence. Recall from basic electrodynamics that

$$\vec{B} = \mu_o(\vec{H} + \vec{M}); \quad \vec{\nabla} \cdot \vec{B} = 0. \tag{10.25}$$

These two equations imply that

$$\vec{\nabla}\cdot\vec{H} = -\vec{\nabla}\cdot\vec{M}. \tag{10.26}$$

Hence, a vanishing magnetization value at the surface results in negative divergence of the auxilliary field $\vec{H}$, directed opposite to the magnetization inside the magnetic medium, see Figure 10.3.6(c). The $\vec{H}$ field manifests itself as a **demagnetizing field**, $\vec{H}_d$, within the sample and outside the sample $\vec{H}$ is known as the stray field. The relationship between $\vec{H}$ and $\vec{M}$ is given by

$$\vec{H}_d = -\mathcal{N}\vec{M}, \tag{10.27}$$

where the unitless factor $\mathcal{N}$ is known as the **demagnetizing factor**. The value of $\mathcal{N}$, in general, depends on different directions. So we often express the demagnetization factors as a **demagnetization tensor** N. In this case, Equation (10.27) should be generalized to

$$\vec{H}_d = -\mathrm{N}\cdot\vec{M}, \tag{10.28}$$

where the dot represents matrix−vector multiplication. Demagnetization factors for some common shapes are listed in Table 10.3.3. In an ellipsoidal sample, if the semiaxes of the ellipsoid coincide with the axes of the coordinate system, the tensor becomes diagonal. Typically, associate three demagnetization factors along the principal axes as $\mathcal{N}_\mathrm{a}$, $\mathcal{N}_\mathrm{b}$, and $\mathcal{N}_\mathrm{c}$. These factors can be arranged in a matrix form known as the demagnetization tensor

$$(\mathrm{N}) = \begin{pmatrix} \mathcal{N}_\mathrm{a} & 0 & 0 \\ 0 & \mathcal{N}_\mathrm{b} & 0 \\ 0 & 0 & \mathcal{N}_\mathrm{c} \end{pmatrix}, \tag{10.29}$$

which obey the general constraint

$$\mathcal{N}_\mathrm{a} + \mathcal{N}_\mathrm{b} + \mathcal{N}_\mathrm{c} = 1. \tag{10.30}$$

Table 10.3.3: Examples of demagnetization factor $\mathcal{N}$ for various geometries.

| Shape | Magnetization direction | $\mathcal{N}$ |
|---|---|---|
| Sphere | Any direction | $\frac{1}{3}$ |
| Long needle | Parallel to the axis | 0 |
| | Perpendicular to the axis | $\frac{1}{2}$ |
| Thin film | Parallel to the plane | 0 |
| | Perpendicular to the plane | 1 |

**Example 10.3.2.2**
Calculate the demagnetization energy of a long needle.
**Solution**
The energy contribution $\mathscr{E}_d$ of the demagnetizing field is defined by

$$\mathscr{E}_d = -\frac{\mu_o}{2}\int_V \vec{M}\cdot\vec{H}_d dV. \tag{10.31}$$

From Table 10.3.3 we can write down the demagnetization tensor for a long needle as

$$\mathrm{N} = \begin{pmatrix} \frac{1}{2} & 0 & 0 \\ 0 & \frac{1}{2} & 0 \\ 0 & 0 & 0 \end{pmatrix}, \tag{10.32}$$

so that with $\vec{M} = M_x\hat{\imath} + M_y\hat{\jmath} + M_z\hat{k}$, $\vec{H}_d = -\frac{1}{2}M_x\hat{\imath} - \frac{1}{2}M_y\hat{\jmath}$, and using Equation (10.31) we can compute the demagnetization energy as

$$\begin{aligned}\mathscr{E}_d &= \frac{\mu_o}{4}\int_V \vec{M}\cdot(M_x\hat{\imath}+M_y\hat{\jmath})dV,\\ &= \frac{1}{4}\mu_o V(M_x^2+M_y^2).\end{aligned} \tag{10.33}$$

Using $M_x = M_s \sin\theta\cos\varphi$, $M_y = M_s\sin\theta\sin\varphi$, and $M_z = M_s\cos\theta$ (see Figure 10.3.7). The azimuthal angle is given by $\varphi$ and polar angle is given by $\theta$. We can write the final form for the energy as

$$\mathscr{E}_d = \frac{1}{4}\mu_o V M_s^2 \sin^2\theta. \tag{10.34}$$

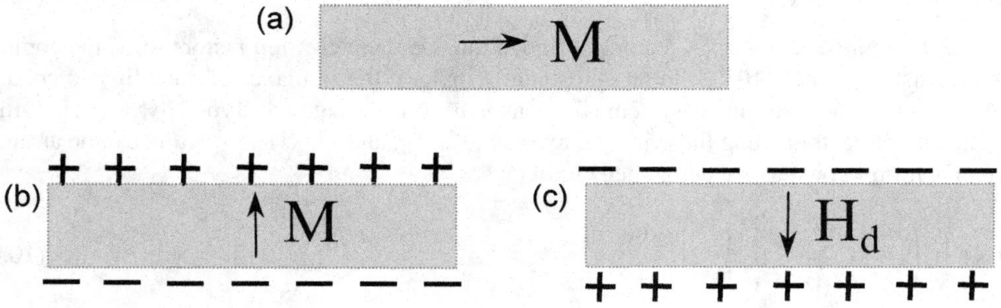


Figure 10.3.6: Demagnetization field effect in an infinite flat plate. (a) When the magnetization lies in the plane of the plate, no magnetic poles are created at the surface. However, when the magnetization $\vec{M}$ is perpendicular to the plane of the plate, as shown in (b), surface magnetic poles are created due to the discontinuity. This sets up an opposing demagnetizing field $H_d$ displayed in (c).

Experimentally, magnetization $M$ is measured in response to an applied field $H_a$. However, a demagnetizing field is set up within the medium in response to the externally applied field giving rise to a net internal field $H_i$. The value of this net internal field and the corresponding magnetic field can vary within the magnet. For the special case of a spherical or ellipsoidal sample, we have $H_i = H_a = -\mathscr{N}M$. Using Equation (10.25) we see that this gives the internal magnetic field as $\vec{B}_i = \mu_o(\vec{H}_i + \vec{M})$.

**Example 10.3.2.3**

Calculate the demagnetization energy of a spheroid.

**Solution**

In this example, we compute the demagnetization energy of a spheroid. To do so, we first define

$$\vec{M} = M_x\hat{\imath} + M_y\hat{\jmath} + M_z\hat{k}. \tag{10.35}$$

Utilizing the demagnetization tensor, Equation (10.29) and Equation (10.35), we can write

$$\vec{H}_d = -\mathscr{N}_a M_x\hat{\imath} - \mathscr{N}_b M_y\hat{\jmath} - \mathscr{N}_c M_z\hat{k}. \tag{10.36}$$

Using Equations (10.35) and (10.36) in the definition for $\mathscr{E}_d$ (Equation (10.31)) and integrating we obtain

$$\mathscr{E}_d = \frac{\mu_o}{2}(\mathscr{N}_a M_x^2 + \mathscr{N}_b M_y^2 + \mathscr{N}_c M_z^2)V, \tag{10.37}$$

where V is the sample volume. In the case of a spheroid with a = b ≠ c, we define $\mathscr{N}_a = \mathscr{N}_b = \mathscr{N}_\perp$ and $\mathscr{N}_c = \mathscr{N}_\parallel$. This implies we can write the expression for the demagnetization energy as

$$\mathscr{E}_d = \frac{\mu_o}{2}[\mathscr{N}_\perp(M_x^2 + M_y^2) + \mathscr{N}_\parallel M_z^2]. \tag{10.38}$$

The above equation can be further rewritten in terms of the angle $\theta$ subtended by the magnetization vector with respect to the easy axis (polar) direction, see Figure 10.3.7. Using $M_x = M_s \sin\theta\cos\varphi$, $M_y = M_s \sin\theta \sin\varphi$, and $M_z = M_s \cos\theta$, we can write

$$\mathscr{E}_d = \frac{\mu_o}{2} M_s^2 [\mathscr{N}_\perp \sin^2\theta + \mathscr{N}_\parallel \cos^2\theta] V, \tag{10.39}$$

where $\mathscr{N}_\parallel$ and $\mathscr{N}_\perp$ are demagnetization factors along the polar and equatorial axis, respectively. For a prolate (elongated) sphere $\mathscr{N}_\perp > \mathscr{N}_\parallel$ and for an oblate (flattened) sphere $\mathscr{N}_\perp < \mathscr{N}_\parallel$. Using these parameters, demagnetization energy can be rewritten as

$$\mathscr{E}_d = \frac{\mu_o}{2} M_s^2 V [\mathscr{N}_\parallel + (\mathscr{N}_\perp - \mathscr{N}_\parallel)\sin^2\theta]. \tag{10.40}$$

From Equation (10.40) we see that for a prolate spheroid when $\theta = 0$, the energy is a minimum, and when $\theta = \frac{\pi}{2}$ it is a maximum. The converse holds true for an oblate spheroid. The dependence of the demagnetization energy on the geometrical shape of the body gives rise to shape anisotropy (Exercise 10.9.12). Shape anisotropy is homogeneous for uniformly magnetized ellipsoidal bodies. For non-ellipsoidal samples, it is position dependent.

*Zeeman energy*

A magnetic body in an external field has an interaction energy given by

$$\mathscr{E}_z = -\mu_o \int_V \vec{M} \cdot \vec{H}_a dV, \tag{10.41}$$

where $\vec{H}_a$ is the applied external field. For example, a dipole moment of magnitude $\mu$ oriented at an angle $\theta$ to the external magnetic field $H_a$ applied along the z-axis, will have a **Zeeman energy** given by -$\mu_o \mu H \cos\theta$, where we have used the definition of magnetization given in Equation (10.11).

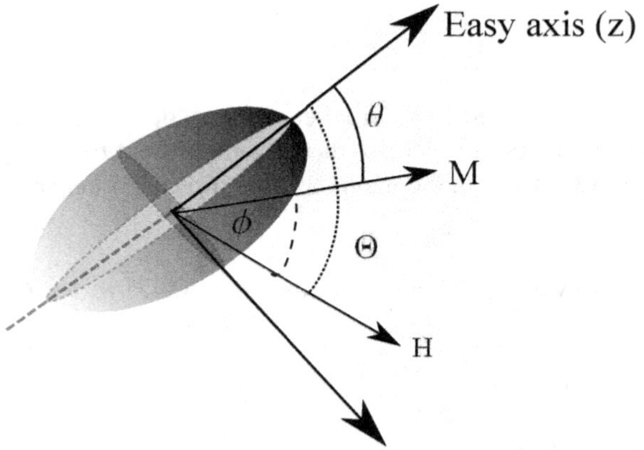


Figure 10.3.7: Ellipsoidal prolate shaped single domain magnetic particle. The magnetization vector $\vec{M}$ makes an angle $\theta$ with the easy axis (taken as the z-axis direction) and an angle $\phi$ with the applied field $H_a$. The angle between the applied field and the easy axis is $\Theta$. Uniaxial anisotropy is a characteristic feature of elongated particles.

*Magnetocrystalline (crystal) anisotropy*

The dependence of magnetic properties on the direction in which they are measured gives rise to **magnetic anisotropy**. Consider a single ferromagnetic domain where the magnetization is oriented at an angle $\theta$ to the z-axis taken to be along the principle axis of crystal symmetry. For crystals with a single axis of high symmetry the **uniaxial anisotropy** energy expression is written as

$$E_{anis} = K_u \sin^2\theta, \tag{10.42}$$

where $\theta$ is the angle between the polar z-axis and the magnetization vector as shown in Figure 10.3.7 and $K_u$ is the uniaxial anisotropy energy constant. If you compare the form of this uniaxial anisotropy energy with Equation (10.34), you will find that they have a similar angular symmetry dependence. In the most general case, the multiplicative anisotropy constant $K_u$ can be either positive or negative. For $K_u > 0$, minimization of $E_{anis}$ with respect to the angle $\theta$ gives zero, which implies that the z-axis is a direction of easy magnetization. In this case the z-axis is known as the easy axis. For $K_u < 0$, $\theta = \pi/2$ the direction of **easy axis** magnetization lies in the xy (basal) plane of the crystal. Such a ferromagnet is known as the easy-plane type. Similarly there are directions along which the magnetization is difficult to align and are known as the **hard axis.** In magnetic domains, the net magnetization direction tends to align along an easy axis. Different materials have different easy axis, see Figure 10.3.8.

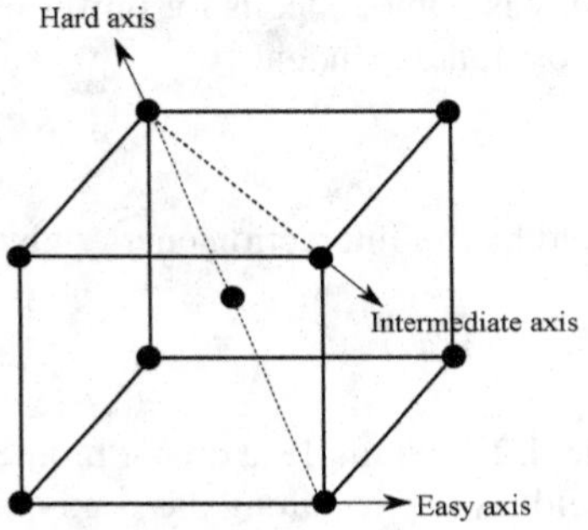


Figure 10.3.8: Easy, intermediate, and hard axis for a unit cell of bcc iron.

**Example 10.3.2.4**
The anisotropy field, $H_{\text{anis}}$, is defined as the applied field needed to saturate the magnetization of a uniaxial crystal in a hard direction. Calculate the anisotropy field for a single-domain ellipsoidal particle as shown in Figure 10.3.7 with the magnetization directed at an angle $\theta$ with respect to the easy axis.
**Solution**
Supplementing our anisotropy energy expression Equation (10.42) with a Zeeman energy term due to the external field $H$ we have

$$E = K_u \sin^2\theta - \mu_o M_s H \cos(\frac{\pi}{2} - \theta), \tag{10.43}$$

where we have chosen $\Theta = \frac{\pi}{2}$ and $\theta$ is the angle made by the magnetization vector with the anisotropy axis. To obtain the anisotropy field, $H_{\text{anis}}$, we first minimize the energy by setting $\partial E/\partial\theta = 0$. This gives

$$H_{\text{anis}} = \frac{2K_u}{\mu_o M_s} \sin\theta. \tag{10.44}$$

Since the hard axis is at $\theta = \pi/2$, we obtain $H_{anis} = 2K/\mu_o M_s$. Typical anisotropy field ranges from less than 2 $kAm^{-1}$ to greater than 20 $MAm^{-1}$.

Magnetocrystalline anisotropy is an intrinsic property of a material dictated by crystal symmetry. Expressions for anisotropy energies can be stated either in terms of the magnetization vector direction cosines ($*\theta_1, \theta_2$, and $\theta_3$) or in terms of a set of orthonormal spherical harmonics $(\theta, \phi)$. Magnetocrystalline anisotropy expression for the hexagonal, tetragonal, and cubic crystal systems are given by

$$\begin{aligned}
\text{Hexagonal}: E_{anis} &= K_0 + K_1 \sin^2\theta + K_2 \sin^4\theta + K_3 \sin^6\theta + K_3' \sin^6\theta \sin 6\phi, \\
\text{Tetragonal}: E_{anis} &= K_1 \sin^2\theta + K_2 \sin^4\theta + K_2' \sin^4\theta \cos 4\phi + K_3 \sin^6\theta + K_3' \sin^6\theta \sin 4\phi, \\
\text{Cubic}: E_{anis} &= K_0 + K_1(\alpha_1^2\alpha_2^2 + \alpha_2^2\alpha_3^2 + \alpha_3^2\alpha_1^2) + K_2(\alpha_1^2\alpha_2^2\alpha_3^2).
\end{aligned}$$

where the angle $\varphi$ is with respect to the x-axis in the basal plane (azimuthal angle), the polar angle $\theta$ is measured with respect to the z-axis, and the explicit expressions for the magnetization direction cosines in terms of spherical harmonic variables are given by $\alpha_1 = \sin\theta\cos\phi$, $\alpha_2 = \sin\theta\sin\phi$, and $\alpha_3 = \cos\theta$. All the anisotropy constants have dimensions of energy per unit volume. In Table 10.3.4 we list the $K_1$ and $K_2$ values for the three common ferromagnetic elements with cubic anisotropy. Values of the other anisotropy parameters may be found in materials science data book.

Table 10.3.4: Anisotropy constant values for Fe, Co, and Ni.

| Anisotropy Constant [$J/m^3$] | Fe(bcc) | Co(hcp) | Ni(fcc) |
|---|---|---|---|
| $K_1$ | 54800 | 760000 | -126300 |
| $K_2$ | 1960 | 100500 | 57800 |

In the absence of any anisotropy energy term, the magnetization vector does not choose any particular direction in space. The energy cost associated with the orientation of the ferromagnetic vector is equal for all possible orientations in space. But, with the introduction of anisotropy the spherical symmetry of the system is broken. In general, the magnetocrystalline energy equation is a complicated function of anisotropy parameters and angles. To concisely display and extract useful information concerning anisotropy, it is often customary to draw what is known as a **magnetization anisotropy energy surface**, for example as shown in Figure 10.3.9. Consider the example shown below.

**Example 10.3.2.5**
Using the expression for cubic magnetic anisotropy energy, write a MATLAB code to display the magnetic anisotropy energy surface of an isotropic cubic crystal in (i) the absence and (ii) the presence of a magnetic field.
**Solution**
Since we are dealing with a cubic system, we use the energy expression below with an appropriate magnetic field term

$$E_{anis} = K_0 + K_1\left(\alpha_1^2\alpha_2^2 + \alpha_2^2\alpha_3^2 + \alpha_3^2\alpha_1^2\right) + K_2\left(\alpha_1^2\alpha_2^2\alpha_3^2\right) + \frac{h}{\sqrt{3}}(\alpha_1 + \alpha_2 + \alpha_3), \tag{10.45}$$

where $\alpha_1 = \sin\theta\cos\phi$, $\alpha_2 = \sin\theta\sin\phi$, and $\alpha_3 = \cos\theta$. Since we are concerned with an isotropic system $K_1 = K_2 = 0$. We choose $K_0 = 1$. The magnetic field parameter $h = 1.5$ the code. The MATLAB code displaying the anisotropy energy surface is given by

```
%copyright by J. E Hasbun and T. Datta
%ch10_magnetic_anisotropy_iso.m
```

```
% ch10_magnetic_anisotropy.m - by J. Hasbun and T. Datta
% code to simulate the magnetic anisotropy energy surface
% for an isotropic cubic crystal - with and without magnetic field

clear;

% declaring the anisotropy constants
% isotropic case
K0 = 1; K1 = 0;

% definining the anisotropy energy surface projections along x,y, and z

ecubic = @(x,y,K0,K1,H) (K0+ K1*(((sin(x).*cos(y)).^2).*...
       ((sin(x).*sin(y)).^2 + ((cos(x)).^2).*((sin(x).*sin(y)).^2)+...
       ((cos(x)).^2).*((sin(x).*cos(y)).^2))+ ...
       (H/sqrt(3))*(sin(x).*cos(y)+sin(x).*sin(y)+cos(x)));

fcubicx = @(x,y,K0,K1,H) (sin(x).*cos(y)).*ecubic(x,y,K0,K1,H);
fcubicy = @(x,y,K0,K1,H) (sin(x).*sin(y)).*ecubic(x,y,K0,K1,H);
fcubicz = @(x,y,K0,K1,H) (cos(x)).*ecubic(x,y,K0,K1,H);

rotate3d on;
colormap(copper);
subplot(1,2,1)
ycubic = ezsurf(@(x,y) fcubicx(x,y,K0,K1,0),@(x,y) ...
   fcubicy(x,y,K0,K1,0),@(x,y) fcubicz(x,y,K0,K1,0),[0 pi],[0 2*pi]);
title('H = 0 (Zero field)');
% view(45,45)

% plot with magnetic field for the simple cubic lattice

rotate3d on;
colormap(copper);
subplot(1,2,2)
ycubich = ezsurf(@(x,y) fcubicx(x,y,K0,K1,1.5),@(x,y) ...
   fcubicy(x,y,K0,K1,1.5),@(x,y) fcubicz(x,y,K0,K1,1.5),[0 pi],[0 2*pi]);
title('H = 1.5 (Strong field)')
% view(45,45)
```

The output from the code is shown below

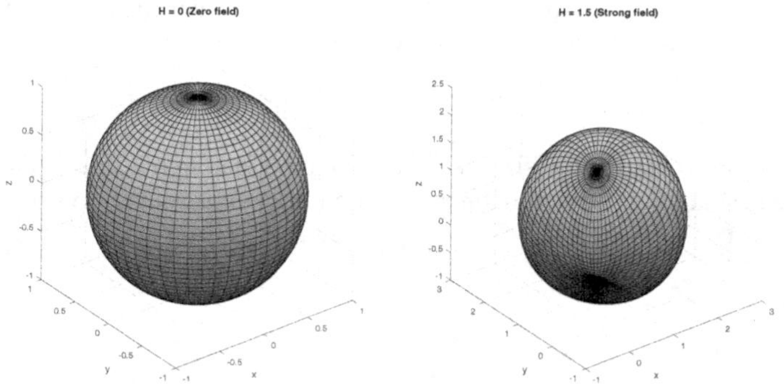


Figure 10.3.9: Magnetic anisotropy energy surface for a cubic crystal. The spherical shape in the absence of anisotropy demonstrates the energy insensitivity of the system to any preferrred mangetization direction. However, with the introduction of a magnetic field, the surface starts to get pinched with the evolution of a preferred orientation axis.

In Problem 10.9.12 we will plot the magnetic anisotropy energy surface for a FCC and a HCP crystal. In Figure 10.3.10 we display the anisotropy energy surface of a BCC cubic crystal.

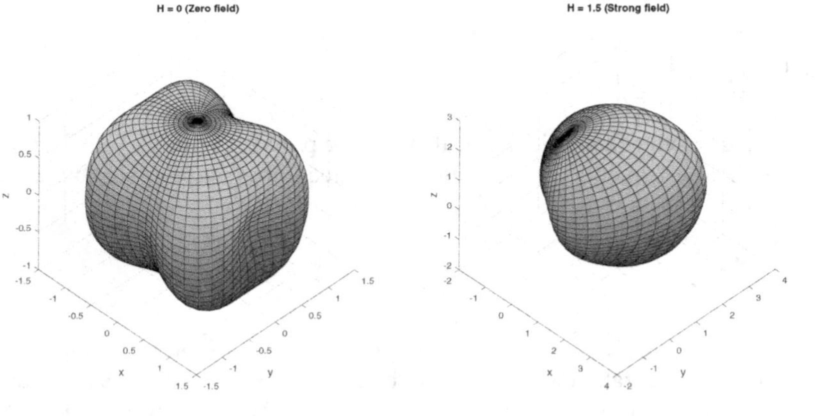


Figure 10.3.10: Magnetic anisotropy energy surface for Fe, a BCC cubic crystal, both in the presence and absence of an external magnetic field H. For a large magnetic field, there is only preferred direction. Parameter choice: $K_0 = 1$, $K_1 = 2$, and $K_2 = 0$.

The microscopic origin of magnetocrystalline anisotropy is either due to crystal field interaction and spin-orbit coupling or due to interatomic dipole–dipole interaction. An intuitive understanding of the spin-orbit coupling effect origin of anisotropy can be obtained by considering the fact that initially all the dipoles are pointed in the z-direction. Upon application of the field, they are eventually rotated to point in the x-direction. Orienting the direction also implies that the electronic orbit has to be rotated. But the orbit is intimately coupled to the lattice. Any attempt to rotate the electron charge cloud implies that these bonds will have to be twisted, which is is met with stiff resistance. For materials in which spin-orbit coupling is weak, twisting the orbits does not cost much. However in rare earth magnets – where spin orbit coupling is substantial – this sets up a strong magnetocrystalline anisotropy. As mentioned earlier, hard magnets require strong anisotropy. Rare earths with their strong magnetocrystalline anisotropy, therefore, make good permanent magnets. Soft magnets have weak anisotropy; they make good temporary magnets.

*Magnetostriction*

Magnetostriction is the change in the size of the ferromagnetic material due to magnetoelastic coupling. The change is small, but present. Consider a point in an unstrained crystal

$$\begin{aligned} x &= x_o + A_{11}x_o + A_{12}y_o + A_{13}z_o, \\ y &= y_o + A_{21}x_o + A_{22}y_o + A_{23}z_o, \\ z &= z_o + A_{31}x_o + A_{32}y_o + A_{33}z_o. \end{aligned} \tag{10.46}$$

We define the direction cosines of the vector in the unstrained crystal by $\beta_1$, $\beta_2$, and $\beta_3$ given by

$$\beta_1 = \frac{x_o}{r_o};\; \beta_2 = \frac{y_o}{r_o};\; \beta_3 = \frac{z_o}{r_o}. \tag{10.47}$$

Using (Equation 10.47), we can rewrite Equation (10.46) as

$$x = r_o\left(\beta_1 + \sum_{j=1}^{3} A_{1j}\beta_j\right);\; y = r_o\left(\beta_2 + \sum_{j=1}^{3} A_{2j}\beta_j\right);\; z = r_o\left(\beta_3 + \sum_{j=1}^{3} A_{3j}\beta_j\right).$$

We can then write the length vector (Exercise 10.9.9) as

$$r = r_o\left(1 + \sum_{ij} 2A_{ij}\beta_i\beta_j\right)^{1/2}, \tag{10.48}$$

where we have neglected higher order terms that involve products of the $A_{ij}$. Thus the approximate change in length in a certain direction in terms of the direction cosines is given by (Exercise 10.9.9)

$$\frac{\delta\mathscr{L}}{\mathscr{L}} \equiv \frac{r - r_o}{r_o} = \sum_{ij} A_{ij}\beta_i\beta_j. \tag{10.49}$$

To proceed further, we need to make the assumption that the crystal strain depends on the direction of the spontaneous magnetization with respect to the crystal axes. This implies that it depends on the direction cosines $\alpha_1$, $\alpha_2$, and $\alpha_3$ of the magnetization vector when it changes from the demagnetized state to saturation along that direction. For our purpose, we will consider the following dependence

$$A_{ii} = a + b\alpha_i^2 \text{ and } A_{ij} = c\alpha_i\alpha_j. \tag{10.50}$$

Substituting these coefficients into Equation (10.49), we obtain

$$\frac{\delta\mathscr{L}}{\mathscr{L}} = a + b(\alpha_1^2\beta_1^2 + \alpha_2^2\beta_2^2 + \alpha_3^2\beta_3^2) + 2c(\alpha_1\alpha_2\beta_1\beta_2 + \alpha_2\alpha_3\beta_2\beta_3 + \alpha_3\alpha_1\beta_3\beta_1), \tag{10.51}$$

where in expanding the summation in Equation (10.49) we have used the fact that $A_{ij} = A_{ji}$. This introduces the factor of two in Equation (10.51). For a demagnetized specimen $\frac{\delta\mathscr{L}}{\mathscr{L}}$=0. Then since $\bar{\alpha_i^2} = \frac{1}{3}$ and $\bar{\alpha_i\alpha_j}$=0, we obtain

$$\frac{\delta\mathscr{L}}{\mathscr{L}} = h_1(\alpha_1^2\beta_1^2 + \alpha_2^2\beta_2^2 + \alpha_3^2\beta_3^2 - \frac{1}{3}) + h_2(\alpha_1\alpha_2\beta_1\beta_2 + \alpha_2\alpha_3\beta_2\beta_3 + \alpha_3\alpha_1\beta_3\beta_1). \tag{10.52}$$

where -3a = b = $h_1$ and $2c$ = $h_2$. The saturation magnetostriction measured along the [100]- and [111]- directions is given by

$$\lambda_{[100]} = \frac{2}{3}h_1 \text{ and } \lambda_{[111]} = \frac{2}{3}h_2. \tag{10.53}$$

Substituting Equation (10.53) into Equation (10.51), we finally obtain

$$\frac{\delta \mathscr{L}}{\mathscr{L}} = \frac{3}{2}\lambda_{[100]}(\alpha_1^2\beta_1^2 + \alpha_2^2\beta_2^2 + \alpha_3^2\beta_3^2 - \frac{1}{3}) + 3\lambda_{[111]}(\alpha_1\alpha_2\beta_1\beta_2 + \alpha_2\alpha_3\beta_2\beta_3 + \alpha_3\alpha_1\beta_3\beta_1) \tag{10.54}$$

The above expression is appropriate for the description of the experimental data for cubic crystal systems. For an isotropic saturation magnetostriction $\lambda_{[100]} = \lambda_{[111]} = \lambda_s$, we have

$$\frac{\delta \mathscr{L}}{\mathscr{L}} = \frac{3}{2}\lambda_s(\cos^2\theta - \frac{1}{3}), \tag{10.55}$$

where $\theta = \cos^{-1}\left(\sum_{ij}\alpha_i\beta_j\right)$ is the angle between the spontaneous magnetization and the direction in which the $\frac{\delta \mathscr{L}}{\mathscr{L}}$ is measured.

### 10.3.3 Stoner-Wohlfarth Model of Hysteresis

In 1948 E. C. Stoner and E. P. Wohlfath developed a hysteresis model for single-domain particles. When the size of the domain is small enough that there are no domain walls, one needs to consider a coherent rotation of the domains only. Referring back to Figure 10.3.7, the energy density $\mathscr{E}$ with uniaxial anisotropy energy and the Zeeman energy is given by (also see Example 10.3.2.4)

$$\mathscr{E} = K_u \sin^2(\Theta - \phi) - \mu_o H M_s \cos\phi, \tag{10.56}$$

where we have used the relation $\Theta = \theta + \phi$. The angle $\Theta$ is defined between the easy axis and the externally applied magnetic field $H$; $\phi$ is the angle between the magnetization vector and the magnetic field as shown in Figure 10.3.7. For a given value of the applied magnetic field, the above equation can be minimized to obtain

$$\tilde{\mathscr{E}}' = \frac{1}{2K_u}\frac{d\mathscr{E}}{d\phi} = \frac{1}{2}\sin[2(\phi - \Theta)] + h\sin\phi = 0, \tag{10.57}$$

where we have used the trigonometric identity $\sin 2A = 2\cos A \sin A$, normalized the energy density expression with respect to the uniaxial anisotropy energy $K_u$, and defined h=$\mu_o HM_s/2K_u$. Solutions to the above equation do not guarantee an energy minimum unless the second derivative of Equation (10.56) is positive. Equation (10.56) is actually a torque equation stating that equilibrium is achieved when the rotational forces due to an external field are balanced out from those caused by the uniaxial anisotropy misalignment. We solve Equation (10.57) and its derivative numerically to obtain the hysteresis loop. The derivative of Equation (10.57) is given by

$$\tilde{\mathscr{E}}'' = \frac{d}{d\phi} = \cos[2(\phi - \Theta)] + h\cos\phi. \tag{10.58}$$

Now, using the MATLAB code provided below, we simultaneously solve Equations (10.57) and (10.58).

```
%copyright by J. E Hasbun and T. Datta
%ch10_hysteresisloop.m
```

```
clear;

%here x=phi, y=theta, z=h
fH =@(x,y,z) 0.5*sin(2*(x-y))+z*sin(x); % define fH(x,y,z)
dfH =@(x,y,z) cos(2*(x-y))+z*cos(x);    % define dfH(x,y,z) (the derivative)
y=pi/4;
z1=-2; z2=2; Nz=40; zs=(z2-z1)/(Nz-1);  %h range to work with
z=z1:zs:z2;
x1=-pi; x2=pi; Nx=20; xs=(x2-x1)/(Nx-1);%seek roots using this range of
    guesses
x=x1:xs:x2;
k=0;
for ic=1:length(z)
  %seek all the aeros of fH and collect all the one that make dfH >=0
  for ix=1:length(x)   %let's sweep through many guesses to find many roots
    xx=fzero(@(x) fH(x,y,z(ic)),x(ix));
    if(dfH(xx,y,z(ic)) >=0)  %collect all the roots for which dfH is >=0
      k=k+1;
      phi(k)=xx;
      h(k)=z(ic);
    end
  end
end
plot(h,cos(phi),'k*')
xlabel('h')
ylabel('Cos(\phi)')
```

Results from the code are displayed in Figure 10.3.11. In the case when the field is applied perpendicular to the direction of the easy axis no hysteresis occurs, Figure 10.3.11(b). When the field is applied along the polar axis the jump in magnetization occurs at $h = 1$.

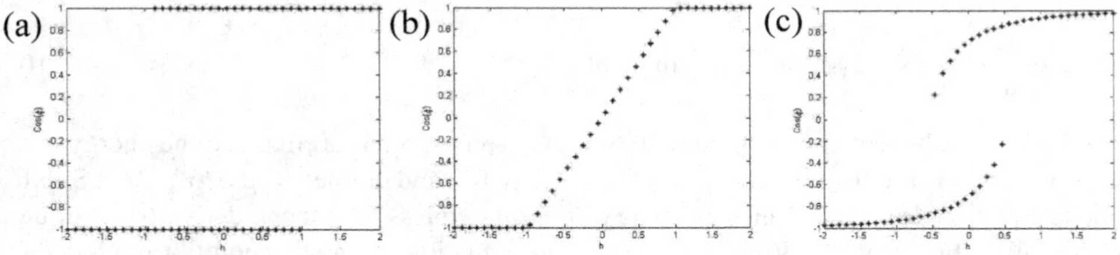


Figure 10.3.11: Hysteresis plot generated from Stoner-Wohlfarth model of hysteresis. (a) $\Theta$=0, (b) $\Theta=\frac{\pi}{2}$, and (c) $\Theta=\frac{\pi}{4}$.

## 10.4 Diamagnetism

Diamagnets are materials that repel any externally applied magnetic field. Diamagnetism is an induced effect and primarily arises due to the response of a magnetic system to an externally applied magnetic field. Classic examples of diamagnets include bismuth, antimony, gold, noble gases, and alkali halides. Almost all organic molecules and many inorganic molecules are diamagnetic. Superconductors, below their critical temperature, have a diamagnetic susceptibility of $\chi = -1$ and are considered to be perfect diamagnets, since they expel magnetic flux from the bulk by a mechanism known as the Meissner effect (see Chapter 12). The expulsion of magnetic flux is characterized by a negative susceptibility. Diamagnetism in insulators (Larmor diamagnetism) is different from that in metals (Landau diamagnetism). In Chapter 11 we will discuss at length the physics of diamagnetism from a quantum physics perspective.

## 10.5 Paramagnetism

In contrast to diamagnetic materials, paramagnets have unpaired electrons which give rise to a net magnetic moment. These moments are weakly coupled to each other and thermal agitation prevents the moments from aligning in the same direction. In the presence of an external magnetic field, the moments can overcome the randomization effects of temperature to line up in the direction of the field and give rise to a net magnetization. The magnitude of the resulting magnetization depends on the strength of the external field. This behavior results in a positive susceptibility typically in the range of $10^{-2}$-$10^{-5}$. The susceptibility varies inversely with temperature, $\chi \sim C/T$, and the dependence is known as Curie's law. In the next section, we will derive an expression for Curie's law and the constant C based on a semi-classical approach. For metallic (Pauli) paramagnets, the susceptibility is temperature independent. Table 10.3.1, as well as Figure 10.8.22, lists examples of paramagnetic materials.

**Example 10.5.0.1**
A quick test to figure out whether a particular element is diamagnetic or paramagnetic is to count the number of unpaired electrons. Based on this rule, predict whether the following elements and ions are diamagnetic or paramagnetic? (a) Cu, (b) $Sc^{3+}$, and (c) Gd (Gadolinium).
**Solution**
(a) From the periodic table in Chapter 1 we find the electronic configuration of Cu atom to be [Ar] $3d^{10}4s^1$. The presence of the unpaired electron makes Cu atom to be paramagnetic. However, if you look at Table 10.3.1 you will find that Cu is listed under diamagnets. Why? This is because we are obtaining the electronic configuration of a copper *atom*. In the case of copper metal, the odd electron is shared with the sea of other electrons forming the metallic bond, giving rise to a diamagnetic contribution, and hence the metal is diamagnetic. $Cu^{2+}$ salts, which have an electronic configuration of $[Ar]3d^9$, are paramagnetic. The lesson is, while it is generally true that counting unpaired electrons will allow you to predict diamagnetic or paramagnetic behavior, one has to be aware of the physical and chemical state of the atom to make the correct prediction.
(b) The electronic configuration of $Sc^{3+}$ is $1s^2\ 2s^2\ 2p^6\ 3s^2\ 3p^6$, which is the same as Argon. All the shells are filled and $Sc^{3+}$ is diamagnetic. No tricks here!
(c) Gadolinium has an electronic configuration of [Xe] $4f^7 5d^1 6s^2$. The unpaired electron suggests that an atom of Gd is a paramagnet. However, in its metallic state, Gd is a paramagnet above 293 K. Below this transition temperature, it undergoes a magnetic phase transition to a ferromagnet. In Sections 10.5.2, 10.7, and 10.7.2, you will study the physics of the paramagnetic–ferromagnetic phase transition. In Section 10.6 you will learn that gadolinium is used as a magnetocaloric material which has applications in magnetic refrigeration technology.

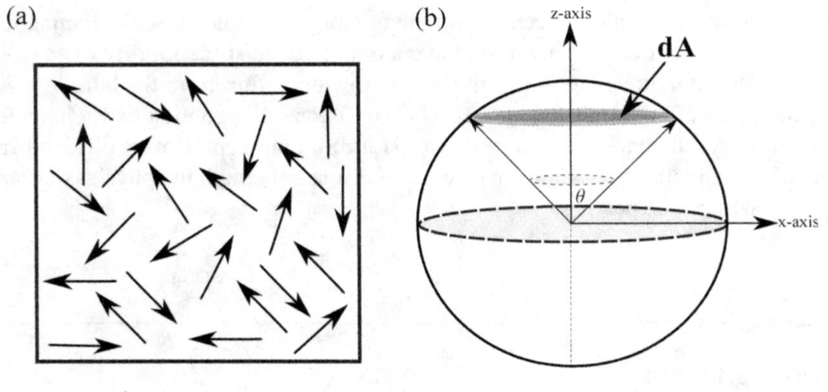


Figure 10.5.12: (a) Randomly oriented magnetic moments of magnitude $\mu$ form a paramagnet, (b) each individual magnetic moment sweeps out a disc of area $dA$ between angle $\theta$ and $\theta$+d$\theta$.

### 10.5.1 Langevin Theory of Paramagnetism

The Langevin theory of paramagnetism explains the temperature dependence of paramagnetic susceptibility by assuming $N$ non-interacting magnetic moments of magnitude $\mu$ in a volume $V$ that are randomly oriented as a result of thermal excitation. In the Langevin formulation, we ignore the quantization of magnetic moments, see Figure 10.5.12. We first calculate the number of moments lying between angles $\theta$ and $\theta+d\theta$ with respect to the field $\vec{H}$, by noting that it is proportional to the fractional surface area surrounding the sphere. We have

$$dA = 2\pi r^2 \sin\theta d\theta. \tag{10.59}$$

This implies the overall probability, p($\theta$), of an atomic magnetic moment making an angle between $\theta$ and $\theta+d\theta$ is

$$p(\theta) = \frac{e^{\mu_o \mu H \cos\theta / k_B T} 2\pi r^2 \sin\theta d\theta}{\int_0^\pi e^{\mu_o \mu H \cos\theta / k_B T} 2\pi r^2 \sin\theta d\theta}. \tag{10.60}$$

where we have used the Zeeman energy expression for a dipole in a magnetic field given by

$$E_{\text{dip}} = -\mu_o \mu H \cos\theta, \tag{10.61}$$

in the Boltzmann factor $\exp(-E_{\text{dip}}/k_B T)$. Also, note that in the denominator we sum over all possible angular orientations ranging from 0 to $\pi$. Since each moment contributes an amount $\mu_z = \mu\cos\theta$ parallel to the magnetic field, the average magnetization of the whole system is

$$\langle \mu_z \rangle = \int_0^\pi \frac{\mu \cos\theta e^{\mu_o \mu H \cos\theta / k_B T} \sin\theta d\theta}{\int_0^\pi e^{\mu_o \mu H \cos\theta / k_B T} \sin\theta d\theta}. \tag{10.62}$$

Carrying out the substitution $y = \mu_o \mu H / k_B T$ and defining $x = \cos\theta$ we have

$$\langle \mu_z \rangle = \frac{\mu \int_{-1}^{1} x e^{yx} dx}{\int_{-1}^{1} e^{yx} dx}. \tag{10.63}$$

Using integration by parts for the numerator and a regular integration for the denominator, we obtain the result (Exercise 10.9.17)

$$\frac{\langle \mu_z \rangle}{\mu} = \coth y - \frac{1}{y} \equiv L(y), \tag{10.64}$$

where $L(y)$ is the Langevin function. If we now make the following approximation

$$\coth y \equiv \frac{1}{y} + \frac{y}{3} + \mathcal{O}(y^3), \tag{10.65}$$

then

$$L(y) = \frac{y}{3} + \mathcal{O}(y^3). \tag{10.66}$$

If N/V denotes the number of magnetic moments per unit volume; then, the magnetization can be written as

$$M = \frac{N}{V}\langle \mu_z \rangle. \tag{10.67}$$

Using the expressions from Equations (10.64) and (10.66) for $\langle \mu_z \rangle$ in the limit of small fields we have

$$M \approx \frac{N\mu y}{3V} = \frac{N\mu_o \mu^2}{3Vk_B T} H, \tag{10.68}$$

which can be rearranged for the susceptibility as

$$\chi = \frac{M}{H} = \frac{N\mu_o \mu^2}{3Vk_B T}. \tag{10.69}$$

So the magnetization is given by

$$M = \frac{N\mu_o \mu^2}{3Vk_B}\frac{H}{T} = C\frac{H}{T}. \tag{10.70}$$

This is known as Curie's law. As mentioned earlier, we find that the susceptibility, M/H, of a paramagnet is inversly proportional to temperature with the Curie constant C = $\frac{N\mu_o \mu^2}{3Vk_B}$. Figure 10.5.13 shows a plot for Curie's law for a paramagnet, ferromagnet, and an antiferromagnet.

**Example 10.5.1.1**
Using the Langevin theory of paramagnetism, compute the Curie constant for a crystal of copper sulfate, $CuSO_4.5H_20$. The density and relative molar mass of pentahydrate $CuSO_4$ is 2286 kg m$^{-3}$ and 249.7 g mol$^{-1}$, respectively.
**Solution**
The Curie constant is given by C = $\frac{N\mu_o \mu^2}{3Vk_B}$. We first compute the number of atoms per m$^3$ n,

$$\begin{aligned} n \equiv \frac{N}{V} &= \frac{(6.023 \times 10^{23}\ \text{atoms/mol})(2.286 \times 10^{-6}\ \text{g cm}^{-3})}{249.7\ \text{g mol}^{-1}}, \\ &= 5.51 \times 10^{27} \text{atoms m}^{-3}. \end{aligned} \tag{10.71}$$

With the value of $n$ from above, setting $\mu$ equal to $\mu_B$ and using the expression for the Curie constant we have C = 0.014 K.

### 10.5.2 Curie-Weiss Law

Paramagnets can arise from ferromagnetic materials above a critical temperature known as the Curie temperature, $T_c$. Thermal energy can disrupt the alignment of the magnetic moments leading to a destruction of net magnetization. The Curie law as derived above is not obeyed; rather, a modified version known as the **Curie-Weiss law** is used

$$\chi = \frac{C}{T - \theta_{CW}}, \tag{10.72}$$

where $\theta_{CW}$ could be positive (ferromagnet) or negative (antiferromagnet) based on the type of magnetic ordering. The Curie-Weiss law was obtained by Pierre Weiss in 1906 to explain the origins of ferromagnetism and to describe the ferromagnetic-paramagnetic phase transition where the Curie constant $C = 9.8 \times 10^{-5}$ $m^3$ K $mol^{-1}$. To explain the unusually large transition temperatures, he postulated the existence of an internal interaction energy between localized magnetic moments which he termed to be the molecular field, $\vec{H}_m$. He assumed what amounts to a **mean-field approximation** in which any one magnetic moment feels an identical average field generated by all the other neighboring moments. Within such a scheme, Weiss considered the internal molecular field responsible for orienting the dipoles to be directly proportional to magnetization

$$\vec{H}_m = C_m \vec{M}, \tag{10.73}$$

where $C_m$ is the Weiss coefficient (internal molecular field constant). Weiss's assumption of the internal molecular field was not non-sensical, rather quite intuitive. In Chapter 11 you will learn that his assumption can be attributed to the interplay of the interatomic Coulomb interaction and quantum mechanics (Pauli exclusion principle) which give rise to magnetic exchange interactions. More on this later. For now we consider the total field $\vec{H}_t$ acting on and in the magnet as

$$\vec{H}_t = \vec{H} + \vec{H}_m. \tag{10.74}$$

From Curie's law we have

$$\chi = \frac{M}{H} = \frac{C}{T}, \tag{10.75}$$

in which if we replace the $H$ with $H_t$ we have

$$\frac{M}{H + C_m M} = \frac{C}{T}. \tag{10.76}$$

Solving for the new $M/H$ ratio, we obtain the susceptibility as

$$\chi = \frac{M}{H} = \frac{C}{T - T_c}, \quad T_c = CC_m, \tag{10.77}$$

where the critical temperature $T_c \equiv \theta_{CW}$ signifies the paramagnetic-ferromagnetic transition as shown in the susceptibility plot of Figure 10.5.13(b). A positive value of $\theta_{CW}$ corresponds to the molecular field helping to align the moments in the same direction as the field. This is the case in ferromagnets. A negative value corresponds to the opposite case where the moments are discouraged from aligning in the same direction. This is the case in antiferromagnets where there are two oppositely directed magnetic sublattices as shown in Table 10.3.1. Curie-Weiss temperatures for ferromagnetic elements Fe, Ni, and Co are 1043 K, 627 K, and 1388 K, respectively. Similar to the Curie-Weiss temperature, we can define the Néel temperature, $T_N$, as the temperature below which an antiferromagnetic material becomes paramagnetic. Typical examples of Néel temperature for antiferromagnets are NiO (525K) and MnO (116K). Within the scope of the Weiss molecular field theory $\theta = -T_N$. In practice typical values of Curie-Weiss temperature for antiferromagnets are larger. For example, we have NiO (2000K) and MnO (610K). The discrepancy arises due to a neglect of further neighbor interactions.

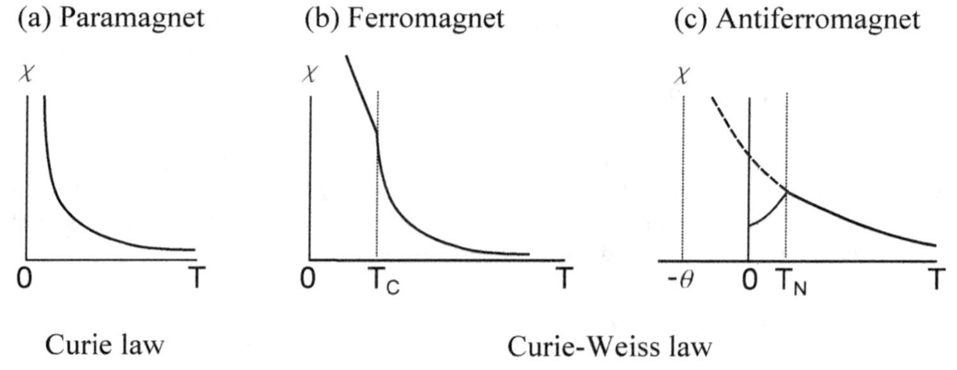


Figure 10.5.13: Susceptibility ($\chi$) versus temperature (T) plots for paramagnet, ferromagnet, and antiferromagnet. $T_c$ refers to the Curie temperature for a paramagnet-ferromagnet transition. $T_N$ refers to the paramagnet-antiferromagnet Néel transition temperature, named after Louis Néel.

## 10.6 Magnetocaloric Effect

The **magnetocaloric effect** is the heating or cooling of magnetic materials when subjected to an increase or decrease in magnetic field. Magnetic refrigeration technology relies on the magnetocaloric effect, originally discovered in iron by Emil Warburg in 1881. Early research on the magnetocaloric effect in paramagnets was carried out because of the drive to reach ultra-low temperature by adiabatic demagnetization cooling. Using adiabatic demagnetization, temperatures as low as 1K to 1mK were reached in electron paramagnets and in the micro Kelvin range in nuclear paramagnets. In 1976, Gerald V. Brown developed a magnetic refrigerator using metallic gadolinium as a magnetic refrigerant. Renewed interest in the magnetocaloric effect came in 1997 when V. Pecharsky and K. Gschneidner, Jr., discovered the giant magnetocaloric effect in the compound $Gd_5Si_2Ge_2$ and its related $Gd_5(Si_{1-x}Ge_x)_4$ alloys. Below you will learn the theory behind the adiabatic demagnetization cooling process.

**a. Isothermal magnetization**: The first step in a typical experiment utilizing a paramagnetic salt such as cerium magnesium nitrate is to isothermally magnetize an already cooled sample. The sample is surrounded by helium which provides a means to extract the energy away and keep the process isothermal (constant temperature). This step is represented by $a \rightarrow$ b on the S-T diagram in Figure 10.6.14. From the TdS equation (see Exercise 10.9.11) for a magnetic system we have

$$TdS = C_H dT + \mu_o T \left(\frac{\partial M}{\partial T}\right)_H dH, \tag{10.78}$$

where $C_H$ is the heat capacity at constant field $H$. For an isothermal process, dT = 0, we then have

$$TdS = \Delta Q = \mu_o T \left(\frac{\partial M}{\partial T}\right)_H \Delta H. \tag{10.79}$$

Since paramagnetic salts obey Curie's law

$$\chi = \frac{C}{T}, \tag{10.80}$$

the slope of $\chi$ with temperature is always negative and we have using Equation (10.14)

$$\Delta Q = \mu_o TH \left(\frac{\partial \chi}{\partial T}\right)_H \Delta H < 0. \tag{10.81}$$

The above change in temperature is the phenomenon of magnetocaloric effect.

**b. Adiabatic demagnetization**: In the second step, b $\rightarrow$ c on the S-T diagram, the system is thermally insulated by extracting the helium. For this isentropic (constant entropy) process dS=0. Using the TdS Equation (10.78) for the adiabatic process we have

$$C_H dT \quad = \quad -\mu_o T \left(\frac{\partial M}{\partial T}\right)_H dH, \tag{10.82}$$

or,

$$\Delta T \quad = \quad -\frac{TH}{\mu_o C_H}\left(\frac{\partial \chi}{\partial T}\right)_H \Delta H. \tag{10.83}$$

Negative slope of Curie's law implies that in this step as the external field is dialed down, the temperature drops, as is clearly evident from Equation (10.82). As the field is turned down, the entropy of the moments increase since they begin to point in random directions. But this increase is balanced by a decrease in the entropy of the lattice vibrations (phonons), thereby keeping the overall process isentropic. Conversely, if the field were to be adiabatically increased, the temperature would increase. Such a coupling between thermal and magnetic properties gives rise to the magnetocaloric effect.

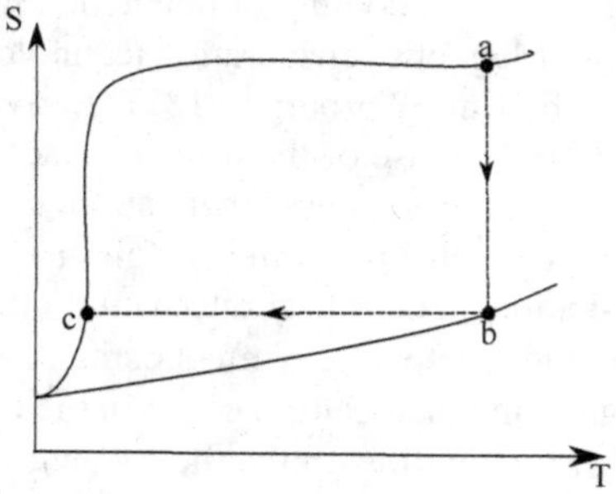


Figure 10.6.14: Entropy (S) versus temperature (T) diagram demonstrating the isothermal magnetization (step a $\rightarrow$ b) and the adiabatic demagnetization (step b $\rightarrow$ c) processes utilized in the magnetocaloric effect.

**Example 10.6.0.1**

Gadolinium sulfate, $Gd_2(S0_4)_3.8H_20$, was used by American chemist William F. Giauque and his graduate student D.P. MacDougall in 1933 to reach temperatures below 1 K. Consider gadolinium sulftate obeying Curie's law present in a magnetic field of $0.64 \times 10^6$ A/m at a temperature of 1.5 K. What is the final temperature if the field is reduced reversibly and adiabatically to zero? The Curie constant is $9.80 \times 10^{-5}$ m$^3$Kmol$^{-1}$ and the heat capacity at constant magnetization could be modeled as $C_M$=A/T$^2$ where A = 2.91 J K$^{-1}$ mol$^{-1}$.

**Solution**:
The TdS equation for a magnetic system (see Exercise 10.9.11) can be written as

$$TdS = C_M dT - T\mu_o \left(\frac{\partial H}{\partial T}\right)_M dM, \tag{10.84}$$

where $C_M$ is the heat capacity at constant magnetization. Based on Curie's law, Equation (10.70) we have

$$\left(\frac{\partial H}{\partial T}\right)_M = \frac{M}{C}. \tag{10.85}$$

Using the expression for $C_M$, Curie's law, and setting dS = 0 in Equation (10.84) for an adiabatic process we have

$$0 = \frac{A}{T^2}dT - T\mu_o \frac{M}{C}dM,$$

or,

$$A\int_{T_i}^{T_f} \frac{dT}{T^3} = \frac{\mu_o}{C}\int_{M_i}^{M_f} MdM,$$

where the subscripts $i$ and $f$ denote the initial and final states, respectively. Performing the integration and using Curie's law, we have

$$\frac{1}{T_f^2} - \frac{1}{T_i^2} = \frac{\mu_o C}{A}\left(\frac{H_i^2}{T_i^2} - \frac{H_f^2}{T_f^2}\right), \tag{10.86}$$

from which rearranging for the final to initial temperature ratio gives

$$\frac{T_f}{T_i} = \sqrt{\frac{\mu_o(C/A)H_f^2 + 1}{\mu_o(C/A)H_i^2 + 1}} \tag{10.87}$$

Now when the field is reversibly and adiabatically reduced to zero, we can set $H_f = 0$, to obtain

$$\left(\frac{T_i}{T_f}\right)^2 = 1 + \frac{\mu_o C}{A}H_i^2. \tag{10.88}$$

Inserting the appropriate numbers, we find that for an initial temperature of 1.5 K, the final temperature will be 0.35 K. Giaque and McDougall in their 1933 experiment reached a final temperature of 0.25 K.

## 10.7 Ferromagnetism

At room temperature, metals such Fe, Ni, and Co are spontaneously magnetic. These materials are different from the paramagnets that we studied in Section 10.5 where magnetization does not survive in the absence of a field. Table 10.3.1 lists examples of ferromagnetic materials. In ferromagnets the magnetic dipoles all line up in the same direction and a residual magnetic moment survives even after the removal of the external applied magnetic field. However, thermal effects can cause a ferromagnet to transition to a paramagnet at the Curie-Weiss point.

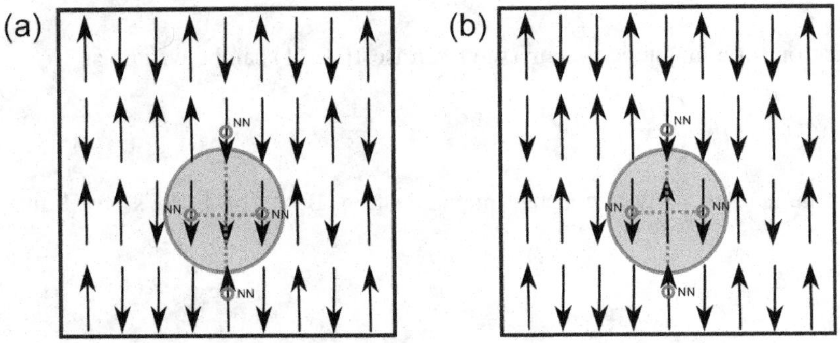


Figure 10.7.15: (a) Ising model magnetic ordering pattern. (b) Hypothetical spin flip in the circled region. In both cases, the nearest neighbors (NN) magnetic moments are shown by the dashed line.

To analyze the ferromagnetic phase, one needs an appropriate model. The **Ising model**, proposed by Wilhelm Lenz and solved by his student Ernst Ising, is the simplest model of ferromagnetism. The Ising model is to the study of magnetism what the simple harmonic oscillator is to quantum mechanics and classical mechanics problems. The model consists of binary variables utilized to represent the two possible orientations, ↑ (+1) and ↓ (-1), of magnetic dipole moments. In the simplest nearest neighbor version, see Figure 10.7.15, the Ising Hamiltonian is given by

$$\mathscr{H} = -J \sum_{\langle i,j \rangle} S_i^z S_j^z, \tag{10.89}$$

where $S_i^z$ is the classical dimensionless Ising variable at site i taking the values $\pm$ 1. It is inherently a *classical* model. The sign on the energy J of the interacting dipoles (which you will learn in Chapter 11 is called the exchange energy) decides the type of magnetic ordering. For $J > 0$, ferromagnetic interaction, all the magnetic moments line up in the same direction to give rise to a ferromagnet. For $J < 0$, antiferromagnetic interaction, the magnetic moments anti-align to give rise to an antiferromagnetic order. Some authors prefer not to include the overall negative sign. In that case, a negative (positive) value of J will give rise to a ferromagnet (antiferromagnet). In this chapter, we will focus on the ferromagnetic model. You will later learn in Chapter 11 that the $z$ superscript in the model stands for the z-component of the electronic spin. The Ising model can be solved exactly in only one and two dimensions using the transfer matrix method. In this chapter, we will focus on solving the model using a mean field description (analytical) and a Monte-Carlo simulation (computational) approach. In Exercise 10.9.19 you will analyze the model using another mean field-like approach called the Landau theory of phase transition.

### 10.7.1 Mean Field Theory of Ising Model

The central idea behind a mean field description is to assume an average interaction experienced by any and every magnetic moment due to their surrounding magnetic dipoles. Mean field theories are the simplest type of theory that can be constructed to study phase transitions. Results from mean field theories should be taken with caution since they ignore correlations and fluctuations which become important near $T_c$ the critical transition temperature. To carry out a mean field analysis let us first define $\bar{M}$, as the net magnetization of the lattice. The Hamiltonian can be rewritten in terms of the magnetization deviation at each site of the lattice as $\delta S_i^z = S_i^z - \bar{M}$

$$\mathscr{H} = -J \sum_{\langle i,j \rangle} \{\bar{M} + (S_i^z - \bar{M})\}\{\bar{M} + (S_j^z - \bar{M})\}. \tag{10.90}$$

Expanding we have

$$\mathscr{H} = -J\sum_{\langle i,j\rangle}\{\bar{M}^2 + \bar{M}(S_j^z - \bar{M}) + \bar{M}(S_i^z - \bar{M}) + (S_i^z - \bar{M})(S_j^z - \bar{M})\}. \tag{10.91}$$

Notice that the last term $\delta S_i^z \delta S_j^z$ is the square of the fluctuation variable (which is already small) and can be neglected. Mathematically, neglecting these contributions constitutes the mean field approximation. We then have,

$$\mathscr{H} = J\sum_{\langle i,j\rangle}\{\bar{M}^2 - \bar{M}(S_i^z + S_j^z)\}. \tag{10.92}$$

Here are a few words about how to carry out the summation, though. Note, you will have to perform a summation over the nearest-neighbor bonds. To do so, realize that

$$\sum_{\langle i,j\rangle} = \sum_i \sum_{\vec{\delta}}, \tag{10.93}$$

where $\vec{\delta}$ are the nearest neighbor vectors taking values $\pm\hat{i}$ and $\pm\hat{j}$. Breaking down the summation this way, we are essentially saying that we will consider a lattice site – one at a time – count the energy contribution from all its nearest neighbors and repeat this process till every site on the lattice has been accounted for. To carry out the summation we define a nearest neighbor coordination number, z, for the lattice. For a square lattice, the coordination number is 4. Then using the strategy for the summation just mentioned, we have

$$\mathscr{H} = J\sum_i\sum_{\vec{\delta}}\bar{M}^2 - J\sum_i\sum_{\vec{\delta}}\bar{M}(S_i^z + S_j^z).$$

Since $i$ and $j$ are dummy variables, in the second summation we can replace $j$ with $i$ to obtain

$$\begin{aligned}
\mathscr{H} &= J\sum_i\sum_{\vec{\delta}}\bar{M}^2 - 2J\bar{M}\sum_i\sum_{\vec{\delta}}S_i^z,\\
\mathscr{H} &= \frac{J}{2}\bar{M}^2Nz - 2J\bar{M}z\sum_i S_i^z,
\end{aligned}$$

where N is the number of sites and the $\frac{1}{2}$-factor in the first term removes any double counting. The purpose of embarking on a mean field description was to understand the origin of the Curie temperature. The mean field approximation provides a simplistic view on how to analyze the ferromagnetic transition. To proceed any further, we now need to obtain an average value of the magnetization at any temperature T. To obtain the average magnetization, we use the standard definition from statistical mechanics

$$\bar{M} = \frac{1}{N}\sum_{\{S_i^z\}}\frac{e^{-\beta\mathscr{H}}\sum_i S_i^z}{\sum_{\{S_i^z\}}e^{-\beta\mathscr{H}}}, \tag{10.94}$$

where the notation $\sum_{\{S_i^z\}}$ implies that the summation has to be carried out for $S_i^z = \pm 1$ in the Hamiltonian. If you compare Equations (10.94) and (10.62), you can see that we are using the same definition for computing the average magnetization. The only difference is in the angular orientation of the spins, which in the case of the Ising model includes only two choices - up and down, hence discrete; whereas, in the semi-classical Langevin model, the spins were free to point in any direction and was continuous.

$$\begin{aligned}
\bar{M} &= \frac{1}{N}\sum_{\{S_i^z\}}\frac{e^{-\beta\frac{J}{2}\bar{M}^2Nz - 2J\bar{M}z\sum_i S_i^z}\sum_i S_i^z}{\sum_{\{S_i^z\}}e^{-\beta\frac{J}{2}\bar{M}^2Nz - 2J\bar{M}z\sum_i S_i^z}},\\
&= \frac{1}{N}\sum_{\{S_i^z\}}\frac{e^{2\beta J\bar{M}z\sum_i S_i^z}\sum_i S_i^z}{\sum_{\{S_i^z\}}e^{2\beta J\bar{M}z\sum_i S_i^z}}.
\end{aligned} \tag{10.95}$$

Now, using the fact that the sum of exponentials can be written as a product, we can rewrite the above equation as

$$\bar{M} = \frac{1}{N} \sum_{\{S_i^z\}} \frac{\prod_i e^{2\beta J\bar{M}zS_i^z} \sum_i S_i^z}{\sum_{\{S_i^z\}} \prod_i e^{2\beta J\bar{M}zS_i^z}}. \tag{10.96}$$

If we define the parameter $\xi = 2\beta J\bar{M}z$, then the partition function can be written as

$$Z = \sum_{\{S_i^z\}} \prod_{i=1}^{N} e^{\xi S_i^z}, \tag{10.97}$$

and the magnetization expression as

$$\bar{M} = \frac{1}{N} \sum_{\{S_i^z\}} \frac{\prod_i e^{\xi S_i^z} \sum_i S_i^z}{\sum_{\{S_i^z\}} \prod_i e^{\xi S_i^z}}. \tag{10.98}$$

Inspecting Equations (10.97) and (10.98), we find the following relation

$$\bar{M} = \frac{1}{N} \frac{\partial \ln Z}{\partial \xi}. \tag{10.99}$$

Hence, if we can obtain an expression for the partition sum, we can compute the magnetization. To do so, observe that $\{S_i^z\} = \pm$. Therefore,

$$\begin{aligned} Z &= \sum_{\{S_i^z\}} \prod_{i=1}^{N} e^{\xi S_i^z}, \\ &= \prod_{i=1}^{N} (e^{\xi} + e^{-\xi}), \qquad (10.100) \\ &= [2\cosh(2\beta J\bar{M}z)]^N. \qquad (10.101) \end{aligned}$$

where we have used the identity $\cosh(x) = (\exp(x) + \exp(-x))/2$. Now using Equations (10.99) and (10.100) above we obtain the self-consistency equation for $\bar{M}$ as

$$\bar{M} = \tanh\left(\frac{T_c}{T}\bar{M}\right) = \tanh\left(\frac{\bar{M}}{\bar{T}}\right), \tag{10.102}$$

where we define $T_c = 2Jz/k_B$ with $k_B$ as the Boltzmann constant. Also $\bar{T} = T/T_c$. A solution to the above transcendental equation can be obtained graphically as shown in Figure 10.7.16. In Exercise 10.9.18(b), you will learn how to solve this equation using an iterative (computational) approach, see Figure 10.7.17 for the final solution.

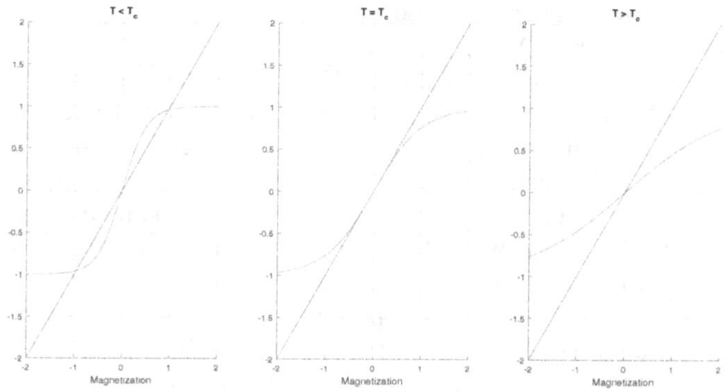


Figure 10.7.16: Graphical solution of the Ising model mean field self-consistency Equation (10.102). At low temperature when $T < T_c$ there are three solutions. Discarding the trivial zero magnetization solution, we find that the system is magnetized. As the temperature is increased to reach the critical value at $T = T_c$ and higher $T > T_c$ there is only one solution at the origin indicating that the system has lost its magnetization. Hence there are two regions, the low temperature ferromagnetic phase and the high temperature paramagnetic phase. In Exercise 10.9.18(a) you will write a MATLAB code to verify these solutions.

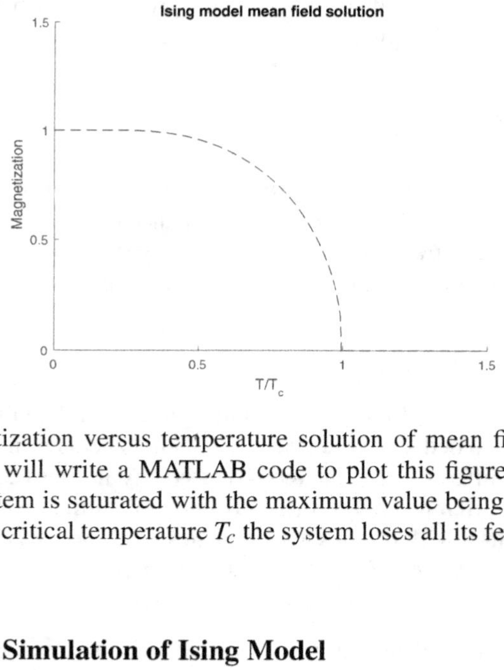


Figure 10.7.17: Magnetization versus temperature solution of mean field Equation (10.102). In Exercise 10.9.18(b) you will write a MATLAB code to plot this figure. At low temperatures the magnetization of the system is saturated with the maximum value being attained at $T = 0$. At temperatures higher than the critical temperature $T_c$ the system loses all its ferromagnetism to transition into a paramagnet state.

### 10.7.2 Monte Carlo Simulation of Ising Model

In the previous two sections, we learned how a ferromagnet can undergo a phase transition to a paramagnet and vice versa. The technique that we utilized to study the transition was analytical. In this section we will discuss a popular computational technique used to study the magnetic phase transition. You will learn about the **Monte Carlo method** using the Metropolis algorithm to simulate the Ising model. The Metropolis algorithm was introduced in 1953 by Nicholas Metropolis and his co-workers in their paper on the simulation of hard-sphere gases. The method relies on two very important concepts: Markov chain and detailed balance. But before we get into those details, let us first understand what is meant by a Monte Carlo method.

The name Monte Carlo originates from the famous European gambling center Monte Carlo Casino located in Monte Carlo, Monaco. The method used to be known as statistical sampling and the idea behind the technique is much older than the name or even before its use in a computer simulation. The use of a random number to decide the next move (similar to tossing a coin to make a decision) makes it a Monte Carlo technique. Initially it was a numerical method for estimating integrals which could not be performed by other analytical or standard numerical integration schemes. Use of the idea was made popular by its application to the hard sphere problem.

The conceptually tricky and challenging part of performing a Monte Carlo simulation is the generation of an appropriate random set of microstates according to the Boltzmann probability distribution. Metropolis and his co-workers introduced the idea of a **Markov chain** to generate the states. In their approach each new state is directly generated from the preceding state. In a Monte Carlo simulation a Markov process is repeatedly utilized to generate a Markov chain of states. The process is chosen specially so that when it is run for a long enough time starting from any state of the system, it will eventually produce a succession of states which appear with probabilities that are given by the Boltzmann distribution. In order to achieve this, two further conditions will have to be placed on the Markov process: ergodicity and detailed balance. **Ergodicity** is the idea that all possible microstates of the system should be attainable. The Markov chain of states which are being used to generate the states should allow for one state of the system to be reached from another. For the Metropolis algorithm the single-spin flip dynamics ensures that the algorithm obeys ergodicity since one can go from any one state to another on a finite lattice by flipping each of the spins, one at a time. The other condition of **detailed balance** ensures that the Boltzmann probability distribution is the equilibrium distribution for our problem; that is, when the system has reached equilibrium. The most important definition is the fact that

$$\sum_i \mathscr{P}_i \mathscr{W}(i \to j) = \sum_i \mathscr{P}_j \mathscr{W}(j \to i) \tag{10.103}$$

where i, j denote the states, $\mathscr{W}$ denotes the transition probabilities between the states, and $\mathscr{P}_i$ is the probability of the state $i$ at time t. This is the detailed balance equation. All it is saying is that the rate at which the system makes a transition into and out of any state $i$ must be equal. The probability of the $i^{th}$ state with energy $E_i$ at temperature T is given by Boltzmann factor

$$\mathscr{P}_i = \frac{e^{-E_i/k_BT}}{\mathscr{Z}}, \tag{10.104}$$

where $\mathscr{Z}$ is the partition function. In a typical simulation, this probability is not exactly known. However, this difficulty is avoided since the states are being generated in a Markovian process. In that case, only the energy difference $\Delta E$ matters. That is why in Steps 3a & 3b of the algorithm (shown below) the energy difference plays a role. To proceed further, we will call each magnetic dipole a spin. The single spin-flip Metropolis algorithm described below is implemented in the Ising model code ch10_ising.m, in the MATLAB GUI, see Figure 10.7.18, and summarized in the flow chart in Figure 10.7.19. The algorithm is outlined as follows:

**Step 1.** Start with any state of the lattice, choose a lattice site randomly (a random number generator can be used for this purpose), and consider a hypothetical flip of a single-spin.
**Step 2.** Compute the energy difference $\Delta E$ for this *hypothetical* single-spin flip (Example 10.7.2.1).
**Step 3a.** If $\Delta E \leq 0$, accept the hypothetical flip.
**Step 3b. Metropolis algorithm**: If the $\Delta E > 0$ for the hypothetical flip, then the system's energy increases. In this case we need to decide whether to flip the spin or not by comparing the Boltzmann factor $e^{-\frac{\Delta E}{k_BT}}$ to a random number, r, generated in the interval $[0,1]$. If $r < e^{-\frac{\Delta E}{k_BT}}$ we

accept the move and flip the spin. If r $> e^{-\frac{\Delta E}{k_B T}}$ we do not make any changes to the current state of the system and move on to Step 4.

**Step 4.** Steps 1, 2, 3a, & 3b are repeated until the entire lattice with N points is swept through once, giving every spin an equal opportunity to flip at least once. This completes one Monte Carlo sweep and is known as a Monte Carlo step per site (MCSS). This is an artifical time in a Monte Carlo simulation and is considered as a standard measure of time.

**Step 5.** Upon completion of one MCSS, the data for that step is saved, the loop is advanced to the next MCSS step, Step 4 repeated, and the simulation continued till a predecided number of MCSS steps have been completed.

**Step 6.** For a given temperature and a set of thermodynamic variables affecting the system, the system takes a while before equilibration is reached. Typically the choice of MCSS is large enough so that the system reaches equilibrium before any measurements are made of the thermodynamics variables of interest, as shown in Figure 10.7.20. The average over the magnetization (or energy) is computed over the last 10% or 20% of the accumulated data to obtain the average energy and magnetization.

**Step 7.** The temperature loop is then incremented and Steps 1 - 7 repeated.

**Step 8.** Finally, the data points are plotted as shown in Figure 10.7.20.

The MATLAB code which simulates the Ising model is provided below.

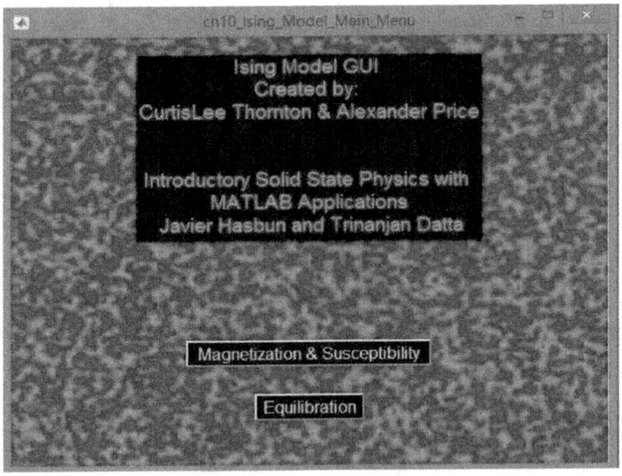


Figure 10.7.18: MATLAB interface for Ising model GUI. By clicking either *Magnetization & Susceptibility* or *Equilibration*, we can run the simulation. The Monte Carlo simulation code (without the GUI) is provided in the chapter. The GUI code is available for download separately from the publisher's weblink. In Exercises 10.9.20 and 10.9.21 you will utilize the GUI to study equilibration and finite size effects.

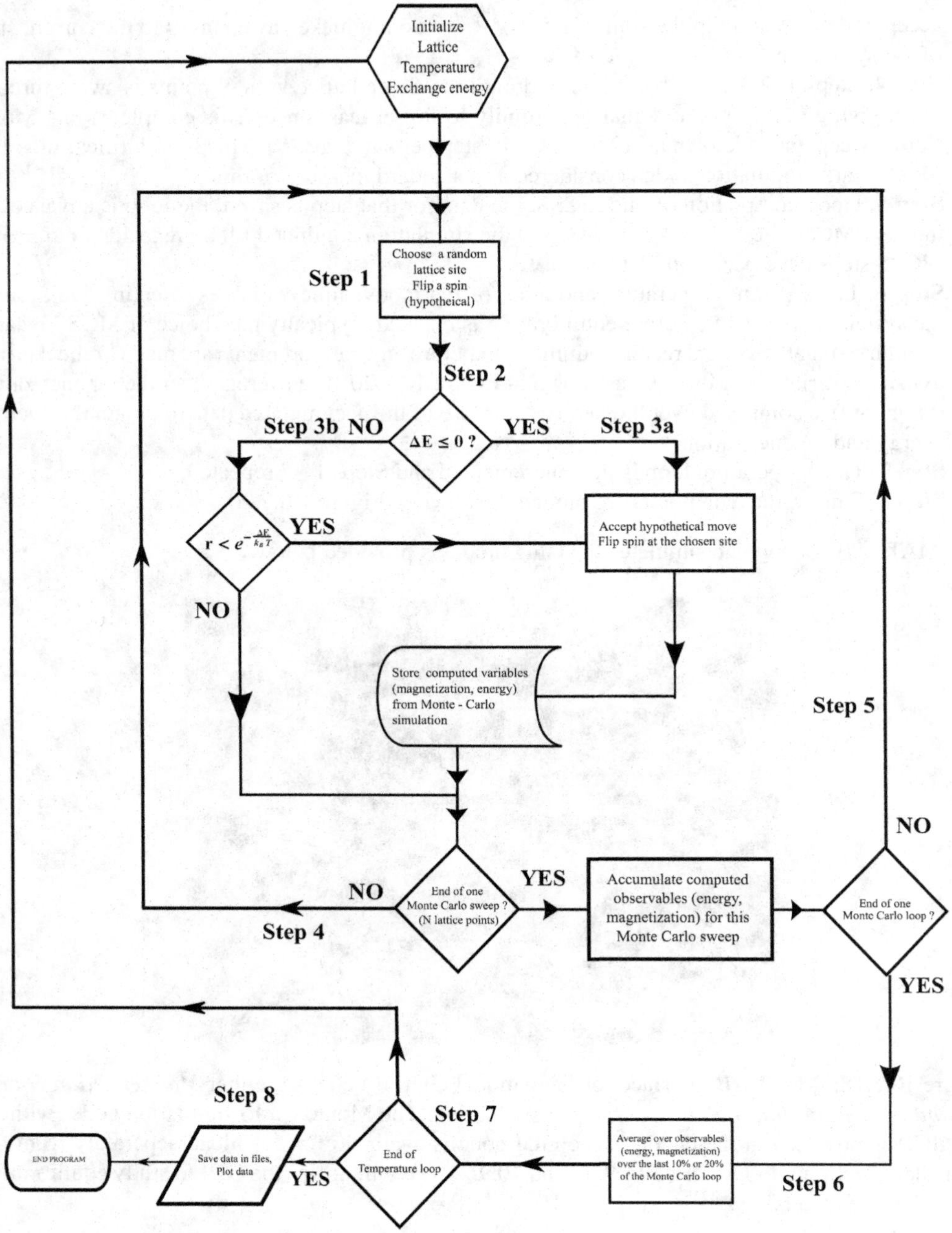


Figure 10.7.19: Metropolis algorithm for Monte Carlo simulation of 2nd nearest neighbor Ising model.

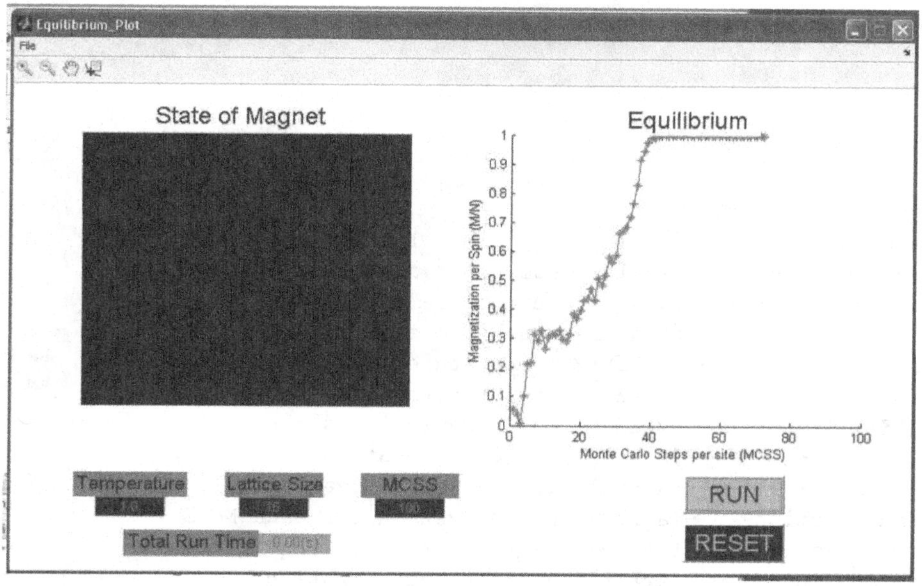


Figure 10.7.20: Initial fluctuations of the magnetization stabilize as higher MCSS steps are reached. The final state of the magnet where all the spins point in the same direction is displayed on the left square panel. The settings allow the user to choose the intial and final temperature, temperature step size, lattice size, and Monte Carlo Step per site (MCSS).

**Example 10.7.2.1**

In the algorithm for simulating the Ising model, an important step is to carry out various tests on the energy change $\Delta E$ after carrying out the hypothetical spin move, see steps 3 & 4. One can obtain a simple and nice expression for this $\Delta E$. To do so, consider the spin arrangement shown in Figure 10.7.15(a) with the orientation of the circled spin flipped as shown in Figure 10.7.15(b). For this two spin arrangement we can define an initial energy

$$E_{initial} = -Js_i(s_{i+1,j} + s_{i-1,j} + s_{i,j+1} + s_{i,j-1}), \tag{10.105}$$

and a final energy after the hypothetical spin flip as

$$E_{final} = Js_i(s_{i+1,j} + s_{i-1,j} + s_{i,j+1} + s_{i,j-1}). \tag{10.106}$$

Therefore the change in energy, $\Delta E = E_{final} - E_{initial}$, between the two configurations is given as

$$\Delta E = 2Js_i(s_{i+1,j} + s_{i-1,j} + s_{i,j+1} + s_{i,j-1}). \tag{10.107}$$

This is precisely the expression we use in our Monte Carlo simulation code to decide whether or not a spin flips.

The MATLAB code which simulates the Ising model is below.

```
%copyright by J. E Hasbun and T. Datta
% ch10_ising.m - by William D. Baez, CurtisLee Thornton, and T. Datta
% GUI implementation by -  CurtisLee Thornton and Alexander Price

% Monte Carlo GUI code is available from the publisher separately
```

```
% Ising model code

clearvars;
close all

%{
VARIABLE DEFINITIONS:
    latsize-----------(scalar) Length of lattice edges
    N-----------------(scalar) Number of sites in the Lattice
    initT-------------(scalar) Initial Temperature
    finT--------------(scalar) Final Temperature
    Tdt---------------(scalar) Temperature Step Size
    Temp--------------(array)  Temperature Array
    MCSS--------------(scalar) Monte Carlo Steps per Site
    MCS---------------(scalar) Monte Carlo Steps
    M-----------------(array)  Initial magnetization (all up)
    magnetization-----(array)  Magnetization for each MCSS
    Emagnetization----(array)  Magnetization for Equilibrium
    average_mag-------(array)  Average Magnetization
    susceptibility----(array)  Susceptibility Array
    Ediff-------------(scalar) Total Energy Contribution from
                               Nearest-neighbor
    expect_mag--------(scalar) Expectation Value for Magnetization
    expect_magsqr-----(scalar) Expectation Value for Magnetization Squared
    mag_sumsqr--------(scalar) Sum of the Squared Magnetization
    initH-------------(scalar) Magnitude of External Magnetic Field

NOTE: Comment or uncomment as need for magnetization and susceptibility or
equilibration study.
 %}

% % % % User Input Variables % % % %
latsize = input('Enter 2-D Lattice Dimension:');
MCSS    = input('Enter the number of Monte Carlo Steps per Site:');
initH   = input('Enter the Magnetic Field Strength:');
initT   = input('Enter the initial Temperature:');
finT    = input('Enter the final Temperature:');
Tdt     = input('Enter the Temperature Step Size:');

% % % % Generates initial lattice with all spins up % % % %
M = ones(latsize);

% % Random Initial State (USED FOR EQUILIBRATION STUDY ONLY) % % % %
% Comment this block if not running equilibration study else uncomment.
for ix = 1:latsize
    for iy = 1:latsize
        if rand < 0.5
            M(ix,iy) = 1;
        else
            M(ix,iy) = -1;
        end
```

```
    end
end

% % % % Temperature Array % % % %
Temp = initT:Tdt:finT;

% % % % Total Number of Lattice Sites % % % %
N = latsize*latsize;

% % % % Total Number of Monte Carlo Steps % % % %
MCS = MCSS*N;

% % % % Initilize Arrays % % % %
magnetization = zeros(1,MCSS);
Emagnetization = zeros(1,MCS);
average_mag = zeros(1,length(Temp));
susceptibility = zeros(1,length(Temp));

% % % % Start Stop Watch % % % %
tstart = tic;

% % % % Temperature Step % % % %
for t = 1:length(Temp)

    % % % % Prints the Current Temperature at Every Step % % % %
    fprintf('Current Temp: %g\n',Temp(t))

    % % % % Starts the counter % % % %
    count = 0;
    countE = 0;

    % % % % Montecarlo Step % % % %
    for montecarlosteps = 1:MCS

        % % % % Select a Random Lattice Site % % % %
        a = floor(rand*latsize + 1); % row number
        b = floor(rand*latsize + 1); % column number

        % % % % Energy for the Spin ABOVE the Lattice Point % % % %
        if a == 1
            Edifftop = M(latsize,b);
        else
            Edifftop = M(a-1,b);
        end

        % % % % Energy for the Spin BELOW the Lattice Point % % % %
        if a == latsize
            Ediffbottom = M(1,b);
        else
            Ediffbottom = M(a+1,b);
        end
```

```
        % % % % Energy for the Spin LEFT of the Lattice Point % % % %
        if b == 1
            Edifflleft = M(a,latsize);
        else
            Edifflleft = M(a,b-1);
        end

        % % % % Energy for the Spin RIGHT of the Lattice Point % % % %
        if b == latsize
            Ediffright = M(a,1);
        else
            Ediffright = M(a,b+1);
        end

        % % % % Total Energy Difference % % % %
        Ediff = 2*M(a,b)*(Edifftop + Ediffbottom + Edifflleft + Ediffright)...
                + (2*M(a,b)*initH);

        % % % % Spin Flip Decision % % % %
        if Ediff <= 0
            M(a,b) = -M(a,b);
        elseif rand < exp(-Ediff/Temp(t))
            M(a,b) = -M(a,b);
        end

        % % % % Magnetization Tracker per FLIP % % % %
        countE = countE + 1;
        Emagnetization(countE) = abs((sum(sum(M))))/N;

        % % % % Magnetization Tracker per SWEEP % % % %
        if mod(montecarlosteps,N) == 0
            count = count + 1;
            magnetization(count) = abs((sum(sum(M))));
        end
    end

    % % % % Calculations % % % %
    average_mag(t)= abs(((sum(magnetization(((0.90)*MCSS+1):MCSS))/
        (0.10*MCSS))))/N;

    expect_mag = sum(magnetization)/length(magnetization);
    expect_magsqr = expect_mag^2;

    mag_sumsqr = sum(magnetization.^2)/length(magnetization);

    susceptibility(t) = ((mag_sumsqr - expect_magsqr)/Temp(t))/N;
end

% % % % End Stop Watch % % % %
telapsed = toc(tstart);
```

```
fprintf('Total Calculation Time: %g seconds\n',telapsed)

% % % % Plots % % % %

% USED FOR MAGNETIZATION AND SUSCEPTIBILITY STUDY ONLY
% Comment block if not running equilibration study, uncommented for
% magnetization and susceptibility study
% Select the block of code below and use Ctrl + R to comment
% Select the block of code below and use Ctrl + T to uncomment

figure('Name','Plots','NumberTitle','off')
subplot 121
title('Magnetic Phase Diagram')
plot(Temp(2:length(Temp)),average_mag(2:length(Temp)),'-bo')
xlabel('Temperature (K)');
ylabel('Average Magnetization per Spin (M/N)');
subplot 122
title('Susceptibility');
plot(Temp(2:length(Temp)),susceptibility(2:length(Temp)),'-bo')
xlabel('Temperature (K)');
ylabel('Susceptibility per Spin (S/N)');

% USED FOR EQUILIBRATION STUDY ONLY
% Comment block if not running equilibration study, else uncomment
% Select the block of code below and use Ctrl + R to comment
% Select the block of code below and use Ctrl + T to uncomment

% figure('Name','Equilibration','NumberTitle','off')
% plot(Emagnetization,'-b*')
% title('Equilibration')
% ylim([0 1])
% xlabel('MCS (Monte Carlo Steps)')
% ylabel('Magnetization per Spin (M/S)')
```

In addition to the Metropolis algorithm, there are several other algorithms called Glauber, Wolff, Swendsen-Wang, and Wang-Landau each with its own pros and cons which incorporate both single spin-flip and cluster-flip processes to simulate the thermal physics system. For further details, refer to the excellent textbooks listed in References [46, 47].

## 10.8 Bohr-van Leeuwen Theorem

In Section 10.2 you learned that orbital motion of electrons gives rise to a microscopic current loop, and in turn, a magnetic dipole moment. Intuitively one may envision the collection of these dipoles to add up to give rise to a net magnetic moment. Unfortunately, this physical picture is incorrect as shown by Niels Bohr and Hendreka J. van Leeuwen. The **Bohr-van Leeuwen theorem**, based on classical statistical mechanics, states that at any finite temperature and in all finite electric and magnetic fields, the net magnetization of a collection of electrons in thermal equilibrium vanishes identically. Therefore it is impossible for electrons within the realm of classical physics to give rise to magnetism. To gain some intuition into the contents of the theorem, realize that in the absence of an applied electric field, an electron moves in a straight line. However, with a magnetic field present, the electrons move in curved cyclotron orbits as shown in Figure 10.8.21.

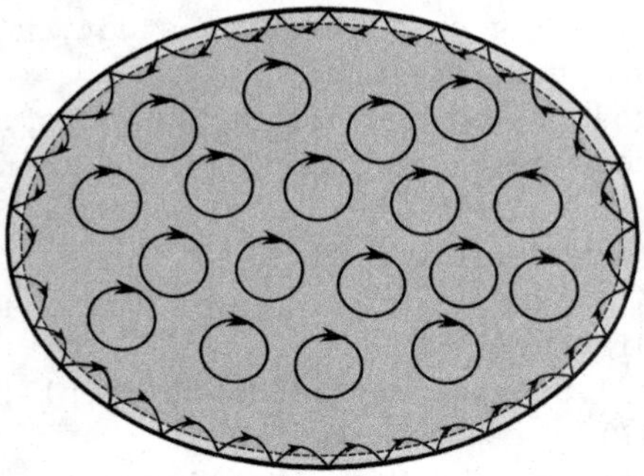

Figure 10.8.21: Cancellation of edge and bulk orbital currents associated with the motion of electrons in a magnetic field. The Bohr-van Leeuwen theorem forbids the existence of magnetism within a classical physics description.

The orbital currents generated from the partially complete skipping orbit at the edges cancel the orbital currents generated inside the bulk of the material. If the net orbital current is cancelled, there can be no magnetization! The conceptual idea discussed in this section can be justified by computing the effects of the magnetic field on the free energy, see Exercise 10.9.23. Hence, magnetism is inherently a quantum phenomenon and can be accounted for only using quantum mechanics.

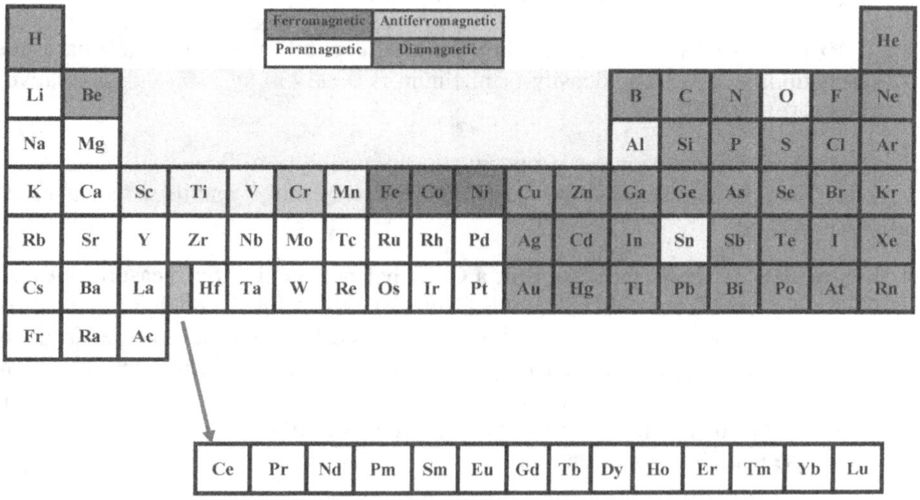


Figure 10.8.22: Periodic table of magnetic elements. At room temperature only Fe, Co, and Ni are ferromagnetic. Note, Cr is antiferromagnetic. The bulk of the elements are either paramagnets or diamagnets.

## 10.9 Chapter 10 Exercises

10.9.1. Similar to the Bohr magneton, the unit of nuclear magnetism is the nuclear magneton $\mu_N = \frac{e\hbar}{2m_p}$ where $m_p$ is the mass of the proton. Calculate the value of $\mu_N$?

10.9.2. In Example 10.3.1.1 we computed the mass susceptibility of platinum. What is the molar susceptibility of Pt? The density of platinum is $\rho$=21450 kg m$^{-3}$ and its relative atomic mass is 0.19509 kg mol$^{-1}$.

10.9.3. Oxygen has a dimensionless paramagnetic susceptibility of $0.19 \times 10^{-5}$. Convert this to $\chi_{mol}$. The relative atomic mass of oxygen is 0.016 kg mol$^{-1}$ and its gaseous phase density is 1.429 kg m$^{-3}$.

10.9.4. Figure 10.9.23 shows the set-up for a Gouy balance method for measuring susceptibility devised by the French physicist Louis Georges Gouy. A finely powdered sample contained in a cylindrical quartz tube is partially suspended between the poles of a magnet with homogeneous magnetic field. The bottom portion resides in the presence of the magnetic field, $H_{xb}$, and the top portion is in a region beyond the poles where the field $H_{xt}$ is weak. This creates a magnetic field gradient of $dH_x/dz$. Show that the sample experiences a force, $F_z$, whose magnitude is given by

$$F_z = \frac{1}{2}\mu_o \chi A(H_{xb}^2 - H_{xt}^2), \tag{10.108}$$

where A is the cross-section of the sample in the z-direction and $\chi$ is the magnetic susceptibility.

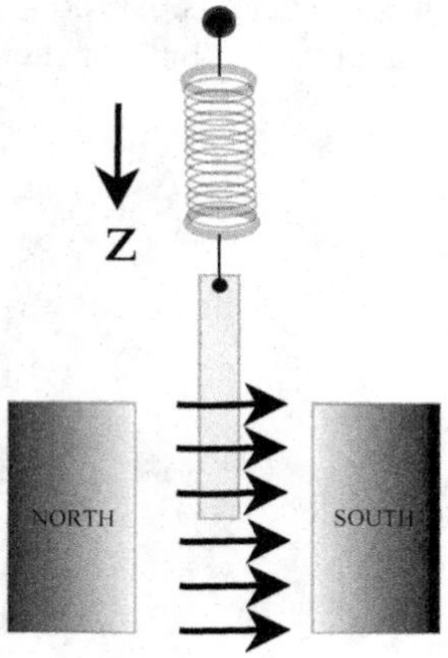


Figure 10.9.23: Gouy balance. Magnetic sample is supsended between the poles of a magnet creating a homogeneous magnetic field. The top of the sample is in a magnetic field free zone, the bottom is not. This creates a gradient in the field in the z-direction, hence a force.

10.9.5. Calculate (a) the saturation magnetization $M_s$ and (b) the corresponding saturation magnetic field for Ni which has 0.60 $\mu_B$. The room temperature density is $\rho$=8.908 g/cm$^3$ and atomic weight $A_{Ni}$ is 58.69 g/mol.

10.9.6. (a) Show that the theoretical upper bound for the energy product is given by $\frac{1}{4}\mu_o M_s^2$ where $M_s$ is the saturation magnetization. (b) Using the saturation magnetization value from Problem 10.9.5 compute the energy product value for Ni.

10.9.7. What are the expressions for $\mathrm{H}_i$ and $\mathrm{B}_i$ in terms of $\mathrm{H}_a$ and M for a sphere which has $\mathcal{N} = 1/3$?

10.9.8. Show that the shape anisotropy constant of a prolate spheroid is given by

$$K_{sh} = \frac{1}{4}\mu_o M_s^2(1 - 3\mathcal{N}_{\|}) \tag{10.109}$$

10.9.9. In Section 10.3.2 we obtained an expression for the fractional change in length, Equation (10.54). Verify Equations (10.48) and (10.49) used as part of the derivation.

10.9.10. Show that if the magnetization is measured along the direction of magnetization we have for the change in length

$$\frac{\delta\mathcal{L}}{\mathcal{L}} = \frac{2}{3}h_1 + 2(h_2 - h_1)(\alpha_1^2\beta_1^2 + \alpha_2^2\beta_2^2 + \alpha_3^2\beta_3^2), \tag{10.110}$$

where $h_1$ and $h_2$ are related to the magnetostriction constants.

10.9.11. Derive (a) Equation (10.78) and (b) Equation (10.84) used in Section 10.6.

10.9.12. Write a MATLAB code to plot the generic magnetic anisotropy energy surface for a (i) BCC (*e.g.*, Fe, choose $\mathrm{K}_0 = 1$, $\mathrm{K}_1 > 0$, and $\mathrm{K}_2 = 0$) , (ii) FCC (*e.g.*, Ni, choose $\mathrm{K}_0 = 1$, $\mathrm{K}_1 < 0$, and $\mathrm{K}_2 = 0$), and (iii) HCP (*e.g.*, Co, choose $\mathrm{K}_0 = 0.1$, $\mathrm{K}_1 = 2$ or $\mathrm{K}_0 = 1.1$, $\mathrm{K}_1 = -1$) crystal lattice. For the HCP crystal lattice, identity the anisotropy constant parameter combination which produces *easy axis* and *easy plane* anisotropy.

10.9.13. The application of a stress $\sigma$ to a ferromagnetic material can change the direction of spontaneous magnetization. This is inverse of the magnetostriction effect we discussed in Section 10.3.2 and is known as the *Villari effect*. Using thermodynamics, show that
(a) magnetostriction and stress are related by the following relation

$$\left(\frac{\partial\lambda}{\partial H}\right)_\sigma = \left(\frac{\partial M}{\partial\sigma}\right)_H \tag{10.111}$$

(b) Using the definition of the applied stress energy

$$\mathcal{E}_{stress} = -\sigma\int d\lambda, \tag{10.112}$$

show that,

$$\mathcal{E}_{stress} = -\frac{3}{2}\lambda_s\sigma\cos^2\theta, \tag{10.113}$$

where $\lambda_s$ is the magnetostriction constant. Is the stress energy a minimum or a maximum when $\lambda_s < 0$.

10.9.14. In Example (10.3.2.4) we computed the anisotropy field required to saturate the magnetization of a uniaxial crystal in a hard direction. In this problem you will compute the dependence of the magnitude of this critical field on the orientation of the applied magnetic field. Using Equation (10.56) show that

$$\sin 2\theta = \frac{1}{\bar{H}^2}\left(\frac{4 - \bar{H}^2}{3}\right)^{3/2}, \tag{10.114}$$

where $\bar{H} = H_o/(K_u/\mu_o M_s)$. Write a MATLAB code to make a plot of the critical field ($\bar{H}$) versus the orientation of the applied field ($\theta$).

10.9.15. Consider a single-domain ellipsoidal particle as shown in Figure 10.3.7. Compute the $\mathscr{M}_s$ versus $H_a$ curve when the field is applied at 10° to the polar axis. What is the value of the coercive field in this case.

10.9.16. Based on the number of unpaired electrons, predict whether the following elements and ions are diamagnetic or paramagnetic? (a) Au, (b) Xe, (c) $Cu^+$,(d) $Ti^{4+}$, (e) $Gd^{3+}$, and (f) $Fe^{3+}$.

10.9.17. Derive the Langevin function $L(y)$ given in Equation (10.64).

10.9.18. (a) Write a MATLAB code to validate the graphical solution of the transcendental Ising model mean field equation shown in Figure 10.7.16. Consider both the low ($T_c > T$), critical ($T_c = T$), and high ($T_c < T$) temperature limit. (b) Write a MATLAB code to reproduce the magnetization versus temperature plot from the mean field solution Equation (10.102) as shown in Figure 10.7.17.

10.9.19. Lev Davidovich Landau was a prominent Soviet physicist who made fundamental contributions to many areas of theoretical condensed matter physics. One of his contributions was to construct a mean field-like approach to study the ferromagnetic-paramagnetic phase transition. He constructed a free energy functional which depends on the order parameter (magnetization), $\bar{M}$, and can be expressed as,

$$F(M,T) = F_0(T) + a(T)M^2 + \frac{1}{2}bM^4 + \cdots - HM, \tag{10.115}$$

where $a(T) = a_o(T - T_c)$ with $a_o > 0$ and $b > 0$. Using the condition that

$$\frac{\partial F}{\partial M_s} = 0, \tag{10.116}$$

show that the critical exponents in zero field is given by

$$M_s \approx (T_c - T)^{-1/2}, \tag{10.117}$$

and for the susceptibility

$$\chi \approx (T - T_c)^{-1}, \tag{10.118}$$

where $T_c$ is the critical temperature. Since the order parameter is small near a phase transition, to a good approximation the free energy of the system can be approximated by the first few terms as shown above. Based on mean field theory approximation the critical exponents are

Table 10.9.5: Mean field critical exponents of the ferromagnetic Ising model.

| Physical quantity | Exponent |
|---|---|
| Specific heat, $C_M \sim \|T - T_C\|^{\alpha}$ | $\alpha = 0$ |
| Magnetization, M $\sim (T - T_C)^{\beta}$ | $\beta = \frac{1}{2}$ |
| Susceptibility, $\chi \sim (T - T_C)^{-\gamma}$ | $\gamma$=1 |
| Critical isotherm, M $\sim H^{-1/\delta}$ | $\delta$=1/3 |

(a) Derive the mean field exponents for magnetization and susceptibility and plot the quantities versus temperature. (b) Plot the mean field solution of magnetization and susceptibility versus temperature within the Landau free energy approach.

10.9.20. Recall from our Monte Carlo simulation discussion that for a given temperature and set of thermodynamic variables, the system takes a while before equilibration is reached. This is similar to inserting a thermometer and waiting for a while before you take the measurement. In a Monte Carlo simulation where we are trying to mimic thermal effects of a system, we need to allow for this process. In this exercise you will utilize the MATLAB GUI to learn about how equilibration is achieved. To do so, download the GUI code from the publisher's website and click on the file named ch10_Ising_Mode_Main_Menu.m. This should open up the interface as shown in Figure 10.7.18. On this interface, click on the *Equilibration* tab to display the interface as shown in Figure 10.7.20. Now make the following choices: Temperature = 0.3, lattice size L = 8, and MCSS = 100. Then *hit run*.

(a) In words, explain what you observe, both for the state of the magnet and in the magnetization per spin versus MCSS plot?
(b) Now keeping the temperature and MCSS fixed, change the lattice size to L = 16, 32, 48, and 64, successively. Before running each set of new trial, click on the *Reset* to clear the GUI plots and calculations. For each choice of the run explain in words what you observe? Do you reach equilibrium for all the lattices for the MCSS choice of 100? If not, which parameter can you change to get the desired state?

*Note*: If you do not have access to the GUI, you can still carry out the exercise by running the ch10_isingmontecarlo.m code. Make sure to uncomment the equilibration portion of the code.

10.9.21. In this exercise you will utilize the MATLAB GUI to observe finite size effects in lattice simulation. Following the instructions in Exercise 10.9.20 open up the interface and click on the *Magnetization & Susceptibility* tab. In the box for choices successively choose $L = 8, 16, 20, 24, 32$ and 64 with MCSS = 5000. Set the initial temperature to 0.01 and the final temperature to 8.0. Keep in mind that the temperature is in scaled units, that is, $T/J$. With a choice of $J = 1$, when we choose the temperature in the GUI simulation to be 3.0, we are actually choosing $T/J = 4.0$. Before running each set of new trial, click on the *Reset* to clear the GUI plots and calculations. In words explain how the shape of both the magnetization and susceptibility curve changes. With larger lattice sizes, the simulation will run somewhat slower, so be patient.

*Note*: If you do not have access to the GUI you can still carry out the exercise by running the ch10_isingmontecarlo.m code. Make sure to uncomment the magnetization and susceptibility portion of the code.

10.9.22. Macroscopic systems contain a large number of degrees of freedom. This is true in the case of an Ising model also where each site has two possible states (up or down). Thus on a lattice of size $L$ we have $2^L$ states, which with increasing lattice size becomes larger. To simulate such systems we typically choose a lattice which is much smaller in size than the actual model and/or is discretized. This approximation introduces systematic errors which are called **finite size effects**. In Exercise 10.9.21 you saw how finite lattice size affects the outcome of the magnetization and susceptibility. In order to reliably predict the exact nature of the phase transition point, we have to develop a systematic procedure which will allow us to extrapolate the finite size data to that of an infiinte system. Such an analysis is termed **finite size scaling**. From thermal physics we know that a true phase transition occurs only in the $L \to \infty$ limit.

Near the critical transition point, spins of an Ising model bunch up together to form clusters of a typical size $\xi$ called the correlation length. The correlation length typically diverges as the temperature approaches the critical value of $T_c$. Let us define a dimensionless parameter, $t$, called the reduced temperature which measures the deviation from $T_c$,

$$t = \frac{T - T_c}{T_c}. \tag{10.119}$$

For the Ising model we can safely assume that the correlation length varies as $\xi \sim |t|^{-\nu}$ where $\nu$ is called a **critical exponent**. The absolute value of $|t|$ ensures that the reduced temperature expression can be utilized both above and below the critical temperature region. A remarkable fact about the critical exponents is that this is a property of the Ising model itself and is independent of the details (e.g., coupling strength or the type of lattice that is being studied, square or triangular, in a particular dimension). This property is known as **universality**. Within a given universality class, all models have the same critical exponents. We can also define a critical exponent for the magnetic susceptibility as $\chi \sim |t|^{-\gamma}$ and $m \sim |t|^{\beta}$. Since critical exponents carry information on the nature of critical phenomena and phase transition, the question is how we should measure these exponents. In this problem you will learn about the **finite size scaling** strategy which will be used to measure the exponent $\gamma$.

The finite size scaling method is a way of extracting values for critical exponents by observing how measured quantities vary as the size L of the system changes. To implement the method, let us express the susceptibility in terms of the correlation length by eliminating temperature, as $\chi \sim \xi^{\gamma/\nu}$ (in the vicinity of a phase transition). In a Monte Carlo simulation where simulations are performed on a finite lattice of size $L$, the correlation length is cut off as it approaches the system size. This implies that the susceptibility $\chi$ will be cut off. Mathematically, this cutoff can be expressed as

$$\xi \sim L; \;\; \chi \sim L^{\gamma/\nu}, \tag{10.120}$$

where as long as $\xi \ll L$, the value of $\chi$ should be the same as that for the infinite system. For the Ising model $\nu$ is equal to one. Thus $\chi \sim L^{\gamma}$. At the critical transition temperature, the susceptibility attains a maximum value, $\chi_{\text{max}}$, and we can choose this value to make a plot of $\chi_{\text{max}}$ versus $\ln L$, the slope of which is the exponent $\gamma$.

Using the concept of finite size scaling and the Ising model code (ch10_isingmontecarlo.m) obtain a value for the susceptibility exponent $\gamma$. In two dimensions the exact value is known to be 7/4, which is 1.75.

10.9.23. Use the ideas of statistical mechanics to prove the Bohr-van Leeuwen theorem. Keep in mind that in the presence of a magnetic field $\vec{p}$ is replaced with the canonical momentum $\vec{p} - q\vec{A}$.

# 11

# *Magnetism II*

**Contents**



## 11.1 Introduction

Magnetism can arise both from *localized* and *delocalized* electrons. While the delocalized electron picture is appropriate when describing the effects of magnetism in metals, alloys and their conducting compounds, the localized models are more suitable for describing the magnetism of insulating ionic $3d$ transition metal compounds and $4f$ electrons from the rare-earth series. In the localized limit, strong Coulomb interaction prevents the transfer of electrons from one atomic site to another. This is the regime of insulating magnets. In the opposite limit, weak Coulomb correlations allow the electrons to be delocalized with kinetic energy as the most dominant energy term and Coulomb interactions acting as a perturbation. This is the metallic regime. In this chapter you will learn the basics of both types of magnetism. First, we will begin with the atomic description.

## 11.2 The Building Blocks of Atomic Magnetism

Magnetic properties of localized electrons in an atom or a crystalline solid are dictated by the nature of orbitals (shells) occupied by the electrons around the nucleus. A permanent magnetic moment in atoms (ions) arise from incompletely filled shells. The quantum state of an electron in an atom is specified by a set of four quantum numbers:

- $n$: Principal quantum number – defines the energy levels of the atomic orbital. Electrons with the same $n$ are said to be in the same shell. $n$ can take values of $1, 2, 3, \cdots$
- $l$: Orbital quantum number – defines the angular momentum of the orbital. This number divides the shells into subshells with $l$ ranging from 0 to $n-1$. It determines the shape of an orbital and its angular distribution. Electrons with $l$ values of $0, 1, 2, 3, ...$ are termed to be in $s, p, d, f, ...$ orbitals.
- $m_l$: Magnetic quantum number – specifies the orientation in space of a given orbital. For a given subshell number $l$, there are $2l+1$ orbitals which range from $-l$ to $l$.
- $m_s$: Component of the spin quantum number specifying the orientation of the intrinsic angular momentum $s$ – with values ranging from $-s, -s+1, \cdots, s-1, +s$.

**Example 11.2.0.1**
For $n = 3$ what are the possible (a) $l$ values and (b) the corresponding $m_l$ values?
**Solution**
(a) For $n = 3$, $l = 0, 1, 2$. These are the $s, p$, and $d$ orbitals.
(b) For a given $l$ value, $m_l$ ranges from -$l$ to +$l$. So we have

$$l = 0 \Longrightarrow m_l = 0; \quad (11.1)$$
$$l = 1 \Longrightarrow m_l = -1, 0, 1; \quad (11.2)$$
$$l = 2 \Longrightarrow m_l = -2, -1, 0, 1, 2. \quad (11.3)$$

A given quantum state is typically denoted by the abstract $\langle bra|$ and $|ket\rangle$ notation. In this terminology an atomic state can be specified by $|n, l, m_l, s, m_s\rangle$. For example, $|3, 2, 0, \frac{1}{2}, -\frac{1}{2}\rangle$ corresponds to an electron in a $n = 3$ principal quantum number shell, with orbital angular momentum $l = 2$, an orbital projection of $m_l = 0$, for an electron spin $\frac{1}{2}$ in a -$\frac{1}{2}$ (down) projection state.
The atomic orbitals are populated with electrons in the order of increasing energy. For example atomic iron, with atomic number $Z = 26$, has an electronic configuration given by [Ar]$3d^6 4s^2$. The 4s shell has two paired electrons (up and down spin states) leading to a net cancellation of the spin quantum number. The 3d shell has four unpaired electrons giving rise to a non-zero spin quantum number of $2\hbar$. The unpaired electron spins cause iron to possess an overall magnetic moment. Recall from Chapter 10 that not all unpaired electron systems are spontaneously magnetic. For example, paramagnetic systems have weakly cooperating interaction mechanisms and possess a magnetization only in the presence of an applied external magnetic field.

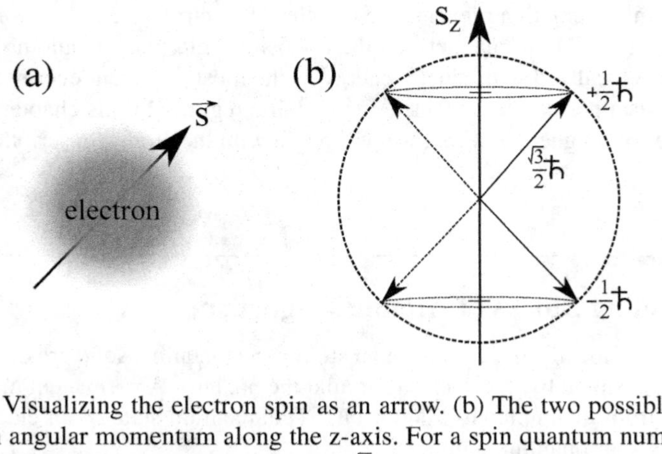


Figure 11.2.1: (a) Visualizing the electron spin as an arrow. (b) The two possible orientation projections of the spin angular momentum along the z-axis. For a spin quantum number state with s = $\frac{1}{2}$ the magnitude of the spin angular momentum is $\sqrt{3}\hbar/2$ (see Equation (11.8)).

### 11.2.1 Spin, Orbital, and Total Angular Momentum

A fundamental characteristic of an electron, besides its mass and charge, is its inherent magnetic moment – a purely quantum mechanical attribute which has no classical analogue. This intrinsic magnetic moment gives rise to an angular momentum which is termed the electron spin. In the language of quantum mechanics, spin angular momentum is a vector operator denoted by the symbol $\hat{\vec{s}}$, where the *hat* denotes the operator nature. The existence of the electron spin was established by the Stern-Gerlach experiment and justified by Paul Dirac's relativistic quantum theory. The eigenvectors and eigenvalues for a given spin state are

$$\hat{s}^2|s,m_s\rangle = s(s+1)\hbar^2|s,m_s\rangle, \tag{11.4}$$
$$\hat{s}_z|s,m_s\rangle = m_s\hbar|s,m_s\rangle. \tag{11.5}$$

For an electron which has a spin angular momentum of $s = 1/2$ and a spin projection of $s_z = -1/2$ we can write

$$\hat{s}^2|\frac{1}{2},-\frac{1}{2}\rangle = \frac{3}{4}\hbar^2|\frac{1}{2},-\frac{1}{2}\rangle, \tag{11.6}$$
$$\hat{s}_z|\frac{1}{2},-\frac{1}{2}\rangle = -\frac{1}{2}\hbar|\frac{1}{2},-\frac{1}{2}\rangle. \tag{11.7}$$

The magnitude of the spin angular momentum, $S$, is given by the expression

$$S = \sqrt{s(s+1)}\hbar = \sqrt{3}\hbar/2, \tag{11.8}$$

as shown in Figure 11.2.1. The quantum numbers $m_s$=$+\frac{1}{2}$ (spin−up state) or $m_s$=$-\frac{1}{2}$ (spin−down state) are denoted with up (down) arrows ↑ (↓), respectively. This is the reason why we used arrows in the previous chapter to denote magnetic dipoles. Similar to the spin angular momentum, one can define the eigenstates and eigenvalues of the orbital angular momentum $\hat{\vec{l}}$ as

$$\hat{l}^2|l,m_l\rangle = l(l+1)\hbar^2|l,m_l\rangle, \tag{11.9}$$
$$\hat{l}_z|l,m_l\rangle = m_l\hbar|l,m_l\rangle. \tag{11.10}$$

In Chapter 10, based on a semiclassical derivation, we arrived at the following expression relating the orbital magnetic dipole moment $\hat{\vec{\mu}}_l$ with the orbital angular momentum of an electron

$$\hat{\vec{\mu}}_l = -\frac{e}{2m_e}\hat{\vec{l}}, \tag{11.11}$$

where we have explicitly included the operator notation on the angular momentum symbol. The above equation is almost correct except that we need to include the appropriate orbital g-factor

$$\hat{\vec{\mu}}_l = -g_l\mu_B\frac{\hat{\vec{l}}}{\hbar} \equiv \gamma_l\hat{\vec{l}}; \tag{11.12}$$

where we recall that $\mu_B = \frac{e\hbar}{2m_e}$ is the Bohr magneton and $\gamma_l = -\frac{g_l\mu_B}{\hbar}$ is the orbital gyromagnetic ratio. The gyromagnetic ratio is essentially the magnetic moment per unit angular momentum. It turns out that the $g_l$ factor is exactly equal to one. So fortunately the semiclassical and the quantum expressions agree. In general, the g-factor is a dimensionless proportionality constant that relates the observed magnetic moment $\vec{\mu}$ of a particle to the appropriate angular momentum. In a similar spirit, we may write down an expression relating the spin magnetic dipole moment and the spin angular momentum of an electron by

$$\hat{\vec{\mu}}_s = -g_s\mu_B\frac{\hat{\vec{s}}}{\hbar} \equiv \gamma_s\hat{\vec{s}}, \tag{11.13}$$

where we have introduced the spin g-factor $g_s$ and $\gamma_s = -\frac{g_s\mu_B}{\hbar}$ is the spin gyromagnetic ratio . The spin g-factor is not equal to one, rather $g_s$= 2.002319304. Since spin is an intrinsic quantum object, we cannot expect a semi-classical derivation to yield the correct answer. This is reflected in the g-value not being equal to one. The above equation implies that the $z$-component of the electron's spin magnetic moment is given by

$$\mu_{s_z} = -g_s\mu_B\frac{s_z}{\hbar} = -g_s\mu_B m_s = -\frac{1}{2}g_s\mu_B, \tag{11.14}$$

where $m_s = \frac{1}{2}$ for a spin-up state and similarly

$$\mu_{s_z} = -g_s\mu_B m_s = +\frac{1}{2}g_s\mu_B, \tag{11.15}$$

with $m_s$ =-$\frac{1}{2}$ for a spin-down state.
The gyromagnetic ratio and the g-factor can be measured using the *Einsten–de Haas effect* or using ferromagnetic resonance experiments. In Exercise 11.11.1 you will derive the equation which governs this experiment.

**Example 11.2.1.1**
What is the numerical value of the spin magnetic dipole moment?
**Solution**
Considering the spin-up state we have

$$\mu_{s_z} = -g_s\mu_B m_s = -\frac{1}{2}(2.002319304)(9.274\times10^{-24})\ \mathrm{Am}^2 = -9.285\times10^{-24}\ \mathrm{Am}^2. \tag{11.16}$$

Note if we had approximated $g_s \approx 2$, we would have gotten $-\mu_B$ for the up-state. The negative charge on the electron causes the magnetic moment to be antiparallel to the angular momentum, hence the flipped sign in Equation (11.14).

In addition to the spin and the orbital magnetic dipole moments, one can define a total angular momentum $\hat{\vec{j}}$ given by,

$$\hat{\vec{j}} = \hat{\vec{l}} + \hat{\vec{s}}, \tag{11.17}$$

with a corresponding total magnetic dipole moment of an electron given by

$$\hat{\vec{\mu}}_j = \hat{\vec{\mu}}_l + \hat{\vec{\mu}}_s = -\frac{\mu_B}{\hbar}(g_l\hat{\vec{l}} + g_s\hat{\vec{s}}) = -\frac{\mu_B}{\hbar}(\hat{\vec{l}} + 2\hat{\vec{s}}), \tag{11.18}$$

where we have taken $g_s \approx 2$ and let $g_l = 1$ as stated above. Similar to $m_l$ and $m_s$, the components of the total angular momentum, $m_j$, range from $-j,\ -j+1,\ \cdots, j-1,\ +j$. The $2j+1$ degeneracy of the energy states can be lifted in the presence of a magnetic field and is known as the Zeeman effect. The eigenstates and eigenvalues of the $\hat{\vec{j}}^2$ and $\hat{j}_z$ operator can be defined as

$$\hat{\vec{j}}^2|j,m_j\rangle = j(j+1)\hbar^2|j,m_j\rangle, \tag{11.19}$$

$$\hat{j}_z|j,m_j\rangle = m_j\hbar|j,m_j\rangle. \tag{11.20}$$

Henceforth, it will be understood that the angular momentum variables defined here are operators, so in all the discussions in the remaining sections we will suppress the *hat* notation.
In Chapter 10, Section 10.3.2.2, we discussed how the bulk magnetization $\vec{M}$ interacts in the presence of an applied external magnetic field $\vec{H}_a$. In this section, we consider the Zeeman interaction from the perspective of the quantum angular momentum. The Zeeman energy, for a magnetic field applied along the $z$-axis, using Equation (11.13) is given by

$$E_z = -\vec{\mu}\cdot\vec{B} = \left(\frac{g_s\mu_B s_z}{\hbar}\right)B_z = g_s\mu_B m_s B_z. \tag{11.21}$$

The Zeeman interaction is another source for splitting atomic orbital degeneracy as shown in Figure 11.2.2. The Zeeman effect was first observed in 1896 by the Dutch physicist Pieter Zeeman as a broadening of the yellow D-lines of sodium in a flame held between strong magnetic poles. The **Zeeman splitting** is given by $\Delta_z$.

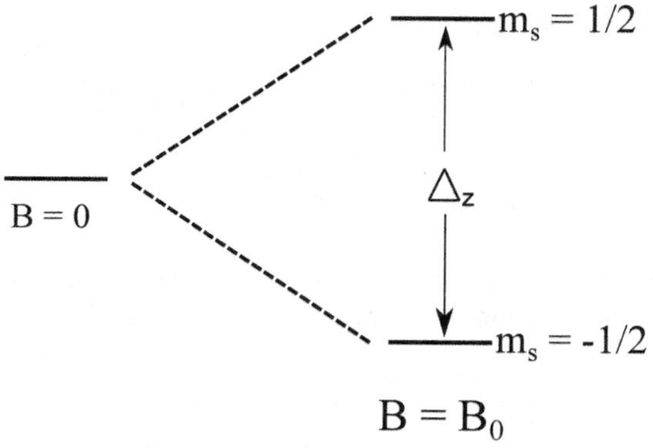


Figure 11.2.2: Zeeman interaction energy of an electron in an external magnetic field.

**Example 11.2.1.2**
Obtain an expression for the Zeeman splitting $\Delta_z$ as shown in Figure 11.2.2. Compute (a) the level splitting energy $\Delta_z$ in eV and in Kelvin, (b) the frequency of electromagnetic radiation required to excite an electron from the lower energy level to the upper one in the presence of an external magnetic field of 0.5 Tesla.
**Solution**
Using Equation (11.21) and the spin projection components $m_s = \pm\frac{1}{2}$, we have

$$\Delta_z = \frac{1}{2} g_s \mu_B B_z - \left(-\frac{1}{2}\right) g_s \mu_B B_z = g_s \mu_B B_z. \tag{11.22}$$

(a) Using the numerical values for $g_s, \mu_B, B_z$ and the conversion between joules and eV we have

$$\Delta_z^{\mathrm{eV}} = (2.002319304)(9.274 \times 10^{-24})(0.5)/(1.60217657 10^{-19}) = 0.12 \ \text{meV}. \tag{11.23}$$

In the Kelvin temperature scale we have,

$$\Delta_z^{\mathrm{K}} \equiv \Delta_z^{\mathrm{K}}/k_B = (2.002319304)(9.274 \times 10^{-24})(0.5)/(1.3807\, 10^{-23}) = 0.67 \ \text{K}, \tag{11.24}$$

where $k_B$ is the Boltzmann constant = $1.3807 \times 10^{-23}$ J/K.
(b) To obtain a numerical value of the exciting electromagnetic radiation we equate $\Delta_z = h\nu$, where $h$ is the Planck constant and $\nu$ is the frequency. We then have $\nu \approx 14$ GHz. Inducing transitions from the lower to the upper electronic level in the presence of an external magnetic field is a well-established method in the field for electron paramagnetic resonance (EPR). The computed frequency in this example lies in the microwave range and is a typical choice in EPR experiments. Two common frequencies used lie in the X-band frequency range (9.5 GHz) or the Q-band frequency range (35 GHz). In Section 11.9, we will discuss in further details the basic concepts of magnetic resonance.

### 11.2.2 Atomic Orbitals

In general atomic orbitals participate in chemical bonding. The resulting coordination chemistry dictates the ensuing chemical and physical properties, including magnetism. Before we discuss the effect of atomic orbitals on the origins of magnetism, it is worthwhile to visualize the shape of the orbitals participating in the bonding process. Spherical harmonic solutions of the angular part of the hydrogen atom problem provides the $Y_l^{m_l}(\theta,\phi)$ functions. Below we list the solution for the wavefunction

$$\psi_{nlm}(r,\theta,\phi) = R_{nl}(r)Y_l^m(\theta,\phi), \tag{11.25}$$

where the normalized radial wavefunction for the bound states of one-electron atom is written as,

$$R_{nl}(r) = -\sqrt{\left(\frac{2Z}{na_0}\right)^3 \frac{(n-l-1)!}{2n[(n+l)!]^3}}e^{-\rho/2}\rho^l L_{n+l}^{2l+1}(\rho), \tag{11.26}$$

where $Z$ is the atomic number, $a_o$ is the Bohr radius, $\rho = 2Zr/na_0$, and $L_{n+l}^{2l+1}(\rho)$ is the associated Laguerre polynomial. The orthonormal spherical harmonic angular wavefunction $Y_l^m(\theta,\phi)$ of degree $l$ and order $m$ is given by

$$Y_l^m(\theta,\phi) = (-1)^{(m+|m|)/2}\sqrt{\frac{(2l+1)}{4\pi}\frac{(l-|m|)!}{(l+|m|)!}}P_l^m(\cos\theta)e^{im\phi}, \tag{11.27}$$

where $P_l^m(\cos\theta)$ is the associated Legendre polynomial of degree $l$ and order $m$. In Tables 11.10.4 and 11.10.5, we list the explicit expressions for the radial and angular part of the wavefunction up to $n = 4$. Combining the radial and the angular solutions, we obtain

$$\psi_{nlm}(r,\theta,\phi) = -\sqrt{\left(\frac{2Z}{na_0}\right)^3 \frac{(n-l-1)!}{2n[(n+l)!]^3}}e^{-\rho/2}\rho^l L_{n+l}^{2l+1}(\rho)Y_l^m(\theta,\phi), \tag{11.28}$$

with the normalization defined by

$$\int_0^\infty r^2 dr \int_0^\pi \sin\theta d\theta \int_0^{2\pi} d\phi \;\; \psi^*_{n_1l_1m_1}(r,\theta,\phi)\psi_{n_2l_2m_2}(r,\theta,\phi) = \delta_{n_1,n_2}\delta_{m_1,m_2}\delta_{l_1,l_2}. \tag{11.29}$$

In practice, to visualize the atomic orbitals we often use a spherical harmonics real basis. These real basis wavefunctions are defined in terms of the linear combination of $Y_l^{m_l}(\theta,\phi)$ functions as

$$Y_{lm} = \begin{cases} \frac{i}{\sqrt{2}}\left(Y_l^m - (-1)^m Y_l^{-m}\right) & \text{if } m < 0, \\ Y_l^0 & \text{if } m = 0, \\ \frac{1}{\sqrt{2}}\left(Y_l^{-m} + (-1)^m Y_l^m\right) & \text{if } m > 0. \end{cases} \tag{11.30}$$

**Example 11.2.2.1**
Write a MATLAB code to generate the radial wavefunction and radial distribution function for $n = 3$, as shown in Figure 11.2.3.
**Solution**
Using the exressions for the radial wavefunction from Table 11.10.4 we can write the following MATLAB code

```
%copyright by J. E Hasbun and T. Datta
% ch11_dorbital_radial.m
```

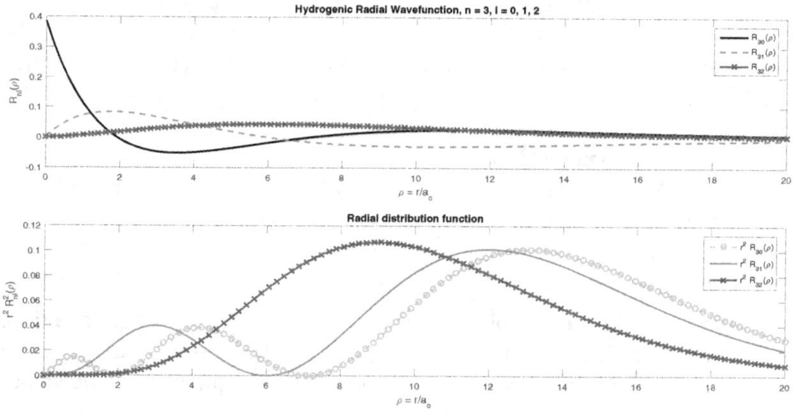


Figure 11.2.3: Radial hydrogenic wavefunctions and radial distribution function for n = 3 orbitals. The MATLAB code used to produce this figure is shown in Example 11.2.2.1.

```
% Hydrogenic radial wavefunction code

% Creating the grid points
rho = linspace(0,20,100);

% Defining the expressions for the radial wavefunction using Table 11.10.4
% in Chapter 11
% Note: Radial wavefunctions scaled by rho = r/a_o
% Note: Atomic nunmber Z is set to one

Rthreezero = ((2/(3^(3/2))))*(1 - (2/3)*rho + ...
    (2/27)*(rho.*rho)).*exp(-rho/3);                % R_30(r)(3s)
Rthreeone = (4*sqrt(2)/9)*((1/(3^(3/2))))*...
    (1 - rho/6).*rho.*exp(-rho/3);                  % R_31(r)(3p)
Rthreetwo = (4/(27*sqrt(10)))*((1/(3^(3/2))))*rho.*rho...
            .*exp(-rho/3);                          % R_32(r)(3d)

% Radial distribution function. Note the multiplicative r^2 factor in front
% of Rnl(r).

Rthreezerosq = rho.*rho.*Rthreezero.*Rthreezero;          % r^2 R^2_30(r)
Rthreeonesq = rho.*rho.*Rthreeone.*Rthreeone;             % r^2 R^2_31(r)
Rthreetwosq = rho.*rho.*Rthreetwo.*Rthreetwo;             % r^2 R^2_32(r)

% Plotting the n = 3 radial wavefunction
subplot(2,1,1);
plot(rho,Rthreezero,'-k',rho,Rthreeone,'--r',rho,Rthreetwo,'-xb',
    'LineWidth',2);
xlabel('\rho = r/a_{o}')
ylabel('R_{nl}(\rho)')
legend('R_{30}(\rho)','R_{31}(\rho)','R_{32}(\rho)')
title('Hydrogenic Radial Wavefunction, n = 3, l = 0, 1, 2')
```

```
% Plotting the radial distribution functions
subplot(2,1,2)
plot(rho,Rthreezerosq,'o--g',rho,Rthreeonesq,'-r',rho,Rthreetwosq,'-xb',...
    'LineWidth',2);
xlabel('\rho = r/a_{o}')
ylabel('r^2 R^{2}_{nl}(\rho)')
legend('r^2 R_{30}(\rho)','r^2 R_{31}(\rho)','r^2 R_{32}(\rho)');
title('Radial distribution function')
```

**Example 11.2.2.2**

Using the expressions for the $d$ orbital spherical harmonics from Table 11.10.5 and the definition of the conversion from the complex to the real basis, Equation (11.30), derive (a) the real form expressions for the five $d$-orbitals and (b) convert the real form to their equivalent Cartesian expressions.

**Solution**

For each of the choice of orbital combinations, we will first derive the real form expression and then use the conversion from polar to Cartesian form

$$x = r\sin\theta\cos\phi;\ \ y = r\sin\theta\sin\phi;\ \ z = r\cos\theta, \tag{11.31}$$

to derive the Cartesian expressions for the orbitals. We will also use the Euler identity $e^{i\theta} = \cos\theta + i\sin\theta$ to express the exponentials in terms of cosines and sines as

$$\cos\theta = \frac{e^{i\theta}+e^{-i\theta}}{2};\ \ \sin\theta = \frac{e^{i\theta}-e^{-i\theta}}{2i}. \tag{11.32}$$

1. $l = 2;\ m = -2$

$$\begin{aligned} d_{l=2}^{m=-2} &\equiv \frac{i}{\sqrt{2}}\left(Y_2^{-2} - Y_2^{2}\right) = \frac{i}{\sqrt{2}}\sqrt{\frac{15}{32\pi}}\sin^2\theta\left(e^{-2i\phi} - e^{2i\phi}\right), \\ &= \frac{i}{\sqrt{2}}\sqrt{\frac{15}{32\pi}}\sin^2\theta(-2i\sin 2\phi), \\ &= \sqrt{\frac{15}{16\pi}}\sin^2\theta\sin 2\phi. \end{aligned} \tag{11.33}$$

Now, to convert to Cartesian form, we write using Equation (11.31)

$$d_{l=2}^{m=-2} = 2\sqrt{\frac{15}{16\pi}}\sin^2\theta\sin\phi\cos\phi = \sqrt{\frac{15}{4\pi}}\frac{xy}{r^2} \equiv d_{xy}. \tag{11.34}$$

2. $l = 2;\ m = -1$

$$\begin{aligned} d_{l=2}^{m=-1} &\equiv \frac{i}{\sqrt{2}}\left(Y_2^{-1} + Y_2^{1}\right) = \frac{i}{\sqrt{2}}\sqrt{\frac{15}{8\pi}}\sin\theta\cos\theta\left(e^{-i\phi} - e^{i\phi}\right), \\ &= \frac{i}{\sqrt{2}}\sqrt{\frac{15}{8\pi}}\sin\theta\cos\theta(-2i\sin\phi), \\ &= \sqrt{\frac{15}{4\pi}}\sin\theta\sin\phi\cos\theta. \end{aligned} \tag{11.35}$$

Now, to convert to Cartesian form, we write using Equation (11.31)

$$d_{l=2}^{m=-1} = \sqrt{\frac{15}{4\pi}}\sin\theta\sin\phi\cos\theta = \sqrt{\frac{15}{4\pi}}\frac{yz}{r^2} \equiv d_{yz}. \tag{11.36}$$

3. $l = 2;\ m = 1$

$$
\begin{aligned}
d_{l=2}^{m=1} &= \frac{1}{\sqrt{2}}\left(Y_2^{-1} - Y_2^1\right) = \frac{1}{\sqrt{2}}\sqrt{\frac{15}{8\pi}}\sin\theta\cos\theta\left(e^{-i\phi} + e^{i\phi}\right), \\
&= \frac{1}{\sqrt{2}}\sqrt{\frac{15}{8\pi}}\sin\theta\cos\theta(2\cos\phi), \\
&= \sqrt{\frac{15}{4\pi}}\sin\theta\cos\phi\cos\theta. \qquad (11.37)
\end{aligned}
$$

Now, to convert to Cartesian form, we write using Equation (11.31)

$$
d_{l=2}^{m=1} = \sqrt{\frac{15}{4\pi}}\sin\theta\cos\phi\cos\theta = \sqrt{\frac{15}{4\pi}}\frac{xz}{r^2} \equiv d_{xz}. \qquad (11.38)
$$

4. $l = 2;\ m = 2$

$$
\begin{aligned}
d_{l=2}^{m=2} &\equiv \frac{1}{\sqrt{2}}\left(Y_2^2 + Y_2^{-2}\right) = \frac{1}{\sqrt{2}}\sqrt{\frac{15}{32\pi}}\sin^2\theta\left(e^{2i\phi} + e^{-2i\phi}\right), \\
&= \frac{1}{\sqrt{2}}\sqrt{\frac{15}{32\pi}}\sin^2\theta\,(2\cos 2\phi), \\
&= \sqrt{\frac{15}{16\pi}}\sin^2\theta\cos 2\phi. \qquad (11.39)
\end{aligned}
$$

Now, to convert to Cartesian form, we write using Equation (11.31)

$$
\begin{aligned}
d_{l=2}^{m=2} &\equiv \sqrt{\frac{15}{16\pi}}\sin^2\theta\left(2\cos^2\phi - 1\right), \\
&= \sqrt{\frac{15}{16\pi}}(\sin^2\theta\cos^2\phi - \sin^2\phi\sin^2\theta) = \sqrt{\frac{15}{16\pi}}\frac{(x^2 - y^2)}{r^2}, \\
&\equiv d_{x^2-y^2}. \qquad (11.40)
\end{aligned}
$$

5. $l = 2;\ m = 0$

$$
\begin{aligned}
d_{l=2}^{m=0} \equiv Y_{20} &= \sqrt{\frac{5}{16\pi}}\left(3\cos^2\theta - 1\right), \\
&= \sqrt{\frac{5}{16\pi}}\left(3\cos^2\theta - \sin^2\theta - \cos^2\theta\right), \\
&= \sqrt{\frac{5}{16\pi}}\frac{2z^2 - x^2 - y^2}{r^2}, \\
&= \sqrt{\frac{5}{16\pi}}\frac{3z^2 - x^2 - y^2 - z^2}{r^2}, \\
&\equiv d_{3z^2-r^2}, \qquad (11.41)
\end{aligned}
$$

where we have used $r^2 = x^2 + y^2 + z^2$ above.

In Section 11.5 you will learn how the shape of these atomic orbitals and their relative orientations play a crucial role in lifting the degeneracies of atomic levels and lead to crystal field splitting. For now, consider the MATLAB code below which generates the 3d atomic orbital. In Figure 11.2.4 we have shown the boundary surface (isosurface) representation of the five 3*d* orbitals. The boundary surface method of representing orbitals assumes that the surface encloses some substantial portion, say 90%, of the total electron density for the orbital. Consider the example shown below.

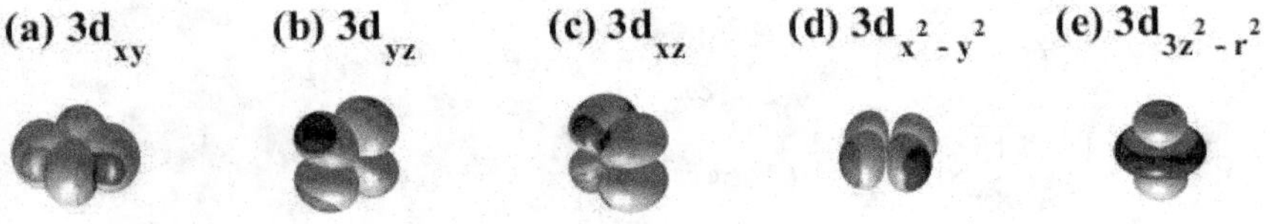


Figure 11.2.4: Boundary surface representation of atomic $3d_{xy}$, $3d_{yz}$, $3d_{xz}$, $3d_{x^2-y^2}$, and $3d_{3z^2-r^2}$ orbitals in real basis. The MATLAB code used to generate these orbitals is given in Example 11.2.2.3.

**Example 11.2.2.3**
Write a MATLAB code to generate the atomic $3d_{xy}$, $3d_{yz}$, $3d_{xz}$, $3d_{x^2-y^2}$, and $3d_{3z^2-r^2}$ orbitals shown in Figure 11.2.4.
**Solution**

```
%copyright by J. E Hasbun and T. Datta
% ch11_dorbitals.m

% Boundary (isosurfaces) of atomic d orbitals
% A substantial portion (say 90%) of the total electron density of the
% orbital is enclosed within the boundary surface

% Wavefunctions defined below

% Normalized wave functions of hydrogen atom
% NOTE: Z =1, a0 = 1, 0<= l < n %-l<= m <=l

% d orbital
% ylm choice for d orbitals
 ylm_20 = @(theta,phi) (3*cos(theta).^2 - 1.0)*sqrt(5/pi)/4;
 ylm_21 = @(theta,phi) sqrt(2).*sin(theta).*cos(theta).*cos(phi)...
          *sqrt(15/pi)/2;
 ylm_22 = @(theta,phi) (sin(theta).^2).*(cos(2*phi))*sqrt(15/pi)/4;

% 3d
 psi_3dz2 = @(r,theta,phi) exp(-(1/3)*r).*(r.^2).*(3*cos(theta).^2 - 1.0)...
          /(81*sqrt(6*pi));
 psi_3dxz = @(r,theta,phi) exp(-(1/3)*r).*(r.^2).*sqrt(2).*sin(theta)...
          .*cos(theta).*cos(phi)/(81*sqrt(pi));
 psi_3dyz = @(r,theta,phi) exp(-(1/3)*r).*(r.^2).*sqrt(2).*sin(theta)...
          .*cos(theta).*sin(phi)/(81*sqrt(pi));
 psi_3dx2y2 = @(r,theta,phi) exp(-(1/3)*r).*(r.^2).*(sin(theta).^2)...
          .*(cos(2*phi))/(81*sqrt(2*pi));
 psi_3dxy = @(r,theta,phi) exp(-(1/3)*r).*(r.^2).*(sin(theta).^2)...
          .*(sin(2*phi))/(81*sqrt(2*pi));

 %configuring the range
 [x, y , z] = meshgrid(-30:0.5:30,-30:0.5:30,-30:0.5:30);
```

```
% Cartesian to spherical coordinates conversion
R=sqrt(x.^2+y.^2+z.^2);
THETA=acos(z./R);
PHI=atan2(y,x);

% Plotting orbtials

figure

subplot(1,5,1);
colors = ylm_22(THETA,PHI); psi = psi_3dxy(R,THETA,PHI); psisq = psi.^2;
set(gcf,'color',[1 1 1]);
daspect([1 1 1]); axis off; view(3);
camlight('left'); camzoom(0.75); lighting phong;
axis vis3d; colormap jet; rotate3d on; brighten(1);
isosurface(psisq,1E-5,colors);
title('(a) 3d_{xy}','FontName','Times','FontSize',12)

subplot(1,5,2)
colors = ylm_21(THETA,PHI); psi = psi_3dyz(R,THETA,PHI); psisq = psi.^2;
set(gcf,'color',[1 1 1]);
daspect([1 1 1]); axis off; view(3);
camlight('left'); camzoom(0.75); lighting phong;
axis vis3d; colormap jet; rotate3d on; brighten(1);
isosurface(psisq,1E-5,colors);
title('(b) 3d_{yz}','FontName','Times','FontSize',12)

subplot(1,5,3);
colors = ylm_21(THETA,PHI); psi = psi_3dxz(R,THETA,PHI); psisq = psi.^2;
set(gcf,'color',[1 1 1]);
daspect([1 1 1]); axis off; view(3);
camlight('left'); camzoom(0.75); lighting phong;
axis vis3d; colormap jet; rotate3d on; brighten(1);
isosurface(psisq,1E-5,colors);
title('(c) 3d_{xz}','FontName','Times','FontSize',12)

subplot(1,5,4)
colors = ylm_22(THETA,PHI); psi = psi_3dx2y2(R,THETA,PHI); psisq = psi.^2;
set(gcf,'color',[1 1 1]);
daspect([1 1 1]); axis off; view(3);
camlight('left'); camzoom(0.75); lighting phong;
axis vis3d; colormap jet; rotate3d on; brighten(1);
isosurface(psisq,1E-5,colors);
title('(d) 3d_{x^2 - y^2}','FontName','Times','FontSize',12)

subplot(1,5,5)
colors = ylm_20(THETA,PHI); psi = psi_3dz2(R,THETA,PHI); psisq = psi.^2;
set(gcf,'color',[1 1 1]);
daspect([1 1 1]); axis off; view(3);
camlight('left'); camzoom(0.75); lighting phong;
```

```
axis vis3d; colormap jet; rotate3d on; brighten(1);
isosurface(psisq,1E-5,colors);
title('(e) 3d_{3z^2 - r^{2}}','FontName','Times','FontSize',12)
```

## 11.3 Spin-orbit Interaction

Spin-orbit interaction plays an important role in dictating the fate of the atomic ground state in addition to being the key microscopic reason for several interesting phenoma in magnetism such as magnetocrystalline anisotropy and magnetostriction. To learn the basic concepts of spin-orbit coupling, let us begin by asking the question: *What is the connection between an electric field and a magnetic field*?

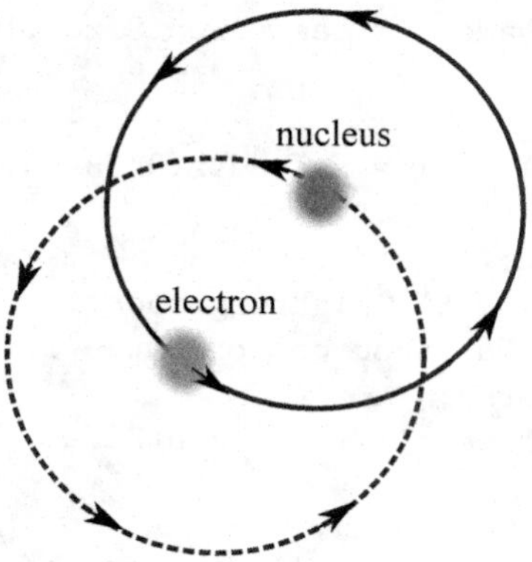


Figure 11.3.5: Spin orbit coupling interaction.

A field is perceived either as *electric* or *magnetic* based on the motion of the reference frame relative to the sources of the field. As shown in Figure 11.3.5, in an atom both the nucleus (which is positively charged) and the electron (which is negatively charged) are in relative motion to each other. Since the electron is the system of interest, we choose a coordinate system attached to the inertial rest frame of the electron. The orbiting nucleus generates at the position of the electron both an electric field $\vec{E}_n$ and a magnetic field (in *SI* units)

$$\vec{B}_e = -\frac{\vec{v}_e \times \vec{E}_n}{c^2}, \tag{11.42}$$

where $\vec{v}_e$ is the velocity of the electron (relative to the nucleus) and the subscript $n(e)$ stands for nucleus (electron). The innocent looking relationship relating $\vec{B}_e$ with $\vec{E}_n$ expresses a profound idea in physics: The Biot-Savart law for the magnetic field of a moving point charge is nothing other than the Coulomb electric field of a stationary point charge transformed into a moving frame of reference. The speed of light squared factor $c^2$ in $\vec{B}_e$ is a consequence of the Lorentz transformation equations between the two frames of reference.

The Zeeman Hamiltonian of an electron magnetic dipole moment $\vec{\mu}_s$ interacting with the magnetic field created by the nucleus at the position of the electron is given by

$$\mathscr{H}_{so} = -\vec{\mu}_s \cdot \vec{B}_e, \tag{11.43}$$

where $\mathscr{H}_{so}$ is termed the spin-orbit interaction energy. Using Equation (11.13) with $\mu_B = \frac{e\hbar}{2m_e}$ and $g_s \sim 2$ for the net spin angular momentum and Equation (11.42) we have

$$\mathscr{H}_{so} = -\left(-\frac{|e|}{m_e}\vec{s}\right)\cdot\left(-\frac{\vec{v}_e \times \vec{E}_n}{c^2}\right), \quad (11.44)$$

$$\mathscr{H}_{so} = -\frac{|e|\vec{s}\cdot\left(\vec{v}_e \times \vec{E}_n\right)}{m_e c^2} \quad (11.45)$$

where $m_e$ is the mass of the electron. To proceed further we need to utilize an expression for the electric field. In general the electric field can be expressed as $\vec{E} = -\vec{\nabla}V(r)$, where $V(r)$ is the electric potential. For a solid if we make the central field approximation then the electrostatic potential is spherically symmetric and is only a function of radius. We can then write

$$\vec{E}_n = -\vec{\nabla}V_n(\vec{r}) = \frac{1}{|e|}\frac{\partial U(r)}{\partial r}\frac{\vec{r}}{r}, \quad (11.46)$$

where we utilized $U(r) = -|e|V_n(r)$. $V_n(r)$ is the Coulomb potential of the nucleus and $U(r)$ is the Coulomb potential energy of interaction between the electron and the nucleus. For hydrogen-like atoms with $Z$-protons we can write

$$U(r) = -\frac{Z|e|^2}{4\pi\varepsilon_o}\frac{1}{r} \quad (11.47)$$

Utilizing Equation (11.46) in Equation (11.45), we obtain

$$\mathscr{H}_{so} = -\frac{1}{m_e c^2 r}\frac{\partial U(r)}{\partial r}\vec{s}\cdot(\vec{v}_e \times \vec{r}), \quad (11.48)$$

$$\mathscr{H}_{so} = \frac{1}{m_e^2 c^2 r}\frac{\partial U(r)}{\partial r}(\vec{r}\times\vec{p}_e)\cdot\vec{s}, \quad (11.49)$$

$$\mathscr{H}_{so} = \frac{1}{m_e^2 c^2 r}\frac{\partial U(r)}{\partial r}\vec{l}\cdot\vec{s} \quad (11.50)$$

The above equation is almost correct except for the inclusion of an overall $\frac{1}{2}$ factor. Even though we considered the electron to be stationary, in reality it is *not*. We derived our spin-orbit interaction in the *instantaneous* rest frame of the electron. The rotation of the electron rest frame has kinematic consequences since any instantaneous rest frame is obtained from the previous one by a non-collinear Lorentz boost. When the kinematics of the electron motion is properly accounted for using the special theory of relativity, it leads to the Thomas precession $\frac{1}{2}$ factor. We then obtain

$$\mathscr{H}_{so} = \frac{1}{2m_e^2 c^2 r}\frac{\partial U(r)}{\partial r}\vec{l}\cdot\vec{s}. \quad (11.51)$$

For the special case of a spherically symmetric Coulomb potential, Equation (11.47), we can write down the following expression for spin-orbit interaction,

$$\mathscr{H}_{so} = \frac{1}{4\pi\varepsilon_o}\frac{Z|e|^2}{2m_e^2 c^2 r^3}(\vec{l}\cdot\vec{s}) = \xi_{so}(r)\vec{l}\cdot\vec{s}, \quad (11.52)$$

where $\xi_{so}(r) = \frac{1}{4\pi\varepsilon_o}\frac{Z|e|^2}{2m_e^2 c^2 r^3}$ is the one-electron spin-orbit coupling.

**Example 11.3.0.1**

Obtain an expression for the spin-orbit coupling energy and the spin-orbit coupling constant.

Consider taking the quantum mechanical average over Equation (11.52). We then have

$$E_{so} = \langle \mathscr{H}_{so} \rangle = \frac{1}{4\pi\varepsilon_o} \frac{Z|e|^2}{2m_e^2c^2} \left\langle \frac{\vec{l}\cdot\vec{s}}{r^3} \right\rangle. \tag{11.53}$$

For a given quantum state $\vec{l}$ and $\vec{s}$ are well defined. To obtain the expectation value, we square the total angular momentum operator $\vec{j}$ as

$$\vec{j}^{\,2} = \vec{l}^{\,2} + \vec{s}^{\,2} + 2\vec{l}\cdot\vec{s}, \tag{11.54}$$

so the scalar product of $\vec{l}\cdot\vec{s}$ may be written as

$$\vec{l}\cdot\vec{s} = \frac{1}{2}\left[\vec{j}^{\,2} - \vec{l}^{\,2} - \vec{s}^{\,2}\right] = \frac{1}{2}\left[j(j+1) - l(l+1) - \frac{3}{4}\right]\hbar^2. \tag{11.55}$$

for $s = \frac{1}{2}$. The spin-orbit energy is thus given by

$$E_{so} = \frac{1}{2}\zeta_{so}\left[j(j+1) - l(l+1) - \frac{3}{4}\right], \tag{11.56}$$

where the **spin-orbit coupling constant** $\zeta_{so} \equiv \hbar^2\langle \xi_{so}(r)\rangle$, or

$$\zeta_{so} = \frac{1}{4\pi\varepsilon_o} \frac{Z|e|^2\hbar^2}{2m_e^2c^2} \left\langle \frac{1}{r^3} \right\rangle. \tag{11.57}$$

A detailed quantum mechanical evaluation of $\left\langle \frac{1}{r^3} \right\rangle$ is beyond the scope of the present textbook. However, calculations show that $\left\langle \frac{1}{r^3} \right\rangle \sim Z^3$. This implies that the spin-orbit interaction energy depends on the fourth power of the atomic number $Z$. Hence, the heavier the element, the stronger is the spin-orbit coupling effect. We will explore the consequences of this statement in the next section on Hund's rules. In summary, spin-orbit interaction is an intrinsic effect generated by the interaction between spin and the orbital degrees of freedom of an electron.

## 11.4 Ground State of an Ion: Hund's Rules

In contrast to the hydrogen atom or a hydrogen-like ion (e.g., $Be^{3+}$) which has a single electron, atoms are multi-electron systems. The presence of multiple electrons cause several interaction energy terms: Coulomb interaction, spin-orbit interaction, and crystal-field effects, to compete in dictating the final ground state configuration. Partially filled electron shells which give rise to a net magnetic moment can be found in the first row $3d$ transition metals (Sc $\rightarrow$ Zn), second row $4d$ transition metals (Y $\rightarrow$ Cd), the $4f$ lanthanides (La $\rightarrow$ Lu), and the $5f$ actinides (Ac $\rightarrow$ Lr). The $3d$, $4f$, and $5f$ electrons can give rise to localized magnetic moments. Interestingly, for the rare earths ($4f$) and the actinides ($5f$), even though the $f$-electrons could be in an insulating state, these materials can be in a metallic state.

To obtain the ground state **term**, **level**, and **state** of an ion, we begin with the assumption that spin-orbit coupling is a small perturbation. In this case the net values of orbital and spin angular momentum are given by the sum total of the individuals

$$L = \sum_i l_i; \;\; M_L = \sum_i m_{li}; \;\; S = \sum_i s_i; \;\; M_S = \sum_i m_{si}; \tag{11.58}$$

The above way of combining the orbital and spin angular momentum is known as the Russell-Saunders scheme or $LS$ coupling scheme (after astronomers Henry Norris Russell and Frederick

Albert Saunders). However, for the heavier elements such as the third row transition metals and the actinides, we need to introduce a $jj$-coupling scheme where the $J$ values are computed first since neither $L$ nor $S$ are good quantum numbers. The expression for the total angular momentum $J$ is given by

$$\vec{J} = \vec{L} + \vec{S}, \tag{11.59}$$

where $M_J$ values range from $-J, -J+1, ..., J-1, J$.
In the vector model of the atom, the addition of $\vec{L}$ and $\vec{S}$ is represented by a vector summation as shown in Figure 11.4.6. From the figure we can clearly observe that the presence of the spin g-factor causes the net magnetic dipole moment, $\vec{\mu}$, to have an orientation different from $\vec{J}$. Nevertheless we can ask the following question: is it possible to express the net magnetic dipole moment $\vec{\mu}$ as proportional to $\vec{J}$? To do so, we consider the following equation

$$\vec{\mu}_J = -g_J \mu_B \frac{\vec{J}}{\hbar}, \tag{11.60}$$

where $g_J$ is the Landé $g$-factor . An explicit expression for the Landé $g_J$ factor can be obtained by taking the dot product with $\vec{J}$ on both sides of Equation (11.60). This leads to

$$g_J = -(\vec{\mu}_J \cdot \vec{J}/\mu_B)/(|\vec{J}|^2/\hbar). \tag{11.61}$$

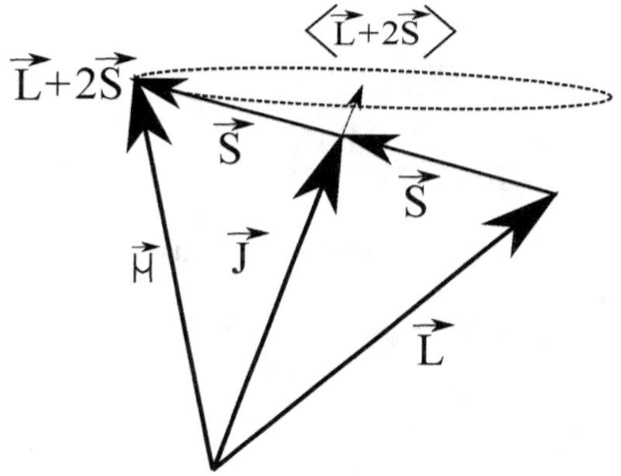


Figure 11.4.6: Vector model of atom displaying the addition of orbital angular momentum ($\vec{L}$), spin angular momentum ($\vec{L}$), the net angular momentum ($\vec{J}$), and the net magnetic dipole moment ($\vec{\mu}$).

The orbital and spin angular momenta can combine in $(2L+1)(2S+1)$ ways. This is the total number of available choices of the $z$ component of the $\vec{L}$ which ranges from $-L, -L+1, ..., L-1, L$ multiplied by the (2S+1) choices for spin projection. Recalling Equation (11.18), the total magnetic dipole moment is written as

$$\vec{\mu}_J = \sum_i \vec{\mu}_i = -\frac{\mu_B}{\hbar} \sum_i (\vec{l}_i + 2\vec{s}_i),$$

which utilizing the definition of the total angular momentum variables $\vec{L}$ and $\vec{S}$, shown in Equation (11.58), gives

$$\vec{\mu}_J = -\frac{\mu_B}{\hbar}(\vec{L} + 2\vec{S}). \tag{11.62}$$

In the next step, invoking the definitions for $\vec{J}$ and $\hat{\vec{\mu}}_J$ from Equations (11.59) and (11.62), respectivley we can compute

$$
\begin{aligned}
\hat{\vec{\mu}}_J \cdot \vec{J} &= -(\mu_B/\hbar)[(\vec{L}+2\vec{S})\cdot(\vec{L}+\vec{S})], \\
\hat{\vec{\mu}}_J \cdot \vec{J} &= -(\mu_B/\hbar)[(\vec{L}+2\vec{S})\cdot(\vec{L}+\vec{S})], \\
\hat{\vec{\mu}}_J \cdot \vec{J} &= -(\mu_B/\hbar)[(\vec{L}^2+3\vec{L}\cdot S+2\vec{S}^2)], \\
\hat{\vec{\mu}}_J \cdot \vec{J} &= -(\mu_B/\hbar)[(\vec{L}^2+(3/2)(\vec{J}^2-\vec{L}^2-\vec{S}^2)+2\vec{S}^2)], \\
\hat{\vec{\mu}}_J \cdot \vec{J} &= -(\mu_B/\hbar)[(3/2)J(J+1)-(1/2)L(L+1)+(1/2)S(S+1)].
\end{aligned} \tag{11.63}
$$

To complete the derivation, we insert Equation (11.63) into the expression for $g_J$ (Equation (11.61)), and noting that $|\vec{J}|^2 = J(J+1)\hbar^2$, the final expression for the Landé $g_J$ factor becomes

$$
g_J = \frac{3}{2} + \{S(S+1) - L(L+1)\}/\{2J(J+1)\}. \tag{11.64}
$$

**Example 11.4.0.1**

Using the values of L, S, and J from Tables 11.4.1 and 11.4.2 compute the Landé $g_J$ factor and $\mu_{\text{eff}}$ for the following ions (a) $Ni^{2+}$ (b) $Nd^{3+}$. Also compute, only the Landé $g_J$ factor for an ion which has (c) S = 0 and (d) L = 0.

**Solution**

(a) From Table 11.4.1 for $Ni^{2+}$ we have $S = 1, L = 3$, and $J = 4$. Using Equation (11.64) and inserting numbers we have

$$
\begin{aligned}
g_J &= \frac{3}{2} + \frac{\{(1)(1+1)-3(3+1)\}}{\{2(4)(4+1)\}} = \frac{5}{4} \\
\text{and } \mu_{\text{eff}} &= \left(\frac{3}{2} + \frac{\{(1)(1+1)-3(3+1)\}}{\{2(4)(4+1)\}}\right)\sqrt{4(4+1)} = 5.59
\end{aligned} \tag{11.65}
$$

where we used the definition $\mu_{eff} = g_J\sqrt{J(J+1)}$ (in $\mu_B$ units). Note, this value agrees exactly with the number reported in Table 11.4.1.

(b) From Table 11.4.2 for $Nd^{3+}$ we have $S = \frac{3}{2}, L = 6$, and $J = \frac{9}{2}$. Using Equation (11.64) and inserting numbers we have

$$
\begin{aligned}
g_J &= \frac{3}{2} + \frac{\{(\frac{3}{2})(\frac{3}{2}+1)-6(6+1)\}}{\{2(\frac{9}{2})(\frac{9}{2}+1)\}} = \frac{8}{11}, \\
\text{and } \mu_{\text{eff}} &= \left(\frac{3}{2} + \frac{\{(\frac{3}{2})(\frac{3}{2}+1)-6(6+1)\}}{\{2(\frac{9}{2})(\frac{9}{2}+1)\}}\right)\sqrt{\left(\frac{9}{2}\right)\left(\frac{9}{2}+1\right)}
\end{aligned} \tag{11.66}
$$

$$
= 3.618 = 3.62 \text{ (in } \mu_B \text{ units)}. \tag{11.67}
$$

Once again, note, this value agrees exactly with the number reported in Table 11.4.2.

(c) When S = 0, J = L. This implies $g_J = 1$. This is the result one would expect for the *purely orbital* case.

(d) When L = 0, J = S. This implies $g_J = 2$. This is the result one would expect for the *purely spin* case.

Table 11.4.1: $3d$ transition metal ions with ground state term symbols. For shells which are less than half-filled we compute $J$ as $J = |L - S|$ and if they are greater than half-filled, $J = |L + S|$.

| ion | 3d$^n$ | S | L | J | Term symbol | $\mu_{eff} = g_J\sqrt{J(J+1)}$ (in $\mu_B$ units) | $\mu_{exp}$ | $\mu_{eff} = g\sqrt{S(S+1)}$ (in $\mu_B$ units) |
|---|---|---|---|---|---|---|---|---|
| $Ti^{3+}$,$V^{4+}$ | 3d$^1$ | $\frac{1}{2}$ | 2 | $\frac{3}{2}$ | $^2D_{3/2}$ | 1.55 | 1.70 | 1.73 |
| $V^{3+}$ | 3d$^2$ | 1 | 3 | 2 | $^3F_2$ | 1.63 | 2.61 | 2.83 |
| $Cr^{3+}$, $V^{2+}$ | 3d$^3$ | $\frac{3}{2}$ | 3 | $\frac{3}{2}$ | $^4F_{3/2}$ | 0.77 | 3.85 | 3.87 |
| $Mn^{3+}$, $Cr^{2+}$ | 3d$^4$ | 2 | 2 | 0 | $^5D_0$ | 0 | 4.82 | 4.90 |
| $Fe^{3+}$, $Mn^{2+}$ | 3d$^5$ | $\frac{5}{2}$ | 0 | $\frac{5}{2}$ | $^6S_{5/2}$ | 5.92 | 5.82 | 5.92 |
| $Fe^{2+}$ | 3d$^6$ | 2 | 2 | 4 | $^5D_4$ | 6.70 | 5.36 | 4.90 |
| $Co^{2+}$ | 3d$^7$ | $\frac{3}{2}$ | 3 | $\frac{9}{2}$ | $^4F_{9/2}$ | 6.63 | 4.90 | 3.87 |
| $Ni^{2+}$ | 3d$^8$ | 1 | 3 | 4 | $^3F_4$ | 5.59 | 3.12 | 2.83 |
| $Cu^{2+}$ | 3d$^9$ | $\frac{1}{2}$ | 2 | $\frac{5}{2}$ | $^2D_{5/2}$ | 3.55 | 1.83 | 1.73 |
| $Zn^{2+}$ | 3d$^{10}$ | 0 | 0 | 0 | $^1S_0$ | 0 | 0 | 0 |

Table 11.4.2: $4f$ lanthanide ions with ground state term symbols. For shells which are less than half-filled we compute $J$ as $J = |L - S|$ and if they are greater than half-filled, $J = |L + S|$.

| ion | 4f$^n$ | S | L | J | Term symbol | $\mu_{eff} = g_J\sqrt{J(J+1)}$ (in $\mu_B$ units) | $\mu_{exp}$ |
|---|---|---|---|---|---|---|---|
| $Ce^{3+}$ | 4f$^1$ | $\frac{1}{2}$ | 3 | $\frac{5}{2}$ | $^2F_{5/2}$ | 2.54 | 2.51 |
| $Pr^{3+}$ | 4f$^2$ | 1 | 5 | 4 | $^3H_4$ | 3.58 | 3.56 |
| $Nd^{3+}$ | 4f$^3$ | $\frac{3}{2}$ | 6 | $\frac{9}{2}$ | $^4I_{9/2}$ | 3.62 | 3.3 – 3.7 |
| $Pm^{3+}$ | 4f$^4$ | 2 | 6 | 4 | $^5I_4$ | 2.68 | – |
| $Sm^{3+}$ | 4f$^5$ | $\frac{5}{2}$ | 5 | $\frac{5}{2}$ | $^6I_{5/2}$ | 0.85 | 1.74 |
| $Eu^{3+}$ | 4f$^6$ | 3 | 3 | 0 | $^7F_0$ | 0.0 | 3.4 |
| $Gd^{3+}$ | 4f$^7$ | $\frac{7}{2}$ | 0 | $\frac{7}{2}$ | $^8S_{7/2}$ | 7.94 | 7.98 |
| $Tb^{3+}$ | 4f$^8$ | 3 | 3 | 6 | $^7F_6$ | 9.72 | 9.77 |
| $Dy^{3+}$ | 4f$^9$ | $\frac{5}{2}$ | 5 | $\frac{15}{2}$ | $^6H_{15/2}$ | 10.65 | 10.63 |
| $Ho^{3+}$ | 4f$^{10}$ | 2 | 6 | 8 | $^5I_8$ | 10.61 | 10.4 |
| $Er^{3+}$ | 4f$^8$ | $\frac{3}{2}$ | 6 | $\frac{15}{2}$ | $^4I_{15/2}$ | 9.58 | 9.5 |
| $Tm^{3+}$ | 4f$^9$ | 1 | 5 | 6 | $^3H_6$ | 7.57 | 7.61 |
| $Yb^{3+}$ | 4f$^{10}$ | $\frac{1}{2}$ | 3 | $\frac{7}{2}$ | $^2F_{7/2}$ | 4.53 | 4.5 |
| $Lu^{3+}$ | 4f$^{10}$ | 0 | 0 | 0 | $^1S_0$ | 0 | 0 |

The angular momentum combinations of $\vec{J}$, $\vec{L}$, and $\vec{S}$ play an important role in determining how the electronic shells are filled up. For multi-electron systems, atomic orbitals are filled up in an order that minimizes the overall energy. An empirical set of rules were devised by the German physicist Friedrich Hund (1896–1997) that obey the minimum energy principle. The three Hund's rules ([HR]), in order of relative importance, are stated as follows:

HR 1: A given electronic configuration of electrons are arranged to maximize the value of $S$. This rule helps to minimize the Coulomb electrostatic interaction consistent with the Pauli exclusion principle.

HR 2: For a given spin multiplicity, $L$ is maximized consistent with $S$. The second rule also arises in an effort to minimize Coulomb interaction effects.

HR 3: The combined effects of $L$ and $S$ are considered by computing the $J$ value. For shells which are less than half-filled $J = |L - S|$ and if they are greater than half-filled $J = |L + S|$. The third rule is an attempt to minimize the effects of spin-orbit coupling energy, given by Equation (11.52) and discussed further below.

Competition between spin-orbit coupling energy and crystal field energy has important implications. Based on their relative strengths, [HR3] will either be obeyed (as in rare-earth ions) or invalidated (as in transition metal ions) where spin-orbit energy is not significant, see Tables 11.4.1 and 11.4.2. For a given atom, once the $S$, $L$, and $J$ values are decided, the ground state **term symbol** is labeled in the form

$$^{2S+1}L_J \tag{11.68}$$

where the spin multiplicity is given by $2S + 1$. The multiplicity number tells us how many atomic spin configurations are energetically degenerate. In the presence of a magnetic field this degeneracy can be broken. In the above spectroscopic terminology the $L$ symbol is written in

| $L$ | 0 | 1 | 2 | 3 | 4 | 5 | 6 | ⋯ |
|---|---|---|---|---|---|---|---|---|
| | $S$ | $P$ | $D$ | $F$ | $G$ | $H$ | $I$ | ⋯ |

capital letters to indicate that we are dealing with the sum total values of the angular momentum. The assignment of the letters $S, P, D, F, \ldots$ is historic. It has its origin in spectroscopists terminology who used **S**harp, **P**rincipal, **D**iffuse, **F**undamental, ... to characterize spectral lines. Let us consider some Hund's rule examples.

**Example 11.4.0.2**
What is the ground state term symbol of $V^{3+}$?
**Solution**
Consider the *free* transition metal ion, $V^{3+}$, which has an outer shell configuration $3d^2$ as shown in Figure 11.4.7. The *free* ion assumption implies that its spectrum and magnetic moment will not be affected by the surrounding atoms, ions, or molecules (compared to the opposite limit, see Section 11.5). The $d$ electrons have $l$=2, so the angular momentum multiplicity is $2l$+1 = 5, where they are all spin-up (maximizing spin angular momentum, [HR 1]). This gives a total spin value of S = $2(\frac{1}{2}) = 1$ with a spin multiplicity of 2(1)+1 = 3. The orbital momentum ranges from $m_l = -2, \cdots, 2$, with the spins occupying $m_l$=2 and 1 (maximizing orbital angular momentum, [HR 2]). Thus L = 2 + 1 = 3 ($F$). Now, for a shell which is *less than half-filled* following rule [HR 3] we get $J = 3 - 1 = 2$. Thus, the ground state term symbol is $^3F_2$ (compare this with the term symbol for $V^{3+}$ in Table 11.4.1).

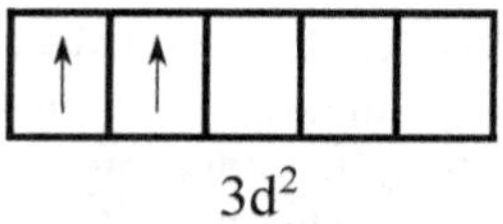


Figure 11.4.7: Electronic configuration of transition metal ion Vanadium 3+ ion. Vanadium has applications in the steel industry where it is used as an additive (ferrovanadium) to increase strength and anti-corrosive properties. Compounds of vanadium were discovered in Mexico by the Spanish-Mexican mineralogist Andrés Manuel del Rio in 1801.

**Example 11.4.0.3**
What is the ground state term symbol of $Gd^{3+}$?
**Solution**
Consider the *free* rare earth ion, $Gd^{3+}$, which has an outer shell configuration $4f^7$ as shown in Figure 11.4.8. The $f$ electrons have $l$=3, so the angular momentum multiplicity is 2$l$+1 = 7, where they are all spin-up (maximizing spin, [HR 1]). This gives a total spin value of S = $\frac{7}{2}$ with a spin multiplicity of 2(7/2)+1 = 8. The electrons do not give rise to a net orbital momentum since the spins give rise to an *exactly half-filled* shell. Thus $m_l = -3, \cdots, 3$ sum up to zero. So, $J = 7/2 + 0$=7/2. This gives the ground state term symbol as $^8S_{7/2}$ (compare this with the term symbol result for $Gd^{3+}$ in Table 11.4.2).

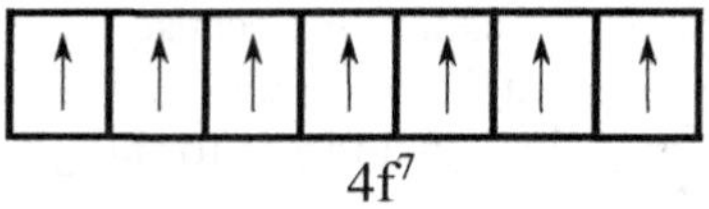


Figure 11.4.8: Electronic configuration of rare-earth Gadolinium 3+ ion. Gadolinium has applications in magnetic refrigeration technology (see Section 10.6 in Chapter 10).

**Example 11.4.0.4**
What is the ground state term symbol of $Dy^{3+}$?
**Solution**
Consider the *free* transition metal ion, $Dy^{3+}$, which has an outer shell configuration $4f^9$. The $f$ electrons have $l$=3, so the angular momentum multiplicity is 2$l$+1 = 7, where they are arranged as shown in Figure 11.4.9 [HR 1, HR 2]. This gives a total spin value of S = $7\left(\frac{1}{2}\right) - 2\left(\frac{1}{2}\right) = 5/2$ with a spin multiplicity of 2(5/2)+1 = 6. The orbital momentum ranges from $m_l = -3, \cdots, 3$, with the spins maximizing the orbital angular momentum to give L = 3(2) + 2(2)+1(1)+0(1)-1(1)-2(1)-3(1) = 5 ($H$). Now, for a shell which is *more than half-filled* following rule [HR 3] we get $J = 5 + 5/2$=15/2. Thus, the ground state term symbol is $^6H_{15/2}$ (compare this with the term symbol for $Dy^{3+}$ in Table 11.4.2).

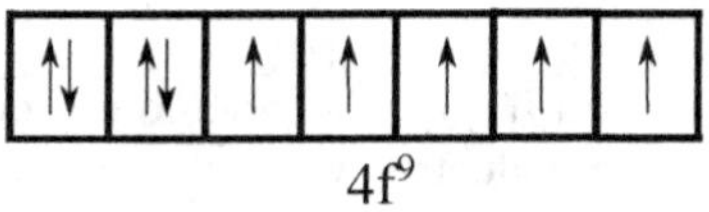


Figure 11.4.9: Electronic configuration of transition metal ion Dysprosium 3+ ion. Dysprosium has applications in control rods in nuclear reactors for its high thermal neutron absorption cross-section. It also finds applications in data storage industry for its high magnetic susceptibility. Dysprosium was first identified in 1886 by the French chemist Paul Émile Lecoq de Boisbaudran.

The net effective magnetic moment per atom can be deduced experimentally from susceptibility measurements. Tables 11.4.1 and 11.4.2 summarize the comparison between theoretical and experimental data. In Section 11.7 we will utilize the concepts and ideas developed on atomic orbitals, net orbital and spin angular momentum, and Hund's rules on evaluating the ground state of atoms and ions to explain the paramagnetic properties and compute the magnetic susceptibility based on a quantum mechanical approach. For now, let us consider another example.

**Example 11.4.0.5**

Utilizing Hund's rule, in Example 11.4.0.3, we obtained only the ground state term symbol of a free ion for a multi-electron configuration (seven to be precise). In this example, you will learn how to obtain all the possible term symbol configurations for an ion with multiple electrons. To keep the discussion tractable, we choose the case of a carbon atom in its excited state whose electronic configuration is given by $1s^2 2s^2 2p^1 3d^1$ configuration.

First, note that the filled shells have zero contribution to both the total $L$ and $S$ values since all their $m_l$ and $m_s$ values add up to zero. Second, the two unpaired electrons in the $p$ and $d$ shells contribute individually an angular momentum value of $l_1 = 1$ and $l_2 = 2$. Using the vector model discussed in Equation (11.58) the resultant $l$-values can be combined as

$$L = l_1 + l_2, l_1 + l_2 - 1, \cdots, |l_1 - l_2|, \tag{11.69}$$

$$L = 3, 2, 1. \tag{11.70}$$

Therefore, the total orbital angular momentum combination can give rise to term symbols $L = P, D, F$. Now, corresponding to each of these angular momentum values there is an associated total spin angular momentum. In this particular example, because the two electrons are occupying the $p$ and the $d$ orbitals separately they can combine with all possible combinations of $s = \pm\frac{1}{2}$ values. The Pauli exclusion principle does not block them from forming a particular quantum state. Hence, the resultant values range from

$$S = s_1 + s_2, s_1 + s_2 - 1, \cdots, |s_1 - s_2|, \tag{11.71}$$

$$S = 1, 0. \tag{11.72}$$

The spin-multiplicity associated with each of the total $S$ values is 2(1) + 1 = 3 and 2(0) + 1 =1. The corresponding term symbols are $^1$P, $^1$D, $^1$F, $^3$P, $^3$D, and $^3$F. Now, a word of caution: *Hund's rules do not predict the excited states or how close the excited states are to the ground state. The rule is useful only to obtain the ground state term symbol.* While we can be fairly certain that the ground state term symbol is given by $^3$F, the hierarchy of states as shown in Figure 11.4.10 will not be experimentally accurate.

Similar to the above example, we can also pose the following question: what are all the possible term symbols for electrons occuyping the same orbital, say in a $2p^2$ , $3d^4$, or $4f^3$ state? In such cases, we cannot simply combine all possible total angular momentum states with all possible spin angular momentum states. The Pauli principle will forbid certain configurations in which the electrons share the same quantum numbers.

**Example 11.4.0.6**

Assuming a jj-coupling scheme, obtain all the term symbols for $\mathrm{n_1} s^1 \mathrm{n_2} p^1$, where $\mathrm{n_1}$ and $\mathrm{n_2}$ represent the level numbers.

In the case of jj-coupling, we will use $(j_1, j_2)_{\mathscr{J}}$ as notation to denote the term symbols. For a s electron we have $j = 1/2$ . While for a p electron we have $j = 1/2, 3/2$. If $j_1 = 1/2$ and $j_2 = 1/2$, the possible $\mathscr{J}$ values are $\mathscr{J} = 0, 1$. But, for $j_1 = 1/2$ and $j_2 = 3/2$ the allowed values of $\mathscr{J}$ are $\mathscr{J} = 1, 2$. Hence, the possible term symbols are

$$(1/2, 1/2)_0, \ (1/2, 1/2)_1, \ (1/2, 3/2)_1, (1/2, 1/2)_2 \tag{11.73}$$

Note, both the $LS$- and the $jj$-coupling represent extreme cases. Many atoms are described by some of form of intermediate coupling; for example, Ge and Sn.

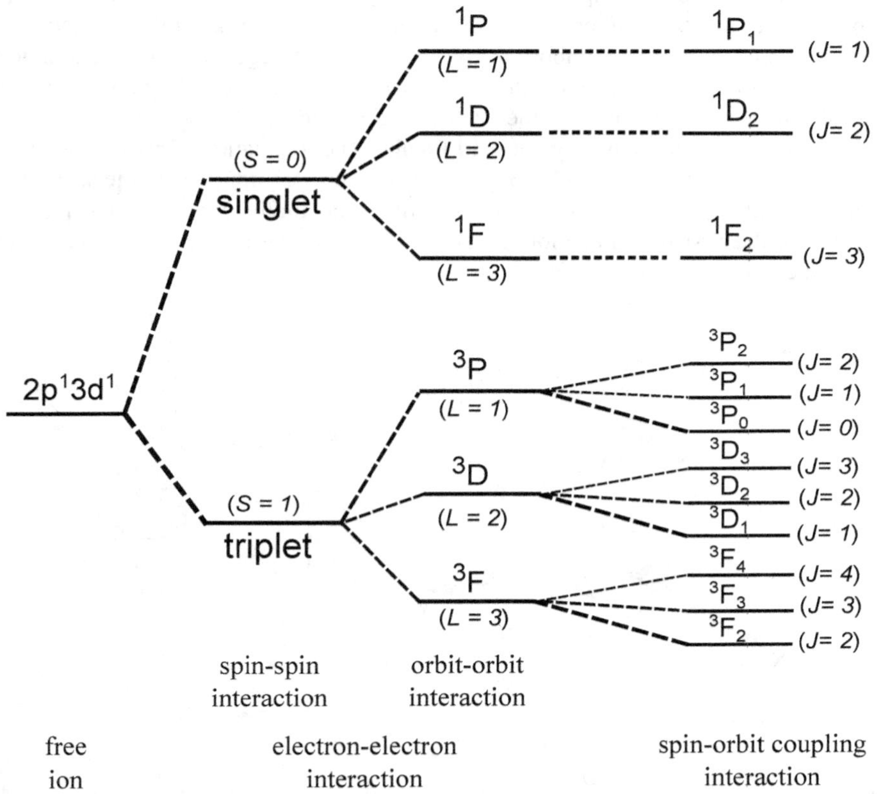


Figure 11.4.10: Schematic illustration of the level splitting of a $2p3d$ electronic configuration for the excited state of carbon in Example 11.4.0.5. To begin with, the free-ion is in a spherically symmetric environment. With the inclusion of electron−electron interactions, the energy levels split based on the singlet $(S = 0)$ and triplet $(S = 1)$ spin configurations and the total orbital momentum combinations $(P, D, F)$. When the effects of spin-orbit interaction are included, further splitting occurs. Those split energy levels are distinguished by the $J$ angular momentum quantum numbers in the term symbol, for example, ${}^3F_2$. Therefore, corresponding to a ${}^3F$ *term* we have a ${}^3F_2$ *level*. These energy *levels* can be further split into states. For example, corresponding to $J = 2$ and $M_J = 2$ we have the state ${}^3F_2$. The arrangement of the energy levels shown in the figure is based on Hund's rule (comparison of actual energy level arrangement with actual experiments will show deviation). Also note, the size of the energy level splitting is not to scale.

## 11.5 Crystal Field Theory

The term symbols in the previous section were obtained for a metal ion in a free environment. By a free environment we mean a situation where the surrounding ions or molecules have no influence on the metal ion, or the atom is simply not surrounded by any charges. In reality, when a transition metal ion participates in a bonding process with an ion or a molecule (known as a ligand) the effect of electrostatic repulsion plays an important role in lifting the degeneracy of the atomic orbitals for a given local geometrical arrangement of the surrounding ligands. The purely ionic effect of the electrostatic field of the ligands on the transition metal ion is the crystal field effect and the theory describing it is known as the **crystal field theory**. The foundation of crystal field theory was developed by physicists Hans Bethe (1906–2005) and John Hasbrouck van Vleck (1899–1980). The historic importance of this theory was in its ability to qualitatively explain the spin only value paramagnetism of the first row transition metal ions and also the ability to explain the colors of transition metal complexes.

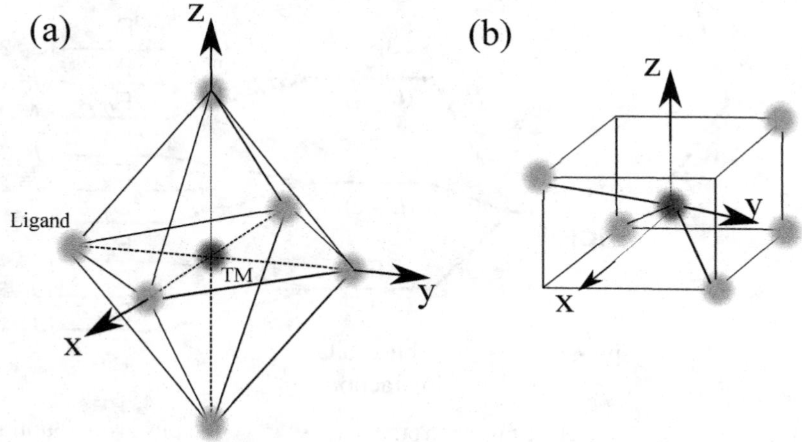


Figure 11.5.11: (a) A transition metal ion, such as $Cu^{2+}$, is located at the central position surrounded by six oxygen ligand atoms. (b) A transition metal ion, such as $X^{n+}$, is located at the central position surrounded by ions.

First, let us consider Figure 11.5.11(a) to understand the basic principle. We have a transition metal ion surrounded by six oxygen atoms (ligands) arranged in an octahedral arrangement. As the transition metal ions and ligands approach each other, electrostatic repulsion causes some of the d orbitals ($d_{z^2}$ and $d_{x^2-y^2}$) which have a strong overlap with the $p$ orbitals, Figure 11.5.12(b), to be raised in energy. This is different from the – $d_{xy}$, $d_{yz}$, and $d_{xz}$ orbitals which lie midway as seen in Figure 11.5.12(a). In this way, the degeneracies of the $d$-orbitals are broken leading to **crystal field splitting** as shown in Figure 11.5.13. For an octahedral environment, the splitting is denoted as $\Delta \equiv 10Dq$. The split $d$-orbitals can be divided into two classes and labeled as – $t_{2g}$: $d_{xy}$, $d_{yz}$, and $d_{xz}$ and $e_g$: $d_{z^2}$ and $d_{x^2-y^2}$ orbitals.

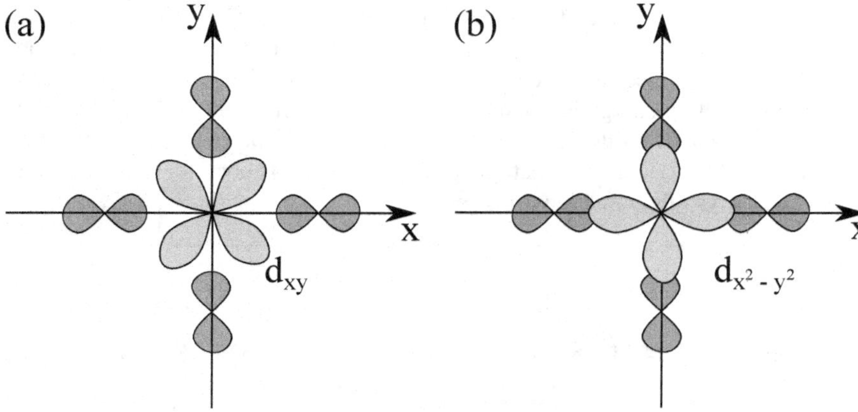


Figure 11.5.12: In the $d_{xy}$ arrangment of orbitals, shown in (a), the d-orbitals lie midway leading to minimal repulsion with the p-orbitals. However, the $d_{x^2-y^2}$ orbital, shown in (b), directly overlaps with the p-orbitals thereby increasing the interaction energy. Eventually this leads to crystal field splitting as shown in Figure 11.5.13.

Crystal field theory comes with a caveat – the interaction between the metal ion and the ligand is assumed to be purely electrostatic (ionic) in nature. The ligands are treated as point charges and are credited with producing a steady crystalline field only. There is no admixture of the transition metal and ligand wavefunction. The mixing of orbitals is treated within *ligand field theory* which is beyond the scope of the present textbook.

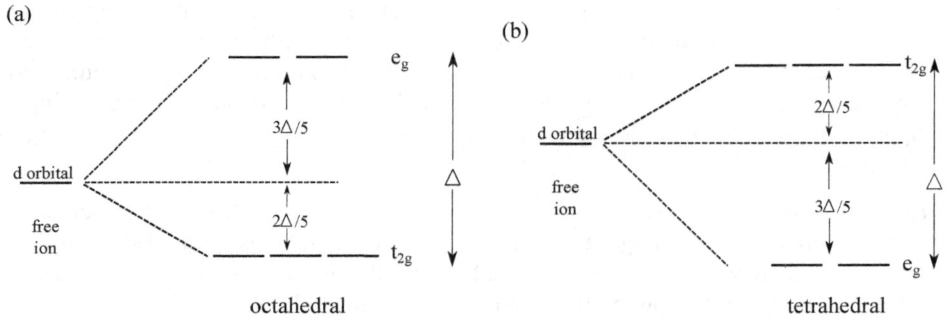


Figure 11.5.13: (a) Crystal field splitting arrangement of $e_g$ and $t_{2g}$ orbitals in (a) octahedral arrangment and (b) tetrahedral arrangement.

We can write down a mathematical model of the neighboring atoms or ions as negative point charges and qualitatively discuss some features. The Hamiltonian can be written as

$$\mathcal{H} = \mathcal{H}_{ion} + \mathcal{H}_{ee} + \mathcal{H}_{so} + \mathcal{H}_{cf}, \tag{11.74}$$

$$\mathcal{H} = \sum_{i=1}^{n} \left[ \frac{\vec{p_i}^2}{2m_e} - \frac{Ze^2}{4\pi\varepsilon_o} \frac{1}{r_i} \right] + \sum_{i<j} \frac{e^2}{4\pi\varepsilon_o} \frac{1}{r_{ij}} + \sum_i \xi_{so}(\vec{r}_i)\vec{l}_i \cdot \vec{s}_i + \sum_{i=1}^{n} e\mathcal{V}(\vec{r}_i), \tag{11.75}$$

Table 11.5.3: Typical energy values for competing interactions in magnetic solids containing $3d$ and $4f$ ions in eV scale. The energy contribution arising from the Coulomb interaction between electron–nuclei ($\mathscr{H}_{ion}$) and electron–electron ($\mathscr{H}_{ee}$) is denoted by $\mathscr{H}_{Coul}$, where $\mathscr{H}_{Coul} = \mathscr{H}_{ion} + \mathscr{H}_{ee}$. The spin-orbit coupling energy is given by $\mathscr{H}_{so}$, the crystal field energy by $\mathscr{H}_{cf}$, and finally the Zeeman energy by $\mathscr{H}_z$. The Hamiltonian, which considers these important interaction contributions, is given in Equation (11.74). See the section on Crystal Field Theory for a full account of how electrons behave in a crystalline field environment.

| | $\mathscr{H}_{Coul}$ | $\mathscr{H}_{so}$ | $\mathscr{H}_{cf}$ | $\mathscr{H}_z$ |
|---|---|---|---|---|
| $3d$ | 0.86 – 4.31 | $8 \times 10^{-3}$ – 0.086 | 0.862 – 8.625 | $8 \times 10^{-5}$ |
| $4f$ | 0.86 – 5.71 | 0.86 – 4.31 | $\approx 0.026$ | $8 \times 10^{-5}$ |

where the first bracketed term is the ionic part, the second term is the electron–electron interaction, the third term is spin-orbit coupling, and the fourth term is the contribution of the crystal fields. The summation is taken over all the electrons and $\mathscr{V}(\vec{r}_i)$ is the crystal field potential produced by the ligands at the site $\vec{r}_i$. The electron-electron part of the Hamiltonian dictates the exchange splittings into the triplet and singlet states, with competition from the spin orbit and crystal field interactions. The kinetic energy and nuclear potential energy is $\sim 1 - 10$ eV. Typical energy range estimates for crystal field strength and spin-orbit interaction is stated in Table 11.5.3. In general, solving the above many-particle Hamiltonian is a formidable task. Nevertheless, we can distinguish three broad cases:

*Case I: Weak crystal field effect* – exchange splittings ($\mathscr{H}_{ion} + \mathscr{H}_{ee}$) > spin-orbit coupling ($\mathscr{H}_{so}$) > crystal fields ($\mathscr{H}_{cf}$). This situation arises in the $4f$ electrons of rare-earth based solids. Electron-electron repulsion, which is the dominant interaction, dictates the splitting of the energy levels, but spin-orbit coupling mixes up the orbital and spin angular momentum such that $J$ is the valid quantum number. In such a situation, $j - j$ coupling is the appropriate way to combine the total angular momentum and Hund's third rule becomes important.

*Case II: Intermediate crystal field effect* – exchange splittings ($\mathscr{H}_{ion} + \mathscr{H}_{ee}$) > crystal fields ($\mathscr{H}_{cf}$) > spin-orbit coupling ($\mathscr{H}_{so}$). The $3d$ transition metal ions are a good example of this case. $L$ and $S$ are valid quantum numbers. In this situation we have to determine how a given $LS$ coupling scheme is split by the octahedral or tetrahedral field.

*Case III: Strong crystal field effect* – crystal fields ($\mathscr{H}_{cf}$) $\gtrsim$ exchange splittings ($\mathscr{H}_{ion} + \mathscr{H}_{ee}$) > ($\mathscr{H}_{so}$). In this situation, which is possible in $4d$ and $5d$ transition metal ions the crystal field strength is comparable to or larger than exchange splittings. There is substantial mixing of the orbitals such that the ionic treatment of crystal field theory is no longer a good approximation. In this case the covalency of the participating orbitals becomes important, which is within the scope of ligand field theory.

The size of the crystal field splitting plays a crucial role in distinguishing between two different magnetic ground states – **high spin** and **low spin** states of transition metal complexes. Intuitively, the high spin and the low spin states can be explained by comparing the pairing energy $U$ and crystal field splitting strength $\Delta$. The pairing energy is the electrostatic cost of repulsion from placing two spins together. When $U > \Delta$, it is favorable for the spins to stay as far apart as possible and they do so by occupying the orbitals in an all-up state, see Figure 11.5.14(a), before filling the lower $t_{2g}$ state. On the other hand for $U < \Delta$, it is beneficial for the system to pair up in the orbital states rather than occuping the higher $e_g$ orbitals to form a low spin state as shown in Figure 11.5.14(b).

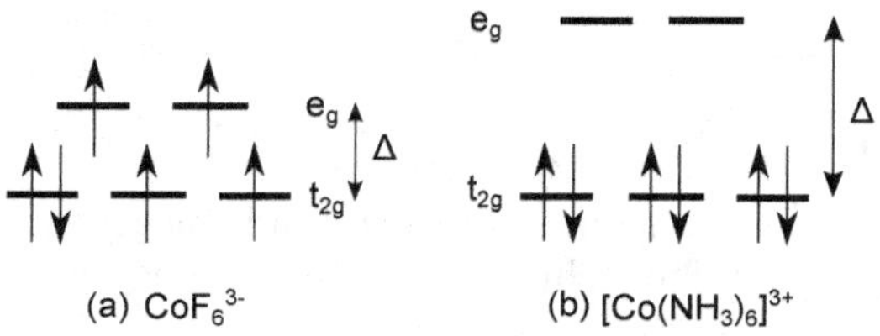


Figure 11.5.14: (a) High spin and (b) low spin configurations of $Co^{3+}$ ion. Both are experimentally observed.

**Example 11.5.0.1**
Compute the electromagnetic frequency range required to probe crystal field splitting energy which lies in the range from 12.5 meV to 2.5 eV.

**Solution**
Using the conversion 1 eV = $1.6 \times 10^{-19}$ J, $E = h\nu$, where $h = 6.626 \times 10^{-34}\text{m}^2\text{kg/s}$ is the Planck's constant and $\nu$ is the frequency we can compute the required electromagnetic frequency. Thus we have

$$\begin{aligned}
\nu &= \frac{(12.5 \times 10^{-3})(1.6 \times 10^{-19})}{6.626 \times 10^{-34}} = 3.02 \times 10^{12}\text{Hz} \equiv 3.02\ \text{THz}, \quad \text{(Far Infrared)}, \\
\nu &= \frac{(2.5)(1.6 \times 10^{-19})}{6.626 \times 10^{-34}} = 6.04 \times 10^{14}\text{Hz} \equiv 604\ \text{THz}, \quad \text{(Near Infrared)}.
\end{aligned}$$

Before concluding this section, let us revisit the question: why is it possible to describe the electronic state of the 3d transition metal ions in terms of the spin-only contribution? To do so, we must first realize that in $3d$ transition metal oxides, the electronic states are well localized due to strong Coulomb interactions. The localized electronic states can be described in terms of the crystal-field states, such as the triply degenerate $t_{2g}$ $(xy, yz, zx)$ orbitals and the doubly degenerate $e_g$ $(x^2 - y^2, 3z^2 - r^2)$ orbital states for the octahedral environment. The energy scale of the level splitting of the $t_{2g}$ $(e_g)$ orbital states is typically $\sim 0.1$ eV, which is much larger than that of spin-orbit coupling ($\sim$ 20 meV). Therefore, in $3d$ transition metal oxides, the orbital angular momentum is completely quenched by degeneracy lifting, and the electronic state can be described in terms of the spin-only Hamiltonian. This is opposite to the $4f$ rare-earths where the electrons are screened by inner shell electrons, such that crystal field effects are not important, although Sm and Eu do show a discrepancy.

## 11.6 Diamagnetism

The starting point will be the Hamiltonian of an ion in a homogenous magnetic field. Ignoring relativistic effects such as spin-orbit coupling and assuming the ion is in a spherically symmetric ligand environment

$$\mathscr{H} = \mathscr{H}_{ion} + \mathscr{H}_{ee} + \mathscr{H}_z, \tag{11.76}$$

$$\mathscr{H} = \frac{1}{2m_e}\sum_{i=1}^{N_e}\left[\vec{p}_i + e\vec{A}(\vec{r}_i)\right]^2 - \frac{Ze^2}{4\pi\varepsilon_o}\sum_{i=1}^{N_e}\frac{1}{r_i} + \frac{e^2}{4\pi\varepsilon_o}\sum_{i<j}\frac{1}{r_{ij}} + 2\frac{\mu_B}{\hbar}\vec{S}\cdot\vec{B}, \tag{11.77}$$

where the total spin $\vec{S} = \sum_{i=1}^{n} s_i$ and $N_e$ is the number of electrons. The kinetic energy term has been modified, from Equation (11.74), to include the effects of electromagnetic field through the replacement of the momentum operator with its canonical version $p_i \to p_i + e\vec{A}(\vec{r}_i)$. The last term represents the Zeeman interaction contribution in the presence of the external magnetic field. In order to perform any calculations with the above Hamiltonian, we first need to make a choice of the vector potential $\vec{A}$. To do so, we pick the Coulomb gauge which is defined as

$$\vec{B} = \vec{\nabla}\times\vec{A};\ \ \vec{\nabla}\cdot\vec{A} = 0. \tag{11.78}$$

A solution to the above equation is given by

$$\vec{A}(\vec{r}) = \frac{1}{2}\vec{B}\times\vec{r}. \tag{11.79}$$

For a magnetic field directed along the $z-$ axis $\vec{B} = (0,0,B)$, using Equation (11.79), we have $\vec{A} = \frac{B}{2}(-y,x,0)$. Let us now focus on the kinetic energy term $\mathscr{K}$ and expand

$$\hat{\mathscr{K}} = \frac{1}{2m_e}\sum_{i=1}^{N_e}\left[\vec{p}_i + e\vec{A}(\vec{r}_i)\right]^2, \tag{11.80}$$

$$\hat{\mathscr{K}} = \frac{1}{2m_e}\sum_{i=1}^{N_e}\left[\vec{p}_i^{\,2} + e\left\{\vec{p}_i\cdot\vec{A}(\vec{r}_i) + \vec{A}(\vec{r}_i)\cdot\vec{p}_i\right\} + e^2\vec{A}(\vec{r}_i)^2\right]. \tag{11.81}$$

where we have displayed the *hat* to be explicity clear that the kinetic energy expression above is a quantum mechanical operator. The second line needs an explanation. First, both $\vec{p}$ and $\vec{A}$ are quantum mechanical operators, so we must be careful in how they are ordered. Second, these operators will act on a wavefunction. Utilizing the vector identity

$$\vec{\nabla}\cdot\left(\psi\vec{A}\right) = \psi\vec{\nabla}\cdot\vec{A} + \vec{A}\cdot\vec{\nabla}\psi \tag{11.82}$$

we have in the Coulomb gauge

$$\vec{p}_i\cdot\vec{A}(\vec{r}_i)\psi = \frac{\hbar}{i}\nabla_i\cdot\left(\vec{A}(\vec{r}_i)\psi\right) = \frac{\hbar}{i}\left(\psi\vec{\nabla}_i\cdot\vec{A}(\vec{r}_i) + \vec{A}(\vec{r}_i)\cdot\vec{\nabla}_i\psi\right) = \vec{A}(\vec{r}_i)\cdot\vec{p}_i\psi, \tag{11.83}$$

where we have used $\vec{p}_i = -i\hbar\vec{\nabla}_i$ with the subscripted $i$ on the gradient operator $\vec{\nabla}$ denoting the electron index. In operator notation, the above equation implies

$$\vec{p}_i\cdot\vec{A}(\vec{r}_i) \equiv \vec{A}(\vec{r}_i)\cdot\vec{p}_i. \tag{11.84}$$

The kinetic energy is re-written as

$$\hat{\mathscr{K}} = \frac{1}{2m_e}\sum_{i=1}^{N_e}\vec{p}_i^{\,2} + \frac{e}{m_e}\sum_{i=1}^{N_e}\vec{A}(\vec{r}_i)\cdot\vec{p}_i + \frac{e^2}{2m_e}\sum_{i=1}^{N_e}\vec{A}(\vec{r}_i)^2. \tag{11.85}$$

The second term can be further rewritten if we notice that

$$\vec{A}(\vec{r}_i)\cdot\vec{p}_i = \frac{B}{2}(-y_i p_{xi} + x_i p_{yi}) = \frac{1}{2}Bl_{iz} = \frac{1}{2}\vec{B}\cdot\vec{l}_i. \tag{11.86}$$

where we have used the definition of angular momentum $\vec{L} = \sum_i \vec{r}_i \times \vec{p}_i = \sum_i \vec{l}_i$. Thus Equation (11.85) becomes

$$\mathscr{H} = \frac{1}{2m_e}\sum_{i=1}^{N_e}\vec{p}_i^{\,2} + \frac{\mu_B}{\hbar}\vec{L}\cdot\vec{B} + \frac{e^2}{2m_e}\sum_{i=1}^{N_e}\vec{A}^2(\vec{r}_i). \tag{11.87}$$

Combining Equations (11.76) and (11.87) we have

$$\mathscr{H} = \frac{1}{2m_e}\sum_{i=1}^{N_e}\left[\vec{p}_i + e\vec{A}(\vec{r}_i)\right]^2 - \frac{Ze^2}{4\pi\varepsilon_o}\sum_{i=1}^{N_e}\frac{1}{r_i} + \frac{e^2}{4\pi\varepsilon_o}\sum_{i<j}\frac{1}{r_{ij}} + 2\frac{\mu_B}{\hbar}\vec{S}\cdot\vec{B}, \tag{11.88}$$

$$\mathscr{H} = \frac{1}{2m_e}\sum_{i=1}^{N_e}\left[\vec{p}_i^{\,2} - \frac{Ze^2}{4\pi\varepsilon_o}\frac{1}{r_i}\right] + \frac{e^2}{4\pi\varepsilon_o}\sum_{i<j}\frac{1}{r_{ij}} + \frac{\mu_B}{\hbar}\left(\vec{L}+2\vec{S}\right)\cdot\vec{B} + \frac{e^2}{2m_e}\sum_{i=1}^{N_e}\vec{A}^2(\vec{r}_i), \tag{11.89}$$

$$\mathscr{H} = \frac{1}{2m_e}\sum_{i=1}^{N_e}\left[\vec{p}_i^{\,2} - \frac{Ze^2}{4\pi\varepsilon_o}\frac{1}{r_i} + \frac{1}{2}\frac{e^2}{4\pi\varepsilon_o}\frac{1}{r_{ij}}\right] + \frac{\mu_B}{\hbar}\left(\vec{L}+2\vec{S}\right)\cdot\vec{B} + \frac{e^2B^2}{8m_e}\sum_{i=1}^{N_e}(x_i^2+y_i^2), \tag{11.90}$$

where the half factor in the third term avoids double counting. The first three terms are independent of the magnetic field and constitute the unperturbed Hamiltonian. The fourth and fifth terms are dependent on the external magnetic field. The magnetic moment can be evaluated by taking the derivative of Equation (11.90) with respect to B. This leads to

$$\vec{\mu} = -\frac{\partial\mathscr{H}}{\partial B} = -\frac{\mu_B}{\hbar}\left(\vec{L}+2\vec{S}\right) - \frac{e^2}{4m_e}\sum_{i=1}^{N_e}(x_i^2+y_i^2)\vec{B}. \tag{11.91}$$

Equation (11.91) perfectly summarizes the conceptual aspects of magnetism. The dipole moment operator is composed of two pieces – one containing the angular momentum operators and the other the effects of an external magnetic field (the diamagnetic term). In atoms, ions, or molecules which have a partially filled shell the total orbital and spin angular momentum operators do not vanish. In these cases, we have a permanent magnetic moment which can give rise to paramagnetism, ferromagnetism, antiferromagnetism, or ferrimagnetism. However, for completely filled shells the orbital and spin terms vanish since $\vec{L} = 0$ and $\vec{S} = 0$. In this situation, the diamagnetic term containing the effects of an induced magnetic moment become important and we have a diamagnet. Note the presence of the negative sign in front of the diamagnetic contribution. This is what causes the susceptibility to become negative. Since diamagnetism is a small effect, it is observable only when the other forms of magnetism are not present. To make further progress, we will focus solely on the diamagnetic term from Equation (11.90)

$$\mathscr{H}_{dia} = \frac{e^2B^2}{8m_e}\sum_{i=1}^{N_e}(x_i^2+y_i^2). \tag{11.92}$$

In a spherically symmetric crystal field environment

$$\langle 0|x_i^2|0\rangle = \langle 0|y_i^2|0\rangle = \langle 0|z_i^2|0\rangle = \frac{1}{3}\langle 0|r_i^2|0\rangle \equiv \frac{1}{3}\langle r_i^2\rangle, \tag{11.93}$$

where $|0\rangle$ is the ground state wave function. Utilizing Equations (11.91) and (11.93) and the fact that magnetization is the magnetic moment per unit volume we obtain an expression for the magnetization $M_z$ as

$$\begin{aligned} M_z &= \frac{N}{V}\langle 0|\mu_z|0\rangle, \\ M_z &= -\frac{Ne^2B}{4m_eV}\sum_{i=1}^{N_e}\langle 0|(x_i^2+y_i^2)|0\rangle, \\ M_z &= -\frac{Ne^2B}{4m_eV}\sum_{i=1}^{N_e}\frac{2}{3}\langle 0|r_i^2|0\rangle = -\frac{Ne^2B}{6m_eV}\sum_{i=1}^{N_e}\langle r_i^2\rangle. \end{aligned} \tag{11.94}$$

By differentiating Equation (11.94) with respect to the magnetic field (at constant temperature T), we obtain the quantum mechanical expression for the diamagnetic susceptibility

$$\chi^{dia} = \left(\frac{\partial M_z}{\partial H}\right)_T = \mu_o\left(\frac{\partial M_z}{\partial B}\right)_T = -\frac{\mu_o Ne^2}{6m_eV}\sum_{i=1}^{N_e}\langle r_i^2\rangle. \tag{11.95}$$

Diamagnetism is present in all materials, either as a weak effect which can be ignored or as a small correction to a larger effect.

**Example 11.6.0.1**
Calculate the dimensionless diamagnetic susceptibility in the 1s state of the hydrogen atom with a number density of $10^{27}$ m$^{-3}$ .
**Solution**
To compute the diamagnetic susceptibility in the 1s state we first compute $\sum_{i=1}^{N_e}\langle r_i^2\rangle$ in Equation (11.95). In this summation $N_e = 1$ since there is only one electron in a hydrogen atom. Note, for atoms with $N_e$ electrons the summation would have yielded $N_e\langle r_i^2\rangle$. However, for our problem $\sum_{i=1}^{N_e}\langle r_i^2\rangle = \langle r^2\rangle$. The expectation value for the 1s state wavefunction is computed as follows. Using Equation (11.25) we have

$$\psi_{100}(r,\theta,\phi) = R_{10}(r)Y_0^0(\theta,\phi). \tag{11.96}$$

Inserting the expressions for $R_{10}(r)$ and $Y_0^0(\theta,\phi)$ from Tables 11.10.4 and 11.10.5, respectively, gives

$$\psi_{100}(r,\theta,\phi) = \left(2\left(\frac{1}{a_o}\right)^{3/2}\exp(-Zr/a_o)\right)\left(\frac{1}{\sqrt{4\pi}}\right) = \frac{1}{\sqrt{\pi a_o^3}}\exp(-r/a_o). \tag{11.97}$$

where we have used $Z = 1$ for the atomic number. Now to compute the expectation value of $r^2$ we write for the single electron present in the hydrogen atom

$$\begin{aligned} \langle r^2\rangle &= \langle\psi_{100}(r,\theta,\phi)|r^2|\psi_{100}(r,\theta,\phi)\rangle, \\ \langle r^2\rangle &= \int_0^\infty\int_0^\pi\int_0^{2\pi}\psi_{100}^*(r,\theta,\phi)r^2\psi_{100}(r,\theta,\phi)r^2\sin\theta drd\theta d\phi, \\ \langle r^2\rangle &= \frac{1}{\pi a_o^3}\int_0^\infty r^4\exp(-2r/a_o)dr\int_0^\pi\sin\theta d\theta\int_0^{2\pi}d\phi, \end{aligned}$$

Carrying out the angular integrations give a factor of $4\pi$. We then write

$$\langle r^2\rangle = \frac{4\pi}{\pi a_o^3}\int_0^\infty r^4\exp(-2r/a_o)dr. \tag{11.98}$$

To compute the above integral we make the substitution $w = 2r/a_o$. With this change of variables we can rewrite the above integral as

$$\langle r^2 \rangle = \frac{4\pi}{\pi a_o^3} \int_0^\infty \left(\frac{a_o w}{2}\right)^4 \exp(-w) \frac{a_o}{2} dw,$$

$$\text{where using } dw = \frac{2}{a_o} dr \text{ we have,}$$

$$\langle r^2 \rangle = \frac{4\pi}{\pi a_o^3} \frac{a_o^5}{32} \int_0^\infty w^{5-1} \exp(-w) dw.$$

$$\text{Thus } \langle r^2 \rangle = \frac{a_o^2}{8} \Gamma(5) = \frac{a_o^2}{8} 4! = 3a_o^2. \tag{11.99}$$

In the above derivation we used the definition of the Gamma function, $\Gamma(n) = (n-1)!$, when $n$ is an integer. Note, our final answer confirms our intuition in Equation (11.93). Thus the expression for the diamagnetic susceptibility is given by

$$\chi^{dia} = -\frac{\mu_o N e^2}{6 m_e V} 3a_o^2 = -\frac{N}{V} \frac{\mu_o e^2 a_o^2}{2 m_e}. \tag{11.100}$$

Now inserting $N/V = 10^{27} m^{-3}$, $a_o = 0.53 \times 10^{-10}$ m, $\mu_o = 4\pi \times 10^{-7} \text{A/m}^2$, $m_e = 9.1 \times 10^{-31}$ kg, and $e = 1.6 \times 10^{-19}$ C yields a value of $\chi^{dia} = 4.97 \times 10^{-8}$. Note, the computed dimensionless diamagnetic susceptibility is in the range quoted in Table 10.1 of Chapter 10.

## 11.7 Quantum Theory of Paramagnetism

The semi-classical theory of paramagnetism was discussed at length in Chapter 10. In this section we will refine the theoretical approach to a quantum version. The splitting of energy levels into its constituent non-degenerate energy levels can be used to build the theory of quantum paramagnetism. The experimental evidence for energy level splitting comes from Electron Spin Resonance experiments (ESR), discussed in Section 11.9.

### 11.7.1 S=$\frac{1}{2}$

In this section we wish to compute the thermal average of the net magnetic moment given by $\langle \mu_J \rangle = \langle g_J \mu_B m_J \rangle$ where $g_J \mu_B m_J$ is the net magnetic dipole moment, see Equation (11.60). Assuming $L = 0$, we have J = $\frac{1}{2}$ and $g_J = 2$. Therefore, the Zeeman energy is given by

$$E_z = g_J \mu_B m_J B = 2\mu_B \left(\pm\frac{1}{2}\right) B = \pm \mu_B B, \tag{11.101}$$

where the plus (minus) refers to the up (down) electronic spin state. Note $\mu_{J\uparrow} = -\mu_B$ (for $m_s = \frac{1}{2}$) and $\mu_{J\downarrow} = +\mu_B$ (for $m_s = -\frac{1}{2}$). From the basic definition of average quantities in statistical mechanics we have

$$\langle \mu_J \rangle = \sum_\nu \frac{\mu_{J\nu} P_\nu}{\sum_\nu P_\nu}, \tag{11.102}$$

where $P_v = e^{-E_v/k_BT}$; i.e., the Boltzmann factor. Using Equation (11.101) in the above equation we have

$$\begin{aligned}\langle\mu_J\rangle &= \frac{(\mu_{J\uparrow})e^{-E_{zJ\uparrow}B/k_BT} + (\mu_{J\downarrow})e^{-E_{zJ\downarrow}B/k_BT}}{e^{-E_{zJ\uparrow}B/k_BT} + e^{-E_{zJ\downarrow}B/k_BT}}, \\ &= \frac{(-\mu_B)e^{-\mu_BB/k_BT} + (\mu_B)e^{+\mu_BB/k_BT}}{e^{-\mu_BB/k_BT} + e^{+\mu_BB/k_BT}}, \\ &= \mu_B \tanh\left(\frac{\mu_BB}{k_BT}\right). \end{aligned} \tag{11.103}$$

For $N/V$ electrons per unit volume, assuming a non-interacting theory, we have the total magnetization $M = \frac{N}{V}\langle\mu\rangle$ given by

$$\begin{aligned} M &= \frac{N}{V}\langle\mu_J\rangle, \\ M &= \frac{N}{V}\mu_B \tanh\left(\frac{\mu_BB}{k_BT}\right), \\ M &= M_s \tanh(y), \end{aligned} \tag{11.104}$$

where in the last step we defined $M_s = \frac{N}{V}\mu_B$ and $y = \mu_BB/k_BT$. In the limit of small magnetic fields, $\tanh(y) \approx y$. Therefore we obtain

$$M = \frac{N}{V}\frac{\mu_BB}{k_BT} = \frac{N}{V}\frac{\mu_B^2B}{k_BT} = \frac{N}{V}\frac{\mu_o\mu_B^2H}{k_BT}. \tag{11.105}$$

Thus, from the above equation we obtain the susceptibility as

$$\chi = \frac{\partial M}{\partial H} = \frac{N}{V}\frac{\mu_o\mu_B^2}{k_BT}. \tag{11.106}$$

### 11.7.2 General J Value

The goal in this section is to compute $\langle\mu_J\rangle = \langle g_J\mu_Bm_J\rangle = g_J\mu_B\langle m_J\rangle$ in the case of any general J value. In this case the partition function sum is given by

$$Z = \sum_{m_J=-J}^{J} e^{-m_Jg_J\mu_BB/k_BT}. \tag{11.107}$$

Introducing the variable $x = g_J\mu_BB/k_BT$, we can write $\langle m_J\rangle$ as

$$\langle m_J\rangle = \frac{\sum_{m_J=-J}^{J} m_Je^{-m_Jx}}{\sum_{m_J=-J}^{J} e^{-m_Jx}} = \frac{1}{Z}\frac{\partial Z}{\partial x}. \tag{11.108}$$

The net magnetization is given by

$$M = \frac{N}{V}g_J\mu_B\langle m_J\rangle = \frac{N}{V}\frac{g_J\mu_B}{Z}\frac{\partial Z}{\partial B}\frac{\partial B}{\partial x} = \frac{N}{V}k_BT\frac{\partial \ln Z}{\partial B}, \tag{11.109}$$

where we have used

$$\frac{\partial B}{\partial x} = \frac{k_BT}{g_J\mu_B}. \tag{11.110}$$

The partition sum Z can be written as a geometric series using the well-known formula

$$a + ar + ar^2 + \cdots + ar^{M-1} = \sum_{j=1}^{M} ar^{j-1} = \frac{a(1-r^M)}{1-r} \tag{11.111}$$

Now, comparing this with the partition function we can identify $a = e^{-Jx}$ and $r = e^x$. Substituting and simplifying yields

$$Z = \frac{\sinh[(2J+1)\frac{x}{2}]}{\sinh[\frac{x}{2}]}. \tag{11.112}$$

If we define a variable $y = xJ = g_J \mu_B JB/k_BT$, we can rewrite the magnetization as

$$M = M_s B_J(y), \tag{11.113}$$

with $M_s = \frac{N}{V} g_J \mu_B J$ and where we have now introduced the Brillouin function $B_J(y)$ given by

$$B_J(y) = \frac{2J+1}{2J} \coth\left(\frac{2J+1}{2J} y\right) - \frac{1}{2J} \coth\frac{y}{2J}. \tag{11.114}$$

In the limit $J \to \infty$, $B_\infty(y) = L(y)$, thus, recovering the Langevin function introduced in Chapter 10. This makes sense, because in the limit that $J \to \infty$ we are approaching the classical limit. In the opposite quantum limit when $J = \frac{1}{2}$ it reduces to the tanh function

$$B_{1/2}(y) = \tanh(y) \tag{11.115}$$

where we have used the identity

$$\tanh(2y) = \frac{2\tanh(y)}{1+\tanh^2(y)}. \tag{11.116}$$

**Example 11.7.2.1**
Estimate a value for the variable $y$ introduced in the above derivation for a hydrogen atom at room temperature in the presence of earth's magnetic field. Assume the earth's field to be $5 \times 10^{-5}$ T.
**Solution**
The hydrogen atom has a single electron in a s-orbital. Thus $J = 0 + 1/2 = 1/2$ [using Equation (11.17)] and $g_J = 2$ [using Equation (11.64)]. Inserting values for $\mu_B = 9.27 \times 10^{-24}\ A \cdot m^2$ and $k_B = 1.38 \times 10^{-23} J/K$ we have

$$y = g_J \mu_B JB/k_BT = \frac{2(9.27 \times 10^{-24})(0.5)(5 \times 10^{-5})}{2(1.38 \times 10^{-23})} = 1.12 \times 10^{-8}, \tag{11.117}$$

which is much smaller than 1. Except at low temperatures, large magnetic fields, or both it is appropriate to use Equation (11.119). Of course, to be absolutely sure about the validity of the approximation you should explicitly compute the value of $y$ as we have done in this example.

For small $y$ we have an expansion, $\coth(y) \approx 1/y + y/3$. Using this leads to the following approximate expression for the Brillouin function (see Exercise 11.11.17)

$$B_J(y) = \frac{(J+1)y}{3J} + \mathcal{O}(y^3), \tag{11.118}$$

$$\text{Thus, } \chi = \frac{M}{H} = \frac{\mu_o M}{B} \approx \frac{N}{V} \frac{\mu_o \mu_{eff}^2}{3k_BT}, \tag{11.119}$$

with the effective value of the magnetic moment defined as

$$\mu_{eff} = g_J \mu_B \sqrt{J(J+1)}, \tag{11.120}$$

with the Landé g-value given by Equation (11.64).

**Example 11.7.2.2**
Using Equations (11.64) and (11.114) write a MATLAB code to generate the plot for Brillouin function versus $B/T$ for the combination of $L$ and $S$ values shown in Figure 11.7.15. The choice of $L$ and $S$ is made keeping in mind that we want to study the Brillouin function's trend in the limit that $J \to \infty$. Here $B$ represents magnetic field and $T$ is the temperature.
**Solution**
The MATLAB script which reproduces the plot is below.

```
%copyright by J. E Hasbun and T. Datta
% ch11_brillouin.m

% This script plots the Brillouin function for different values of
% the total angular momentum J. The paramagnetic magnetization follows
% the Brillouin function curve.

% Constants
mub = 9.27*10^-24; kb = 1.38*10^-23;

% defining x = B/T (magnetic field/temperature ratio)

% Function definitions
landeg = @(L,S) 1.5 + (S*(S+1) - L*(L+1))/(2*(L+S)*(L+S+1));
brillouinJ = @(x,L,S)((2*(L+S)+1)/(2*(L+S)))*coth(((2*(L+S)+1)/...
    (2*(L+S)))*landeg(L,S)*(mub/kb)*(L+S)*x)-(1/(2*(L+S)))
    *coth((landeg(L,S)*...(mub/kb)*(L+S)*x)/(2*(L+S)));

% Printing Landeg values
fprintf('landeg(0.0,0.5)= %0.3f\n',landeg(0.0,0.5));
fprintf('landeg(1.0,0.5)= %0.3f\n',landeg(1.0,0.5));
fprintf('landeg(2.0,0.5)= %0.3f\n',landeg(2.0,0.5));
fprintf('landeg(3.0,0.5)= %0.3f\n',landeg(3.0,0.5));
fprintf('landeg(3.0,3.5)= %0.3f\n',landeg(3.0,3.5));
fprintf('landeg(6.0,2.0)= %0.3f\n',landeg(6.0,2.0));
fprintf('landeg(5.0,10.0)= %0.3f\n',landeg(5.0,10.0));


% Plotting the Brillouin function curves for different J values
xmax = 4;
hold on;
fplot(@(x)brillouinJ(x,0.0,0.5),[0,4],'LineWidth',2,'-k');
fplot(@(x)brillouinJ(x,1.0,0.5),[0,4],'--r');
fplot(@(x)brillouinJ(x,2.0,0.5),[0,4],'-g');
fplot(@(x)brillouinJ(x,3.0,0.5),[0,4],'--b');
fplot(@(x)brillouinJ(x,3.0,3.5),[0,4],'-c');
fplot(@(x)brillouinJ(x,6.0,2.0),[0,4],'--m');
```

```
fplot(@(x)brillouinJ(x,5.0,10.0),[0,4],'LineWidth',5,'-k');
xlabel('B/T');ylabel('Brillouin Function');
legend('J=0.5','J=1.5','J=2.5','J=3.5','J=6.5','J=8.0','J=15.0');
hold off;
```

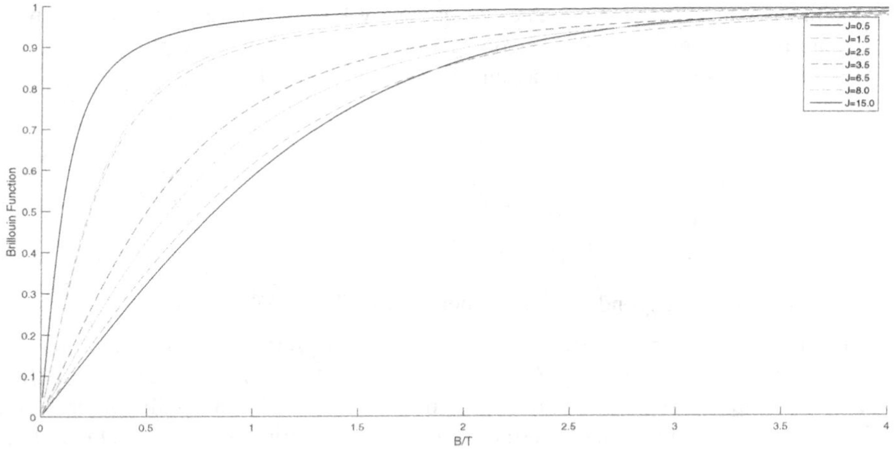


Figure 11.7.15: Brillouin function plots, $B_J(y)$. For magnetic systems with large values of J, the system saturates more rapidly in response to an external magnetic field. Note, the horizontal axis represents the ratio of B/T and the vertical axis is the Brillouin function. From Equation (11.113) we see that the magnetization is proportional to $B_J(y)$. Thus these curves also display the trend in magnetization variation as the magnetic field to temperature ratio is changed.

## 11.8 Exchange Interaction

In Section 10.8 of Chapter 10 we mentioned the Bohr van Leeweun theorem which suggested that magnetism is purely a quantum phenomenon. Until now, even though we have discussed at length the building blocks of atomic magnetism we have not provided an explanation for the microscopic (quantum mechanical) origins of the interactions which generate spontaneous magnetization. To understand the issue at a deeper level, consider the following example.

**Example 11.8.0.1**
Calculate the interaction energy of (a) two magnetic dipoles of strength 1 $\mu_B$ lying along the x-axis and separated by 1 Å, (b) two electrons separated by 1 Å.
**Solution**
(a) From basic electrodynamics, we know that two magnetic dipoles interact via the dipole interaction energy via the equation

$$E = \frac{\mu_o}{4\pi r^3}\left[\vec{\mu}_1 \cdot \vec{\mu}_2 - \frac{3}{r^2}(\vec{\mu}_1 \cdot \vec{r})(\vec{\mu}_2 \cdot \vec{r})\right], \tag{11.121}$$

where $\vec{r}$ is the separation between the dipoles. Since the dipoles point along the x-axis and are of the same magnitude we can write

$$\begin{aligned} E &= \frac{\mu_o}{4\pi r^3}\left[\mu_B^2 - \frac{3}{r^2}\mu_B^2 r^2\right], \\ E &= -2\frac{\mu_o}{4\pi r^3}\mu_B^2. \end{aligned} \tag{11.122}$$

Inserting numbers, $\mu_B = 9.274\times10^{-24} A\cdot m^2$ and $r$= 1 Å, we have $|E| = 1.7\times10^{-23}$ J. Dividing by the Boltzmann constant the temperature scale is given by 1.25 K. This energy scale is low compared to the Curie transition temperatures of ferromagnets such as Fe, Ni, and Co!
(b) The Coulomb interaction energy of a pair of electrons is given by

$$\begin{aligned} U &= \frac{1}{4\pi\varepsilon_o}\frac{q^2}{r}, \\ U &= \frac{1}{4\pi(8.85\times10^{-12})}\frac{(1.6\times10^{-19})^2}{1\times10^{-10}}, \\ U &= 2.3\times10^{-18}\text{J}. \end{aligned} \tag{11.123}$$

The temperature scale corresponding to this energy is $1.67\times10^5$K!

So what does the above example teach us? It clearly demonstrates that magnetic dipole–dipole interactions cannot explain the very high Curie-Weiss temperatures of Fe, Ni, and Co of 1043 K, 1400 K, and 627 K. Rather, consider the following scenario where two atoms with unpaired electrons interact with each other. If the spins of these two electrons are antiparallel to each other, the electrons can approach each other as close as possible, thereby raising the Coulomb interaction energy. However, if the states are parallel, the Pauli exclusion principle causes the electrons to stay as far away as possible leading to a reduction of the Coulomb interaction energy. The order of magnitude of this Coulomb interaction was estimated to be at $10^5$ K in the above example. So even for a small change of 1%, the cost associated with the process would be of the order of 1000 K, an order of magnitude similar to the Curie-Weiss temperature of the aforementioned ferromagnetic materials. Thus, the key to explaining magnetism is the combined effect of Coulomb interaction energy and Pauli exclusion principle. These two basic ideas give rise to what is known as the **exchange interaction**. The exchange interaction was introduced by the german theoretical physicist Werner Heisenberg in 1926 to interpret the origin of the large Curie-Weiss transition temperature.
To derive a quantum mechanical expression for the exchange interaction energy, consider two electrons, for example, in the 3d$^2$ state. The state of each electron is specified by the product of the orbital wavefunction $\psi(\vec{r})$ and spin wavefunction (spinor, $\chi(\vec{s})$). It will be assumed that the one-electron problem has been solved and the orbital wavefunctions of the two electrons are given by the two orthonormal eigenstates $\psi_a(\vec{r})$) and $\psi_b(\vec{r})$). For the $\uparrow$ ($\downarrow$) spin electron state, we choose the normalized spinor wavefunctions as $\alpha(s)(\beta(s))$. The Hamiltonian for the two-electron problem is given by

$$\mathscr{H}_{el} = \mathscr{H}_0(\vec{r}_1) + \mathscr{H}_0(\vec{r}_2) + \frac{e^2}{|\vec{r}_1 - \vec{r}_2|}, \tag{11.124}$$

consisting of two spin independent one-electron Hamiltonian $\mathscr{H}_0(\vec{r})$ and the Coulomb interaction. The one-electron eigenenergies $E_a$ and E$_b$ are given by

$$\mathscr{H}_0\psi_a(\vec{r}) = E_a\psi_a(\vec{r}), \tag{11.125}$$

$$\mathscr{H}_0\psi_b(\vec{r}) = E_b\psi_b(\vec{r}). \tag{11.126}$$

To proceed further we will diagonalize the two-electron Hamiltonian $\mathscr{H}_{el}$ in the subspace of the 3d$^2$ state where orbital character is fixed to d-orbitals, but the spins can vary. We then have four possible wavefunction combinations. Two with parallel spins and two with antiparallel spins. Now, quantum mechanics informs us that for electrons, which are fermions, the overall wavefunction must be antisymmetric in order to satisfy the Pauli principle. The easiest way to ensure this condition is to construct the **Slater determinant**. When both spins point up, we can write

$$\Psi_{\uparrow\uparrow} = \frac{1}{\sqrt{2}}\begin{vmatrix}\psi_a(\vec{r}_1)\alpha(s_1) & \psi_a(\vec{r}_2)\alpha(s_2)\\ \psi_b(\vec{r}_1)\alpha(s_1) & \psi_b(\vec{r}_2)\alpha(s_2)\end{vmatrix}, \tag{11.127}$$

$$\Psi_{\uparrow\uparrow} = \frac{1}{\sqrt{2}}\alpha(s_1)\alpha(s_2)\left[\psi_a(\vec{r}_1)\psi_b(\vec{r}_2) - \psi_a(\vec{r}_2)\psi_b(\vec{r}_1)\right]. \tag{11.128}$$

With both spins pointing down, we can write

$$\Psi_{\downarrow\downarrow} = \frac{1}{\sqrt{2}}\begin{vmatrix}\psi_a(\vec{r}_1)\beta(s_1) & \psi_a(\vec{r}_2)\beta(s_2)\\ \psi_b(\vec{r}_1)\beta(s_1) & \psi_b(\vec{r}_2)\beta(s_2)\end{vmatrix}, \tag{11.129}$$

$$\Psi_{\downarrow\downarrow} = \frac{1}{\sqrt{2}}\beta(s_1)\beta(s_2)\left[\psi_a(\vec{r}_1)\psi_b(\vec{r}_2) - \psi_a(\vec{r}_2)\psi_b(\vec{r}_1)\right]. \tag{11.130}$$

For the antiparallel combinations we have

$$\Psi_{\downarrow\uparrow} = \frac{1}{\sqrt{2}}\begin{vmatrix}\psi_a(\vec{r}_1)\beta(s_1) & \psi_a(\vec{r}_2)\beta(s_2)\\ \psi_b(\vec{r}_1)\alpha(s_1) & \psi_b(\vec{r}_2)\alpha(s_2)\end{vmatrix}, \tag{11.131}$$

$$\Psi_{\downarrow\uparrow} = \frac{1}{\sqrt{2}}\left[\psi_a(\vec{r}_1)\psi_b(\vec{r}_2)\beta(s_1)\alpha(s_2) - \psi_b(\vec{r}_1)\psi_a(\vec{r}_2)\alpha(s_1)\beta(s_2)\right], \tag{11.132}$$

and

$$\Psi_{\uparrow\downarrow} = \frac{1}{\sqrt{2}}\begin{vmatrix}\psi_a(\vec{r}_1)\alpha(s_1) & \psi_a(\vec{r}_2)\alpha(s_2)\\ \psi_b(\vec{r}_1)\beta(s_1) & \psi_b(\vec{r}_2)\beta(s_2)\end{vmatrix}, \tag{11.133}$$

$$\Psi_{\uparrow\downarrow} = \frac{1}{\sqrt{2}}\left[\psi_a(\vec{r}_1)\psi_b(\vec{r}_2)\alpha(s_1)\beta(s_2) - \psi_b(\vec{r}_1)\psi_a(\vec{r}_2)\beta(s_1)\alpha(s_2)\right], \tag{11.134}$$

We now diagonalize the $\mathscr{H}_{el}$ in the subspace of

$$|\Psi\rangle = |\Psi_{\uparrow\uparrow}, \Psi_{\downarrow\uparrow}, \Psi_{\uparrow\downarrow}, \Psi_{\downarrow\downarrow}\rangle \tag{11.135}$$

Using the above definition we construct the 4 × 4 matrix obtained by taking the expectation value of the two-electron Hamiltonian, $\mathscr{H}_{el}$

$$E_{el} = \langle\Psi_{\uparrow\uparrow}, \Psi_{\downarrow\uparrow}, \Psi_{\uparrow\downarrow}, \Psi_{\downarrow\downarrow}|\mathscr{H}_{el}|\Psi_{\uparrow\uparrow}, \Psi_{\downarrow\uparrow}, \Psi_{\uparrow\downarrow}, \Psi_{\downarrow\downarrow}\rangle \tag{11.136}$$

We now need to compute each and every possible combination of matrix element. Since it is a 4 × 4 matrix, we have sixteen possible choices. In Exercise 11.11.19 you will derive Equation (11.137) and the solutions for the singlet ($E_s$) and triplet ($E_{tr}$) eigenenergy states. The singlet state is characterized by a state with only one eigenenergy (since total $S = 0$, $m_S$ can take only one value) and the triplet state by three (since total $S = 1$, $m_S$ can take three values 0, $\pm 1$), hence their names. For the moment we will simply state the solution of the two-electron energy matrix and focus on the physical interpretation of the solution. The solution is given by the expression

$$E_{el} = (E_a + E_b)\begin{pmatrix}1 & 0 & 0 & 0\\ 0 & 1 & 0 & 0\\ 0 & 0 & 1 & 0\\ 0 & 0 & 0 & 1\end{pmatrix} + \begin{pmatrix}K_{ab} - J_{ab} & 0 & 0 & 0\\ 0 & K_{ab} & -J_{ab} & 0\\ 0 & -J_{ab} & K_{ab} & 0\\ 0 & 0 & 0 & K_{ab} - J_{ab}\end{pmatrix}, \tag{11.137}$$

with $E_s$ and $E_{tr}$ given by

$$E_s = E_a + E_b + K_{ab} + J_{ab}, \tag{11.138}$$

$$E_{tr} = E_a + E_b + K_{ab} - J_{ab}. \tag{11.139}$$

In the above equations, we have introduced the definitions

$$E_a = \langle \Psi | \mathscr{H}_0(\vec{r}_1) | \Psi \rangle = \langle \Psi_{\uparrow\uparrow}, \Psi_{\downarrow\uparrow}, \Psi_{\uparrow\downarrow}, \Psi_{\downarrow\downarrow} | \mathscr{H}_0(\vec{r}_1) | \Psi_{\uparrow\uparrow}, \Psi_{\downarrow\uparrow}, \Psi_{\uparrow\downarrow}, \Psi_{\downarrow\downarrow} \rangle, \tag{11.140}$$

$$E_b = \langle \Psi | \mathscr{H}_0(\vec{r}_2) | \Psi \rangle = \langle \Psi_{\uparrow\uparrow}, \Psi_{\downarrow\uparrow}, \Psi_{\uparrow\downarrow}, \Psi_{\downarrow\downarrow} | \mathscr{H}_0(\vec{r}_2) | \Psi_{\uparrow\uparrow}, \Psi_{\downarrow\uparrow}, \Psi_{\uparrow\downarrow}, \Psi_{\downarrow\downarrow} \rangle, \tag{11.141}$$

$$K_{ab} = \int\int d\vec{r}_1 d\vec{r}_2 |\psi_a(\vec{r}_1)|^2 \frac{e^2}{|\vec{r}_1 - \vec{r}_2|} |\psi_b(\vec{r}_2)|^2, \text{ (Coulomb integral)} \tag{11.142}$$

and

$$J_{ab} = \int\int d\vec{r}_1 d\vec{r}_2 \psi_a^*(\vec{r}_1) \psi_b^*(\vec{r}_2) \frac{e^2}{|\vec{r}_1 - \vec{r}_2|} \psi_b(\vec{r}_1) \psi_a(\vec{r}_2). \text{ (Exchange integral)} \tag{11.143}$$

**Example 11.8.0.2**
Show that the energy for the singlet and triplet state can be combined into a single expression as

$$E = \frac{E_s + E_{tr}}{2} \pm \frac{E_s - E_{tr}}{2}. \tag{11.144}$$

**Solution**
Adding the two eigenenergy expressions we get

$$E_a + E_b + K_{ab} = \frac{E_s + E_{tr}}{2}, \tag{11.145}$$

and subtracting the two eigenenergies we have

$$J_{ab} = \frac{E_s - E_{tr}}{2}. \tag{11.146}$$

Thus we can write

$$E = E_a + E_b + K_{ab} \pm J_{ab} = \frac{E_s + E_{tr}}{2} \pm \frac{E_s - E_{tr}}{2}. \tag{11.147}$$

Inspecting the singlet and triplet energies we observe that since $J_{ab} > 0$, the energy of the triplet energy state will always be lower than the singlet. For orthogonal orbitals participating in a direct exchange process, we find that a $S = 1$ state is favored over $S = 0$. Direct exchange operates between spins which are close enough to have sufficient overlap of their wavefunctions. It gives a strong but short range coupling which decreases rapidly as the ions are separated. What does this mean? It implies that ferromagnetism is always the choice of ground state. Intuitively, this makes sense if we keep in mind that two parallel electrons because of the Pauli exclusion principle will try to avoid each other as much as possible to try and reduce the Coulomb repulsion between them. However, with antiparallel spins, such as in the singlet state, quantum mechanics does not forbid the electron spins to get closer to each other thereby raising the Coulomb repulsion energy. There is a technical name for this – it is called the **exchange hole effect**. Note, that with non-orthogonal orbitals either the singlet or the triplet state can be favored based on the degree of overlap. In fact, in nature antiferromagnetism (anti-aligned spin configuration) is seen more often than the ferromagnetic arrangement we just discussed.

**Example 11.8.0.3**
Show that for two electron spin operators $\vec{S}_1$ and $\vec{S}_2$ we have

$$2\vec{S}_1 \cdot \vec{S}_2 + \frac{1}{2} = \left(\vec{S}_1 + \vec{S}_2\right)^2 - 1. \tag{11.148}$$

Evaluate the above expression for the case of (i) a singlet and (ii) a triplet state.
**Solution**
We start off by noting that S = 1/2 for an electron. For two electrons the total spin is given by $\vec{S} = \vec{S}_1 + \vec{S}_2$. Squaring both sides gives

$$\vec{S}_1^2 + \vec{S}_2^2 + 2\vec{S}_1 \cdot \vec{S}_2 = \left(\vec{S}_1 + \vec{S}_2\right)^2. \tag{11.149}$$

Now, using Equation (11.9) for both the spin 1/2 operators we have

$$\begin{aligned} 2S(S+1) + 2\vec{S}_1 \cdot \vec{S}_2 &= \left(\vec{S}_1 + \vec{S}_2\right)^2, \\ 2\frac{1}{2}\left(\frac{1}{2}+1\right) + 2\vec{S}_1 \cdot \vec{S}_2 &= \left(\vec{S}_1 + \vec{S}_2\right)^2, \\ \Rightarrow 2\vec{S}_1 \cdot \vec{S}_2 + \frac{1}{2} &= \left(\vec{S}_1 + \vec{S}_2\right)^2 - 1. \end{aligned} \tag{11.150}$$

For a singlet the total S = 0. Thus we have

$$2\vec{S}_1 \cdot \vec{S}_2 + \frac{1}{2} = \left(\vec{S}_1 + \vec{S}_2\right)^2 - 1 = -1. \tag{11.151}$$

For a triplet state $S = 1$. Thus

$$\begin{aligned} 2\vec{S}_1 \cdot \vec{S}_2 + \frac{1}{2} &= \left(\vec{S}_1 + \vec{S}_2\right)^2 - 1 \\ &= S^2 - 1 = S(S+1) - 1 = 1(1+1) - 1 = 1. \end{aligned} \tag{11.152}$$

Let us pause here for a moment and make the following observations. From Example 11.8.0.2 we know that the two-electron energy can be combined into a single expression involving the total singlet and triplet energies and the relative singlet and triplet energies, with a $\pm 1$ factor in between. But note from Example 11.8.0.3 we learned we could generate a $\pm$ simply out of the space of two spins interacting with each other. Thus we could in principle mimic the energy expression simply out of a set of spin operators and not worry about the space components. This remarkable connection was put forward by Paul A. M. Dirac (1902–1984). His brilliant insight allows us to combine all these facts into a single Hamiltonian expression dependent only on spin operators and now popularly known in the magnetism community as the **Heisenberg exchange Hamiltonian** $\mathscr{H}_{\text{ex}}$. The explicit expression for a pair of electron spins is given by

$$\mathscr{H}_{12} = \frac{E_s + E_{tr}}{2} - \frac{E_s - E_{tr}}{2}\left(2\vec{S}_1 \cdot \vec{S}_2 + \frac{1}{2}\right), \tag{11.153}$$

$$= \text{const.} - 2J_{12}\vec{S}_1 \cdot \vec{S}_2. \tag{11.154}$$

Finally, dropping the constant energy term, absorbing the factor of two in the definition of the exchange constant, and generalizing the interaction to act between any two electron pairs in the magnetic solid we have

$$\mathscr{H}_{\text{ex}} = -\sum_{i,j} J_{ij}\vec{S}_i \cdot \vec{S}_j, \tag{11.155}$$

where $J_{ij}$ is the exchange constant generalized to include the interaction between any two pairs of spins in the system. In practice the interactions are usually truncated at the nearest neighbor or next-nearest neighbor level. The value of the exchange constant can be computed using the exchange integral Equation (11.143) for orthonormal orbitals (for non-orthogonal orbitals the expression needs to be modified). Typical values of exchange energies range from meV to eV.

**Example 11.8.0.4**
Assuming that exchange interaction acts only between any two pairs of nearest neighbor z component of spin on a lattice and a constant J, write down an expression for the magnetic Hamiltonian of the system.
**Solution**
With only the z component of the nearest neighbor interacting spin and a constant J, the Heisenberg exchange Hamiltonian reduces to

$$\mathscr{H} = -J \sum_{\langle i,j \rangle} S_i^z S_j^z. \tag{11.156}$$

The expression above should be familiar to you from the ferromagnetism section of Chapter 10. It is the celebrated Ising Hamiltonian! So, it appears that everything has come full circle. While in Chapter 10 we simply postulated a classical magnetic model, in this chapter using quantum mechanics we were able to derive the general magnetic Heisenberg exchange Hamiltonian. We were also able to show how it reduces to the special case of an Ising model and the Ising Hamiltonian!

Yet, in another version of the model, if we completely ignore the z component of spin interaction and preserve only the x and y components we end up with what is known as the $XY$ model. Interestingly enough, phase transition studies of the *classical XY* model in two lattice dimensions led by Vadim Berezinskii, John M. Kosterlitz, and David J. Thouless predicted a vortex binding–unbinding transition. Studies of topological excitations such as vortices and topological phases of matter were recognized by the 2016 Nobel Prize in physics. The award went to David J. Thouless (1934–), F. Duncan M. Haldane (1951–) and J. Michael Kosterlitz (1943–) "for theoretical discoveries of topological phase transitions and topological phases of matter".

## 11.9 Magnetic Resonance

From Section 11.5 we know that the local crystalline environment and applied external magnetic field can affect the energy levels of a magnetic system. To probe excitations, experimentalists often use parmagnetic resonance spectroscopy. While the electronic version is known as electron spin resonance (ESR), the nuclear variety is termed nuclear magnetic resonance (NMR). Magnetic resonance phenomena come in a variety of other flavors: nuclear quadrupole resonance (NQR), ferromagnetic resonance (FMR), spin wave resonance (SWR), antiferromagnetic resonance (AFMR), and conduction electron spin resonance (CESR).
The characteristic frequencies employed in the resonance experiments are known as Larmor frequencies. Conceptually, from a semi-classical perspective, this frequency is the same as that of a magnetic moment (originating from either the electron or the proton) experiencing a torque in the presence of a magnetic field. The magnetic field can originate either due to the presence of an external field or due to its crystalline environment. The frequency of precession is the Larmor frequency. ESR transitions among the energy levels can be detected by monitoring the power absorbed from an alternating magnetic field, just as ordinary atomic transitions are detected by absorption of light. Comparing the observed transitions with model calculations allow us to deduce features of

the environment around the magnetic moment. ESR experiments have several applications. They are used to study free radicals, ions of the transition metals, solid bodies with defects, metals and semiconductors.

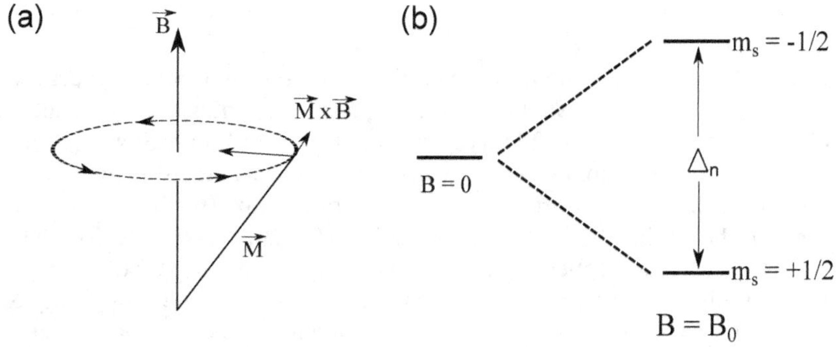


Figure 11.9.16: (a) Precession of a magnetic moment in an external magnetic field. (b) Two-state energy level diagram for a nucleus of spin $I = \frac{1}{2}$. The hydrogen atom would be an example of a two-state system. Note, the relative difference in the relative location of the up ($\frac{1}{2}$) and down (-$\frac{1}{2}$) spin nuclear state in comparison to the Zeeman split energy level diagram of an electronic spin shown in Figure 11.2.2. The difference arises from the sign on the charge of the electron and the proton.

Similar to ESR, in NMR experiments two different energy states with nuclear spin, say $I = \frac{1}{2}$, arise from the alignment of the nuclear magnetic moments relative to the applied field, see Figure 11.9.16(a). However, subsequent transitions between the levels are induced by electromagnetic waves in the radio (Larmor) frequency range. This difference is due to the lower gyromagnetic ratio of the proton relative to an unpaired electron. In analogy with the electronic Zeeman effect, the nuclear Zeeman interaction energy is written as

$$E_z^n = -\vec{\mu}_n \cdot \vec{B}_a, \tag{11.157}$$

where $\vec{B}_a$ is the applied external magnetic field and $\vec{\mu}_n$ is the nuclear magnetic moment. To proceed further, we need an expression for the nuclear magnetic moment $\vec{\mu}_n$. If we recognize that the electronic moment should be negative since the electron is negatively charged, then the nuclear magnetic moment must be positively charged since the proton has a positive charge. Hence, with a field applied along the z-direction and using $\vec{I}$ as the symbol for angular momentum of the nuclear spin, we have

$$\vec{\mu}_n = \gamma_n \vec{I} = g_n \mu_N \frac{\vec{I}}{\hbar}. \tag{11.158}$$

The interaction energy can be written as

$$E_z^n = -\mu_n^z B_z = -g_n \mu_N \frac{I_z}{\hbar} B_z = -g_n \mu_N m_I B_z, \tag{11.159}$$

where $g_n$ is the nuclear g-factor, $\mu_N$ is the nuclear magneton (see Chapter 10, Exercise 10.1), and the nuclear spin $I_z = m_I \hbar$ admits projections $m_I = -I, I-1, ..., I$. The energy level splitting for $I = \frac{1}{2}$ is given by

$$\Delta_n = -\left(-\frac{1}{2} g_n \mu_N B_z\right) - \left(-\frac{1}{2} g_n \mu_N B_z\right) = g_n \mu_N B_z. \tag{11.160}$$

The above expression implies that the nuclear down spin projection of $m_I = -\frac{1}{2}$ occupies a higher energy level compared to $m_I = \frac{1}{2}$ in the presence of an external field, as shown in Figure 11.9.16. In Example 11.2.1.2, we will compute the Larmor frequency for ESR from a quantum mechanical perspective.

**Example 11.9.0.1**
In 1939 Isidor I. Rabi and collaborators detected the NMR phenomenon in hydrogen molecules in the Stern–Gerlach setup. For his studies on the magnetic properties of atomic nuclei, Rabi was awarded a Nobel Prize in 1944. In 1952 Felix Bloch and Edward M. Purcell were awarded the Noble Prize in physics for their development of new methods for NMR precision measurements. While Purcell *et al.* measured the NMR absorption spectra in paraffin wax, Bloch *et al.* investigated NMR in water. At present, the National High Magnetic Field Laboratory (NHMFL) located in Tallahassee, Florida uses a 21.1 Tesla magnet to peform NMR research and magnetic resonance imaging (MRI) scanning. For this magnetic field strength compute (a) the level splitting energy $\Delta_n$ in eV and Kelvins, (b) the frequency of electromagnetic radiation required to excite transitions between the two-state energy levels of a hydrogen proton.
**Solution**
(a) Using Equation (11.160), $g_n = 5.59$ (for a proton), $\mu_N = 5.051 \times 10^{-27}$ Am$^2$, B$_z$= 21.1 T and the conversion between joules and eV we have

$$\Delta_n^{\mathrm{eV}} = (5.59)(5.051 \times 10^{-27})(21.1)/(1.60217657\,10^{-19}) = 3.7 \ \mu\mathrm{eV}. \tag{11.161}$$

In the Kelvin temperature scale we have,

$$\Delta_n^{\mathrm{K}} = \frac{\Delta_n}{k_B} = (5.59)(5.051 \times 10^{-27})(21)/(1.3807\,10^{-23}) = 43 \ \mathrm{mK}. \tag{11.162}$$

Compared to room temperature, the 43 mK splitting energy is tiny. Hence, thermal effects will overpower the alignment energy of the external field and randomize the orientation of the nuclear spins to destroy any net magnetization.
(b) To obtain a numerical value of the exciting radio frequency (rf) we equate

$$\nu = \Delta_n/h = ((5.59)(5.051 \times 10^{-27})(21))/(6.626 10^{-34}) = 894 \ \mathrm{MHz} \approx 900 \ \mathrm{MHz}, \tag{11.163}$$

where $h$ is the Planck constant and $\nu$ is the frequency. Within the NMR community it is typical to refer to a 21 T magnet NMR set-up as a 900 MHz NMR magnet. Other typical magnetic field range includes 12–15 T fields for which the rf range is between 500–650 MHz. The MHz frequency lies in the radio wave range. By supplying the nucleus with the appropriate energy, it is possible to flip the proton from one orientation to the other. The flipping of the proton from one magnetic alignment to the other by the radio waves is known as the resonance condition.

For spin$-\frac{1}{2}$ nuclei at thermal equilibrium the energy difference between the up and the down states prevents these states from being equally populated. Similar to the S=$\frac{1}{2}$ case in the quantum theory of paramagnetism, Section 11.7, we can utilize the Boltzman factors to write down an expression for either the up- or the down-states

$$\frac{N_{m_I}}{N} = \frac{\exp\left(-E_{m_I}/k_BT\right)}{\sum\limits_{m_I=-I}^{m_I=+I} \exp\left(-E_{m_I}/k_BT\right)}, \tag{11.164}$$

where $N_{m_I}$ is the population number for the spins in state $m_I$ and $E_{m_I}$ the corresponding energy. Using the above equation, the relative population of the down- to the up-states can be written as

$$\frac{N_\downarrow}{N_\uparrow} = \exp\left(-\frac{\Delta_n}{k_BT}\right) = \exp\left(-\frac{g_n\mu_N B_z}{k_BT}\right). \tag{11.165}$$

**Example 11.9.0.2**

Calculate the population imbalance ratio between the down to the up spins at 300 K for (a) 14.1 T (600 MHz) and (b) 21.1 Tesla (900 MHz) NMR setup.

**Solution**

Using Equation (11.165) for both cases we have

$$\frac{N_\downarrow}{N_\uparrow} = \exp\left(-\frac{((5.59)(5.051\times10^{-27})(14.1))}{(300)(1.3807\times10^{-23})}\right) = 0.999904 \tag{11.166}$$

$$\frac{N_\downarrow}{N_\uparrow} = \exp\left(-\frac{((5.59)(5.051\times10^{-27})(21.1))}{(300)(1.3807\times10^{-23})}\right) = 0.999856. \tag{11.167}$$

Since NMR transition energies are very small, the up and the down states are nearly equally populated. From quantum mechanics we know that for net absorption of radiation to occur there must be more particles in the lower-energy state than in the higher one. If no net absorption is possible, a condition called saturation is attained. If there are equal numbers of spin up and spin down nuclear moments, the net rf absorption will be zero because equal numbers of rf photons will be absorbed and emitted. The population difference is accentuated by working with the highest possible magnetic field. This small difference in parts per million is sufficient to cause an NMR signal.

Originally, NMR experiments were utilized for accurate measurement of nuclear magnetic moments in condensed matter systems. Subsequently NMR found widespread applications in the studies of chemical structure and dynamics. Presently, a popular application of NMR is in the domain of magnetic resonance imaging (MRI) utilized primarily for medical diagnosis. Magnetic resonance imaging has become an essential diagnostic tool worldwide due to its ability to non-invasively depict and distinguish soft tissues within the body. At the most basic level, it utilizes an induced magnetic field and a pulsed radio frequency wave to create detailed images of a patient. The human body is composed primarily of hydrogen atoms (63%) and most of our tissues contain roughly 75% water. MRI machines that are currently used for clinical diagnostic purposes make use of this fact through what is known as the **chemical shift**. The chemical shift is defined as the difference in resonant frequency between isolated hydrogen and its value when bound to a specific site within a molecule. Different tissues will contain diverse chemical compositions/environments, and therefore different chemical shifts, allowing for discernible contrast between them. In the next few paragraphs, we focus on deriving the basic equations for NMR spectroscopy.

To understand the time evolution of the magnetic moments in NMR, the first step is to write down the equation of motion describing the torque experienced by the $i^{th}$ individual nuclear magnetic moment $\vec{\mu}_{n,i}$ in the presence of an applied magnetic field $\vec{B}_a$. The torque induces precession at the Larmor frequency appropriate to the $i^{th}$ nuclear moment and is related to the rate of change of its angular nomentum as

$$\frac{d\vec{I}_i}{dt} = \vec{\mu}_{n,i}\times\vec{B}_a, \text{ or } \frac{d\vec{\mu}_{n,i}}{dt} = \gamma_n\vec{\mu}_{n,i}\times\vec{B}_a, \tag{11.168}$$

where we have used the relation (see Equation 11.158)

$$\vec{\mu}_{n,i} = \gamma_n\vec{I}_i. \tag{11.169}$$

The gyromagnetic ratio $\gamma_n$ is a constant. For a bulk sample, we sum over all the individual magnetic moments $\vec{\mu}_{n,i}$ to obtain the nuclear magnetization

$$\vec{\mathsf{M}} = \sum\vec{\mu}_{n,i}, \tag{11.170}$$

in a unit volume. The torque equation then becomes

$$\frac{d\vec{\mathsf{M}}}{dt} = \gamma\vec{\mathsf{M}}\times\vec{B}_a. \tag{11.171}$$

Further advancement of NMR theory came in 1946 when Felix Bloch formulated a set of equations that described the behavior of nuclear spin in the presence of both a static field and an oscillating radio-frequency (rf) field. Bloch modified Equation (11.171) to account for the observation that nuclear spins relax to equilibrium values following the application of rf pulses. **Relaxation** refers to the process by which the spins return to equilibrium due to interactions with their surroundings. Bloch assumed that the spin relaxation rates along the z-axis and the x-y plane occur at different rates, but following first-order kinetics. He introduced the $T_1$ time scale called the longitudinal or spin-latice relaxation that originates from the interaction between spins and the lattice. Additionally, the spin-spin relaxation that arises from the interaction between different parts of the spin system is modeled by $T_2$. If we combine both the effects of thermal equilibrium and torque, we have the set of **Bloch equations** shown below in vector form

$$\frac{d\vec{\mathsf{M}}(t)}{dt} = \underbrace{\gamma_n(\vec{\mathsf{M}}(t)\times\vec{B}(t))}_{\text{precession}} - \underbrace{\frac{1}{T_2}(\mathsf{M}_x\hat{\imath}+\mathsf{M}_y\hat{\jmath})}_{\text{spin-spin relaxation}} - \underbrace{\frac{1}{T_1}(\mathsf{M}_z-\mathsf{M}_{z0})\hat{k}}_{\text{spin-lattice relaxation}}, \tag{11.172}$$

or, in component form

$$\frac{d\mathsf{M}_x}{dt} = \gamma_n(\vec{\mathsf{M}}\times\vec{B}_a)_x - \frac{\mathsf{M}_x}{T_2}, \tag{11.173}$$

$$\frac{d\mathsf{M}_y}{dt} = \gamma_n(\vec{\mathsf{M}}\times\vec{B}_a)_y - \frac{\mathsf{M}_y}{T_2}, \tag{11.174}$$

$$\frac{d\mathsf{M}_z}{dt} = \gamma_n(\vec{\mathsf{M}}\times\vec{B}_a)_z - \frac{\mathsf{M}_z-\mathsf{M}_{z0}}{T_1}. \tag{11.175}$$

Let us consider the solution of the Bloch equations in the presence of a static external magnetic field $\vec{B}_a = B_0\hat{k}$ and initial condition $\vec{\mathsf{M}}(t=0) = (\mathsf{M}_x(0), \mathsf{M}_y(0), \mathsf{M}_z(0))$. Using Equations (11.173) – (11.175) and expanding the vector cross-product, we then obtain

$$\frac{d\mathsf{M}_x}{dt} = \omega_o\mathsf{M}_y - \frac{\mathsf{M}_x}{T_2}, \tag{11.176}$$

$$\frac{d\mathsf{M}_y}{dt} = -\omega_o\mathsf{M}_x - \frac{\mathsf{M}_y}{T_2}, \tag{11.177}$$

$$\frac{d\mathsf{M}_z}{dt} = \frac{1}{T_1}(\mathsf{M}_{z0}-\mathsf{M}_z), \tag{11.178}$$

where $\omega_o = \gamma_n B_0$ and $\mathsf{M}_{z0}$ is the equilibrium value which obeys the Curie law. We will solve the above system of differential equations beginning with Equation (11.178). We rearrange the equation to obtain

$$\int_{\mathsf{M}_z(0)}^{\mathsf{M}_z} \frac{d\mathsf{M}_z}{\mathsf{M}_{z0}-\mathsf{M}_z} = \int_0^t \frac{dt}{T_1}. \tag{11.179}$$

Integrating both sides of the equation and keeping in mind that the saturation magnetization $\mathsf{M}_{z0}$ is greater than $\mathsf{M}_z(t)$ or $\mathsf{M}_z(0)$ we obtain

$$-\ln|\mathsf{M}_{z0}-\mathsf{M}_z|\Bigg|_{\mathsf{M}_z(0)}^{\mathsf{M}_z(t)} = \frac{t}{T_1}, \tag{11.180}$$

or

$$-\ln\left|\frac{\mathsf{M}_{z0}-\mathsf{M}_z(t)}{\mathsf{M}_{z0}-\mathsf{M}_z(0)}\right| = \frac{t}{T_1}, \tag{11.181}$$

which can be rearranged to give the final solution as

$$M_z(t) = M_z(0)\exp(-t/T_1) + M_{z0}[1 - \exp(-t/T_1)] \tag{11.182}$$

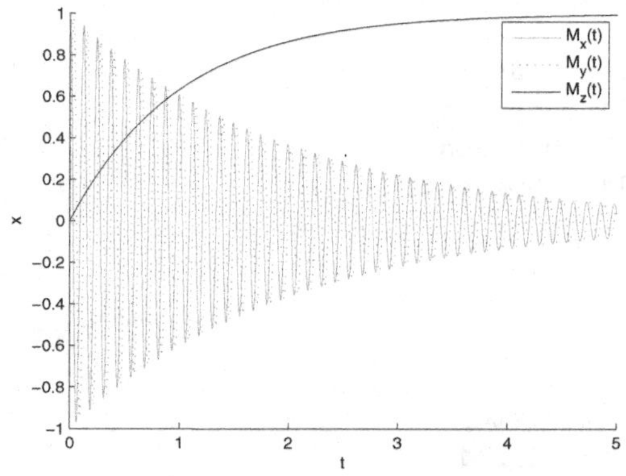


Figure 11.9.17: Decay of the transverse magnetization components in the x-y plane and relaxation of $M_z$ component of magnetization to the z-axis.

To solve Equations (11.176) and (11.177), we will employ a trick that will eliminate the relaxation terms. Let us assume the solutions are of the form

$$M_x(t) = -m_x \exp(-t/T_2); \; M_y(t) = m_y \exp(-t/T_2), \tag{11.183}$$

where $m_x$ and $m_y$ are functions of time. With the above choice, it is easy to show (Exercise 11.11.20) that

$$\frac{d^2 m_x}{dt^2} + \omega_o^2 m_x = 0, \tag{11.184}$$

$$\frac{d^2 m_y}{dt^2} + \omega_o^2 m_y = 0. \tag{11.185}$$

Explicit expressions for $m_x(t)$ and $m_y(t)$ which solve Equations (11.184) and (11.185) are given by

$$M_x(t) = -e^{-t/T_2}(M_x(0)\cos\omega_o t - M_y(0)\sin\omega_o t), \tag{11.186}$$

$$M_y(t) = e^{-t/T_2}(M_y(0)\cos\omega_o t + M_x(0)\sin\omega_o t), \tag{11.187}$$

$$M_z(t) = M_z(0)\exp(-t/T_1) + M_{z0}[1 - \exp(-t/T_1)], \tag{11.188}$$

where we have assumed an initial condition of $\vec{M}(0) = (M_x(0), M_y(0), M_z(0))$.

In Figures 11.9.17 and 11.9.18 we display the solutions of the Bloch equations obtained using a computational approach. As expected, the computational solution agrees well with the analytical solutions obtained above in Equations (11.186) - (11.188). From these equations we observe that the general time-dependent solution for the transverse components, $M_x(t)$ and $M_y(t)$, have a damped solution. The damping is created by the presence of the exponential decay term multiplying the sinusoidal component. The sinusoidal terms correspond to the precessional motion of the magnetic dipole moment. Note, the magnitude of $|\vec{M}|$ is not fixed.

**Example 11.9.0.3**

By executing the Bloch equations simulation code below, can you explain why the z-component of magnetization is called the relaxation term?

Below is the MATLAB code for simulating the Bloch equations.

```
%copyright by J. E Hasbun and T. Datta
% ch11_blocheqsim.m

% This script simulates the Bloch equations in the presence of
% longitudinal and transverse relaxation

% Solving the Bloch equations
% dMx_dt(1) = write all components
% dMy_dt(2) = write all components
% dMz_dt(3) = write all components

%initial conditions
Xo = [1;0;0];
%timespan
tspan = [0,5];
% defining simulation parameters
omega= 50;tone=1;ttwo=1;mzero=1;
%set an error for solving coupled ODE
options=odeset('RelTol',1e-6);
magparam = @(t,x)mag(t,x,omega,tone,ttwo,mzero);
%call the solver
[t,X] = ode45(magparam,tspan,Xo,options);
%plot the results
figure(1)
hold on;
plot(t,X(:,1),'r');plot(t,X(:,2),':');plot(t,X(:,3),'k-');
legend('M_x(t)','M_y(t)','M_z(t)');ylabel('x');xlabel('t')
figure(2)
plot(X(:,1),X(:,2));

% Create a separate function file ch11_mag.m

function dm = mag(~,m,omega,tone,ttwo,mzero)
%a function which returns a rate of change vector
dm = zeros(3,1);
dm(1)=  omega*m(2)- m(1)/ttwo;
dm(2)= -omega*m(1)- m(2)/ttwo;
dm(3)=  (mzero - m(3))/tone;
```

As mentioned earlier, relaxation refers to the process by which the spins return to equilibrium by interacting with their surroundings. Changing the initial condition in the Bloch simulation code shows that the z-component, irrespective of the choice of the initial magnetization value, always returns to the initial value. A similar behavior is observed in Figures 11.9.17 and 11.9.18 also for an initial condition of $\vec{M}(t=0) = (1,0,0)$.

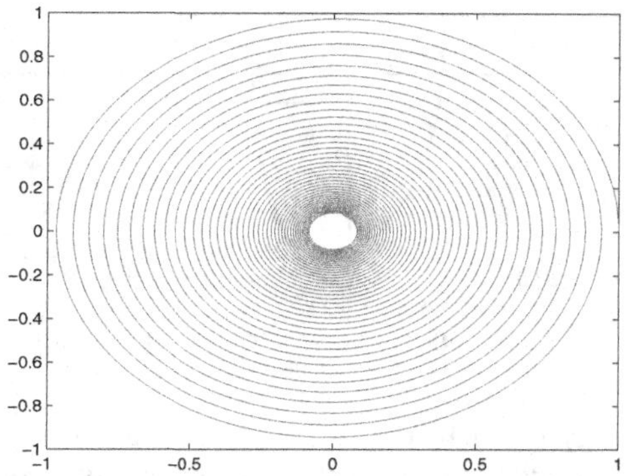


Figure 11.9.18: Dephasing of the transverse magnetization components $\mathsf{M}_x(t)$ and $\mathsf{M}_y(t)$ displaying the damped nature of the solution.

## 11.10 Pauli Paramagnetism

We will finish this chapter by briefly discussing the effects of magnetic field on the spin polarization of an itinerant electron system – the free electron gas (see Chapter 5 for an introduction). Since the electron possesses a permanent magnetic moment, a classical electron gas should exhibit a response to an external magnetic field. Consider the free electron band structure in the absence of a magnetic field shown in Figure 11.10.19(a). With an applied external magnetic field $B_0$ acting along the z axis, the electron subbands $n_\uparrow$ and $n_\downarrow$ separate out as shown in Figure 11.10.19(b). Thus electrons in one half band find themselves in states lying above the empty states, and the others below. This population imbalance gives rise to magnetization

$$M_{el} = \mu_B[N(\uparrow) - N(\downarrow)] \tag{11.189}$$

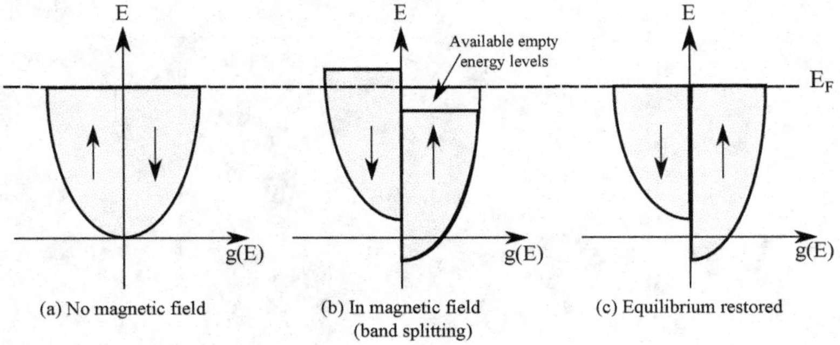


Figure 11.10.19: Plot of energy $E$ versus density of states $g(E)$. Arrows represent magnetic moments. The dashed line corresponds to the highest occupied energy level (in the absence of a field) known as the Fermi energy. (a) In the absence of a magnetic field the electron energy levels (bands) are degenerate. (b) In the presence of an external field, splitting occurs with a relative shift between the up and down magnetic moments. (c) Eventually, equilibirum is restored when some of the down magnetic moments flip their orientation to occupy the available empty energy levels shown in (b). Finally, the up band population is in excess of the down giving rise to a magnetization.

The subbands are displaced by the Zeeman interaction energy of $-\mu_B B_0$. If we define the density of states as $g(E)$, then we can rewrite the magnetization expression $M_{el}$ as

$$M_{el} = \frac{\mu_B}{2}\left[\int_0^{E_F+\mu_B B_0} g(E)dE - \int_0^{E_F-\mu_B B_0} g(E)dE\right] \tag{11.190}$$

Since $\mu_B B_0 \ll E_F$, we can write

$$\int_0^{E_F\pm\mu_B B_0} g(E)dE = \int_0^{E_F} g(E)dE \pm \mu_B B_0 g(E_F). \tag{11.191}$$

Using the above we finally obtain for the magnetization

$$M_{el} = \mu_B^2 g(E_F) B_0 \tag{11.192}$$

We see from the above equation that the susceptibility is independent of temperature and proportional to the density of states. This is **Pauli paramagnetism**.

Table 11.10.4: Radial wavefunction.

| n | l | Radial Wavefunction $R_{nl}(r)$ |
|---|---|---|
| 1 | 0 | $R_{10}(r) = 2\left(\frac{Z}{a_o}\right)^{3/2} \exp(-Zr/a_o)$ |
| 2 | 0 | $R_{20}(r) = 2\left(\frac{Z}{2a_o}\right)^{3/2}\left(1-\frac{Zr}{2a_o}\right)\exp(-Zr/2a_o)$ |
| | 1 | $R_{21}(r) = \frac{1}{\sqrt{3}}\left(\frac{Z}{2a_0}\right)^{3/2}\left(\frac{Zr}{a_0}\right)\exp(-Zr/2a_0)$ |
| 3 | 0 | $R_{30}(r) = 2\left(\frac{Z}{3a_o}\right)^{3/2}\left(1-\frac{2Zr}{3a_o}+\frac{2Z^2r^2}{27a_o^2}\right)\exp(-Zr/3a_o)$ |
| | 1 | $R_{31}(r) = \frac{4\sqrt{2}}{9}\left(\frac{Z}{3a_0}\right)^{3/2}\left(\frac{Zr}{a_0}\right)\left(1-\frac{Zr}{6a_0}\right)\exp(-Zr/3a_0)$ |
| | 2 | $R_{32}(r) = \frac{4}{27\sqrt{10}}\left(\frac{Z}{3a_o}\right)^{3/2}\left(\frac{Zr}{a_o}\right)^2\exp(-Zr/3a_o)$ |
| 4 | 0 | $R_{40}(r) = 2\left(\frac{Z}{4a_o}\right)^{3/2}\left(1-\frac{3Zr}{4a_o}+\frac{Z^2r^2}{8a_o^2}-\frac{Z^3r^3}{192a_o^3}\right)\exp(-Zr/4a_o)$ |
| | 1 | $R_{41}(r) = \frac{5}{2\sqrt{15}}\left(\frac{Z}{4a_o}\right)^{3/2}\left(1-\frac{Zr}{4a_o}+\frac{Z^2r^2}{80a_o^2}\right)\left(\frac{Zr}{a_o}\right)\exp(-Zr/4a_o)$ |
| | 2 | $R_{42}(r) = \frac{1}{8\sqrt{5}}\left(\frac{Z}{4a_o}\right)^{3/2}\left(1-\frac{Zr}{12a_o}\right)\left(\frac{Zr}{a_o}\right)^2\exp(-Zr/4a_o)$ |
| | 3 | $R_{43}(r) = \frac{1}{96\sqrt{35}}\left(\frac{Z}{4a_o}\right)^{3/2}\left(\frac{Zr}{a_o}\right)^3\exp(-Zr/4a_o)$ |

Table 11.10.5: Spherical Harmonics for $l = 0, 1, 2, 3, 4$.

| l | m | Spherical Harmonic $Y_{lm}(\theta, \phi)$ |
|---|---|---|
| 0 | 0 | $Y_0^0(\theta,\phi) = \frac{1}{(4\pi)^{1/2}}$ |
| 1 | 0 | $Y_1^0(\theta,\phi) = \left(\frac{3}{4\pi}\right)^{1/2} \cos\theta$ |
| | $\pm1$ | $Y_1^{\pm1}(\theta,\phi) = \mp\left(\frac{3}{8\pi}\right)^{1/2} \sin\theta e^{\pm i\phi}$ |
| 2 | 0 | $Y_2^0(\theta,\phi) = \left(\frac{5}{16\pi}\right)^{1/2} (3\cos^2\theta - 1)$ |
| | $\pm1$ | $Y_2^{\pm1}(\theta,\phi) = \mp\left(\frac{15}{8\pi}\right)^{1/2} \sin\theta\cos\theta e^{\pm i\phi}$ |
| | $\pm2$ | $Y_2^{\pm2}(\theta,\phi) = \left(\frac{15}{32\pi}\right)^{1/2} \sin^2\theta e^{\pm 2i\phi}$ |
| 3 | 0 | $Y_3^0(\theta,\phi) = \left(\frac{7}{16\pi}\right)^{1/2} (5\cos^3\theta - 3\cos\theta)$ |
| | $\pm1$ | $Y_3^{\pm1}(\theta,\phi) = \mp\left(\frac{21}{64\pi}\right)^{1/2} \sin\theta(5\cos^2\theta - 1)e^{\pm i\phi}$ |
| | $\pm2$ | $Y_3^{\pm2}(\theta,\phi) = \left(\frac{105}{32\pi}\right)^{1/2} \sin^2\theta\cos\theta e^{\pm 2i\phi}$ |
| | $\pm3$ | $Y_3^{\pm3}(\theta,\phi) = \mp\left(\frac{35}{64\pi}\right)^{1/2} \sin^3\theta e^{\pm 3i\phi}$ |
| 4 | 0 | $Y_4^0(\theta,\phi) = \left(\frac{9}{256\pi}\right)^{1/2} (35\cos^4\theta - 30\cos^2\theta + 3)$ |
| | $\pm1$ | $Y_4^{\pm1}(\theta,\phi) = \mp\left(\frac{45}{64\pi}\right)^{1/2} \sin\theta(7\cos^3\theta - 3\cos\theta)e^{\pm i\phi}$ |
| | $\pm2$ | $Y_4^{\pm2}(\theta,\phi) = \left(\frac{45}{128\pi}\right)^{1/2} \sin^2\theta(7\cos^2\theta - 1)e^{\pm 2i\phi}$ |
| | $\pm3$ | $Y_4^{\pm3}(\theta,\phi) = \mp\left(\frac{315}{64\pi}\right)^{1/2} \sin^3\theta\cos\theta e^{\pm 3i\phi}$ |
| | $\pm4$ | $Y_4^{\pm4}(\theta,\phi) = \left(\frac{315}{512\pi}\right)^{1/2} \sin^4\theta e^{\pm 4i\phi}$ |

## 11.11 Chapter 11 Exercises

11.11.1. The *Einstein–de Haas effect* experiment can be used to measure the gyromagnetic ratio and the g-factor. The basic idea behind the experiment is to use conservation of angular momentum. Consider a magnetized iron bar, in the presence of an external magnetic field, suspended from a thin elastic fiber and free to rotate. If the orientation of the magnetization is changed by reversing the direction of the external magnetic field, then the associated magnetic moments must also change direction. However, the system is isolated mechanically. Therefore the total angular momentum must be conserved. To do so, the crystal lattice must rotate to compensate for the change in angular momentum of the magnetic atoms. For this experimental set-up show that

$$\vec{M}_{bar} = \frac{g\mu_B}{\hbar}\vec{L} \tag{11.193}$$

where $\vec{M}_{bar}$ is the magnetization of the bar and $\vec{L}$ the angular momentum of the electron. By measuring $\vec{M}_{bar}$ and angular momentum $\vec{L}$ from the torsion vibrations of the string, the gyromagnetic ratio and the g-factor can be obtained. Typically, $\vec{L}$ is measured using an optical set-up based on deflection of light as shown in the experimental set-up in Figure 11.11.20.

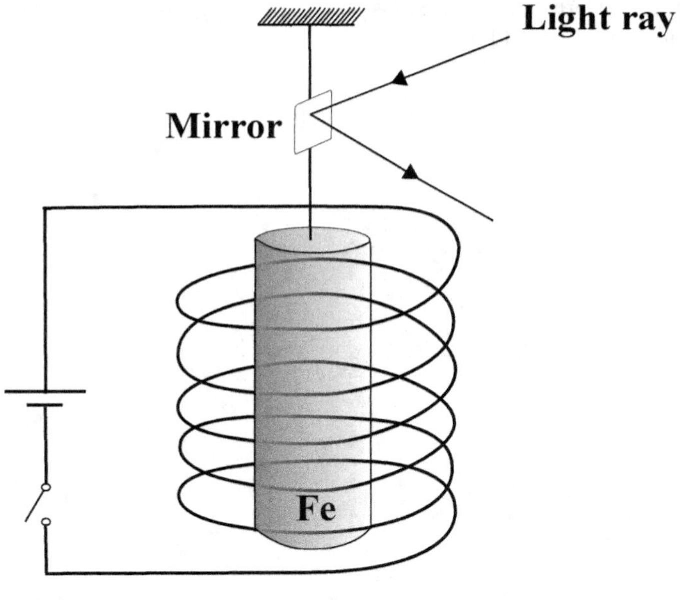


Figure 11.11.20: Einstein-de Haas effect experimental set up.

11.11.2. Using the expressions for the $p$ orbital spherical harmonics from Table 11.10.5 and the definition of the conversion from the complex to the real basis, Equation (11.30), derive (a) the real form expressions for the three $p$-orbitals, (b) convert the real form to their equivalent Cartesian expressions, and (c) write a MATLAB code to generate the 2p orbitals shown in Figure 11.11.21.

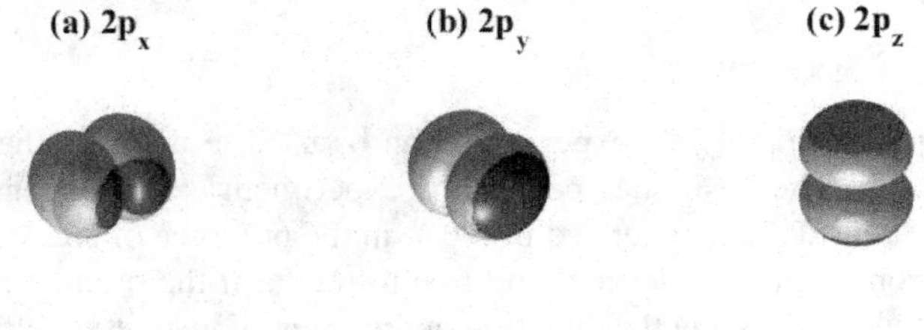


Figure 11.11.21: Boundary surface electron-density representation of atomic $2p_x$, $2p_y$, and $2p_z$ orbitals in real basis.

11.11.3. The Lanthanides (atomic numbers 57–71) and Actinides (atomic numbers 89–103) belong to the $f$-block of elements where the electrons occupy the $4f$ and $5f$ electron shells, respectively. The Lanthanides and Actinides are also referred to as Rare Earth metals. As mentioned in Chapter 10, rare earth neodymium (NdFeB) and samarium–cobalt (SmCo) magnets form good quality permanent magnets (which you can buy at Lowes or Home Depot). Using the expressions for the $f$ orbital spherical harmonics from Table 11.10.5 and the definition of the conversion from the complex to the real basis, Equation (11.30), derive (a) the real form expressions for the seven $f$-orbitals, (b) convert the real form to their equivalent Cartesian expressions, and (c) write a MATLAB code to generate the 4f orbitals shown in Figure 11.11.22.

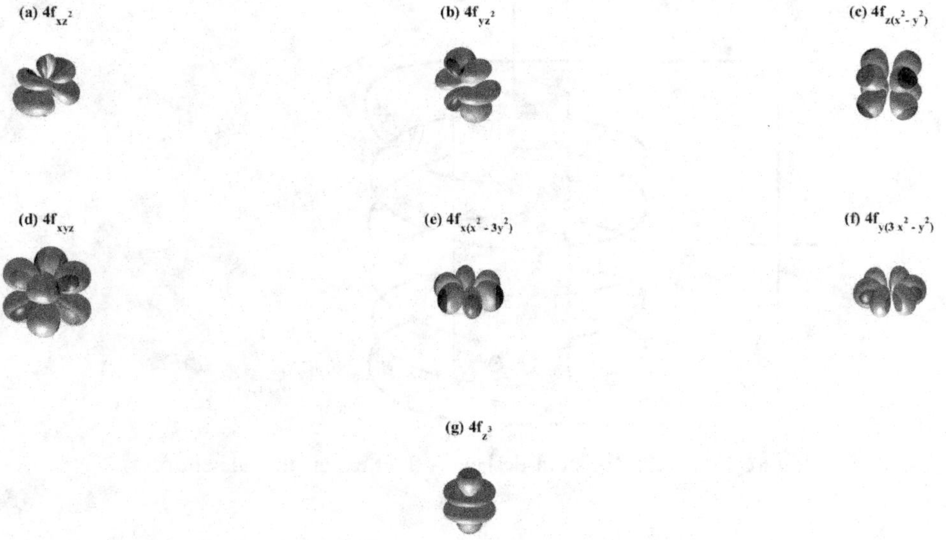


Figure 11.11.22: Boundary surface representation of atomic $4f_{xz^2}$, $4f_{xz^2}$, $4f_{yz^2}$, $4f_{y(3x^2-y^2)}$, $4f_{x(x^2-3y^2)}$, $4f_{xyz}$, and $4f_{z(x^2-y^2)}$ orbitals in real basis.

11.11.4. By calculating the sum of squares of the wavefunctions given by $m_l = 0, \pm 1, \pm 2$, and $\pm 3$ for the $4f^7$ $f$-orbital electron shell show that it is spherical.

11.11.5. Derive the Cartesian expressions for the spherical harmonics $Y_4^4, Y_4^0$, and $Y_4^{-4}$ to show that

$$\begin{aligned} Y_4^{\pm 4} &= \sqrt{\frac{315}{512\pi}} \frac{(x \pm iy)^4}{16r^4}, \\ Y_4^0 &= \sqrt{\frac{9}{256\pi}} \frac{(3r^4 - 30r^2z^2 + 35z^4)}{r^4}. \end{aligned} \tag{11.194}$$

11.11.6. Confirm the following ground state term symbols: (a) $Cu^{2+}$, ${}^2D_{5/2}$ ; (b) $Co^{2+}$, ${}^4F_{9/2}$; (c) $Sm^{3+}$, ${}^6H_{5/2}$; (d) $Ho^{3+}$, ${}^5I_8$

11.11.7. Using the values of L, S, and J from Tables 11.4.1 and 11.4.2, compute the Landé $g$-factor and $\mu_{eff} = g\sqrt{J(J+1)}$ (in $\mu_B$ units) for the following ions (a) $V^{4+}$, (b) $Fe^{2+}$, (c) $Gd^{3+}$, (d) $Dy^{3+}$.

11.11.8. Assuming the *LS* (Russell–Saunders) coupling scheme, obtain all the possible term symbols for electrons occupying orbitals (a) $2p^1 3p^1$ and (b) $2p^5$?

11.11.9. Assuming $jj$-coupling scheme, obtain all the possible term symbols for electrons occupying orbitals $np^1$ $nd^1$?

11.11.10. The mineral perovskite has a general stoichiometry of $ABX_3$ where $A$ and $B$ represent the cations and $X$ is the anion. The original sample was supplied to Gustav Rose by the Chief Mines Inspector August Alexander Kammerer of the Russian Empire. Subsequently, Rose determined its physical properties and chemical composition. On the Chief Mines Inspector's suggestion Rose named the mineral after the Russian politician and mineralogist Count Lev Alekseyevich Perovski (Perovskite: Name Puzzle and German-Russian Odyssey of Discovery, Eugene Katz, *Helv. Chim. Acta* 2020, 103, e2000061). Present interest in these materials originates from their wide range of applicability ranging from solid-state ionics, sensors, fuel cells, electrooptical devices to memory devices (RAM), amplifiers, high temperature superconductors, and multiferroic materials. In Figure 11.5.11 we show a typical cubic perovskite crystal structure arrangement. An ideal cubic perovskite is realized in $SrTiO_3$ which has applications in microelectronics technology due to its high charge storage capacity (large dielectric constant). In the octahedral arrangement, eight Ti atoms reside at the cube corners, one Ti atom sits in the centre of the cube, and six oxygen atoms are located at the centre of the faces. In this problem, you will derive the crystal field potential experienced by the Ti ion at the center of the octahedra. For this purpose, consider equal point charges of value $eZ$, where $eZ$ is the effective ligand charge, placed on each of the six corners of an octahedron. Choosing the origin of the Cartesian coordinates to be at the centre of the octahedron, show that the potential close to the centre is given by

$$V_{cf}^{oct} = \frac{eZ}{4\pi\varepsilon_o a}\left[6 + \frac{35}{4a^4}\left(x^4 + y^4 + z^4 - \frac{3}{5}r^4\right) + \mathcal{O}\left(\frac{r^6}{a^6}\right)\right], \tag{11.195}$$

where $e$ is the magnitude of each charge and $a$ is the distance between the origin and each charge. The constant term is the contribution if we treated each charge as point-like objects. The spatial part is the crystal field potential contribution.

11.11.11. The Legendre polynomials $P_l(\cos\theta))$ can be used to generate a **multipole expansion** in electrodynamics which relates the potential at point $\mathbf{r}$ created by a unit point charge located at $\mathbf{r}'$ by

$$\frac{1}{|\mathbf{r} - \mathbf{r}'|} = \sum_{l=0}^{\infty} \frac{r_<^l}{r_>^{l+1}} P_l(\cos\gamma), \tag{11.196}$$

where $r_<(r_>)$ is the smaller (larger) distance of $\mathbf{r}$ and $\mathbf{r}'$. Equation (11.196) then can be taken a step further if we utilize the addition theorem for spherical harmonic functions which can relate two coordinate vectors $\mathbf{r}$ and $\mathbf{r}'$ with spherical coordinates $(r, \theta, \phi)$ and $(r', \theta', \phi')$ respectively, which have an angle $\gamma$ in between by

$$\frac{1}{|\mathbf{r}-\mathbf{r}'|} = \sum_{l=0}^{\infty} \sum_{m=-l}^{l} \frac{4\pi}{2l+1} \frac{r_<^l}{r_>^{l+1}} Y_l^{m*}(\theta', \phi') Y_l^m(\theta, \phi). \tag{11.197}$$

The mathematical advantage of the above equation is that it completely factorizes the

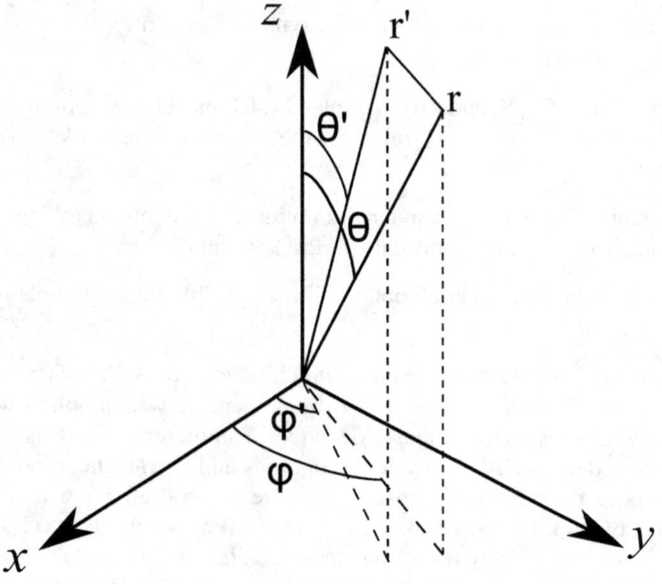


Figure 11.11.23: Spherical harmonics addition theorem.

source (primed variable) and environment (unprimed variable) charge coordinates.

(i) In Figure 11.5.11(a) an octahedral arrangement of ligand ions of charge $Ze$ is shown surrounding a central transition metal ion where $eZ$ is the effective ligand charge and $e$ is the electronic charge. By considering only the $l = 0$ and $l = 4$ terms in Equation (11.197), show that the octahedral crystal field potential $V_{cf}^{oct}(r, \theta, \phi)$ can be written in terms of the spherical harmonics $Y_m^l(\theta, \phi)$ as

$$V_{cf}^{oct}(r, \theta, \phi) = \frac{eZ}{4\pi\varepsilon_o a} \left[ 6 + \frac{7\sqrt{\pi}}{3} \frac{r^4}{a^4} \left\{ Y_4^0 + \sqrt{\frac{5}{14}} \left( Y_4^4 + Y_4^{-4} \right) \right\} \right]. \tag{11.198}$$

*Hint*: To solve the problem, identity the spherical polar coordinates of the ligands. Note, in your problem $r_< = r$ and $r_> = a$, where $a$ is the distance of the ligand ion along the axis in any of the three directions of the octahedral arrangement.

(ii) Show that the above form reduces to the Cartesian expression derived in Equation (11.195).

11.11.12. The ideal cubic perovskite structure in the previous problem is usually distorted resulting in an orthorhombic perovskite crystal structure, as seen for example in Lanthanum Manganite ($LaMnO_3$) – a compound which exhibits colossal magnetoresistance effect.

Show that for an orthorhombic structure the crystal field potential is given by

$$V_{cf}^{ortho} = \frac{eZ}{4\pi\varepsilon_o}\left[\frac{2}{a}+\frac{2}{b}+\frac{2}{c}+x^2\left(\frac{2}{a^3}-\frac{1}{b^3}-\frac{1}{c^3}\right)+y^2\left(\frac{2}{b^3}-\frac{1}{a^3}-\frac{1}{c^3}\right) + z^2\left(\frac{2}{c^3}-\frac{1}{a^3}-\frac{1}{b^3}\right)\right], \tag{11.199}$$

where $e$ is the magnitude of each charge and $a, b$, and $c$ is the distance between the origin and each charge. The constant term is the contribution if we treated each charge as point-like objects. The spatial part is the crystal field potential contribution.

11.11.13. As mentioned in Section 11.5, it is customary to denote the crystal field splitting value as $\Delta = 10Dq$. In this problem we will use concepts from quantum mechanics to obtain an expression for the $10Dq$ splitting. Since the crystal field splitting is an energy difference, we will compute the expectation value of the crystal field Hamiltonian (energy), in the complex basis. Beginning with Equation 11.198, derived in Exercise 11.11.11,

(a) Show that the crystal field perturbation can be written as

$$H_{m,m'} = \int d\vec{r}\psi^*_{nlm}(\vec{r})(eV_{cf}^{oct})\psi_{nlm} = \begin{pmatrix} \mathrm{Dq} & 0 & 0 & 0 & 5\mathrm{Dq} \\ 0 & -4\mathrm{Dq} & 0 & 0 & 0 \\ 0 & 0 & 6\mathrm{Dq} & 0 & 0 \\ 0 & 0 & 0 & -4\mathrm{Dq} & 0 \\ 5\mathrm{Dq} & 0 & 0 & 0 & \mathrm{Dq} \end{pmatrix}. \tag{11.200}$$

where $D = \frac{35e}{4a^5}$ and $q = \frac{2Ze}{105}\langle R_{32}|r^4|R_{32}\rangle$.

(b) Diagonalize the matrix in Equation 11.200 to obtain the eigenvalues. By inspecting the eigenvalues, can you explain the origin of the $t_{2g}$ and the $e_g$ orbital levels in Figure 11.5.11?

11.11.14. Show that the susceptibility computed in Example 11.6.0.1 is dimensionless.

11.11.15. If U is the internal energy of a statistical mechanical system at temperature T and Z the partition sum thenshow that

$$U = -\frac{d\ln Z}{d\beta}, \tag{11.201}$$

where $\beta = 1/\mathrm{T}$.

11.11.16. Derive the approximate expression for the Brillouin function given in Equation (11.112).

11.11.17. Prove the expression for the partition function Z used in Equation (11.118).

11.11.18. In Section 11.7 we introduced the idea of a two-level quantum system. Considering the splitting to be given by $\Delta$, compute (a) the heat capacity $C_V$, (b) heat capacity, (c) Helmholtz-free energy, and (d) entropy for this system. For each quantity, express them in terms of suitably scaled dimensionless variable and plot the temperature variation as shown in Figure 11.11.24.

11.11.19. Derive the two-electron energy matrix, Equation (11.137), and the eigenenergy solution, Equations (11.138) and (11.139), for the singlet and triplet state. In the process of your derivation, also introduce the appropriate definition of the Coulomb integral and the Exchange integral expressions. Below are some useful hints to help you with the derivation.

(a) When considering two orbitals, we can have both the spins to be the same or different.

(b) Within different spin combinations, we could have the case where $\Psi_{\uparrow\downarrow}$ can overlap with $\Psi_{\uparrow\downarrow}$ or $\Psi_{\downarrow\uparrow}$.

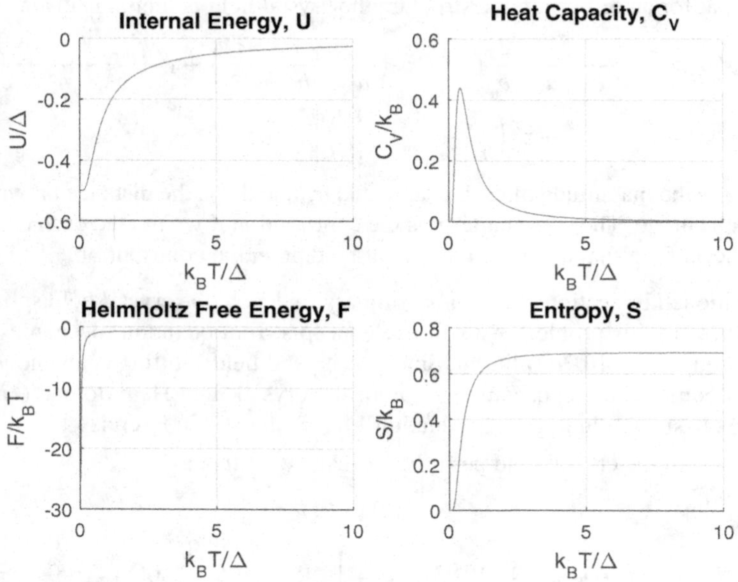


Figure 11.11.24: Plot of internal energy, heat capacity, Helmholtz-free energy, and entropy.

(c) The rest of the two-electron energy matrix elements are zero.

(d) The wavefunctions that we need to consider are stated in Equations (11.128) – (11.134).

11.11.20. In the section on magnetic resonance, you have learned how the Bloch equations can simulate magnetic resonance phenomena. Utilizing the proposed solution in that section, Equation (11.183), derive the differential equation system shown in Equations (11.184) and (11.185).

11.11.21. In magnetic materials, a disturbance of a spin excitation is carried in the form of spin waves. Ferromagnetic resonance (FMR), is a standard spectroscopic technique which is utilized to probe the magnetization of ferromagnetic materials, gyromagnetic ratio, spin waves, and spin dynamics. Historically, FMR was discovered (unknowingly) in 1911 by V. K. Arkadýev when he observed the absorption of ultra high frequency (UHF) radiation (300 MHz to 3 GHz) by ferromagnetic materials. Subsequently, in a 1935 paper published by Lev Landau and Evgeny Lifshitz, a prediction of ferromagnetic resonance of the Larmor precession was made. The prediction of this theory was confirmed independently by experiments on FMR by J. H. E. Griffiths and E. K. Zavoiskij in 1946. Considering an ellipsoidal sample with demagnetization factors ($\mathscr{N}_x, \mathscr{N}_y, \mathscr{N}_z$), obtain the shape effects of specimen on the resonance frequency. Show that in the absence of any relaxation processes within the sample the equation for resonance frequency is given by

$$\omega_o^2 = \gamma^2 \left[B_0 + (\mathscr{N}_y - \mathscr{N}_z)\mu_o M\right]\left[B_0 + (\mathscr{N}_x - \mathscr{N}_z)\mu_o M\right], \tag{11.202}$$

where $B_0$ is the static external field and $\omega_o$ is called the frequency of the uniform mode. In the uniform mode, all the magnetic moments precess together in phase with the same amplitude. Shape effects in FMR were studied by Charles Kittel in 1948.

# 12

## *Superconductivity*

**Contents**



### 12.1 Introduction

In 1908 Dutch physicist Heike Kamerlingh Onnes (1853–1926) was the first person to achieve helium liquefaction through a series of compression and expansion cycles. Onnes's primary motivation to liquefy helium was driven by a fundamental question of how metals behave when cooled to near absolute zero. Using liquid helium as a refrigerant, he was able to lower the temperature down to 1.5 K. In the process of experimenting with various metals at the Kamerlingh Onnes Laboratory, on April 8$^{\text{th}}$ 1911, it was observed that when the temperature was lowered below 4.2 K the resistance of mercury suddenly vanished and was restored when warmed back up. Kamerlingh Onnes wrote in his notebook, "*The temperature measurement was successful. [The resistivity of] Mercury practically zero*". He had discovered the phenomenon of superconductivity (although he preferred to call it supraconductivity). Later, in 1913, he was awarded a Nobel Prize in physics. The Nobel Prize citation read, "his investigations on the properties of matter at low temperatures which led, inter alia, to the production of liquid helium". For a historical account of the history of superconductivity read reference [52]. Following Kamerlingh Onnes discovery, several other elemental metals were found to display a vanishing resistance when the temperature was lowered below a characteristic temperature, called the **critical temperature** $T_c$.

Twenty-two years following the discovery of superconductivity, in 1933 Fritz W. Meissner (1882–1974) and Robert Ochsenfeld (1901–1993) found that a bulk superconductor in a weak magnetic field will act as a perfect diamagnet, with a zero magnetic induction in the interior. Similar to the phenomenon of a Faraday cage which acts as a shield to block electric field lines, Meissner–Ochsenfeld observed that magnetic field lines are expelled from the interior by a superconducting sample of tin. The physical principle underlying the Meissner–Ochsenfeld effect is the reason behind magnetic levitation, see Figure 12.1.1. An explanation of the flux exculsion effect and perfect diamagnetic behavior of superconductors soon followed in the form of the London equations. In 1935, using electrodynamics, brothers Fritz London (1900–1954) and Heinz London (1907–1970) proposed a phenomenological theory which explained both perfect conductivity and flux exclusion.

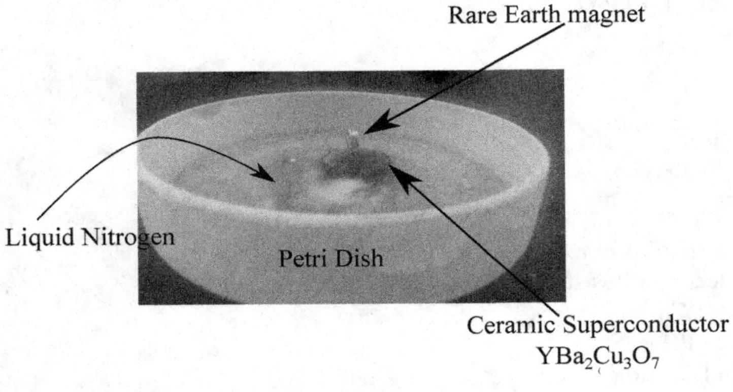


Figure 12.1.1: A levitated neodymium magnet on top of a ceramic high temperature superconductor Yttrium barium copper oxide ($YBa_2Cu_3O_7$ or YBCO). The critical temperature of YBCO is 93 K, thus allowing liquid nitrogen to be used as the cooling agent. The above picture was taken at the *2016 Department of Physics Augusta University Science Show*.

Inspired by London, in 1950, Lev Landau (1908–1968) and Vitaly Lazarevich Ginzburg (1916–2009) formulated another phenomenological approach to understand superconductivity. The approach based on thermodynamic arguments constructed the Ginzburg–Landau free energy functional. They introduced the idea of a superconducting order parameter and using their theoretical approach were able to explain the behavior of a superconductor depending on the temperature and the magnetic field. Further application of the Ginzburg–Landau approach prompted Alexei Abrikosov (1928–2017 [Source: Wikipedia]) in 1952 to predict the existence of a periodic lattice structure of magnetic flux in a class of superconductors called Type II. In such superconductors there is a range of magnetic fields in which the system is permeable to magnetic field lines in tubes of circulating supercurrents known as **vortices**. Abrikosov predicted that these vortices would arrange themselves in triangular networks in 1957, a fact which was experimentally confirmed twelve years later. Simultaneously, in 1957 John Bardeen (1908–1991), Leon Cooper (1930–), and John Robert Schrieffer (1931–) proposed the celebrated microscopic theory of superconductivity now popularly known as the BCS theory. They introduced the idea of a Cooper pair and subsequent condensation of these pairs to provide the first successful comprehensive quantum (microscopic) theory of low temperature superconductors. Now, for more than half a century, BCS theory has been successfully applied to most low temperature superconductors. In 1972 Bardeen, Cooper, and Schrieffer won the Nobel Prize in physics.

Progress in superconductivity has not stalled since BCS theory, rather, the field has been bolstered by several key discoveries listed in Table 12.1.1. In Table 12.2.2 we list some of the popular superconducting materials discovered to date. In this chapter you will learn about the basic equations and physical concepts which explain low temperature superconductivity. As the search for an ultimate room temperature superconductor continues, hopefully, after reading this chapter you will be inspired to pursue one of the most challenging unsolved issues of condensed matter physics – "A comprehensive understanding of the physical mechanism which gives rise to high temperature superconductivity" and "the quest for a room temperature superconductor".

Table 12.1.1: Milestones in experimental and theoretical discoveries in superconductivity. Conventional superconductors are materials that display superconductivity as described by the BCS theory or its extensions. However, non-BCS (unconventional) superconductors such as cuprates cannot be described by BCS theory. Conventional superconductors can be of either Type I or Type II. Niobium and vanadium are Type II, while most other elemental superconductors are Type I. The most commonly used conventional superconductor in applications is a niobium-titanium alloy; this is a type-II superconductor with a superconducting critical temperature of 11 K. The highest critical temperature so far achieved in a conventional superconductor is 39 K (-234degC) in magnesium diboride.

| | |
|---|---|
| 1908 | Liquefaction of $^4$He at 4.2 K |
| 1911 | Superconductivity discovered in mercury (Hg) at 4.15 K |
| 1933 | Meissner–Ochsenfeld effect |
| 1935 | Shubnikov phase in Type II superconductivity observed |
| 1950 | Ginzburg–Landau theory of superconductivity |
| 1957 | Bardeen Cooper Schrieffer (BCS) theory of low temperature superconductivity |
| 1957 | Abrikosov flux lattice |
| 1962 | Josephson effect |
| 1963/1964 | Anderson–Higgs mechanism |
| 1979 | Heavy fermion superconductivity discovered in the magnetic material $CeCu_2Si_2$ |
| 1980 | Organic superconductivity discovered |
| 1986 | Cuprate high temperature superconductivity (HTS) discovered by Bednorz–Müller (LBCO, 35 K) |
| 1991 | Superconductivity in alkali metal fullerides discovered |
| 2001 | Superconductivity of $MgB_2$ discovered, $T_c = 39K$ |
| 2006 | Pnictide superconductivity discovered in LaFePO, $T_c = 4K$, (Hosono group) |
| 2008 | Pnictide superconductivity discovered in LaFeAsO, $T_c = 26K$, (Hosono group) |

## 12.2 Basic Properties of a Superconductor

Much like magnetism, superconductivity is a distinct phase of matter. Most superconductors satisfy certain basic properties. Below we summarize these features.

1. **Zero electrical resistance:** An obvious and popular characteristic of a superconductor is vanishing DC electrical resistance at zero magnetic field in the superconducting state. Above the critical temperature $T_c$, the DC resistivity is finite in zero magnetic field, see Figure 12.2.2.

Table 12.2.2: Historically important low (1–4) and high-temperature (5–23) superconductors with their critical transition temperature at ambient pressure. Material 5 is an unconventional Heavy fermion superconductor and material 6 is an organic superconductor. Cuprate high temperature superconductors are listed from 7 to 11. Material 13 is an example of alkaline fulleride superconductor. The newly discovered iron based superconductors are listed from 15 to 20.

| | **Material** | **$T_c$ (Kelvin)** |
|---|---|---|
| 1. | Mercury (Hg) | 4.15 |
| 2. | Aluminum (Al) | 1.1 |
| 3. | Lead (Pb) | 9.25 |
| 4. | Niobium (Nb) | 9.25 |
| 5. | $CeCu_2Si_2$ | 0.7 |
| 6. | $(TMTSF)_2PF_6$ Bechgaard salt | 1.1 |
| 7. | Lanthanum Barium Copper Oxide $La_{2-x}Ba_xCuO_4$ (LBCO) | 35 |
| 8. | Lanthanum Strontium Copper Oxide $La_{2-x}Sr_xCuO_4$ (LSCO) | 37 |
| 9. | Yttrium Barium Copper Oxide $YBa_2Cu_3O_7$ (YBCO) | 91 |
| 10. | Bismuth Strontium Calcium Copper Oxide $Bi_2Sr_2CaCu_2O_{8+x}$ (BSCCO) | 89 |
| 11. | Hg–Ba–Ca–Cu–O $HgBa_2Ca_2Cu_3O_8$ | 134 |
| 12. | Barium Bismuthate $Ba_{1-x}K_xBiO_3$ | 30 |
| 13. | $K_3C_{60}$ | 19 |
| 14. | $MgB_2$ | 39 |
| 15. | LaFePO | 4 |
| 16. | $LaFeAsO_{0.89}F_{0.11}$ | 26 |
| 17. | $SmFeAsO_{0.9}F_{0.1}$ | 55 |
| 18. | $Ba_{1-x}K_xFe_2As_2$ | 38 |
| 19. | LaFeAs | 18 |
| 20. | FeSe | 8 |

2. **Persistent currents:** A sensitive test to display the absence of any resistance in a superconducting state is to generate a current in a superconducting ring and observe if there is any decay or not. Such a set-up is shown in Figure 12.2.3. For example, a metallic ring in the presence of an external magnetic field will enclose a quantized trapped flux when cooled below its superconducting transition temperature. If the field is now decreased to zero, the trapped flux remains and is maintained by a **persistent current** which flows around the ring (read Section 12.4 for a mathematical explanation). Experiments on persistent currents, also known as supercurrents, show that both the magnetic field and the persistent superconducting current is stable for several years. The supercurrent persists with zero applied voltage. The resistivity of a superconductor based on such measurements is shown to be less than $10^{-26}$ Ω m. Just for comparison, note that copper has a resistivity value of $1.68\times10^{-8}$ Ωm.

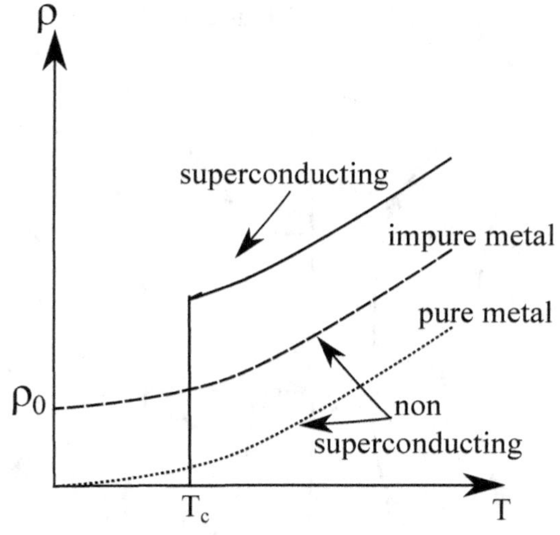


Figure 12.2.2: Schematic plot comparing resistivity versus temperature of a regular metal and a superconductor. At zero temperature, a non-superconducting normal metal attains a finite residual resistivity value $\rho_o$ controlled by the concentation of impurities. For a pure metal, the resistivity vanishes at zero temperature. But, for a superconductor, the resistivity abruptly disappears below the critical transition temperature $T_c$ even if the metallic sample may have impurities. In 1911, Kamerlingh Onnes observed such an effect in mercury.

3. **Meissner–Ochsenfeld effect (flux expulsion):** In 1933 Meissner and Ochsenfeld discovered that magnetic fields in superconducting tin (Sn) and lead (Pb) do not penetrate into the bulk of a superconductor, rather the current is confined to a surface layer of thickness $\lambda$, called the **London penetration depth**. The penetration depth is typically on the scale of tens to hundreds of nanometers.

4. **Critical fields:** Superconductivity can be destroyed by a magnetic field. The breakdown of superconductivity can happen in two possible ways. In a **Type I superconductor**, the Meissner effect exists only for $T < T_c$ with the applied field being smaller than the **critial magnetic field** $H_c(T)$. When the applied field is greater, $H > H_c(T)$, the system abruptly goes to the normal state. Examples of type I materials are Hg, Pb, Nb, and Sn. In a **type II superconductor**, there are two critical fields, $H_{c1}(T)$ and $H_{c2}(T)$. For $H < H_{c1}$, we have flux expulsion and the system is in the Meissner phase. For $H > H_{c2}$ we have uniform flux penetration and the system is in a normal state. For $H_{c1}(T) < H < H_{c2}(T)$, the system is in a *mixed state* in which quantized vortices of flux $\Phi$ penetrate the sytem. In Figure 12.2.4 we display the two possible scenarios in a field versus temperature diagram. In Table 12.3.3 we list the critical temperature and critical field values of some superconducting materials. The critical magnetic field is temperature dependent, decreasing with increasing temperature.

5. **Specific heat jump:** There is a jump in the specific heat at the critical transition temperature $T_c$. This jump is a deviation from the linear temperature dependence contribution expected from an ideal gas. Rather, there is an exponential suppression of heat capacity at low temperature (Type I superconductors). The jump is consistent with a second order (or continuous) phase transition. The presence of the exponential heat capacity behavior $C \approx \exp(-\Delta/T)$ points to the existence of an energy gap $E_g$ in the excitation spectrum. In fact, the energy

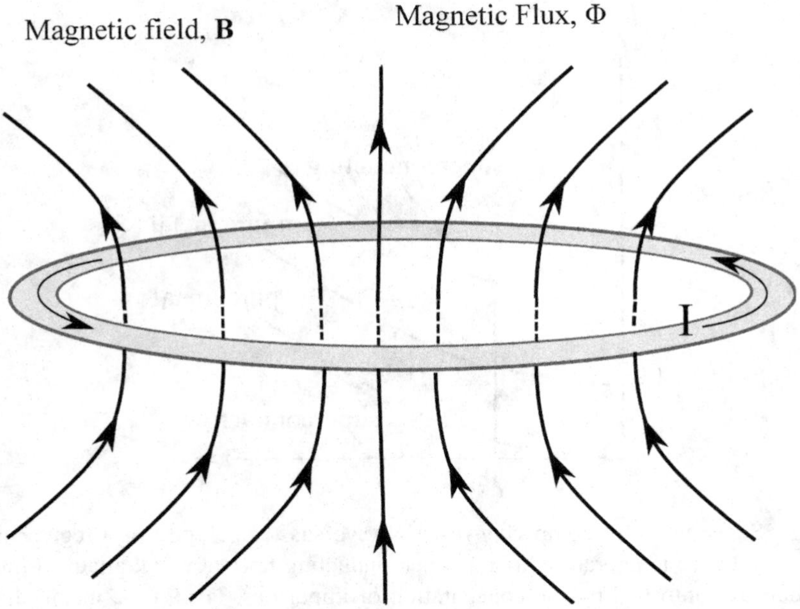


Figure 12.2.3: Persistent current in a superconducting ring.

gap is twice as large as $\Delta$. Energy gaps in Type I superconductors are of the order of meV, compared to electron volts in semiconductors or metals.

6. **Tunneling and Josephson effect:** Ivar Giaever (Norway, 1929–) and Brian Josephson (Great Britain, 1940 –) studied the effects of tunneling and superconductivity. Tunneling is a quantum mechanical phenomenon that allows particles to access classically forbidden regions. Using the idea of tunneling, Giaever measured the energy gap in superconductors. Developing upon the ideas of Giaever, Josephson, during his PhD in Cambridge at the age of only 22, proposed that supercurrents can also appear in the tunnel barrier. The tunnel barrier is the thin piece of insulation that appears in between two metals, a metal and a superconductor, or even two superconductors. Josephson predicted that in the case of a weak link between two superconductors, current can flow at zero bias voltage, a situation known as the **Josephson effect**. These effects have now been experimentally verified and led to the creation of the superconducting quantum interference device (SQUID).

## 12.3 Zero Electrical Resistance

In Figure 12.2.2 we displayed the resistivity behavior difference of a metal and a superconductor. In general the electrical resistivity of metals and alloys decreases when cooled. Upon cooling a specimen of the metal or alloy, the lattice vibrations subside and the conduction electrons experience less scattering. For a perfectly pure sample, the resistivity should approach zero as the temperature tends towards zero Kelvin. *Note*, this is not the phenomenon of superconductivity. In reality, there are no perfect samples and most specimens will have some impurities. In this case the electrons will be scattered by impurities leading to a residual resistivity as shown in Figure 12.2.2. However, there

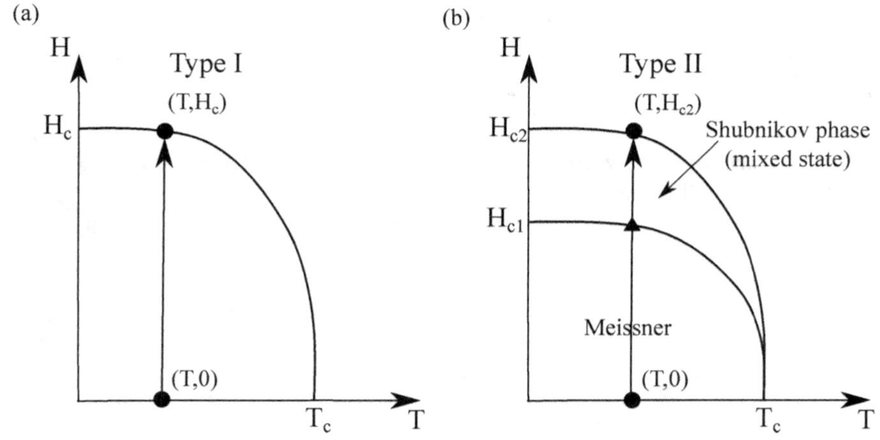


Figure 12.2.4: Field versus temperature plot of type I and type II superconductor.

are certain metals which display the remarkable behavior that their resistivity first decreases and then completely vanishes below a critical temperature value. The phase transition to the superconducting state occurs even if the metal is impure, which from our previous discussion would have resulted in a large residual resistivity. Thus, the superconducting state is not just a metal with zero resistance, but, is a distinct electronic phase of matter.

Table 12.3.3: Critical temperatures and critical fields ($H_c(0)$ or $H_{c2}(0)$) at T = 0 K for some elemental Type I and Type II superconductors. For Type II superconductors, the upper critical field value is stated. *Source*: [57]

| **Superconductor** | **Type** | **$T_c$ (Kelvin)** | **Critical field (MA $m^{-1}$)** |
|---|---|---|---|
| Hg | I | 4.15 | 0.033 |
| Pb | I | 7.19 | 0.064 |
| Nb | I | 9.25 | 0.158 |
| NbTi | II | 9.6 | 11.94 |
| $Nb_3Al$ | II | 18.7 | 25.78 |
| $Nb_3(AlGe)$ | II | 21 | 35.01 |

## 12.4 Persistent Current

The most convincing evidence that superconductors have zero resistivity is the observation of a persistent current. In experiments carried out in a coil of $Nb_{0.75}Zr_{0.25}$ with magnetically induced persistent currents and observed via NMR, it was estimated that the decay time is greater than $10^5$ years! To provide a description of persistent currents based on the laws of electrodynamics, consider the closed loop of superconducting wire shown in Figure 12.2.3. The flux $\Phi$ for the system is defined by the surface integral

$$\Phi = \int \vec{B} \cdot d\vec{S}, \tag{12.1}$$

where $d\vec{S}$ is an infinitesimal area vector element enclosed by and perpendicular to the plane of the ring. Using one of Maxwell's equation

$$\vec{\nabla} \times \vec{E} = -\frac{\partial \vec{B}}{\partial t}, \tag{12.2}$$

in Stoke's theorem

$$\int \left(\vec{\nabla} \times \vec{E}\right) \cdot d\vec{S} = \oint \vec{E} \cdot d\vec{r}, \tag{12.3}$$

we can write

$$\int \left(-\frac{\partial \vec{B}}{\partial t}\right) \cdot d\vec{S} = \oint \vec{E} \cdot d\vec{r}.$$

Now, using Equation (12.1) and interchanging the order of integration and the time derivative we have

$$-\frac{\partial \Phi}{\partial t} = \oint \vec{E} \cdot d\vec{r}. \tag{12.4}$$

The line integral in the above equation around any closed path is equal to the negative rate of change of the magnetic flux $\Phi$ through the loop. Since $\vec{E} = 0$ everywhere inside the superconductor, the integral over any closed path circling inside the superconductor is zero. Thus the integral over any closed path is equal to zero. Thus we have

$$\begin{aligned} \frac{\partial \Phi}{\partial t} &= 0, \\ \Rightarrow \Phi &= \text{const.}, \end{aligned} \tag{12.5}$$

thereby demonstrating the fact that the magnetic flux in the supeconductor is constant with time. This implies that the magnetic field must also remain constant inside the superconductor.

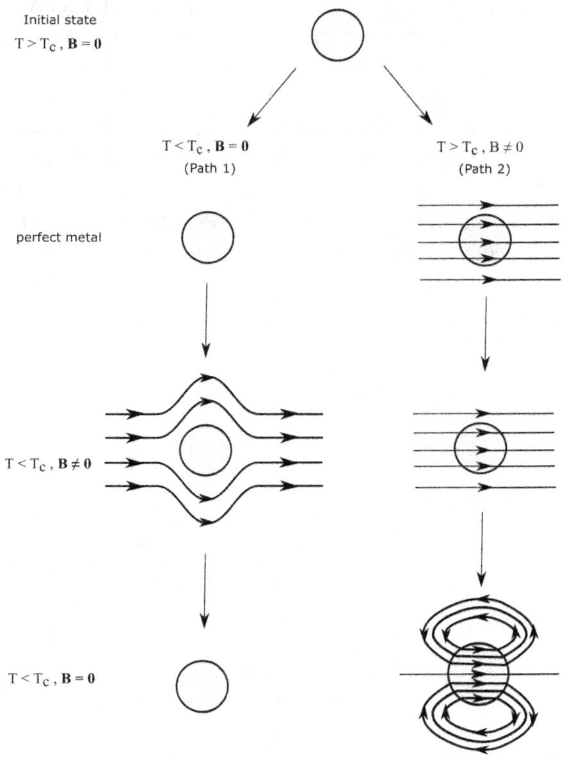


Figure 12.4.5: A perfect metal is sensitive to the presence or absence of a magnetic field as it is cooled across its transition temperature. In one scenario, path 1, it returns to its unmagnetized state. Whereas in another case, path 2, it traps the magnetic field inside it.

## 12.5 Meissner Effect

In Section 12.4 we argued that persistent currents around a closed loop of a superconducting sample ensure that flux remains constant with time. In this section we will expand upon this effect further by drawing analogy between a perfect conductor and a superconductor. By highlighting the difference between a perfect conductor and a superconductor, we can truly realize that this is indeed a true new phase of matter.

Consider the situation depicted in Figure 12.4.5. Let us assume that a sample of a metal loses its resistance in the absence of any magnetic field. We call such a specimen a perfect metal. Subsequently a magnetic field is applied (path 1). Since the metal is resistanceless, the flux density in the metal cannot change. Thus it must remain zero even after the application of the magnetic field. Thus field lines travel around the metal without penetrating it. Physically what happens is that the application of magnetic field induces resistanceless currents which circulate on the surface of the specimen. These surface currents generate a flux density that exactly cancels the flux density of the applied magnetic field everywhere inside the metal. These surface currents are often referred to as **screening currents**. If we now switch off the applied magnetic field to zero, the specimen is left in its original unmagnetized condition, as shown in the last step of path 1 in Figure 12.4.5.

Now, consider a different sequence of events as shown in path 2 of Figure 12.4.5. In path 2 we expose the sample to a magnetic field above its transition temperature, then cool the sample down so that it loses its resistance, before finally removing the applied magnetic field. It turns out this sequence of operations has a different effect on the final state of the metal. As shown in the last step of path 2, this brings about a different outcome. Since, the flux density inside the perfectly conducting metal cannot change, and persistent currents are induced on the specimen to maintain the flux inside, this leaves the specimen behind as being permanently magnetized. We conclude that the state of magnetization of a perfect conductor is not uniquely determined by the external conditions but depends on the sequence by which these conditions were arrived at.

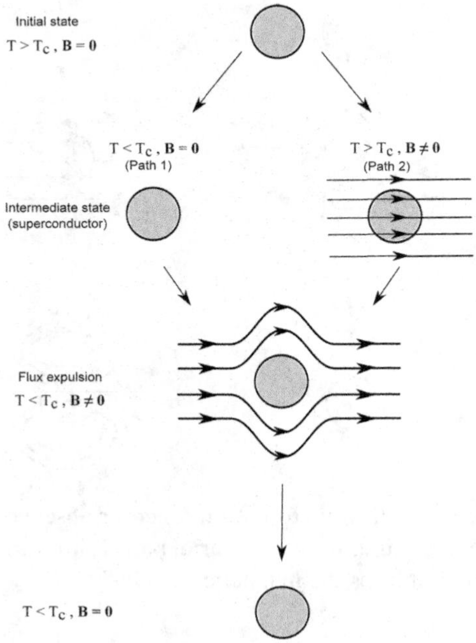


Figure 12.5.6: Demonstration of Meissner effect in a superconductor. A superconductor always expels flux irrespective of when it was exposed to the applied external magnetic field. Additionally, upon removing the field, it does not trap any magnetic field lines inside it.

However, in 1933 Meissner and Ochsenfeld discovered something remarkable when they repeated the same experiments with a superconducting material. They found that no matter what the path was, at the transition temperatures the specimens spontaneously became perfectly diamagnetic, expelling all flux inside, as in Figure 12.5.6. This experiment conclusively demonstrated that superconductors are more than just perfect metals! To compare the two systems, as shown in Figure 12.5.7, we can succinctly state that a perfect metal is a flux conserving system, whereas a superconductor is a flux expelling system. A superconductor is more than a perfect conductor, it is a perfect diamagnet.

As described above in the Meissner effect in superconductors, magnetic fields cannot be frozen in, rather, they are expelled as soon as the conductor transitions from the normal to the superconducting state. Magnetic fields can decay due to resistive losses or dissipation, both of which are irreversible processes. However, in a superconductor the screening currents flow without resistance and there is no dissipation. Furthermore, the Meissner effect in a superconductor is reversible.

The Meissner effect has an important consequence: **magnetic levitation**. Note, a permanent magnet has magnetic field lines emerging from the north pole to the south pole and loops around to enter the magnet. We know from our science show demo picture in Figure 12.1.1 that a magnet hovering over a superconducting material will levitate. The reason of course being that its magnetic field lines are repelled away from the superconducting material because the material is diamagnetic (flux expulsion). In Figure 12.5.8, we provide a step-by-step graphical explanation of the Meissner effect. In situation (a) we simply have a bar magnet in free space. This magnet is then placed on a superconducting material which is above its critical temperature and thus not superconducting. Upon cooling, by pouring liquid nitrogen on top of the YBCO sample, since the field lines emerging from the north pole of the permanent magnet cannot penetrate the diamagnetic superconductor to complete the field line loop, the magnet is forced to rise above the superconducutor to allow for the magnetic field lines to return to the south pole. That is it! It is this effect that leads to magnetic levitation. Wonderful practical technological applications in the form of superconducting magnetic levitation technology, which can minimize heating issues, to run high speed trains is in fact a reality. So, now you can see why we should learn about solid state physics and its properties. You never know when a material becomes useful for the next technological advancement.

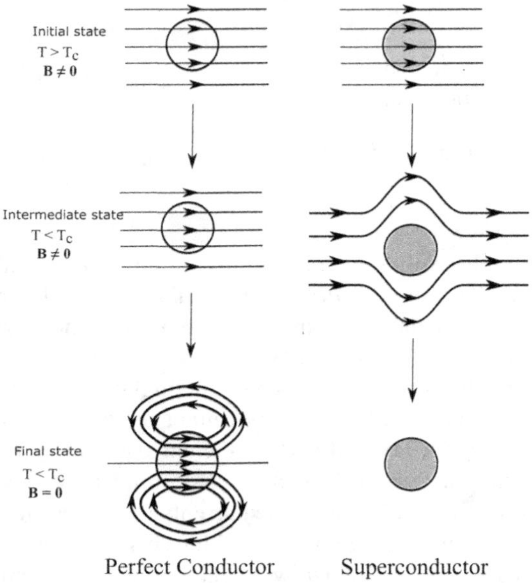


Figure 12.5.7: Differences between a perfect metal and a superconductor. The two paths yield strikingly different results. The perfect conductor is a flux preserving medium, but, the superconducting is not!

In order to preserve a zero magnetic field inside the sample, the screening currents flowing around the edges of the sample flow in a direction so as to create a magnetic field opposite to the applied external field, leaving zero total field. From Chapter 10, Equation (10.13), we know that

$$\vec{B} = \mu_o \left( \vec{H} + \vec{M} \right). \tag{12.6}$$

Now imposing the Meissner condition $\vec{B} \doteq 0$ in the above equation we have $\vec{M} = -\vec{H}$. Thus using the definition from Chapter 10, the magnetic susceptibility $\chi$ for a superconductor is given by

$$\chi = \frac{dM}{dH}|_{H=0} = -1. \tag{12.7}$$

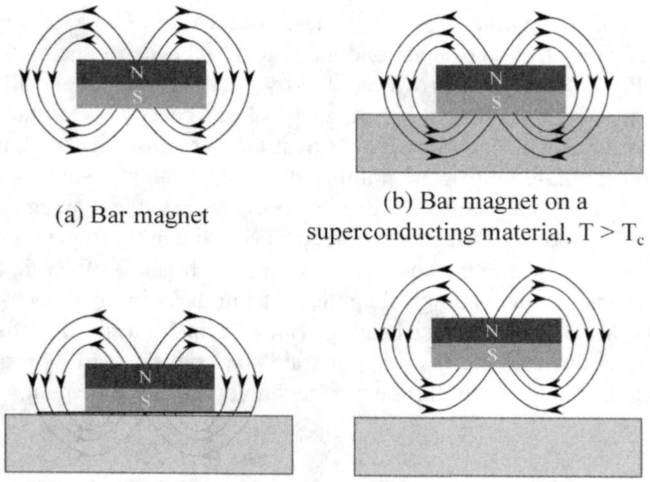


Figure 12.5.8: Meissner effect explanation.

Now recall from Chapter 10 that solids with a negative value of $\chi$ are called perfect diamagnets. From the above, we find that superconductors are **perfect diamagnets**. Note, it is possible to destroy the state of superconductivity with an applied magnetic field called the thermodynamic critical field $H_c$. A stronger applied field $H > H_c$ will destroy superconductivity.
In its perfectly diamagnetic flux expelling superconducting state, electric currents cannot flow through the body of the material. But, the currents cannot be confined entirely to the surface either. With a zero thickness sheet or layer of surface the current density would seem to be infinite, which is unphysical. In reality currents flow within a very thin surface layer whose thickness is of the order of $\mu$m - nm. This depth within which the currents flow is termed the **penetration depth**. It is typically denoted by the symbol $\lambda$. Simply speaking, it is the depth to which the flux of the applied magnetic field appears to penetrate. Drawing analogy with electrodynamics, this phenemon is similar to the concept of a "skin depth" whereby a high frequency alternating field is unable to penetrate the bulk of a metal, thereby residing only on a thin layer of the surface. The penetration depth in a superconducting metal depends on the purity, with the depth increasing as the metal becomes more impure. The existence of this flux leakage depth ensures that the flux density does not abruptly fall to zero at the boundary of the metal, but, dies away within the region where the screening currents are flowing. In Figure 12.5.9 we display this concept pictorially via a decaying magnetic field profile. In Example 12.6.0.1 you will calculate a mathematical expression for this field profile using one of the London equations.

## 12.6 London Equation

We have learned about two important properties of a superconductor. First, it is a perfect conductor. Second, it expels magnetic flux. In 1935, using electrodynamics, brothers Fritz London and Heinz London proposed a phenomenological theory of superconductivity which explained both the perfect conductivity and the flux exclusion. Their theory, inspired by the two-fluid model of superfluid $^4$He,

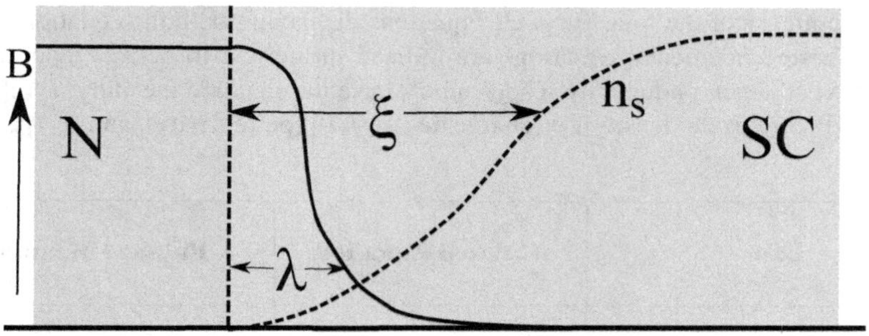


Figure 12.5.9: Variation of magnetic field strength (B) and number density of superconducting electrons ($n_s$) in the region of a boundary between normal and superconducting regions. The symbols N and SC stand for normal and superconducting, respectively. The penetration depth ($\lambda$) and the coherence length ($\xi$) concepts are explained in the text.

postulated the existence of a (i) normal fluid with concentration $n_n$ which has a finite resistivity and (ii) a superelectron fluid of concentration $n_s$. The theory also assumed that as temperature increases up to the critical value, the superfluid density drops to zero. In Figure 12.5.9 we show a profile of how the superelectron density gradually (not abruptly) changes inside a superconductor. This fact was later recognized by the British physicist Sir Alfred Brian Pippard (1920–2008) by analyzing the shortcomings of the London theory. The length scale associated with the gradual buildup of this superelectron density has a special name, it is called the **coherence length** and is denoted by the symbol $\xi$. Note, the coherence length in a superconductor is temperature dependent. It typically is of the order of $1-100$ nm.

Initally the idea of the superfluid superelectron density was not well understood conceptuallly. It was difficult to explain how electrons could combine to form a superfluid. But, regardless of that fact the London brothers' intuition put forward a theory which was successful in explaining (with some shortcomings) the basics of the observed superconducting phenomena. If we define the corresponding normal and superfluid current densities as $\vec{j}_n$ and $\vec{j}_s$, respectively, then the total number density $n$ and the total current density $\vec{j}$ are written as

$$n = n_n + n_s, \tag{12.8}$$
$$\vec{j} = \vec{j}_n + \vec{j}_s. \tag{12.9}$$

The normal fluid which carries an ohmic current is governed by

$$\vec{j}_n = \sigma_n \vec{E}, \quad \sigma_n = \frac{e^2 n_n \tau}{m}. \tag{12.10}$$

The normal fluid conductivity $\sigma_n$ is given by the Drude formula as shown in Equation (12.10). Now recall from Chapter 5, the Drude equation for conductivity

$$m\frac{d\vec{v}}{dt} = -e\vec{E} - m\frac{\vec{v}}{\tau}. \tag{12.11}$$

where $m$ is the mass of the electron (or the effective mass of the charge carriers), $e$ the charge, $\tau$ is the mean free collision time, and $\vec{E}$ is the electric field. In a superconductor, there is no scattering term. Thus ($\tau \to \infty$) makes the second term on the right-hand side of the Drude equation drop off to zero. We can then replace the velocity of the charge carriers with a superfluid velocity $v_s$. The equation becomes

$$m\frac{d\vec{v}_s}{dt} = -e\vec{E}, \tag{12.12}$$

Table 12.6.4: Summary of the four Maxwell Equations displaying both the equation and its physical meaning. These fundamental equations are utilized in the text to derive important formulae within the context of superconductivity. The symbols have the standard meaning: $\vec{E}$ (electric field), $\vec{B}$ (magnetic field), $\vec{J}$ (current density), $\rho$ (charge density), $\varepsilon_0$ (permittivity), and $\mu_o$ (permeability).

| Law | Maxwell Equation | Physical Meaning |
|---|---|---|
| Gauss's Law (Electric field) | $\vec{\nabla} \cdot \vec{E} = \frac{\rho}{\varepsilon_0}$ | Electric flux is proportional to charge inside volume. |
| Gauss's Law (Magnetic field) | $\vec{\nabla} \cdot \vec{B} = 0$ | Magnetic flux is zero in an enclosed volume. |
| Faraday's Law of Induction (Maxwell–Faraday Law) | $\vec{\nabla} \times \vec{E} = -\frac{\partial \vec{B}}{\partial t}$ | Induced electric field is proportional to changing magnetic flux. |
| Ampére's law | $\vec{\nabla} \times \vec{B} = \mu_o \left( \vec{J} + \varepsilon_o \frac{\partial \vec{E}}{\partial t} \right)$ | Magnetic field is proportional to electric and displacement currents. |

or,

$$\frac{d\vec{v}_s}{dt} = -\frac{e}{m}\vec{E}. \tag{12.13}$$

Assuming that the superfluid electron density $n_s$ is constant in space and time, we multiply both sides of Equation (12.12) by $(-n_s e)$ to recast the equation in terms of current density (for superfluid electrons) defined by

$$\vec{j}_s = -n_s e \vec{v}_s, \tag{12.14}$$

to obtain

$$\begin{aligned} \frac{d(-n_s e \vec{v}_s)}{dt} &= \frac{n_s e^2}{m}\vec{E}, \\ \frac{\partial \vec{j}_s}{\partial t} &= \frac{n_s e^2}{m}\vec{E}. \end{aligned} \tag{12.15}$$

Equation (12.15) is referred to as the *first* **London equation** which provides a phenomenological explanation of perfect conductivity in superconductors.

The *second* **London equation** follows from the perfect diamagnetism condition, or the Meissner effect. To derive this equation, take the curl of Equation (12.15) to obtain

$$\frac{\partial(\vec{\nabla} \times \vec{j}_s)}{\partial t} = \frac{n_s e^2}{m}\vec{\nabla} \times \vec{E}, \tag{12.16}$$

which using the third Maxwell equation (Faraday's Law of Induction) from Table 12.6.4 becomes

$$\frac{\partial(\vec{\nabla} \times \vec{j}_s)}{\partial t} = -\frac{n_s e^2}{m}\frac{\partial \vec{B}}{\partial t}. \tag{12.17}$$

Integrating the above with respect to time yields

$$\vec{\nabla} \times \vec{j}_s = -\frac{n_s e^2}{m}\vec{B} + \vec{C}(\vec{r}), \tag{12.18}$$

where the last term represents a constant of integration at each point $\vec{r}$ inside the superconductor. For a superconducting sample in zero applied magnetic field, we have $\vec{j}_s = \vec{0}$ and $\vec{B} = \vec{0}$. This implies $\vec{C}(\vec{r}) = 0$. For the Meissner-Ochsenfeld effect to be explained we have to consider the case when the body becomes superconducting in a non-zero applied field. However, within the first London equation we have assumed a constant superfluid density. To account for flux expulsion, the London theory postulated that $\vec{C}(\vec{r}) = 0$ regardless of the history of the system. Thus we have

$$\vec{\nabla} \times \vec{j}_s = -\frac{n_s e^2}{m}\vec{B}, \tag{12.19}$$

which gives the second London equation.
Combining Equation (12.19) and the fourth Maxwell relation for the case of static fields (that is ignoring the displacement current term) we have

$$\vec{\nabla} \times (\vec{\nabla} \times \vec{B}) = \mu_o \vec{\nabla} \times \vec{j}_s = -\frac{\mu_o n_s e^2}{m}\vec{B} = -\frac{1}{\lambda^2}\vec{B}, \tag{12.20}$$

where $\lambda$ has the dimensions of length and is called the **London penetration depth** defined by

$$\lambda = \sqrt{\frac{m}{\mu_o n_s e^2}}. \tag{12.21}$$

Now using the vector identity

$$\vec{\nabla} \times (\vec{\nabla} \times \vec{B}) = \vec{\nabla}\left(\vec{\nabla} \cdot \vec{B}\right) - \vec{\nabla}^2 \vec{B}, \tag{12.22}$$

and the second Maxwell equation $\vec{\nabla} \cdot \vec{B} = 0$, we obtain a simple form of the second equation,

$$\vec{\nabla}^2 \vec{B} = \frac{1}{\lambda^2}\vec{B}. \tag{12.23}$$

Now consider the example below as an application of the second London equation.

**Example 12.6.0.1**
Let us consider a slab (see Figure 12.6.10) of superconductor filling the half space $x > 0$. A magnetic field $\vec{B}_a$ is applied parallel to the surface. Using the second London equation, obtain a profile for the field versus spatial dependence.
**Solution**
To obtain the magnetic field profile inside the superconductor, we apply the London Equation (12.23) in the x-direction. We then have the one-dimensional form of the London equation as

$$\frac{\partial^2 B(x)}{\partial x^2} = \frac{1}{\lambda^2}B(x), \tag{12.24}$$

where $B(x)$ is the magnetic field inside the superconducting metal. Let us assume that the solution is of the form

$$B(x) = C_1 \exp(C_2 x). \tag{12.25}$$

Then differentiating and substituting into the one-dimensional form of the London equation, we have

$$\left(C_2^2 - \frac{1}{\lambda^2}\right) C_1 \exp(C_2 x) = 0. \tag{12.26}$$

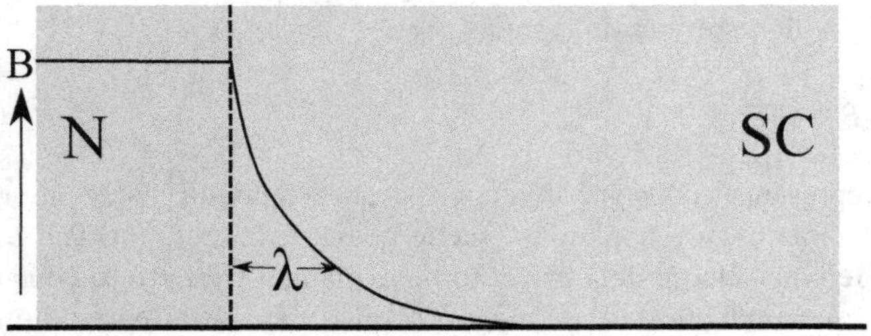


Figure 12.6.10: A superconducting slab with an external field applied parallel to its surface. N represents the normal outside region. SC refers to superconducting.

Since the exponential cannot be zero, we must set the term inside the bracket equal to zero to obtain

$$C_2 = \pm\frac{1}{\lambda}. \tag{12.27}$$

Thus the two possible solutions are

$$B(x) = C_1 \exp(x/\lambda) \ \text{ or } \ B(x) = C_1 \exp(-x/\lambda). \tag{12.28}$$

We know from our knowledge of Meissner effect that the magnetic field inside the superconductor does not increase. Thus the first expression cannot be a solution. Hence, the correct form of the solution must be $B(x) = C_1 \exp(-x/\lambda)$. To obtain the unknown constant $C_1$, note the boundary condition $B(0) = B_a$. Thus, the magnetic field profile inside the superconductor must be of the form

$$B(x) = B_a \exp(-x/\lambda). \tag{12.29}$$

The above solution agrees with our intuition of flux expulsion. Furthermore, it clearly demonstrates that the magnetic penetration depth length is set by $\lambda$. The London equation predicts, therefore, an exponential decay of the flux density at the surface of a superconductor. The London penetration depth can also be rewritten in another fashion where we can express it as the ratio, that is,

$$\lambda = \sqrt{\frac{m\varepsilon_o c^2}{n_s e^2}} = \frac{c}{\omega_{ps}}, \tag{12.30}$$

where we have used $c = \frac{1}{\sqrt{\varepsilon_o \mu_o}}$ and defined the plasma frequency in the superconducting state $\omega_{ps} = \sqrt{\frac{n_s e^2}{m\varepsilon_o}}$. For electron concentrations of a typical metal, the penetration depth is of the order of $10^2$-$10^3$ Å.

The London equation can also be rewritten in terms of the magnetic vector potential $\vec{A}$ defined by

$$\vec{B} = \vec{\nabla} \times \vec{A}. \tag{12.31}$$

Thus we can write

$$\begin{aligned} \vec{\nabla} \times \vec{j}_s &= -\frac{n_s e^2}{m}\vec{B}, \\ \vec{\nabla} \times \vec{j}_s &= -\frac{n_s e^2}{m}\left(\vec{\nabla} \times \vec{A}\right), \end{aligned}$$

or,

$$\vec{j}_s = -\frac{n_s e^2}{m}\vec{A} = -\frac{1}{\mu_o \lambda^2}\vec{A}, \tag{12.32}$$

where we have used Equation (12.21). The above equation works only if we make an appropriate choice of **gauge**. A gauge is similar to an unrestricted degree of freedom. For example, from Equation (12.31) we observe that any choice of $\vec{A} = \vec{A} + \vec{\nabla}\theta$ would equally well satisfy $\vec{B}$ since the curl of a gradient is zero. Thus $\vec{A}$ is not uniquely defined and the $\vec{\nabla}\theta$ (the gauge degree of freedom) would allow for different possible functions to satisfy the magnetic field expression. We conclude that $\vec{A}$ *is not* gauge invariant. But, note that the current density *is* gauge invariant. It obeys the continuity equation given by

$$\frac{\partial \rho}{\partial t} + \vec{\nabla}\cdot\vec{j} = 0, \tag{12.33}$$

where $\rho$ is the charge density. Thus using Equation (12.32) the only valid choice of a gauge is one that satisfies the condition

$$\vec{\nabla}\cdot\vec{A} = \mu_o \lambda^2 \frac{\partial \rho}{\partial t}. \tag{12.34}$$

In the case of a time-independent situation the charge density is independent of time and we have the familiar form of the **London gauge**

$$\vec{\nabla}\cdot\vec{A} = 0. \tag{12.35}$$

## 12.7 Thermodynamics of Superconductors

This section is devoted to understanding the normal to superconducting phase transition region and the specific heat jump in superconductors. We begin the discussion by writing the first law of thermodynamics of a magnetic system system as

$$dU = TdS + \mu_o VHdM, \tag{12.36}$$

where $V$ is the volume, $M$ is the magnetization, $S$ is the entropy, and $H$ is the auxilliary field. As in Chapter 10 we define the Helmholtz and Gibbs-free energy of a superconductor in a magnetic field as

$$F(T,M) = U - TS, \tag{12.37}$$

$$G(T,H) = U - TS - \mu_o VHM. \tag{12.38}$$

Since there is no entropy meter to measure entropy change (dS), directly we cannot conveniently analyze internal energy to study the nature of the superconducting phase transition. Also, in a circuit the free current establishes the auxilliary field over which we have direct control and not magnetization. Thus, amongst the thermodynamic variables, temperature and auxilliary field can be controlled. We thus focus on Gibbs free energy and its change given by

$$dG = -SdT - \mu_o VMdH. \tag{12.39}$$

To compute the free energy difference between the superconducting and the normal state we begin by considering the energy difference in the superconducting state itself. For a Type I superconductor which exhibits the Meissner–Ochsenfeld effect (flux expulsion) we know that $M = -H$. Using the condition for perfect diamagnetism, we integrate Equation (12.39) along the vertical path shown in Figure 12.2.4(a). Since dT = 0, using Equation (12.39), we have in the superconducting phase

$$\begin{aligned}\int_{G_s(T,0)}^{G_s(T,H_c(T))} dG &= -\int_0^{H_c(T)} \mu_o V M dH,\\ G_s(T,H_c(T)) - G_s(T,0) &= -\int_0^{H_c(T)} \mu_o V(-H)dH,\\ G_s(T,H_c(T)) - G_s(T,0) &= \frac{1}{2}\mu_o [H_c(T)]^2 V.\end{aligned} \tag{12.40}$$

Now, in the normal state magnetization is zero (ignoring any negligible weak metallic paramagnetism and diamagnetism contribution). Thus, if we had carried out the above analysis for the normal state below the critical field, we would have obtained

$$G_n(T,H_c) - G_n(T,0) = 0. \tag{12.41}$$

Additionally, note that at the critical field the normal state and the superconducting state are in thermodynamic equilibrium. Thus

$$G_s(T,H_c) = G_n(T,H_c). \tag{12.42}$$

Combining Equations (12.40), (12.41), and (12.42), we find that at zero applied external field

$$G_s(T,0) - G_n(T,0) = -\frac{1}{2}\mu_o [H_c(T)]^2 V. \tag{12.43}$$

The Gibbs energy for the superconducting state is lower at zero field making it the more stable phase. The quantity $\frac{1}{2}\mu_o [H_c(T)]^2 V$ is referred to as the **condensation energy**. The condensation energy is a measure of the gain in free energy in the superconducting state compared with the normal state at the same temperature.

**Example 12.7.0.1**
Calculate the condensation energy per unit volume of mercury (Hg) at T = 0 K.
**Solution**
The formula for condensation energy per unit volume at T = 0 K is given by $\frac{1}{2}\mu_o [H_c(0)]^2$. From Table 12.3.3, we find that the critical field is given by 0.033 M A m$^{-1}$. Thus, inserting values for the permeability $\mu_o = 4\pi \times 10^{-7}$NA$^{-2}$ into the formula we find 0.67 KJ m$^{-3}$.

The concept of condensation energy can be extended to include Type II superconductors. However, as evident from Figure 12.2.4, the field $H_c$ has no meaning since the real transitions occur at $H_{c1}$ and $H_{c2}$. In the remainder of this section, we will derive an expression for the heat capacity jump which was earlier listed as one of the basic properties.

**Example 12.7.0.2**
Using Equation (12.39), derive an expression for the difference in entropy between the normal and the superconducting state of a Type I superconductor at the critical field $H_c(T)$ phase boundary.
**Solution**
Using Equation (12.39) in the Type I superconducting phase for dT = 0 and integrating up to a field

$H$ less than $H_c$ we have

$$\int_{G_s(T,0)}^{G_s(T,H)} dG = -\int_0^H \mu_o VMdH,$$

$$G_s(T,H) - G_s(T,0) = -\int_0^H \mu_o V(-H)dH,$$

$$G_s(T,H) - G_s(T,0) = \frac{1}{2}\mu_o H^2 V. \tag{12.44}$$

Similarly, for the normal state we can write as in Equation (12.41), $G_n(T,H) = G_n(T,0)$. Now, using the expression for $G_s(T,0)$ from Equation (12.43) in Equation (12.44) we have

$$G_s(T,H) = G_n(T,0) - \frac{1}{2}\mu_o[H_c(T)]^2 V + \frac{1}{2}\mu_o H^2 V, \tag{12.45}$$

which can be rearranged to give

$$G_s(T,H) - G_n(T,H) = \frac{1}{2}\mu_o V\left\{H^2 - [H_c(T)]^2\right\}, \tag{12.46}$$

where we have used the condition $G_n(T,H) = G_n(T,0)$.
Finally, to derive the expression for entropy, we resort back to Equation (12.39) from which we have

$$S = -\left(\frac{\partial G}{dT}\right). \tag{12.47}$$

Using the above entropy formula at $H = H_c(T)$ and applying to Equation (12.46) we have

$$-\left(\frac{\partial}{dT}[G_s(T,H) - G_n(T,H)]\right)_{H=H_c(T)} = -\frac{1}{2}\mu_o V\frac{\partial}{dT}\left\{H^2 - [H_c(T)]^2\right\},$$

$$S_s(T,H_c) - S_n(T,H_c) = \mu_o VH_c(T)\frac{dH_c(T)}{dT},$$

$$S_n(T,H_c) - S_s(T,H_c) = -\mu_o VH_c(T)\frac{dH_c(T)}{dT}. \tag{12.48}$$

Since $\frac{dH_c}{dT}$ is negative (see Figure 12.2.4), the entropy of the normal state is always greater than that of the superconducting state. This implies that the there is more ordering in the superconducting phase than in the normal state.

Using Example 12.7.0.2, we can compute the difference $\Delta C$ in heat capacity in the superconducting and normal states to be given by

$$\Delta C = C_s - C_n = T\frac{d}{dT}(S_n - S_s) = \mu_o VTH_c\frac{d^2H_c}{dT^2} + \mu_o VT\left[\frac{dH_c}{dT}\right]^2, \tag{12.49}$$

which for $T = T_c$ and $H_c = 0$ we have

$$\Delta C = \mu_o VT_c\left(\frac{dH_c}{dT}\right)^2. \tag{12.50}$$

Note, the above equation holds for zero applied field. From the entropy expression, Equation (12.48), we find that at $H_c = 0$ there is no latent heat of transition, that is $\Delta S = 0$. From Equation (12.50) we find that there is a discontinuity in the heat capacity. Thus the transition at $T = T_c$ (with $H_c = 0$) is of second order, but away from $T_c$, the phase transition has a latent heat and is a first-order phase transition.

## 12.8 Type I and Type II Superconductors

In Section 12.2 we stated that there are some superconductors called Type II where in the intermediate state the system allows partial penetration of the magnetic field. That such materials existed was known experimentally for a while although without proper theoretical understanding. The first inkling towards the existence of such a class of superconductors was provided by the behavior of superconductivity, in alloys and impure metals, which could not be explained within the traditional Type I theory. In 1957 Alexei A. Abrikosov (1928–2017) published a theoretical proposal precisely predicting the existence of a mixed state superconductor with inherent physical properties now classified as Type II superconductivity. After Abrikosov's insightful theory, it was realized that the anomalous properties of these second-class superconductors is not just a trivial impurity effect. In the next few paragraphs using our understanding of the thermodynamics equations in the superconducting state (see Section 12.7), the concept of coherence length, and the idea of penetration depth, we will formulate a conceptual understanding of how and why exactly such a mixed state can occur. The key to explaining the occurrence of Type I or Type II superconductivity lies in examining how the free energy varies at the surface between the normal and the superconducting state. We know from our discussions in Sections 12.5 and 12.6 that the system does not abruptly change from being normal to fully superconducting when the magnetic field starts to enter the sample. Rather, the field leaks into the material over a distance governed by the penetration depth and the corresponding superconducting electron density builds up in the region over a distance spanning the coherence length. Thus, there is a delicate energy balance at the surface dictated by the rate of growth of the coherence length $\xi$ and the penetration depth $\lambda$. Hence the surface free energy $\mathscr{F}_s$ is given by

$$\mathscr{F}_s = \frac{1}{2}\mu_o[H_c(T)]^2(\xi - \lambda)V, \tag{12.51}$$

where the condensation energy contribution Equation (12.43) is multiplied with the spatially dependent $\xi$. Furthermore, the magnetic energy contribution is modulated by $\lambda$. Note, if superelectrons are formed close to the surface, then an amount of energy corresponding to the condensation energy is gained by the superconducting system. However, if normal electrons are present, the system loses

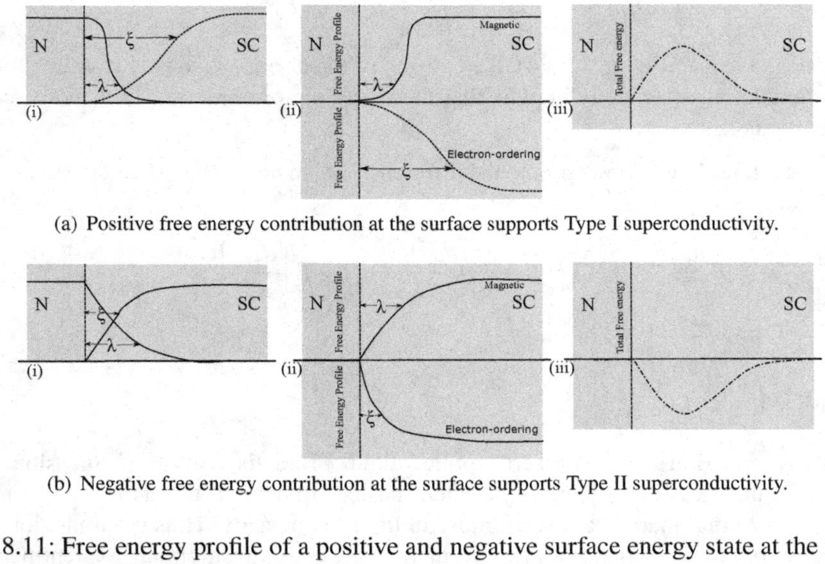


(a) Positive free energy contribution at the surface supports Type I superconductivity.

(b) Negative free energy contribution at the surface supports Type II superconductivity.

Figure 12.8.11: Free energy profile of a positive and negative surface energy state at the normal (N) and superconductor (SC) boundary.

this energy to the surface which corresponds to a gain in the surface energy term. Thus the first expression in Equation (12.51) has a positive contribution. On the other hand, the magnetic field contribution leads to a lowering of the condensation energy at the interface. Thus it is subtracted. The difference

$$\delta = \xi - \lambda, \tag{12.52}$$

is called the **wall-energy parameter**.

Table 12.8.5: Coherence length $\xi$ and penetration depth $\lambda$ for various types of superconductors. Data compiled from chapter 9 of *Handbook of Superconductivity* [58].

| **Superconductor** | **$T_c$ (Kelvin)** | **$\xi$ (nm)** | **$\lambda$ (nm)** | **$\kappa$ (GL parameter)** |
|---|---|---|---|---|
| Al | 1.18 | 1550 | 45 | 0.03 |
| Sn | 3.72 | 180 | 42 | 0.23 |
| Nb | 9.25 | 39 | 52 | 1.3 |
| NbTi | 9.6 | 3.8 | 130 | 27 |
| $UPt_3$ | $< 1$ | 11.1 | 600 | 54 |
| $K_3C_{60}$ | 19.4 | 2.8 | 240 | 92 |
| $La_{2-x}Sr_xCuO_4$ | 37 | 2.0 | 200 | 100 |
| $YBa_2Cu_3O_{7-\delta}$ | 91 | 1.65 | 156 | 95 |
| $Bi_2Sr_2CaCu_2O_{8+x}$ | 89 | 1.8 | 250 | 139 |

In Figure 12.8.11(a) we show the case when the coherence length builds up slowly compared to the penetration length. From the figure it is clearly evident that there is a positive free energy contribution at the surface when $\xi > \lambda$. Thus under this condition if a normal state was created inside the superconducting sample, the free energy would increase. Hence, a normal region is not possible. But, there exists another scenario where $\xi < \lambda$, in which case the free energy becomes negative as shown in Figure 12.8.11(b). Under this scenario the surface free energy is negative and the system can gain energy by allowing for the existence of the normal state within the superconducting volume. Thus by applying an external magnetic field, energy would be released when the interfaces are formed and the magnetic flux could penetrate the material. This is what gives rise to Type II superconductivity with a mixed state, see Figure 12.8.12. The former condition gives a Type I superconductor. Typical examples of Type I materials are Hg, Al, Sn, and In. Type II superconductors include $Nb_3Sn$, NbTi, all high $T_c$ cuprates, fullerenes, $MgB_2$, and iron-based superconductors. In Table 12.8.5 we list a collection of Type I and Type II superconductors.

In most pure metals, the coherence length is quite large compared to the penetration depth. Thus, they are of Type I. However, impurities in a metal can reduce the coherence range making it substantially less than the penetration depth, thereby creating the negative surface energy possibility. Before Abrikosov's prediction, the possibility of a negative surface energy normal superconductor boundary was never appreciated. Alloys or sufficiently impure metals are thus typically Type II superconductors. The arguments presented above were mainly conceptual, derived from a purely thermodynamic perspective. A more comprehensive approach, beyond the scope of this textbook, to understanding the origins of the two different classes of superconductivity is provided by the Ginzburg-Landau theory of superconductivity. You may recall from Chapter 10, Problem 10.19, about the Landau theory of phase transition. The Ginzburg-Landau theory is a more advanced application of the same concept. The theory introduces a **Ginzburg-Landau parameter** $\kappa$ defined as

$$\kappa = \lambda/\xi, \tag{12.53}$$

which classifies the two types of superconductors as

$$\text{Superconductor Type}: \begin{cases} \kappa < 1/\sqrt{2}, & \text{Type I} \\ \kappa > 1/\sqrt{2}, & \text{Type II} \end{cases} \tag{12.54}$$

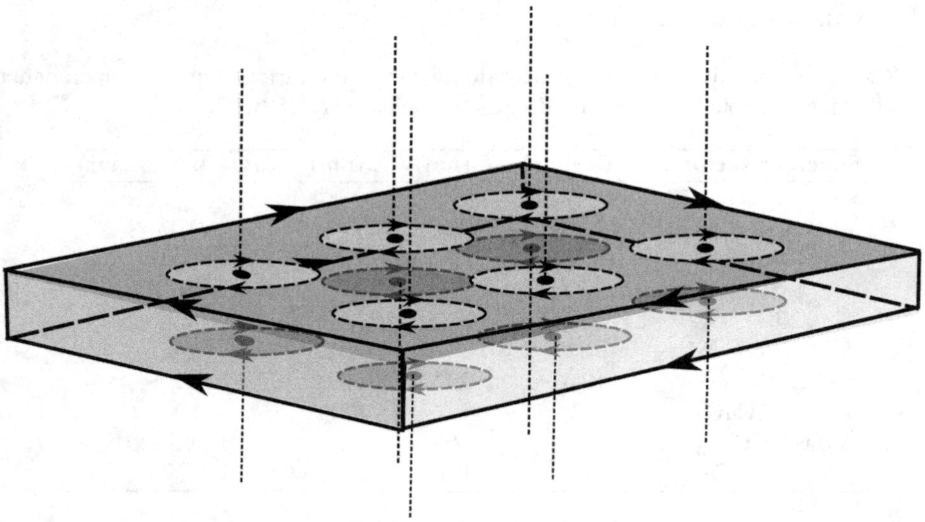

Figure 12.8.12: Mixed state in a Type II superconductor. The superconducting state is penetrated by an array of vortices. Around the vortex core, the system is in its normal state allowing for flux penetration. Beyond a certain length, the system expels flux as in the Meissner state.

The physical effects of magnetic field penetration and its subsequent effects on magnetization of the superconducting material is shown in the magnetization (M) versus field (H) plots in Figure 12.9.13. For a Type I superconductor the slope of the line is negative one, displaying (perfect) diamagnetic behavior. In reality a long, thin needle-like specimen, which has a zero demagnetization factor (recall from Chapter 10), with an applied field parallel to its axis can display this perfectly diamagnetic behavior. The magnetization can be tracked experimentally using a pickup coil. But, for a Type II superconductor, the Meissner state exists only up to a lower critical field value of $H_{c1}$. Beyond this point as the magnetic field lines begin to partially penetrate the sample, the magnetization is reduced before being completely destroyed above the upper critical field of $H_{c2}(T)$. The mixed normal-superconducting state that exists between $H_{c1}$ ¡ $H$ ¡ $H_{c2}$ is called the *Shubnikov phase* in honor of the Russian physicist Lev V. Shubnikov (1901–1937), who along with his collaborators discovered this phenomenon experimentally in 1935.

## 12.9 Vortices in Superconductors

We have learned that in the mixed state of a Type II superconductor a mixture of normal and superconducting zones is energetically favored due to the formation of a negative surface energy, see Figure 12.8.12. These normal regions allow the magnetic field penetration in the form of thin filaments usually called flux lines, fluxons, fluxoids or **vortices**. The vortices, also known as Abrikosov vortices, which can be visualized as a swirl of electrical current associated with this state. Within the vortex state the material surrounding these normal regions can have zero resistance and have partial flux penetration. In the center of each vortex, superconducting behavior does not exist. As the strength of the external field increases, the number of vortex filaments increase until the field reaches the upper critical value $H_{c2}$. At this transition point, the filaments crowd together and amalgamate so that the entire material goes into a normal state.

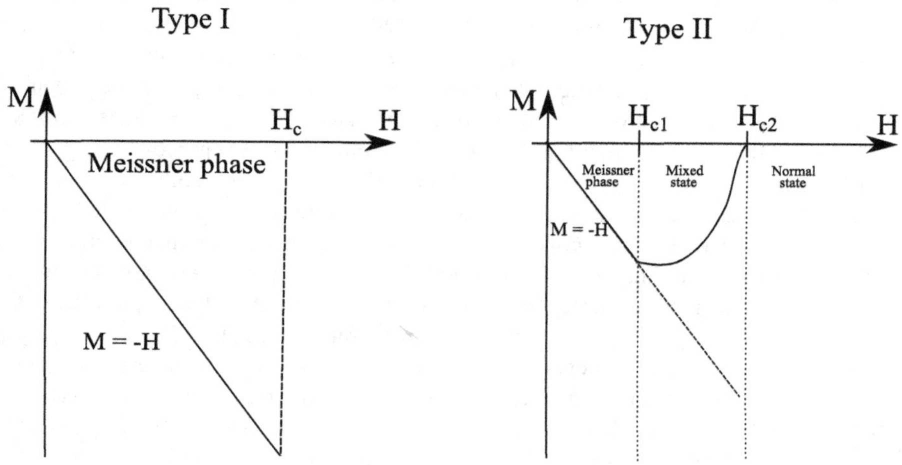


Figure 12.9.13: (a) Type I and (b) Type II superconductors.

Abrikosov predicted that each vortex or flux line carries a single fixed quantized flux number given by

$$\Phi_0 = h/2e = 2.07 \times 10^{-15}\ \mathrm{T\ m^2}, \tag{12.55}$$

where $h$ is the Planck's constant and $e$ is the magnitude of the electronic charge. In a homogeneous superconductor, these vortices are typically arranged in the form of a triangular lattice. Magnetic flux lines can be imaged using neutron diffraction, magneto-optical methods, and decoration techniques. In fact, the very first experimental detection of the fluxoid pattern was done by Essmann and Träuble using a decoration technique.

## 12.10 Technological and Scientific Applications of Superconducting Materials

To create a strong magnetic field which permeates a large volume of space such as in superconducting wires and does not decay in space, we need to use electromagnets with large values of current. But, a high value of current in the coil comes with a price – loss of energy due to Joule heating. One way to avoid this problem, for example, is to cool the wire with water, but, this approach is expensive and not very convenient. However, the use of a superconducting wire which does not resist the flow of current and hence curbs joule dissipation is the perfect way to tackle the issue of heating. The superconducting wires are often made from niobium and titanium alloys (NbTi) or niobium and tin ($Nb_3Sn$). Using coils with several thousand turns of superconducting wires, plunged in liquid helium, strong magnetic fields can be created which are used for Magnetic Resonance Imaging (MRI) devices in hospitals; for Nuclear Magnetic Resonance (NMR) experiments in chemistry and in physics; in particle accelerators to make particles deflect; or even on board a magnetically levitated train called a Maglev.

One of the most obvious application of superconductivity is for low-loss electrical power transmission. Since a substantial amount of electrical energy is lost during transmission of current through normal conductors, by replacing the cables with superconductors, in principle all the dc losses could be eliminated. This is one of the pursuits that still keeps research in the area of superconductivity alive. The concept of magnetic levitation has been used to test superconducting Maglev (*mag*netic *lev*itation) trains. At present, the fastest train in the world is found in Japan, which does not touch the tracks and uses superconductors. The Maglev train levitates a few inches above the tracks based on technology that uses strong magnetic fields created by superconducting coils located and cooled on board. Although not marketed yet, the prototype has been tested with a record speed of 581 km/h. Economic issues limit the development and wide-scale deployment of superconducting Maglevs.

As mentioned earlier, a SQUID is an electronic system that uses a superconducting ring in which one or two small insulating layers have been inserted. This device based on the Josephson effect in the superconductor-insulator-superconductor sandwich and on the flux quantization in the ring makes it ultra-sensitive to any magnetic field. Squids are hence the most efficient systems to measure magnetic fields with great accuracy, even the weaker ones. For instance, squids are sensitive enough to measure the magnetic activity of the human brain in real time, enabling very precise magnetoencephalographic measurements. Squids are used whenever very powerful magnetometers are needed: in physics, archaeology, and geology. The exceptional properties of Josephson junction circuits might some day replace silicon transistors, enabling computers to reach processing rates of 100 GHz. There are theoretical proposals to use superconductivity to build a quantum computer, enabling massively parallel processing.

Superconductivity has had a profound effect in the field of medical applications. Currently, there are a couple of devices, Magnetic Resonance Imaging (MRI) and MagnetoEncephaloGraphy (MEG) which are based on the use of superconducting properties. Using MRI, it is possible to view inside the human body with virtually no ill effect, contrary to x-ray scanners for instance. At present, MRIs are routinely used to diagnose tumours, sclerosis and oedemas. To do that, the *Nuclear* MRI machine is helped by the NMR effect described in detail in Chapter 10 of this book. To generate strong magnetic fields, with minimal heating effects, the coil is made of a superconducting wire plunged in a very cold liquid such as helium. There is no electric resistance, hence no heating. In addition to that, once the magnetic field has been created, the coil can be closed. The current (and hence the magnetic field) keeps flowing since there is no resistance. In addition to visualizing the organs, superconductors can assist with mapping brain activity. Due to the electrical activity of the brain, it generates magnetic fields of the order of picotesla. However, it is possible to use a MEG to measure and probe these fields to obtain "real time" brain activity data. In order to achieve that, SQUIDS which are ultrasensitive magnetic sensors are placed on the surface of the patient's brain. MEGs can be used to study normal and abnormal behaviours of the brain with a time resolution of about a millisecond. Similar to MRI, it is a non-invasive and harmless approach to medical diagnoses. It can be used to diagnose normal and abnormal areas of the brain, thereby helping to treat, for example, epilepsy.

The electric field that accelerates beams of charged particles (electrons, protons, ions) in a particle accelerator is produced by radio-frequency resonant cavities. The magnetic field that guides and focuses the charged objects is produced by electromagnets. Since superconducting cables cut down on transmission lose the large hadron collider (LHC) of the CERN in Geneva uses several thousand superconducting magnets spread on the 27-km circumference to produce a magnetic field four times higher than classical electromagnets, with an electric intake ten times smaller (considering the power consumed by the cryogenic cooling device). At present, superconductivity has become a key technology for particle accelerators. In addition, superconductors are often used in radiation sensors called **bolometers.** These sensors work at a very low temperature and are very sensitive tools to study extremely weak radiations, such as the background radiation of the Universe at 3 K, for instance. Bolometers are used in many astrophysics and astroparticle physics experiments. Finally, experiments conducted in search of dark matter in the universe make use of thermometers made of superconducting materials.

## 12.11 Chapter 12 Exercises

12.11.1. In words, explain the concepts of critical temperature and critical field as applied to superconductors.

12.11.2. What does the London penetration depth in a superconductor mean?

12.11.3. What is the fundamental difference between a perfect conductor and superconductor both of which can have zero resistance?

12.11.4. What is the physical property of a superconductor that causes magnetic levitation?

12.11.5. The first London equation provides a phenomenological explanation of which property of a superconductor?

12.11.6. The superelectron density inside a superconductor builds up gradually rather than abruptly at the normal-superconductor interface. What is the name given to this length scale?

12.11.7. In Example 12.6.0.1 we calculated the magnetic field profile of a superconductor occupying half the space. Using the expression for the calculated field, Equation (12.29), obtain an expression for the surface supercurrent. Such currents are also known as **Meissner currents.**

12.11.8. Calculate (a) the magnetic field profile and (b) the Meissner current of a superconducting slab of finite thickness $d$, see Figure 12.11.14. Recall, in Example 12.6.0.1 we calculated the magnetic field profile of a semi-infinite superconductor occupying half the space.

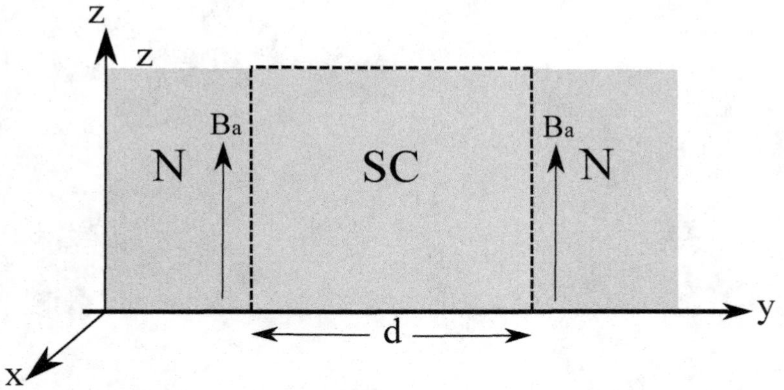


Figure 12.11.14: A superconducting slab of finite thickness d surrounded by a non-superconducting region. The magnetic field is applied along the z axis. The slab is oriented along the y direction.

12.11.9. Why does a Type II superconductor allow flux penetration?

12.11.10. Experiments on lead (Pb) suggest that it has a $\xi = 87$ nm and a $\lambda = 39$ nm. Is Pb a Type I or Type II superconductor?

# 13

# *Optical Properties of Solids*

**Contents**



## 13.1 Introduction

Have you ever wondered why metals are opaque but glass is transparent when we shine a light on it? Why are some objects translucent? How does a **light-emitting diode (LED)** which uses **electroluminescence** work? In this chapter we will provide a conceptual framework to understand the response of a material to visible light, which is part of the **electromagnetic spectrum (EMS)**, see Figure 13.1.1. You will also learn how optical phenomena can be utilized for technological benefit. The EMS spans a wide range of radiation ranging from $\gamma$-rays, through x-rays, to ultraviolet, visible, infrared, microwaves, and finally to radio waves.

Visible light occupies a narrow part of the EMS with wavelength that ranges from 0.4 $\mu$m to 0.7 $\mu$m. When light is incident on an optical medium, metallic or nonmetallic, it can undergo **reflection**, **propagation**, or **transmission**, see Figure 13.1.2. The propagating part of the light within the medium undergoes further interaction with potential processes including **refraction**, **absorption**, **luminescence**, and **scattering**.

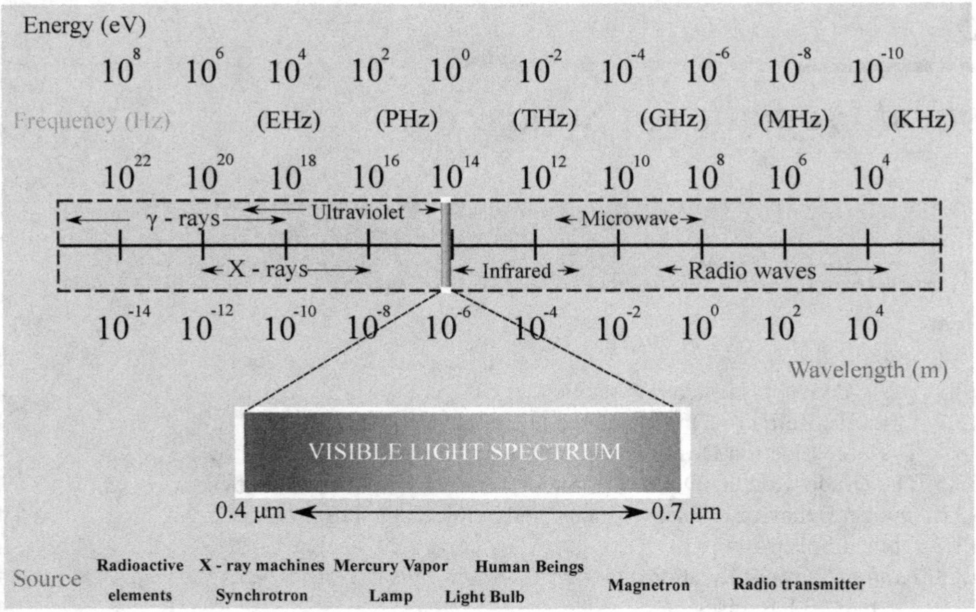


Figure 13.1.1: Electromagnetic spectrum.

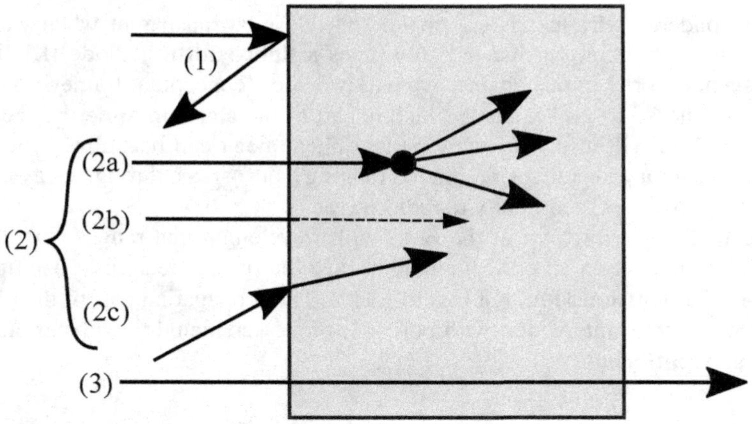


Figure 13.1.2: The three different processes (1) reflection, (2) propagation, and (3) transmission by which light can interact with an optical medium. The propagating light beam could further undergo scattering (process 2a), absorption/ luminescence (process 2b), and refraction (process 2c).

Electromagnetic radiation consists of fluctuating electric and magnetic fields. In the frequency range of the visible light spectrum, the electrical field component interacts with the charge cloud surrounding each atom to displace and separate the positive and negative charge centers to create **electronic polarization**. The polarized medium then subsequently either absorbs some of the energy or leads to a retardation of the light wave velocity as it passes through the medium, thereby causing refraction. Absorption by electronic polarization is important only at frequencies in the vicinity of the relaxation frequency of the constituent atoms.

Besides polarization, **electron energy transition** is another important interaction that allows us to explain optical properties of solid state matter. When a photon (packet of light energy) is absorbed by an atom, the atom gains the energy of the photon, and electrons may jump to a higher energy level. The atom is then said to be in an excited state. Absorption occurs when the photon's energy, $h\nu$, matches the energy difference, $\Delta E$, between two energy levels (see Figure 13.1.3). Thus, we have

$$\Delta E = h\nu, \tag{13.1}$$

where the frequency $\nu$, the wavelength $\lambda$, and the speed of light $c$ are related by

$$c = \nu\lambda. \tag{13.2}$$

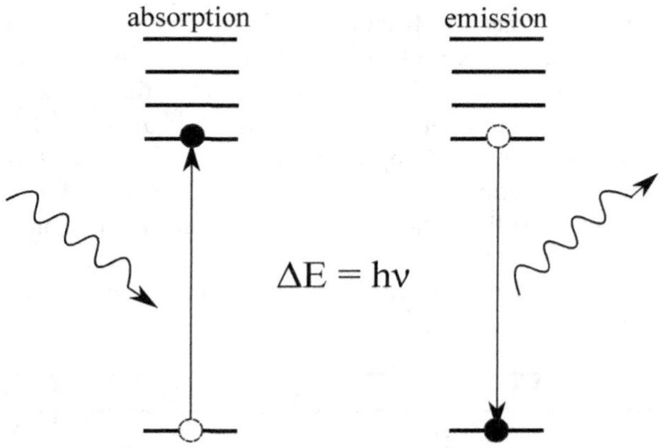


Figure 13.1.3: Atomic level description of electromagnetic radiation **absorption** and **emission**. In an absorption process the incident radiation $h\nu$, where h is the Planck's constant and $\nu$ is the frequency, can be tuned to match the energy level difference $\Delta E$ to promote an electron from a lower energy level to a higher one. When the atom re-emits light via spontaneous emission, known as luminescence, those processes can be further classified into **fluorescence** and **phosphorescence** based on the timescale of deexcitation.

When the electron deexcites to a lower energy level, the energy of the atom is lowered, and the excess energy of the electron is emitted in the form of a photon. Such absorption–emission processes are routinely exploited in spectroscopic techniques which experimentally measure optical properties of materials. Extending this atomic level description to the domain of solid state physics, we can say that an electron from the nearly filled valence band transitions across the band gap into an empty state within the conduction band. In Section 13.6 we use such mechanisms to qualitatively explain the optical properties of glass, metals, and other non-metallic materials. In Section 13.7 we list the various spectroscopic and technological applications that use optical properties.

**Example 13.1.0.1**
The acronym **LASER**, **L**ight **A**mplification by **S**timulated **E**mission of **R**adiation, was coined by the American physicist Gordon Gould (1920-2005). He introduced this common worldwide household name as a graduate student at Columbia university. Although controversial, the construction of the first laser in 1960 is credited to Theodore H. Maiman (1927-2007). A LASER is a device that emits light which is both spatially and temporally coherent. Spatial coherence allows for the laser to be focused in a tight region allowing for technological applications such as laser pointer, laser cutting,

and lithography. The temporal coherence aspect enables the device to emit monochromatic light or light restricted to a narrow spectrum. The red laser operates at a wavelength of about 650 nm. What is the frequency corresponding to this?

**Solution**

Using Equation (13.2) we have

$$3 \times 10^8 = \nu(350 \times 10^{-9}). \tag{13.3}$$

Thus we get $\nu = 8.57 \times 10^{14}$ Hz. Comparing with Figure 13.1.1, we find that the answer correctly lies in the visible light region.

In Table 13.1.1 we list the frequency range description within the optical part of the EMS. The typical wavelength in the optical zone varies between a nanometer to a millimeter, which is usually larger than the atomic dimensions of the underlying crystal lattice being probed. We can study the optical properties of a solid state material from two different physical perspectives. In the classical treatment of electromagnetic radiation, a phenomenological description using Maxwell equations considers the solid to be a dielectric media characterized by optical constants. To provide a microscopic description, the optical constants can be related to the dielectric constant and the optical conductivity using a quantum mechanical approach (beyond the scope of this chapter). In an even more advanced treatment, the radiation field is considered to be quantized and their interaction with the elementary excitations in the solids (phonons, magnons) needs to be considered.

Table 13.1.1: Frequency ($\nu$), wavelength ($\lambda$), and energy (eV) range of different regions in the optical part of the electromagnetic spectrum.

| **Region** | **$\nu$ (THz)** | **$\lambda$ ($\mu$m)** | **Energy (eV)** |
|---|---|---|---|
| Far infrared (FIR) | 0.1 – 10 | 25 – 1000 | $5\times10^{-4} - 0.05$ |
| Mid infrared (MIR) | 10 – 120 | 2.5 – 25 | 0.05 – 0.5 |
| Near infrared (NIR) | 120 – 400 | 0.8 – 2.5 | 0.5 – 1.6 |
| Visible | 400 – 800 | 0.4 – 0.8 | 1.6 – 3 |
| Ultraviolet (UV) | 800 – 32000 | 0.010 – 0.4 | 3 – 120 |

Optical methods form an important branch of experimental techniques that can be used to probe the internal structure of solid state materials. At the beginning of the $20^{\text{th}}$ century, the study of interactions between light with matter (blackbody radiation) laid the foundations of quantum theory. The traditional use of optical materials for windows, antireflection coating, lenses, and mirrors have been extended to include fiber optics technology and optoelectronic devices. Further applications such such as waveguides, photodetectors, lasers, LEDs, and flat-panel displays have had an important technological societal impact.

Table 13.1.2: The origins of the various spectral regions can be generally classified as shown in the table above. To probe the corresponding transitions at the microscopic level, an appropriate form of spectroscopy needs to be applied. For example, the resonant and recoil-free emission and absorption of gamma radiation by atomic nuclei bound in a solid is called **Mössbauer** effect. It is a physical phenomenon discovered by the German physicist Rudolf Mössbauer (1929-2011) in 1958 and its main application is in **Mössbauer** spectroscopy. In physics this particular spectroscopic technique has been used to demonstrate gravitational red-shift, measurement of gamma-ray line width, test of lattice dynamical models, and hyperfine interactions. It also finds widespread use in the fields of geology and bioinorganic chemistry.

| **Radiation** | **Type of Transition** |
|---|---|
| Gamma rays | Nuclear |
| X-rays | Inner electrons |
| Ultraviolet light | Outer electrons |
| Visible light | Outer electrons |
| Infrared light | Vibration |
| Microwaves | Rotation |
| Radiowaves | Spin flips |

The optical response can be described by a set of quantities called the **optical constants**. In this chapter we will be studying and exploring the various interconnected relationships amongst these constants (which can vary based on wavelength) and highlight their importance. Some of the important quantities that you will learn about include the dielectric constant ($\varepsilon$), the refractive index ($n$), the extinction coefficient ($\kappa$), the electromagnetic skin depth ($\delta$), and the surface impedance ($Z$). Knowledge of optical constants and interaction is useful for designing lasers, LEDs, non-linear optical crystals, photovoltaic cells, and materials characterization of electronic band structure, impurity levels, lattice vibrations, and magneto-optical behavior. In the next section, we will introduce some of the basic concepts in Maxwell's equations related to our study of optical phenomena.

## 13.2 Basic Concepts in Electrodynamics

A **dielectric** material is by nature electrically insulating and exhibits a structure composed of an **electric dipole**. An electric dipole constitutes a spatially separated positive and negative charge center at the molecular or atomic level. The bonding in an electric dipole originates from the Coulomb attraction between the opposite charges. A dipole could be created by a temporary short-lived distortion (vibration of a molecule) of an otherwise symmetrically shaped molecule or it could be a permanent dipole originating from an asymmetrical arrangement of charge centers. Irrespective of its origins, when a dielectric material is exposed to an external electric field, its behavior can be characterized by three macroscopic vectors – the electric field strength ($\vec{E}$), the polarization ($\vec{P}$), and the electric displacement vector ($\vec{D}$).

The microscopic response of a dielectric substance is determined primarily by its polarization. Polarization is defined as the net electric dipole moment per unit volume. The application of an electric field tends to polarize symmetric charge centers to create polar molecules which produce microscopic dipoles aligned parallel to the direction of the external field. This generates an overall net electric dipole moment within the material and hence an electric polarization. Assuming an isotropic medium where the polarization response to the electric field is linear, we can write

$$\vec{P} = \varepsilon_o \chi_{el} \vec{E}, \tag{13.4}$$

where $\varepsilon_o$ is the **electric permittivity** of free space and $\chi_{el}$ is the **electric susceptibility** of the medium. The assumption of isotropic medium and linear response does not hold in real materials. Using the above relation, we can write the following expression betwen the electric displacement $\vec{D}$ of the medium, the electric field $\vec{E}$, and the polarization $\vec{P}$ of the medium as

$$\vec{D} = \varepsilon_o \vec{E} + \vec{P}, \tag{13.5}$$

We can combine Equations (13.4) and (13.5) to write

$$\vec{D} = \varepsilon_o \varepsilon_r \vec{E} = \varepsilon \vec{E}, \tag{13.6}$$

where $\varepsilon_r = 1 + \chi_{el}$ is the **relative dielectric constant** of the medium. The parameter $\varepsilon_r$ provides us with insight on how light propagates through dielectrics. The **electric permittivity** of the medium is given by $\varepsilon$. The electric displacement vector is the systems response to an external electric field which causes the bound charges in the material to separate. This induces a local electric dipole moment leading to the polarization $\vec{P}$ term. Conceptually, it is important to recognize that the electric displacement vector is related to the free charge density of the medium, whereas the electric field includes the response of both the bound and free charges.

Changing the topic of discussion to magnetic fields, recall from Chapter 10, that in a magnetic solid the relation between $\vec{B}$ and $\vec{H}$ is modified due to the magnetization, $\vec{M}$, of the medium in a spirit much similar to Equation (13.6). In the case of a linear, isotropic, and homogeneous magnetic media, we had defined

$$\vec{B} = \mu_o(\vec{H} + \vec{M}). \tag{13.7}$$

The magnetization in this case is given by

$$\vec{M} = \chi_{mag} \vec{H}, \tag{13.8}$$

where $\chi_{mag}$ is the dimensionless magnetic susceptibility and the magnetization $\vec{M}$ is in units of $\mathrm{Am}^{-1}$. From the above we have

$$\vec{B} = \mu_o(1 + \chi_{mag})\vec{H} = \mu_o \mu_r \vec{H} = \mu \vec{H}, \tag{13.9}$$

where $\mu_r = 1 + \chi_{mag}$ is the **relative magnetic permeability** and $\mu$ is the magnetic permeability of the medium. Using the above definitions we can rewrite Maxwell's equations of electromagnetism in a medium as displayed in Table 13.2.3. Using the redefined expressions, it is possible to show that wave-like solutions are consistent in a medium with no free charges ($\rho_f = 0$) or currents ($\vec{J}_f = 0$). To derive the equations for light that propagates in a uniform optical medium, we have to further satisfy the following conditions:

- **uniform (homogeneous)**: the permittivity and permeability have the same values at all points in space,
- **isotropic**: the permittivity and permeability do not depend on the direction of propagation,
- **non-dispersive**: permeability and permittivity are frequency independent.

Using the above-stated assumptions and Equations (13.6) and (13.9), we can write

$$\vec{\nabla} \times \vec{E} = -\mu_o \mu_r \frac{\partial \vec{H}}{\partial t}, \tag{13.10}$$

$$\vec{\nabla} \times \vec{H} = \varepsilon_o \varepsilon_r \frac{\partial \vec{E}}{\partial t}. \tag{13.11}$$

Table 13.2.3: Summary of the four Maxwell Equations displaying both the equation and its physical meaning. These fundamental equations are utilized in the text to derive important formulae within the context of superconductivity. The symbols have the standard meaning: $\vec{D}$ (displacement vector), $\vec{B}$ (magnetic field), $\vec{J}_f$ (free current density), and $\rho_f$ (free charge density).

| **Law** | **Maxwell Equation (medium)** | **Physical Meaning** |
|---|---|---|
| Gauss's Law (Electric field) | $\vec{\nabla}\cdot\vec{D}=\rho_f$ | Electric flux is proportional to free charge inside volume. |
| Gauss's Law (Magnetic field) | $\vec{\nabla}\cdot\vec{B}=0$ | Magnetic flux is zero in an enclosed volume. |
| Faraday's Law of Induction (Maxwell–Faraday Law) | $\vec{\nabla}\times\vec{E}=-\frac{\partial\vec{B}}{\partial t}$ | Induced electric field is proportional to changing magnetic flux. |
| Ampére's Law | $\vec{\nabla}\times\vec{H}=\vec{J}_f+\frac{\partial\vec{D}}{\partial t}$ | Magnetic field is proportional to electric and displacement currents. |

Now consider taking the curl of Equation (13.10) and eliminate the auxilliary field using Equation (13.11) to obtain

$$\vec{\nabla}\times\left(\vec{\nabla}\times\vec{E}\right)=-\mu_o\mu_r\varepsilon_o\varepsilon_r\frac{\partial^2\vec{E}}{\partial t^2}. \tag{13.12}$$

Next, we use the vector identity

$$\vec{\nabla}\times\left(\vec{\nabla}\times\vec{E}\right)=\vec{\nabla}\left(\vec{\nabla}\cdot\vec{E}\right)-\vec{\nabla}^2\vec{E}, \tag{13.13}$$

to simplify the left-hand side of the equation. Furthermore, the zero-free charge condition tells us that $\vec{\nabla}\cdot\vec{E}=0$, thus yielding

$$\vec{\nabla}^2\vec{E}=\mu_o\mu_r\varepsilon_o\varepsilon_r\frac{\partial^2\vec{E}}{\partial t^2}=\mu\varepsilon\frac{\partial^2\vec{E}}{\partial t^2}. \tag{13.14}$$

To proceed further, we draw comparison between Equation (13.14) and the general wave equation given by the expression

$$\frac{\partial^2 y}{\partial x^2}=\frac{1}{v^2}\frac{\partial^2 y}{\partial t^2}, \tag{13.15}$$

where $v$ is the propagation speed which can be identified as

$$\frac{1}{v^2}=\mu_o\mu_r\varepsilon_o\varepsilon_r=\mu\varepsilon. \tag{13.16}$$

Note, in free space $\varepsilon_r = \mu_r = 1$. Thus we obtain for the speed of electromagnetic radiation (which includes light) in vacuum as

$$v = \frac{1}{\sqrt{\mu_o \varepsilon_o}} = 3 \times 10^8 \text{m/s} \equiv \text{c (speed of light)}. \tag{13.17}$$

To obtain an expression for the velocity of light in a medium, we use Equation (13.16) to obtain

$$v = \frac{1}{\sqrt{\mu_r \varepsilon_r}} \text{c}, \tag{13.18}$$

which can be used to define the **refractive index** $n$ (or slowness) of the medium as

$$n = \frac{c}{v}. \tag{13.19}$$

Microscopically, the electric dipoles are excited by the incoming electric field component and radiate back with retardation. This results in a permittivity which is different from unity. However, the magnetic dipoles in the medium interact extremely weakly with the oscillating magnetic field component of the EMS spectrum. Hence we set the relative magnetic permeability equal to one to obtain

$$v = \frac{1}{\sqrt{\mu_r \varepsilon_r}} \text{c} = \frac{1}{\sqrt{\varepsilon_r}} \text{c}. \tag{13.20}$$

Thus, using Equation (13.19) we can write

$$n = \sqrt{\varepsilon_r}. \tag{13.21}$$

Equation (13.21) is worth reflecting upon. Starting with the fundamental laws of electromagnetism we are able to relate the refractive index to the dielectric constant, which is pretty remarkable. Now combining all the above concepts, we can write the wave equation, Equation (13.14), for the electric field as

$$\vec{\nabla}^2 \vec{E} = \frac{1}{v^2} \frac{\partial^2 \vec{E}}{\partial t^2}. \tag{13.22}$$

Eliminating $\vec{E}$ instead of $\vec{B}$ from Maxwell's equations, we obtain the magnetic field wave equation (Exercise 13.9.6)

$$\vec{\nabla}^2 \vec{B} = \frac{1}{v^2} \frac{\partial^2 \vec{B}}{\partial t^2}. \tag{13.23}$$

Equations (13.22) and (13.23) belong to a class of partial differential equations named the **Helmholtz equations** after the German physicist Hermann Ludwig Ferdinand von Helmholtz (1821-1894). Solution to the electric or magnetic field wave equations plays an important role in how we analyze light–matter interaction such as reflection, refraction, or propagation through a medium. The simplest solution, called the plane wave solution, is given by the expression

$$\vec{E}(\vec{r},t) = \vec{E}_o \exp\left(i\vec{k} \cdot \vec{r} - i\omega t\right), \tag{13.24}$$

where $\vec{E}_o$ is the wave amplitude, $\vec{k}$ is the propagation wave vector, $\vec{r}$ is the position vector, and $\omega$ is the angular frequency. Inserting the solution into the wave equation and evaluating the spatial derivative gives us

$$\begin{aligned} \nabla^2 \vec{E} &= \vec{E}_o \nabla^2 \exp\left(i\vec{k} \cdot \vec{r} - i\omega t\right), \\ &= \vec{E}_o \vec{\nabla} \cdot \left\{ +i\vec{k} \exp\left(i\vec{k} \cdot \vec{r} - i\omega t\right) \right\}, \\ &= -k^2 \vec{E}_o \exp\left(i\vec{k} \cdot \vec{r} - i\omega t\right) = -k^2 \vec{E}. \end{aligned} \tag{13.25}$$

Computing the time derivative we find

$$\mu\varepsilon\frac{\partial^2\vec{E}}{\partial t^2} = -\omega^2\mu\varepsilon\vec{E}. \tag{13.26}$$

Equating the time and space derivatives, Equations (13.25) and (13.26), we now have a couple of important relations

$$k = \omega\sqrt{\mu\varepsilon} = \frac{\omega}{v} = \frac{2\pi}{\lambda}, \tag{13.27}$$

First, the above equation relates the **wave vector** to frequency and speed of propagation as

$$\omega = vk, \tag{13.28}$$

which is known as the **dispersion** relation. Second, we have the relationship

$$k = \frac{2\pi}{\lambda}, \tag{13.29}$$

where we have used the fact that

$$v = \nu\lambda = \frac{\omega}{2\pi}\lambda, \tag{13.30}$$

where $\omega$ is the angular frequency measured in rad $s^{-1}$. This equation relates the propagation wave vector with the wavelength of light. Just the way the frequency counts the number of oscillations based on time period, $f = 1/T$, the wave vector (or wave number) counts how many waves of wavelength $\lambda$ fit in an interval of "$2\pi$", where the $2\pi$ originates due to the cyclic nature of the oscillations. Further information on the properties of wave propagation (in free space) can be extracted from Maxwell's equation. For example,

$$\vec{\nabla}\cdot\vec{E} = 0. \tag{13.31}$$

Taking the divergence we obtain

$$\begin{aligned} \vec{\nabla}\cdot\left\{\vec{E}_o\exp\left(i\vec{k}\cdot\vec{r} - i\omega t\right)\right\} &= \vec{E}_o\cdot\nabla\left\{\exp\left(i\vec{k}\cdot\vec{r} - i\omega t\right)\right\}, \\ &= -i\vec{k}\cdot\vec{E}_o\exp\left(i\vec{k}\cdot\vec{r} - i\omega t\right). \end{aligned} \tag{13.32}$$

Equating the last line of the above equation to zero

$$\vec{k}\cdot\vec{E} = 0, \tag{13.33}$$

we obtain the condition that the electric field is transverse to the direction of propagation since the zero dot condition implies orthogonality. A similar equation holds for the magnetic field. On the other hand, if we take the curl of the electric field, it can be shown (Exercise 13.9.7) that

$$\vec{k}\times\vec{E} = \omega\vec{B}. \tag{13.34}$$

Thus, based on Equation (13.34) we can conclude that $\vec{k}, \vec{E}$, and $\vec{B}$ are mutually perpendicular to each other. The magnitude of the magnetic field can be related to the electric field as

$$|\vec{B}| = \frac{|\vec{k}|}{\omega}|\vec{E}| = \frac{n}{c}|\vec{E}|, \tag{13.35}$$

where we have used Equation (13.27). Thus we can consider either the electric or the magnetic field equation. From Equation (13.35) we can infer that

$$k = \frac{n}{c}\omega, \tag{13.36}$$

which is consistent with Equation (13.27). Often it is useful to consider the auxilliary field $\vec{H}$ (because it relates to the free current) in which case the equation becomes

$$\frac{|\vec{E}|}{|\vec{H}|} = \mu \frac{|\vec{E}|}{|\vec{B}|} = \sqrt{\frac{\mu}{\varepsilon}} = Z. \tag{13.37}$$

The quantity Z has units of Ω and is called the impedance of the medium.

**Example 13.2.0.1**
What is the impedance of free space?
**Solution**

$$Z_o = \sqrt{\frac{\mu_o}{\varepsilon_o}} = 377\ \Omega \tag{13.38}$$

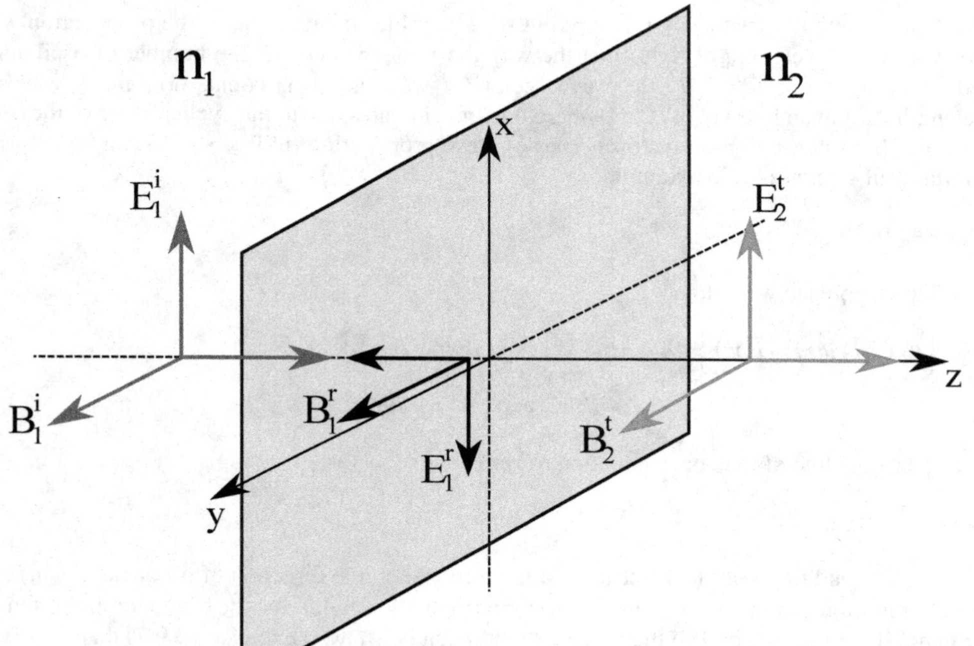


Figure 13.2.4: Reflection and transmission of electromagnetic wave at normal incidence.

In the next few paragraphs we will apply our knowledge of plane wave solutions to obtain expressions for the reflection and transmission coefficients at the boundary of two non-conducting media with normal incidence. Consider Figure 13.2.4 where an electromagnetic wave strikes the interface of two media with refractive indices $n_1$ and $n_2$. We consider the incident electromagnetic waves

$(\vec{E}_1^i, \vec{H}_1^i)$ in medium 1, the reflected electromagnetic waves $(\vec{E}_1^r, \vec{H}_1^r)$ in medium 1, and the transmitted electromagnetic waves $(\vec{E}_2^t, \vec{H}_2^t)$ in medium 2. The electromagnetic waves are assumed to propagate along the $z$ axis with the waves impinged on the x–y plane. Assuming linearly polarized waves in the x direction, we can define

$$\vec{E}_1^i = E_{1x}^i \exp(k_1 z - \omega t)\,\hat{\imath}, \tag{13.39}$$
$$\vec{E}_1^r = -E_{1x}^r \exp(-k_1 z - \omega t)\,\hat{\imath}, \tag{13.40}$$
$$\vec{E}_2^t = E_{2x}^t \exp(k_2 z - \omega t)\,\hat{\imath}, \tag{13.41}$$

where the incident and transmitted propagation vectors have the same magnitude, but the reflected wave vector has a negative sign. Using the relations

$$k_1 = \frac{n_1}{c}\omega; \quad k_2 = \frac{n_2}{c}\omega \tag{13.42}$$

and combining with Equations (13.34) and (13.36) we obtain the relation

$$\vec{B} = \frac{n}{c}\hat{\hat{k}} \times \vec{E}, \tag{13.43}$$

where $\hat{\hat{k}} = \hat{k}$ for the incident and transmitted waves and $\hat{\hat{k}} = -\hat{k}$ for the reflected wave. Since the magnetic field is constrained to obey the above equation, we can write for the magnetic field components in the above process

$$c\vec{B}_1^i = n_1 E_{1x}^i \exp(k_1 z - \omega t)\,\hat{\jmath}, \tag{13.44}$$
$$c\vec{B}_1^r = n_1 E_{1x}^r \exp(-k_1 z - \omega t)\,\hat{\jmath}, \tag{13.45}$$
$$c\vec{B}_2^t = n_2 E_{2x}^t \exp(k_2 z - \omega t)\,\hat{\jmath}, \tag{13.46}$$

For the case of normal incidence that we are considering here, the electric and magnetic components of all three electromagnetics waves are parallel to the boundary surface between the two dielectric media. Thus, the boundary conditions for the total fields which apply in this case at z = 0 are given by

$$E_{\parallel,1} = E_{\parallel,2}, \tag{13.47}$$
$$H_{\parallel,1} = H_{\parallel,2}. \tag{13.48}$$

Since we are considering non-magnetic media $\mu_1 = \mu_2 = \mu_o$, the second condition for the total fields can be rewritten as

$$E_{\parallel,1} = E_{\parallel,2}, \tag{13.49}$$
$$B_{\parallel,1} = B_{\parallel,2} \tag{13.50}$$

Application of the boundary conditions from Equations (13.49) and (13.50) and keeping in mind Equations (13.39-13.41) and (13.44-13.45) implies

$$E_1^i - E_1^r = E_2^t, \tag{13.51}$$
$$n_1\left(E_1^i + E_1^r\right) = n_2 E_2^t \tag{13.52}$$

The above equations can now be solved to obtain

$$E_1^r = \frac{n_2 - n_1}{n_2 + n_1} E_1^i, \tag{13.53}$$
$$E_2^t = \frac{2n_1}{n_1 + n_2} E_1^i. \tag{13.54}$$

We now introduce the **Fresnel coefficients** $r_{12}$ and $t_{12}$ for normal incidence reflection and transmission, respectively. We have

$$r_{12} = \frac{n_2 - n_1}{n_2 + n_1}, \tag{13.55}$$

$$t_{12} = \frac{2n_1}{n_1 + n_2}. \tag{13.56}$$

The Fresnel conditions (or equations) were derived by the French civil engineer and physicist Augustin-Jean Fresnel (1788-1827). These equations capture the reflected and transmitted ampltiudes of the light wave. Since amplitude can be related to power, an alternate way of capturing the optical behavior at the interface is by using the concept of **reflectance** $R$ for normal incidence as

$$R = r_{12}^2 = \left(\frac{E_1^r}{E_1^i}\right)^2 = \left(\frac{n_2 - n_1}{n_2 + n_1}\right)^2, \tag{13.57}$$

and **transmittance** $T$, defined as

$$T = \frac{n_2}{n_1} t_{12}^2 = \frac{n_2}{n_1}\left(\frac{E_1^t}{E_1^i}\right)^2 = \frac{n_2}{n_1}\left(\frac{2n_1}{n_2 + n_1}\right)^2. \tag{13.58}$$

Reflectance is the fraction of the incident power that is reflected from the interface. Transmittance is the fraction of transmitted power through the interface. Using Equations (13.57) and (13.58) we note that $R + T = 1$ for any pair of non-conducting media which is essentially an expression of energy conservation at the interface.
What are the real-life implications of the $R$ and the $T$ values? For example, if we consider an air–glass interface with $n_1 = 1$ (air–incidence) and $n_2 = 1.5$ (glass–transmittance) we have, using Equations (13.57) and (13.58), R = 4% and T = 96%, respectively. The 4% reflectance show up as lens flare in photography. Fresnel equations also play a role in crime solving. One-way mirrors used by police become partial reflectors due to a very thin layer of aluminum coating applied to one side of the mirror in the interrogation room. With a brightly lit interrogation room the suspect only sees their reflection in the mirror. But, the investigating crime officers waiting outside in a dimly lit room that does not allow transmission of enough light. Other popular applications are antireflective coatings applied to prescription lenses, optical surfaces, and photographic lenses to reduce reflection.

## 13.3 Complex Refractive Index and the Dielectric Constant

We will now venture beyond the description of propagating electromagnetic waves in free space to develop an understanding of how light interacts within a conducting medium. Maxwell's equations provide a definition of the fields that are generated by currents and charges in matter. But, they do not describe how these currents and charges are generated. Thus, to find a set of self-consistent solutions to the electromangetic field, Maxwell's equations need to be supplemented by the behavior of the material under the influence of the fields. These material–field relationship equations are known as the **constitutive relations**. For a linear and isotropic medium they are given by

$$\vec{D} = \varepsilon_o \varepsilon_r \vec{E} = \varepsilon \vec{E}, \quad (\vec{P} = \varepsilon_o \chi_{el} \vec{E}) \tag{13.59}$$

$$\vec{B} = \mu_o \mu_r \vec{H} = \mu \vec{H}, \quad (\vec{M} = \chi_{mag} \vec{H}) \tag{13.60}$$

$$\vec{J} = \sigma \vec{E}. \tag{13.61}$$

Note, Equations (13.59) and (13.60) have been reported earlier. In a conducting medium with no free charges, we can use the third constitutive Equation (13.61) in Maxwell's equations to obtain the wave equation (Exercise 13.9.12)

$$\nabla^2\vec{E} = \mu\varepsilon\frac{\partial^2\vec{E}}{\partial t^2} + \mu\sigma\frac{\partial\vec{E}}{\partial t}. \tag{13.62}$$

Next, inserting the plane wave solution, Equation (13.24), into Equation (13.62) we obtain

$$\nabla^2\vec{E}(\vec{r}) + \omega^2\mu\left(\varepsilon + \frac{i\sigma}{\omega}\right)\vec{E}(\vec{r}) = 0. \tag{13.63}$$

We can define the complex dielectric constant given by

$$\tilde{\varepsilon} = \varepsilon + \frac{i\sigma}{\omega}, \tag{13.64}$$

where both $\varepsilon$ and $\sigma$ are real within the current formulation. From the above we have the equation

$$\nabla^2\vec{E}(\vec{r}) + \tilde{k}^2\vec{E}(\vec{r}) = 0, \tag{13.65}$$

where the **complex wave vector** can be expressed in terms of the **complex refractive index** $\tilde{n}$ given by

$$\tilde{k} = \omega\sqrt{\mu\left(\varepsilon + \frac{i\sigma}{\omega}\right)} = \frac{\tilde{n}\omega}{c}. \tag{13.66}$$

Note, Equation (13.66) is a complex generalization of the previously introduced wave vector Equation (13.27). We can further rewrite $\tilde{n}$ as

$$\tilde{n} = n + i\kappa, \tag{13.67}$$

where $n$ is the real part of the refractive index and the imaginary part $\kappa$ is called the **extinction coefficient**. From Equation (13.21) we know that it is possible to relate the dielectric constant with the refractive index. Thus if $n$ is the **complex refractive index** the dielectric constant must also be complex and is defined as

$$\tilde{\varepsilon}_r = \varepsilon_1 + i\varepsilon_2, \tag{13.68}$$

where $\tilde{\varepsilon}_r = \tilde{\varepsilon}/\varepsilon_o$ is the **complex relative dielectric constant**. Thus we can write

$$\tilde{n}^2 = \tilde{\varepsilon}_r. \tag{13.69}$$

In the above we have set the relative permeability $\mu_r = 1$ since we are concerned with optical frequencies. Using the definition of $\tilde{n}$, $\tilde{\varepsilon}$, and after some simple algebra equating the real and imaginary parts we can obtain the following relations (Exercise 13.9.13)

$$\varepsilon_1 = n^2 - \kappa^2, \tag{13.70a}$$
$$\varepsilon_2 = 2n\kappa. \tag{13.70b}$$

It is possible to solve Equations (13.70a) and (13.70b) to obtain explicit expressions for $n$ and $\kappa$ as (Exercise 13.9.14)

$$n = \frac{1}{\sqrt{2}}\left(\varepsilon_1 + (\varepsilon_1^2 + \varepsilon_2^2)^{1/2}\right)^{1/2}, \tag{13.71a}$$
$$\kappa = \frac{1}{\sqrt{2}}\left(-\varepsilon_1 + (\varepsilon_1^2 + \varepsilon_2^2)^{1/2}\right)^{1/2}. \tag{13.71b}$$

The choice of positive square roots ensures that both $n$ and $\kappa$ themselves are real and positive as required from physical consideration. The introduction of a complex refractive index has implications. Consider a plane wave propagating in the $z > 0$ direction in a medium described by a complex refractive index. Introducing the **skin depth** parameter $1/\delta$ we have

$$\vec{E}(\vec{r}) = \vec{E}_o \exp\left(+i\tilde{k}z\right) = \vec{E}_o \exp\left(+in\frac{\omega}{c}z\right) \exp\left(-\frac{z}{\delta}\right) \tag{13.72}$$

This shows that a non-zero extinction leads to an exponential decay of the wave in the medium. The real part of the complex refractive index determines the phase velocity of the wave front. From Equation (13.66) we can identifty the real and imaginary parts of the complex wave vector $\tilde{k}$. Then using the expressions for the real and imaginary parts in Equations (13.71a) and (13.71b) we can write

$$\tilde{k} = \omega\sqrt{\frac{\mu\varepsilon}{2}}\left[\left(1+\sqrt{1+\frac{\sigma^2}{\omega^2\varepsilon^2}}\right)^{1/2} + i\left(\sqrt{1+\frac{\sigma^2}{\omega^2\varepsilon^2}}-1\right)^{1/2}\right]. \tag{13.73}$$

Thus the propogation vector $k$ and the attenuation factor $\alpha$ is given by

$$k = \omega\sqrt{\frac{\mu\varepsilon}{2}}\left(\sqrt{1+\frac{\sigma^2}{\omega^2\varepsilon^2}}+1\right)^{1/2}, \tag{13.74}$$

$$\alpha = \omega\sqrt{\frac{\mu\varepsilon}{2}}\left(\sqrt{1+\frac{\sigma^2}{\omega^2\varepsilon^2}}-1\right)^{1/2}. \tag{13.75}$$

The ratio of $\sigma/\omega\varepsilon$ determines whether a material is a good conductor or otherwise. For a good conductor we have $\sigma \gg \omega\varepsilon$. For this case, we have

$$k = \alpha = \sqrt{\frac{\omega\mu\sigma}{2}}. \tag{13.76}$$

The speed of electromagnetic wave is given by

$$v = \frac{\omega}{k} = \sqrt{\frac{2\omega}{\sigma\mu}}. \tag{13.77}$$

The electric field amplitude diminishes exponentially with distance as $\exp(-\alpha z)$. The distance to which the field penetrates before its amplitude diminishes by a factor $e^{-1}$ is known as the skin depth (mentioned earlier), which is given by the equation

$$\delta = \frac{1}{\alpha} = \sqrt{\frac{2}{\omega\mu\sigma}}. \tag{13.78}$$

Thus we find that the wave does not penetrate much inside the conductor. This effect is known as the **skin effect**. From a commercial perspective, this has financial implications. It is evident from the skin depth equation that with increasing frequency, currrent conduction is limited exclusively to the outer surface since the penetration depth starts to decrease. As a cost-saving measure, when skin depth is shallow, a solid conductor can be replaced with a hollow tube with no loss of performance, thereby saving hundreds of dollars in production and manufacturing cost.

**Example 13.3.0.1**
Compare the skin depths of copper and titanium at 1 GHz. The resistivity $\rho$ of copper and titanium are $1.69 \times 10^{-8}$ $\Omega$–m and $54 \times 10^{-8}$ $\Omega$–m at 20°C, respectively.

**Solution**
Using Equation (13.78), $\rho = 1/\sigma$, the resistivity values, and the conversion 1GHz = $10^9$ Hz we find for the skin depth of each metal

$$\delta_{\mathrm{Cu}} = 2.07\ \mu\mathrm{m} \quad \text{and} \quad \delta_{\mathrm{Ti}} = 11.7\ \mu\mathrm{m}. \tag{13.79}$$

In general Equations (13.70a) and (13.70b) are complicated to analyze. But, it is possible to create three categories of approximations which allow us to infer useful information. The real and imaginary parts of the complex dielectric constant $\tilde{\varepsilon}$ can then be used to analyze the propagation of electromagnetic waves in insulators and conductors.

(i) **Insulator**: $\omega \gg \sigma/\varepsilon$; $\varepsilon_2 \ll |\varepsilon_1|$, $\varepsilon_1 > 0 \Rightarrow n \approx \sqrt{\varepsilon_1}$, $\kappa = \varepsilon_2/2n \ll n$.

(ii) **Metals (in upper infrared)**: $\omega \gg \sigma/-\varepsilon$; $\varepsilon_2 \ll |\varepsilon_1|$, $\varepsilon_1 < 0 \Rightarrow \kappa \approx \sqrt{-\varepsilon_1}$, $n = \varepsilon_2/2k \ll k$.

(iii) **Metals (at or below microwave frequencies)**: $\omega \ll \sigma/|\varepsilon|$; $\varepsilon_2 \gg |\varepsilon_2|$, $n \approx \kappa \approx \sqrt{\varepsilon_2/2}$.

The first condition above is for a weakly absorbing medium where $\kappa$ is very small. We can conclude that the refractive index is determined by the real part of the dielectric constant. But, the absorption is mainly determined from the imaginary part.

## 13.4 The Free Electron Drude Theory of Optical Properties

The Drude free electron theory of metals is based on a theoretical construct that treats the valence electrons of the atoms as if they are free. Since free electrons are not bound to a particular atom, there is no resonant restoring force. But, in the presence of an applied electric field, the free electrons are subjected to an electrical force. Subsequently, these electrons undergo a collision with a scattering time given by $\tau$. Taking into account the scattering processes, the Drude model for conductivity can be generalized to include alternating-current (ac) effects to study how metals and doped semiconductors interact with light. Applying Newton's second law gives

$$m\left(\frac{d\vec{v}}{dt} + \frac{v}{\tau}\right) = -e\,\vec{E}(t) = -e\,\vec{E}_o e^{-i\omega t}, \tag{13.80}$$

where we have introduced a velocity dependent damping term and $\tau$ is a relaxation time. We introduce a complex notation to describe the time dependent electric field, with the understanding that only the real part of this expression has a direct physical meaning. Assuming there is a steady state solution of the form

$$\vec{v}(t) = \vec{v}_o \exp(-i\omega t), \tag{13.81}$$

we can show (Exercise 13.9.15) that the amplitude is given by

$$\vec{v}_o = -\frac{e}{m}\left(\frac{\vec{E}_o}{-i\omega + 1/\tau}\right). \tag{13.82}$$

Next, we define the time dependent current density vector as

$$\vec{J}(t) = \vec{J}_o \exp(-i\omega t). \tag{13.83}$$

Combining the above definition with current density we can write

$$\vec{J}_o = -ne\vec{v}_o, \tag{13.84}$$

$$= \sigma(\omega)\vec{E}_o, \tag{13.85}$$

where the frequency dependent **AC conductivity** can be identified with

$$\sigma(\omega) = \frac{\sigma_o}{1 - i\omega\tau}, \tag{13.86}$$

where

$$\sigma_o = \varepsilon_o \omega_p^2 \tau. \tag{13.87}$$

In the above we have introduced the definition of the **plasma frequency** as

$$\omega_p^2 \equiv \frac{ne^2}{m\varepsilon_o}. \tag{13.88}$$

The expression in Equation (13.88) represents the oscillation frequency of a neutral gas of charged particles known as **plasma**. Both metals and doped semiconductors can be physically modeled as an equal collection of positive ions and free electron charges, i.e., a plasma. For ordinary metals, typical plasma energy values range from 3 – 17 eV. The relaxation times are typically of the order of $10^{-14}$ s and the damping constant arising from the relaxation time is of the order of 0.1 eV. Equation (13.87) for the complex conductivity expression can be realized by multiplying the numerator and denominator with the complex conjugate $1 + i\omega\tau$ to obtain

$$\sigma(\omega) = \frac{ne^2}{m}\frac{\tau}{1 - i\omega\tau}, \tag{13.89}$$

$$= \frac{ne^2}{m}\frac{\tau}{1 - i\omega\tau}\frac{1 + i\omega\tau}{1 + i\omega\tau}, \tag{13.90}$$

$$= \left(\frac{ne^2\tau}{m}\right)\frac{1 + i\omega\tau}{1 + \omega^2\tau^2}. \tag{13.91}$$

Notice from Equation (13.91) that as $\omega \to 0$, the $\Re\mathrm{e}[\sigma(\omega)]$ reduces to the Drude conductivity (or DC conductivity) formula

$$\sigma = \frac{ne^2\tau}{m}. \tag{13.92}$$

Thus, in the low frequency limit, $\Im\mathrm{m}[\sigma(\omega)] \to 0$ as $\omega \to 0$. The complex dielectric constant and the complex AC conductivity are related to each other. This can be shown by using the fourth Maxwell equation (see Ampére's Law, Table 13.2.3) and substituting the expression for the frequency dependent electric field in the right-hand side of the equation to give

$$\begin{aligned}
\vec{\nabla} \times \vec{B} &= \mu_o\left(\vec{J} + \varepsilon_o \frac{\partial \vec{E}}{\partial t}\right), \\
&= \mu_o\left(\sigma(\omega)\vec{E}_o e^{-i\omega t} + (-i\omega)\varepsilon_o \vec{E}_o e^{-i\omega t}\right), \\
&= \mu_o\left(\varepsilon_o + \frac{i\sigma(\omega)}{\omega}\right)(-i\omega)\vec{E}_o e^{-i\omega t}, \\
&= \mu_o \frac{\partial}{\partial t}\varepsilon(\omega)\vec{E}(t),
\end{aligned} \tag{13.93}$$

where we have identified

$$\varepsilon(\omega) = \varepsilon_o + \frac{i\sigma(\omega)}{\omega}. \tag{13.94}$$

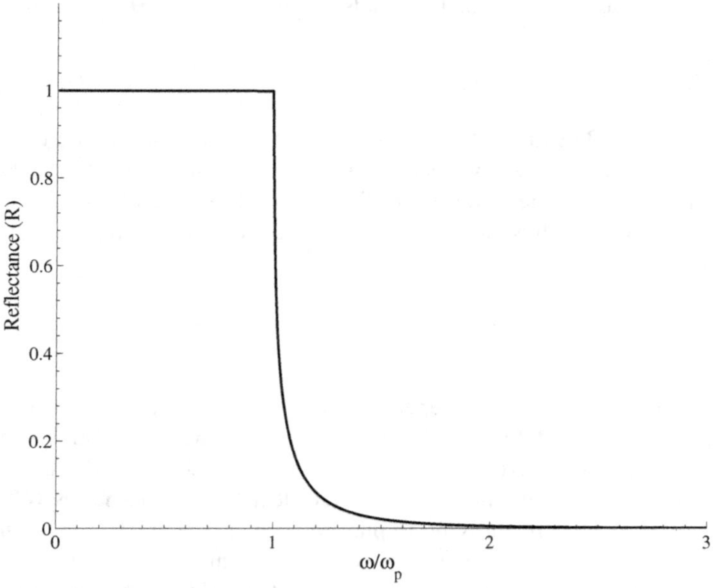


Figure 13.4.5: Reflectance variation against frequency for a metal.

We can conclude from Equation (13.94) that optical measurements of $\varepsilon(\omega)$ are equivalent to measurements of AC conductivity. Now substituting Equation (13.89) in Equation (13.94) we derive the frequency-dependent relative dielectric function $\varepsilon_r(\omega)$

$$\begin{aligned}
\varepsilon_r(\omega) &= \frac{\varepsilon(\omega)}{\varepsilon_o}, \\
&= \frac{1}{\varepsilon_o}\left(\varepsilon_o + \frac{i\sigma(\omega)}{\omega}\right), \\
&= \frac{1}{\varepsilon_o}\left(\varepsilon_o + \frac{i}{\omega}\frac{ne^2}{m}\frac{\tau}{1-i\omega\tau}\right), \\
&= 1 + \frac{i}{\omega}\frac{ne^2\tau}{m\varepsilon_o}\frac{1}{(1-i\omega\tau)}, \\
&= 1 + \frac{i}{\omega}\frac{ne^2\tau}{m\varepsilon_o}\frac{1}{(-i\tau)}\frac{1}{(\omega+i/\tau)}, \\
&= 1 - \frac{\omega_p^2}{\omega(\omega+i/\tau)}.
\end{aligned} \tag{13.95}$$

For future analysis we will extract the real and imaginary parts from Equation (13.95); that is,

$$\Re\mathrm{e}[\varepsilon_r(\omega)] = 1 - \frac{\omega_p^2}{\gamma^2+\omega^2}, \tag{13.96}$$

$$\Im\mathrm{m}[\varepsilon_r(\omega)] = \frac{\omega_p^2\gamma}{\omega(\gamma^2+\omega^2)}, \tag{13.97}$$

where $\gamma \equiv 1/\tau$ is the damping frequency. Let us analyze Equation (13.95) in the case of a weakly damped system whereby we set $1/\tau \to 0$. This leads to the following expression

$$\varepsilon_r(\omega) = 1 - \frac{\omega_p^2}{\omega^2}. \tag{13.98}$$

We know from Equation (13.69) that the complex dielectric function and the complex refractive index are related. In Equation (13.57) we had derived a relationship for reflectance. It is possible to apply this equation to an air-metal interface. We note in this case $n_1 = 1$ (real, air) and $n_2 = \tilde{n} = n + i\kappa$ (imaginary, metal). Thus, substituting into Equation (13.57) we can define the **plasma reflectivity** expression $\mathscr{R}$ as

$$\mathscr{R} = \left| \frac{\tilde{n} - 1}{\tilde{n} + 1} \right|^2, \tag{13.99}$$

where $\tilde{n} = \sqrt{\varepsilon_r}$. Inserting the frequency dependent refractive index expression into $\mathscr{R}$ we obtain Figure 13.4.5. Let us now analyze this plot. When $\omega < \omega_p$, $\tilde{n}$ is imaginary. But for $\omega > \omega_p$ it is real and positive. At $\omega = \omega_p$ the refractive index is zero. For $\omega < \omega_p$, $\mathscr{R} = 1$ and decreases for $\omega > \omega_p$. In simple terms this implies that the reflectivity of a free electron gas (plasma) is 100% up to plasma frequencies. Thus, most metals are shiny up to plasma frequency and will reflect visible light and infrared. This is the reason why metals such as silver and aluminum have been utilized for centuries to make mirrors. Beyond the plasma frequency some of the light can be transmitted through the metal. This implies that metals will eventually become transparent for frequencies which extend into the ultraviolet for which $\omega > \omega_p$. This phenomenon is known as the **ultraviolet transparency of metals**. In passing, we should note that some metals such as gold and copper appear colored since they selectively absorb particular wavelength ranges. This effect is in addition to the plasma reflectivity effects.

**Example 13.4.0.1**

Using Equation (13.98) for the frequency dependent dielectric function for a weakly damped system write a MATLAB code to generate Figure 13.4.5. Scale the expression with respect to the plasma frequency $\omega_p$.

**Solution**

The MATLAB code to generate the figure is below.

```
%copyright by J. E Hasbun and T. Datta
% ch13_metalR.m

% This script plots the reflectance from a metal below, at, and above
% the plasma frequency using the Drude free electron theory for a weakly
% damped system.

% Function definitions
% Equation (13.98)
% The imaginary part is zero in the limit of weak damping.
% The function is scaled w.r.t to the plasma frequency.

epsr = @(w) (1 - 1/w^2);

% Refractive index, Equation (13.69)

nrefr = @(w) sqrt(epsr(w));
```

```
% Reflectance, Equation (13.99)
refl = @(w) (abs((nrefr(w)-1)/(nrefr(w)+1)))^2;

% Plots

figure1 = figure;
axes1 = axes('Parent',figure1,'YMinorTick','on',...
    'XTick',[0 1 2 3 4 5 6 7 8 9 10 11 12],...
    'XMinorTick','on');
ylim(axes1,[0 1.2]);
hold(axes1,'all')

hold on;
fplot(@(w)refl(w),[0,3],'LineWidth',2,'-k');

% Create xlabel
xlabel('\omega/\omega_{p}',...
    'FontSize',16,'FontName','Times New Roman');

% Create ylabel
ylabel('Reflectance (R)','FontSize',16,...
    'FontName','Times New Roman');

hold off;
```

**Example 13.4.0.2**
Write a MATLAB script to generate Figure 13.4.6 which plots the refractive index ($n$), extinction coefficient $\kappa$, and reflectance derived within the Drude theory for the case when damping is not negligible.

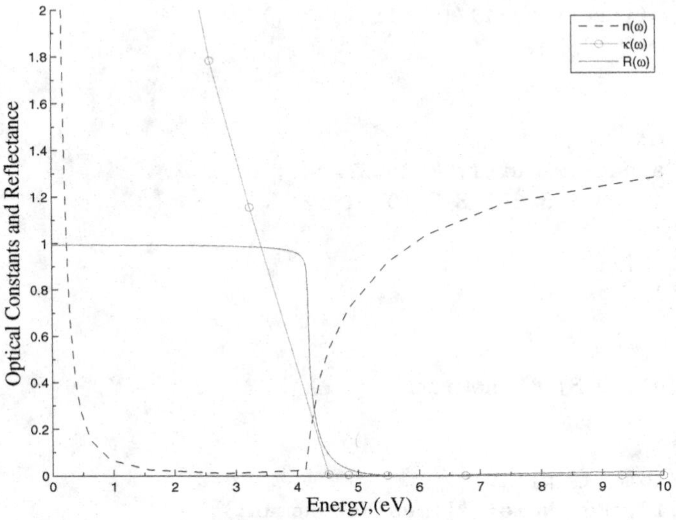


Figure 13.4.6: Frequency dependent refractive index, extinction coefficient, and reflectance variation of a metal with damping.

**Solution**
The MATLAB code to generate the figure is below.

```
%copyright by J. E Hasbun and T. Datta
% ch13_nwkappawmetal.m

% This script plots the refractive index (n), extinction coefficient
% kappa, and the reflectance derived within the Drude theory in the case
% when damping is not negligible. The plasma frequency and damping
% parameters are for Na (metal). Data sourced from Reference [13.3],
% Table 11.2.

% Constants
omp = 5.914;    % Na plasma frequency (eV)
hgamma = 0.0198512; % Na damping frequency (eV)

% Function definitions
% Frequency dependent real part of the dielectric function.
% Compare with Equation (13.96).
realpart = @(w) 1 - omp^2/(hgamma^2 + w^2);

% Frequency dependent imaginary part of the dielectric function.
% Compare with Equation (13.97).
impart = @(w) (omp^2)*hgamma/((w)*(w^2 + hgamma^2));

% Refractive index (n) and extinction coefficient (\kappa)
nw = @(w) sqrt(0.5*(sqrt((1+realpart(w))^2 + impart(w)^2) + 1 ...
    + realpart(w)));
```

```
kappaw = @(w) sqrt(0.5*(sqrt((1+realpart(w))^2 + impart(w)^2)...
    - 1 - realpart(w)));

% Reflectance, Equation (13.99)

refl = @(w) ((1 - nw(w))^2 + kappaw(w)^2)/((1 + nw(w))^2 + kappaw(w)^2);

% Plots

figure1 = figure;
axes1 = axes('Parent',figure1,'YMinorTick','on',...
    'XTick',[0 1 2 3 4 5 6 7 8 9 10 11 12],...
    'XMinorTick','on');
ylim([0,2]);
hold(axes1,'all')

hold on;

fplot(@(w)nw(w),[0,10],'LineWidth',2,'--k');
fplot(@(w)kappaw(w),[0,10],'LineWidth',2,'-or');
fplot(@(w)refl(w),[0,10],'LineWidth',2,'-b');

% Create xlabel
xlabel('Energy,(eV)',...
    'FontSize',16,'FontName','Times New Roman');

% Create ylabel
ylabel('Optical Constants and Reflectance', 'FontSize',16,...
    'FontName','Times New Roman');

% % Create legend
legend1 = legend('n(\omega)','\kappa(\omega)','R(\omega)');
set(legend1,'FontSize',14,'FontName','Times New Roman','show');

hold off;
```

To study the optical properties of the free electron carriers in metals, we can classify the frequency response into three regimes I, II, and III.

- **Regime I:** Non-relaxation Hagen–Rubens region ($0 < \omega \ll \tau^{-1}$). In this regime Equation (13.95) can be rewritten as

$$\varepsilon_r(\omega) = 1 - \frac{\omega_p^2}{\omega(\omega + i/\tau)} \approx 1 + i\frac{\omega_p^2\tau}{\omega}, \tag{13.100}$$

  thus the real and imaginary parts are given by

$$\varepsilon_1 = \Re e[\varepsilon_r(\omega)] = 1; \;\; \varepsilon_2 = \Im m[\varepsilon_r(\omega)] = \frac{\omega_p^2\tau}{\omega} \tag{13.101}$$

  In the non-relaxation Hagen–Rubens region because $\omega\tau \ll 1$, the dielectric function becomes using Equations (13.71a) and (13.71b) we obtain the frequency dependent refractive index as

$$n(\omega) \approx \frac{1}{\sqrt{2}}\left[\left\{1+\left(\frac{\omega_p^2\tau}{\omega}\right)^2\right\}^{1/2}+1\right] \approx \frac{1}{\sqrt{2}}\left\{1+\left(\frac{\omega_p^2\tau}{\omega}\right)^2\right\}^{1/2} \approx \left(\frac{\omega_p^2\tau}{2\omega}\right)^{1/2} \tag{13.102}$$

Similarly from Equation (13.95), we write for the absorption part

$$\kappa(\omega) \approx \left(\frac{\omega_p^2\tau}{2\omega}\right)^{1/2} \tag{13.103}$$

Thus in this regime $n(\omega) \approx \kappa(\omega)$. We thus calculate the reflectivity using Equation (13.99) as

$$\begin{aligned}
R(\omega) &= \frac{\{1-n(\omega)\}^2+\kappa^2(\omega)}{\{1+n(\omega)\}^2+\kappa^2(\omega)}, \\
&= \frac{\left\{1-(\omega_p^2\tau/2\omega)^{1/2}\right\}^2+\omega_p^2\tau/2\omega}{\left\{1+(\omega_p^2\tau/2\omega)^{1/2}\right\}^2+\omega_p^2\tau/2\omega}, \\
&\approx \frac{\omega_p^2\tau/\omega-2\left(\omega_p^2\tau/2\omega\right)^{1/2}}{\omega_p^2\tau/\omega+2\left(\omega_p^2\tau/2\omega\right)^{1/2}}, \\
&\approx 1-4\left(\omega/2\omega_p^2\tau\right)^{1/2}, \\
&= 1-\left(8\omega/\omega_p^2\tau\right)^{1/2}
\end{aligned} \tag{13.104}$$

- **Regime II:** Relaxation regime ($\tau^{-1} \ll \omega \ll \omega_p$). In this regime Equation (13.95) can be rewritten as

$$\varepsilon_r(\omega) \doteq 1+i\frac{\omega_p^2\tau}{\omega}\frac{1}{(1-i\omega\tau)} \approx 1-\frac{\omega_p^2}{\omega^2}\frac{1}{(1+i/\omega\tau)} \approx -\frac{\omega_p^2}{\omega^2}\left(1-\frac{i}{\omega\tau}\right), \tag{13.105}$$

thus the real and imaginary parts are given by

$$\varepsilon_1 = \Re e[\varepsilon_r(\omega)] = -\frac{\omega_p^2}{\omega^2};\ \varepsilon_2 = \Im m[\varepsilon_r(\omega)] = \frac{\omega_p^2}{\omega^3\tau} \tag{13.106}$$

In this regime because $\omega\tau \gg 1$ we have $\Im\mathrm{m}[\varepsilon_r(\omega)] \ll |\Re\mathrm{e}[\varepsilon_r(\omega)]|$. Thus from Equation (13.95) $n(\omega)$ and $\kappa(\omega)$ is approximately given by

$$n(\omega) \approx \frac{1}{\sqrt{2}}\left[\frac{\omega_p^2}{\omega^2}\left(1+\frac{1}{\omega^2\tau^2}\right)^{1/2}-\frac{\omega_p^2}{\omega^2}\right] \approx \frac{\omega_p}{2\omega^2\tau}, \tag{13.107}$$

$$\kappa(\omega) \approx \frac{\omega_p}{\omega}. \tag{13.108}$$

Thus, the reflectivity $R(\omega)$ from Equation (13.98) is then approximately given by

$$\begin{aligned}
R(\omega) &\approx \frac{\left(1-\omega_p/2\omega^2\tau\right)^2+\omega_p^2/\omega^2}{\left(1+\omega_p/2\omega^2\tau\right)^2+\omega_p^2/\omega^2}, \\
&\approx \frac{-\omega_p/\omega^2\tau+\omega_p^2/\omega^2}{\omega_p/\omega^2\tau+\omega_p^2/\omega^2} = \frac{1-1/\omega_p\tau}{1+1/\omega_p\tau} = 1-\frac{2}{\omega_p\tau}
\end{aligned} \tag{13.109}$$

- **Regime III:** High frequency regime ($\omega \gg \omega_p$) and $\omega\tau \gg 1$, thus

$$\varepsilon_r(\omega) = 1 + i\frac{\omega_p^2\tau}{\omega}\frac{1}{(1-i\omega\tau)} \approx 1 - \frac{\omega_p^2}{\omega^2}\frac{1}{(1+i/\omega\tau)} \approx 1 - \frac{\omega_p^2}{\omega^2}\left(1 - \frac{i}{\omega\tau}\right), \tag{13.110}$$

thus the real and imaginary parts are given by

$$\varepsilon_1 = \Re e[\varepsilon_r(\omega)] = 1 - \frac{\omega_p^2}{\omega^2}; \quad \varepsilon_2 = \Im m[\varepsilon_r(\omega)] = \frac{\omega_p^2}{\omega^3\tau} \tag{13.111}$$

In this case $\Im\mathrm{m}[\varepsilon_r(\omega)] \ll \Re\mathrm{e}[\varepsilon_r(\omega)]$ and we can approximate the refractive index and absorption by

$$n(\omega) \approx \frac{1}{\sqrt{2}}\{2\Re\mathrm{e}[\varepsilon_r(\omega)]\} \approx 1 - \frac{\omega_p^2}{2\omega^2}, \tag{13.112}$$

$$\kappa(\omega) \approx \frac{\omega_p^2}{2\omega^3\tau}. \tag{13.113}$$

The reflectivity $R(\omega)$ is then approximately given by (Exercise 13.9.17)

$$R(\omega) = \left(\frac{\omega_p}{2\omega}\right)^4 \tag{13.114}$$

## 13.5 The Drude–Lorentz Dipole Oscillator Theory of Optical Properties

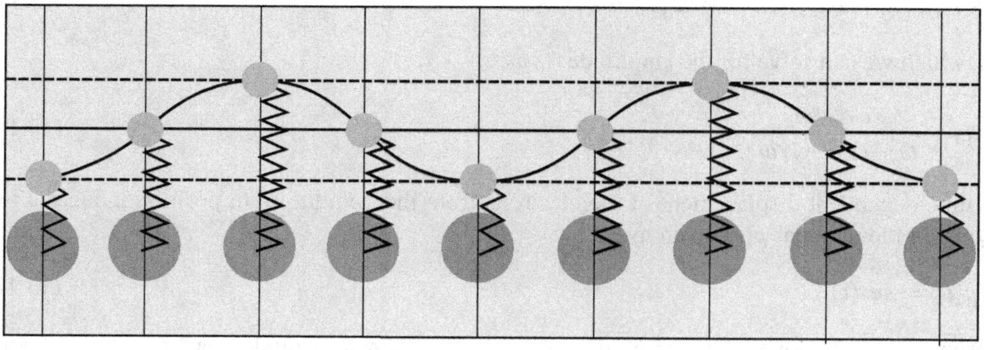

Figure 13.5.7: Drude–Lorentz oscillator model of the optical properties of an insulator.

In the previous section, we developed a theory of optical properties for metals where the participating electrons are itinerant (unbound to any particular atom and free to move). In this section we will discuss a generalization of the free electron Drude theory, called the Drude–Lorentz theory of a dipole oscillator that explains reasonably well the optical properties of an insulator where electrons are bound to an atom. For a realistic model of an insulator, the Drude–Lorentz model needs to take into consideration the fact that an insulator is a collection of Lorentz oscillators with different frequencies. But, we proceed with a simplifed model that considers the interaction between light waves and an atom of a single resonant frequency $\omega_o$ due to the bound electrons, see Figure 13.5.7. Since dipoles can lose energy by collisional processes, in a realistic scenario, we must also consider damping. This effect is treated using a damped harmonic oscillator system. In solids thermally excited phonons can give rise to such losses. The oscillating electric field of the light wave (part of the electromagnetic wave) induces forced oscillations of the atomic dipole through the driving forces exerted on the electrons. Thus, we have a damped driven harmonic oscillator system that is typically studied as part of a standard theoretical mechanics course, see Reference [64]. We can ignore the motion of the nucleus by applying the Born-Oppenheimer approximation. We write for the time dependent displacement $x(t)$ of the electron the equation

$$\underbrace{m_e \frac{d^2x}{dt^2}}_{\text{acceleration}} + \underbrace{m_e \gamma \frac{dx}{dt}}_{\text{damping}} + \underbrace{m_e \omega_o^2 x}_{\text{restoring force}} = -e\,\mathscr{E}(t), \tag{13.115}$$

where $m_e$ is the electron mass, $e$ is the magnitude of the electron charge, $\gamma$ is the damping rate, and $\mathscr{E}$(t) is the time dependent electric field of the incident light wave. Monochromatic light wave of angular frequency $\omega$ can be mathematically described as

$$\mathscr{E}(t) = \mathscr{E}_o \cos(\omega t + \Omega) = \mathscr{E}_o \Re e \left[\exp(-i\omega t - \Omega)\right], \tag{13.116}$$

where $\mathscr{E}_o$ is the amplitude and $\Omega$ is the phase of the light. For convenience of solving the differential equation, we choose a complex form for the electric field. The AC electric field will drive oscillations at its own frequency $\omega$. Thus using the standard approach of solving differential equations we adopt the displacement solutions of the time-dependent type given by

$$x(t) = \mathscr{X}_o \Re e \left[\exp(-i\omega t - \Omega)\right]. \tag{13.117}$$

Now substituting Equations (13.116) and (13.117) into Equation (13.115), and cancelling the common complex phase factors we obtain an expression of the form

$$-m_e \omega^2 X_o - i m_e \gamma \omega X_o + m_e \omega_o^2 X_o = -e\,\mathscr{E}_o, \tag{13.118}$$

from which we can solve for the amplitude $X_o$ as

$$X_o = \frac{-e\,\mathscr{E}_o / m_e}{\omega_o^2 - \omega^2 - i\gamma\omega}. \tag{13.119}$$

The time-dependent displacement of the electrons from their equilibirium position induces a time varying dipole moment $p(t)$ given by

$$p(t) = -ex(t). \tag{13.120}$$

A resonant contribution to the macroscopic polarization, defined as dipole moment per unit volume, of the medium is given by

$$\begin{aligned} P_{res} &= np(t), \\ &= -ne\,x(t), \\ &= \left(\frac{ne^2}{m_e}\right) \frac{1}{\omega_o^2 - \omega^2 - i\gamma\omega} \mathscr{E}, \end{aligned} \tag{13.121}$$

where $n$ represents the number of atoms per unit volume and $P_{res}$ is the resonant polarization. As it is clearly evident from Equation (13.121), the polarization amplitude response is small unless $\omega \approx \omega_o$; hence, the use of the qualifier resonant.

As we have learned in previous sections, optical properties are intimately related to dielectric constants. Equation (13.121) can be used to obtain the complex relative dielectric constant $\varepsilon_r$. The electric displacement $\vec{D}$ of the medium is related to the electric field $\vec{\mathscr{E}}$ and polarization $\vec{\mathscr{P}}$ through (see Equation 13.5)

$$\vec{D} = \varepsilon_o \vec{\mathscr{E}} + \vec{P}. \tag{13.122}$$

Since we are interested in resonant phenomenon, we can split the polarization into two contributing parts, referred to as bound ($\vec{P}_{\text{b}}$) and resonant ($\vec{P}_{\text{res}}$). Using Equation (13.121) we can write

$$\begin{aligned} \vec{D} &= \varepsilon_o \vec{\mathscr{E}} + \vec{P}_{\text{b}} + \vec{P}_{\text{res}}, \\ &= \varepsilon_o \vec{\mathscr{E}} + \varepsilon_o \chi \vec{\mathscr{E}} + \vec{P}_{\text{res}}, \\ &= \varepsilon_o \varepsilon_{\text{r}} \vec{\mathscr{E}}, \end{aligned} \tag{13.123}$$

where $\varepsilon_{\text{r}}(\omega)$ is defined as

$$\begin{aligned} \varepsilon_{\text{r}}(\omega) &= 1 + \chi + \left(\frac{ne^2}{\varepsilon_o m}\right) \frac{1}{\omega_o^2 - \omega^2 - i\gamma\omega}, \\ &= 1 + \chi + \chi(\omega), \end{aligned} \tag{13.124}$$

where $\chi(\omega)$ is the frequency-dependent susceptibility. Realizing Equation (13.124), we write the real and imaginary parts as

$$\Re e\left[\varepsilon(\omega)_r\right] \equiv \varepsilon_1(\omega) = 1 + \chi + \left(\frac{ne^2}{\varepsilon_o m}\right) \frac{\omega_o^2 - \omega^2}{(\omega_o^2 - \omega^2)^2 + (\gamma\omega)^2}, \tag{13.125}$$

$$\Im m\left[\varepsilon(\omega)_r\right] \equiv \varepsilon_2(\omega) = \left(\frac{ne^2}{\varepsilon_o m}\right) \frac{\gamma\omega}{(\omega_o^2 - \omega^2)^2 + (\gamma\omega)^2}, \tag{13.126}$$

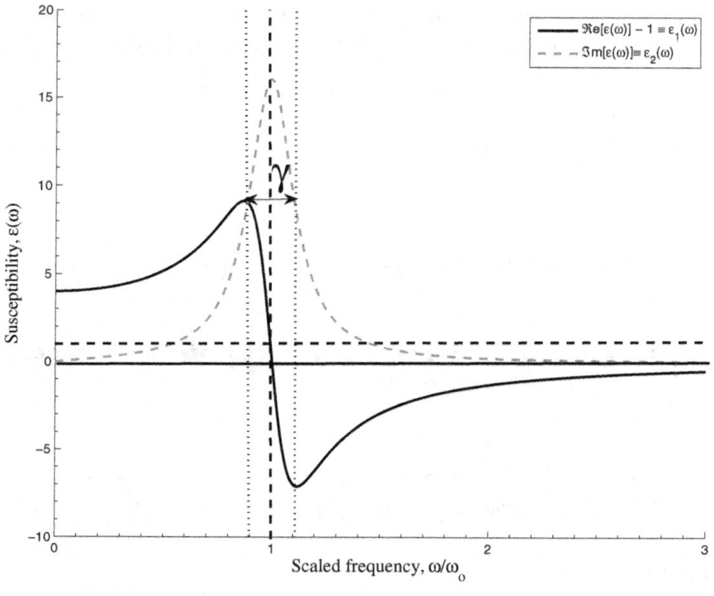


Figure 13.5.8: Frequency dependence of the real and imaginary parts of the dielectric constant.

In Figure 13.5.8 we notice that the real part increases as the frequency is increased. This trend is refered to as **normal dispersion**, where dispersion here refers to the dependence of the refractive index on the wavelength and frequency. But, there is a region near the resonant frequency where the real part decreases with increasing frequency. Historically, this is called **anomalous dispersion**. The anomaly arose due to neglecting the imaginary part of the susceptibility function. This in turn would make the real part singular (infinite) as it approached $\omega_o$. However, experiments did not observe any such trend. This inconsistency between theory and experiment was dubbed *anomalous dispersion*. From a conceptual perspective, the non-inclusion of the imaginary part implies violation of the principle of causality. The complex nature of the susceptibility cures the causality issue making the theory of optical spectroscopy consistent with the principle of relativity. We also notice the following symmetry properties of the susceptibility function

$$\Re e[\chi(-\omega)] = \Re e[\chi(\omega)], \tag{13.127}$$

$$\Re e[\chi(\omega)] = -\Re e[\chi(\omega)]. \tag{13.128}$$

The above relations imply that the real part is symmetric about the origin, but, the imaginary part is asymmetric.

**Example 13.5.0.1**

Using the expressions for the frequency dependent real and imaginary parts of the dielectric function shown in Equations (13.125) and (13.126), write a MATLAB code to generate Figure 13.5.8. Take the static contribution $\chi = 0$ (which produces a constant shift) and ignore the one in Equation 13.125) for simplicity.

**Solution**

The dielectric function involves three different energy scales – resonant frequency $\omega_o$ of the material, the plasma frequency $\omega_p$, and the damping rate $\gamma$. When multiple energy scales are involved, it is typically a good idea to scale them with respect to each other. This allows for an easy comparison and discussion of how the system behaves since it is the relative competition of all the energy scales that ultimately decides how the material will respond optically. We choose $\omega_o$ as the scale relative to which we plot the optical constants. Why $\omega_o$? Well, it is a natural frequency of the system. The two dimensionless frequencies are

$$\overline{\omega_p} = \frac{\omega_p}{\omega_o}, \tag{13.129}$$

$$\overline{\gamma} = \frac{\gamma}{\omega_o}. \tag{13.130}$$

```
%copyright by J. E Hasbun and T. Datta
% ch13_drudelorentz.m

% This script plots the real and imaginary parts of the susceptibility
% derived within the Drude-Lorentz theory.

% Constants
ompscl = 2;    % scaled plasma frequency, \omega_p/\omega_o
gammascl = 0.25; % scaled damping rate, \gamma/\omega_o

% Function definitions
% Frequency dependent real part of the susceptibility (scaled).
% Compare with Equation (13.125). The frequency w is scaled w.r.t omega_{o}

realpart = @(w) (ompscl^2)*(1 - w^2)/((1 - w^2)^2+gammascl^2*w^2);
```

```
% Frequency dependent imaginary part of the susceptibility (scaled).
% Compare with Equation (13.126).
impart = @(w) (ompscl^2)*(gammascl*w)/((1 - w^2)^2+gammascl^2*w^2);

% Plots

figure1 = figure;
axes1 = axes('Parent',figure1,'YMinorTick','on',...
    'XTick',[0 1 2 3 4 5 6 7 8 9 10 11 12],...
    'XMinorTick','on');
hold(axes1,'all')

hold on;
fplot(@(w)realpart(w),[0,3],'LineWidth',2,'-k');
fplot(@(w)impart(w),[0,3],'LineWidth',2,'--r');

% Create xlabel
xlabel('Scaled frequency, \omega/\omega_{o}',...
    'FontSize',14,'FontName','Times New Roman');

% Create ylabel
ylabel('Susceptibility, \epsilon(\omega)','FontSize',14,...
    'FontName','Times New Roman');

% Create legend
legend('\Ree[\epsilon(\omega)] - 1 \equiv \epsilon_{1}(\omega)',...
    '\Imm[\epsilon(\omega)]\equiv \epsilon_{2}(\omega)');
legend(axes1,'show');

% Create textbox
annotation(figure1,'textbox',...
    [0.378380952380953 0.602947368421053 0.0489999999999994 0.118],...
    'String',{'\gamma'},...
    'FontSize',30,...
    'FontName','Times New Roman',...
    'FitBoxToText','off',...
    'EdgeColor',[1 1 1]);

% Create doublearrow
annotation(figure1,'doublearrow',[0.353571428571429 0.417857142857143],...
    [0.632 0.634]);

% Create line
annotation(figure1,'line',[0.354761904761905 0.361904761904762],...
    [0.924 0.108],'LineStyle',':','LineWidth',2);

% Create line
annotation(figure1,'line',[0.388095238095238 0.383333333333333],...
    [0.107270676691729 0.926],'LineStyle','--','LineWidth',2);

% Create line
```

```
annotation(figure1,'line',[0.416666666666667 0.415476190476191],...
    [0.11 0.924],'LineStyle',':','LineWidth',2);

% Create line
annotation(figure1,'line',[0.131897711978466 0.909825033647376],...
    [0.377594249201278 0.381789137380192],'LineWidth',2);

% Create line
annotation(figure1,'line',[0.129761904761905 0.907142857142857],...
    [0.409 0.414],'LineStyle','--','LineWidth',2);

hold off;
```

When frequencies are close to resonance, we can carry out a **resonance approximation**. Under this condition $\omega \approx \omega_o \gg \gamma$, we can approximate $\omega_o^2 - \omega^2$ by $2\omega_o\Delta\omega$, where the detuning from resonance is defined as $\Delta\omega = (\omega - \omega_o)$. Applying the resonance approximation, we find that

$$\Re e\left[\varepsilon(\omega)\right] \equiv \varepsilon_1(\omega) = 1 + \chi + \frac{\omega_p^2}{2\omega_o}\frac{\omega_o - \omega}{(\omega_o - \omega)^2 + (\gamma/2)^2}, \tag{13.131}$$

$$\Im m\left[\varepsilon(\omega)\right] \equiv \varepsilon_2(\omega) = \frac{\omega_p^2}{2\omega_o}\frac{\gamma/2}{(\omega_o - \omega)^2 + (\gamma/2)^2}, \tag{13.132}$$

where $\gamma$ is the full width at half maximum (FWHM) of the imaginary part of the dielectric function, which has a Lorentzian functional form. In the resonance approximation, note that the imaginary part of the susceptibility function is symmetric about $\omega_o$, but, the real part is antisymmetric. This is opposite of what we found earlier.

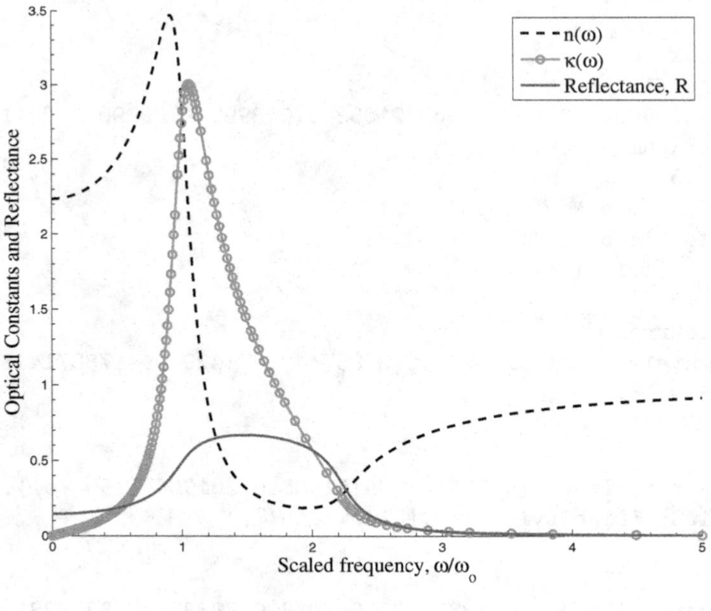


Figure 13.5.9: $n(\omega)$, $\kappa(\omega)$, and reflectance R plot for representative points (see Exercise 20).

The relative dielectric constant in the low and the high frequency limits of $\varepsilon_r(\omega)$ has simple forms. Using Equation (13.124) we have in each limit

$$\varepsilon_r(0) \equiv \varepsilon_{low} = 1 + \chi + \frac{\omega_p^2}{2\omega_o} \frac{\omega_o - \omega}{(\omega_o - \omega)^2 + (\gamma/2)^2}, \tag{13.133}$$

$$\varepsilon_r(\infty) \equiv \varepsilon_{high} = 1 + \chi. \tag{13.134}$$

We have mentioned at the beginning of this section that atoms and molecules have several resonances, not just a single one as in our model. Quantum mechanically, there are multiple electronic resonances and molecules that may exhibit vibrational and rotational levels. If one takes into consideration these effects quantum mechanically it can be shown that the susceptibility becomes a sum over various oscillator strengths

$$\chi(\omega) = \omega_p^2 \sum_j \frac{f_j}{\omega_j^2 - \omega^2 - i\gamma_j\omega}, \tag{13.135}$$

where $\omega_j$ is the frequency for a transition between two electronic states with energy difference $\hbar\omega_j$. $\gamma_j$ is the decay rate for the final state and $f_j$ is the oscillator strength which obeys the **Thomas–Reich–Kuhn sum rule**

$$\sum_j f_j = Z, \tag{13.136}$$

for an atom with Z electrons.This tells us that the total absorption, integrated over all frequencies is dependent only on the atomic number Z.

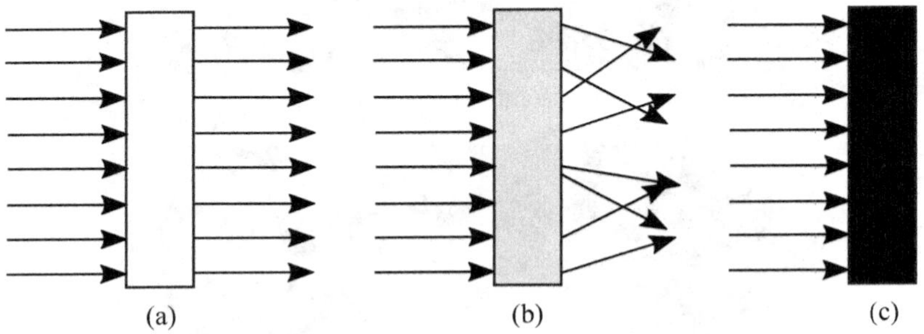


Figure 13.5.10: Light scattering diagram for materials. (a) Transmission, (b) transluscency, and (c) opacity occur as a result of light that is allowed to pass through, be partially scattered, or be completely absorbed.

## 13.6 Optical Behavior of Glass, Metals, and Semiconductors

Light incident on a material can be either reflected, absorbed or transmitted exclusively or a combination of all these phenomena might occur, see Figure 13.5.10. If light passes through the material easily, then the material is capable of transmitting light with minimal absorption or reflection. Such materials are called **transparent**, for example, optical glass as shown in Figure 13.5.10(a). If the material transmits light diffusively, then light is scattered within the material and objects are not clearly visible. Such materials are called **translucent**. For example, frosted glass or thin sheets of plastic, as in Figure 13.5.10(b). Finally, if the light is completely blocked so that the material is impervious to the transmission of visible light, then it is called **opaque**, as in Figure 13.5.10(c). For example, metals and wood. Opacity or transparency of a material is based upon its absorption and transmission properties. The material's composition dictates whether a specific wavelength or a range of wavelength has enough energy to transfer an electron from the ground state to the excited state and in turn for that wavelength to get absorbed. The material thus appears as opaque to that wavelength. If the material transmits most of the incident light, it appears as transparent with respect to the incident light.

The material composition of ordinary glass consists of silica and aluminates. Ultraviolet light possesses adequate energy to excite electrons from the occupied to unoccupied energy levels. Thus incident UV light gets absorbed by glass. However, the energy of visible and infrared light is not enough to excite electrons. In this case most of the incident light gets transmitted. Thus glass appears transparent to visible and infrared light. Transparent materials can appear colored as a consequence of specific wavelength ranges of light that are selectively absorbed; the color discerned is a result of the combination of wavelengths that are transmitted. If the absorption is uniform for all visible wavelengths, the material appears **colorless**. For example high-purity inorganic glasses and high-purity and single-crystal diamonds and sapphire.

(a) Sheesh Mahal (exterior)

(b) Sheesh Mahal ceiling (interior)

Figure 13.6.11: (a) & (b) The house of mirrors or *Sheesh Mahal* is located in the palatial premises of Amer Fort of Jaipur, Rajasthan, India. It was built by Raja Man Singh I in the $16^{th}$ century and completed by 1727. The walls and ceilings of this entire *mahal* or palace are decorated with glass inlaid panels and multi-mirrored ceilings such that even a solitary ray of light can create the effects of thousands of candles. The entire palace is decorated by exquisite mosaics, colored glasses, and fine quality convex-shaped mirrors. Personal collection of author (T. Datta).

The color of a material is determined by the frequency distribution of both transmitted and reemitted light beams resulting from electronic transitions. For example, high-purity single-crystal sapphire ($Al_2O_3$) is colorless. But, if only 0.5-2.0% of $Cr_2O_3$ is added, the material looks red. This red-colored material is popularly known as the ruby gemstone. From a solid state physics perspective, the presence of Cr which substitutes for the Al, introduces impurity levels in the band gap of sapphire. These levels allow for strong absorption between the 560 nm-520 nm (green) to 490 nm-450 nm (blue) range. Thus, only red is transmitted. Glasses have played a vital role in history including one of displaying opulence, see Figure 13.6.11.

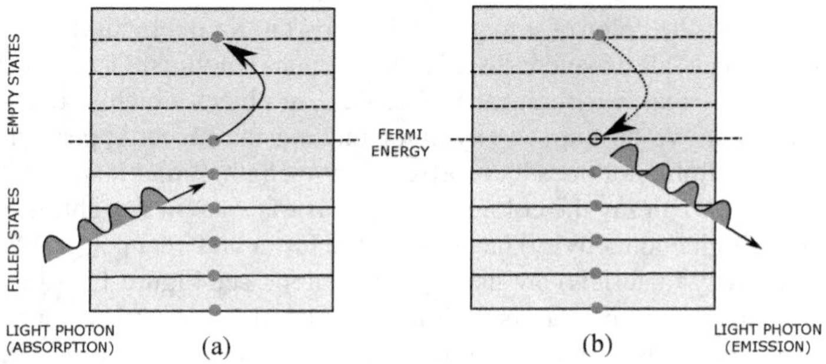


Figure 13.6.12: Schematic light absorption–emission process in metals. The optical properties of light can be explained by considering the fact that metals have a set of available unoccupied energy levels into which the electrons can be excited upon absorption of a light photon. Based on this simple energy level picture, metals thus appear opaque.

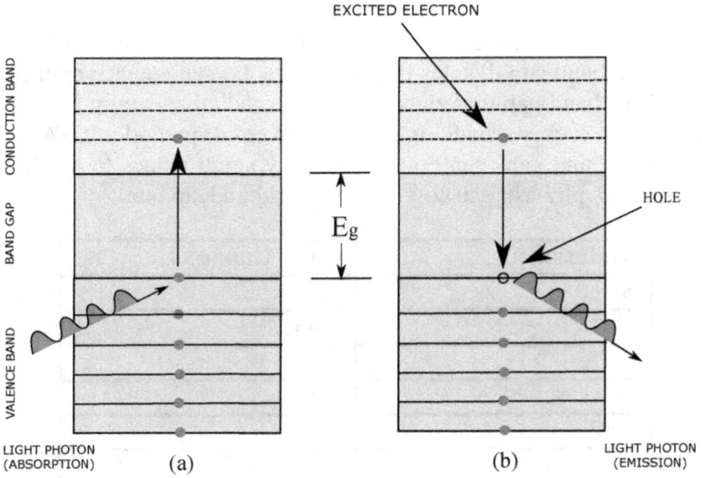


Figure 13.6.13: Schematic light absorption–emission process in insulators. As described in the text, the size of the band gap controls the optical properties.

Metals are opaque because visible light excites electrons into unoccupied energy states above the Fermi energy, as demonstrated in Figure 13.6.12. Total absorption is within a very thin outer layer, usually less than 0.1 $\mu$ thus only metallic films thinner than 0.1 $\mu$m are capable of transmitting visible light. In fact, metals are opaque to all electromagnetic radiation on the low end of the frequency spectrum, from radio waves, through infrared, the visible, and into about the middle of the ultraviolet radiation. Metals are transparent to high-frequency (x- and $\gamma$-ray) radiation. All frequencies of visible light are absorbed by metals because of the continuously available empty electron states, which permit electron transitions. Most of the absorbed radiation is reemitted from the surface in the form of visible light of the same wavelength, which appears as reflected light. Aluminum and silver are two metals that exhibit this reflective behavior. Copper and gold appear red-orange and yellow, respectively, because some of the energy associated with light photons having short wavelengths is not reemitted as visible light. Nonmetallic materials may be opaque or transparent to visible light; and, if transparent, they often appear colored. Every nonmetallic material becomes opaque at some wavelength, which depends on the magnitude of its bandgap.

The visual appearance and the color of a material depends on the interaction of the electrons with light. A white object reflects all wavelengths of light in equal amounts. Our eyes detect this entire range of wavelengths, thus making it appear white. But, an object which is black absorbs all the wavelengths of light. Thus we do not see any color making the object appear black. A colorless object does not reflect the light, nor does it absorb any wavelength, rather it simply allows the wavelengths to pass through. We can say the colourless object is transparent to light. As a general rule of thumb, we find metals (high conductivity) have a metallic luster and are opaque. Insulators (high resistivity or low conductivity materials) are usually transparent, see Figure 13.6.13. Semiconductors which are in between metals and insulators with their conducting properties can be both opaque or transparent. Their color depends on the size of the band gap. If the energy of the incident photon is greater than the band gap, then all the photons will be absorbed. Consequently, it will appear black. If the photon energy is less than the gap, then the photons will be transmitted. For photon energies that lie within the energy gap range, only those with energy higher than the band gap will be absorbed. The rest are allowed to pass through (transmitted). If all the colors are transmitted, then the perceived light is white in color. For example, Si has an energy gap of 1.11 eV, see Table 13.6.4. Given that visible light has photons in the energy range of approximately 1.8 eV to 3.1 eV, all the energy from visible light can be absorbed. Thus, silicon appears black.

Table 13.6.4: The energy gap between the valence and the conduction band is referred to as the band gap. The size of the band gap dictates the potential solid state device application of the semiconducting material. Low band gap materials require less radiative energy for transition across the gap. The process of photon energy transfer is used in light emitting diodes. GaAs and GaPs are used in LEDs. Semiconductor band gaps can range from very small values to all the way up to 6 eV. *Source*: http://hyperphysics.phy-astr.gsu.edu/hbase/Solids/bandgap.html.

| **Material** | **Band Gap (eV)** |
|---|---|
| Si | 1.11 |
| GaAs | 1.43 |
| CdTe | 1.58 |
| AlAs | 2.16 |
| GaP | 2.26 |
| Diamond | 5.5 |

**Example 13.6.0.1**

Diamond, a solid form of carbon with a diamond cubic crystal structure, is created when carbon is subjected to extremely high pressures and temperatures found at the earth's lithosphere (the rigid outer part of the earth, consisting of the crust and upper mantle). The earliest diamonds have been known to be found in India. A majority of these early stones were transported along the network of trade routes that connected India and China, commonly known as the Silk Road. Currently, South Africa, Russia, Botswana, and Australia are some of leading producers of gem and industrial diamond, respectively. Diamond has the highest hardness and thermal conductivity of any bulk material. These properties make it useful in cutting and polishing tools and in scientific applications such as diamond knives and diamond anvil cells. Explain why diamond is transparent?

**Solution**

From Table 13.6.4 we know that diamond has a band gap of 5.5 eV. Since visible light lies between 1.8 eV to 3.1 eV, a diamond due to its large band gap will allow all of visible light to pass through making it transparent.

Depending on the material structure, light may be scattered so that it is not coherently transmitted. Even after the light has entered the material, it will undergo internal scattering. Thus a beam of light will spread out or the image will become blurred. In extreme cases, the material could become opaque due to excessive internal scattering. Scattering can come from obvious causes such as grain boundaries in poly-crystalline materials. In highly pure materials, scattering still occurs and an important contribution comes from Rayleigh scattering. This is due to small, random differences in refractive index from place to place. In amorphous materials such as glass this is typically due to density or compositional differences in the random structure.

In the preceding paragraphs we discussed light scattering through solid state materials. It is possible for light to scatter from molecular aggregates, polymers, colloids, or gases during the process of transmission with subsequent reemission that has no preferential propagation direction. In **Rayleigh scattering** a photon interacts with orbiting electron and is deflected without any change in photon energy (elastic scattering). This is significant for high atomic number atoms and low photon energies. For example, the blue color in the sunlight gets scattered more than other colors in the visible spectrum and thus making the sky look blue. In contrast, in the **Tyndall effect**, the scattering occurs from particles much larger than the wavelength of light thereby giving clouds the whitelike appearance. In contrast to elastic scattering, one can also have inelastic light scattering of light by molecules excited to higher vibrational or rotational energy levels. This form of scattering is called **Raman scattering**. It was discovered by the Indian Nobel Laureate Sir Chandrasekhara Venkata Raman (1888-1970) and his student K. S. Krishnan (1898-1961) who demonstrated the effect to occur in solids, liquids, and vapors. It is also widely accepted that G. Landsberg (1890-1957) and L. Mandelstam (1879-1944) disovered this effect independently. It was predicted theoretically by the Austrian theoretical physicist Adolf Smekal (1895-1959).

## 13.7 Optical Spectroscopy

**Spectroscopy** is the analysis of the interaction between matter and any part of the EMS. For example, visible light passing through a prism produces a rainbow pattern. The decomposition of white light into its constituents allows us to infer the composition of visible light. In this section we are concerned with optical spectroscopy which plays a crucial role in physics, chemistry, and biology. In Table 13.7.5 we list the many commonly applied spectroscopic techniques spanning across the entire EMS.

Table 13.7.5: Common spectroscopic techniques (beyond optical) utilized at present to study interactions in solid state materials. This chapter is focused on a tiny portion of this vast spectrum. UV, visible, and infrared all use optical materials to disperse and focus the radiation. Thus, the term optical spectroscopy is typically associated with spectrometers used to study the interaction of all these three forms of radiation. The frequency, wavelength, and energy of these regions are mentioned in Table 13.1.1.

| Energy | Frequency | Wavelength | Spectroscopy | Information |
|---|---|---|---|---|
| 1 $\mu$eV – 1 meV | 1 GHz – 1 THz | 1 m – 1 mm | NMR<br>ESR | nuclear spin<br>electron spin |
| $\approx$ 1 eV | < 1 PHz<br>(near IR) | $\approx 1\ \mu$m | Raman | molecular vibrations<br>phonon dispersion |
| 1 eV – 10 eV | $\approx$ PHz<br>(VIS – UV) | 1 $\mu$m – 1 nm | Optical | band electrons<br>interband transitions<br>plasmon, magnon |
| $\approx$ 1 keV (soft)<br>> 1 keV (hard) | $\approx$ 1 EHz<br>(X–ray) | $\approx$ 1 – 0.1 nm | XAS, XES<br>EXAFS, RIXS<br>XANES, EXAFS | core electron<br>magnon, bimagnon<br>trimagnon |
| Particle beam<br>0.1 meV<br>– 500 meV | | $\approx$ 0.1 nm | neutron | magnon excitation<br>exchange interaction |
| 100 eV | | | ARPES | band structure |

Spectroscopic approaches can be broadly classified into two categories. In one category, energy is transferred between the sample and the photon. Within this class, in **absorption spectroscopy** a photon is absorbed by an atom or molecule, which undergoes a transition from a lower-energy state to a higher energy, or excited state. An electron inside a solid state material may end up in an excited state by exchanging energy via thermal means, by absorption of a photon, or by a chemical reaction. In all these cases, the number of photons passing through the sample decreases. The measurement of this decrease in photons, which we call **absorbance**, is a useful analytical signal. Based along these lines of reasoning we can derive a formula for the intensity.

Consider a monochromatic electromagnetic radiation passing through an infinitesimally thin layer of sample thickness $dz$. Let us take the intensity $I(z)$, which is the optical power per unit area, to be $z$ dependent. Then the decrease in intensity dI after passing through the dz layer can be related to the thickness and concentration as

$$dI = -\alpha C dz I(z), \tag{13.137}$$

where $\alpha$ is the **absorptivity**, $C$ is the concentration, and $z$ the path length. Integrating the above equation we get the Beer–Lambert (or more commonly Beer's) law to obtain

$$I(z) = I_o e^{-\alpha C z}, \tag{13.138}$$

where $I_o$ is the optical intensity at $z = 0$.

Historically speaking this law was first discovered by the French mathematician Pierre Bouger (1698 – 1758). Later the Swiss polymath Johann Heinrich Lambert (1728-1777) and the German chemist August Beer (1825-1863) reintroduced it in different forms that relate to the sample thickness and concentration, respectively. The modern version of the law, Equation (13.138), incorporates both these effects. A plot of absorbance as a function of the photon's energy is called an **absorbance spectrum**. The optical properties described in the previous paragraphs lead to various optical spectroscopic techniques. Below we list and qualitatively describe a select few of them.

i. **UV–Vis spectroscopy**: The instrument used for UV–Vis spectroscopy is called a **spectrophotometer**. A spectrophotometer is an optical instrument which measures absorbance or transmittance from the UV range to the visible range (sometimes in the NIR also) relative to the wavelength of light. This particular type of spectroscopy reveals information on electronic transitions which can provide valuable clues to the structure of a molecule and molecular properties such as color. The first commercially available UV–Vis spectrophotometer was introduced in 1940 by Arnold O. Beckman, Howard Cary, and colleagues from the National Technologies Laboratories. UV–Vis spectroscopy finds routine use in analytical chemistry and biochemistry.

ii. **Atomic absorption (AA) spectroscopy**: The electronic configuration of each atom is unique to itself. Thus every atom has a distinct energy fingerprint where only specific energies will be absorbed. During the mid-19th century, a German physicist Gustav Robert Kirchoff (1824 – 1887) and a German chemist Robert Wilhelm Eberhard Bunsen (1811-1899) utilized this fundamental concept to introduce atomic absorption as a spectroscopic procedure. Moving forward, in 1953 the Australian physicist Sir Alan Walsh (1916-1998) demonstrated that atomic absorption could be used as a quantitative analytical tool for the determination of single elements in compounds. The first commercial atomic absorption spectrometer was introduced in 1955.

    In AA the sample, which could be a liquid or a solid, is atomized in either a flame or a graphite furnace to create gas–phase atoms. Upon the absorption of ultraviolet or visible light, the free atoms undergo electronic transitions from the ground state to excited electronic states. The incident light beam is attenuated by atomic vapor absorption according to Beer's law (at the most simple level of description). In order to tell how much of a known element is present in a sample, one must first establish a basis for comparison using known quantities. It can be done producing a calibration curve. For this process, a known wavelength is selected, and the detector will measure only the energy emitted at that wavelength. However, as the concentration of the target atom in the sample increases, absorption will also increase proportionally. Thus, one runs a series of known concentrations of some compound, and records the corresponding degree of absorbance, which is an inverse percentage of light transmitted. A straight line can then be drawn between all of the known points. From this line, one can then extrapolate the concentration of the substance under investigation from its absorbance. The uniqueness of energy level arrangement of each atom allows for the qualitative analysis of a pure sample. This particular spectroscopy finds widespread use in nanoscience, biophysical applications, food technology, petrochemical industry and forensic sciences to name a few.

iii. **Infrared spectroscopy**: Infrared was discovered in 1800 by the German astronomer Sir Frederick William Herschel (1738-1822), who discovered a type of invisible radiation in the spectrum lower in energy than red light, by means of its effect on a thermometer. Infrared spectroscopy (IR spectroscopy or vibrational spectroscopy) involves the interaction of infrared radiation with matter. In contrast to UV spectroscopy, the infrared spectrum provides a rich array of absorption bands which can provide a wealth of structural information about a molecule. The information contained in the IR spectrum originates from molecular vibrations. The method or technique of infrared spectroscopy is conducted with an instrument called an infrared spectrometer (or spectrophotometer) which produces an infrared spectrum. As in the case of the other absorption spectroscopy techniques, an IR spectrum can be visualized in a graph of infrared light absorbance (or transmittance) on the vertical axis vs. frequency or wavelength on the horizontal axis. The energies are affected by the shape of the molecular potential energy surfaces, the masses of the atoms, and the associated vibrational modes.

In **emission spectroscopy** an atom, molecule, or electrons inside a solid state material can be promoted to an excited state. When the system deexcites *spontaneously* the excess energy is then released as a photon, a process we call **emission.** In **luminescence** the material absorbs energy and then immediately emits visible or near-visible radiation in the range 1.8 eV $< h\nu <$ 3.1 eV. Emission following the absorption of a photon is also called **photoluminescence**, and that following a chemical reaction is called **chemiluminescence**. Sometimes the reemission of radiation occurs within nanoseconds after the excitation, in which case the luminescence is called **fluorescence**, and if it takes longer than nanoseconds, such as micro- or milliseconds then it is known as **phosphorescence**. If it takes of the order of seconds, then it is called an **afterglow**. Special materials called phosphors have the capability of absorbing high-energy radiation and spontaneously emitting lower-energy radiation (remember the green CRT screens in oscilloscopes?). Sulfides (ZnS), oxides (ZnO with surplus Zn), tungstates, rare earth elements such as $Eu^{3+}$ or Tb, and a few organic materials exhibit phosphorescence. Ordinarily, pure materials do not display these phenomena and, to induce them, impurities in controlled concentrations must be added.

Luminescence can be classified based on the energy source for electron excitation. **Photoluminescence** occurs in fluorescent lamps. Arc between electrodes excites electrons within mercury lamp to higher energy levels. Electrons fall back emitting UV light. Fluorescent lamps consist of a glass housing, coated on the inside with specially prepared tungstates or silicates. Ultraviolet light is generated within the tube from a mercury glow discharge, which causes the coating to fluoresce and emit white light. **Cathode-luminescence** is produced by an energized cathode which generates a beam of high-energy bombarding electrons. Applications of this include electron microscope; cathode-ray oscilloscope; and color television screens. Spontaneous light emission can also occur in candles or incandescent bulbs. In these two cases the electrons are excited to a higher state by heat energy and the process is called **thermoluminescence**.

**Electroluminescence** occurs in devices with p–n rectifying junctions which are stimulated by an externally applied voltage. When a forward biased voltage is applied across the device, electrons and holes recombine at the junction and emit photons in the visible range (monochromatic light, i.e., singe color). These diodes are called light emitting diodes (LEDs). LEDs emit light of many colors, from red to violet, depending on the composition of the semiconductor material used. For example, GaAs, GaP, GaAlAs, and GaAsP are typical materials for LEDs.

The optical properties described in the emission spectroscopy type lead to the **atomic emission spectroscopy**, **fluorescence spectroscopy**, and **phosphorescence spectroscopy**.

In the second broad class of spectroscopic techniques, the electromagnetic radiation undergoes a change in amplitude, phase angle, polarization, or direction of propagation as a result of its refraction, reflection, scattering, diffraction, or dispersion by the sample. X-ray diffraction is a prime example of such a technique. Other techniques in the UV–V is domain include refractometry, nephelometry, turbidimetry, and optical rotatory dispersion. Some of the other forms of spectroscopy that we have not discussed (since they are not optical) include Mössbauer, x-ray absorption, atomic fluorescence, and chemiluminescence spectroscopy.

## 13.8 Kramers-Kronig Relationship

** This section is optional and can be skipped.

Susceptibility is a complex variable, see Equation (13.124). Thus we can apply the concepts of complex analysis to extract useful information about it. Such an approach was formulated by the collective efforts of a Dutch physicist Hendrik Anthony Hans Kramers (1894-1952) and a German American physicist Ralph Kronig (1904-1995). We will utilize two basic theorems of complex analysis to initiate our discussion. The first is the **Cauchy integral theorem** which states that if we define a close path in the complex plane, then for a function $f(z)$ that is analytic everywhere inside the path we have

$$\oint \frac{f(z)}{z-z'} = \begin{cases} 2\pi i f(z'), & \text{if } z' \text{ lies inside the closed path,} \\ 0 & \text{otherwise.} \end{cases} \quad (13.139)$$

The second equation is the **Dirac formula**

$$\frac{1}{x-x'-i\varepsilon} = \mathscr{P}\frac{1}{x-x'} + i\pi\delta(x-x'), \quad (13.140)$$

where $\varepsilon$ is an infinitesimal, real number, and $\mathscr{P}$ is the principal value.
A generic complex form of the susceptibility can be written as

$$\chi(\omega) = C\frac{1}{(\omega_o^2-\omega^2)+i\Gamma\omega} \quad (13.141)$$

The above function is strongly peaked around $\omega = \omega_o$. With the approximation $|\omega-\omega_o| \ll \omega_o$, we can rewrite the above equation as

$$\begin{aligned} \chi(\omega) &= C\frac{1}{(\omega_o-\omega)(\omega_o+\omega)+i\Gamma\omega}, \\ &\approx \frac{1}{2\omega_o}C\frac{1}{\omega_o-\omega+i\Gamma/2}. \end{aligned} \quad (13.142)$$

We find that there is a pole at $\omega = \omega_o + i\Gamma/2$ in the upper half of the complex plane. *Causality* requires that the susceptibility cannot have a pole in the lower half of the complex plane, irrespective of the physical details of the system. It is possible to reason this fact using mathematics as follows. We know that

$$\vec{P}(\omega) = \chi(\omega)\varepsilon_o\vec{E}(\omega). \quad (13.143)$$

Now consider $\vec{E}$ to be given by a $\delta$-function in time which amounts to a short impulse. Since this corresponds to a constant electric field in space, we can write $\vec{E}_i$. If we now take the Fourier transformation of $\vec{P}(\omega)$ to get $\vec{P}(t)$, we obtain

$$\begin{aligned} \vec{P}(t) &= \frac{C\varepsilon_o\vec{E}_i}{4\pi\omega_o}\int_{-\infty}^{\infty} d\omega\, e^{i\omega t}\frac{1}{\omega_o-\omega+i\Gamma/2}, \\ &= \frac{C\varepsilon_o\vec{E}_i}{2\pi\omega_o} i\Theta(t)e^{i(\omega_o-\Gamma/2)t}, \end{aligned} \quad (13.144)$$

where we used the contour integral definitions as in Equation (13.139). The Heavyside (step) function $\Theta(t)$ is defined as

$$\Theta(t) = \begin{cases} 1 & t > 0, \\ 0 & t > 0. \end{cases} \quad (13.145)$$

Interestingly enough, if the sign on the exponential were negative, rather than a decay we would have a growth of the polarization preceding the impulse, thereby violating the principle of casuality. We observe that the total susceptibility $\chi(\omega)$ of a medium is simply equal to the sum of the susceptibilities of all the individual resonances with different relative weight factors C. Based on the above properties, we can deduce some useful formula. Since causality requires that $\chi(\omega)$ cannot have a pole in the lower half of the complex plane, the function

$$f(\omega) = \frac{\chi(\omega)}{\omega'-\omega+i\varepsilon}, \quad (13.146)$$

is analytic in the entire lower half-plane, where $\varepsilon$ is an infinitesimal real, positive number. Since $f(\omega)$ decreases as $\frac{1}{\omega^2}$, we can use the standard trick in analytical calculus of creating a loop in the complex plane by taking the real axis as one part of the loop and completing the loop somewhere very far away in the lower-halfplane. Then we can use the theorem stated in Equation (13.139) to perform the integral in the real axis to obtain

$$\int_{-\infty}^{\infty} d\omega \, \frac{\chi(\omega)}{\omega' - \omega + i\varepsilon} = 0. \tag{13.147}$$

Now using the Dirac Delta function definition from Equation (13.140), we get

$$\mathscr{P} \int_{-\infty}^{\infty} d\omega \, \frac{\chi(\omega)}{\omega' - \omega} + i\pi \, \chi(\omega') = 0. \tag{13.148}$$

If we interchange the $\omega$ and $\omega'$, we can write

$$\chi(\omega) = \frac{1}{i\pi} \mathscr{P} \int_{-\infty}^{\infty} d\omega' \, \frac{\chi(\omega')}{\omega' - \omega}. \tag{13.149}$$

Since the susceptibility is a complex number, we write

$$\chi(\omega) = \Re e\chi(\omega) + i\Im m\chi(\omega), \tag{13.150}$$

and insert it into Equation (13.149) to obtain the **Kramers–Kronig** relations as

$$\Re\text{e}[\chi(\omega)] = \frac{1}{\pi} \mathscr{P} \int_{-\infty}^{\infty} d\omega' \frac{\Im\text{m}[\chi(\omega')]}{\omega' - \omega}, \tag{13.151}$$

$$\Im\text{m}[\chi(\omega)] = -\frac{1}{\pi} \mathscr{P} \int_{-\infty}^{\infty} d\omega' \frac{\Re\text{e}[\chi(\omega')]}{\omega' - \omega}, \tag{13.152}$$

The relation between the $\Re e\chi(\omega)$ and the $\Im m\chi(\omega)$ is known mathematically as a Hilbert transform, analogous to the Fourier transform but with the $\frac{1}{(x-y)}$ replaced with $e^{xy}$.
The Kramers–Kronig relations can be recast in a useful form if we assume that the susceptibility has the following symmetry properties

$$\Re\text{e}[\chi(-\omega)] = \Re\text{e}[\chi(\omega)], \text{ (real function)}, \tag{13.153}$$

$$\Im\text{m}[\chi(-\omega)] = -\Im\text{m}[\chi(\omega)], \text{ (odd function)}. \tag{13.154}$$

To utilize the above properties, we break Equation (13.149) into two parts, as follows

$$\chi(\omega) = \frac{1}{i\pi} \mathscr{P} \int_{0}^{\infty} d\omega' \, \frac{\chi(\omega')}{\omega' - \omega} + \frac{1}{i\pi} \mathscr{P} \int_{-\infty}^{0} d\omega' \, \frac{\chi(\omega')}{\omega' - \omega}. \tag{13.155}$$

Now using the properties that the $\Re e\chi(\omega)$ is even and $\Im m\chi(\omega)$, we have

$$\begin{aligned} \Re\text{e}[\chi(\omega)] + i\Im\text{m}[\chi(\omega)] &= \frac{1}{i\pi} \mathscr{P} \int_{0}^{\infty} d\omega' \, \Re\text{e} \, \chi(\omega') \left( \frac{1}{\omega' - \omega} - \frac{1}{\omega' + \omega} \right), \\ &+ \frac{1}{\pi} \mathscr{P} \int_{0}^{\infty} d\omega' \, \Im\text{m} \, \chi(\omega') \left( \frac{1}{\omega' - \omega} + \frac{1}{\omega' + \omega} \right), \end{aligned} \tag{13.156}$$

Combining the common terms we have

$$\Re\mathrm{e}[\chi(\omega)] + i\Im\mathrm{m}[\chi(\omega)] = \frac{1}{i\pi}\mathscr{P}\int_0^\infty d\omega'\, \Re\mathrm{e}\,\chi(\omega')\left(\frac{2\omega}{\omega'^2-\omega^2}\right), + \frac{1}{\pi}\mathscr{P}\int_0^\infty d\omega'\, \Im\mathrm{m}\,\chi(\omega')\left(\frac{2\omega'}{\omega'^2-\omega^2}\right). \tag{13.157}$$

Finally, equating the real and imaginary parts we have on each side, we have

$$\Re\mathrm{e}[\chi(\omega)] = \frac{2}{\pi}\mathscr{P}\int_0^\infty d\omega' \frac{\Im\mathrm{m}[\chi(\omega')]\omega'}{\omega'^2-\omega^2}, \tag{13.158}$$

$$\Im\mathrm{m}[\chi(\omega)] = -\frac{2}{\pi}\mathscr{P}\int_0^\infty d\omega' \frac{\Re\mathrm{e}[\chi(\omega')]\omega}{\omega'^2-\omega^2} \tag{13.159}$$

The physical implications of the Kramers–Kronig relations are as follows. With knowledge of the real part of the susceptibility over the entire frequency range, we can compute the imaginary part, or vice versa. This means that if we know the absorption spectrum of a material, we can compute the index of refraction over the entire wavelength range without any additional measurements, and vice versa. Thus even from a mathematical point of view we recover the fact that absorption and refraction are not independent properties of a material, but, rather come from the same underlying physical mechanism.

## 13.9 Chapter 13 Exercises

13.9.1. What is the wavelength and frequency range for the visible part of the electromagnetic spectrum?

13.9.2. Terahertz radiation has applications ranging from medical imaging, national security, to scientific imaging in submillimeter astronomy. What is the wavelength of electromagnetic radiation corresponding to terahertz waves of frequency $10^{12}$ Hz?

13.9.3. What are the three different processes that light can undergo when interacting with an optical medium?

13.9.4. What microscopic process within a material causes light waves to refract? What is the speed of light traveling through a medium of refractive index of 1.33.

13.9.5. The band gaps of some common semiconductors are listed in Table 13.6.4. What frequency of radiation will cause a transition across the band gap for GaAs?

13.9.6. Derive Maxwell's wave Equation (13.23) relating the space and time derivatives of $\vec{B}$.

13.9.7. Derive Equation (13.34).

13.9.8. In Figure 13.9.14 we show electromagnetic wave $\{\vec{E}_i, \vec{B}_i, \vec{k}_i\}$ incident on the interface of two media with refractive indices $n_1$ and $n_2$.

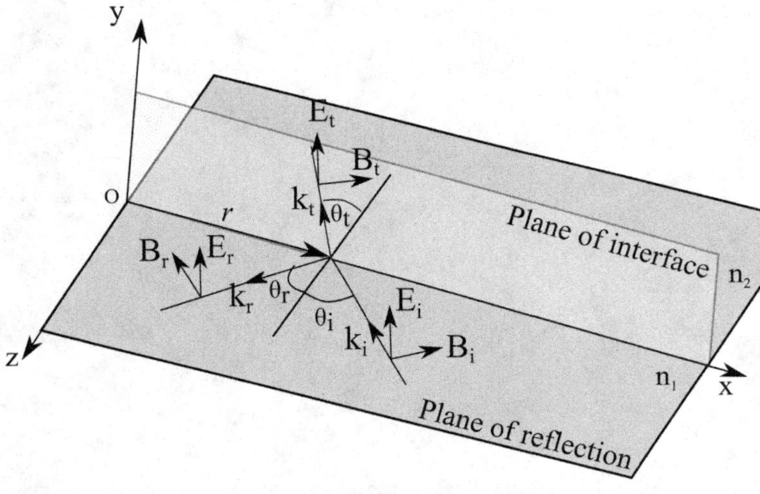


Figure 13.9.14: Transverse electric (TE) mode scattering, also known as $S$ or $\sigma$ scattering. The scattering is of the transverse type because the incident electric field is perpendicular to the plane of incidence. For an electric field parallel to the plane of incidence the scattering is called transverse magnetic (TM), $P$, or $\pi$ type. In the TM case, the magnetic field sticks out of the plane of incidence.

The reflected and transmitted waves are denoted by $\{\vec{E}_r, \vec{B}_r, \vec{k}_r\}$ and $\{\vec{E}_t, \vec{B}_t, \vec{k}_t\}$, respectively. The angle of incidence, reflection, and of the transmitted ray is given by $\theta_i$, $\theta_r$, and $\theta_t$, in that order. At the interface of two media, the parallel and perpendicular components of the electric and magnetic fields obey boundary conditions (in the absence of free charges and free currents) given by

$$\varepsilon_1 E_1^{\perp} = \varepsilon_2 E_2^{\perp}, \tag{13.160a}$$
$$E_1^{\parallel} = E_2^{\parallel}, \tag{13.160b}$$
$$B_1^{\perp} = B_2^{\perp}, \tag{13.160c}$$
$$\frac{1}{\mu_1} B_1^{\parallel} = \frac{1}{\mu_2} B_2^{\parallel}. \tag{13.160d}$$

Using the boundary conditions, derive the law of reflection which states that $\theta_i = \theta_r$.

13.9.9. For the TE wave set-up of Problem 13.9.8 derive the law of refraction which states that $n_i \sin\theta_i = n_t \sin\theta_t$.

13.9.10. For the TE wave set-up of Problem 13.9.8 derive the Fresnel reflection $(r_{TE})$ and transmission $(t_{TE})$ coefficients.

13.9.11. For the TE wave set-up of Problem 13.9.10 show that the result reduces to Equation (13.99) for normal incidence.

13.9.12. Derive Maxwell's wave Equation (13.62) in a conducting medium with no free charges.

13.9.13. Derive Equations (13.70a) and (13.70b) relating the real and imaginary parts of the complex dielectric constant with the refractive index.

13.9.14. Derive Equations (13.71a) and (13.71b).

13.9.15. Derive Equation (13.86).

13.9.16. (a) Calculate the plasma frequency in Hz for (i) Li (ii) Na, and (iii) K. The free electron density for these metals are $n = 4.70 \times 10^{28}, 2.65 \times 10^{28}$, and $1.40 \times 10^{28}$ per $m^3$, respectively. The experimentally observed plasma frequency (in $\times 10^{14}$ Hz) are 14.6, 14.3, and 9.52, see Reference [61].

(b) The **effective number of free electrons,** $N_{eff}$ is a parameter that provides us with information on how many free electrons per atom contribute to the electron gas. This parameter appears in several nonoptical equations such as Hall constant and superconductivity. Thus it is important to get a sense of its numerical value. We can define $N_{eff}$ as

$$N_{eff} = \left( \frac{\omega_p(\text{observed})}{\omega_p(\text{calculated})} \right)^2. \tag{13.161}$$

Calculate $N_{eff}$ for (i) Li (ii) Na, and (iii) K.

13.9.17. Derive the reflectivity Equation (13.114) in the high frequency regime governed by the conditions $\omega \gg \omega_p$ and $\omega\tau \gg 1$.

13.9.18. Gallium phosphide (GaP) is a compound semiconducting material that has been used in the manufacture of low-cost red, orange, and green light-emitting diodes (LEDs) with low to medium brightness since the 1960s. Pure GaP LEDs emit green light at a wavelength of 555 nm. Nitrogen-doped GaP emits yellow-green (565 nm) light, zinc oxide doped GaP emits red (700 nm). Gallium phosphide has applications in optical systems, also. The biological pigments of the cones in the human eye have maximum absorption values at wavelengths of about 420 nm (blue), 534 nm (bluish-green), and 564 nm (yellowish-green). Given that the human eye is more sensitive to yellow than it is to red, what color will we perceive GaP to exhibit?

13.9.19. Write a MATLAB code to reproduce Figure 13.5.9 of the refractive index, kappa, and reflectance within the Drude–Lorentz theory.

# 14

# *Transport Properties of Solids*

**Contents**



## 14.1 Introduction

Our current technological revolution relies heavily on our knowledge of electrical conduction in solids. Electrical conduction is part of a broader class of phenomena known as transport properties. Transport of mass leads to **diffusion**, transport of energy leads to **thermal conduction**, and transport of charge gives rise to **electrical conduction**. Broadly speaking, the electrical conductivity values can be used to classify materials in three different categories: **conductors**, **semiconductors**, and **insulators**. Electrical conductivity can range from $10^{-20}$ $(\Omega\ \text{m})^{-1}$ (insulators) to $10^{4}$ $(\Omega\ \text{m})^{-1}$ (semiconductors), to $10^{7}$ $(\Omega\ \text{m})^{-1}$ (metals). In this chapter we will utilize a Boltzmann transport equation formalism to develop a conceptual framework to treat the electric and thermal conduction behavior. We also discuss the physics of **thermoelectric** behavior such as the **Seebeck effect**, **Thomson effect** and **Peltier effect**.

Transport theory can be formulated from two different perspectives, see Figure 14.1.1. In one conceptual viewpoint, say *framework A*, the electrical current is generated due to an applied electric field. Thus, in this scenario the field *causes* the current *response*. In another approach, say *framework B*, the voltage builds up in *response* to the current flow. The current flux is thus determined by the boundary conditions at the interface of the sample that is being investigated. An inhomogeneous electric field is created across the sample as a consequence of the current flow. These two viewpoints form the conceptual foundations of modern transport theory. In framework A, we have theories such as **Drude conductivity**, **Kubo formalism**, and the **Boltzmann equation** approaches. In the second framework B, we have the **Landauer theory** of transport, a conceptual framework for analyzing transport in *nano-* or *meso-*scopic systems.

In simple metals, resistance to electrical motion arises due to impurities, phonon scattering, or both. At low temperatures (relative to the Debye temperature) the impurity scattering dominates over

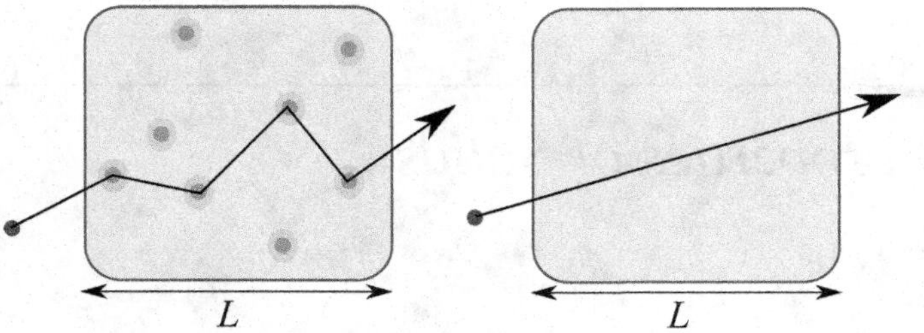


(a) Diffusive (b) Ballistic

Figure 14.1.1: Transport phenomena can exhibit two broad regimes of interest. (a) In the diffusive regime, the electron scatters either elastically or inelastically from the scattering centers as it traverses through the material. The Drude theory or the Boltzmann approach is an appropriate theoretical formulation to describe the associated transport behavior. (b) In the opposite ballistic regime, the charge carrier traverses through the system without any scattering process. All the inelastic processes are in the contacts or leads connected to the ballistic device. The Landauer theory of transport is then the appropriate strategy.

the phonon scattering and at high temperatures (relative to the Debye temperature $\Theta_D$) it is the opposite scattering mechanism. In 1900, the German physicist Paul Karl Ludwig Drude (1863–1906) formulated a theory of electrical conduction in metals where he treated electrons as classical particles, akin to gas molecules which can scatter off the localized ionic cores. The model assumed the following conditions related to electronic collision:

- the electrons collide only with the heavy and relatively stationary ionic cores,
- in between collisions, the electrons do not interact with each other or with other ions. These two approximations are known as the **independent electron approximation** and the **free electron approximation**, respectively,
- the collisions are instantaneous and result in abrupt velocity changes,
- the collision probability per unit time is given by $1/\tau$, also known as the scattering rate,
- electrons achieve thermal equilibrium with their surroundings only through collisions.

Using the above assumptions, Drude was able to explain an empirical fact discovered in 1853 by Wiedemann and Franz, which states that the ratio of electrical and thermal conductivities at a given temperature is the same for all metals, see Equation (14.70) associated with the **Wiedemann-Franz law**. This excellent agreement between theory and experiment was a short-lived success of the Drude theory. It was soon realized that the Drude theory failed to account for the temperature dependence of electrical and thermal conductivities. Furthermore, if one uses Drude's ideas and treats electrons as classical particles and applies the above argument to the electron gas, then the heat capacity per unit volume turns out to be $\frac{3}{2}N_e k_B$ where $N_e$ is the number of electrons in the unit volume, and $k_B$ is the Boltzmann constant. This result implies that the electrons should have a substantial contribution to the heat capacity (on the order of lattice heat capacity, $3Nk_B$, where $N$ is the number of atoms). This electronic contribution was never observed in experiments.

The discrepancies of the Drude-Lorentz theory were resolved by Arnold Sommerfeld (1868–1951) in 1927 and 1928, who analyzed the Drude-Lorentz theory and replaced the Maxwell-Boltzmann distribution with the Fermi-Dirac distribution. His approach immediately led to a successful account of the Lorentz number and the electronic heat capacity. Assuming the existence of a relaxation time $\tau$ as a parameter, Sommerfeld obtained from the Drude-Lorentz approach the correct temperature dependence of the electronic heat capacity and also an excellent agreement with the **Lorentz number**, $L$. The Sommerfeld theory provided the correct order of magnitude for the electronic heat capacity which is negligibly small compared to its classical value. This supported Debye's observation that the heat capacity of metals at ordinary temperatures is predominately due to phonons. While Sommerfeld's theory resolved the electronic heat capacity issue, it did not offer any explanation of the problems associated with the temperature dependencies of electrical and thermal conductivities. The solution of this problem was resolved by Bloch in 1928 when he investigated the scattering of electrons by lattice vibrations based on his Bloch wave approach applied to the Boltzmann transport equation. By considering an electron-phonon scattering mechanism, he correctly explained the electrical conductivity in the high temperature limit (above the Debye temperature). The Wiedemann-Franz law was also correctly reproduced. In 1933, the German physicist Eduard Grüneisen (1877 – 1949) performed a theoretical calculation which explained the temperature dependence of resistivity $\rho$ both above and below the Debye temperature and obtained as

$$\rho = \begin{cases} T & T \gg \Theta_D \\ T^5 & T \ll \Theta_D. \end{cases} \tag{14.1}$$

Improvements to transport theory have been ongoing and the literature is vast. In this chapter we will focus on a couple of basic conceptual frameworks to get you started. These include the **Boltzmann transport equation** and the **Landauer theory** of transport. For further studies, you are recommended to consult the list of excellent references listed at the end of this chapter. To motivate for you why these two regimes can exist, let us consider the various transport length scales and regimes. Transport length scales serve a very useful purpose to classify the various transport regimes by comparing with the dimension $L$ of the sample and the Fermi wavelength $\lambda_F$, see Table 14.1.1. The three important length scales are:

1. **Elastic mean free path**, $l_e$: The distance between elastic scattering events experienced by the carrier, in this case an electron, is called the elastic mean free path. Due to lattice imperfections, such as impurities or dislocations, the electrons can scatter without any loss of energy. This is possible if we imagine an impurity atom which is more massive than an electron. Minimal energy is transferred because of the large mass difference, but the momentum changes drastically. The elastic mean free path can be calculated from the scattering time $\tau$ between successive events as $l_e = \tau v_F$, where $v_F$ is the Fermi velocity. For semiconductors $\tau$ can be obtained from the electron mobility defined in Equation (14.62).

2. **Inelastic mean free path**, $l_{in}$: The lattice irregularties which lead to scattering events could be dynamical or non-stationary. For example, quantized lattice vibrations called phonons can scatter an electron. It is also possible that an electron could excite lattice vibrations and lose energy. Thus, these processes lead to an inelastic scattering event. Further examples of inelastic scattering processes include electron-electron or even electron-magnon interaction.

3. **Phase coherence length**, $l_\phi$: This parameter measures the distance an electron travels before its phase is randomized. Elastic scattering events do not randomize the phase, but inelastic collisions do. The elastic process modifies the phase of the electron by the same amount for the same path. But, in an inelastic process, the nature of the scattering target evolves with time. Thus the phase shift is different with the passage of time.

When the elastic mean free path is less than the sample size, many elastic scattering events are possible. Thus the carrier randomly diffuses through the crystal. This **diffusive** regime of transport phenomena can be described via the Drude theory or the Boltzmann equation approach. The classical or quantum-*ness* of the situation is dictated by the comparison of the coherence length in relation to the elastic mean free path. However, in the opposite end of this spectrum is the limit where the

Table 14.1.1: Transport regime classification based on transport length scales.

| | | |
|---|---|---|
| Diffusive | Classical | $l_e \ll L, l_\phi < l_e$ |
| | Quantum | $l_e \ll L, l_\phi > l_e$ |
| Ballistic | Classical | $\lambda_F \ll L < l_\phi, l_e$ |
| | Quantum | $\lambda_F, L < l_\phi < l_e$ |

elastic mean free path length is larger than the sample length where the electron traverses through the system without any scattering. This is the **ballistic** transport regime. The Landauer theory of transport was formulated to describe this type of phenomenon which is widely prevelant in our nanostructures.

## 14.2 The Boltzmann Transport Equation

In Chapter 5 the Drude model of conduction provided the simplest possible description of electric and heat current propogation in metals. Within the Drude model it was assumed that electrons participate in elastic collision processes where the magnitude of the velocity is unchanged, but they emerge with an average velocity that is randomly oriented. The average velocity assumption implies that there are some electrons which are travelling either faster or slower compared to the average speed, in general. Thus, an appropriate description of the transport problem should involve a **distribution function** which will provide a spatial, momentum, and temporal dependence of how the electrons are distributed. Transport (matter or heat) is inherently a non-equilibrium process. Thus, we wish to find the non-equilibrium distribution function that can describe transport phenomena. Non-interacting electron at equilibrium with an external temperature bath T is described in momentum space $\vec{p}$ by the **Fermi-Dirac distribution** $f^{FD}(\vec{p})$ given by

$$f^{FD}(\vec{p}) = \frac{1}{\exp\left(\varepsilon(\vec{p}) - \mu\right)/k_B T + 1}, \tag{14.2}$$

where $k_B$ is the Boltzmann constant, $\varepsilon(\vec{p})$ is the dispersion relationship, and $\mu$ is the chemical potential. In contrast, phonons which are bosonic would be described by the Bose-Einstein distribution function. The classical explanation of the Drude model was improved upon using the quantum mechanical Sommerfeld model description which even though it is quantum mechanical in nature, it has neglected electron-ion and electron-electron interactions. For strongly correlated electron systems, an exact quantum mechanical description is the most appropriate. But, if we are concerned with material systems in which the mean free path of electrons is much larger than their de Broglie wavelength, then the conduction electrons can be approximated as a semiclassical electron gas. In the next few pages, we wish to derive the Boltzmann transport equation that will allow us to obtain the non-equilibrium distribution function that can describe transport properties.
Consider the six-dimensional phase space of Cartesian coordinates $\vec{r}$ and $\vec{k}$ shown in Figure 14.2.2. Let $f(\vec{r},\vec{k},t)$ be the non-equilibrium distribution function, defined by the relationship

$$f(\vec{r},\vec{k},t)d\vec{r}d\vec{k} = \text{number of particles in } d\vec{r}d\vec{k}, \tag{14.3}$$

where $d\vec{r}d\vec{k}$ is the infinitesimal phase space volume $dV_{\vec{r},\vec{k}}$ around the point $\vec{r}$ and $\vec{k}$ at time t. To determine the non-equilibrium distribution function, we study the evolution of the particles within the volume element $dV_{\vec{r},\vec{k}}$ at a later time $t$ around the region $\vec{r}'$ and $\vec{k}'$ with the volume element $dV_{\vec{r}',\vec{k}'}$. Since conservation of particle number holds, we have

$$f(\vec{r}+\dot{\vec{r}}dt, \vec{k}+\dot{\vec{k}}dt, t)d\vec{r}d\vec{k} = f(\vec{r},\vec{k},t)d\vec{r}d\vec{k}, \tag{14.4}$$

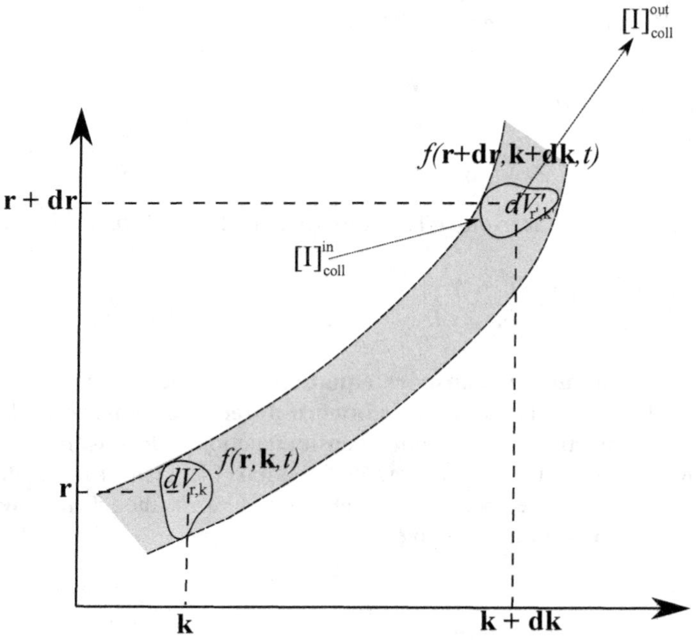


Figure 14.2.2: Evolution of the non-equilibrium distribution function appearing in the Boltzmann transport equation is tracked in the six-dimensional phase space of $\mathbf{r}-\mathbf{k}$ and time. The Liousville theorem guarantees that the volume element is preserved, thus $dV_{\vec{r}',\vec{k}'} = dV_{\vec{r},\vec{k}}$. Here, $[I]^{\text{in,out}}_{\text{coll}}$ represents the collision processes.

where the dots denote the time derivative of position and wavevector. Now according to the Liouville theorem of classical mechanics the phase space volume remains constant, see Figure 14.2.2, implying

$$dV_{\vec{r}',\vec{k}'} = dV_{\vec{r},\vec{k}}. \tag{14.5}$$

Thus using Equation (14.5) in Equation (14.4) we have

$$f(\vec{r}+\dot{\vec{r}}dt,\vec{k}+\dot{\vec{k}}dt,t) = f(\vec{r},\vec{k},t). \tag{14.6}$$

However, if collisions do occur during the *dt* time evolution, particles are scattered in- or out- of the phase-space trajectory. Thus we modify Equation (14.4) with a collision term as

$$f(\vec{r}+\dot{\vec{r}}dt,\vec{k}+\dot{\vec{k}}dt,t) - f(\vec{r},\vec{k},t) = \left(\frac{\partial f(\vec{r},\vec{k},t)}{\partial t}\right)_{\text{coll}} \text{dt}, \tag{14.7}$$

where

$$I[f] \equiv \left(\frac{\partial f(\vec{r},\vec{k},t)}{\partial t}\right)_{\text{coll}} \text{dt} = \left(\frac{\partial f(\vec{r},\vec{k},t)}{\partial t}\right)_{\text{in}} \text{dt} - \left(\frac{\partial f(\vec{r},\vec{k},t)}{\partial t}\right)_{\text{out}} \text{dt}, \tag{14.8}$$

is a functional of the distribution function $f(\vec{r},\vec{k},t)$ known as the **collision integral** or the **scattering operator**. The subscripts coll, in, and out represent net, incoming, and outgoing collision integral terms, respectively. With the understanding that

$$\frac{\partial}{\partial \vec{r}} \equiv \frac{\partial}{\partial x}\hat{\imath} + \frac{\partial}{\partial y}\hat{\jmath} + \frac{\partial}{\partial z}\hat{k}, \tag{14.9}$$

$$\frac{\partial}{\partial \vec{k}} \equiv \frac{\partial}{\partial k_x}\hat{\imath} + \frac{\partial}{\partial k_y}\hat{\jmath} + \frac{\partial}{\partial k_z}\hat{k}, \tag{14.10}$$

for small time differences the linear-order Taylor expansion of the left-hand side equation gives us

$$\frac{\partial f}{\partial t} + \dot{\vec{r}} \cdot \frac{\partial f}{\partial \vec{r}} + \dot{\vec{k}} \cdot \frac{\partial f}{\partial \vec{k}} = \left( \frac{\partial f(\vec{r},\vec{k},t)}{\partial t} \right)_{\text{coll}}, \tag{14.11}$$

which is the famous **Boltzmann transport equation** formulated in 1872 by Ludwig Boltzmann (1844–1906). The above is a **nonlinear integro-differential equation**. The unknown non-equilibrium distribution function is a probability density function in the six-dimensional phase space of particle position and momentum. We can rewrite Equation (14.11) in terms of velocity $\vec{v}$ and force $\vec{F}$. If we ignore collision processes between the times $t$ and $t+dt$, then $\vec{r}$ and $\vec{k}$ will change based on the semiclassical equations of motion given by

$$\dot{\vec{r}} = \vec{v}(\vec{k}); \;\; \hbar\dot{\vec{k}} = \vec{F}(\vec{r},\vec{k}). \tag{14.12}$$

Using the above in Equation (14.11) we then have

$$\frac{\partial f}{\partial t} + \dot{\vec{r}} \cdot \frac{\partial f}{\partial \vec{r}} + \frac{\vec{F}}{\hbar} \cdot \frac{\partial f}{\partial \vec{k}} = \left( \frac{\partial f(\vec{r},\vec{k},t)}{\partial t} \right)_{\text{coll}}, \tag{14.13}$$

where the derivatives on the left-hand side of the equation are known as **drift terms** so that their total sum equals collision term. The Boltzmann transport equation provides a powerful theoretical approach to compute transport properties, but, when the sample dimension scales become of the order of the de Broglie wavelength of the particle distribution the above equation fails, see Table 14.1.1. These shortcomings can be avoided by approaching the transport theory from the Landauer approach described in Section 14.8. The semiclassical Boltzmann transport equation approach described has its quantum generalization called the Quantum Boltzmann Equation. This approach also goes by a different name known as the **non-equilibrium Green's function** (NEGF) formalism initiated by the works of Schwinger, Baym, Kadanoff, and Keldysh. The basic quantity in this theoretical approach is the correlation function (or electronic Green function) which plays the role of the Boltzmann non-equilibrium distribution function, $f(\vec{r},\vec{k},t)$.

## 14.3 Relaxation Time Approximation and Drude Conductivity

The presence of the collision term in the Boltzmann transport equation makes the problem of solving for the non-equilibrium distribution function challenging. However, the **relaxation time approximation** provides a simple approach to recover some relevant information regarding the transport properties. Within this approach, local thermodynamic equilibrium in the system is attained through collisions where

1. the distribution of the electrons after the collision is not affected by the non-equilibrium distribution just prior to it,

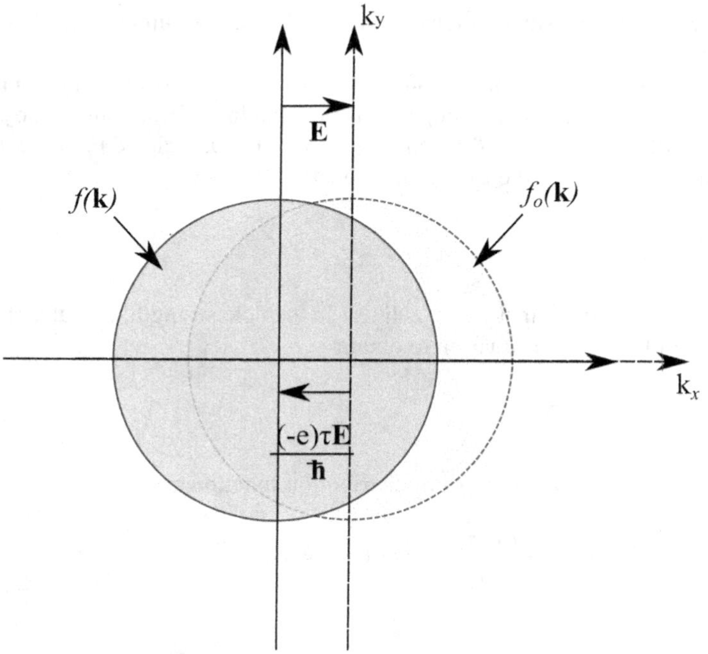


Figure 14.3.3: The presence of an external electric field $\vec{E}$ (here denoted with boldface E) causes the Fermi surface to be displaced. The external perturbation disturbs the equilibrium distribution to change it to the non-equilibrium $f$ function appearing in the Boltzmann formulation. $\tau$ is the relaxation time and $\hbar$ the reduced Planck's constant.

2. the local equilibrium electronic distribution $f_o$ conforms to the local temperature, thereby preventing any change in the form of the distribution function.

Based on these assumptions we can claim that the deviation of $f$ from the thermal equilibrium distribution $f_o$ is small. Thus, based on the relaxation time approximation we can approximate the collision integral as

$$\left[\frac{\partial f}{\partial t}\right]_{\text{coll}} = -\frac{f - f_o}{\tau}, \tag{14.14}$$

where $\tau$ is called the **relaxation time**. In general $\tau = \tau(\vec{k}, \vec{r})$ can depend on the energy $\varepsilon(\vec{k})$ and on position, and is a semi-empirical parameter. Thus, in the relaxation time approximation the Boltzmann equation becomes

$$\frac{\partial f}{\partial \vec{r}} \cdot \vec{v} + \frac{\partial f}{\partial \vec{k}} \cdot \frac{\vec{F}}{\hbar} + \frac{\partial f}{\partial t} = -\frac{f - f_o}{\tau} \tag{14.15}$$

The non-equilibrium distribution function $f$, evaluated with the Boltzmann Equation (14.15) permits the evaluation of several transport phenomena coefficients. For example, electron current density $\vec{J}$ and the energy flux density $\vec{U}$ can be computed from the definitions

$$\vec{J} = \frac{1}{4\pi^3} \int (-e) \vec{v}_{\vec{k}} f d^3\vec{k}, \tag{14.16}$$

$$\vec{U} = \frac{1}{4\pi^3} \int \varepsilon(\vec{k}) \vec{v}_{\vec{k}} f d^3\vec{k}, \tag{14.17}$$

where spin degeneracy and density of allowed points in $\vec{k}$ space per unit volume has been accounted for by using $2/(2\pi)^3$ factor.

The static Drude conductivity of a metal can be derived using the Boltzmann approach within the relaxation time approximation. For a homogeneous material in a uniform and steady electric field $\vec{E}$, the electric field distribution function $f$ depends only on $\vec{k}$. The associated Fermi surface is displaced as shown in Figure 14.3.3. Thus, the Boltzmann equation becomes

$$\frac{1}{\hbar}\frac{\partial f}{\partial \vec{k}}\cdot(-e)\vec{E} = -\frac{f - f_o}{\tau} \tag{14.18}$$

For low electric fields, assuming that $f - f_o$ is linear in the field strength, we can approximate $f \approx f_o$ in the left-hand side of the above equation to obtain

$$f = f_o + \frac{e\tau}{\hbar}\frac{\partial f_o}{\partial \vec{k}}\cdot\vec{E}, \tag{14.19}$$

from which we identify the non-equilibrium distribution function as

$$f_1 \;=\; \frac{e\tau}{\hbar}\frac{\partial f_o}{\partial \vec{k}}\cdot\vec{E} = \frac{e\tau}{\hbar}\frac{\partial f_o}{\partial \varepsilon}\frac{\partial \varepsilon(\vec{k})}{\partial \vec{k}}\cdot\vec{E} = e\tau\frac{\partial f_o}{\partial \varepsilon}\vec{v}\cdot\vec{E}, \tag{14.20}$$

where

$$\vec{v} = \frac{1}{\hbar}\frac{\partial \varepsilon(\vec{k})}{\partial \vec{k}}, \tag{14.21}$$

is the semi-classical expression of the velocity and $\varepsilon(\vec{k})$ is the single electron dispersion. Inserting the above definition of $f_1$ ($f_o = 0$ gives zero contribution) in Equation (14.16), we obtain

$$\vec{J} \;=\; \frac{1}{4\pi^3}\int(-e)\vec{v}f_1 d^3\vec{k} = \frac{e^2}{4\pi^3}\int\tau\left(-\frac{\partial f_o}{\partial \varepsilon}\right)\vec{v}(\vec{v}\cdot\vec{E})d^3\vec{k}. \tag{14.22}$$

For simplicity we will assume that $\vec{J}$ and $\vec{E}$ are parallel to each other such that $\vec{J} = \sigma\vec{E}$. Thus defining the unit vector $\hat{e} = \vec{E}/|\vec{E}|$, dotting both sides of Equation (14.22) with $\hat{e}$, we can extract the static conductivity expression as

$$\sigma_o = \frac{e^2}{4\pi^3}\int\tau(\hat{e}\cdot\vec{v})^2\left(-\frac{\partial f_o}{\partial \varepsilon}\right)d^3\vec{k}. \tag{14.23}$$

Based on our previous understanding, we know that for low temperature the Fermi-Dirac distribution function changes sharply from unity to zero within a small interval around the Fermi level of the order of $k_BT$. The distribution function changes significantly from zero in this same interval. Furthermore, to estimate the conductivity given by Equation (14.23) we can replace $(\hat{e}\cdot\vec{v})^2$ with $v^2/3$ (assuming isotropic contributions from x, y, or z direction) and approximate the change in the negative derivative of the distribution function with respect to $\varepsilon$ with $\delta(\varepsilon(\vec{k}) - \varepsilon_F)$, where $\varepsilon_F$ is the Fermi energy. Thus we have

$$\sigma_o = \frac{e^2}{12\pi^3}\int\tau v^2\delta(\varepsilon(\vec{k}) - \varepsilon_F)d^3\vec{k}. \tag{14.24}$$

We can transform the integration variable in Equation (14.24) from $d^3\vec{k}$ to an integral calculated over the Fermi surface FS. To do so, observe that between two isoenergy surfaces of $\varepsilon$ and $\varepsilon + d\varepsilon$ we can write

$$d\varepsilon = \nabla_{\vec{k}}\varepsilon(\vec{k})\cdot d\vec{k} = |\nabla_{\vec{k}}\varepsilon(\vec{k})|dk. \tag{14.25}$$

Then using Equation (14.25) we can express the volume element $d^3\vec{k}$ in terms of a surface integral and energy integration variable as

$$d^3\vec{k} = dS\, dk = dS\, \frac{d\varepsilon}{|\nabla_{\vec{k}}\varepsilon(\vec{k})|}. \tag{14.26}$$

Using Equation (14.26) and invoking the Delta function energy constraint over the Fermi surface (FS), we can write

$$\begin{aligned} \sigma_o &= \frac{e^2}{12\pi^3}\int \tau v^2 \delta(\varepsilon(\vec{k}) - \varepsilon_F)d\varepsilon\, \frac{dS}{|\nabla_{\vec{k}}\varepsilon(\vec{k})|} \\ &= \frac{e^2}{12\pi^3}\int_{\text{FS}} \tau v^2 \frac{dS}{|\nabla_{\vec{k}}\varepsilon(\vec{k})|} \quad \text{(performing the Delta function integration)} \\ &= \frac{e^2}{12\pi^3\hbar}\int_{\text{FS}} \tau\, v\, dS, \end{aligned} \tag{14.27}$$

where we have used $|\vec{\nabla}_{\vec{k}}\varepsilon(\vec{k})| = \hbar v$. Thus, for a parabolic conduction band $\varepsilon(\vec{k})$ with effective mass $m^*$ the Fermi velocity $v_F$ is computed as

$$v_F = \left.\frac{\partial\varepsilon(\vec{k})}{\partial\vec{k}}\right|_{k=k_F} = \frac{\partial}{\partial\vec{k}}\left.\left(\frac{\hbar^2}{2m^*}k^2\right)\right|_{k=k_F} = \frac{\hbar k_F}{m^*}. \tag{14.28}$$

Finally we perform the Fermi surface integration in Equation (14.27). To do so, we replace $\tau$ and $v$ with $\tau_F$ and $v_F$, substitute the surface area integral over dS with $4\pi k_F^2$, and use the Fermi wavevector's relationship to the electron density $n$ given by

$$k_F^3 = 3\pi^2 n, \tag{14.29}$$

to obtain the static conductivity as

$$\sigma_o = \frac{e^2}{4\pi^3}\tau_F v_F 4\pi k_F^2 = \frac{ne^2\tau_F}{m^*}. \tag{14.30}$$

Recall from Chapter 5 that Equation (14.30) is the well-known Drude formula for DC electrical conductivity!

## 14.4 Boltzmann Equation in Electric Field and Temperature Gradients

Equation (14.13) will allow us to study transport effects due the presence of electric fields at uniform temperature, that is isothermal (constant temperature) conditions. In this section we will further generalize the Boltzmann transport equation derived in the previous section to take into account temperature gradients. Consider a band of energy $E(\vec{k})$ in a crystal in thermal equilibrium at a non-uniform temperature. We define the space and momentum dependent, but time independent (stationary), local equilibrium distribution function $f_o(\vec{r},\vec{k})$ as

$$f_o(\vec{r},\vec{k}) = \frac{1}{\exp\left[\left(\varepsilon(\vec{k}) - \mu(\vec{r})\right)/k_B T(\vec{r})\right] + 1}, \tag{14.31}$$

where the local temperature $T(\vec{r})$ and chemical potential $\mu(\vec{r})$ have spatial dependence. To solve for $f_1$ from Equation (14.13), we need the gradient of $f_o$ with respect to $\vec{k}$ and with respect to $\vec{r}$. These derivatives (see Problem 14.9.5) are given by

$$\frac{\partial f_o}{\partial \vec{r}} = \frac{\partial f_o}{\partial \varepsilon} k_B T \frac{\partial}{\partial \vec{r}} \left[ \frac{\varepsilon(\vec{k}) - \mu}{k_B T} \right] = \frac{\partial f_o}{\partial \varepsilon} \left[ -\frac{\varepsilon(\vec{k})}{T} \frac{\partial T}{\partial \vec{r}} - T \frac{\partial}{\partial \vec{r}} \frac{\mu}{T} \right], \tag{14.32}$$

and

$$\frac{\partial f_o}{\partial \vec{k}} = \frac{\partial f_o}{\partial \varepsilon} \frac{\partial \varepsilon(\vec{k})}{\partial \vec{k}} = \frac{\partial f_o}{\partial \varepsilon} \hbar \vec{v}. \tag{14.33}$$

Within the relaxation time approximation, the Boltzmann equation for the stationary distribution $f(\vec{r},\vec{k})$ in the presence of an electric field $\vec{E}$ and temperature gradient we have

$$\frac{\partial f}{\partial \vec{r}} \cdot \vec{v} + \frac{(-e)}{\hbar} \frac{\partial f}{\partial \vec{k}} \cdot \vec{E} = -\frac{f - f_o}{\tau} = -\frac{f_1}{\tau}, \tag{14.34}$$

where we used $\vec{F} = -e\vec{E}$. Assuming the electric field and temperature gradient are small, which is usually the case, we can substitute $f \approx f_0$ in the left-hand side of Equation (14.34). This implies that

$$\frac{\partial f_0}{\partial \vec{r}} \cdot \vec{v} + \frac{(-e)}{\hbar} \frac{\partial f_0}{\partial \vec{k}} \cdot \vec{E} = -\frac{f_1}{\tau}. \tag{14.35}$$

Note, with this approximation we can solve for $f_1$ as

$$f_1 = -\tau \left\{ \frac{\partial f_0}{\partial \vec{r}} \cdot \vec{v} + \frac{(-e)}{\hbar} \frac{\partial f_0}{\partial \vec{k}} \cdot \vec{E} \right\}. \tag{14.36}$$

Now using Equations (14.32) and (14.33) in Equation (14.36) we obtain for the stationary non-equilibrium distribution function as

$$f_1 = \tau \left( -\frac{\partial f_o}{\partial \varepsilon} \right) \left[ -e\vec{E} - T\vec{\nabla}\left(\frac{\mu}{T}\right) \right] \cdot \vec{v} + \tau \left( -\frac{\partial f_o}{\partial \varepsilon} \right) (-\varepsilon(\vec{k})) \frac{\vec{\nabla} T}{T} \cdot \vec{v}, \tag{14.37}$$

where again $\vec{\nabla} \equiv \frac{\partial}{\partial \vec{r}}$ is the gradient operator acting on the space dependent temperature and chemical potential terms. To proceed further with the derivation, let us assume for simplicity that $\vec{E}$ is along the $x$ direction. Furthermore, assuming an isotropic system implies that $\vec{J}$ and $\vec{U}$ are parallel to the $x$ direction and $\vec{\nabla} \equiv \partial_x$. Thus replacing the expression for $f$ with the above $f_1$ in Equations (14.16) we can write for $J_x$ as

$$
\begin{aligned}
J_x &= 2\frac{(-e)}{2\pi} \int v_{k_x} \tau \left( -\frac{\partial f_o}{\partial \varepsilon} \right) \left\{ \left[ -eE_x - T\partial_x \left(\frac{\mu}{T}\right) \right] v_{k_x} + (-\varepsilon(k_x)) \frac{\partial_x T}{T} v_{k_x} \right\} dk_x, && (14.38)\\
&= 2\frac{(-e)}{2\pi} \tau \int v_{k_x}^2 \left( -\frac{\partial f_o}{\partial \varepsilon} \right) \left\{ \left[ -eE_x - T\partial_x \left(\frac{\mu}{T}\right) \right] + (-\varepsilon(k_x)) \frac{\partial_x T}{T} \right\} dk_x, && (14.39)\\
&= e \left\{ \frac{2}{2\pi} \int \tau v_{k_x}^2 \varepsilon^0(k_x) \left( -\frac{\partial f_o}{\partial \varepsilon} \right) dk_x \right\} \left[ eE_x + T\partial_x \left(\frac{\mu}{T}\right) \right], \\
&+ e \left\{ \frac{2}{2\pi} \int \tau v_{k_x}^2 \varepsilon^1(k_x) \left( -\frac{\partial f_o}{\partial \varepsilon} \right) dk_x \right\} \frac{\partial_x T}{T}, && (14.40)\\
&= eK_0 \left[ eE_x + T\partial_x \left(\frac{\mu}{T}\right) \right] + e\frac{K_1}{T} \partial_x T, && (14.41)
\end{aligned}
$$

where we have identified the expressions within the integral as the kinetic coefficients $K_0$ and $K_1$. Since the derivation is being performed in one dimension, the spin degeneracy and density of allowed points in momentum space is taken to be $2/(2\pi)$ instead of $2/(2\pi)^3$ as in three dimensions. The zero or one subscript for $K$ follows the power on the dispersion relation $\varepsilon(k_x)$. Following a similar strategy, we can derive the expression for $U_x$ along $x$. But, now note the presence of the $\varepsilon(k_x)$ in the definition of Equation (14.17). This will cause the kinetic coefficients to pick up another power of $\varepsilon(k_x)$ to transform $K_0$ to $K_1$ and $K_1$ to $K_2$. Generalizing the above expressions for the current density and the energy density to three dimensions along any arbitrary unit vector $\hat{e}$ of the electric field direction we have

$$\vec{J} = eK_0\left[e\vec{E} + T\vec{\nabla}\left(\frac{\mu}{T}\right)\right] + e\frac{K_1}{T}\vec{\nabla}T, \tag{14.42}$$

$$\vec{U} = -K_1\left[e\vec{E} + T\vec{\nabla}\left(\frac{\mu}{T}\right)\right] - \frac{K_2}{T}\vec{\nabla}T, \tag{14.43}$$

where the kinetic coefficient expression is given by

$$K_n = \frac{1}{4\pi^3}\int \tau(\hat{e}\cdot\vec{v})^2\varepsilon^n(\vec{k})\left(-\frac{\partial f_o}{\partial \varepsilon}\right)d\vec{k}. \tag{14.44}$$

The transport coefficients $K_n$ can be evaluated replacing $(\hat{e}\cdot\vec{v})^2$ by $v^2/3$ as we did with the evaluation of the DC Drude conductivity calculation. Next, we can replace the three-dimensional integral in $d^3\vec{k}$ as a 2D constant energy surface integral and an integration over the energy variable $d\varepsilon$ (same trick as in the Drude conductivity calculation) to have

$$K_n = \frac{1}{12\pi^3}\int\left(-\frac{\partial f_o}{\partial \varepsilon}\right)\varepsilon^n(\vec{k})d\varepsilon \int_{\varepsilon=\text{const}} \frac{\tau v^2}{|\vec{\nabla}_k\varepsilon(\vec{k})|}dS, \tag{14.45}$$

where $|\vec{\nabla}_k\varepsilon(\vec{k})| = \hbar v$. Keeping in mind Equation (14.30), we can define a generalized conductivity $\sigma(\varepsilon)$ as

$$\sigma(\varepsilon) = \frac{e^2}{12\pi^3\hbar}\int_{\varepsilon=\text{const}} \tau\, v\, dS, \tag{14.46}$$

where notice that $\sigma(\varepsilon_F) = \sigma_o$ is the standard conductivity of the metal. Thus we have the form

$$e^2K_n = \int\left(-\frac{\partial f_o}{\partial \varepsilon}\right)\varepsilon^n(\vec{k})\sigma(\varepsilon)d\varepsilon. \tag{14.47}$$

The integrals appearing in the above equation can be calculated using the Sommerfeld expansion which is given by

$$\int_0^\infty\left(-\frac{\partial f_o}{\partial \varepsilon}\right)g(\varepsilon)d\varepsilon = g(\mu) + \frac{\pi^2}{6}k_B^2T^2\left(\frac{d^2g}{d\varepsilon^2}\right)\Bigg|_{\varepsilon=\mu} + \mathcal{O}(T^4), \tag{14.48}$$

with $g(\varepsilon) = \varepsilon^n\sigma(\varepsilon)$. Hence, the transport coefficients to $\mathcal{O}(T^4)$ can then be written as (see Problem 14.9.7)

$$e^2K_0 \approx \sigma(\mu) + \frac{\pi^2}{6}k_B^2T^2\sigma''(\mu), \tag{14.49a}$$

$$e^2K_1 \approx \mu\sigma(\mu) + \frac{\pi^2}{6}k_B^2T^2\left[2\sigma'(\mu) + \mu\sigma''\right], \tag{14.49b}$$

$$e^2K_2 \approx \mu^2\sigma(\mu) + \frac{\pi^2}{6}k_B^2T^2\left[2\sigma(\mu) + 4\mu\sigma'(\mu) + \mu^2\sigma''\right], \tag{14.49c}$$

where the first and second derivatives of $\sigma(\varepsilon)$ are calculated at the Fermi energy $\varepsilon = \mu$.

We conclude this section by rewriting Equations (14.42) and (14.43) in a slightly different way which renders an easier analysis of thermoelectric phenomena. To do so, notice that

$$\vec{\nabla}\left(\frac{\mu}{T}\right) = \frac{1}{T}\vec{\nabla}\mu - \frac{\mu}{T^2}\vec{\nabla}T. \tag{14.50}$$

Next, substituting Equation (14.50) in Equations (14.42)–(14.43) and simplifying we can write the basic transport equations for charge and energy flux as

$$\vec{J} = e^2K_0\left[\vec{E} + \frac{1}{e}\nabla\mu - S(T)\vec{\nabla}T\right], \tag{14.51}$$

$$\vec{U} = -\frac{1}{e}\frac{K_1}{K_0}\vec{J} - k_e\vec{\nabla}T, \tag{14.52}$$

where we can identify $\sigma_o = e^2K_o$ (compare with the integral in the last line of Equation 14.27), introduce the **Seebeck coefficient (absolute thermoelectric power)** definition, discussed further in Section 14.7, to write

$$S(T) = -\frac{1}{eT}\left(\frac{K_1}{K_0} - \mu\right) \approx \frac{\pi^2}{3}\frac{k_B^2T}{(-e)}\frac{\sigma'(\mu)}{\sigma(\mu)}, \tag{14.53}$$

and the **thermal conductivity (electronic)** definition (discussed further in Section 14.6) to have

$$k_e = \frac{1}{T}\left(K_2 - \frac{K_1^2}{K_0}\right) \approx \frac{\pi^2}{3}\frac{k_B^2T}{e^2}\sigma_o. \tag{14.54}$$

In the above, we have kept the leading order term in the temperature expansions of $e^2K_o$ and also neglected the second derivative terms. The explicit temperature dependence of the Seebeck coefficient and the thermal conductivity are evaluated using the Sommerfeld equations listed in Equations (14.49a)–(14.49c), see Problems 14.9.8, 14.9.9, and 14.9.10.

## 14.5 Drift and Diffusion Current

As a first application of Equations (14.51) and (14.52), we consider the electron current density in a metal in isothermal conditions ($\nabla T = 0$), but with a non-uniform carrier concentration $\nabla n \neq 0$, which implies $\nabla\mu \neq 0$. Thus inserting $\nabla T = 0$ in Equation (14.51) we have

$$\vec{J} = \sigma_o\left[\vec{E} + \frac{1}{e}\vec{\nabla}\mu\right], \tag{14.55}$$

from which we can identify two contributions, one for the **drift current**

$$\vec{J}_{\text{drift}} = \sigma_o\vec{E}, \tag{14.56}$$

and the other is the **diffusion current**

$$\vec{J}_{\text{diff}} = \frac{\sigma_o}{e}\vec{\nabla}\mu, \tag{14.57}$$

We can apply the above equation to the case of a free-electron-like conduction band for a metal. We have in this case, the conductivity and the T=0 Fermi energy $\varepsilon_F$ as

$$\sigma_o = \frac{ne^2\tau}{m^*}, \tag{14.58}$$

$$\mu = \frac{\hbar^2}{2m^*}(3\pi^2n)^{2/3} \equiv \varepsilon_F. \tag{14.59}$$

Note, from Equation (14.59) we can obtain the relation

$$\frac{\nabla\mu}{\mu} = \frac{2}{3}\frac{\nabla n}{n}. \tag{14.60}$$

We now rewrite Equation (14.56) as

$$\vec{J}_{\text{drift}} = \sigma_o \vec{E}, \tag{14.61a}$$

$$= \left(\frac{ne^2\tau}{m^*}\right)\vec{E}, \tag{14.61b}$$

$$= ne\mu_e \vec{E}, \tag{14.61c}$$

where we have introduced the **electron mobility** defined as

$$\mu_e = \frac{e\tau}{m^*}. \tag{14.62}$$

Next, we rewrite Equation (14.57). To do so, we insert Equations (14.58), (14.59), and (14.60) into Equation (14.57) to obtain

$$\vec{J}_{\text{diff}} = \frac{\sigma_o}{e}\vec{\nabla}\mu, \tag{14.63a}$$

$$= \left(\frac{ne^2\tau}{m^*}\right)\frac{1}{e}\vec{\nabla}\mu, \tag{14.63b}$$

$$= \left(\frac{ne^2\tau}{m^*}\right)\frac{1}{e}\frac{2}{3}\frac{\vec{\nabla}n}{n}\mu = ne\left(\frac{e\tau}{m}\right)\frac{1}{ne}\frac{2}{3}\varepsilon_F\vec{\nabla}n = eD\vec{\nabla}n, \tag{14.63c}$$

where the **diffusion coefficient**, D, is given by

$$D = \frac{2}{3}\frac{\varepsilon_F}{e}\mu_e. \tag{14.64}$$

Thus we can rewrite the current density in the metal as

$$\vec{J} = ne\mu_e\vec{E} + eD\nabla n \tag{14.65}$$

In the case of the free electron gas which is non-degenerate and following the Boltzmann distribution, we have

$$\varepsilon_F = \frac{3}{2}k_B T, \tag{14.66}$$

based on the equipartition theorem which distributes $\frac{1}{2}k_B T$ amount of energy to each quadratic degree of freedom, in this case the velocities along x, y, and z. Thus the diffusion coefficient can be rewritten as

$$D = \frac{k_B T}{e}\mu_e. \tag{14.67}$$

Equations (14.64) and (14.67) are the **Einstein relations** between the mobility and the diffusion coefficient for the degenerate and non-degenerate electron gas, respectively.

## 14.6 Thermal Conductivity of Metals

Consider a metal in the presence of a uniform temperature gradient $\nabla T$ and in open circuit situation as shown in Figure 14.6.4. In this situation $\vec{J} = \vec{0}$. Thus Equation (14.52) for the energy flux density takes the form

$$\vec{U} = -k_e \vec{\nabla} T \tag{14.68}$$

with $k_e$ given by Equation (14.54),

$$k_e = \frac{\pi^2}{3} \frac{k_B^2}{e^2} T \sigma_o. \tag{14.69}$$

Thus the energy flows in the direction opposite to $\vec{\nabla} T$ as expected, i.e., from the hot to the cold side. From the expression of the electron thermal conductivity $k_e$, we see that the ratio of thermal conductivity to the electrical conductivity is proportional to T (Wiedemann–Franz law), thus one has the ratio

$$\frac{k_e}{T \sigma_o} = L \equiv \frac{\pi^2}{3} \frac{k_B^2}{e^2}, \tag{14.70}$$

which is known as the **Lorentz number** $L = 2.45 \times 10^{-8} (V/K)^2$. The Lorentz number would actually be a universal constant (independent of the specific metal, temperature and relaxation time), provided the approximations made in the transport equations are justified. The most vulnerable part is the relaxation time approximation of the collision term. This approximation is justified above the Debye temperature, where electron–phonon scattering is the dominant process, and also at very low temperature, where the impurity scattering is dominant. In both temperature regimes, the Lorentz number is approximately the same for all metals. At intermediate temperatures, however, significant deviations may occur.

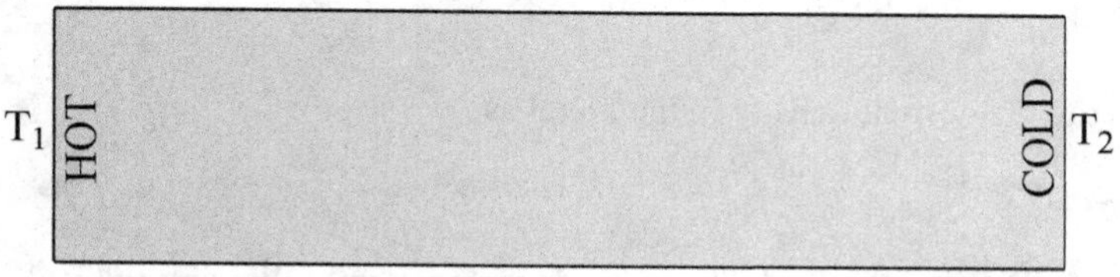


Figure 14.6.4: A bar of homogeneous material with a hot and a cold end (temperature gradient: this gradient of T points from the cold side to the hot side).

## 14.7 Thermoelectric Phenomena

Thermoelectric effects are a manifestation of electrical current and heat flow interaction in materials. The coupling parameter, termed **thermopower**, provides a quantitative sense of the direct conversion of heat into electricity and vice versa. In the reverse process, cooling is achieved by the application of a voltage across a thermoelectric material. Thermoelectric materials have applications in green technologies, where waste heat from industrial applications or car engines can be converted into usable power. Interestingly enough many of the **topological insulators** are excellent thermoelectrics. Examples include $Bi_2Se_3$, $Bi_2Te_3$, and $Sb_2Te_3$.

### 14.7.1 Seebeck Effect and Thermoelectric Power

In 1821 the Baltic German physicist Thomas Johann Seebeck (1770–1831) was credited with the discovery that when two dissimilar metals such as copper and bismuth wires are joined at two ends to form a loop, a voltage is developed in the circuit if the two junctions are kept at different temperatures. The metal pair forms a circuit called a **thermocouple**. This is the Seebeck effect. The existence of the current in the closed circuit can be confirmed in several ways, one of which includes the deflection of a magnetic needle caused by the current flowing in the circuit (this is how Seebeck performed his classic experiment). Although attributed to Seebeck, historical accounts have revealed

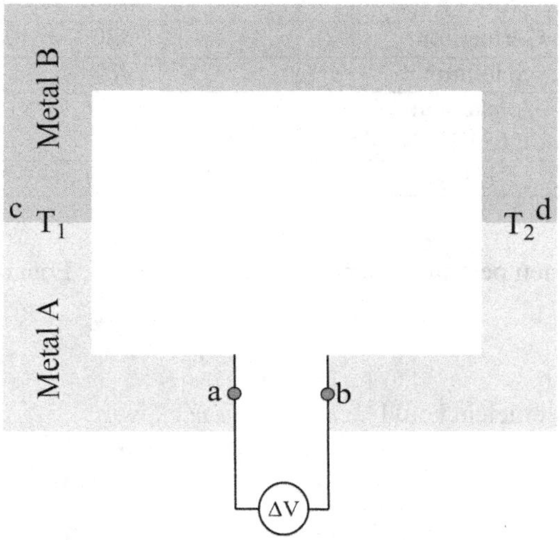


Figure 14.7.5: Standard bimetallic strip geometry used in Seebeck coefficient measurement setup.

that the first experiments on thermoelectricity were performed by the Italian physicist Alessandro Giuseppe Antonio Anastasio Volta (1745–1827) in 1794. Additionally, Seebeck did not fully comprehend the underlying physical mechanism that caused the effect he measured. He called it "thermomagnetische Reihe" or "thermomagnetism". The correct physical explanation was provided by the Danish physicist and chemist Hans Christian Øersted (1777–1851) who recognized that it was the heat that caused the electric current in the circuit. Øersted coined the term thermoelectricity.

The **Seebeck voltage** $\Delta V$ can be measured due to a temperature difference and is related to the Seebeck coefficient via

$$\Delta V = -\int_{T_1}^{T_2} [S_A(T) - S_B(T)]\, dT, \tag{14.71}$$

where $\Delta V$, as shown in Figure 14.7.5, measures the open circuit voltage difference across terminal points $a$ and $b$, $T_1$ ($T_2$) are the temperatures at the two metal junction points $c(d)$, and $S_A(S_B)$ represent the thermoelectric power mentioned before or the Seebeck coefficient for the two metals forming the bimetallic strip. The SI unit of the Seebeck coefficient is volts per kelvin $(V/K)$, although it is customarily reported in $10^{-6}\mu V/K$. The term thermoelectric power is a misnomer though since the Seebeck coefficient does not measure any power. Table 14.7.2 lists the Seebeck coefficient for some standard materials.

Table 14.7.2: The performance of a thermocouple is determined by the Seebeck coefficient of a pair of metals. The chosen standard is platinum, which has a value of $-5\ \mu V/K$. The room temperature Seebeck values, relative to platinum, of some common materials are reported above. In conductors, the Seebeck coefficient is negative for negatively charged carriers (such as electrons), and positive for positively charged carriers (such as holes).

| **Material** | **Seebeck Coefficient** ($\mu V/K$) |
|---|---|
| Bismuth | -72 |
| Nickel | -15 |
| Silicon | 440 |
| Germanium | 330 |
| Selenium | 900 |
| Constantan<br>Alloy - Cu (55%), Ni (45%) | -35 |
| $Bi_2Te_3$ | -270 |

Assuming that the $S(T)$ is temperature independent, we can rewrite Equation (14.71) as

$$S = -\frac{\Delta V}{\Delta T}. \tag{14.72}$$

The sign of the Seebeck coefficient could be positive or negative.

### 14.7.2 Thomson Effect

Lord Kelvin (William Thomson) found that when a current is passed through a wire of single homogeneous material along which a temperature gradient exists, heat must be exchanged with the surroundings in order to preserve the original temperature gradient along the wire. This fact can be understood if we recognize that the hot charge carriers will diffuse towards the cold end. The charge separation sets up an electric field that opposes the motion of the carriers due to the thermal effect. Thus, heat is released or absorbed reversibly at a rate depending on the current density and the material specific properties. One important distinction between the Joule heating effect and Thomson effect is that the latter is reversible. If the direction of the current is reversed, the Thomson effect changes sign, but the irreversible Joule heating process does not. It is possible to show that the **Thomson coefficient**, $K_{TH}$, is related to the absolute thermoelectric power via the relationship

$$K_{TH}(T) = T\frac{dS(T)}{dT}, \tag{14.73}$$

where $T$ is the temperature in Kelvin.

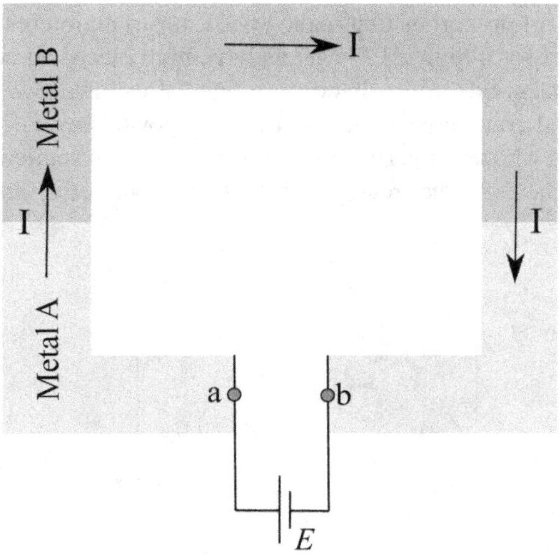


Figure 14.7.6: In additional to Joule heating, heat can also be reversibly released or absorbed when current flows across the junction of two dissimilar metals. The french watchmaker turned physicist Jean Charles Peltier discovered this thermoelectric effect in 1834.

### 14.7.3 Peltier Effect

In 1834 the French physicist Jean Charles Athanase Peltier (1785–1845) discovered that heat is generated reversibly when a current flows in a given homogeneous material when current flows across a junction between two conducting materials, see Figure 14.7.6. In contrast to the Joule effect, but similar to the Thomson effect, this transport property exhibits reversibility. Thus, if the direction of the current changes, the Peltier effect changes sign. The Peltier effect can be thought of as a reverse Seebeck effect. In fact the **Peltier coefficient**, $\Pi$, is connected to the Seebeck coefficient via the simple relationship

$$\Pi(T) = TS(T), \tag{14.74}$$

where $T$ is the temperature in Kelvin.
Applications of the Peltier effect are abundant. **Thermoelectric coolers** operating on the Peltier effect are abundant. Applications range from portable coolers, cooling electronic components to cooling of CCDs in telescopes and spectrometers.

### 14.7.4 Thermoelectric Figure of Merit, Z

The performance of a thermoelectric material is evaluated in terms of a dimensional **figure of merit** $ZT$ defined by

$$ZT = \frac{S^2\sigma}{\kappa_l + \kappa_e}, \tag{14.75}$$

where $S$ is the Seebeck coefficient, $\sigma$ is the electrical conductivity, $\kappa_l$ is the lattice thermal conductivity, and $\kappa_e$ is the electronic thermal conductivity, $T$ is the temperature. The product $S^2\sigma$ is called the **power factor**. A larger $ZT$ leads to a high conversion efficiency. The denominator is mostly determined by lattice thermal conductivity.

The combination of transport properties that make up ZT, raises an interesting question: "What material properties are needed for a high ZT?" Metals have high electrical conductivity, low Seebeck coefficient, and high thermal conductivity. In contrast, insulators have low large Seebeck coefficient, but extremely low electrical conductivity leading to a tiny power factor. Thus, the optimal thermoelectric materials are those whose properties lie in between, that is semiconducting. This region, as highlighted in [70] lies at the crossover region between semiconducting and metallic properties, see Figure 14.7.7.

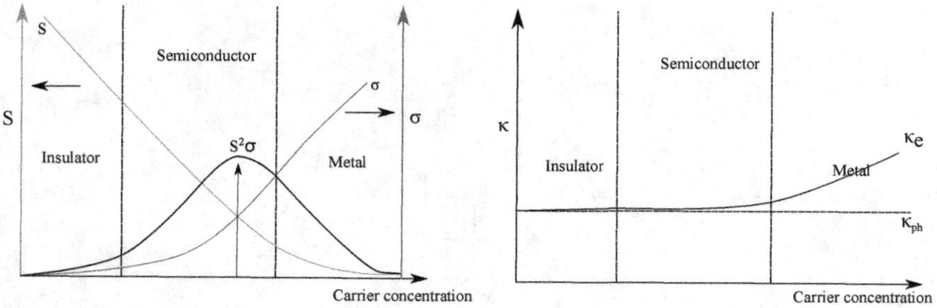


Figure 14.7.7: Carrier concentration variation of the Seebeck coefficient $S$, electrical conductivity $\sigma$, and lattice (electronic) thermal conductivity $\kappa_l(\kappa_e)$. The ratio of the power factor $S^2\sigma$ to $\kappa_l + \kappa_e$ decides the optimal thermoelectrical figure of merit $ZT$. The optimized carrier concentration is around $1 \times 10^{19}$ cm$^{-1}$. $Bi_2Te_3$ is a good thermoelectric with $ZT \approx 1$ and Silicon is poor with $ZT \approx 0.01$. Figure redrawn by the authors based on reference [70].

## 14.8 Landauer Theory of Transport

The modern human society experiences all the benefits of nanotechnology. At the heart of this immense technological revolution lies a fundamental understanding of how electron transport in devices which are really tiny ($\approx 10^{-9}$ m) cannot be described by the diffusive transport equations of the previous sections. Table 14.1.1 is an excellent reminder of the ballistic transport regimes that have come to dominate the landscape of mesoscopic or nanophysics. In the 1980s experiments on mesoscopic systems (those neither at the nano level nor macroscopic) revealed the important role played by contacts.

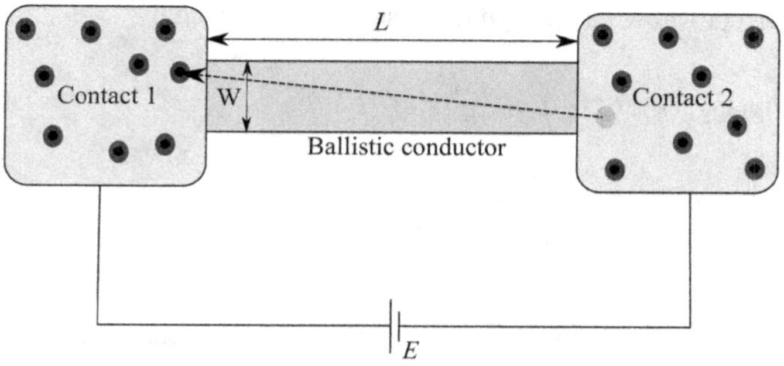


Figure 14.8.8: A ballistic conductor between two contacts is a canonical set-up within which the Landauer thory of transport can be examined. The electron injected into the ballistic conductor moves through it without any scattering events to appear at the other contact.

The transport model introduced by the German-American physicist Rolf William Landauer (1927–1999) provided a successful explanation of the phenomena observed in these devices. Within his approach, Landauer considered that all the irreversibility and the dissipation originate in the contacts, with the conductor itself being free from all interactions. Although this classification seems artifical for large macroscale conductors (say a copper wire carrying current in our homes), at the atomic/nano-scale (IC chips inside our laptop, cell phone) this is precisely the regime the conductors are in! In the next few paragraphs, we will develop the formalism that can adequately treat this situation.

As mentioned earlier, electric current flow is often viewed as an electron (or the relevant charge carrier) responding to an external electric field (or corresponding stimuli). Landauer conceptualized the current flow as a transmission process or a consequence of carrier injection at the contacts and the probability of the carriers to reach the other end. Consider an ideal one-dimensional sample of length L, see Figure 14.8.8. The density of states between $k$ and $k+dk$ including electron spin is given by

$$g(k)dk = 2\frac{1}{L}\frac{L}{2\pi}dk. \tag{14.76}$$

Furthermore, we note that the electron group velocity is given by

$$v = \frac{\hbar k}{m}, \tag{14.77}$$

where $m$ is the mass of the electron. Assuming the right (left) modes R(L) in the conductor are described by the Fermi-Dirac distribution function $f_R(k)(f_L(k))$ occupied upto the chemical potential level $\mu_L(\mu_R)$, the current $I$ flowing through the system is then given by

$$\begin{aligned}
I &= e\int_0^{\infty} v(k)g(k)f_R(k)dk - e\int_0^{\infty} v(k')g(k')f_L(k')dk', \\
&= e\int_0^{\infty} \frac{\hbar k}{m}\frac{1}{\pi}f_R(\varepsilon)\frac{dk}{d\varepsilon}d\varepsilon - e\int_0^{\infty}\frac{\hbar k'}{m}\frac{1}{\pi}f_L(\varepsilon')\frac{dk'}{d\varepsilon'}d\varepsilon', \\
&= e\int_0^{\mu_R}\frac{\hbar k}{m}\frac{1}{\pi}\frac{dk}{d\varepsilon}d\varepsilon - e\int_0^{\mu_L}\frac{\hbar k'}{m}\frac{1}{\pi}\frac{dk'}{d\varepsilon'}d\varepsilon', \text{(invoking the distribution function cut-off)} \\
&= e\int_{\mu_L}^{\mu_R}\frac{\hbar k}{m}\frac{1}{\pi}\frac{m}{\hbar k}d\varepsilon, \text{(reversing the limits on the second integral and combining the two)} \\
&= \frac{2e}{h}(\mu_R-\mu_L) = \frac{2e^2}{h}V, \qquad (14.78)
\end{aligned}$$

where we have related the voltage bias to the chemical potential via the relation $V = e(\mu_R - \mu_L)$. Thus, the **quantum of conductance** $G$ in units of Siemens (S) is given by

$$G = \frac{2e^2}{h} = (12.9\,k\Omega)^{-1} = 77.4\,\mu S, \qquad (14.79)$$

where we have used the well-known relation $I = GV$ in Equation (14.78). Notice, the $G$ replaces the inverse of the resistance in Ohm's law $V = IR$. It is the maximum conductance for a channel (in this case the wire) with one energy level (also known as an energy mode) in the range of interest $\mu_R > E > \mu_L$. Thus, only the most perfect contact can achieve this maximum value of conductance. Additionally, this relationship implies that the contact resistance of a single-moded conductor is not negligible, rather quite large and of the order of $k\Omega$. Fo most samples, the transmission is not ideal at 100%, but dependent on the scattering process. In addition, there could be more than one mode or channel of conduction in the conductor. Thus, we modify our derived formula to rewrite the **Landauer formula** as

$$G = \frac{2e^2}{h}MT, \qquad (14.80)$$

where $M$ represents the number of modes and $T$ is the transmission which gives the average probability an electron injected into the right lead will be transmitted to the left lead. It is assumed that the electron will exit from the conductor into the contacts without any reflection. The conductance quantum $\frac{2e^2}{h}$ can be measured in a **quantum point contact (QPC)** first reported in 1988 by two research groups: Van Wees *et al.* [72] and Wharam *et al.* [73]. A QPC is a narrow constriction between two wide electrically conducting regions, of a width comparable to the electronic wavelength (nano- to micrometer). A QPC can be formed in a two-dimensional electron gas heterostructure such as GaAs/AlGaAs.

## 14.9 Chapter 14 Exercises

14.9.1. In this problem you will generalize the results of the static Drude conductivity $\sigma_o$, Equation (14.24), derived in Section 14.3 to the case of a wavevector ($\vec{q}$) and frequency ($\omega$) dependent conductivity $\sigma(\vec{q},\omega)$. Using Equation (14.15) and assuming the metal is subject to an electric field periodic in both space and time given by the expression

$$\vec{E}(\vec{r},t) = \vec{E}_o e^{i(\vec{q}\cdot\vec{r}-\omega t)}, \tag{14.81}$$

show that

$$\sigma(\vec{q},\omega) = \frac{e^2}{4\pi^3}\int \frac{\tau(\vec{e}\cdot\vec{v})^2}{1-i\tau(\omega-\vec{q}\cdot\vec{v})}\left(-\frac{\partial f_o}{\partial \varepsilon}\right)d\vec{k}, \tag{14.82}$$

where $\varepsilon$ is the single electron dispersion.

14.9.2. Using Equation (14.82) obtain a general expression for the transverse conductivity $\sigma(q,\omega)$ for a spherical energy band. Assume, the impinging electric field to be polarized along the x-direction and propagating along the z-direction, thereby making the conductivity response transverse since the electric field itself is transversely polarized. State the final answer in terms of the static Drude conductivity $\sigma_o$ and the parameter

$$\rho = \frac{iqv_F\tau}{1-i\omega\tau}, \tag{14.83}$$

14.9.3. Show that for $|\rho| \ll 1$ the general expression for $\sigma(\vec{q},\omega)$ derived in the previous problem reduces to the AC conductivity formula derived in Chapter 13, Equation (13.86). Also, show that in the opposite anomalous regime where $|\rho| \gg 1$, the conductivity is real and frequency independent.

14.9.4. Use the Boltzmann equation formalism within the relaxation time approximation to derive the Hall effect current density equations

$$j_x = \frac{E_x - \omega_c\tau E_y}{1+(\omega_c\tau)^2}\sigma_o, \tag{14.84}$$

$$j_y = \frac{E_y + \omega_c\tau E_x}{1+(\omega_c\tau)^2}\sigma_o, \tag{14.85}$$

for an electric and magnetic field uniform in space and steady in time. The applied magnetic field points along the $z$ direction and the electric field along the $x$ direction. Model the non-equilibrium distribution function as $f = f_o + ak_x + bk_y$ (Reference [69]). $\omega_c = eB/m$ is the cyclotron frequency and $\sigma_o$ the static Drude conductivity.

14.9.5. Applying your knowledge of chain rule from calculus, derive Equation (14.32).

14.9.6. Electrical conductivity depends on the scattering time $\tau$, see Equation (14.30). At very low temperatures, assuming impurity scattering only, the scattering is roughly constant or temperature independent. But, electrical conductivity (thus resistivity) can be and is known to be temperature dependent. There are three main independent scattering processes: scattering by impurities (imp), scattering by electron-electron interaction (el-el), and by electron-phonon (el-ph) collisions via which temperature dependence can enter the resistivity formula. For example, well below the Debye temperature threshold the el-ph scattering is $\propto T^5$. But, far above the Debye temperature, it is $T^{-1}$. The el-el scattering scales as $T^2$. Calculation of these lifetimes are beyond the scope of the current textbook.

Inspired by the work of the British chemist and physicist Augustus Matthiessen (1831–1870) it is possible to formulate a very crude empirical additive law of effective scattering, which goes by the name of **Matthiessen's rule**. It states that the effective scattering $\tau_{eff}$ is given by

$$\frac{1}{\tau_{eff}} = \frac{1}{\tau_{imp}} + \frac{1}{\tau_{el-el}} + \frac{1}{\tau_{el-ph}}. \tag{14.86}$$

The rule fails to apply when the scattering processes are correlated with each other or if they are wavevector dependent. Assuming these approximations and using the concept of collision integral, Equation (14.14), restated below

$$\left[\frac{\partial f}{\partial t}\right]_{\text{coll}} = -\frac{f - f_o}{\tau}, \tag{14.87}$$

demonstrates the validity of Matthiessen's rule.

14.9.7. Derive Equations (14.49a)–(14.49c) using the Sommerfeld expansion. *Hint*: You will need to differentiate the functions $\sigma(E), E\sigma(E)$, and $E^2\sigma(E)$ at $E = \mu$.

14.9.8. Using the Sommerfeld expansion coefficients, derive the temperature dependence of the Seebeck coefficient, Equation (14.53). *Hint*: Express $e^2K_1$ in terms of $e^2K_0$ and then simplify, neglecting any temperature dependence above linear order.

14.9.9. Derive the $e^2K_2$ expression with $g(E) = (E-\mu)^2\sigma(E-\mu)$.

14.9.10. Using the Somerfeld $e^2K_2$ expansion coefficient derived in the previous problem, compute the temperature dependence of the electronic thermal conductivity, Equation (14.54). Note, for metals the $K_1^2/K_0 \approx T/T_F$ is typically small and can be neglected. T is the temperature and $T_F$ is the Fermi temperature.

# Appendix A: MATLAB TUTORIAL

**Contents**



## A.1 Introduction

The acronym MATLAB stands for matrix laboratory. MATLAB is a commercial software tool for the purpose of performing numerical computation and data visualization. It can be obtained from http://www.mathworks.com/. There are several versions available. The student version is the least expensive. For the most part, this textbook makes use of the numerical capabilities of the bare bones MATLAB. It is important to mention that there exist toolboxes available for needs beyond the textbook, but at an additional cost. As employed in the text, MATLAB programs or scripts (m-files) are run by means of the "command window." The command window accepts simple calculator type of commands which the MATLAB engine interprets after the user presses "Enter." The scripts are simple text documents written with the MATLAB language and which are run from the command line. In this tutorial, we will give simple examples of how to run MATLAB and how to use the scripts that accompany the text. For further help, refer to the text's commented scripts as well as other tutorials available throughout the web. Much of this tutorial as well as other example scripts can be found in Hasbun's classical mechanics [64] and DeVries-Hasbun's computational physics [18].

## A.2 Tutorial Notation

Within the command window, MATLAB's prompt ">>" indicates the place where the commands are entered. Anything typed at the prompt is user input which the engine will try to act on once the user presses Enter. MATLAB is case sensitive, which allows for many different ways to define variables. For example, here is an example of addition of three different variables.

>> a=2
a =
2
>> B=3
B =
3
>> b=4
b =
4
>> a+b+B
ans =
9

In the above example, we defined the variable and then added them. We could have also added the three numbers together in one stroke as follows:

>> 2+3+4
ans =
9

Both answers are equivalent, but the first case enables the user to use and reuse variables without the need to retype them.

## A.3 General Features

MATLAB allows the use of comments. These are preceded by the percent sign "%" and can be placed anywhere on a line within a MATLAB script or m-file. Here is an example.

>> a=3 %defines a
a =
3

We could have made the same definition and suppressed the output if a semicolon ";" had been placed immediately after the command. Also more than one command can be placed on the same line when separated by a comma or a semicolon. For example, the command

>> a=3; b=4; %define a and b

produces no output, but does define the variables. This can be checked by looking at the defined variables:

>> who
Your variables are:
a b

To check on the value of a variable, just type its name and press Enter. For example,

>> a
a =
3

To clear the value of a variable, such as a, just type "clear a" after the prompt. To clear all the variables, type "clear all" after the prompt. Typing "clear" also works, and typing "clc" and pressing return clears the command window. The keyboard control sequence "ctrl-c" stops the execution of a script (press the ctrl key while simultaneously pressing the c key.) MATLAB has a "workspace window" that shows all the defined variables it recognizes during the current session. To view the workspace window, choose that option under the "view" drop-down menu.

Define a row vector c with components 1,3, and 5, like this:

>> c=[1 2 3]
c =
1 2 3

To define a column vector d with components 2,4, and 7 separate the elements with semicolons, as for example,

```
>> d=[2;4;7]
d =
2
4
7
```

The matrix product of these two vectors is done according to the rules of matrices. For example c*d should produce a single element 1x1, but the product d*c should produce a 3x3 matrix. Let's see,

```
>> c*d
ans =
31
>> d*c
ans =
2 4 6
4 8 12
7 14 21
```

The transpose of this matrix is as follows.

```
>> (d*c)'
ans =
2 4 7
4 8 14
6 12 21
```

You can do the "transpose(d*c)" at the prompt. To learn about the details of the stored variable, type "whos" at the prompt as follows.

```
>> whos
Name Size Bytes Class
ans 3x3 72 double array
c 1x3 24 double array
d 3x1 24 double array
e 1x3 24 double array
Grand total is 18 elements using 144 bytes
```

Yo can also ask for help at any time. For example, typing "help" at the MATLAB prompt gives a list of all help topics available. Typing "help whos" at the prompt outputs the description for the command "whos" and so on. MATLAB's power lies in the way it uses arrays to perform numerical tasks.

MATLAB has a unique multiplication feature. To see this, first, let's create a square matrix.

```
>> A=[1 3 5;2 4 6; 7 9 1]
A =
1 3 5
2 4 6
7 9 1
```

If one desired to square the matrix, one would simply do,

```
>> A^2
ans =
42 60 28
52 76 40
32 66 90
```

which is the simple matrix product of A*A. However, the multiplication operation preceded by a dot has a different result.

```
>> A.*A
```

ans =
1 9 25
4 16 36
49 81 1
This is NOT a matrix product; instead, it is a dot-multiplication. The result is a matrix that contains each of the elements of A squared. The usefulness of this feature in MATLAB is realized when plotting a function. Suppose you wanted a plot of the function $y = 1/x$ on the interval $[2, 6]$. Since $y$ is a function of $x$, then first create an array containing the variable $x$ from 2 to 7 in steps of say 0.25. After that, invert an element of the array at a time and store it in $y$, then plot $y$ versus $x$. MATLAB does this as follows. First, use the "colon" loop idea for the $x$ array
>> x=[2:0.25:6];
where the output has been suppressed, but basically an array for the variable $x$ in the range from 2 to 6 in steps of 0.25 has been created for a total of 136 different values of $x$ or array elements. This is checked by looking at the workspace window or by typing "whos" at the prompt. Next use MATLAB's dot-division, which is similar to the dot-multiplication explained above. Place a "." before the "/" symbol as follows
>> y=1./x;
which is ready for MATLAB's plot command.
>> plot (x,y)
The resulting graph is shown in Figure A.1.

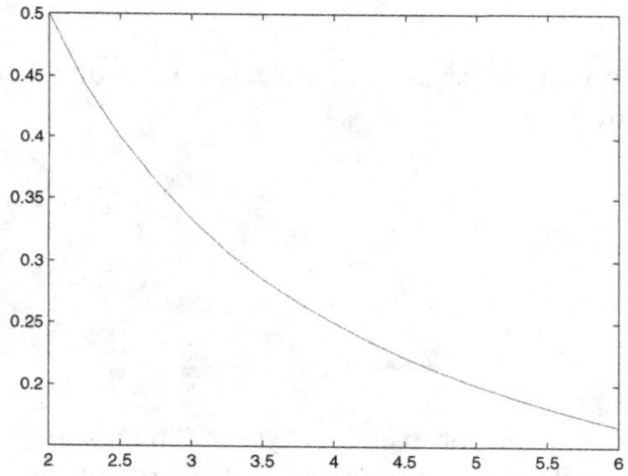


Figure A.1: Plot of $y = 1/x$.

It is possible to place labels on a graph as well, which is shown in many of the text's scripts. The help menu can also be accessed through MATLAB's command window for further details as well. While on the subject of vectors, it is useful to point out the vector dot and cross-products. For example, let's define the following vectors:
>> A=[1,2,3]
A =
1 2 3
>> B=[4,5,6]
B =
4 5 6
Next, let's find the dot and cross-products of A and B,
>> dot(A,B)
ans =

32
>> cross(A,B)
ans =
-3 6 -3
Let's find the angle between A and B and convert it to degrees
>> dot(A,B)/(sqrt(dot(A,A))*sqrt(dot(B,B)))*180/pi
ans =
55.8423
Here, the operation dot(A,A) is identical to sum(A.*A). We can also draw and add these vectors in three dimensions as done below.
>> axis([0,4,0,5,0,6]) %determines the axes ranges to use
>> line([0,1],[0,2],[0,3],'color','r') % the A vector line in red
>> line([0,4],[0,5],[0,6],'color','b') % the B vector line in blue
>> line([1,4],[2,5],[3,6],'color','k') % the connecting vector between A and B in black
>> xlabel('x'),ylabel('y'),zlabel('z') % put labels for the axes
The graph obtained by the above operations is shown in Figure A.2.

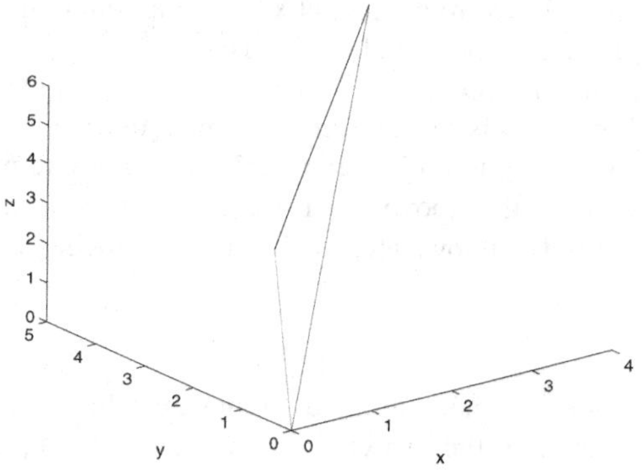


Figure A.2: The addition of two vectors and the resultant.

## A.4 Operands

MATLAB's full range of operands are listed below. These are important when you are writing a script or when you are working in the command window during a MATLAB session.

| Operand | Description | Operand | Description | Operand | Description |
|---|---|---|---|---|---|
| + | addition | ~= | not equal | , | separator |
| - | subtraction | > | greater than | ; | end row |
| * | Scalar multiplication | >= | greater than or equal | () | subscript enclosure, expression precedence |
| / | scalar right division | \| | or | [] | matrix |
| \ | matrix left division | & | and | Ctrl-c | abort |
| ^ | scalar power | ~ | not | | |
| .* | array multiplication | >> | MATLAB prompt | ; | suppress output |
| ./ | array division | ' | transpose | | |
| .^ | array exponentiation | % | comment, format specification | | |
| = | assignment | . | decimal point | | |
| < | less than | ... | line continuation | | |
| <= | less than or equal | \n | new line format specification | | |
| == | logical equal | : | vector generation | | |

## A.5 M-Files and Functions

An m-file is a text file that contains MATLAB commands in a suitable sequence for it to interpret and execute. Lines that do not refer to standard MATLAB language are commented (by preceding a line with the % sign). A simple m-file can be created from the command window by typing the command "diary test.m" followed by several commands wished to be included or tested. The file can then be closed by typing "diary off" and be accessed through the MATLAB editor or any other text editor. With the editor, MATLAB output lines, which are not standard commands, can be deleted to keep just the basic typed commands and the file can be re-saved in the working directory. By invoking the name of the file within the MATLAB command window; that is, by typing, "test", MATLAB's engine will proceed to interpret it and produce the needed output. Any errors in the script will be seen in the command window and can be fixed with the text editor. This textbook contains scripts that are ready to run in the manner explained.

MATLAB has built-in functions as described in the next section; however, a user can also build functions. A user function in MATLAB is also an m-file that a user builds in order to perform operations that are repetitive in nature. For example, suppose that you needed the numerical derivative of the $log(x)$ at $x$; you could create the script file "logder.m", which is a function, as shown here.

```
function [F] = logder(x,Stateplacedel)
% logder.m calculates the derivative of the log(x) at x within an interval [x,x+Stateplacedel]
% where the value of Stateplacedel is provided by the user
F = (log(x+Statedel)-log(x))/Stateplacedel;
```

Once the text file with the above lines is saved with the name lodger.m, it can be invoked through the MATLAB command window or through a line in a script file. For example, let's do the derivative of the $log(x)$ at $x = 5$. Assuming the above file is saved in the working directory, type the line "logder(5,1.e-3)" at the command window. The session would proceed as follows.

```
>> logder(5,1.e-3)
ans =
0.2000
```

Furthermore, typing "help logder" at the command window, produces the output:
"logder.m calculates the derivative of the log(x) at x within an interval [x,x+Stateplacedel]
where the value of placeStatedel is provided by the user",
which are basically the lines that are commented within the user-built script itself. Thus it is important to write comments within the functions to recall their usage as well as to make sure that such functions exist or are present in the working directory. The above is not the only way to create functions. Another way is to create functions within a script with the help of the "inline" command. For example f=inline('exp(x)') defines the function f(x)=exp(x) within a script.

## A.6 Built-in Functions

Some of MATLAB's built-in functions are listed below. These are important when performing commands for mathematical operations in a script or in a command window's working session.

| Function | Description | Function | Description | Function | Description |
|---|---|---|---|---|---|
| sin | trig sine | ceil | round towards positive infinity | diag | extract diagonal of a matrix or make a diagonal matrix |
| cos | trig cosine | max | largest component | triu | upper triangular part of a matrix |
| tan | trig tangent | min | smallest component | tril | lower triangular part of a matrix |
| asin | inverse sine | sign | signum | size | size of a matrix |
| acos | inverse cosine | length | length of a vector | det | determinant of a square matrix |
| atan | inverse tangent | sort | sort in ascending order | inv | inverse of a matrix |
| exp | exponential | sum | sum of elements | rank | rank of a matrix |
| log | natural logarithm | prod | product of elements | rref | reduced row echelon form |
| abs | absolute value | median | median value | eig | eigenvalues and eigenvectors |
| sqrt | square root | mean | mean value | poly | polynomial |
| rem | remainder | std | standard deviation | lu | LU factorization |
| round | round to nearest integer | eye | identity matrix | qr | QR factorization |
| floor | round towards negative infinity | zeros | matrix of zeros | quad | numerical integration |
| mod | modulus | ones | matrix of ones | fzero, roots | zeros of functions and roots of polynomials |
| inline | create a function on the fly | rand | randomly generated matrix | ode23, ode45 | differential equation solvers |

To use these functions, invoke the "help" command. For example, for help on the function "mean", type "help mean" in the command window to get help on that function. In this particular case, the help command says that, for vectors, "mean($x$)" is the mean value of the elements in the $x$ array. An example of this follows.

```
>> mean([1,2,3,4,5,6,7])
ans =
4
```

The help command is very useful in MATLAB and should be used as often as possible, at least until you are familiar with it.

## A.7 Plotting

Some of MATLAB's plotting commands are listed below. These are useful in visualizing a function in two or three dimensions. The textbook contains scripts that make use of most of these commands.

| Command | Description | Command | Description | Command | Description |
|---|---|---|---|---|---|
| plot | 2-dimensional plot | semilogx | use log scale on the x-axis | surfl | 3-dimensional shaded surface with lighting |
| subplot | table of plots | semilogy | use log scale on the y-axis | mesh | 3-dimensional mesh plot |
| loglog | use log-log scales | surf | 3-dimensional shaded surface | grid | Adds grid lines |

Please refer to the scripts in the text for the various uses of these commands.

## A.8 Programming

Programming refers to a set of commands in a script for the express purpose of performing tedious calculations. Programming is most useful when loops and if-else statements are used in conjunction with built-in or user-built functions as well as with command lines within m-files. Below are two examples that involve a loop and an if-else statement.

Loops

A simple loop in MATLAB is as follows:

```
for i=1:1:101
```

```
x(i)=(i-1)*2*pi/(101-1);
y(i)=sin(x(i));
end
```

The above lines create a 101-element array for $x$ between 0 and 2 for which the sin is evaluated and stored in the array for $y$. Of course, in MATLAB there are other ways of doing the same thing, but the above is an example of a loop. You could add an "if-else" statement, say, for the purpose of converting half the above wave into a square box. This can be done as follows.

```
for i=1:1:101
x(i)=(i-1)*2*pi/(101-1);
if x(i)<=pi
y(i)=sin(x(i));
else
y(i)=sign(sin(x(i)));
end
end
```

Finally, you can perform a "plot(x,y)" command to visualize the results.

## A.9 Zeros of Functions

The MATLAB built-in function "fzero" can be used to obtain the variable values at which a function takes a zero value. One example of this is to find the value of x at which x=exp(-x) is satisfied. This can be done as follows:

```
>> fzero('x-exp(-x)',.3)
ans =
0.5671
```

where we have used 0.3 as an initial guess. See the help command for a full function description. You can also use built-in functions as well as user-built functions in place of the expression between the single quotes, as explained before. Another example is to obtain roots of polynomials. The built-in function "roots" can be employed for this purpose. If you needed to know the roots of the cubic polynomial $x^3+3x^2-2x+5$, for example, you could proceed as follows.

```
>> roots([1 3 -2 5])
ans =
-3.8552
0.4276 + 1.0555i
0.4276 - 1.0555i
```

which gives one real and two complex conjugate roots.

## A.10 Numerical Integration

Numerical integration is possible with MATLAB. The built quad function can be used for that purpose. One example for the integration of the sin function on $[0,\pi]$ is as follows.

```
>> quad(@sin,0,pi)
ans =
2.0000
```

You could also replace the "sin" in the above line with the name of a user built-in function. Another example is to create a function on the fly with the "inline" command

```
>> g=inline('2*x.^2+x.^3.*sin(x)');
>> quad(g,0,pi)
ans =
```

32.8276
Here the function $2x^2 + x^3\sin(x)$ was integrated on $[0,\pi]$.

## A.11 Differential Equations

MATLAB has the capability of solving differential equations with the use of the built-in function "ode23" or the "ode45". For example, to solve the problem $y' = (y - t^2)\exp(-t)$ for $y(t)$ on the interval $[0,20]$ with the initial condition that at $t = 0$, $y = 1$, do it as follows.

```
f=inline('(y-t.^2).*exp(-t)');
[t,y]=ode23(f,[0,20],1);
plot(t,y)
```

Other more complicated cases make use of user-built functions. See the textbook's scripts for examples or type "help ode23" within the command window for further details. The "ode45" solver and other more sophisticated methods of tackling differential equations are available as explained in the MATLAB help facility.

## A.12 Movies

Here we show how MATLAB makes movies and we use a traveling wave as an example. First make a space and time array, plot the waves at every time step, save the plot frame into an array M, pause for a short time to see the plot versus position on the screen, play the movie at 15 frames per second, and finally save it as an avi file to be used later as the following demonstrates.

```
x=0:0.01:1; %space array
t=0:0.01:1; %time array
nt=length(t); %time steps
for j=1:nt
y=sin(2*pi*(x-t(j))); %a traveling wave
plot(x,y) %plot for every time step
M(j) = getframe; %make movie frames if desired
pause(0.05); %pause for a short time
end
movie(M,1,15) %play once, at 15 frames per second
%save the movie as wave.avi (at 15 frames per sec without compression)
movie2avi(M,'wave','fps',15,'compression','None');
```

According to the help command for code that is compatible with all versions of MATLAB, including versions before MATLAB Release 11 (5.3), use the following lines in place of the above loop.

```
M = moviein(nt);
for j=1:nt
plot commands here
M(:,j) = getframe;
end
movie(M)
```

## A.13 Publish Code to HTML

It is possible to publish written code on a website. To do so, one needs to convert the script and its output into html format. To do so, let's work with the script "file.m". In the command line, type the following line

```
>>publish('file.m')
```

which runs the script, waits for the output, and automatically converts the script and the output to html code in the subdirectory "html" under the main directory where the file.m is run from. Typing "help publish" gives further information on this command.

## A.14 Symbolic Operations

With the presence of the symbolic toolbox, MATLAB has symbolic functions capability. Typing "help symbolic" within the command line gives a list of available functions if the *Symbolic Math Toolbox* is available with the version of the software you have. A simple example of a symbolic operation is to plot the function $y = 2t$. This is accomplished by typing the following lines:

```
>> syms t %defines the symbolic variable t
>> y=2*t
y =
2*t
>> ezplot(t,y)
```

Another example is to plot the function $y = 2\exp(-at^2)$ and its derivative. This is accomplished as follows:

```
>> syms t y f a
>> y=2*exp(-a*t^2)
y =
2*exp(-a*t^2)
>> f=diff(y,t)
f =
-4*a*t*exp(-a*t^2)
>> a=0.5
a =
0.5000
>> ezplot(t,eval(y),[-5 5])
>> hold on
>> ezplot(t,eval(f),[-5 5])
```

Other examples for symbolic operations are available within the textbook or within the help menu within MATLAB.

## A.15 Toolboxes

In MATLAB basic capabilities can be extended through the use of toolboxes and can be ordered separately. Toolboxes are separate sets of scripts created by www.mathworks.com in order to solve specific sets of problems. Currently, the toolboxes available include:

Parallel Computing
- Parallel Computing Toolbox
- MATLAB Distributed Computing Server

Math, Statistics, and Optimization
- Symbolic Math Toolbox
- Partial Differential Equation Toolbox
- Statistics Toolbox
- Curve Fitting Toolbox
- Optimization Toolbox
- Global Optimization Toolbox
- Neural Network Toolbox
- Model-Based Calibration Toolbox

Control System Design and Analysis
- Control System Toolbox
- System Identification Toolbox
- Fuzzy Logic Toolbox
- Robust Control Toolbox
- Model Predictive Control Toolbox
- Aerospace Toolbox

Signal Processing and Communications
- Signal Processing Toolbox
- DSP System Toolbox
- Communications System Toolbox
- Wavelet Toolbox
- RF Toolbox
- Phased Array System Toolbox

Image Processing and Computer Vision
- Image Processing Toolbox
- Computer Vision System Toolbox
- Image Acquisition Toolbox
- Mapping Toolbox

Test and Measurement
- Data Acquisition Toolbox
- Instrument Control Toolbox
- Image Acquisition Toolbox
- OPC Toolbox
- Vehicle Network Toolbox

Computational Finance
- Financial Toolbox
- Econometrics Toolbox
- Datafeed Toolbox
- Database Toolbox
- Spreadsheet Link EX (for Microsoft Excel)
- Financial Instruments Toolbox
- Trading Toolbox

Computational Biology
- Bioinformatics Toolbox
- SimBiology

Code Generation and Verification
- MATLAB Coder
- HDL Coder
- HDL Verifier
- Filter Design HDL Coder
- Fixed-Point Designer

Application Deployment
- MATLAB Compiler

MATLAB Builder NE (for Microsoft .NET Framework)
MATLAB Builder JA (for Java language)
MATLAB Builder EX (for Microsoft Excel)
Spreadsheet Link EX (for Microsoft Excel)
MATLAB Production Server

Database Connectivity and Reporting
Database Toolbox
MATLAB Report Generator

Simulink Product Family
Simulink

Event-Based Modeling
Stateflow
SimEvents

Physical Modeling
Simscape
SimMechanics
SimDriveline
SimHydraulics
SimRF
SimElectronics
SimPowerSystems

Control System Design and Analysis
Simulink Control Design
Simulink Design Optimization
Aerospace Blockset

Signal Processing and Communications
DSP System Toolbox
Communications System Toolbox
Computer Vision System Toolbox

Code Generation
Simulink Coder
Embedded Coder
HDL Coder
Simulink PLC Coder
Fixed-Point Designer
DO Qualification Kit (for DO-178)
IEC Certification Kit (for ISO 26262 and IEC 61508)

Rapid Prototyping and HIL Simulation
xPC Target
xPC Target Embedded Option
Real-Time Windows Target

Verification, Validation, and Test
- Simulink Verification and Validation
- Simulink Design Verifier
- SystemTest
- Simulink Code Inspector
- HDL Verifier

Simulation Graphics and Reporting
- Simulink 3D Animation
- Gauges Blockset
- Simulink Report Generator

More detailed information on these toolboxes can be obtained through http://www.mathworks.com.

## A.16 MATLAB Websites, other Tutorials, and Clones

In addition to the help facility included within the MATLAB software, the MathWorks website contains excellent sources of information, help, and code example such as:

1. http://www.mathworks.com/support/books/
2. http://www.mathworks.com/academia/student_center/tutorials/index.html?link=body
3. http://www.mathworks.com/access/helpdesk/help/techdoc/index.html?

Often the MATLAB clones's websites are excellent sources of code examples that could be ported into MATLAB.

1. http://www.octave.org/
2. http://www.scilab.org/

The MATLAB clones are open source software that is capable of running MATLAB code with some minor, if any, modifications. The clones may be most useful if much of the code has been written by the user. This is in contrast to the MATLAB dependence on its toolboxes for specific needs.

# Appendix B: Distribution Functions

Contents



## B.1 The Boltzmann Distribution Function

The Boltzmann distribution is the result of studying the equilibrium configuration of an assembly of $N$ distinguishable (classical) particles subject to the constraints

$$\Phi \equiv \sum_j N_j = N, \tag{14.1a}$$

where $N_j$ is the number of particles with single particle energy level $\varepsilon_j$ and

$$\Psi \equiv \sum_j N_i \varepsilon_i = U, \tag{14.1b}$$

with the total number of particles $N$ and total energy $U$ are constants. In these expressions, the sums over $j$ are to be carried out up to $n$ energy levels. The idea is to find the occupation number of each level when the thermodynamic probability is a maximum. For example, the number of ways of selecting $N_1$ particles from a total of $N$ to be placed in the 1st level is

$$\left( \begin{array}{c} N \\ N_1 \end{array} \right) = \frac{N}{N_1!(N-N_1)!}, \tag{14.2a}$$

and if we consider that level 1 has $g_1$ distinct states, then the number of ways to place $N_1$ particles into level 1 containing $g_1$ options is

$$\frac{N! g_1^{N_1}}{N_1!(N-N_1)!}. \tag{14.2b}$$

For the second energy level, the counting is similar but there are only $(N-N_1)$ particles remaining, to write

$$\frac{(N-N_1)! g_2^{N_2}}{N_2!(N-N_1-N_2)!}, \tag{14.2c}$$

etc. Continuing this process for $n$ energy levels and multiplying the results, we get the expression

$$\omega_B(N_1, N_2, \ldots, N_n) = N! \frac{g_1^{N_1} g_2^{N_2} \cdots g_n^{N_n}}{N_1! N_2! \ldots N_n!} = N! \prod_{j=1}^{n} \frac{g_j^{N_j}}{N_j!}, \tag{14.3}$$

for the total number of ways that $N$ distinguishable particles can be arranged if they are divided into $n$ groups, with $N_1$ objects in the first group, $N_2$ in the second, etc., and with level $i$ having $g_i$

available states. Since $\ln(\omega_B)$ is an increasing function of $\omega_B$, maximizing $\ln(\omega_B)$ is equivalent to maximizing $\omega_B$. Applying the method of Lagrange multipliers and maximizing with respect to the number of particles $N_j$, subject to the constraints of Equation 14.1, we have

$$\frac{\partial}{\partial N_j}\ln(\omega_B)+\alpha\frac{\partial\Phi}{\partial N_j}+\beta\frac{\partial\Psi}{\partial N_j}=0, \tag{14.4}$$

where $\alpha$ and $\beta$ are parameters related to the system's properties. Using Equation 14.1, in addition to applying Stirling's formula (for large $N$, $\ln(N_j!)\approx N_j\ln N_j-N_j$) on Equation 14.3, and substituting the results into Equation 14.4, find

$$\frac{N_j}{g_j}=\exp(\alpha+\beta\varepsilon_j)\equiv f_{jB}(\varepsilon_j) \tag{14.5}$$

which is the equilibrium number of particles per quantum state, for every energy level. This is the Boltzmann distribution function. Using entropy considerations ($s=k\ln(\omega)$ where $k$ is Boltzmann's constant and taking $\omega=\omega_B$), in addition to the first law of thermodynamics ($du=Tds-pdV$ where $u$ is energy, $T$ is temperature, $s$ is entropy, and $p$ is pressure), it is possible (see Carter [74] or Reif [75]) to solve for the constants $\alpha$ and $\beta$ to obtain

$$e^{\alpha}=\frac{N}{\sum\limits_j g_j\exp(\beta\varepsilon_j)},\qquad \beta=-\frac{1}{kT}. \tag{14.6}$$

Substituting these expressions back into Equation 14.5 gives

$$f_{jB}(\varepsilon_j)=\frac{N_j}{g_j}=\frac{Ne^{-\varepsilon_j/kT}}{Z} \tag{14.7a}$$

which is the Boltzmann distribution function for discrete states and where $Z$ is the partition function

$$Z=\sum_j g_je^{-\varepsilon_j/kT}. \tag{14.7b}$$

In the case when the energy levels are very close to each other, we can make the replacement $g_j\to g(\varepsilon)d\varepsilon$, which is the number of states in the range of $\varepsilon$ and $\varepsilon+d\varepsilon$. We also make the replacements $\varepsilon_j\to\varepsilon$ and $N_j\to N(\varepsilon)d\varepsilon$. With these changes, Equation 14.7b becomes

$$f_B(\varepsilon)=\frac{N(\varepsilon)}{g(\varepsilon)}=\frac{Ne^{-\varepsilon/kT}}{\int g(\varepsilon)e^{-\varepsilon/kT}d\varepsilon}. \tag{14.8}$$

If in this relation we substitute the free electron density of states from Chapter 5 (see the Free Three-Dimensional Electron Gas Section) for $g(\varepsilon)$; that is, if we let

$$g(\varepsilon)=\frac{V}{2\pi^2}\left(\frac{2m}{\hbar^2}\right)^{3/2}\varepsilon^{1/2}, \tag{14.9a}$$

we find that

$$Z=\int_0^\infty g(\varepsilon)e^{-\varepsilon/kT}d\varepsilon=\frac{V}{4}\left(\frac{2mkT}{\pi\hbar^2}\right)^{3/2}. \tag{14.9b}$$

Substituting this result for the denominator of Equation 14.8 we see that

$$f_B(\varepsilon)=\frac{N(\varepsilon)}{g(\varepsilon)}=\frac{4N}{V}\left(\frac{\pi\hbar^2}{2mkT}\right)^{3/2}e^{-\varepsilon/kT}, \tag{14.10}$$

which is the Boltzmann distribution for continuous free electron energies. From Equation 14.8, notice that for free electrons, the number of particles as a function of energy is given

$$N(\varepsilon) = g(\varepsilon) f_B(\varepsilon) = \frac{2N}{\sqrt{\pi}} \frac{\varepsilon^{1/2}}{(kT)^{3/2}} e^{-\varepsilon/kT}, \tag{14.11}$$

where we have used Equations 14.9a and 14.10 for $g(\varepsilon)$ and $f_B(\varepsilon)$, respectively. Equation 14.11 is known as the Maxwell-Boltzmann particle distribution, normalized to the total number of particles, $N$, as can be seen directly from Equation 14.8. Finally, it should be noted that the exponential term of Equations 14.10 and 14.11; i.e., $e^{-\varepsilon/kT}$, is also known as the Boltzmann factor.

## B.2 The Fermi-Dirac Distribution Function

This distribution is associated with indistinguishable (quantum, identical with $1/2$ spin) particles which obey the Pauli exclusion principle in which no quantum state can accept more that one particle with the same spin. Examples of spin $1/2$ particles are electrons, positrons, protons, neutrons, muons, etc. and are collectively known as fermions. Here an energy state can either be occupied or unoccupied and we cannot have more particles than there are degenerate levels (spin included); thus, for the $j$th level $N_j \leq g_j$ and the problem is similar to a binary system in which if we have $g_j$ states, then there will be $N_j$ states occupied with one particle and $(g_j - N_j)$ unoccupied states. Thus, for the $j$th level, the number of ways to organize $N_j$ particles from $g_j$ available states is

$$\omega_j = \frac{g_j!}{N_j!(g_j - N_j)!}. \tag{14.12}$$

The total number of microstates corresponding to an allowable configuration is the product of the individual factors for all levels, or

$$\omega_{FD}(N_1, N_2, ..., N_n) = \prod_{j=1}^{n} \frac{g_j!}{N_j!(g_j - N_j)!}. \tag{14.13}$$

Here, as in the previous subsection, the idea is to maximize $\ln(\omega_{FD})$ with respect to $N_j$ subject to the constraints of Equation 14.1. Again, the details have been worked out (see Carter [74] or Reif [75]) with the result

$$f_{jFD} \equiv \frac{N_j}{g_j} = \frac{1}{1 + e^{(\varepsilon_j - \mu)/kT}}, \tag{14.14}$$

which is the Fermi-Dirac distribution for a discrete system of energy levels, with chemical potential $\mu$. For a continuous energy spectrum, we make the replacements $f_j \rightarrow f(\varepsilon)$ and $\varepsilon_j \rightarrow \varepsilon$, to write

$$f_{FD} = \frac{1}{1 + e^{(\varepsilon - \mu)/kT}}, \tag{14.15}$$

as the Fermi-Dirac distribution for a continuous energy spectrum.

## B.3 The Bose-Einstein Distribution Function

This distribution refers to particles that do not obey the Pauli exclusion principle, such as photons (spin zero) and other particles with integral spin (1, 2, etc.) which are collectively called bosons. These are indistinguishable particles and any number of which can occupy a given quantum state. For the $j$th energy level, there will be $g_j$ quantum states containing a total of $N_j$ identical particles without any restriction on the number of particles in each state. Here, one pictures an arrangement of

$N_j$ particles among $g_j$ states by placing them into $(g_j - 1)$ partitions; new microstates are obtained by shuffling the partitions and the particles while keeping the numbers $N_j$ and $g_j$ fixed. For the $j$th level, the number of ways to arrange $(N_j + g_j - 1)$ objects (particles and partitions) into $(g_j - 1)$ partitions and $N_j$ particles is

$$\omega_j = \frac{(N_j + g_j - 1)!}{N_j!(g_j - 1)!}. \tag{14.16}$$

Again, the total number of microstates is the product of individual factors for all levels, or

$$\omega_{BE}(N_1, N_2, \ldots, N_n) = \prod_{j=1}^{n} \frac{(N_j + g_j - 1)!}{N_j!(g_j - 1)!}. \tag{14.17}$$

By following the maximization recipe discussed in the two previous subsections (see also Carter [74] or Reif [75] for further details), subject to the constraints of Equation 14.1, the result is

$$f_{jBE} \equiv \frac{N_j}{g_j} = \frac{1}{e^{(\varepsilon_j - \mu)/kT - 1}}, \tag{14.18}$$

which is the Bose-Einstein distribution for a discrete system of energies. As before, we can write the continuous energies analogue as

$$f_{jBE}(\varepsilon) = \frac{1}{e^{(\varepsilon - \mu)/kT - 1}}. \tag{14.19}$$

# *Bibliography*


[1] Transmission Electron Microscopy of Multi-Layer InGaAs Quantum Wires Grown on GaAs (311)A, G. D. Lian, M. B. Johnson, H. Wen, Z. M. Wang, G. Salamo, D. A. Blom, and L. F. Allard Jr. *Microsc. Microanal.* 10(Suppl 2) (2004).

[2] Free-Standing Graphene at Atomic Resolution, Mhairi H. Gass, Ursel Bangert, Andrew L. Bleloch, Peng Wang, Rahul R. Nair, And A. K. Geim, *Nature Nanotechnology*, Vol 3, 676–681 (November 2008); <http://www.nature.com/naturenanotechnology>.

[3] *Introduction to Solid State Physics,* Charles Kittel, 8th Ed. 13-19 (John Wiley, NY 2005).

[4] Optical-beam-deflection Atomic Force Microscopy: The NaCl (001) Surface, G. Meyer, N. M. Amer, *Appl. Phys. Lett.* Vol. 56, p2100 (1990). See also D. Rugar and P. Hansma, Physics Today (October 1990) pp23.

[5] <http://www.almaden.ibm.com/vis/stm/blue.html>.

[6] *Fundamentals of Solid State Physics,* J. Richard Christman (John Wiley, NY 1988).

[7] *Principles of Electronic Materials and Devices,* S. O. Kasap, 3rd Ed. (McGraw Hill, NY 2006).

[8] *Foundations of Crystallography with Computer Applications,* Maureen M. Julian (CRC Press Taylor & Francis Group, Boca Raton, FL 2008).

[9] Electron Diffraction Using Transmission Electron Microscopy, Leonid A. Bendersky and Frank W. Gayle, *J. Res. Natl. Inst. Stand. Technol.* 106, 997–1012 (2001) available online at <http://nvl.nist.gov/nvl3.cfm?doc_id=89&s_id=117#jr>.

[10] *Solid State Physics,* Neil Ashcroft and N. David Mermin, (Saunders College, Philadelphia 1976).

[11] On the Calculation of Certain Crystal Potential Constants, and on the Cubic Crystal of Least Potential Energy, J. E. Jones and A. E. Ingham, *Proc. Roy. Soc.* (London) A107, 636 (1925).

[12] *Fundamentals of Solid State Physics,* J. Richard Christman (John Wiley, NY, 1988).

[13] *Introduction to Quantum Mechanics,* L. Pauling and E. B. Wilson, Jr. (McGraw-Hill, NY, 1935).

[14] The Wave Mechanics of an Atom with a Non-Coulomb Central Field. Part I. Theory and Methods, D. R. Hartree, *Proc. Camb. Phil. Soc.* Vol. 24, pp 89 (1928).

[15] *Computational Physics,* J. M. Thijssen, (Cambridge UP, NY, 2007).

[16] *Electronic Structure and the Properties of Solids,* W. A. Harrison, (W. H. Freeman and Co., San Francisco, 1980).

[17] *A First Course in Computational Physics,* 2nd. Ed., Paul L. DeVries and Javier E. Hasbun (Jones and Bartlett, Sudbury, MA, 2011).

[18] *Introductory Solid State Physics,* 2nd Ed., H. P. Myers (Taylor & Francis, CRC Press, NY, 1997).

[19] *Mathematical Handbook of Formulas and Tables,* Murray R Spiegel, Schaum's Outline Series (McGraw-Hill, New York, 1968).

[20] *Condensed Matter Physics,* Michael P. Marder (John Wiley & Sons, New York, 2000).

[21] A Calculation of the Debye Characteristic Temperature of Cubic Crystals, M.M. Shukla and N.T. Padial, *Revista Brasileira de Fisica*, Vol. 3, No. 1, 39-45 (1973).

[22] *Specific Heat: Nonmetallic Solids, vol.5. of Thermophysical Properties of Matter,* Y. S. Touloukian and E. H. Buyco (IFI/Plenum, New York, 1970).

[23] Measurements of X-Ray Lattice Constant, Thermal Expansivity, and Isothermal Compressibility of Argon Crystals, O. G. Peterson, T D. N. Batchelder, and R. O. Simmons, *Phys. Rev.* vol. 150, No. 2 (1966).

[24] *Solid State Physics: Essential Concepts,* David W. Snoke (Addison-Wesley, San Francisco, CA 2009).

[25] *Mathematical Handbook of Formulas and Tables,* Murray R. Spiegel, Schaum's Outline Series (McGraw-Hill Book Co., New York, NY 1968).

[26] The Quantized Hall Effect, Klaus von Klitzing, *Rev. Mod. Phys.* 58, 519–531 (1986). `http://link.aps.org/doi/10.1103/RevModPhys.58.519`

[27] F. J. Morin and J. P. Maita, Phys. Rev. **94**, 1525 (1954) as presented on p287 of reference [12] from where we extracted the data.

[28] Maximum Concentration of Impurities in Semiconductors, E. F. Schubert, G. H. Gilmer, R. F. Kopf, and H. S. Luftman, *Physical Review B*, **46**, No.23, 15078–15084 (1992-I).

[29] *Physics of Semiconductor Devices,* S. M. Sze, 2nd Ed. (John Wiley, NY 1981).

[30] *Quantum Theory of the Solid State,* Joseph Callaway, 2nd ed. (Academic Press, New York, 1991).

[31] *Green's Functions in Quantum Physics,* E. N. Economou (Springer-Verlag, NY, 1983).

[32] Tight-Binding Models of Amorphous Systems: Liquid Metals, L. M. Roth *Phys. Rev. B* Vol. 7, 4321–4337 (1973).

[33] Simple Brillouin-Zone Scheme for the Spectral Properties of Solids, An-Ban Chen, *Phys'. Rev. B* V16, 3291 (1977).

[34] The Density of States of Some Simple Excitations in Solids, R. Jelitto, *J. Phys. Chem. Solids* V30 609 (1969).

[35] Calculated Valence-Band Densities of States and Photoemission Spectra of Diamond and Zinc-Blende Semiconductors, J. Chelikowsky, D. J. Chadi, and Marvin L. Cohen, *Phys. Rev. B*, Vol. 8, 2786 (1973).

[36] A Semi-Empirical Tight-Binding Theory of the Electronic Structure of Semiconductors, P. Vogl, H.P. Hjalmarson, J.D. Dow *J. Phys. Chem. Solids* 44 (5), 365 (1983).

[37] *The Solid State,* H. M. Rosenberg, 3rd Ed. (Oxford Science Publications, NY 1988).

[38] Coherent-Potential Model of Substitutional Disordered Alloys, P. Soven *Phys. Rev. V156*, 809 (1967); also, Single-Site Approximations in the Electronic Theory of Simple Binary Alloys, B. Velicky, S. Kirkpatrick, and H. Ehrenreich, *Phys. Rev.* V175, 747 (1968).

[39] Theoretical Densities of States of a-Si(1-x)Ge(x):H Alloys. Comparison with X-Ray Spectra, D. A. Papaconstantopoulos, C. Senemaud, and E. Belin, *Europhys. Lett.* V6 (7) 635 (1988).

[40] The Theory and Properties of Randomly Disordered Crystals and Related Physical Systems, R. J. Elliot, J. A. Krumhansl, and P. L. Leath, *Rev. Mod. Phys.* V46 (3), 465 (1974).

[41] Electrical Resistivity of Ten Binary Alloy Systems, C. Y. Ho et al., *J. Phys. Chem. Ref. Data* V12 (2) 183 (1983).

[42] *Magnetism in Condensed Matter,* Stephen Blundell, 1st Ed. (Oxford University Press Inc., Oxford 2001).

[43] *Magnetism and Magnetic Materials,* J. M. D. Coey, 1st Ed. (Cambridge University Press, Cambridge 2010).

[44] *Heat and Thermodynamics,* Mark W. Zemansky and Richard H. Dittman, 6th Ed. (McGraw-Hill Book Co, Singapore 1981).

[45] *Hysteresis in Magnetism,* Giorgio Bertotti, 1st Ed. (Academic Press, 1998).

[46] *A Guide to Monte Carlo Simulations in Statistical Physics,* David P. Landau & Kurt Binder, $4^{th}$ Ed. (Cambridge University Press, 2014).

[47] *Monte Carlo Methods in Statistical Physics,* M. E. J. Newman & G. T. Barkema, $1^{st}$ Ed. (Clarendon Press, 1999).

[48] *Ligand Field Theory and Its Applications,* Brian N. Figgis & Michael A. Hitchman, 1st Ed. (Wiley–VCH, 1999).

[49] *Multiplets of Transition-Metal Ions in Crystals,* Satoru Sugano, Yukito Tanabe, & Hiroshi Kamimura, (Academic Press, 1970).

[50] *Principles of Magnetic Resonance (Springer Series in Solid-State Sciences)* Charles P. Slichter, $3^{rd}$ Ed. (Springer, 1990).

[51] *Lecture Notes on Electron Correlation and Magnetism,* Patrik Fazekas, (World Scientific, 1999).

[52] The Discovery of Superconductivity, Dirk van Delft, *Phys. Today* **63**, 38 (2010).

[53] *Introduction to Superconductivity*, A. C. Rose-Innes and E. H. Rhoderick , $2^{nd}$ Ed. (Pergamon Press Inc., 1978).

[54] *Superconductivity, Superfluids, and Condensates*, James F. Annett, 1st Ed. (Oxford University Press, Oxford 2004).

[55] *Introduction to Superconductivity*, Michael Tinkham, $2^{nd}$ Ed. (Dover Publications, 2004).

[56] *Superconductivity*, J. B. Ketterson and S. N. Song, 1st Ed. (Cambridge University Press, Cambridge 1999).

[57] *Superconductivity Supplement for Physics for Scientists and Engineers and Physics for Scientists and Engineers with Modern Physics*, Raymond A. Serway, (Saunders College Publishing, 1988).

[58] *Handbook of Superconductivity*, Edited by: Charles P. Poole, Jr., (Academic Press, 2000).

[59] For more general information on superconductivity visit, `http://superconductors.org/`.

[60] *Optical Properties of Solids,* Mark Fox, $2^{nd}$ Ed. (Oxford University Press Inc., Oxford 2001).

[61] *Electronic Properties of Materials,* Rolf E. Hummel, $3^{rd}$ Ed. (Springer-Verlag New York, Inc. 2001).

[62] *Optical Properties of Solids,* Frederick Wooten, $1^{st}$ Ed. (Academic Press, 1972).

[63] *Electrodynamics of Solids: Optical Properties of Electrons in Matter,* Martin Dressel and George Grüner, $1^{st}$ Ed. (Cambridge University Press, 2002).

[64] *Classical Mechanics with MATLAB Applications,* Javier E. Hasbun, $1^{st}$ Ed. (Jones & Bartlett Learning, 2008).

[65] *Solids State Physics,* Giuseppe Grosso and Giuseppe Pastori Parravicini, (Academic Press, NY 2000).

[66] *Electronic Transport in Mesoscopic Systems,* Supriyo Datta, 1st Ed. (Cambridge University Press, 1995).

[67] *Electrical Transport in Nanoscale Systems,* Massimiliano Di Ventra, 1st Ed. (Cambridge University Press Inc., Oxford 2008).

[68] *Electronic Transport Theories,* Navinder Singh, 1st Ed. (CRC Press Taylor & Francis Group, 2017).

[69] *Transport Phenomena,* H. Smith and H. Hojgaard Jensen, (Clarendon Press, Oxford 1989).

[70] *Semiconductor Thermoelements and Thermoelectric Cooling Information,* A. F. Ioffe, 1st Ed. (Infosearch, ltd., 1957).

[71] Materials for Thermoelectric Energy Conversion, C. Wood, *Rep. on Prog. in Physics* **51**, 459 (1988).

[72] Quantized Conductance of Point Contacts in a Two-dimensional Electron Gas, B. J. van Wees, H. van Houten, C. W. J. Beenakker, J. G. Williamson, L. P. Kouwenhoven, D. van der Marel, and C. T. Foxon, *Phys. Rev. Lett.* **60**, 848 (1988).

[73] One-dimensional Transport and the Quantisation of the Ballistic Resistance, D A Wharam, T J Thornton, R Newbury, M Pepper, H Ahmed, J E F Frost, D G Hasko, D C Peacock, D A Ritchie and G A C Jones, *J. Phys. C* **21**, L209 (1988).

[74] *Classical and Statistical Thermodynamics,* Ashley Carter, (Prentice Hall, Upper Saddle River, NJ 2001).

[75] *Fundamentals of Statistical and Thermal Physics,* F Reif, (McGraw-Hill, NY 1965).

[76] See this reference for further concepts in thermodynamics: *Thermal Physics,* Charles Kittel and Herbert Kroemer, (W. H. Freeman, NY 1980).

# *Index*

Printed in the United States
by Baker & Taylor Publisher Services